S0-ATB-955

CONCRETE CONSTRUCTION HANDBOOK

CONCRETE CONSTRUCTION HANDBOOK

Joseph J. Waddell, P.E.

Joseph A. Dobrowolski, P.E.

Third Edition

McGRAW-HILL, INC.

New York San Francisco Washington, D.C. Auckland Bogotá
Caracas Lisbon London Madrid Mexico City Milan
Montreal New Delhi San Juan Singapore
Sydney Tokyo Toronto

Library of Congress Cataloging-in-Publication Data

Concrete construction handbook / [edited by] Joseph J. Waddell, Joseph A. Dobrowolski. — 3rd ed.
p. cm.
Includes index.
ISBN 0-07-067666-6
1. Concrete construction. I. Waddell, Joseph J. II. Dobrowolski, Joseph A.
TA681.C745 1993
6245.1′834—dc20 92-35397
CIP

1 2 3 4 5 6 7 8 9 0 DOC/DOC 9 8 7 6 5 4 3 2

ISBN 0-07-067666-6

The sponsoring editor for this book was Larry S. Hager, the editing supervisor was Joseph Bertuna, and the production supervisor was Pamela A. Pelton. It was set in Times Roman by The Clarinda Company.

Printed and bound by R. R. Donnelley & Sons Company.

This handbook is dedicated to Raymond E. Davis Sr., Joe W. Kelly, Lewis H. Tuthill, Paul Klieger, and Milos Polivka who always had time to give encouragement, advice, and respect to young engineers launching careers in the field of concrete. They were and are pioneers and giants in concrete research and education.

CONTENTS

Contributors xi
Preface xiii

Part 1 Materials for Concrete

Chapter 1. Cement 1.3
Chapter 2. Aggregates 2.1
Chapter 3. Steel for Reinforcement and Prestressing 3.1
Chapter 4. Admixtures, Curing Mediums, and Other Materials 4.1

Part 2 Properties of Concrete

Chapter 5. Workability 5.3
Chapter 6. Strength and Elastic Properties 6.1
Chapter 7. Durability 7.1
Chapter 8. Permeability and Absorption 8.1
Chapter 9. Volume Changes 9.1
Chapter 10. Thermal and Acoustic Properties 10.1

Part 3 Proportioning Mixes and Testing

Chapter 11. Proportioning Mixes 11.3
Chapter 12. Testing Concrete 12.1
Chapter 13. Inspection and Laboratory Service 13.1

Part 4 Formwork and Shoring

Chapter 14. Formwork Objectives and Types 14.3
Chapter 15. Design, Application, and Care of Forms 15.1

Part 5 Batching, Mixing, and Transporting

Chapter 16. Plant Layout 16.3
Chapter 17. Batching Equipment 17.1
Chapter 18. Concrete Mixers 18.1
Chapter 19. Materials Handling and Storage 19.1
Chapter 20. Heating and Cooling 20.1
Chapter 21. Plant Operation 21.1
Chapter 22. Plant Inspection and Safety 22.1
Chapter 23. Transporting Concrete 23.1

Part 6 Placing Concrete

Chapter 24. Methods and Equipment 24.3
Chapter 25. Preparation for Placing Concrete 25.1
Chapter 26. Placing Concrete 26.1

Part 7 Finishing and Curing

Chapter 27. Finishing 27.3
Chapter 28. Joints in Concrete 28.1
Chapter 29. Curing 29.1

Part 8 Special Concretes and Techniques

Chapter 30. Roller-Compacted Concrete 30.3
Chapter 31. Lightweight Concrete 31.1
Chapter 32. Heavyweight Concrete 32.1

Part 9 Building Construction Systems

Chapter 33. On-Site Precasting 33.3
Chapter 34. Slipform Construction of Buildings 34.1
Chapter 35. Lift-Slab and Precast Tiltup Construction 35.1

Part 10 Specialized Practices

Chapter 36. Pumped and Sprayed Concrete and Mortar 36.3
Chapter 37. Vacuum Processing of Concrete 37.1

Chapter 38. Preplaced-Aggregate Concrete 38.1
Chapter 39. Portland Cement Plaster 39.1
Chapter 40. Concrete Masonry 40.1

Part 11 Precast and Prestressed Concrete

Chapter 41. Prestressed Concrete 41.3
Chapter 42. Precast Concrete 42.1

Part 12 Architectural Concrete

Chapter 43. Materials for Architectural Concrete 43.3
Chapter 44. Construction of Architectural Concrete 44.1

Part 13 Repair and Grouting of Concrete

Chapter 45. Repair of Concrete 45.3
Chapter 46. Grouting of Concrete 46.1

Index I.1

CONTRIBUTORS

Robert F. Adams *Construction Concrete Engineer, Sacramento, Calif.* (CHAP. 24, 25, 26)

Joseph F. Artuso *Assistant to the Vice President, Pittsburgh Testing Laboratory, Pittsburgh, Pa.* (CHAP. 11, 12, 13)

Joseph A. Dobrowolski *Civil Engineer-Concrete Consultant, Altadena, Calif.* (CHAP. 27, 28, 35–39, 41–45)

Anthony L. Felder *Assistant Technical Director, Concrete Reinforcing Steel Institute, Shaumburg, Ill.* (CHAP. 3)

Jerome H. Ford *Symons Corp., Des Plaines, Ill.* (CHAP. 14, 15)

David P. Gustafson *Technical Director, Concrete Reinforcing Steel Institute, Schaumburg, Ill.* (CHAP. 3)

John K. Hunt *President, Hunt Control Systems Inc., Urbana, Ill.* (CHAP. 19–23)

Oswin Keifer, Jr. *U.S. Army, CDE-NoPacDiv, Portland, Oreg.* (CHAP. 30)

F.E. Legg, Jr. *Associate Professor of Construction Materials, University of Michigan, Ann Arbor, Mich.* (CHAP. 2)

Robert E. Philleo *Chief, Concrete Branch, Office of Chief of Engineers, Washington, D.C.* (CHAP. 46)

Charles E. Proudley *Deceased* (CHAP. 29)

John M. Scanlon *Senior Consultant, Wiss, Janney, Elstner Assoc. Inc., Northbrook, Ill.* (CHAP. 31, 32)

Raymond J. Schutz *Materials Consultant, Waupun, Wis.* (CHAP. 4)

Dean E. Stephan, Jr. *Vice President, Pankow Co. Ltd., Altadena, Calif.* (CHAP. 33, 34)

Robert W. Strehlow *Professional Engineer, New Berlin, Wis.* (CHAP. 16–23)

Claude B. Trusty, Jr. *CBR Cement Corp., San Mateo, Calif.* (CHAP. 1)

Joseph J. Waddell *Consulting Engineer, Construction Materials and Methods, Los Osos, Calif.* (CHAP. 35–42)

George W. Washa *Professor of Engineering Mechanics, University of Wisconsin, Madison, Wis.* (CHAP. 5–10)

PREFACE TO THE THIRD EDITION

Since publication of the second edition, economics, environmental requirements, and new technologies have driven concrete suppliers and constructors to develop new materials and new methods. This third edition was written to provide information about the new developments, the continuing practices, and a glimpse of the future.

The last decade has seen tremendous changes in concrete construction due to environmental requirements. They, in turn, created transient problems from admixtures required to be chemically revised to an increased use of flyash, and to the introduction of silica fume from power generating plants. High-range water reducing (superplasticizing) admixtures, the use of gang forming, new types of form liners, steel reinforcement congestion, high-strength concrete, and the onsite precasting of architectural and structural units are revolutionizing concrete constructions.

This third edition of the handbook will be a most useful tool for the engineer and constructor interested in cement and concrete. Extensive revisions have been made in the sections on materials and cement handling, building construction systems, pumping, repairs, formwork, and joints in concrete. The new chapters added include roller compacted concrete and materials, and construction of architectural concrete. All standards and specifications have been updated from current editions.

The Coeditors wish to thank all of the new and the past authors who added their expertise to this edition. Appreciation is given to those sources of technical information such as the American Concrete Institute, the American Society for Testing and Materials, the Precast/Prestressed Institute, the National Ready Mix Concrete Association, the Portland Cement Association, and the National Masonry Association. It is due to their efforts that the concrete industry remains technically vital.

Joseph J. Waddell, P.E.
Joseph A. Dobrowolski, P.E.

CONCRETE CONSTRUCTION HANDBOOK

PART • 1

MATERIALS FOR CONCRETE

CHAPTER 1
CEMENT

Claude B. Trusty, Jr.

Portland cement is a hydraulic cement: that is, when mixed in the proper proportions with water, it will harden under water (as well as in air). Its name is derived from the similarity in appearance, in the hardened state, of Portland cement to the Portland Stone of England. This quarried stone was used in much of the building construction during the early nineteenth century, when the first cement of this type was made. In addition, portland-pozzolan and portland-slag blended cements are produced by blending or intergrinding fly ash, pozzolans, or slag with portland cement clinker.

PRODUCTION

1.1 Manufacture

Raw Materials. The basic ingredients for portland cement consist of (1) lime-rich materials, such as limestone, seashells, marl, and chalk that provide the calcareous components; (2) clay, shale, slate, fly ash, or sand to provide the silica and alumina; and (3) iron ore, mill scale, or similar material to provide the iron or ferriferous component. The number of raw materials required at any one plant depends on the chemical composition of these materials and the types of cement being produced. A typical blend of raw materials might be proportioned so that the clinker, after burning, would consist of approximately 65% CaO, 21% SiO_2, 5% Al_2O_3, and 3% Fe_2O_3, with lesser amounts of the other minor constituents. To accomplish the proper uniform blend, the raw materials are continually sampled and analyzed, and adjustments are made to the proportions while they are being blended.

Processing. After excavation (Fig. 1.3), the first processing operation is crushing (Figs. 1.1 and 1.2). Quarry stone is passed through the primary crusher, usually a gyratory crusher. It accepts rock fragments as large as 5 ft in diameter, crushing them to a maximum size of 6 in. The crusher product then passes to the secondary crusher (a higher-speed gyratory or hammer mill), which reduces it to about ½ in in size. Soft materials bypass this step, but may use a jaw or roll mill if necessary for size reduction. At this point, blending of the ingredients is accomplished usually in circular or linear storage piles. Samples are taken and analyzed usually by an automatic sampler and transfer sys-

STONE IS FIRST REDUCED TO 5-IN. SIZE, THEN 3/4-IN., AND STORED

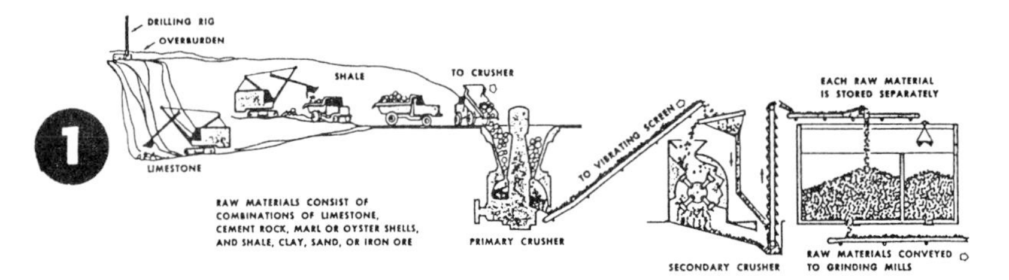

RAW MATERIALS ARE GROUND TO POWDER AND BLENDED

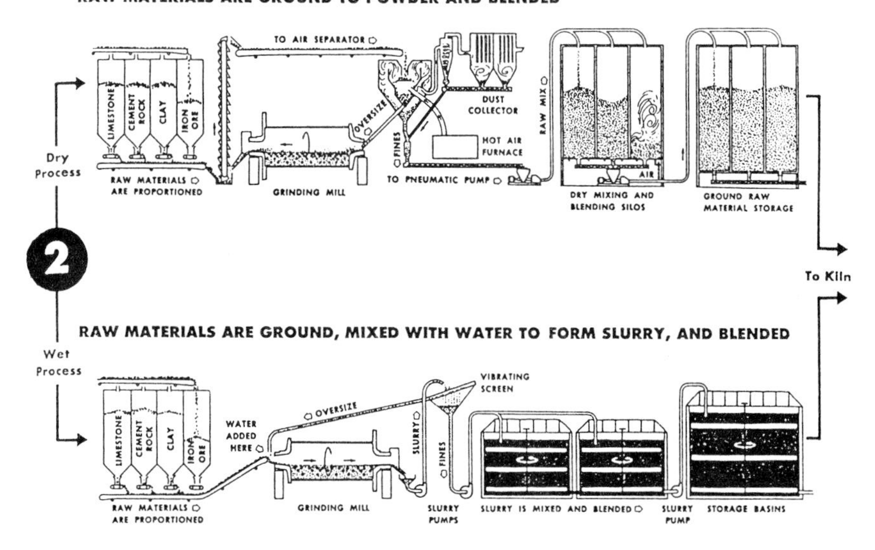

BURNING CHANGES RAW MIX CHEMICALLY INTO CEMENT CLINKER

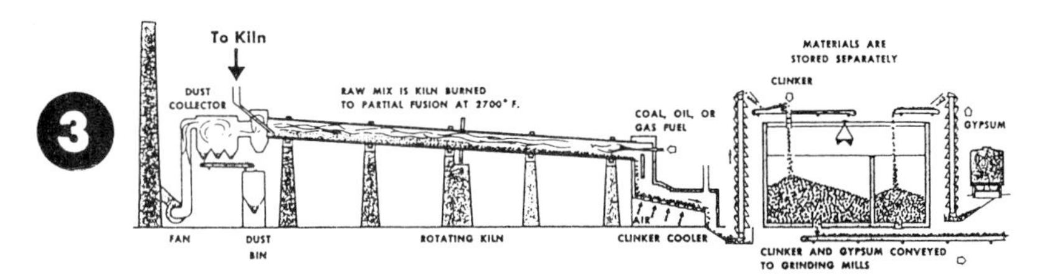

CLINKER WITH GYPSUM ADDED IS GROUND INTO PORTLAND CEMENT AND SHIPPED

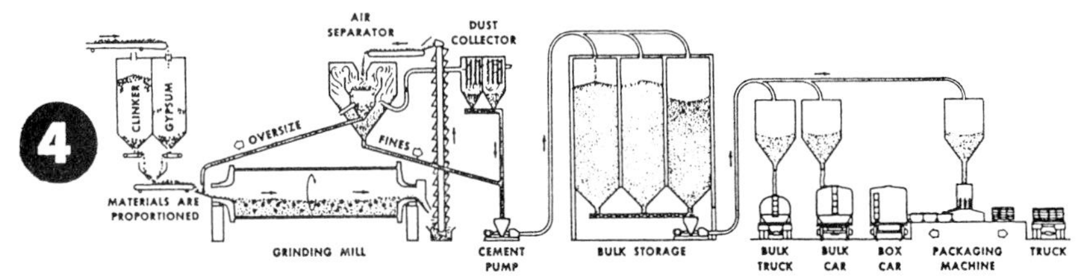

FIGURE 1.1 Schematic flowchart of cement production. (*Courtesy of Portland Cement Assoc.*)

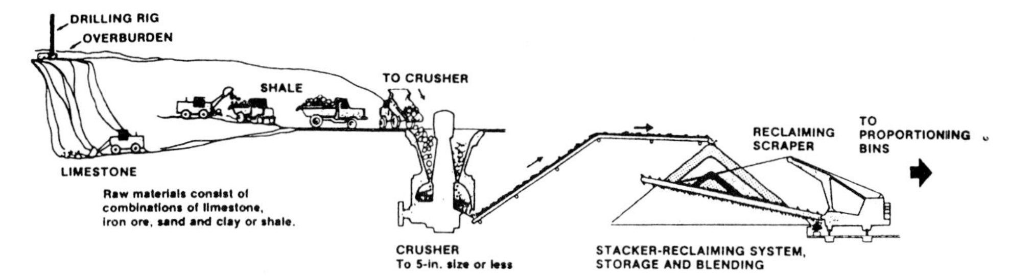

1 Quarrying and blending of raw materials.

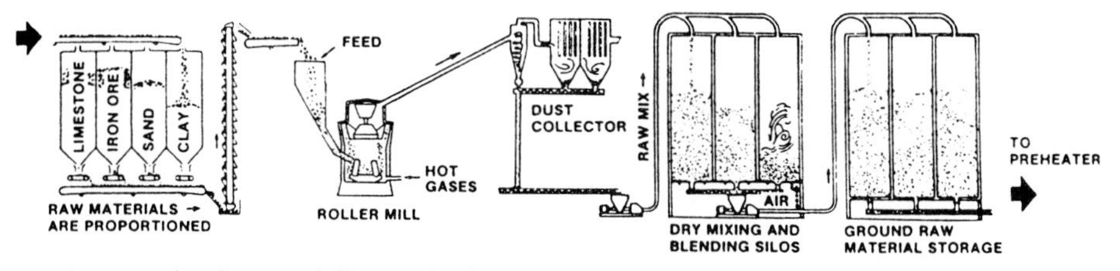

2 Proportioning and fine grinding of raw materials.

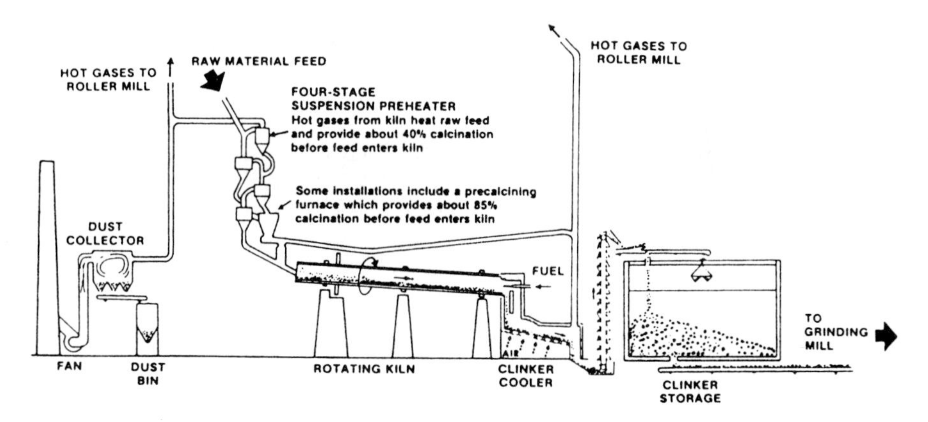

3 Kiln system. Preheating, burning, cooling and clinker storage.

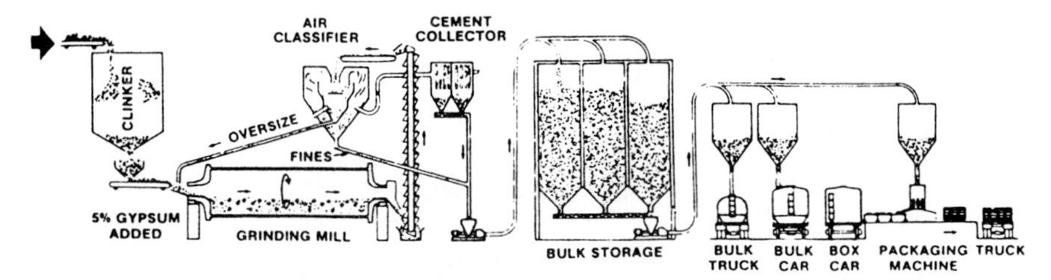

4 Finish grinding and shipping.

FIGURE 1.2 Steps in the manufacture of cement by the dry process. (*Courtesy of Portland Cement Assoc.*)

FIGURE 1.3 Blasting operation in limestone quarry. (*Courtesy of CBR Cement Corp.*)

tem to the analytical equipment, or by in-line or off-line analyzers which make continuous analyses. The data is usually fed automatically into a digital computer controlling the composition of the stored and blended raw feed. A stacker is generally used for blending the material in storage piles from which it is removed by a reclaiming machine. A few plants use belt conveyers and/or cranes for blending and reclaiming.

At this point in the process, there is a divergence in the methods, depending on whether a preheater-precalciner or a dry or wet process is being used. About 45 percent of North America's production is divided about equally between wet and dry processes, while 55 percent uses the preheater-precalciner process. Figure 1.4 is an aerial view of a preheater-precalciner process plant.

Dry Processes

Preheater-Precalciner Process. This process is similar to the long dry process described in the following paragraph (see Fig. 1.4). A vertical roller mill which may be preceded by a roll press and deagglomerator is used to dry and reduce the material to 80 to 90 percent passing the 200-mesh screen. The roll press and vertical roller mill involve the most efficient use of electricity. Raw materials are fed to the table to be crushed between the rolls and the table. Hot gases from the kiln enter the mill around the table. As the material is forced into the hot-gas stream it is dried and carried upward to separators which return the coarse material to the mill. Newer installations use dynamic separators to increase mill efficiency. A few plants use modified ball or tube mills when harder raw materials would result in excessive wear problems with the use of a roller mill.

Long Dry Process. Materials preparation in this dry process is accomplished first by transferring material to mill feed bins. Material that has not been preblended may be blended on a belt by proportional weigh feeders before transfer. Drying may be done if

FIGURE 1.4 Overall view of cement plant using the preheater-calciner method of cement manufacture. (*Courtesy of CBR Cement Corp.*)

necessary in rotary-drum dryers similiar to short horizontal kilns, where the material receives only enough heat to drive off free moisture. From the storage bins the material is fed to rotating ball mills operating in closed circuit with an air separator, where the raw meal is reduced to 80 to 90 percent passing the 200-mesh screen. If the moisture content of the material is low enough, the material may be dried by hot gases from the kiln circuit, or with an auxiliary furnace. Dry-process mills may be tube mills, ball mills, or roller mill or roller press mills. Sometimes the feed goes first through a ball mill, then through a tube mill for final grinding. A ball mill is a horizontal cylinder rotating at 15 to 20 r/min (revolutions per minute) containing a charge of steel balls 4 to 6 in in diameter. As the cylinder revolves, the steel balls continuously cascade with the feed material, pulverizing the latter. A tube mill is similar to a ball mill, except that it has a smaller diameter, contains smaller balls (1 to 2 in in diameter), and does finer grinding. A compartment mill consists of two or three compartments separated by perforated-steel-plate bulkheads. Each compartment has different-sized balls, graduated from large balls in the feed end to small ones in the last compartment. In this manner the mill accomplishes the same grinding as a ball-and-tube mill combination circuit.

The ground material is carried to the separator, where the fines are separated and transferred to blending or storage, while the coarse material is returned to the mill for further grinding.

Wet Process. In the wet process, the raw feed is transferred from the raw storage piles to the grinding mills, which are essentially the same as the ball, tube, or compartment mills used for dry grinding. Water is introduced into the mill along with the feed. Mineral processing additives such as lignins and soda ash are added to lower viscosity and increase solids content. The mills are usually operated in closed circuit with classifying equipment such as hydroseparators, screws, or rake classifiers that separate the fines from the coarse material. The fine slurry is then pumped to the thickeners, which may remove much of the water, and then into blending and storage tanks. The coarse material is returned to the mill for additional grinding. The storage tanks are kept agitated to prevent settlement or segregation of the slurry before it is fed to the kilns.

Burning: Preheater/Precalciner. This process is illustrated in Figs. 1.5 through 1.7. Dry kiln feed is withdrawn from storage and transferred to the top of the preheater tower where it is fed into the first-stage vessel. A preheater kiln consists of a series of one to six stages in series (each stage has one or more vessels) in which raw feed meal spirals downward through a swirling hot-gas stream rising from the kiln or other stage. Very efficient heat transfer occurs, enabling clinker to be produced at 2.7 million Btu (British thermal units) per short ton. The preheater with precalciner has an additional vessel in the tower where up to 60 percent of the total fuel is added. Here most of the CO_2 is removed from the limestone to form quicklime, a necessary step in the reaction process. The preheater may have either an air-through or air-separate precalciner. An air-through does not have a separate vessel and is generally limited to 20 percent of total fuel use. An air-separate has a tertiary air duct and can normally burn up to 60 percent of total fuel use, thus permitting a smaller horizontal kiln diameter. The horizontal rotary kiln section is much smaller than in the dry and wet processes, typically running 200 to 275 ft long and 12 to 18 ft in diameter. This is due to greater thermal efficiency and the addition of most of the fuel in the back end, with the precalciner. The drive for greater fuel efficiencies, higher product quality, and lower emissions from kiln gases has resulted in sophisticated process control systems using fuzzy-logic-type software to optimize operations. These computerized systems, using input from O_2, CO, NO_x, and temperature sensors combined with feed and fuel flow rates, make process adjustments automatically as required. This results in higher production and higher-quality product using less fuel and producing lower emissions.

In addition to natural gas, oil, and coal, most kilns are capable of using coke, rubber tires, rice hulls, wood waste, and other supplemental fuels.

At the discharge end the walnut-sized clinker drops into a rotary, planetary, or grate cooler. This cools the clinker rapidly to "freeze" the chemical reaction while at same time preheating the air entering the kiln to improve fuel efficiency.

Long Dry and Wet Processes. Dry kiln feed or slurry is drawn from storage and fed into the horizontal rotary kiln. The temperature rises in the material moving toward the heat source, and clinker lumps are formed as the iron, alumina, silica, and lime react to become new cement compounds. The material is in the kiln for 2.5 to 4 h, where it cascades and tumbles as it moves toward the exit and the heat source.

The kiln, which may be as large as 18 ft in diameter and 600 ft long, is inclined about $^1/_2$ to $^3/_4$ in per foot to facilitate the movement of raw meal and clinker through it. Burners at the discharge end burn the previously listed fuels. The kiln has a refractory lining to protect the steel shell and to provide insulation against heat loss. The rotational speed is about 1 to 1.5 r/min. Wet-process kilns require a greater length than do dry-process ones of similar capacity and usually have more elaborate heat exchangers (chain garlands) to increase the amount of slurry exposed to combustion gases. The most efficient dry kilns use 4.5 million Btu per ton of clinker, while efficient wet kilns use almost 6 million Btu.

Control Systems. Most installations now have centralized process control in which a computer keeps a continuous record of a kiln instrumentation data. Software programs facilitate operational changes to provide optimal operating efficiency at all times. Closed-circuit color television enables the operator to keep the kiln firing operation constantly in

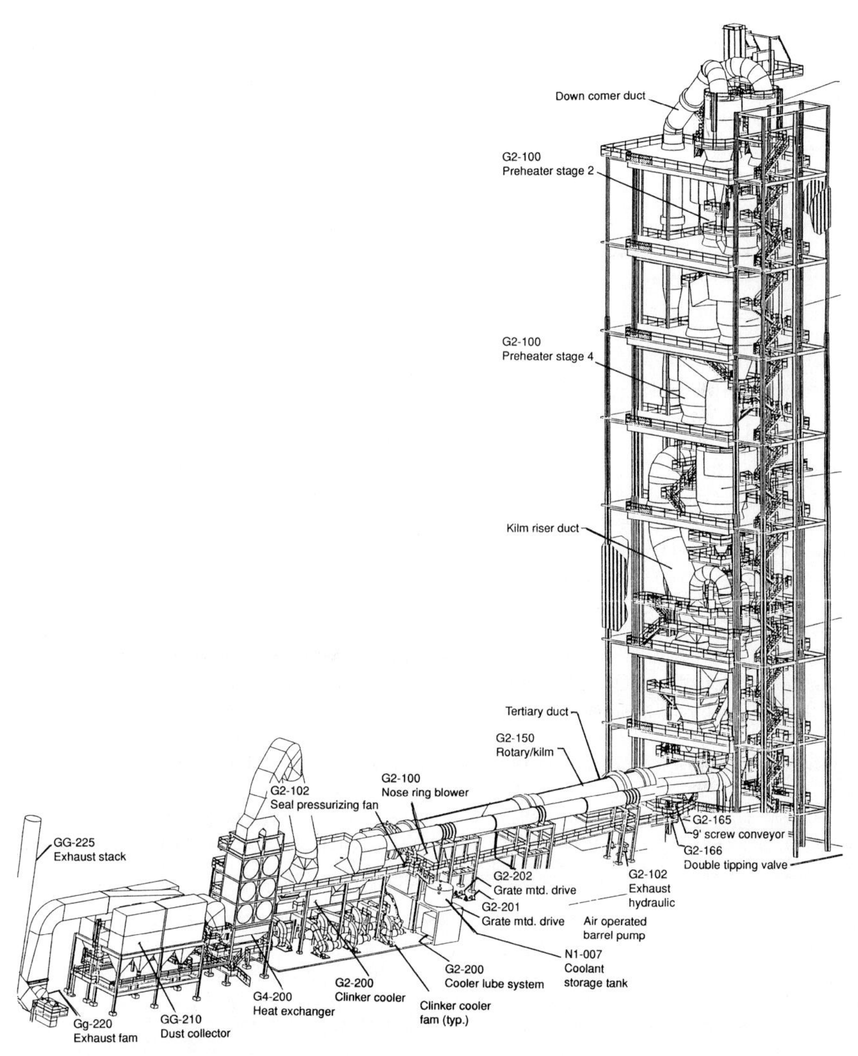

FIGURE 1.5 Plan preheater-calciner plant.

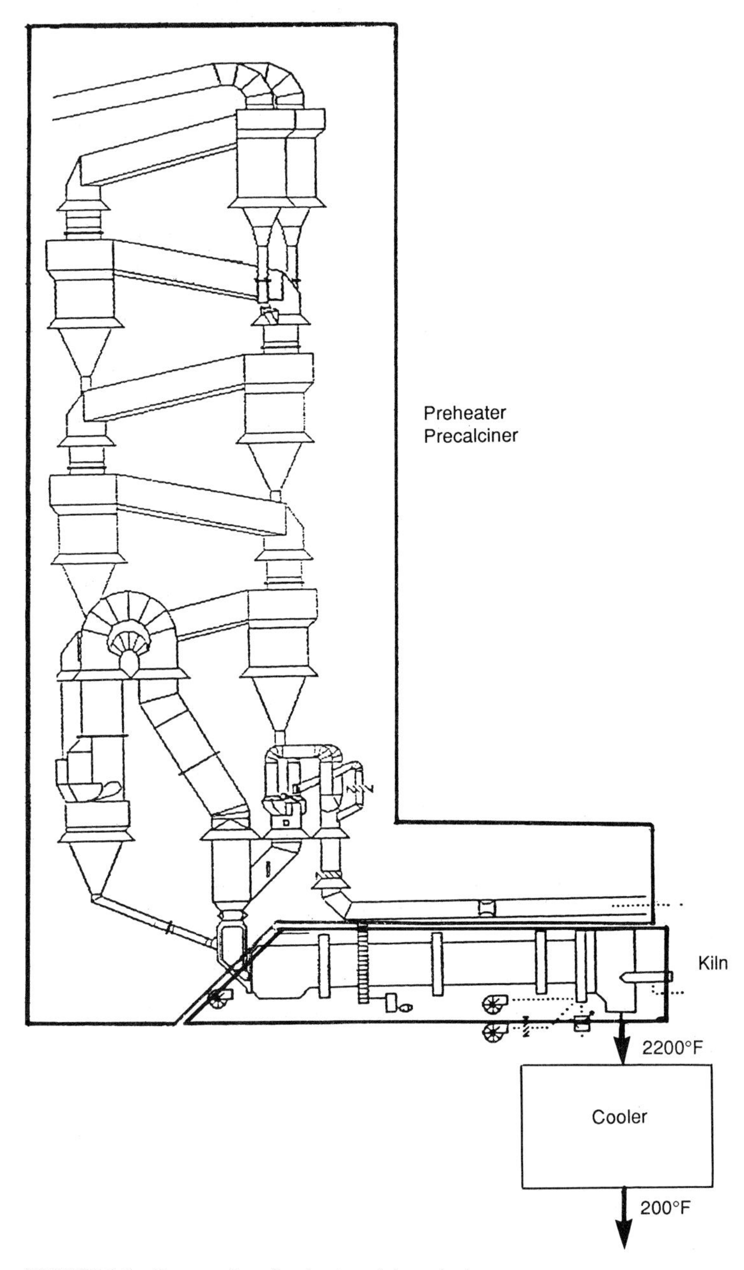

FIGURE 1.6 Cross section of preheater-calciner plant.

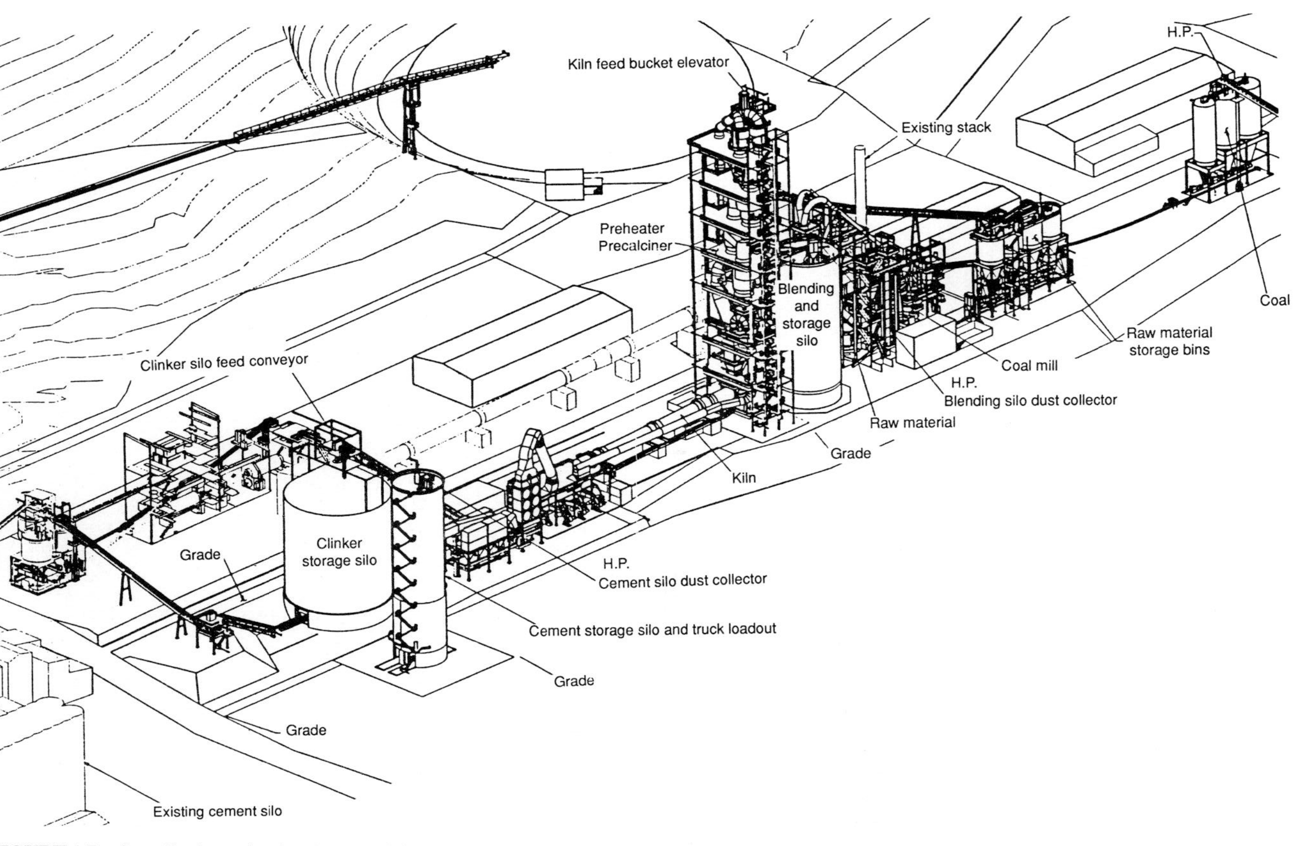

FIGURE 1.7 Overall schematic of preheater-calciner cement plant.

view. The most modern centralized process control systems have integrated data acquisition and supervision using four to six color cathode-ray tubes (CRT's) with multiple process screens and statistical graphics (see Fig. 1.8).

Finish Grinding. Because of environmental concerns and quality requirements, clinker is stored inside in halls or silos before finish grinding. From clinker storage clinker is transferred to finish-grinding bins. Clinker is interground with 5 to 7% gypsum to control setting time and improve strength and volume change characteristics. Additions such as grinding aid–pack set inhibitor are interground to improve flowability characteristics. The finish-grinding mills are generally one- to two-compartment ball mills in closed circuit with a dynamic air separator. A number of plants have recently installed high-efficiency separators which minimize the coarse and superfine fractions and increase mill out-put, while improving electrical efficiency. Several plants have also installed roller presses ahead of the finish mills to improve mill capacity and efficiency. The cement is transferred from the finish mill to storage silos or shipping silos. Shipments are predominantly in bulk using specially designed units for truck and rail transport. Packaged product is also available in 94 lb (U.S.) and 45 kg (99 lb) (Mexico, Canada) kraft paper bags (Figs. 1.9 and 1.10).

COMPOSITION OF CEMENT

1.2 Chemical Composition

Chemical Analysis. Portland cement cannot be shown by a chemical formula as it is a complex mixture of four major compounds, as shown in Table 1.1, and a number of minor compounds. There are four compounds, known as "Bogue compounds," computed from the oxide analysis of portland cement, which for all practical purposes represent the cement according to ASTM (American Society for Testing and Materials) C150. Minor

FIGURE 1.8 Control room of a modern cement plant. *(Courtesy of CBR Cement Corp.)*

FIGURE 1.9 Sacked cement being shipped from plant. (*Courtesy of·CBR Cement Corp.*)

FIGURE 1.10 Typical storage silos and view of preheater. (*Courtesy of CBR Cement Corp.*)

TABLE 1.1 Analysis of a Typical Type II Cement

Chemical analysis, %		Physical properties	
SiO_2	21.0	Fineness	
Al_2O_3	4.6	Blaine, m^2/kg	380
Fe_2O_3	3.2	Minus 325-mesh, %	92.5
CaO	64.5	Compressive strength, psi	
MgO	2.0	1-day cube	1750
SO_3	2.9	3-day cube	3200
Ignition loss	1.5	7-day cube	4200
Insoluble residue	0.3	28-day cube	5650
Free CaO	1.1	Setting time, h:min	
Na_2O	0.3	Initial	1:40
K_2O	0.24	Final	3:15
Total alkalies	0.46	Air content, %	8.0
Compound composition, %		Autoclave expansion, %	0.03
C_3S	59.2	Specific gravity	3.15
C_2S	15.5		
C_3A	6.8		
C_4AF	9.7		

constituents play their part as described below. The main compounds and their formulas and symbol abbreviations are as follows:

Tricalcium silicate	$3CaO \cdot SiO_2 = C_3S$
Dicalcium silicate	$2CaO \cdot SiO_2 = C_2S$
Tricalcium aluminate	$3CaO \cdot Al_2O_3 = C_3A$
Tetracalcium aluminoferrite	$4CaO \cdot Al_2O_3 \cdot Fe_2O_3 = C_4AF$

These compounds or "phases," as they are called, are not true compounds in the chemical sense; however, the computed proportions of these compounds provide valuable information in predicting the properties of the cement. The strength-developing characteristics of the cement depend primarily on C_3S and C_2S, which constitute about 75% of the cement. The C_3S hardens rapidly and has a major influence on setting time and early strength; therefore, a high proportion of C_3S results in high early strength and a higher heat of hydration. On the other hand, C_2S hydrates more slowly and contributes to strength gain after 7 days; C_3A contributes to early strength and high heat of hydration, but results in undesirable properties of the concrete, such as poor sulfate resistance and increased volume change. The use of more iron oxide in the kiln feed helps lower the C_3A, but leads to the formation of C_4AF, a product which acts like a filler with little or no strength. However, it is necessary as a flux to lower the clinkering temperature.

Compound composition is determined from the chemical analysis according to ASTM C150 as follows:

$$\text{Tricalcium silicate} = (4.071 \times CaO) - (7.600 \times SiO_2) - (6.718 \times \%\ Al_2O_3) - (1.430 \times \%\ Fe_2O_3) - (2.852 \times \%\ SO_3)$$

$$\text{Dicalcium silicate} = (2.867 \times \%\ SiO_2) - (0.7544 \times \%\ C_3S)$$

Tricalcium aluminate = $(2.65 \times \% \, Al_2O_3) - (1.692 \times \% \, Fe_2O_3)$

Tetracalcium aluminoferrite = $3.043 \times \% \, Fe_2O_3$

When the alumina/ferric oxide ratio is less than 0.64, a calcium aluminoferrite solid solution [expressed as ss ($C_4AF + C_2F$)] is formed. Contents of this solid solution and of tricalcium silicate are calculated by the following formulas:

$ss(C_4AF + C_2F) = (2.1 \times \% \, Al_2O_3) + (1.702 \times \% \, Fe_2O_3)$

Tricalcium silicate = $(4.071 \times \% \, CaO) - (7.6 \times \% \, SiO_2) - (4.479 \times \% \, Al_2O_3) - (2.859 \times \% \, Fe_2O_3) - (2.852 \times \% \, SO_3)$

No tricalcium aluminate will be present in the cements of this composition. Dicalcium silicate is calculated as shown previously.

Some producers use a "net" C_3S which is calculated by subtracting 4.07 times the percentage of free CaO from the "gross" C_3S as calculated above.

A small amount of gypsum, up to roughly 7% $CaSO_4 \cdot 2H_2O$, is added to the clinker while it is being ground, to control the setting time of the cement by its action in retarding the hydration of tricalcium aluminate. The optimum SO_3 content in concrete is higher by 0.5 to 1.0% than ASTM C563 optimum SO_3 using Ottawa sand. Further, some of the SO_3 present in the cement is not soluble. For this reason present practice is to control the SO_3 maximum for improved cement performance in concrete using ASTM C1038. Excessive amounts of free lime, CaO, can result from underburning of the clinker in the kiln. This may cause expansion and disintegration of the concrete. Free lime should not exceed 2%. Conversely, very low amounts of free lime reduce fuel efficiency, and produce a hard-grinding clinker which reacts slower. Magnesium oxide is limited by specifications to 6%, since it results in unsound expansion in concrete due to delayed hydration, especially in a moist environment.

Prehydration and carbonation of the cement result from moisture absorption and exposure to the air, as measured by ignition loss, should not exceed 3 percent. The alkalies, Na_2O and K_2O, are important minor constituents, as they can cause expansive deterioration when reactive types of siliceous aggregates are used for the concrete. Low-alkali cement is specified in areas where these aggregates are present. Low-alkali cement contains not more than 0.60% total alkalies, computed as percentage Na_2O plus 0.658 times the percentage of K_2O. However, the total alkali percent in the concrete must be controlled, since alkali may enter the concrete mix from ingredients other than cement such as water, aggregates, and admixtures.

False set or premature stiffening is manifested by a rapid stiffening of concrete shortly after mixing. If it results from dehydration of gypsum during the grinding process because of excessive temperature in the finish-grinding mill, it will usually disappear with additional mixing. If it results from cement–admixture interaction, both additional water and mixing may be required to mitigate the problem.

Heat of Hydration. The reaction of cement with water is exothermic; that is, heat is generated in the reaction. The average is about 120 cal/g during complete hydration of the cement. In normal construction, structural members have relatively high surface/volume ratios such that dissipation of the heat generated is not a problem. By insulation of the forms, this heat can be used to advantage during cold weather to maintain proper curing temperatures. However, for dams and other mass concrete structures, measures must be taken to reduce or remove heat by proper design and construction methods. This may involve circulating cold water in embedded pipe coils or other cooling means. Another method in controlling heat evolution is to reduce the percentage of compounds with high heat of hydration, such as C_3A and C_3S, and use a coarser cement fineness to produce a Type IV cement. Since Type IV is generally no longer available in most locations, Type II and pozzolans or slag are being used as a substitute. The use of large (≤ 6 in) aggregate also helps reduce the cement requirement and consequent heat, by reducing the water demand hence less cement at the same water/cement ratio.

The literature Ref. 1 gives the following values (in calories per gram) for the total quantity of heat evolved during complete hydration of the cement:

Tricalcium silicate	120
Dicalcium silicate	62
Tricalcium aluminate	207
Tetracalcium aluminoferrite	100

If we consider the amount of heat generated during the first 7 days of hydration for Type I cement to be 100%, then

Type II, moderately sulfate-resistant	85–94%
Type II, moderate heat of hydration	75–85%
Type III, high early strength	≤150%
Type IV, low heat of hydration	40–60%
Type V, sulfate-resistant	60–90%

PHYSICAL PROPERTIES OF CEMENT

Fineness. Cement particles, because of their small size, cannot be characterized by screen sieves: thus other methods of measuring particle size are necessary. The most commonly used method designates specific surface with particles considered as spheres. (Actually, the particles resemble crushed rock under the microscope.) The surface area is designated in square meters per kilogram of cement.

The Blaine air-permeability method is the present ASTM standard. This method depends on the rate of airflow through a prepared bed of cement in the apparatus cell. The rate of flow is a function of the size and number of pores, which is a function of particle size. Several automatic analyzers measure particle size distribution from 0.1 to 90 μm. Present results between the methods are not comparable. However, they are suitable for research work.

Fineness affects the strength-gain properties, especially up to 7 days' age. For this reason, Type III cement is ground finer than other types. While ASTM specifications designate a minimum fineness, most cements exceed this minimum by 20 to 40 percent. Some customers specify an upper limit on fineness in an effort to minimize drying shrinkage of concrete.

Strength. Present specifications use 2-in mortar cubes at a constant 0.485 water/cement ratio for compressive-strength tests. Mortar for the tests consists of 1 part cement and 2.75 parts graded standard sand mixed with water. Comparable strength development for the five types of cement are as follows, where Type I strength is assumed to be 100 percent at each age listed:

The rate of cement hardening depends on the chemical and physical properties of the cement and the curing conditions, such as temperature and moisture during curing. The

Relative Compressive-Strength Values, %

Cement Type	Age of Specimens, days 3	7	28	90
I	100	100	100	100
II	85	89	96	100
III	195	120	110	100
IV	—	36	62	100
V	67	79	85	100

Source: *Portland Cement Assoc. Bull.* EB 1.12.

water/cement ratio affects the value of the ultimate strength, based on the effect of the water on the paste porosity. A high water/cement ratio produces a paste of high porosity and low strength.

Soundness. Excess free lime or magnesia in cement results in ultimate expansion and disintegration of concrete made with that cement. Soundness is determined by testing 1 × 1 × 10-in neat cement specimens in an autoclave at a steam pressure of 295 psi (pounds per square inch) for 3 hr. ASTM specifications limit the increase in length of the specimen to 0.80 percent. Present specifications limit the amount of magnesia to 6.0%.

Stickiness. Occasionally bulk users encounter reluctance of the cement to flow, or poor-flowing cement. This stickiness, or "pack set," has no effect on the concrete-making properties of cement. Usually the problem is moisture, improperly designed handling facilities, or allowing the cement to set too long without moving it.

Pack set may occur in railroad cars or bottom dump trucks where vibration in transit has removed most of the air surrounding cement particles. A similar situation can occur in storage silos. Usually the use of air jets will fluff the cement enough to allow it to flow. Sometimes it may be necessary to increase the air pressure by 4 to 8 psi. The use of cement grinding aids has significantly reduced flow problems. Modern aeration systems with water-vapor collectors, proper bin vibrators, and correctly designed bins and silos experience little, if any, problems. Round bins with conical bottoms are much better than square or rectangular bins with pyramidal bottoms.

"Warehouse set" can occur when cement is in storage too long—generally sacks, but rarely in bulk. It results from loss of air, surface hydration of the cement by absorbed water, and carbonation. This is partially mitigated in wet climates by a plastic liner between paper layers of the bags. However, stocks should be rotated on a first-in, first-out basis. After 6 to 8 weeks in moist climate, problems can occur.

TYPES OF CEMENT

There are five types of ASTM C150 Portland Cement, eight types of ASTM C595 Blended Hydraulic Cement, three types of ASTM C91 Masonry Cement, two types of plastic cement, three types of expansive cement, and a number of special portland or blended cements for block, pipe, and other product applications. There are now available a number of rapid-setting, rapid-strength-gain cements, some of which meet C595 specifications. In addition, there are high-alumina and magnesite or Sorel cements. Some cements

Typical Compound Composition of Portland Cements

Cement Type	Compound						Ignition loss	Free CaO
	C_3S	C_2S	C_3A	C_4AF	MgO	SO_3		
I	55	19	10	7	2.8	2.9	1.0	1.0
II	51	24	6	11	2.9	2.5	0.8	1.0
III	57	19	10	7	3.0	3.1	1.0	1.6
IV	28	49	4	12	1.8	1.9	0.9	0.8
V	38	43	4	9	1.9	1.8	0.9	0.8

Source: *PCA Bull.* IS 004.09T

are available in limited local markets, but Types I, II, and III are normally readily available in all markets.

1. Many Types I and II cements meet the requirements of both specifications. Some Type II cements meet Type V performance specifications.
2. These are typical analyses for relative value comparison only. Actual values from any producer may vary somewhat from these.

Standard portland cements are listed in the following paragraphs.

Type I, ASTM C150 Common Cement. This type is generally supplied unless another type is specified. It frequently meets Type II specifications.

Type II, ASTM C150 Modified; Moderate Sulfate Resistance. Moderate heat of hydration may be specified. The maximum content of C_3A is 8%. This type of cement is used in concrete exposed to seawater.

Type III, ASTM C150; High Early Strength. This cement is obtained by finer grinding and higher percentage of C_3A and C_3S. Concrete has 3-day compressive strength approximately equal to the 7-day compressive strength for Types I and II, and 7-day compressive strength almost equal to the 28-day compressive strength for Types I and II. However, ultimate strength is about the same or less than that of Types I and II.

Type IV, ASTM C150; Low Heat. Percentages of C_2S and C_4AF are relatively high. While those of C_3A and C_3S are low. Heat of hydration is less than that for other types, and heat develops more slowly. Strength development is much slower. This cement is used for massive concrete structures with low surface/volume ratios and is available only on special order for very large tonnages with long lead times, if at all. It requires much longer curing than other types.

Type V, ASTM C150; Sulfate-Resistant. This cement has a very low C_3A content ($\leq$5%). An alternate C452 limit for sulfate-resistant performance may be specified in lieu of those for $C_3A + C_4AF$. This type is used in concrete exposed to soil alkali sulfates, groundwater sulfates, and for seawater exposure. It is normally available from a few mills, but may require special advance order.

Types IA, IIA, and IIIA, ASTM C150 Air-Entraining. These types are similar in composition as Types I, II, and III except that an air-entraining agent is interground during manufacturing. This is a poor approach to obtaining air entrainment since agent dosage cannot be varied for other factors affecting air in concrete. These types normally are available only in the eastern United States.

Blended Cements ASTM C595. These cements consist of interground blends of clinker and fly ash, natural or calcined pozzolan, or slag within the constituent percent limits specified. They may also consist of slag lime and pozzolan lime blends. Generally, but not necessarily, these cements provide increased resistance to alkali–aggregate reaction, sulfate attack, and seawater attack, but require longer curing and tend to be less resistant to freeze-thaw and deicing salt damage. They provide lower heat liberation and may gain strength more slowly, especially at low temperatures.

Masonry ASTM C91. This cement is produced as Types N, S, and M; Type M has the highest strength for masonry mortars. This type is also used for plaster. This type usually contains a finely interground limestone and an air-entraining plasticizer.

White ASTM C150. This type meets the requirements of Type I and/or Type III. It uses low-iron and low-manganese raw materials and special quenching to produce a pure-white color.

Plastic and Gun Plastic Cement. Plastic cement is produced by intergrinding a mineral plasticizing agent with portland cement clinker that meets the requirements of ASTM Type I or II cement. The Uniform Building Code (UBC) permits plasticizing agents to be added provided they do not exceed 12% of the total volume. Plastic cement meets the requirements of C150 except for insoluble residue, air entrainment, and additions subsequent to calcination as well as UBC special requirements. Plastic cement is developed for portland cement plaster and stucco. Because of the high amount of air entrainment, it is not recommended for concrete.

Oil-Well API Spec. 10. This type consists of several classes and is designed to meet high pressure and temperature conditions encountered in oil-well grouting. This type produces a low-viscosity, slow-setting slurry that remains as liquid as possible to ease pumping pressure in deep wells. It is low in C_3A, is coarse-ground, and cannot contain a grinding aide.

Expansive Types S. K. and M. These types are used to inhibit shrinkage of concrete and minimize cracking. They have low sulfate resistance and are available only on special order.

High-Alumina Cement. This type contains calcium aluminates instead of calcium silicates. It has high early strength (24 h) and refractory properties. It can experience a 40 percent strength retrogression on drying in concrete over a 6-month period, if the concrete is not kept cool during the first 24 hours after mixing and placement.

REFERENCE

1. Lea, F.M.: The Chemistry of Cement and Concrete, 3rd ed., Chemical Publishing Co., New York 1971.

CHAPTER 2
AGGREGATES

F. E. Legg, Jr.*

Approximately three-fourths of the volume of conventional concrete is occupied by aggregates consisting of such materials as sand, gravel, crushed rock, or air-cooled blast-furnace slag. It is inevitable that a constituent occupying such a large percentage of the mass should contribute important properties to both the plastic and hardened product. Additionally, in order to develop special lightweight, thermal-insulating, or radiation-shielding characteristics, aggregates manufactured specifically to develop these properties in concrete are often employed. Characteristics of such special aggregates and concretes are given in Chapter 11.

PURPOSE OF AGGREGATES

2.1 Aggregates in Fresh, Plastic Concrete

When concrete is freshly mixed, the aggregates really are suspended in the cement-water-air bubble paste. Behavior of this suspension (i.e., the fresh, plastic concrete), for instance, ease of placement without segregation causing rock pockets or sand streaks, is importantly influenced by selection of the amount, type, and size gradation of the aggregate. Depending on the nature of the aggregates employed, a fairly precise balance between the amount of fine- and coarse-sized fractions may have to be maintained to achieve the desired mobility, plasticity, and freedom from segregation, all lumped under the general term "workability." Selection of mixture proportions is aimed to achieve optimum behavior of the fresh concrete consistent with developing desired properties of the hardened product. This selection is covered in Chap. 11.

2.2 Aggregates in Hardened Concrete

The aggregates contribute many qualities to the hardened concrete. The strength-giving binding material holding concrete together results from the chemical union of the mixing

*Revised by Joseph A. Dobrowolski.

water and cement and is, of course, the basic ingredient. This hardened cement-water-air bubble paste would, by itself, be a very unsatisfactory building material, not to speak of its high cost. Indicative of the cost is the observation that if such paste were used alone at water contents of average concrete, it would contain from about 19 to 26 cwt (hundred weight) of cement per cubic yard. The paste, subsequent to initial hardening, unless restrained by contained aggregates, undergoes an intolerable amount of shrinkage on drying. The exposed portions of such pastes dry out first, and differential shrinkage between the outside and inside portions often results in cracking. The presence of aggregates provides an enormous contact area for intimate bond between the paste and aggregate surfaces. Rigidity of the aggregates greatly restrains volume change of the whole mass. Figure 2.1 shows the amount of shrinkage of concrete relative to paste as the amount of aggregate in concrete is increased. At the usual aggregate content of about 75%, by absolute volume, shrinkage of concrete is only one-tenth that of paste. Thus, the aggregate is not only cost-conserving but is really essential.

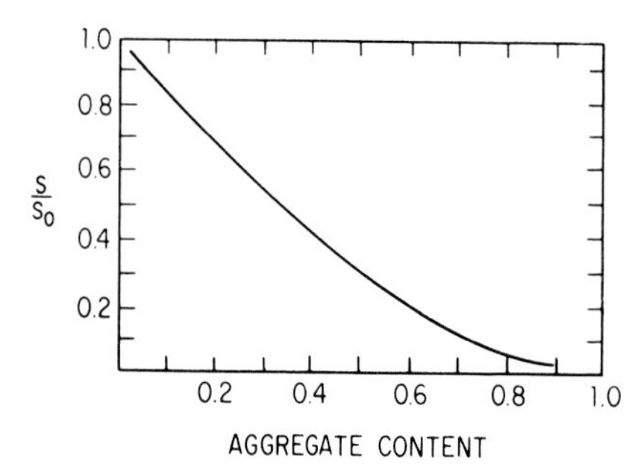

FIGURE 2.1 Ratio of shrinkage of concrete S to shrinkage of paste S_0 as a function of aggregate content.[1]

In some cases, cement grouts containing little or no aggregate are employed successfully for underground work where severe drying is not expected. Expanding agents are also used to compensate for shrinkage where drying is anticipated such as in grouting tendons in posttensioned prestressed members (see Chap. 41).

The conclusion becomes inescapable that aggregates are not simply fillers used to dilute the expensive water-cement paste and thus make a cheaper product. Economics are important, but significant improvements in the workability of the fresh concrete are contributed by proper choice of aggregates. Such choice influences highly important properties of the hardened concrete as well, such as volume stability, unit weight, resistance to destructive environment, strength, thermal properties, and pavement slipperiness. The latter factors are discussed more fully later in this chapter.

DEFINITIONS

The great variety of granular material incorporated in concrete makes formulation of an entirely satisfactory definition of "aggregate" very difficult. For instance, some finely divided pozzolans such as fly ash may simultaneously act as a supplement to the fine aggregate and also contribute cementitious properties binding the mass together. Similarly, finely divided air bubbles characteristic of those in air-entrained concrete may contribute workability to fresh concrete akin to that achieved by adding very fine sand. Extremely lightweight cellular concrete is made with the "aggregate" constituent consisting partly or entirely of air bubbles. Such special concretes and those made containing manufactured lightweight aggregates are treated in Chap. 31. The American Concrete Institute defines aggregate as "Granular material, such as sand, gravel, crushed stone, crushed hydraulic-cement concrete or iron blast-furnace slag used with a hydraulic cement medium to produce either concrete or mortar.

2.3 Coarse Aggregate

(1) Aggregate predominantly retained on the No. 4 (4.75-mm) sieve; or (2) that portion of an aggregate retained on the No. 4 (4.75-mm) sieve.*

2.4 Fine Aggregate

(1) Aggregate passing the ⅜-in (9.5-mm) sieve and almost entirely passing the No. 4 (4.75-mm) sieve and predominantly retained on the No. 200 (75-μm) sieve; or (2) that portion of an aggregate passing the No. 4 (4.75-mm) sieve and predominantly retained on the No. 200 (75-μm) sieve.*

2.5 Gravel

(1) Granular material predominantly retained on the No. 4 (4.75-mm) sieve and resulting from natural disintegration and abrasion of rock or processing of weakly bound conglomerate; or (2) that portion of an aggregate retained on the No. 4 (4.75-mm) sieve and resulting from natural disintegration and abrasion of rock or processing of weakly bound conglomerate.*

2.6 Sand

(1) Granular material passing the ⅜-in (9.5-mm) sieve and almost entirely passing the No. 4 (4.75-mm) sieve and predominantly retained on the No. 200 (75-μm) sieve, and resulting from natural disintegration and abrasion of rock or processing of completely friable sandstone; or (2) that portion of an aggregate passing the No. 4 (4.75-mm) sieve and predominantly retained on the No. 200 (75-μm) sieve, and resulting from natural disintegration and abrasion of rock or processing of completely friable sandstone.*

2.7 Crushed Stone

The product resulting from the artificial crushing of rocks, boulders, or large cobblestones, substantially all faces of which possess well-defined edges and have resulted from the crushing operation.

2.8 Air-Cooled Blast-Furnace Slag

The nonmetallic product, consisting essentially of silicates and aluminosilicates of calcium and other bases, that is developed in a molten condition simultaneously with iron in a blast furnace.

*The definitions are alternatives to be applied under differing circumstances. Definition (1) is applied to an entire aggregate either in a natural condition or after processing. Definition (2) is applied to a portion of an aggregate. Requirements for properties and grading should be stated in specifications. Fine aggregate produced by crushing rock, gravel, or slag is known as "manufactured" sand.

2.9 Crushed Gravel

The product resulting from the artificial crushing of gravel with a specified minimum percentage of fragments having one or more faces resulting from fracture.

SOURCES OF AGGREGATES

As indicated by the previous definitions, conventional concrete aggregates considered here, and relative to their source, may be divided into four major classifications:

1. Sand, gravel, and crushed gravel
2. Crushed stone
3. Air-cooled blast-furnace slag
4. Crushed hydraulic-cement concrete[8]

2.10 Sources of Sand, Gravel, and Crushed Gravel

Geologic history dictates the character and extent of aggregates derived directly from the earth of which humans can make use as a concrete ingredient. Considering first sand and gravel, both are the result of being eroded from parent rock, and they usually have then been washed and sorted over the ages by action of flowing water. Wind-deposited sand dunes and loess areas are not generally useful to the concrete technologist. The origin of cobbles and boulders from which crushed gravel or crushed stone may be made is similar, but because of either their inherent greater resistance to breakdown or their less rigorous environment, they have not been reduced to such small sizes as the sands and gravels.

Natural deposits of sand-gravel-cobble materials of interest to the construction engineer may be broadly classified into those of glacial and nonglacial origin.

FIGURE 2.2 Map of geographical area where glaciation occurred.

Glacial Origin. Much of the northern part of the United States and all of Canada was covered at one time or another with sheets of glacial ice, in some places reaching great thicknesses. Over the past million years, the climate apparently changed at least four times so as to alter the extent of these ice coverings. Prolonged, and sometimes torrential, volumes of water flowed during the recessions of these ice fronts, either on top, underneath in fissures or crevasses in the ice, or at the edges. In addition, movement of the ice created enormous grinding action on the underlying rock. The eroded material was deposited in snakelike ridges by the flowing water into formations called *eskers* or conical mounds called *kames.* At the outfall, the water spread the deposits out into *outwash plains.* Additionally, rainfall must have occurred to augment the meltwater to spread out the eroded material occasionally on *terraces* even at higher elevations. Figure 2.2 gives areas where glacial action occurred. However, granular deposits of value to the concrete engineer are not necessarily found throughout the glaciated region.

Nonglaciated Deposits. Water-deposited supplies of sand and gravel occur in many regions south of the glaciated areas. These deposits were derived from either ancient rivers no longer existing or from present streams and are then known as *river deposits.* Sandy-soil mixtures may be deposited far from the stream source. Nearer the mountains from which the material originated, the more gravelly material may be laid down as *alluvial fans, bars, terraces,* and so on.

2.11 Sources of Crushed Stone

Crushed stone is derived primarily from quarry operations. Bedrock occurs in many areas in the country either as small outcroppings, or with a sufficiently thin soil overburden that it can be economically quarried, or as mountains. "Granite rocks, used for the production of mineral aggregates, are to be found in widely scattered sections of the country. Trap rock, as well as other igneous material, is quarried extensively in the Connecticut Valley, New Jersey, several sections of the Piedmont in the East, the Columbia Plateau, and certain sections of the western ranges. Certain metamorphic rocks, such as quartzites, are found in South Dakota, Wisconsin, and Minnesota; other desirable metamorphic materials are found in the Piedmont sections in the East and in many of the mountainous regions of the country. Limestone is the most important sedimentary rock used for aggregate production. It occurs extensively throughout the Middle West, Kentucky, Tennessee, the Ozark Plateau, and widely scattered areas in the Great Plains and some small areas of the Far West."

2.12 Air-Cooled Blast-Furnace Slag

Blast-furnace slag is a by-product of blast-furnace operation and therefore occurs only in the areas where pig iron is produced. For the year 1969, approximately 65% of the slag marketed was produced in Indiana, Illinois, Michigan, Ohio, and Pennsylvania. Substantial quantities were also produced in Alabama, California, Colorado, Kentucky, Louisiana, Maryland, Minnesota, New York, Texas, Utah, and West Virginia.

Because of the increasing practice of iron-ore concentration prior to entering the blast furnace, production of blast-furnace slag has increased slowly in recent years despite greater pig-iron production. As little as 335 lb of slag has been produced per ton of pig iron by carefully selecting the furnace charge.

In its production the molten slag is dumped into cooling pits immediately after withdrawing from the blast furnace. Stratification of the layers in the pit results from pouring hot molten slag over partially cooled slag, causing cracking so that upon further cooling it can be readily picked up with power shovels. Water quenching is employed to aid fragmentation and to hydrate possible incomplete fused pieces of flux stone which might produce spalls or popouts after the concrete has set. After cooling (to 200°F or lower) the slag is transported to a plant for crushing and screening to sizes appropriate for concrete use.

Granulated slag ("popcorn slag") used for lightweight block or as a constituent of portland blast-furnace-slag cement is chilled much more rapidly than air-cooled slag by complete immersion in water. Expanded slag used in lightweight concrete and masonry units is chilled by aid of controlled amounts of water or steam, yielding a product of intermediate density and strength between air-cooled and granulated slag.

Most blast-furnace slags will fall within the following range of chemical analyses:

	%		%
Silica (SiO_2)	33–42	Sulfur (S)*	1–3
Alumina (Al_2O_3)	10–16	Iron oxide (FeO)	0.3–2
Lime (CaO)	36–45	Manganese oxide (MnO)	0.2–1.5
Magnesia (MgO)	3–12		

*In form of calcium sulfide or calcium sulfate.

ROCK TYPES AND CHARACTERISTICS

Each rock of interest to the concrete technologist, whether in the form of a sand grain, pebble, cobble, or crushed ledge rock fragment, generally consists of an identifiable assembly of "minerals." Minerals thus constitute the elementary building blocks from which rocks are made and each has a definite chemical formula, specific crystalline structure, and characteristic molecular structure. Thus the mineral quartz (silicon dioxide) predominates as a constituent of such rocks as sandstone and quartzite. On the other hand, the silicon dioxide may be chemically combined with aluminum, potassium, sodium, calcium, or magnesium compounds in such a manner as to furnish a series of very complicated minerals demanding the assistance of a professional to make positive identification.

2.13 Rock Classification

Table 2.1 gives a general classification of rocks relative to their origin: igneous, sedimentary, or metamorphic. The *igneous* rocks are formed from molten material; if cooling was slow, the mineral grains are relatively large and the rocks are classified as *intrusive,* and if cooling was rapid, the grains are small and the fine-grained rocks are classified as *extrusive.* Two rocks both of igneous origin, such as granite and rhyolite, may have identical mineral composition, but their names and engineering properties may be quite different depending on grain size (see Table 2.2). As implied by the name, the *sedimentary* rocks are formed from sediments deposited in water or by wind. Limestones and dolomites so frequently used as concrete aggregates are of sedimentary origin and contain a predominance of the mineral calcite (calcium carbonate) in the case of *limestone* or the mineral dolomite (54% calcium carbonate, 46% magnesium carbonate) in the case of the rock *dolomite.* As in the case of so many rock types, calcite and dolomite may be intermixed in the same rock, leading to the designation dolomitic limestone or calcitic dolomite. The sedimentary rocks containing silica, particularly shale and chert, deserve careful consideration before using as concrete aggregate (see Sec. 2.26).

Prolonged heat and/or pressure over geologic times may eventually alter one rock type into another type giving rise to the *metamorphic* class. This alteration may be beneficial, such as transforming sandstone into the harder and tougher quartzite. The *foliated* metamorphic rocks possess a more or less parallel layered structure of mineral grains such as does schist or slate; the *nonfoliated* rocks, for example, quartzite and marble, are more random in structure.

TABLE 2.1 General Classification of Rock

Class	Type	Family
Igneous	Intrusive (coarse-grained)	Granite† Syenite† Diorite† Gabbro Peridotite Pyroxenite Hornblendite
	Extrusive (fine-grained)	Obsidian Pumice Tuff Rhyolite*,† Trachyte*,† Andesite*,† Basalt† Diabase
Sedimentary	Calcareous	Limestone Dolomite
	Siliceous	Shale Sandstone Chert Conglomerate‡ Breccia‡
Metamorphic	Foliated	Gneiss Schist Amphibolite Slate
	Nonfoliated	Quartzite Marble Serpentinite

*Frequently occurs as a porphyritic rock.

†Included in general term *felsite* when constituent minerals cannot be determined quantitatively.

‡May also be composed partially or entirely of calcareous materials.

Tables 2.3 and 2.4 give some physical and engineering properties of the principal types of rocks

Additional geological and mineralogical terms can be found in ASTM C294 Standard Description Nomenclature for Constituents of Natural Mineral Aggregates.

PROSPECTING AND EXPLORATION

Exploration or prospecting for material suitable for use as concrete aggregate may take one or more of several avenues depending on the amount of material needed, location, and

TABLE 2.2 Mineral Composition of Rock

Name of rock	No. of samples tested	Essential mineral composition, %[a]											
		Quartz	Orthoclase, microcline	Plagioclase	Augite	Hornblende	Mica	Calcite	Chlorite	Kaolin	Epidote	Iron ore	Remainder
Igneous rocks													
Granite	165	30	45	(8)	—	—	6	—	—	(6)	—	—	5
Biotite granite	51	27	41	9	—	—	11	—	—	(7)	—	—	5
Hornblende granite	20	23	34	12	—	13	4	—	—	(10)	—	—	4
Augite syenite	23	(4)	52	7	8	—	4	—	(3)	(11)	(3)	(4)	4
Diorite	75	8	7	30	—	27	(4)	—	(3)	(8)	(5)	(3)	5
Gabbro	50	—	—	44	28	9	—	—	(3)	(6)	—	—	10
Rhyolite	43	32	45	(3)	—	—	(5)	—	(4)	(3)	—	(4)	4
Trachyte	6	(3)	42	—	—	6	—	(3)	(3)	(14)	(8)	(7)	5[b]
Andesite	67	—	—	48	14	3	—	—	(6)	—	(3)	(8)	6[b]
Basalt	70	—	—	36	35	—	—	—	—	—	—	(3)	5[b]
Altered basalt	196	—	—	32	31	—	—	—	(9)	(4)	—	(4)	8[b]
Diabase	29	—	—	44	46	—	—	—	—	—	—	(4)	6
Altered diabase	231	—	—	35	26	—	—	—	(15)	(9)	—	(4)	11
Sedimentary rocks													
Limestone	875	(6)	—	—	—	—	—	83[c]	—	—	—	—	3
Dolomite	331	(5)	—	—	—	—	—	11[c]	—	—	—	—	2
Sandstone	109	79	(5)	—	—	—	—	—	—	(4)	—	(9)	3
Feldspathic sandstone	191	35	26	—	—	—	—	(3)	(3)	(22)	—	(4)	7
Calcareous sandstone	53	46	(3)	—	—	—	—	42	—	—	—	(3)	6
Chert	62	93	—	—	—	—	—	—	—	—	—	—	7[d]
Metamorphic rocks													
Granite gneiss	169	34	35	(4)	—	—	20	—	—	—	—	—	7
Hornblende gneiss	18	10	16	15	(3)	45	(4)	—	—	—	—	—	7
Mica schist	59	36	14	(1)	—	—	40	—	—	—	—	—	9
Chlorite schist	23	11	—	10	—	(5)	—	—	39	—	28	(4)	3
Hornblende schist	68	10	(3)	12	—	61	—	—	—	—	(7)	—	7
Amphibolite	22	(3)	—	8	—	70	—	—	—	—	12	—	7
Slate	71	29	(4)	—	—	—	55	—	—	—	—	(5)	7
Quartzite	61	84	(3)	—	—	—	(4)	—	—	—	—	—	9
Feldspathic quartzite	22	46	27	—	24	—	(7)	—	(3)	(10)	—	—	7
Pyroxene quartzite	11	29	19	15	—	—	—	—	—	—	—	(5)	8[e]
Marble	61	(3)	—	—	—	—	—	96	—	—	—	—	1

[a]Values shown in parentheses indicate minerals other than those essential for the classification of the rock.
[b]Includes 10 to 20% rock glass.
[c]Limestone contains 8% of the mineral dolomite; the rock dolomite contains 82% of this mineral.
[d]Includes 3% opal.
[e]Includes 3% garnet.

TABLE 2.3 Average Values for Physical Properties of the Principal Types of Rocks

Type of rock	Bulk specific gravity	Absorption,* %	Loss by abrasion, % Los Angeles†
Igneous			
Granite	2.65	0.3	38
Syenite	2.74	0.4	24
Diorite	2.92	0.3	
Gabbro	2.96	0.3	18
Peridotite	3.31	0.3	
Felsite	2.66	0.8	18
Basalt	2.86	0.5	14
Diabase	2.96	0.3	18
Sedimentary			
Limestone	2.66	0.9	26
Dolomite	2.70	1.1	25
Shale	1.8–2.5		
Sandstone	2.54	1.8	38
Chert	2.50	1.6	26
Conglomerate	2.68	1.2	
Breccia	2.57	1.8	
Metamorphic			
Gneiss	2.74	0.3	45
Schist	2.85	0.4	38
Amphibolite	3.02	0.4	35
Slate	2.74	0.5	20
Quartzite	2.69	0.3	28
Marble	2.63	0.2	47
Serpentine	2.62	0.9	19

*After immersion in water at atmospheric temperature and pressure.
†ASTM C535.

knowledge already acquired or readily procurable. In heavily populated areas where demand for concrete is large, exhaustion of present supplies of aggregate may occur, requiring major and continued attention by large suppliers in efforts to keep up with demand. Zoning restrictions, acquirement of property or mineral rights, investigating suitable haul roads, and attendant economic considerations make this an extremely complex undertaking even after successful location of suitable deposits. Avoidance of legal entanglements with property owners and various public agencies may be of overriding importance to all other considerations. Figure 2.3 shows an artificial lake developed by a producer subsequent to a gravel operation; good public relations result from such activities.

Clear distinction is to be made between methods for *searching out* deposits of granular materials or ledge rock and the methods used for *evaluating the suitability* of such materials as concrete aggregates. The desired qualities are treated in Secs. 2.21 through 2.29. Devices or procedures capable of simultaneously evaluating both aspects are not yet available, and until both tasks are successfully accomplished undue optimism with respect to a potential aggregate source may be unwarranted.

TABLE 2.4 Summary of Engineering Properties of Rocks

Type of rock	Mechanical strength	Durability	Chemical stability	Surface characteristics	Presence of undesirable impurities	Crushed shape
Igeneous						
Granite, syenite, diorite	Good	Good	Good	Good	Possible	Good
Felsite	Good	Good	Questionable	Fair	Possible	Fair
Basalt, diabase, gabbro	Good	Good	Good	Good	Seldom	Fair
Peridotite	Good	Fair	Questionable	Good	Possible	Good
Sedimentary						
Limestone, dolomite	Good	Fair	Good	Good	Possible	Good
Sandstone	Fair	Fair	Good	Good	Seldom	Good
Chert	Good	Poor	Poor	Fair	Likely	Poor
Conglomerate, breccia	Fair	Fair	Good	Good	Seldom	Fair
Shale	Poor	Poor	—	Good	Possible	Fair to poor
Metamorphic						
Gneiss, schist	Good	Good	Good	Good	Seldom	Good to poor
Quartzite	Good	Good	Good	Good	Seldom	Fair
Marble	Fair	Good	Good	Good	Possible	Good
Serpentinite	Fair	Fair	Good	Fair to poor	Possible	Fair
Amphibolite	Good	Good	Good	Good	Seldom	Fair
Slate	Good	Good	Good	Poor	Seldom	Poor

2.14 Existing Geologic Maps and Surveys

Much useful data pertaining to location of potential aggregate sources may be procured from federal or state geologic or agricultural agency maps. For instance, study of surface-geology maps available from the U.S. Geological Survey, U.S. Department of Interior, Washington, D.C., or more detailed maps from state geologists may be rewarding as to location, for instance, of outwash plains and eskers where granular materials could be expected to be found. Similarly, state geology surveys are sometimes able to provide bedrock-formation maps which may give a clue as to potential quarry sources.

2.15 Aerial Surveys

Air-photography interpretation, as contrasted with precise horizontal and vertical measurements derived from photogrammetry, is truly a specialized art, particularly with ref-

FIGURE 2.3 Attractive 80-acre artificial lake, part of land-rehabilitation project after gravel operation. (*Courtesy of American Aggregates Corp.*)

erence to searching for granular materials or rock deposits. Competence derives from intensive study of the geologic and soil sciences, and in the hands of an expert a remarkable amount of information can be derived. Generally such information is interpreted from stereo pairs taken with black-and-white film but single, overlapping aerial photographs may also be employed. Color aerial photography is also being employed and appears to offer some advantages despite the greater cost of the film.

The interpreter of aerial photographs looks for air-photography pattern elements involving vegetation, drainage patterns, topography, and erosion as well as presence of the type of artificial features.

The entire United States has been covered by aerial black-and-white photographs at a scale between 1:15,000 and 1:25,000 and such photographs may be procured from agencies of the federal government at nominal cost, particularly from the Soil Conservation Service of the Department of Agriculture. The usefulness of such aerial photographs to the concrete engineer will depend entirely on the experience and proficiency of the specialist engaged to interpret the photographs.

2.16 Geophysical Exploration

Portable instruments are available which provide information, if coupled with supplementary knowledge, that may be useful to the prospector for mineral aggregates. These instruments operate on either of two principles: (1) electrical resistivity of the different underlying materials enabling distinguishing between bedrock, sand, gravel, groundwater level, or soil; and (2) seismic exploration by which the velocities of subsurface seismic or shock waves through these materials are measured or the waves may be observed to be reflected from underlying strata. Generally, dense materials such as bedrock or dry granular materials will offer greater resistance to passage of electric current employed in resistivity measurements, whereas shock waves will travel faster through dense rock or water.

Resistivity Surveys. In the resistivity method, four equally spaced brass electrodes (usually about ¾ in diameter, 20 in long) are driven in the ground and electric current is made to flow between the two outer electrodes (Fig. 2.4). Voltage drop is then measured

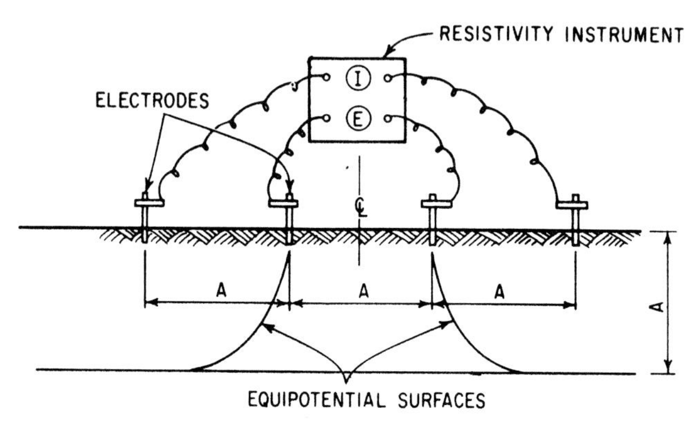

FIGURE 2.4 Schematic diagram of resistivity-survey method.

between the two inner electrodes, from which the resistivity of the underlying material is measured according to the Wenner general equation

$$\rho = 191\frac{AE}{I}$$

where ρ = resistivity of the underlying layer, Ω-cm
A = depth of layer and distance between electrodes, ft
E = voltage drop across the two inner electrodes, V (volts)
I = current flow between two outer electrodes, A (amperes)

It is noted that the depth of material investigated A is numerically equal to the distance between electrodes; and by changing the amount of electrode separation, investigation of the electrical resistivity of the material at different depths can be made. Depending on the character of the underlying material and sensitivity of the particular instrument, soundings up to depths of 50 to 100 ft are readily obtained.

Skill, experience, and familiarity with the geology of the terrain are recognized necessities for successful predictions from resistivity surveys. For instance, moisture in the ground, such as occurs in early spring, is considered an asset to distinguish between sand, gravel, and loamy sand or silt. In dry seasons, resistivity values of such materials tend to merge closer together. Typical resistivity values for different materials are as follows:

Material	Resistivity, Ω-cm
Clay and silt	0–10,000
Sand-clay mixtures	10,000–50,000
Sand	50,000–150,000
Gravel	150,000–400,000
Rock	≥ 400,000

Seismic Surveys. Seismic exploration employs devices by which the speed of travel of shock waves traveling through the subsurface material can be accurately measured. The shock waves are induced either by striking a sledge hammer on a steel plate resting on the ground or by detonating a small explosive charge lowered in an auger hole. Electrically connected to the sledge hammer or explosive charge is an electronic device capable of measuring intervals of time as small as 1/4 ms (0.00025 s). Also connected to the timing instrument is a pickup geophone which also rests on the ground a selected distance away from the impact or detonating point and responds to incoming signals transmitted through the subsurface material. The timing device measures the interval necessary for the induced shock wave to reach the geophone. By progressively moving the impact point farther away from the geophone, a time–distance graph can be plotted. Such a plot is generally neither a straight line nor a smooth curve but is observed to be a series of straight lines of diminishing slope indicating swifter pathways for the shock waves in the deeper and denser materials as distance increases between the impact point and pickup unit.

SAMPLING PROSPECTIVE AGGREGATE SOURCES

After a potential source of concrete aggregate is located, it is necessary to determine if the desired quality and sizes of aggregates are present, or can be produced, so as to qualify the operation as an economical one. Failure to give proper attention to this step is foolhardy since substantial investment may be wasted upon a venture which could have been proved unwise in the first place if the proper investigational techniques had been utilized.

The reasons for sampling an undeveloped deposit are basically twofold: (1) to ensure the prospective operators that the deposit is of sufficient magnitude and contains the required sizes of aggregate or, if not, that the sizes needed can be acquired by crushing, so as to justify investment in processing equipment; and (2) to ascertain if the deposit contains aggregates which, if properly processed, will meet the quality requirements of the major consuming agencies within the market area. Public agencies or authorities will normally have the strictest quality requirements, and it is desirable to invite such agencies to participate in the testing of preliminary samples. The next-best step is to submit samples to a testing organization whose qualifications for such work are acceptable to the public agency. Proved ability to furnish aggregate meeting stringent requirements of a respected agency in the area is a distinct business advantage.

Sampling of prospective aggregate sources is conveniently divided into the methods used for sand-gravel deposits and those used for prospective rock-quarrying operations. In either case, provision should be made for carefully identifying and preserving the samples in clean bags or tight boxes. Record should be made of both horizontal and vertical locations at which the samples were taken so that a detailed cross section can later be constructed.

2.17 Sampling Prospective Quarry Sites

Quarries which have been once abandoned or exposed rock in mountainous regions may be potential sources of concrete aggregates, in which case samples should be procured from the exposed faces.

SAMPLING PROCESSED AGGREGATE

It is highly recommended that ASTM Designation D75, Practice for Sampling Aggregates, be studied for general instructions regarding sampling. This method is equally applicable to sampling aggregates for general concrete use. Too much emphasis cannot be placed upon the necessity for careful and thoughtful sampling if the subsequent tests of the aggregate are to be meaningful.

Sampling should be entrusted only to those possessing the specialized knowledge regarding the amount of material needed for the schedule of tests to be performed and to those fully aware of the false indications which may derive from a sample not taken in such a manner as to be truly representative. It may well be that the sampler needs to exercise judgment superior to that of the personnel performing the more routine tests.

2.18 Sampling During Production

Sampling and testing aggregates during production has several important advantages: (1) the producer, and often also the consumer, are immediately informed as to the character of the production, i.e., whether it fails the specifications or is borderline or is well within the tolerance limits; (2) corrective measures can be undertaken at once by the producer before a large quantity of aggregate is produced which may otherwise have to be disposed of or reprocessed, or before unacceptable material contaminates already approved material in a stockpile or bin; (3) even more importantly, opportunity is provided to obtain periodic representative samples from the production stream in a manner which best characterizes the product.

Whenever possible, samples should be procured from the final conveyor belt or chute by passing a suitable container through the product stream so as to intercept the full cross section briefly. If it is feasible to stop the belt momentarily, an excellent method is to place across the belt two suitably spaced templates shaped to fit the belt convexity and then scrape and brush the entire belt load between the templates into a clean container.

Temptation to take samples which are too small must be resisted at all costs. They must be sufficiently ample in size so that chance inclusion or exclusion in the sample of one or two particles in the larger sizes does not make it appear that the material seriously departs from the grading limits. Coarse-aggregate samples to be examined for grading size, only, should be at least as large as the following:

Maximum particle size, in	Minimum mass of sample, lb (kg)
3/8	25 (10)
1/2	35 (15)
3/4	55 (25)
1	110 (50)
1 1/2	165 (75)
2	220 (100)
2 1/2	275 (125)
3	330 (150)
3 1/2	385 (175)

If the sample actually obtained chances to be considerably larger than prescribed above, it can be appropriately reduced in size by quartering or use of a mechanical splitter following techniques of Practice for Reducing Field Samples of Aggregate to Testing Size, ASTM C702. If tests other than grading are contemplated, the sampler should inform himself of the increased amount needed.

Freshly produced moist sand is much less subject to segregation and consequent sampling errors, and samples weighing 25 lb. are adequate for routine testing and can be obtained in practically any convenient manner such as by intercepting the flow off a belt or chute. Again, if a larger schedule of tests is contemplated, larger samples may be needed.

Opinion is divided, and statistical theory is not well developed, as to whether it is better to composite several samples taken at shorter intervals or whether it is better to take one "grab" sample less frequently to represent a certain amount of production. For instance, some agencies require one test per 100 tons of production. If the plant is producing 100 tons per hour, three samples taken at 20-min intervals could be composited and the results be compared with a single grab sample taken sometime during the same hour. Some favor the latter since an examination of, say, 10 samples over a 10-h day would give indication of the magnitude of the extremes in production whereas the compositing procedure tends to smooth out the results of each hour's production.

Whatever the frequency of sampling chosen, or whether "grab" or composite samples are elected to be taken, both producer and consumer benefit from choosing a rigorous method wherein opportunity for personal choices of the sampler is minimized. Sampling, splitting, and subsequent testing must be strictly devoid of personal whim. Complete randomness of sampling must be preserved wherein every grain of sand or coarse-aggregate particle has equal chance of being found in the selected test sample. This is the goal to be sought.

2.19 Sampling After Production

Sampling subsequent to production and at the destination has great advantage from the standpoint of best representing the aggregate actually being incorporated in the concrete. For instance, possibility of contamination, segregation of sizes, or generation of fines during handling may be suspected. Aggregates strictly conforming to specifications when produced may be subject to improper stockpiling procedures causing segregation or breakage, or may be contaminated by mud or clay scraped up or tracked onto the pile by bulldozers, or be handled so much as to generate fines. However, the obstacles to procuring thoroughly representative samples after delivery are many. Obtaining truly random samples is almost an insurmountable problem whether the aggregate be in trucks, railroad cars, or a stockpile. Practical considerations of the amount of manual labor involved preclude the probability that interior aggregate in the load or stockpile will be truly randomly sampled unless extraordinary effort is expended. It is almost mandatory that the aid of power equipment such as a front-end loader, power shovel, or clam be enlisted to aid sampling from large masses of aggregate.

Stockpiles. Coarse-aggregate stockpiles, particularly if coned or tent-shaped, are subject to segregation of sizes unless formed in relatively thin layers. The larger particles tend to roll down to the toe of the slope. Therefore, whenever possible, sampling should be accomplished either as the pile is being formed or during loading out. If neither alternative is feasible, and it is recognized that such is the case in many instances, then some idea of the content of the pile can be obtained in the following manner: Individual shovelsful of coarse aggregate should be selected from the top, middle, and bottom of the pile and composited to form a single sample. Several such composite samples may be needed to represent adequately the material in the exterior surface of the pile. It is advisable to push a board into the pile just above the point of taking each shovelful so as to prevent segregation during sampling. If a fine-aggregate pile is being sampled, the exterior dry sand should first be scraped away before the individual shovelful is taken.

Railroad Cars. Railroad cars of aggregate are manually sampled by first digging at least three equally spaced trenches across the car with the bottom of each trench at least 1 ft below the surface of the aggregate at the side of the car and the trench at least 1 ft wide at the bottom. Each trench is recommended to be sampled at three equally spaced points by pushing the shovel downward into the trench without scraping horizontally.

Trucks, etc. Trucks should be similarly sampled by trenching as prescribed above for cars, but only one or two trenches may be needed, depending on the size of the load. Barges and other conveyances may require more trenches to provide adequately representative samples.

2.20 Identification of Samples

Figure 2.5 is a reproduction of a form successfully used to identify individual test samples. Persons submitting samples should be instructed to fill out meticulously such a form, or equivalent, to provide proper identification.

AGGREGATES CHARACTERISTICS

Certain tests performed on concrete aggregates are for the purpose of establishing that minimum intrinsic *quality* requirements are fulfilled; others are more related to determin-

SAMPLE IDENTIFICATION

Project ______

Pur. Order No. ______

Date Sampled ______

Lab. No. ______ Date Rec'd. ______

Name of Material ______

Source ______ Manufacturer ______

Address ______ Address ______

Sampled from ______ Pit Name ______

Give car number and initial, if rail shipment. If sampled at pit, give 1/4 section, town line and range.

Quantity of Material Represented by Sample ______

Consigned to: ______

If sampled at source, state to whom and where material is to be shipped.

Sampled by ______
Name Title

Submitted by ______
Name Title Address

Intended Use ______

Specification ______ Sender's Identification ______

Remarks ______

Consign Sample to: (Give specific address)

FIGURE 2.5 Suggested form to be filled out to accompany test samples.

ing characteristics useful for selecting proportions for concrete, and still others may be a much abbreviated group of tests to assure routinely compliance with the job requirements. The first category includes such basic desirable qualities as toughness, soundness, and abrasion resistance, whereas specific gravity and absorption are included in the second category. Clear-cut distinction cannot always be made in a given case. For instance, the absorption value is a necessary item for the job engineer in calculating water/cement ratio of the concrete mix, but it may also, in some cases, reflect pore structure affecting freeze-thaw resistance of concrete in which the aggregate is placed. In most cases, tests applied to aggregates give an *index* to predicted behavior in concrete rather than evaluating a truly basic attribute.

2.21 Surface Texture

Satisfactory concrete is made containing aggregate of a great variety of surface characteristics ranging from very smooth to very rough and honeycombed. As a consequence, only recently have studies been initiated making rigorous examination of the matter.[4, 5, 6, 9, 10] Table 2.5 gives surface textures typical of a selected group of aggregates. Some specifications presently limit the amount of "glassy" pieces in slag coarse aggregate to a negligible amount, thus recognizing the poor bond between cement paste and such extremely smooth particles. Further discussion of surface texture is contained in Ref. 11.

2.22 Aggregate Shape

As is the case of surface texture, satisfactory concrete has been made with aggregate consisting of particles of a great variety of individual shapes. Natural aggregate particles which have been subjected to wave and water action over geologic history may be essentially spherical; others broken by crushing may be cubical or highly angular with sharp

TABLE 2.5 Surface Texture of Typical Aggregates

Group	Surface texture	Characteristics	Examples
1	Glassy	Conchoidal fracture	Black flint, vitreous slag
2	Smooth	Water-worn, or smooth due to fracture of laminated or fine-grained rock	Gravels, chert, slate, marble, some rhyolites
3	Granular	Fracture showing more or less uniform rounded grains	Sandstone, oolite
4	Rough	Rough fracture of fine- or medium-grained rock containing no easily visible crystalline constituents	Basalt, felsite, porphyry, limestone
5	Crystalline	Containing easily visible crystalline constituents	Granite, gabbro, gneiss
6	Honeycombed and porous	With visible pores and cavities	Brick, pumice, foamed slag, clinker, expanded clay

Source: Reproduced by permission of the British Standards Institution.

corners. Of interest to the concrete technologist is that such changes in shape, without compensating changes in particle size, will be influential in altering the void characteristics of the aggregate. A highly angular coarse aggregate possessing larger void content will demand a greater amount of sand to provide a workable concrete. Conversely, well-rounded coarse aggregate tending toward spherical particles will require less sand. It is interesting to note, however, that concretes made with a great disparity in particle shapes at a given cement content per cubic yard of concrete will frequently have about the same compressive strength. Efforts at placing a numerical scale on particle shape so as to be able to characterize it better other than with the words "rounded," "subangular," "angular," resulted in ASTM Test Method D3398.[12] The latter test involves compacting the aggregate into a cylindrical mold.

Particle shape and texture of fine aggregate have also been measured by an orifice flow test[13] and by other means. Research indicates particle shape and surface texture of the fine aggregate may more importantly influence concrete strength than does the coarse aggregate.[14] The sand significantly influences the water required to provide a given concrete slump.

In view of the present uncertainty of the role of aggregate-particle shape in concrete technology, few specifications for coarse aggregate prescribe special requirements except possibly to limit the amount of thin or elongated particles to a maximum of about 10 to 15% by weight, to minimize harsh concrete mixtures whose surfaces may tear during finishing operations. Such particles are defined as those whose ratio of greatest dimension of a circumscribing rectangular prism to the least dimension is greater than 5. Exposed aggregate concrete sometimes uses entirely crushed or entirely rounded particles for pleasing aesthetic effects.

Specifications sometimes inadvertently influence particle shape as the result of controlling quality of gravel by requiring crushing of oversize. Decision may be made that the desired quality occurs only in the large-sized material in a gravel deposit. In this case, confirmation is made of crushing by observing that each particle has at least one fractured face resulting from the crushing process.

Summarizing, particle shape of coarse aggregate in concrete has not proved to be an important problem if increased and content is chosen so as to compensate for aggregates tending otherwise to make harsh mixes such as can result from use of entirely crushed stone aggregate or blast-furnace slag. Information developed indicates particle shape of the fine-aggregate fraction is more consequential than heretofore thought. A standard test has been adopted for directly evaluating particle shape of either fine or coarse aggregate, ASTM Standard Test-Method D3398.

2.23 Structural Strength

High-strength concrete cannot be made containing aggregates which are structurally weak. For instance, insulating concrete containing vermiculite aggregate, which is itself a soft and friable material, rarely exceeds 750 psi compressive strength at 28 days, whereas carefully proportioned and cured concrete containing high-strength crushed limestone, crushed traprock, or quartzite gravel can be made to exceed 10,000 psi. Despite the seemingly obvious relation between concrete strength and aggregate strength, at least in extreme cases, other factors such as particle shape, surface texture, grading, and water/cement ratio of the concrete conspire against precise evaluation of the contribution of the structural strength of the aggregate itself. This is the case despite much research effort. For instance, compressive strengths of various rocks are shown in Table 2.6. The variability of compression values even for similar rocks does not give much encouragement for predicting concrete strengths. Elastic-modulus tests have likewise not been suc-

TABLE 2.6 Compression Strength of Rocks Commonly Used as Concrete Aggregates

Type of rock	No. of samples*	Compressive strength, psi		
		Average†	After deletion of extremes‡	
			Maximum	Minimum
Granite	278	26,200	37,300	16,600
Felsite	12	47,000	76,300	17,400
Trap	59	41,100	54,700	29,200
Limestone	241	23,000	34,900	13,500
Sandstone	79	19,000	34,800	6,400
Marble	34	16,900	35,400	7,400
Quartzite	26	36,500	61,300	18,000
Gneiss	36	21,300	34,100	13,600
Schist	31	24,600	43,100	13,200

*For most samples, the compressive strength is an average of 3 to 15 specimens.
†Average of all samples.
‡Of all samples tested with highest or lowest values 10 percent have been deleted as not typical of the material.
Source: Ref. 15.

cessful in making predictions. The effect of shape, texture, and modulus of elasticity on concrete strength can be found in Table 2.7.

Tests with high-performance concretes of compressive strengths of 13,900 to 15,200 psi at 56 days indicated that granite was low and strengths increased progressively in this order of the four aggregates tested: crushed granite, round and smooth siliceous gravel, crushed limestone, and crushed diabase. The modulus of elasticity in the concrete increased at the same age with aggregate from low to high granite, gravel, diabase, and limestone. The fine-grained coarse aggregates tended to be better for the high-performance concretes.[10] Further research is needed to determine whether elastic properties of aggregates must be considered for high-performance concretes.

The test most often used in the United States to assess overall structural quality of coarse aggregate is the Los Angeles abrasion test (Fig. 2.6). In this procedure a carefully graded and weighed sample of the aggregate is placed in a hollow steel revolving drum along with a charge of steel balls. A shelf inside the rotating drum picks up the charge of balls and aggregate each revolution and drops them as the shelf approaches the high point

TABLE 2.7 Proportional Contribution of Shape, Texture, and Modulus of Elasticity to Concrete Strength

	Relative effect (%) of		
	Shape	Surface Texture	Modulus of elasticity
Flexural strength	31	26	43
Compressive strength	22	44	34

Source: Ref. 6.

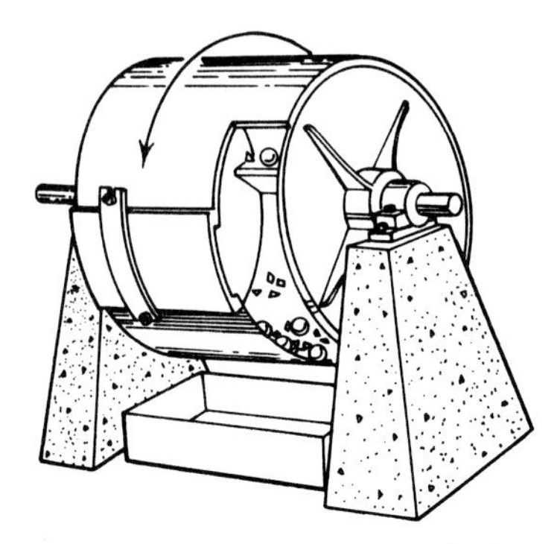

FIGURE 2.6 Los Angeles abrasion machine. *(From T.D. Larson, "Portland Cement and Asphalt Concretes," McGraw-Hill Book Company, New York, 1963, used by permission.)*

of its travel. Thus, the aggregate experiences some scrubbing and tumbling action and considerable impact during the specified 500 revolutions of the drum. The aggregate is reexamined after the expiration of the required number of drum revolutions to determine the amount broken down finer than the No. 12 sieve. Except in the case of blast-furnace slag, the test appears to give a useful index of overall structural integrity of the aggregate as evidenced by the fact that so many organizations make use of it in their specifications. Uniquely enough, similar evidence is lacking with respect to the applicability of the test for blast-furnace slag; correlation of slag abrasion with concrete strength is poor or nonexistent.

Opinion is prevalent that the flexural strength of concrete used in design of pavements is importantly influenced by the structural quality of the coarse aggregate. No single routine test, or group of tests, is presently capable of reliably predicting development of high or low flexural strengths, and there is apparently no substitute for actually making trial batches of concrete from which flexural-strength specimens are tested.

2.24 Specific Gravity and Absorption

Specific gravity expresses the weight of an aggregate particle relative to that of an equal volume of water. As an example, granite was listed in Table 2.3 as having an average specific gravity of 2.65. Thus, since water weighs 62.4 pcf, a cubic foot of solid granite weighs approximately $2.65 \times 62.4 = 165$ lb. However, all aggregates are porous to a degree, allowing entrance of water into the pore or capillary spaces when placed in the concrete mixture or, as is more usual, they are already wet when entering the concrete. Careful definition of specific gravity must therefore properly account for both the weight and volume of the portion of water contained *within* the particles (see Fig. 2.7). Free water on the exterior surfaces of wet aggregate does not enter into computation of specific gravity but does contribute to water/cement ratio of the concrete. It is also noted that specific gravity by its definition disregards void spaces *between* aggregate particles, and average granite, for instance, will have the same specific gravity regardless of whether it consists of angular material quarried from rock or of rounded particles from a gravel deposit. The same is true for absorption, and both specific gravity and absorption are therefore basic properties of the rock and are not influenced by the method of processing.

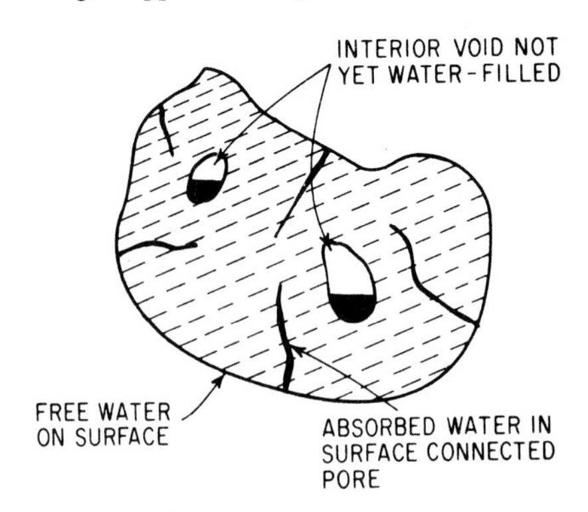

FIGURE 2.7 Moist-aggregate particle showing distribution of exterior and interior water.

Absorption of an aggregate is arbitrarily expressed in terms of the water that enters the pores or capillaries during a soaking period of 24 h and is calculated on the basis of the weight of the oven-dry aggregate as follows:

$$\text{Absorption, \%} = \frac{B - A}{A} \times 100$$

where A = weight, g of oven-dry sample in air
B = weight, g of saturated surface-dry sample in air

Average absorption values of rock types were shown in Table 2.3. It is observed that some of the softer, more porous sedimentary rocks typically have higher absorption values. Aggregates whose absorption of water has just been satisfied but whose surfaces are not visibly wet are said to be in a "saturated surface-dry" condition, frequently abbreviated "SSD."

Different agencies are not unanimous as to whether the quantities of ingredients for a concrete batch are most advantageously computed in terms of oven-dry or saturated surface-dry aggregates, which leads, in part, to two definitions of specific gravity (sp gr), i.e., "bulk specific gravity" or "bulk specific gravity (saturated surface-dry basis)." The differences can be exemplified by considering a coarse aggregate which has been soaked in water and then brought to a surface-dry condition, then weighed in air and again weighed when suspended in water:

$$\text{Bulk sp gr} = \frac{A}{B - C}$$

and

$$\text{Bulk sp gr}_{\text{SSD}} = \frac{B}{B - C}$$

where A = weight, g of oven-dry sample in air
B = weight, g of saturated surface-dry sample in air
C = weight, g of saturated sample in water

The buoyant force $B - C$ in the denominator is the same in both cases and in the metric system can be considered equivalent to the volume of the aggregate in cubic centimeters as if each particle were enclosed in an infinitely thin sheath surrounding the bulk volume including the penetrable voids. It follows that, if the absorption value of the aggregate is known, bulk specific gravity (saturated surface-dry basis) can be calculated from the bulk specific gravity (sometimes the latter is referred to as "bulk specific gravity, dry basis" to reduce confusion) by the following relation:

$$\text{Bulk sp gr}_{\text{SSD}} = \text{bulk sp gr}\left(1 + \text{absorption}\right)$$

where absorption is expressed as a decimal fraction. Thus, average dolomite is listed in Table 2.5 as having a bulk specific gravity of 2.70 and absorption value of 1.1 percent, from which it is calculated that its bulk specific gravity (saturated surface-dry basis) is $2.70(1 + 0.011) = 2.73$. The above equation can, of course, be used to solve for the bulk specific gravity by dividing the bulk specific gravity (SSD basis) by (1 + absorption), where again the absorption is expressed as a decimal fraction. Chapter 11 gives further details regarding use of specific gravity and absorption in concrete-mix computations.

Laboratory procedures for determining specific gravity and absorption are contained in Specific Gravity and Absorption of Coarse Aggregate, ASTM Designation C127, and Specific Gravity and Absorption of Fine Aggregate, ASTM Designation C128.

Specific gravity is *not necessarily related* to aggregate behavior. However, reference to Table 2.3 indicates three individual rock types, shale, sandstone, and chert, which may display poor performance in concrete, particularly in exposed concrete in northern climates. They are observed in the table to be of somewhat lower specific gravity than the others. It is for this reason that heavy-media beneficiation plants which sort materials on a specific-gravity basis are proving successful in many areas. An important exception to the rule that low-gravity materials are necessarily suspected is the case of air-cooled blast-furnace slag, which may have a specific gravity as low as 2.2 because of the high porosity of individual particles. Although not technically classified as a lightweight aggregate, concrete made containing such slag will average roughly 10 pcf lower than that containing natural aggregates.

2.25 Voids and Grading

Voids. The amount of space *between* aggregate particles, or voids, will be importantly influenced by the amount of compaction, aggregate shape, and surface texture, and by the amounts of the respective particle sizes, i.e., the grading of the aggregate. A graded aggregate is one which contains appropriate amounts of the progressively finer-size particles to fill in the apertures between the larger sizes and thus reduce the void content. However, excellently graded aggregate such as to provide minimum voids has not been found fundamental to acceptable concrete. In fact, "gap-graded" aggregates deficient in one or more sieve sizes have been successfully used and are even encouraged by some.

Voids in aggregates can be determined from the relation

$$\text{Voids, \%} = \left(1 - \frac{M}{S \times 62.3}\right) \times 100$$

where S = bulk specific gravity (dry basis)
M = unit weight of the aggregate, pcf or kg/cm^3

The value of M will depend upon the compactive effort expended to consolidate the aggregate. The "dry loose" and "dry rodded" are two standard conditions frequently used in concrete technology and are defined in Standard Method of Test for Unit Weight of Aggregate, ASTM Designation C29. Void contents of typical concrete aggregates will range from about 30 to 50%. Mixtures of coarse and fine aggregate will provide lower void content than either constituent measured separately because of intermingling of the sizes.

Grading. Concrete aggregates after excavation or quarrying are almost always subjected to a screening process to provide the proper sizes. Confirmation that the desired sizes are present in the product is made by the "mechanical analysis" or sieving test wherein a weighed sample of the aggregate is placed in the top of a set of nested testing sieves with progressively smaller openings from top to bottom. The nested set is then shaken by hand or by a mechanical shaker until practical refusal, i.e., when continued agitation causes essentially no more particles to pass through the respective sieves. The individual size fractions are then weighed and computation is made of the percentages retained or passing. ASTM Method C136 gives details of the testing procedure.

Tables 2.7 and 2.8 display data from typical fine- and coarse-aggregate mechanical analyses, respectively. Computation has been made in the fourth column of each table of the fineness modulus of the aggregate, a number found convenient to characterize the overall "coarseness" or "fineness" of the aggregate. Definition of the fineness modulus requires that the sum of the cumulative percentages retained on a definitely specified set of sieves be determined and the result be divided by 100. The sieves specified to be used in determining fineness modulus (and no other) are No. 100, No. 50, No. 30, No. 16, No. 8, No. 4, ⅜-in, ¾-in, 1½-in, 3-in, and 6-in. Note in Table 2.8 that the sieves not in this particular series have been omitted in calculating fineness modulus. Fineness modulus of fine aggregate is often used as a uniformity requirement wherein successive ship-

TABLE 2.7 Mechanical (Sieve) Analysis of a Concrete Fine Aggregate

Sieve	Fraction retained g	Fraction retained %	Cumulative retained %	Cumulative passing %	ASTM Specification C33 grading requirements % passing
¾-in	0	0	0	100	100
No. 4 (4.75 mm)	22	4.0	4.0	96	95–100
No. 8 (2.36 mm)	65	12.0	16.0	84	80–100
No. 16 (1.18 mm)	103	19.0	35.0	65	50–85
No. 30 (600 μm)	119	21.9	56.9	43	25–60
No. 50 (300 μm)	157	29.0	85.9	14	10–30
No. 100 (150 μm)	60	11.1	97.0	3.0	2–10
Pan	16	3.0			
Total	542	100.0	294.8		
Fineness modulus			$\frac{294.8}{100} = 2.95$		

ments are required to not deviate from a base fineness modulus by more than 0.2. This ensures that variations in grading will not be so large as to require change of concrete-mix proportions. It is often helpful to remember that coarser, larger-sized aggregate will have a larger numerical value of fineness modulus.

The last column in Tables 2.7 and 2.8 includes the grading limits given by ASTM Specification C33 for concrete sand and for size 57 coarse aggregate. Coarse-aggregate

TABLE 2.8 Mechanical (Sieve) Analysis of a Concrete Coarse Aggregate, Size 57

Sieve	Fraction retained, lb	Fraction retained, %	Cumulative retained, %	Cumulative passing, %	ASTM Specification C33 grading requirements, size 57, % passing
1½ in	0	0	0	100	100
1 in	1.2	4	—	96	95–100
¾ in	9.3	30	34		
½ in	6.8	22	—	44	25–60
⅜ in	4.3	14	70		
No. 4 (4.75 mm)	8.4	27	97	3	0–10
No. 8 (2.36 mm)	0.6	2	99	1.0	0–5
No. 16 (1.18 mm)	0.3	1	100		
No. 30 (600 μm)	0	—	100		
No. 50 (300 μm)	0	—	100		
No. 100 (150 μm)	0	—	100		
Pan	0				
Total	30.9	100.0	700		
Fineness modulus			$\frac{700}{100} = 7.00$		

grading requirements for most classes needed for concrete are shown in Table 2.9. Although specifications of local agencies sometimes differ slightly from the ASTM C33 gradings, it is debatable whether such deviations are really justified. Concerted efforts are being made to encourage standardization to these gradings so as to reduce the necessity for producers to stock a multiplicity of sizes which differ only in insignificant details.

It is sometimes advantageous to plot aggregate gradings on special paper made for this purpose to enable rapid visualization of the grading. The coarse aggregate and sand in Tables 2.7 and 2.8 have been so plotted in Fig. 2.8 with shaded areas showing the allowable limits of grading for the respective aggregates as specified in ASTM C33. Such plotting of gradings often reveals trends difficult to estimate from tabulated data. For instance, the plot clearly reveals that the coarse aggregate in Table 2.8 is very close to failing the intent of size 57 in the amount passing the 3/8-in sieve since the actual grading curve almost falls outside the shaded area.

The horizontal spacings in Fig. 2.8 are proportional to the logarithm of the sieve opening. Paper is also available wherein the spacings are proportional tc the sieve opening raised to the 0.45 power. The latter paper is often advantageous in plotting aggregate gradings for dense-graded aggregates for road bases or bituminous mixtures. Such gradings plotted on this paper will often approximate a straight line, making visual estimates of deviations in grading very easy.

A small amount of oversize must be allowed for in commercial production, and it is noted that the *maximum size* designated for the particular aggregate in the second column of Table 2.9 is always one size smaller than that through which 100 percent of the material is required to pass. Such a definition of maximum size becomes consequential in selecting proportions for concrete consistent with mixing-water requirements, form dimensions, and reinforcing-bar spacing, as further detailed in Chap. 11. Maximum size may also be a consequential item as to whether or not a particular granular deposit is really an economical one to operate. If major projects in the market area demand 1½-in maximum-size aggregate of which the deposit is found deficient, then expensive importation

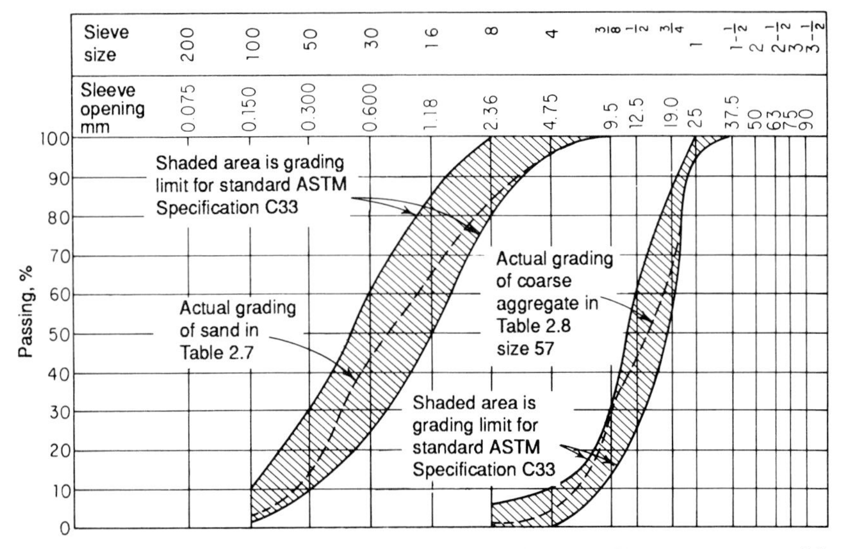

FIGURE 2.8 Aggregate gradings plotted. Sand grading on left and size 57 coarse aggregate on right (data from Tables 2.7 and 2.8).

TABLE 2.9 Grading Requirements for Coarse Aggregates (Simplified Practice Recommendations)

Size No.	Nominal size sieves with square openings)	Amounts finer than each laboratory sieve (square openings), % by weight												
		4 in (100 mm)	3½ in (90 mm	3 in (75 mm)	2½ in (63 mm)	2 in (50 mm)	1½ in (37.5 mm)	1 in (25.0 mm)	¾ in (19.0 mm)	½ in. (12.5 mm)	⅜ in (9.5 mm)	No. 4 (4.75 mm)	No. 8 (2.36 mm)	No. 16 (1.18 mm)
1	3½ in to 1½ in	100	90–100	—	25–60	—	0–15	—	0–5					
2	2½ in to 1½ in	—	—	100	90–100	35–70	0–15	—	0–5					
357	2 in to No. 4	—	—	—	100	95–100	—	35–70	—	10–30	—	0–5		
467	1½ in to No. 4	—	—	—	—	100	95–100	—	35–70	—	10–30	0–5		
56	1 to ⅜ in	—	—	—	—	—	100	90–100	40–85	10–40	0–15	0–5		
57	1 in to No. 4	—	—	—	—	—	100	95–100	—	25–60	—	0–10	0–5	
67	¾ in to No. 4	—	—	—	—	—	—	100	90–100	—	20–55	0–10	0–5	
3	2 to 1 in	—	—	—	100	90–100	35–70	0–15	—	0–5				
4	1½ in ¾	—	—	—	—	100	90–100	20–55	0–15	—	0–5			
5	1 to ½ in	—	—	—	—	—	100	90–100	20–55	0–10	0–5			
6	¾ to ⅜ in	—	—	—	—	—	—	100	90–100	20–55	0–15	0–5		
7	½ to No. 4	—	—	—	—	—	—	—	100	90–100	40–70	0–15	0–5	
8	⅜ in to No. 8	—	—	—	—	—	—	—	—	100	85–100	10–30	0–10	0–5

of suitable larger sizes may be mandatory. On the other hand, if the deposit contains many good-quality cobbles and boulders, crushing down to the appropriate maximum size may entail additional expense. The latter eventuality may be offset by the sweetening effect of the larger sizes to remedy a deposit of poorer quality in the small sizes.

Blending of aggregates is undertaken for a variety of purposes, for instance, to "sweeten" an aggregate with that of better quality so as to make the combined aggregate acceptable, or to remedy deficiencies in grading. If interest centers on the fineness modulus of the blend or its resultant grading, both can be calculated if the characteristics of the component aggregates are known. If two aggregates, designated as A and B, are mixed together having fineness moduli of FM_A and FM_B, respectively, the resultant blend will have the following fineness modulus:

$$FM_{\text{blend}} = FM_A \times \frac{P_A}{100} + FM_B \times \frac{P_B}{100}$$

Where P_A and P_B are the percentages, by weight, of aggregates A and B in the blend. As an example, if the sand in Table 2.7 (fineness modulus 2.95) is blended with the coarse aggregate in Table 2.8 (fineness modulus 7.00) in the ratio of 40% sand to 60% coarse aggregate, the blend will have the fineness modulus

$$FM_{\text{blend}} = 2.95 \times 40/100 + 7.00 \times 60/100 = 5.38$$

If it is desired to determine in what proportion to combine materials A and B to achieve a certain fineness modulus of the blend, the amount of material A to be used can be calculated from

$$P_A = \frac{FM_{\text{blend}} - FM_B}{FM_A - FM_B} \times 100$$

Assume it was desired to determine how to combine the previously mentioned coarse and fine aggregates to achieve a blend with fineness modulus of 5.20; then

$$P_A = \frac{5.20 - 2.95}{7.00 - 2.95} \times 100 = 55\% \text{ coarse aggregate}$$

and the amount of fine aggregate, by weight, is 100 − 55 = 45%.

A problem often arising is that of determining in what proportion to blend two or more materials to meet a certain grading. For example, consider two hypothetical sands identified below as "fine" and "coarse," respectively. Their individual gradings below are compared with the requirements of concrete sand in ASTM C33.

Sieve	% passing		
	Fine	Coarse	ASTM C33 requirements
⅜ in	—	100	100
No. 4	—	95	95–100
No. 8	100	55	80–100
No. 16	98	30	50–85
No. 30	75	15	25–60
No. 50	40	5	10–30
No. 100	15	1	2–10

Inspection of the gradings reveals that both sands seriously depart from C33 requirements, and question arises as to whether blending will be a successful remedy. Some prefer to attack the problem by first plotting the individual gradings on paper such as that used in preparing Fig. 2.8. Geometric intuition is then used to guess how hard to the right or left it is necessary to pull the two lines to fall within the acceptable grading band.

Another method used is one of trial and error without plotting the data. For example, attention might be first given to the amount passing the No. 50 sieve in the above tabulation since many concrete technicians consider this amount as importantly influencing the workability of the concrete. A 50–50 blend might be first considered so as to provide (40 + 5)/2 = 22.5% passing, a value well within the 10 to 30% limit. However, it is quickly revealed that this is not an acceptable ratio for the No. 8 sieve since (100 + 55)/2 = 77.5%, which violates the 80 to 100% requirement. A 60–40 ratio could then be tried, providing a little more of the fine sand, yielding the following:

Sieve		% passing	
		60–40 blend	ASTM C33 requirement
⅜-in	$0.6 \times 100 + 0.4 \times 100 =$	100	100
No. 4	$0.6 \times 100 + 0.4 \times 95 =$	98	95–100
No. 8	$0.6 \times 100 + 0.4 \times 55 =$	82	80–100
No. 16	$0.6 \times 98 + 0.4 \times 30 =$	71	50–85
No. 30	$0.6 \times 75 + 0.4 \times 15 =$	51	25–60
No. 50	$0.6 \times 40 + 0.4 \times 5 =$	26	10–30
No. 100	$0.6 \times 15 + 0.4 \times 1 =$	9	2–10

The 60–40 blend successfully meets the C33 grading requirements, and inspection likewise reveals that only a very little more of the fine sand could be used since the amount passing the No. 100 sieve is already close to the upper limit.

2.26 Deleterious Substances

Aggregates may contain mineral particles which in some exposure conditions of the concrete undergo excessive volume change, causing rupture of the concrete surface, or they may create sufficient interior stress so as to cause cracking and impair the structural integrity of the concrete. In other environments, these same mineral types may have negligible influence. Wetting and drying or substantial water saturation simultaneous with freezing and thawing will be destructive to some rock types. The latter is particularly true for lightweight, porous cherts, highly argillaceous limestones, and some shales. Water expands by about 10 percent on freezing and can be highly destructive when contained in aggregate pores unable to accommodate such expansion. Very soft particles such as ocher are detrimental if close to concrete surfaces subjected to abrasion since the thin mortar covering over the particles will be dislodged and the underlying soft fragments worn away, causing pitting of the surface.

Adherent clay coatings on aggregate particles which persist during the concrete-mixing process may impair bond with the cement paste, and specifications limit the amount of the very fine material (passing No. 200 sieve) which is removable by washing. Clay-like materials, whether occurring as coatings or dispersed in such rocks as argillaceous limestones, are objectionable since the volume of the rock is then responsive to changes in moisture content. Shrinking and swelling are detrimental to the concrete. Shrinkage cracking of concrete in slabs or large masses or in otherwise restrained members, particularly that caused by some aggregates in the southwestern United States area, has been observed due to rocks of this type. Drying shrinkage of concrete is used as an acceptance test for eliminating such moisture-sensitive rocks for use.

Coal and lignite are detrimental to concrete. Lignite, a brownish-black substance intermediate between peat and coal, is particularly objectionable since it will cause unsightly brown stains on the concrete surface when it disintegrates.

Tables 2.10 and 2.11 give the deleterious substances permitted to be present in fine and coarse aggregates provided by Standard Specifications for Concrete Aggregates, ASTM C33.

Still another class of rocks react deleteriously with alkalies of the cement, giving rise to the "alkali–aggregate reaction" or "alkali–carbonate rock reaction." Severe and highly destructive expansions in the first category have occurred for concretes in moist environments when the aggregates contain sufficient amounts of opal, chalcedony, tridymite, cristobalite, and certain rhyolites, andesites, or dacites. Freezing and thawing, although sometimes a complicating factor for a given structure, are not necessary to initiate this destructive action. Figure 2.9 gives locations in the United States known to have structures affected by alkali-aggregate reaction.[7,16] Limestones have also been found which react adversely with the cement alkalies. Those limestones which exhibit destructive reactions have a characteristic texture in which large crystals of dolomite are scattered in a fine grained matrix of calcite and clay.[17]

Assesment of the suitability of an aggregate which demonstrates susceptibility to either the aggregate-alkali reaction or the alkali-carbonate rock reaction for a given use requires a high degree of engineering judgment and careful examination of past performance under similar exposure. ASTM Specification C33 gives useful criteria for evaluating potential alkali–aggregate reactivity and criteria for the alkali–carbonate rock reaction.[7]

Positive identification and assignment of the degree of destructiveness of rock types which contribute to freeze-and-thaw destruction of concrete is probably the most troublesome matter presently facing both the aggregate producer and consumer. Specifications of highway departments, etc., in a given locality frequently limit the percentages of cer-

TABLE 2.10 Limits for Deleterious Substances in Fine Aggregate for Concrete*

Item	Maximum % by weight of total sample
Clay lumps and friable particles	3.0
Material finer than No. 200 (75-μm) sieve	
Concrete subject to abrasion	3.0*
All other concrete	5.0*
Coal and lignite	
Where surface appearance of concrete is of importance	0.5
All other concrete	1.0

*In the case of manufactured sand, if the material finer than the No. 200 sieve consists of the dust of fracture, essentially free from clay or shale, these limits may be increased to 5 and 7%, respectively.

Source: ASTM C33.

TABLE 2.11 Limits for Deleterious Substances in Coarse Aggregate for Concrete*

Item	Maximum % by weight of total sample
Clay lumps and friable particles	5.0
Soft particles*	5.0
Chert as an impurity† that will disintegrate in 5 cycles of the soundness test, or 50 cycles of freezing and thawing (0 to 40°F);‡ or that has a specific gravity, saturated surface-dry, of less than 2.35	
Severe exposure	1.0
Mild exposure	5.0
Material finer than No. 200 sieve	1.0§
Coal and lignite	
Where surface appearance of concrete is of importance	0.5
All other concrete	1.0

*This limitation applies only when softness of individual coarse aggregate particles is critical to performance of the concrete, e.g., in heavy-duty floors or other exposures where surface hardness is especially important.

†These limitations apply only to aggregates in which chert appears as an impurity. They are not applicable to gravels that are predominantly chert. Limitations on soundness must be based on service records in the environment in which they are used.

‡Disintegration is considered to be actual splitting or breaking as determined by visual examination.

§In the case of crushed aggregates, if the material finer than the No. 200 sieve consists of dust of fracture, essentially free from clay or shale, percentage may be increased to 1.5.

Source: ASTM C33.

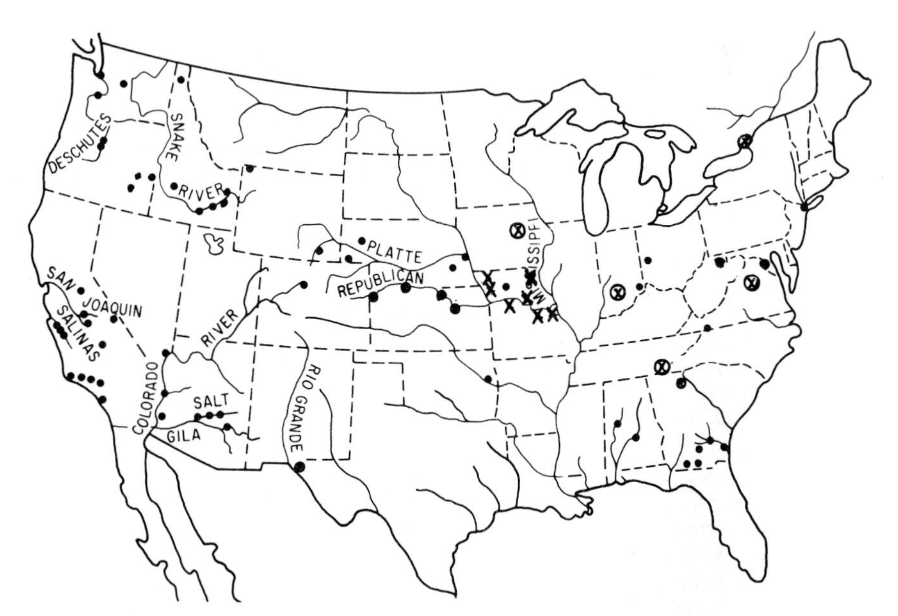

FIGURE 2.9 Locations where alkali–aggregate reaction has been reported.[7, 16] Key: × sources of reactive crushed stone; ● sources affected by alkali–silica reaction; ⊗ sources affected by alkali–carbonate reaction.

tain rock types—e.g., chert, shale, argillaceous limestone—allowed in the aggregate for exposed concrete based upon past performance. It is often necessary to obtain local interpretation of such specifications in order to determine just what the impact on production will be. Efforts to establish criteria and tests of nationwide application have had success, and the following are being used: (1) Sulfate Soundness Test, ASTM C88; (2) Freeze-Thaw Tests in Concrete, ASTM C666; (3) Potential Volume Change of Cement-Aggregate Combinations, ASTM C342; (4) Practice for Evaluation of Frost Resistance of Coarse Aggregates in Air-Entrained Concrete by Critical Dilation Procedures, ASTM C682; (5) Carbonate Reactivity ASTM C586.

2.27 Organic Impurities

Organic impurities sometimes occur in natural fine aggregates which impair hydration of the cement and consequent strength development of the concrete. Such impurities are normally avoided by proper stripping of the deposit to remove topsoil completely and by vigorous washing of the sand. Problems with high-organic-content sand seem to be largely alleviated by the efficient washing equipment now provided by manufacturers. Detection of high organic content in sand is readily determined by the sodium hydroxide colorimetric test, ASTM C40. Some impurities in sand may give indication of high organic content but not be really injurious. Assessment of this possibility can be made by ASTM Test C87.

2.28 Bulking of Moist Sand

Sand which is moist such as fresh from a washing plant or even after prolonged stockpile storage has a considerably greater gross volume than that in the dry state because of the moisture films surrounding each particle. The thin films of moisture inhibit sliding of the particles on each other to achieve a compact condition. This effect is the reverse of what the uninitiated tends to suspect would be the case; one mistakenly believes moisture to be an aid to compaction. Figure 2.10 shows typical volumes of a concrete sand plotted against moisture content. Most concrete sands will demonstrate maximum volume at about 5 to 6% moisture content. The sensitivity of sand volume to small changes in moisture content and resultant uncertainties as to how much sand is actually contained in a given volume of moist sand caused abandonment of volume proportions of job concrete. Likewise, the basis of purchase of sands has changed from a volume basis to weight basis in almost all areas.

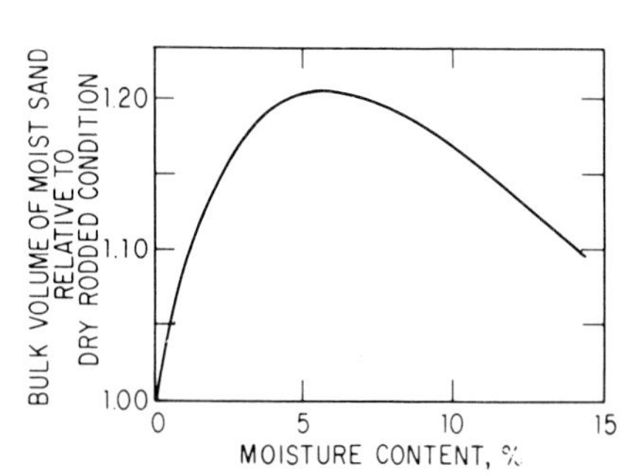

FIGURE 2.10 Loose volume of concrete sand versus moisture content.

2.29 Additional Aggregate Properties

Concrete projects sometimes require considerations of conditions not normally encountered such as unusually high- or low-temperature environments, potentially destructive alkalies or acids, or sulfates. Ice-removal salts have been known to contaminate aggregate supplies inadvertently. The role of aggregates in such circumstances may be questioned or it may be desired to find authoritative information. Reference 18 is particularly recommended for comprehensive coverage under such circumstances. Reference 19 may be helpful and gives many additional references.

AGGREGATE PROCESSING

Processing the raw product of a pit or quarry into an aggregate useful for concrete may be quite simple or very involved depending on several factors. The reasons for choice of specific processing equipment acquired by the deposit operator do not lend themselves to precise engineering analysis because of variations in the deposit, experience of the operator, variations in equipment among different manufacturers, and a host of less definable factors. Economic considerations sometimes lead operators to continue using less efficient equipment simply to amortize their investment. "Package" installations wherein all individual units are assembled into a prearranged pattern for installation at the plant are becoming more frequent for concrete aggregates (Fig. 2.11). Preassembled portable gravel plants have long been common, for instance, for processing road-base aggregates.

Concrete-aggregate-processing plants are often field-designed and assembled with a great variety of possible solutions. The individual processing operations shown in Fig. 2.11 are often physically separated some distance or even housed in separate structures.

2.30 Stripping

Undesired overburden consisting of vegetation, trees, topsoil, clay, etc., must be removed to expose the desired material prior to beginning excavation. The processes do not differ essentially regardless of whether ledge rock or granular deposits are being developed.

FIGURE 2.11 Rotary screen "package" aggregate plant with screw classifier for sand. Plant feed *(A)* first goes to scrubber *(B)*. Outer rotary screen *(C)* furnishes sand to screw classifier *(D)*. Finished sand stored in bin *(E)*. Pea gravel screened off to bin *(F)*. Inner coarse-aggregate screens *(G)* provide three sizes with oversize stored in bin *(H)*. (*Courtesy of Pioneer Engineering.*)

Stripping should be accomplished by means that do not require rehandling the material. Power shovels or draglines loading into trucks are often used, with the trucks disposing of the material into the spent portion of the pit or quarry. With a small amount of overburden and a deep deposit, this may be only an intermittent operation to just keep ahead of the excavation and may be done with bulldozers. With shallow deposits and relatively more overburden, stripping may be a continuous operation. In some cases, lower costs are claimed for single-operator, push-pull, twin-engine, rubber-tired scrapers of 24 cu yd or more capacity (Fig. 2.12). These versatile machines can maintain their own haul roads and travel at high speeds under favorable conditions.

The depth of stripping that can be tolerated for an economical operation cannot be stated except to observe that as much as 50 ft has been removed in rare cases to expose desired material. Some gravel plants are able to utilize a portion of the clay overburden in making road-base material and reserve the deeper, clean gravel for manufacture of concrete aggregates.

Crawler-mounted rippers, with or without bulldozer blades, are sometimes used to break up softer rock or frozen ground in stripping operations.

2.31 Blasting Rock Quarries

Most rock in quarries is initially dislodged by blasting with explosives in order to reduce it to sizes which can be handled for subsequent crushing.

Spacing of drill holes and choice of explosive must be matched with the hardness of the rock. Such choices are made with close liaison between the explosive manufacturer and quarry operator. Rapid developments in drilling techniques and explosives behavior preclude formulating hard-and-fast rules, but the following general observations appear justified:

1. Blasting energy derives only from rapid expansion of *confined* gases. It is for this reason that charges initiated at the bottom of a hole with detonation progressing upward are more effective. Similar advantage derives from using a more powerful explosive at the hole bottom.
2. Smaller-diameter closely spaced holes at a given distance back from the working face, or "burden," are most effective, but staggered rows are also used, with delayed blasting caps giving better fragmentation.

FIGURE 2.12 Gravel-stripping operation using twin-engine scraper. Struck capacity 24 yd^3, heaped capacity 30 $yd.^3$ Average haul distance at this pit 1500-ft or 3,000-ft round trip. Daily production based on double shifting (16 h) is 3500 yd^3. (*Courtesy of American Aggregates Corp.*)

3. Harder rock requires closer spacing of drill holes.
4. Greater face heights enable greater spacing of drill holes. Spacings of about 10 to 30 ft are prevalent.
5. Blast holes are drilled by churn or well drills, percussion drills, rotary drills, and jet-piercing process. The well drill exerts a pile-hammer action on a steel shaft by alternately raising and dropping the shaft. The tip of the shaft is fitted with cutting edges. This technique is, of course, applicable only to vertical holes. Carbide-insert cutting heads are often used on percussion drills and compressed air is employed for removing cuttings. Rotary drills in the softer rocks are frequently used. Percussion and rotary drills can penetrate the rock at any desired angle assuming a suitable drill rig is used. A great variety of drill rigs are available, usually crawler-mounted with hydraulic-operated booms for accurate control. Hole sizes drilled are of the order of 4 to 8 in diameter. Jet-piercing drills employ an intense source of heat to shatter and fuse the rock, the heat deriving from a mixture of petroleum fuel and oxygen. Steam is used to aid the shattering process and blow the hole free of rock fragments.
6. A great variety of explosives are available for rock blasting. These explosives vary as to strength, rate of detonation, mode of placement in the hole, and sensitivity.

In summary, blasting is a complex, specialized problem for the quarry manager and requires decision considering the following variables:

1. Height of face
2. Hardness and uniformity of rock
3. Dip of beds
4. Prevalence of bedding seams and joints
5. Dip of quarry floor
6. Size and depth of drill holes
7. Arrangement and spacing of drill holes
8. Number of holes shot at one time
9. Sixe of charge
10. Position of charges in drill holes
11. Type of explosive used
12. Method of firing shots
13. Method of loading rock
14. Size of shovel
15. Size of crusher

2.32 Excavation

Power Shovels. Broken rock blasted from the quarry face is normally picked up by power shovels of from about 1- to 10-yd^3 dipper capacity. Electric power shovels appear to be gaining popularity for the larger-capacity dippers and require current furnished through flexible cables. Rock has been reported to have been loaded using as little as 0.22 kWh per ton with electric shovels. High-voltage alternating current is fed to the shovel and ac/dc converters drive dc motors for the individual operations. With a 2½-yd dipper and 90° swing, rock loading is reported to progress at 200 to 275 yd^3 h. The larger dippers do not appear to have commensurate increase in loading rate but obviously have greater power and reach for conditions where such is needed.

Sand-gravel deposits with a relatively shallow face are commonly excavated with power shovels. Loading rates are much higher than for rock and the corresponding $2^1/_2$-yd dipper as mentioned above for rock may load as much as 400 yd^3 h of the more easily handled gravel material. Higher working faces make shovels impractical because of danger of cave-in which may bury the shovel, and draglines are then used.

Draglines. Draglines (Fig. 2.13) with ever longer booms and larger scraper buckets are becoming available; 120-ft booms and 15-yd buckets are not uncommon. Some operators find it advantageous to drag *up* the bank, others *down.* Dragging down the bank may be desirable for a completely dry operation, and the operator can achieve blending of the material in the entire bank if such is desired—or he may exercise some selection. If part of the deposit worked is under water, dragging upward is mandatory. There is the attendant possibility of undermining the machine when dragging up the bank.

Dragline distances can be greatly increased by use of slackline techniques with a deadman. In one method, the scraper bucket rides on a sheave, or sheaves, attached to a

FIGURE 2.13 Excavating raw material with walking dragline; 9-yd^3 bucket on 165-ft boom. Complete electric power, 4100 V ac with 115-V dc conversion. 750 tons per hr capacity. Total weight of machine 500 tons. (*Courtesy of American Aggregates Corp.*)

cable which terminates at a fixed anchor some distance away and downgrade from the machine. To take a bite, the cable is tightened and the bucket rides down the cable by gravity a selected distance. The support cable is then slackened and the bucket drops down to the deposit. The separate load cable attached directly to the bucket then pulls the bucket toward the machine for loading. Repositioning the deadman is necessary for digging other than a single deep trench, or the machine can be moved radially around the deadman. By use of another cable with a sheave at an elevated deadman, dragging upgrade can be accomplished, or even excavated toward the deadman if desired. Slack-line techniques allow a great variety of possibilities for loading or formation of surge piles and can cover considerable distances vertically and horizontally.

Dragline excavation under water is accomplished with perforated scraper buckets to release water from the load as it is brought up.

Hydraulic Excavation. Underwater deposits frequently lend themselves to dredging operations (Fig. 2.14). The dredging equipment is mounted on a floating barge or ship and either the dredged material is pumped ashore for processing or, in some cases, the same or adjacent barges house the processing machinery. Suction is supplied by centrifugal pumps specially lined for abrasion resistance, usually driven by 200-hp or frequently much larger engines.

The suction line, varying from 8 to 28 in diameter, is lowered to the bottom by an A-frame arrangement with hoist and winch. In soft, sandy deposits suction alone is often adequate to bring up the material. In harder deposits containing gravel, mechanical means may be advantageous to dislodge the material (Fig. 2.15). Such mechanical devices employ rotary cutters of various types or chain-type digging ladders with special arrangement at the nozzle to prevent boulders from obstructing the suction line. The large amounts of water pumped simultaneously with the granular material tend to wash away

FIGURE 2.14 Excavating raw material with hydraulic dredges. (*Courtesy of American Aggregates Corp.*)

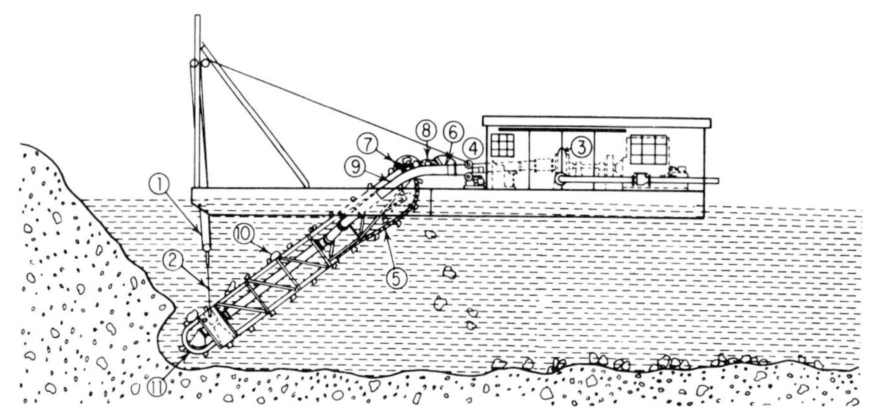

FIGURE 2.15 Schematic view of "ladder" dredge. Front or nozzle end suspended on hoist line (1) and bail (2). Centrifugal dredge pump (3) is independently powered. Traveling-screen chain (5) driven by motor (4) through pulley (6), sprocket (7), and gear train (8). Feed enters nozzle (11) and is carried up through suction line (9). Boulder (10) too large for nozzle is carried up and dropped out of the way behind nozzle. (*Courtesy of Eagle Iron Works.*)

fines desired in concrete sand. Construction of dikes and dewatering the enclosed area have been used with excavation proceeding in the dry.

Barge-mounted dipper shovels and clamshells are also used to excavate materials underwater.

2.33 Transportation of Materials

Aggregate production is typically a high-volume business requiring handling large tonnages sometimes over quite a distance even within the plant area. Each cubic yard of aggregate weighs roughly 1½ tons (over 2 tons for solid ledge rock), and search for more economical means of handling these large tonnages is a continuing effort. Plants use a variety of means for conveying materials, ranging from railways, tramways, trucks, barges, and belts to open sand flumes. Others have chosen belts for practically all operations.

Railways. Some pit-and-quarry operations, particularly where the haul to plant exceeds about 1 mile, use diesel-electric locomotives with a variety of car types in use. Additional track is laid as excavation proceeds farther and farther from the plant, and this method seems to provide economical transportation despite necessary track laying and maintenance. Side-dump cars of various designs are most prevalent, enabling rapid discharge of their load at the plant, but bottom-dump cars are also used. Car capacities range up to 50 tons, but much smaller ones are also employed, particularly with narrow-gage track.

Motor Trucks. A great variety of sizes and types of dump bodies is available ranging up to 85-ton payloads without semitrailer. Better maneuverability seems to favor the single unit, without trailer, type of vehicle for intraplant hauling. Too, public-highway loading restrictions do not apply, and to the extent that plant haul roads can be maintained in good condition, these seem to provide economical transport. Major emphasis is given to condition of haul roads by many operators in view of vulnerability to injury of the large-sized, expensive rubber tires used on these vehicles. Driver fatigue and general maintenance costs are also reduced by maintaining excellent haul roads. Some plants operate a blade grader almost continuously to ensure smooth surfaces free of loose rocks.

Side-dump bodies are used but end-dump types are more prevalent, probably for reason of real or imagined greater stability.

Belt Conveyors. Materials are transported almost universally by belt conveyors (Fig. 2.16), Belt conveyors can conveniently transport materials upward, downward, or on the level but individual flights do not normally exceed 20° from the horizontal. Excessive angle causes large rocks to roll back down the belt.

Raw feed from a quarry or pit is not recommended for belt conveyors without first removing oversize because of possible injury to the belt. This is often achieved by scalping and preliminary crushing at the deposit close to the point of excavation.

Belts lend themselves to a great variety of applications such as stacking of materials, loading cars, bins, and trucks, and forming surge piles. Belts are sometimes loaded at the center of the flight, and by reversing direction, material is piled at either end. Short flights (up to about 60 ft) are also mounted on wheels for portability and can quickly be transferred around the plant for a variety of loading purposes.

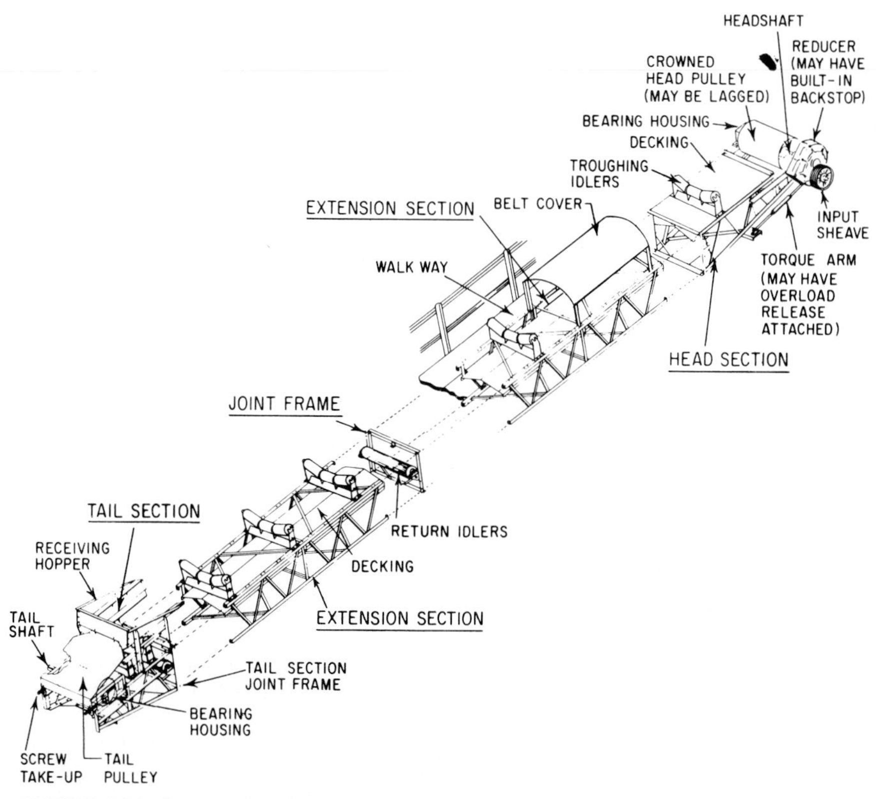

FIGURE 2.16 Cutaway view of belt conveyor. (*Courtesy of Pioneer Engineering.*)

Devices are available for weighing the load traversing a belt in a given time interval using conventional weighing systems with special integrators and more recently one using nuclear-measurement principles has been developed.

2.34 Crushing

Practically all quarried rock, slag, and oversize in a gravel deposit must be reduced to usable sizes for concrete by crushing. Such crushing is accomplished by a variety of commercially constructed machines classified as to type as follows: (1) jaw, (2) gyratory or cone, (3) roll (single, double, and triple), (4) hammer mill, and (5) impact. Diagrammatic sketches of three types are shown in Fig. 2.17.

Choice of the type of crusher appropriate to a given application should consider at least the following:

1. Size and design should be such as to handle the largest-sized rock in the feed.
2. Capacity should be such as to handle the peak rate of feed contemplated.
3. Presence of fines in the feed should not jam and stall the crusher.
4. Wearing parts should have liners and be readily replaceable.
5. Provision should be made for stray iron or other uncrushable material which occasionally finds its way into the feed. Breakdown of the crusher should not result from such materials entering the crusher, but stalling or jamming is unavoidable.

Over the years of development of crushers, much effort has been expended attempting to evaluate the character of the product, i.e., whether it typically shatters into long, splintery

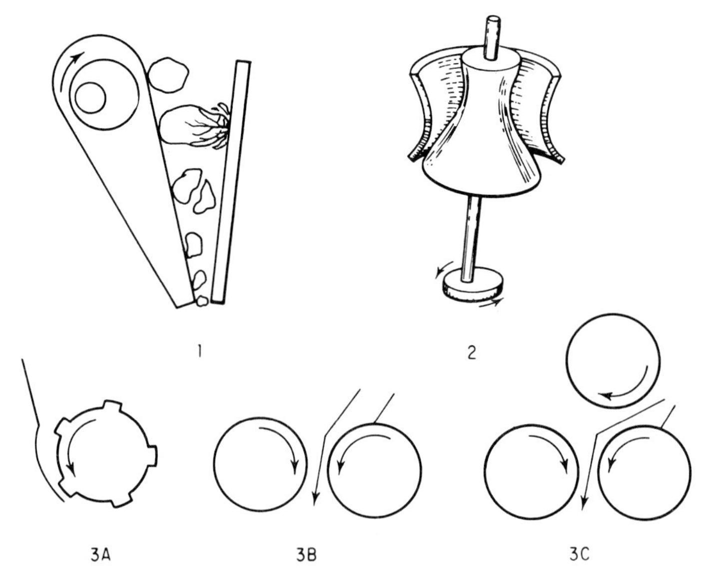

FIGURE 2.17 Schematic sketches of (1) jaw crusher, (2) gyratory or cone crusher, and (3) roll crushers—(3*A*) single roll, (3*B*) double roll, and (3*C*) triple roll.

pieces or into cubical, blocky fragments. The latter-shaped fragments are more desirable in concrete since the void content will be lower and less cement-sand-water mortar will be required. Opinion is prevalent that crushers depending more upon impact for breakage, such as the hammer-mill or impact types, will provide a more cubical product than those depending upon pure compression such as the jaw or double-roll type. It must be acknowledged, however, that thinly laminated stone, for instance, will tend to break along the laminations irrespective of the type of crusher employed, and superiority of performance of the crusher must be measured relative to the character of the particular stone itself. Gyratory crushers are considered intermediate in their tendency to impart pure compression and impact and perhaps for this reason, and for their compactness, have become very popular in the aggregate-production industry.

Jaw crushers are widely used in primary crushing of ledge rock particularly because of the large jaw openings available, up to 66 × 86 in. Replaceable special steel jaw liners, curved, smooth, or corrugated, are used and the different model crushers impart somewhat differing types of movement to the jaws. One popular model has an eccentric-mounted jaw such that movement at the top opening is quite large with relatively little movement at the bottom discharge. Jaw crushers invariably use a large flywheel to store energy for succeeding "bites." There is some opinion, if excessive reduction in one stage of crushing is not attempted, that jaw crushers tend to generate fewer fines than other types of crushers.

Gyratory crushers consist of a vertical shaft carrying a cone or "mantle." The cone does not rotate but is given a gyrating motion by an eccentric bearing arrangement on the lower end of the shaft. Movement of the cone is thus largest at the bottom discharge end (large end of cone) and minimum at the top. Gyrations of the cone occur within a bowl which is also cone-shaped and the space between the cone and bowl is therefore always wedge-shaped. Crushing action is quite similar to the jaw crusher except that curvature of the cone and bowl introduces some bending and shearing stresses to the stone. Stone is introduced through two semicircular openings in the top of the crusher which, because of the shaft housing and frame supports, do not accommodate stone larger than 72-in. size in the largest model now available.

Roll crushers are available in a great variety of types varying from single-to triple-roll types. Single-roll crushers with a studded roll, sometimes called a "slugger roll," are occasionally used as primary crushers. Roll crushers, in general, are not widely used in the concrete-aggregate industry, however, but are advantageous in being able to break down aggregate to a fine size in one pass. This may be desirable in processing agricultural stone or in the portland cement industry.

Crushing of aggregate for concrete is almost always done stepwise; that is, following primary crushing the undersize is removed and the oversize recirculated to a secondary crusher. Size reduction in a single crusher does not normally exceed about 1:4 for minimum production of fines.

2.35 Screening

As discussed in Sec. 2.25, aggregates for use in concrete are furnished in particle sizes appropriate to the use of the concrete. This requires screening in commercial production to provide the necessary sizes.

Except for initial sand removal from bank material, the term "screening" in the aggregate industry applies exclusively to processing the coarser sizes, i.e., larger than about No. 10 (2-mm) sieve. Separation of the larger aggregate particle sizes is accomplished by various types of screens having square, round, diamond, or rectangular slotted openings. Sand-size materials are almost entirely processed by hydraulic classifying means since size separation of these finer materials by this method is so greatly facilitated. Sizing of the finer materials by sieves is confined to laboratory testing.

A few general comments apply to all screening operations:

1. Every particle traversing the screen must be given several chances to drop through if it is sufficiently small to do so. Irregularly shaped fragments may require several opportunities before their least dimension is so oriented as to allow dropping through.
2. It then follows that too steeply sloped screens or those given too violent agitation will cause the particles to hop rather than drop through, particularly if the screen is underloaded. Rolling action, not hopping, is desired.
3. If the screen is overloaded and agitation is insufficient, it may become "blinded" with particles slightly too large to pass through, thus blocking the openings.
4. Rounded gravel particles are easier to screen than angular crushed stone or slag.
5. Production rate in tons per hour is greater for screening materials having a considerable disparity in particle sizes. It is much easier to screen a material 90 percent of which passes the screen than one which is so closely sized that only 10 percent passes.
6. Wet materials screen more easily than dry material, particularly for the smaller gravel sizes. Dry screening is frequently used, however, in the crushed-stone and slag industries.

Grizzlies. Raw feed excavated from the pit or quarry is usually fed to a grizzly, which consists of a series of parallel bars, the spacing between which is adjusted to accommodate the largest rock which will be fed directly to the plant for further processing. The retained portion too large to pass the bars is fed to crushers to bring it down to usable sizes. The grizzly bars may be stationary or vibrating as the nature of the feed dictates. If the feed readily flows without sticking or hanging up on the bars, then vibration is not needed. Vibrating, sloped grizzlies act similarly to commercially manufactured mechanical *feeders* in that they level the material down and distribute it to a belt, crusher, or screen at a uniform rate.

Some gravel-pit grizzlies are job-made and simply consist of a hopper which is topped with an open horizontal grid of railroad-track iron to catch oversize material or clods (see Fig. 2.13). The latter are sledged down as necessary or wasted if only a few oversize boulders occur.

Sand-gravel operators sometimes feed directly by truck or shovel on a stationary grizzly opening and the material builds up, thus serving as a surge pile. Belt conveyors, trucks, etc., carry the material away at the bottom of the sloped bin underneath. Care must be exercised, of course, that the grizzly screen effectively prevents blocking at the bin discharge.

Revolving Screens. Rotating or revolving screens consist of wire or perforated-plate open-end cylinders revolving on a slightly inclined axis. Material is fed into the high end of the cylinder and if sufficiently small drops through the openings in the downhill tumbling action imparted by the rotating screen. Particles too large to pass the openings drop out the end of the cylinder into a chute, and oversize is crushed down and recirculated through the screens. By placing three cylindrical screens, for instance, concentrically, with successively smaller sieve openings in the outer cylinders, four sizes of materials can be obtained, i.e., oversize for the inner screen, that retained by each of the two outer screens, and that passing through the outer screen. With dry materials subject to abrasion, this may be a dusty operation, and water spray is introduced to remedy the situation. In other installations where the coarse aggregate has adherent clay coatings, larger volumes of spray water fed by nozzles inside the inner screen are introduced, and the screening operation acts simultaneously as a scrubber.

Rotating screens may be used at various points in the flow of materials through the plant, for instance, in sand-gravel operations sometimes immediately following removal of sand by vibrating screens and in quarry operation sometimes immediately ahead of primary crushing to remove oversize passed by the grizzly. In Fig. 2.11, however, initial separation of fine and coarse aggregate is accomplished by rotary screening.

Deck Screens. Flat, vibrating screens are widely used either singly or stacked with up to three decks, for provision of closely sized materials. Different manufacturers impart vibratory action to the screens by various eccentric or unbalanced weight drives. Some drives are so arranged as to induce a forward flow of material across the screen even when it is not sloped. Most screens are sloped a few degrees (about 20° is common) to provide gravity flow. Slopes up to 45° are used in rare cases for fine materials. The screens are steel-spring- or air-cushion-mounted to absorb vibration. A variety of screening areas are available for single screens ranging up to about 6 ft wide by 20 ft long. Likewise, the screening surface may be wire, bars, or perforated plate. Most wire screens require backup stiffeners to keep them from depressing from the superimposed load, and a 2-in opening supporting perforated plate is also prescribed for the more fragile wire screens.

Computation of the tons per hour that a deck screen can handle presents uncertainties which are difficult to assess. One manufacturer suggests computation be based upon a certain "basic" capacity in tons per hour per square foot of screen to which a "correction factor" is applied whose value depends upon the product of four coefficients involving: (1) number of screen decks above screen considered, (2) percentage of feed smaller than one-half screen opening, (3) presence or absence of water spray directly on screen, and (4) percentage of oversize fed to screen. The values of the individual coefficients are influenced in the following manner:

1. Efficiency of the lower decks is somewhat reduced from that of the top screen.
2. Greater percentages of feed whose size is less than one-half screen opening increase screening capability by a factor of 3 when percentage of small size increases from 10 to about 60%.
3. Water spray on screens increases screening capability almost three times as much for a $\frac{3}{16}$-in-opening screen as for a 1-in.-opening screen.
4. Screening capacity is about halved for 90% oversize as compared with 20% oversize.

2.36 Washing and Sand Production

Most, but not all, aggregate production is carried on utilizing generous amounts of water. The purpose of the water may be for any one or a combination of the following: (1) to remove undesired adherent coatings such as clay, silt, or dust of fracture; (2) to lubricate the stone sizes to hasten screening operations; (3) to reduce dust nuisance created by dry screening and/or crushing or other handling; and (4) to accomplish sizing, or water "classification" of sand.

Some crushing and screening operations of slag or extremely hard stone such as quartzite are carried on entirely dry except for slight fogging to control dust. In such cases, proprietary wetting agents are sometimes added to the water for greater effectiveness. Such minimum amounts of treated water have been found desirable, particularly when making manufactured sand by crushing, in that balling or clumping is inhibited when roll-mill or pan-mill crushing is employed.

Washing of aggregates may consist of thorough drenching with high-pressure water by means of spray bars equipped with nozzles suspended over deck screens or inside rotary screens or may involve complete inundation with vigorous scrubbing by means of log washers or scrubbers. Stickiness of the coatings needed to be removed and other characteristics of the aggregates will dictate which method of washing is most advantageous. Additionally, many sand-gravel operations use both methods, for instance, spray-bar washing during preliminary screening followed later by log-washer scrubbing for only the coarse aggregate.

Log Washers and Scrubbers. Scrubbers are used to wash aggregates when the nature of the adherent clay coatings is such that the more vigorous treatment of log washers is not needed. Scrubbers consist of cylindrical containers rotating on a horizontal or slightly inclined axis into which the aggregate is fed at one end along with large volumes of water. Lifters fastened to the inside of the drum bring up the aggregate and drop it at each revolution and churn the material violently. Suspended fines are washed away in the water. Some scrubber-screen combinations incorporate wet screening of the aggregate by having an interior cylindrical screen concentric with an outer solid jacket, thus making a separation of the coarse and fine aggregate simultaneously with scrubbing. Baffles impede flow through the scrubber so as to give a more thorough washing action than would be imparted by rotary wet screening alone.

Log washers derive their name from an old practice of putting wood logs inside a rotating drum along with the water-aggregate mixture to augment scrubbing action. Large-sized durable boulders were also used for a similar purpose, making the device akin to a ball mill. Present-day log washers more resemble pugmills and consist of a slightly inclined chamber into which aggregate is fed at the low end along with generous quantities of water. Screws or rotating shafts, similarly inclined, carrying hard-steel blades provide a violent churning action and work the material toward the high end for discharge. Adherent coatings are effectively removed from the aggregate, and the abrading action is sufficient to break down softer fragments such as soft sandstones, ocher, and soft shale. The suspended clay and abraded stone dust are carried away in the wash water. Figure 2.18 illustrates a blade-type log washer.

Like so many other items of aggregate-processing equipment, choice of whether a scrubber, log washer, or both will be needed depends largely on the nature of the deposit. Operators sometimes experiment several seasons to determine just what equipment best fits their needs and to establish optimum operating characteristics. If the deposit is of a nature that these treatments will both wash and successfully reduce the deleterious materials down to an acceptably low percentage, the operator is indeed fortunate and may not be required to install more sophisticated and expensive beneficiation treatments such as jigs or heavy-media separation.

Hydraulic Sand Classification. Requirements for sand sizes for concrete were given in Sec. 2.25. It almost invariably occurs that water classification of sand is needed in order to provide such graded sand and to furnish sand with an acceptably low amount finer than the No. 200 sieve. Many sand suppliers also stock clean, finer sand for mortar and plaster.

Hydraulic classifiers operate on the basic principle that larger-sized particles fall faster in a water suspension than do the smaller sizes. Based on the idealized assumption of spheres freely falling in a viscous medium, Stokes' law states the velocity of fall is proportional to the square of the radius of the sphere. Sand-classifying devices approximate

FIGURE 2.18 Double-shaft log washer used for removing adherent coatings and for disintegrating soft particles. (*Courtesy of Eagle Iron Works.*)

these "ideal" conditions by various types of horizontal and rising-current classifiers such as screws, drags, rakes, wheels, and cyclones.

If a water suspension of sand is introduced into one end of a long, horizontal box or trough and the water overflows the far end, progressively finer and finer particles will settle out toward the overflow (weir) end. If the trough is lengthened and/or the water flow reduced, even finer particles will be deposited at the far end since retention time is increased. The same concepts apply to rising-current classifiers in that the progressively finer particle size will be raised to greater heights in the pool, assuming lack of turbulence.

Commercial installations use the above principles in a variety of ways. Perhaps the simplest process is shown in Fig. 2.11, where the screw classifier is acting primarily to dewater sand before placing it in the delivery bin. The amount of water overflowing the weir at the low end of the screw classifier controls the amount of fines retained. The slowly turning helical screw tends to dewater the product as it raises the sand out of the pool. Control of even this apparently simple operation may be critical to reserve the desired fines. Some operators whose deposit runs low in fines or who use large volumes of water ahead of sand classification such as results from dredging, thus tending to lose fines, may experience difficulty in maintaining the product within the 10 to 30 percent requirement passing the No. 50 sieve.

Rakes accomplish essentially the same purpose as screw classifiers, but in this case the helical screw is replaced by an endless-chain arrangement with attached baffles which drag the sand up the slope of the container.

Figure 2.19 shows a variation of the use of a combined screw-type device and log washer to remove roots and other vegetation from the gravel. Such materials have proved very troublesome if of the proper size to clog screens. Others have used special cyclones to accomplish the same purpose, the principles of which are explained later.

Still other commercial sand classifiers use the principle of the horizontal long trough explained above but have as much as nine bins in the bottom of the trough to draw off the

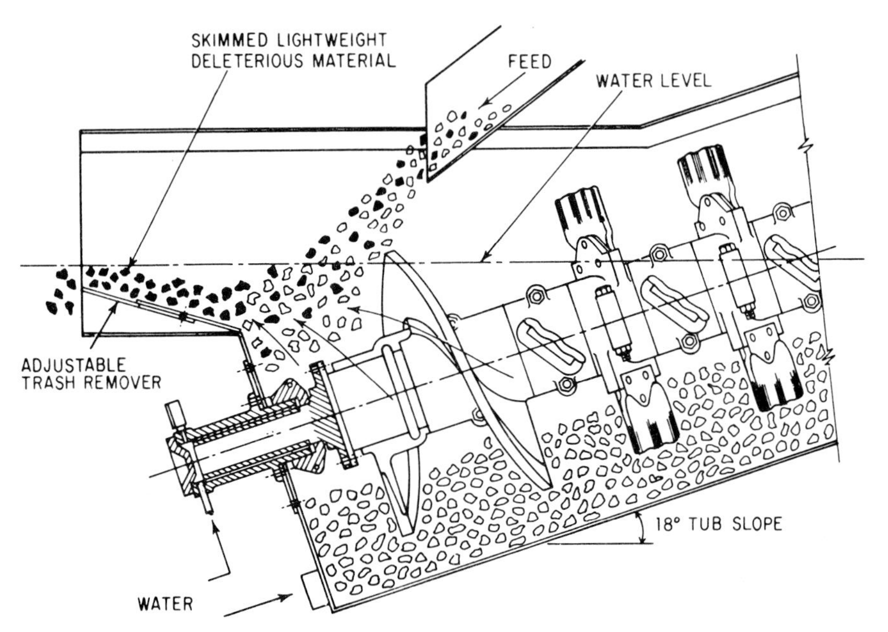

FIGURE 2.19 Coarse-material screw washer-dewaterer removing vegetation from gravel. (*Courtesy of Eagle Iron Works.*)

different sand size which settle out along the tank length (Fig. 2.20). A variety of sand sizes can this be provided by blending these products. Such blending is achieved by wet blending under water by recirculating back through a screw classifier or rake. By blending of fine aggregate such as may be done with coarse aggregate does not successfully intermingle the sizes.

Another device used in concrete sand processing is the cyclone. This apparatus has found particular use in processing water from dredging operations where fines tend to be lost because of the large volume of water handled. Figure 2.21 illustrates the principles of operation of the cyclone. Spinning motion imparted to the feed tends to settle out the larger sand particles to the outside of the cone. By adjusting three flow rates, that of the slurry pumped into the cyclone, the amount of very fine material overflowing the top, and the amount of fines recovered at the bottom of the cone, the latter valuable fines can be recovered to blend back into the sand. Minus-50-mesh material has been reported to be successfully dewatered and recovered with this device, as much as 10 tons/h with a 10-in cyclone and a slurry feed of 300 gal/min.

Dry Classification of Sand. In extremely arid regions with a scarcity of water, dry sizing of fine aggregate is undertaken using air separators. Crushed-stone-manufactured sand and slag sand are also sometimes dry-processed. Some air separators act similarly to the cyclone mentioned just previously except that air replaces water as the carrying medium. These are used more often for removal of the very fine material resulting from handling or the dust of fracture and resemble the cyclone classifiers used in portland cement manufacture (see Chap. 1). In another device, the sand is projected out horizontally with a centrifugal impeller and subjected to a downward blast of air. Divider plates below catch the deflected sand thus separated as to size, centrifugal force casting coarse sizes

FIGURE 2.20 Water scalping-sand classifying tank. Different-sized sand is collected at seven stations and blended in appropriate amounts into collecting flumes below. Remote-control metering panels are available to control valves to provide uniform blends despite changes in feed. (*Courtesy of Eagle Iron Works.*)

the farthest. As in all other sizing operations (except laboratory sieving to refusal), clean-cut separation is not made or, in fact, necessary. Blending is later used to remedy deficiencies of grading.

2.37 Aggregate Beneficiation

Gradual depletion of existing aggregate sources, zoning restrictions preventing development of desired new sources, and tightening of specification requirements by some agencies caused the need to develop and install special equipment to upgrade the quality of aggregate from present deposits or to develop sources heretofore considered to contain aggregate of marginal quality. Some equipment used for beneficiation has proved ineffectual except under very unusual circumstances, and new techniques are constantly being proposed and undergoing trial. The following paragraphs will treat only those successful beneficiation techniques of which there are a substantial number of commercial installations presently in the United States.

Beneficiation techniques presently employ two basic principles for removal of deleterious materials: (1) specific-gravity separation and (2) mechanical impact or abrasion to break up softer fragments. Jigs and heavy-media plants separate the desired from the unwanted materials on the basis of specific gravity. Cage mills, roll mills with rubber liners, and to some extent log washers, break down the unwanted fragments by impact or abrasion. Obviously the nature of the deposit will dictate which of the beneficiation tech-

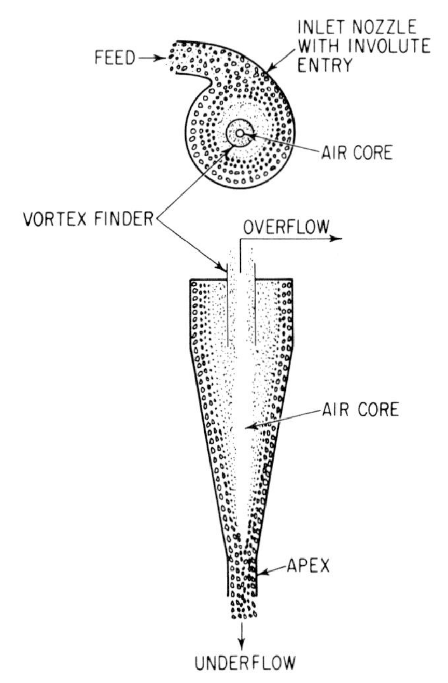

FIGURE 2.21 Cyclone separator. Often used for recovering fines from large volumes of water resulting from dredging operations. Centrifugal force drives large particles to outside. Fires are caught in center overflow. (*Courtesy of Eagle Iron Works.*)

niques is most advantageous in a given case. If the offending particles are predominantly soft, then impact beneficiation may be sufficient. On the other hand, if the deleterious particles are variously hard and soft, or predominantly hard as would be the case with chert, but are characterized by having a slightly lower specific gravity, then more expensive jig or heavy-media separation may be required. The latter seems true for many deposits in the North Central United States since installations are frequent in Ohio, Pennsylvania, Minnesota, and Michigan.

Cage Mill. Aggregate is fed into the center of the cage and drops out of the bottom of the housing by gravity. Cage mills are also available with two and three concentric cages which rotate in opposite direction, thus imparting even greater disintegration action to the product.

Cage mills operating at high speed are used primarily for crushing with the claim that a highly cubical product is produced. However, some operators slow down the rotational speed, thereby reducing the crushing action but still giving sufficient impact to break down soft particles. Rotational speeds of about 600 rpm for a 3-ft diameter cage have been employed for soft-stone reduction. In some installations, the product from the cage mill is then processed through a log washer for removal of the disintegrated fragments; in any event, the broken stone thus reduced to dust must be removed from the product by appropriate means.

As in all beneficiation procedures, too much emphasis cannot be placed on the necessity for prior careful study of the effectiveness of the cage mill in improving the product of a given deposit before installation is made. One of the best means of accomplishing this is to truck a load or two to the nearest cage-mill installation where appreciable quantities can be processed to determine benefits of the processing under full-scale operating conditions.

Jig Beneficiation. Jig beneficiation uses the principle of "hindered settlement" wherein particles of different specific gravities tend to settle out in horizontal layers when placed in water subjected to rapid vertical pulsations. Heavy particles will sink faster than lightweight (low-specific-gravity) particles. If a directed continuous flow of water is maintained in the pulsating suspension within an enclosure with an inclined bottom, the heavy particles will tend toward the bottom of the tank at the discharge end and the lightweight particles are effectively skimmed off in the water overflowing a dam whose height is adjusted to accomplish optimum beneficiation.

Figure 2.22 shows a photograph of one type of jig. Pulsations in this case are introduced by a mechanical eccentric drive; the lower, or hutch, section is attached to the sloping stationary tank by a rubber diaphragm. The tank bottom consists of perforated steel plate whose openings are restricted by a bed of loose steel balls. The balls serve to inhibit the materials from dropping into the hutch section and still allow transmission of water pulsations throughout the tank. Some jigs replace the mechanical drive with pulsating air.

Adjustment of the water-flow rate, rapidity and magnitude of jigging action. Feed rate to the jig, and water overflow are critical and must be brought into careful balance by experimentation to establish optimum operation. Uniform rate of feed to the jig is essential to best operation to maintain essential "fluffing" action of the bed; surges are accompanied by loss of beneficiation effectiveness. Opinion is not unanimous as to how jigging efficiency is influenced by particle shape and size but both coarse and fine aggregates have been treated by the process.

FIGURE 2.22 Mechanically driven jig used for beneficiation to float off undesired lightweight shale, etc. Manifold valves allow adjustment of water flow for optimum operation. Speed and amplitude of jigging motion is also adjustable. (*Courtesy of Eagle Iron Works.*)

Jig beneficiation has been most effective in gravel operations where there is considerable disparity in specific gravity between the wanted and unwanted material such as separating low-gravity wood, shale, or chert of less than 2.30 specific gravity, for example, from the desirable gravel having a gravity greater than about 2.60. Jigging does not provide entirely clean-cut gravity separations and some heavy material is bound to overflow the weir along with the lightweight material; similarly, some lightweight material will be trapped and end up inthe product.

Heavy-Media Separation. Heavy-media separation (HMS), despite its high initial cost, roughly five times that of a jig of equal capacity, has been adopted by the majority of operators where need for beneficiation is indicated for the reason that a precise specific-gravity "cut" can be made. In this process, aggregate is fed into a water suspension of finely divided magnetite and ferrosilicon. Aggregate particles whose specific gravity is lower than the media will float and those of higher gravity than the media will sink. By precise adjustment of the ratio of magnetite to ferrosilicon and the water dilution of the mixture, media gravity can be adjusted to an accuracy of about 0.02 anywhere in the range from about 2.00 to 3.00. Actual specific gravities used in gravel operations usually range from about 2.40 to 2.60.

Figure 2.23 shows a flow sheet of a typical heavy-media plant. The actual separation of sink and float takes place within the "separatory vessel." The latter may variously be of a stationary-tub type, rotating drum, spiral screw, or cone. In the tub type, the sink mate-

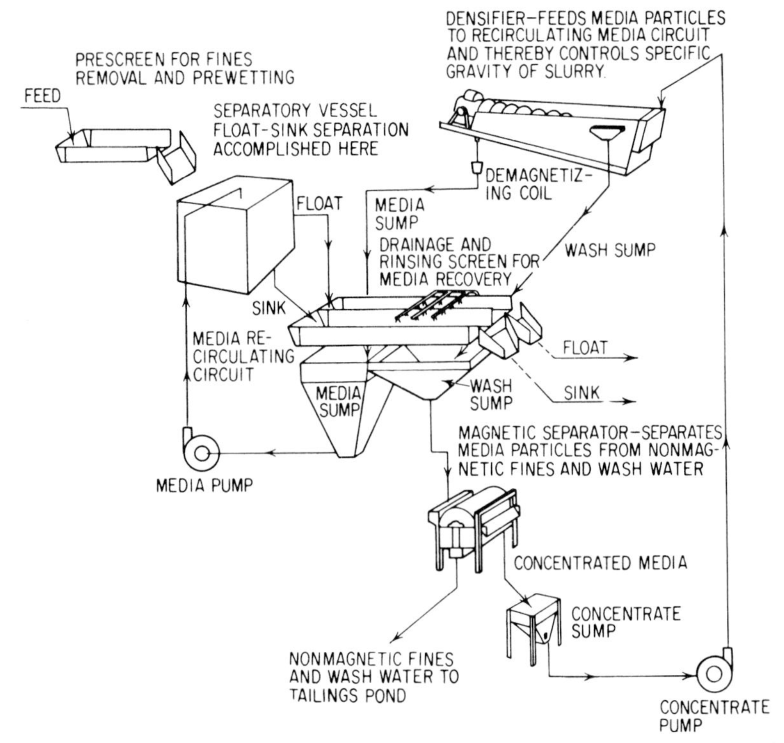

FIGURE 2.23 Schematic view of heavy-media separation (HMS) beneficiation plant. (*Courtesy of Eagle Iron Works.*)

rial is swept up from the bottom of the tank by a reciprocating paddle with the float material overflowing a weir along with the continuously circulating media. The drum type resembles a stationary mounted concrete mixer and has lifters attached to the inside perimeter of the drum which elevate the sink material from the bottom as the drum revolves and deposits it on a chute leading out of the drum. Again, float material overflows the lip of the drum along with the continuously circulating media. The spiral-screw type operates similarly to the screw dehydrator of Sec. 2.36 except that heavy medium replaces water. Float material overflows a weir and sink gravel is carried up the incline by the spiral screw. The cone type uses a deep pool to accomplish separation with the sink product air lifted up the center from the bottom apex of the cone. Advantages are claimed for each of the four types of separating devices and all have been successfully used in gravel or coarse-sand fraction beneficiation.

The remaining devices in Fig. 2.23 are all concerned with reclaiming the relatively expensive heavy medium and returning it for reuse. Both the sink and float aggregate particles are coated with the black adhering media material when leaving the separating device. Some of the medium drains off at once and can be returned for immediate reuse. The adhering medium which is washed off by water sprays is magnetically recovered, screw-dewatered, and then demagnetized to inhibit clumping of the media particles. Both magnetite and ferrosilicon are magnetic, and a properly operating plant loses a minimum of these materials in the wash water despite their extreme fineness. Media losses amounting to 3 to 4 cents per ton of product have been reported. Total estimated cost of media beneficiation has ranged from 35 to 50 cents per ton of product.

The HMS plant does not distinguish deleterious aggregate particles as such but separates the material quite successfully into two specific-gravity groups, i.e., higher than the media gravity and lower than the media gravity. Insofar as the deleterious particles are of lower gravity, they will be successfully removed. However, some acceptable aggregate will inevitably also be of lower gravity and be wasted by the process. Similarly, high-gravity material such as limonite will not be removed. Thus, although the process is capable of making a clean-cut gravity separation, its success is dependent upon the character of the deposit and the type of particles necessary to remove to make an acceptable product. HMS gravel plants currently in use reject about 20 percent of the feed. It is not unusual to find that about one-half of this reject is acceptable material under prevailing specifications.

Disposal of the reject material has been a troublesome problem for some installations considerably removed from a metropolitan area. In the latter case, it can usually be sold for driveway gravel, porous backfill around drains, and such uses.

AGGREGATE TESTS AND SPECIFICATIONS

Large projects, particularly those supported by public funds, operate under written "specifications" in the contract document. The word "specifications" is loosely applied to (1) a document which lists the properties and appropriate numerical limits for each measured characteristic or (2) a document giving detailed procedures by which a particular test to evaluate a given property is to be conducted. Sometimes both are contained in a single "specification" but most organizations now attempt to separate the two. The contractor, inspector, or engineer should be fully aware of this distinction and be familiar with both test methods and truly designated "specifications" for the job. The unwary may discover, for instance, that despite unchanged specification limits, a really different material is being asked for because of a recent change in test method. Such change may be not too obvious in the contract document. Conversely, development of an improved testing technique may necessitate change of test limits for a certain aggregate when no revision of basic quality is really intended.

2.40 Specifying Agencies

Public agencies such as cities, counties, state highway departments, and federal agencies provide specifications and, in some cases, test methods for concrete aggregates. Consulting engineering firms who prepare plans and specifications for public or private work may also furnish special requirements for concrete aggregates leading to even greater complexity in the overall number of provisions of which the aggregate producer must be aware in order to furnish materials successfully for such projects. The specifications of local agencies will naturally reflect requirements applicable to locally available aggregates, local climate, and prevalent construction practices. On the other hand, aggregates are sometimes specified appropriate for a given locality but inappropriate for another.

Three organizations of national scope which undertake preparation of aggregate specifications are (1) The American Association of State Highway Officials (AASHO), (2) The American Society for Testing and Materials (ASTM), and (3) federal government agencies. As the name implies, the first listed organization confines its attention strictly to highway construction and maintenance. The test methods as well as many material specifications prescribed by AASHO are often identical to those of ASTM.

American Association of State Highway Officials. This organization currently publishes its specifications in two volumes:

Part 1. Standard Specifications for Highway Materials

Part 2. Standard Methods of Sampling and Testing

In addition, annual "Interim Specifications" are published which are ultimately provided in the formal publications. These publications may be obtained from American Association of State Highway Officials, 341 National Press Building, Washington, DC 20004.

American Society for Testing and Materials. This is a broadly based organization made up of producers, consumers, and "general-interest" groups. Society regulations of ASTM prohibit any particular one of these groups from exercising predominant voting influence in establishing the provisions of a test method or material specification. Additionally, ASTM establishes "Recommended Practices" which, because of the nature of the information provided, are not intended for insertion in a contract document. One of these, enumerated later, is of particular interest to the field of concrete aggregates.

Many of the test methods and specifications enumerated below can be purchased from ASTM as separate reprints at nominal cost. All are contained in the annually published book of ASTM Standards, Section 4, procurable from The American Society for Testing and Materials, 1916 Race St., Philadelphia, PA 19103.

1. Specifications for Concrete Aggregates, ASTM C33. Provides specifications for fine and coarse aggregate, other than lightweight aggregate, for use in concrete. Several provisions of this specification were given in Secs. 2.25 and 2.26.
2. Specifications for Wire-Cloth Sieves for Testing Purposes ASTM: E 11. Gives detailed requirements for laboratory testing sieves.

3. Method of Test for Resistance to Abrasion of Large Size Coarse Aggregate by Use of the Los Angeles Machine, ASTM C535. This gives method for testing coarse aggregate larger than ¾ in. See Sec. 2.23 for additional details.
4. Method of Test for Resistance to Abrasion of Small Size Coarse Aggregate by Use of Los Angeles Machine, ASTM C131. Gives method of testing coarse aggregate smaller than 1½ in. See Sec. 2.23 for further details. Aggregate otherwise identical except for size may not give the same abrasion value when tested by this method and Method C535 above. The latter procedure is inappropriate for concrete aggregates for most uses.
5. Method of Test for Clay Lumps and Friable Particles in Aggregates, ASTM C142. This method provides a means, after soaking the sample, of determining the content of particles so soft and friable that they can be broken with the fingers.
6. Materials Finer than No. 200 Sieve in Mineral Aggregate by Washing, ASTM C117. This test evaluates the content of fine material passing No. 200 sieve which is either brought into suspension or dissolved when the aggregate is vigorously agitated with water. Has also been termed "elutriation test" or "loss by washing."
7. Lightweight Pieces in Aggregates, ASTM C123. This is a method by which the amount of material lighter than a selected specific gravity is determined by floating in heavy liquids. The coal and lignite determination, for instance, uses a liquid whose specific gravity is 2.00.
8. Organic Impurities in Sands for Concrete, ASTM C40. Potentially detrimental organic impurities in sands are detected by observing color, developed by the supernatant liquid when the sand is inundated in a 3% solution of sodium hydroxide. Glass color-comparison standards are available to facilitate assignment of color of liquid.
9. Effect of Organic Impurities in Fine Aggregate on Strength of Mortar, ASTM C87. Sands suspected of containing injurious amounts of organic matter detected by Method ASTM C40 above are evaluated for development of compressive strength in mortar by comparison with mortar strength of the same sand when thoroughly washed with sodium hydroxide solution to remove organic matter.
10. Potential Alkali Reactivity of Cement-Aggregate Combinations (Mortar Bar Method), ASTM C227. This is a long-time test wherein length changes are observed of mortar bars containing the sand under test when stored up to 1 year or longer. Bars are usually made with different cements having a range of alkali contents to establish sensitivity of the aggregate to different alkali-content cements. Warm, moist storage of the bars is used (100°F over water). Duration of the test over such long periods presents a serious obstacle to routine acceptance work but the test seems to be the most reliable now available to predict injurious expansions in concrete.
11. Potential Reactivity of Aggregates (Chemical Method), ASTM C289. This is the so-called "quick chemical method" for determining aggregates which may exhibit detrimental volume changes in concrete with high-alkali cements. The results of this test are not completely reliable but may be helpful when combined, for instance, with petrographic examination.
12. Potential Volume Change of Cement-Aggregate Combinations, ASTM C342. This is another long-time test involving exposing mortar bars to a series of moist storage and different temperature environments for 1 year or longer, to determine susceptibility of the particular aggregate-cement combination to detrimental volume change.
13. Potential Alkali Reactivity of Carbonate Rocks for Concrete Aggregates (Rock Cylinder Method), ASTM C586. This is a research screening method to determine

observable length changes occurring during immersion of carbonate rocks in sodium hydroxide (NaOH), indicating general level of reactivity and whether tests should be made to determine the effect of aggregate prepared from the rocks upon volume change of concrete.

14. Practice for Sampling Aggregates, ASTM D75. This method gives helpful material on sampling aggregates for concrete.

15. Method for Sieve Analysis of Fine and Coarse Aggregates, ASTM: C136. Gives detailed procedures for conducting sieve analyses of aggregates.

16. Standard Test Method for Index of Aggregate Particle Shape and Texture, ASTM D3398. This method gives particle index as an overall measure of particle shape and texture characteristics.

17. Test Method for soundness of Aggregates by Use of Sodium Sulfate or Magnesium Sulfate, ASTM: C88. The test is designed to simulate the destructive action of freezing and thawing to which some aggregates are vulnerable when water-soaked. In this case, the aggregates are alternately soaked in a saturated solution of either sodium or magnesium sulfate and then ovendried to drive off the water of crystallization. Reimmersion causes expansive action in the rock pores because of hydration of the desiccated crystals and is similar to the destructive action of formation of ice during freezing. Five cycles of the sulfate test is considered equivalent to many cycles of freezing and thawing.

18. Practice for Evaluation of Frost Resistance of Coarse Aggregates in Air-Entrained Concrete by Critical Dilation Procedures, ASTM C682. This details means of evaluating the frost resistance of coarse aggregate in air-entrained concrete using the slow-freezing method, ASTM C671, Test Method for Critical Dilation of Concrete Specimens Subjected to Freezing. Nonlinear length changes occurring in a concrete specimen as it is slowly cooled through the freezing point indicate that expansion caused by the freezing water cannot be accommodated and that the contained aggregate is causing the concrete to be vulnerable to frost attack.

19. Test Method for Specific Gravity and Absorption of Coarse Aggregate, ASTM C127. These methods were briefly discussed in Sec. 2.24. Two precautions in technique are advised regarding these tests: (1) In order to attain a surface-dry condition following a 24-h soak of the coarse aggregate, it is necessary to surface-dry each particle with a towel. It follows that it is a tedious operation to prepare more than a few hundred grams to a surface-dry condition. Furthermore, the surface-dry condition does not persist long unless the aggregate is sealed in a tight container to prevent further drying. (2) The specific-gravity determination requires that the aggregate be weighed when suspended in a basket in water. The buoyant force in water will be importantly influenced by air bubbles adhering to the basket and aggregate particles. Such bubbles must be dislodged by appropriate agitation to avoid a fictitiously low value of specific gravity.

20. Test Method for Specific Gravity and Absorption of Fine Aggregate, ASTM C128. The same precautions as noted above for coarse aggregate apply to sand but are even more difficult to remedy. Some operators aid dislodgment of entrapped air in the volumetric flask by application of vacuum. Others prefer use of clean, dry, white kerosene as the liquid in the flask rather than water. Although such deviations from standard ASTM procedures are not recommended for continued use, a new operator will do well to check his techniques by such methods. Fictitiously low values of specific gravity are not unusual for an inexperienced operator.

21. Test Method for Unit Weight and Voids in Aggregate ASTM C29. This method gives a precise method of determining the unit weight of dry aggregates under three standard conditions of compaction: (1) rodding, (2) jigging, and (3) loose (by shoveling).

22. Standard Descriptive Nomenclature for Constituents of Natural Mineral Aggregates, ASTM C294. This method provides brief description of minerals in aggregates and terms to describe them.

23. Standard Practice for Petrographic Examination of Standard Aggregates for Concrete, ASTM: C 295. This is an important "standard" as distinguished from a formal contract document, for the concrete engineer as well as the trained petrographer. The petrographer is required to have had formal scholastic training and experience, but the recommendations and techniques in this document are aimed specifically at the field of concrete aggregates. Likewise, the engineer required to examine the results of petrographic examinations will derive much benefit from study of this recommended practice to alert him more fully to the strengths and limitations of such examinations.

24. Terminology Relating to Concrete and Concrete Aggregates, ASTM C125.

25. Descriptive Nomenclature of Constituents of Natural Mineral Aggregates, ASTM C294. This gives a highly authoritative description of the minerals composing natural aggregates.

26. Practice for Reducing Field Samples of Aggregate to Testing Size, ASTM C702. Describes techniques for reducing samples to size appropriate for testing using mechanical splitter or quartering. Suggests caution that, when only a few large particles are present, any reduction may impair proper representation.

Federal Government Agencies. Projects supported by funds from Federal agencies operate under a variety of specifications. The Federal Housing Authority relies quite heavily on ASTM specifications for aggregates. Some of the test methods employed by the U.S. Bureau of Reclamation are given in Ref. 3. The U.S. Corps of Engineers publishes the "Handbook for Cement and Concrete" with quarterly supplements available from the U.S. Army Engineer Waterways Experiment Station, Vicksburg, Miss. Some Federal agencies may use specifications procurable from the nearest district office of the General Service Administration. Other agencies rely entirely on specifications prepared by the architect or engineering firm engaged to design the structure. The U.S. Bureau of Public Roads issues Standard Specifications for Construction of Roads and Bridges on Federal Highway Projects for construction supported solely by federal funds.

NATIONAL AGGREGATE INDUSTRY ASSOCIATIONS

The three national organizations listed below are concerned with aggregate production, testing, specifications, and similar matters pertaining to their individual industries. Additionally, most of them actively support laboratories which undertake studies of interest to producers, consumers, and general-interest groups. Printed reports of such studies and reprints of those published in the technical literature are generally available to qualified persons on request. Much of the progress in use of concrete aggregates derives from the devoted efforts of these organizations and their highly competent staffs.

1. National Stone Association, 1415 Elliot Place, N.W., Washington, DC 20007–2599

2. National Aggregates Association, 900 Spring St., Silver Springs, MD 20910

3. National Slag Association, 300 S. Washington, Alexandria, VA 22314

REFERENCES

1. Powers, T. C.: Causes and Control of Volume Change, *J. Res. Develop. Lab.,* PCA, Vol. 1, No. 1, adapted from Fig. 4, Jan. 1959, p. 37.
2. The American Society for Testing and Materials: Standards, Sec. 4, 1991.
3. "Concrete Manual," 8th ed., U.S. Bureau of Reclamation, revised 1981.
4. Alexander, K. M.: Strength of the Cement-Aggregate Bond, Proc. ACI, Vol. 56, Nov. 1959, pp. 377–390.
5. Schmitt, James W.: Effects of Mica, Aggregate Coatings and Water Soluble Impurities on Concrete, ACI Concrete International, Dec. 1990.
6. Ozol, M. A.: Shape, Surface Texture, Surface Areas and Coatings, ASTM STP 169B, 1978.
7. Mielenz, R. C.: Petrographic Examination, ASTM 169B, 1978, Chap. 33.
8. Buck, Alan D.: Recycled Concrete as a Source of Aggregate, *ACI Journal,* May 1977, pp. 212–219.
9. Kaplan, M. F.: Flexural and Compressive Strength of Concrete Affected by Properties of Coarse Aggregates, *ACI Journal,* Vol. 30, No. 11, May 1959, pp. 1193–1208.
10. P. C. Ailcin and P. K. Mehta: Effect of Coarse Aggregate Characteristics on Mechanical Properties of High Strength Concrete, March–April 1990, pp. 103–107, *ACI Materials Journal.*
11. Neville, A. M.: "Properties of Concrete," Wiley, New York, 1963.
12. Huang, Eugene Y.: A Test for Evaluating the Geometric Characteristics of Coarse Aggregate Particles, *ASTM Proceedings,* Vol. 62, 1962, pp. 1223–1242.
13. Rex, H. M., and R. A. Peck: A Laboratory Test to Evaluate the Shape and Texture of Fine Aggregate Particles, *Public Road Magazine,* Dec. 1956.
14. Gaynor, R. D.: Aggregate Properties and Concrete Strength, paper presented at 49th Annual Convention, National Sand and Gravel Association, 1965.
15. Woolf, D. O.: Significance of Tests and Properties of Concrete and Concrete Making Materials, ASTM STP 169, 1956 and ASTM STP 169A, 1966.
16. Proceedings of the Fourth International Symposium on the Chemistry of Cement: Washington D.C., National Bureau of Standards, Monograph 43, Vol. 2, p. 754, 1960.
17. Symposium on Alkali-Carbonate Rock Reactions, Highway Research Brd. Record 45, 1964, pp. 1–244.
18. Lea, F. M.: "The Chemistry of Cement and Concrete," Chemical Publishing Co., New York, 1971.
19. ACI Committee 221 Report, Guide for Use of Normal Weight Aggregates in Concrete, ACI Comm. 221R–1984.

CHAPTER 3
STEEL FOR REINFORCEMENT AND PRESTRESSING

David P. Gustafson
and Anthony L. Felder

Concrete is an inherently brittle material, strong in compression but weak in tension and lacking ductility. Small steel bars, on the other hand, while strong in tension and quite ductile, cannot support sizable compressive loads. In 1861, Coignet set forth the concept of embedding metallic reinforcement through the tensile areas of a concrete structural member, so that the reinforcement carried the tensile loads while the concrete carried the compressive loads. Before the turn of the century, many investigators had provided a rational theoretical basis for reinforced concrete design. This concept of reinforced concrete construction led to structural members that were not only vastly stronger than those made with plain concrete, but also possessed the ductility that was lacking in the plain concrete members.

The widespread use of structural concrete in the twentieth century is, in large measure, due to the use of reinforcing steel acting with the concrete to create a versatile, viable, and cost-effective building material, reinforced concrete. With the capability of being cast in an endless variety of shapes and forms that cannot be obtained using any other common building material, reinforced concrete offers many advantages in creating structures that are esthetically and economically superior in comparison with other structural materials.

REINFORCING STEEL

Reinforced concrete's unlimited variety in shape and form can be safely and economically achieved only through the use of standardized materials. In the earlier days of reinforced concrete, an extremely wide variety of proprietary reinforcing material was available, but obvious advantages have led to a high degree of standardization in modern reinforcing materials. In the United States, the American Society for Testing and Materials (ASTM) has produced standards that govern both the form and materials of modern reinforcing steel.

The principal forms that standard concrete reinforcement takes are deformed bars, welded-wire fabric, and strands, wires, and bars for prestressing. These types of reinforcement are described in this chapter. Steel fibers are another form of steel reinforcement. Discussion of steel fibers is outside the scope of this chapter. Readers are referred to the guide by ACI Committee 544 on Fiber Reinforced Concrete.*

3.1 Deformed Bars

Standard reinforcing bars are rolled with protruding lugs or deformations. A typical deformed steel reinforcing bar is shown in Fig. 3.1. These deformations can serve to increase the bond and eliminate slippage between the bars and the concrete. They act much as the tread on a tire. The use of these deformed bars is an American invention, and in the early days of concrete construction there were literally thousands of different types of reinforcing bars available. By the 1930s the modern deformed reinforcing bar as we know it, was established. With the publication of ASTM Specification A305 in the 1940s, standardization of the sizes and deformations of modern reinforcing bars was essentially completed. This ASTM specification standardized the sizes of deformed bars and set certain specific minimum requirements for deformation spacing, height, and permissible gaps in deformations. The actual pattern of the deformations was not prescribed in the ASTM specification, however, and each producer was free to use its own distinctive pattern as long as the specification requirements were met. Figure 3.2 shows a few of the patterns used. Specification A305 is now obsolete, since the size and deformation requirements have been incorporated into the newer ASTM reinforcing-bar specifications A615, A616, A617, and A706.

The sizes of the standard deformed reinforcing bar are designated by bar numbers. The bar number is based on the number of eighths of an inch in the nominal bar diameter. The nominal bar diameter of a deformed bar is the diameter of a plain round bar having the same weight per foot as the deformed bar. The actual maximum diameter is always larger than the nominal diameter. This increase is always neglected in design, except for the cases of sleeves or couplings that must fit over the bar when the actual maximum diameter, of course, must be used. Table 3.1 shows the nominal specification dimensions for deformed reinforcing bars. The large bars, Nos. 14 and 18, are not commonly stocked and are normally encountered only as column reinforcing on projects where large quantities of these bars are used. In most construction, where isolated instances require only a few items of very heavy reinforcing, building codes permit the use of "bundles" of several smaller bars. The proper method of designating the size of a standard deformed bar is by its "bar number." On a drawing, bill of material, invoice, bar tag, etc., the bar number is preceded by the conventional number symbol (#). When more than one bar of the same size is indicated, the number of bars precedes the size marking; thus "6-#3" indicates 6

*Guide for Specifying, Mixing, Placing, and Finishing Steel Fiber Reinforced Concrete (ACI 544.3R), "ACI Manual of Concrete Practice," Pt. 5.

FIGURE 3.1 A typical deformed steel reinforcing bar. (*Courtesy of Ceco Corporation.*)

FIGURE 3.2 Typical deformed reinforcing bars showing identification marks. (*Courtesy of Concrete Reinforcing Steel Institute.*)

deformed bars of size 3 (approximately ⅜ in. diameter), and "12-#8" would refer to 12 deformed bars of size number 8 (approximately 1 in diameter).

Plain round steel bars, which were the first form of reinforcement, are presently used as column spirals, as expansion-joint dowels, and in the fabrication of bar mats. The requirements for welded plain bar or rod mats are prescribed by ASTM specification A704. The AASHTO Bridge Specifications* permit the use of plain bars for ties.

*"Standard Specifications for Highway Bridges," 14th ed., American Association of State Highway and Transportation Officials, Washington, D.C., 1989.

TABLE 3.1 ASTM Standard Reinforcing Bars

Bar-size designation No.	Weight, lb/ft	Nominal dimensions, round sections		
		Diameter, in	Cross-sectional area, in^2	Perimeter, in
3	0.376	0.375	0.11	1.178
4	0.668	0.500	0.20	1.571
5	1.043	0.625	0.31	1.963
6	1.502	0.750	0.44	2.356
7	2.044	0.875	0.60	2.749
8	2.670	1.000	0.79	3.142
9	3.400	1.128	1.00	3.544
10	4.303	1.270	1.27	3.990
11	5.313	1.410	1.56	4.430
14	7.65	1.693	2.25	5.32
18	13.60	2.257	4.00	7.09

3.2 Grade

Reinforcing bars are hot-rolled from a variety of steels in several different strength grades. Most reinforcing bars are rolled from new steel billets, but some are rolled from used railroad-car axles or railroad rails that have been cut into rollable shapes. Several strengths of reinforcing bars are available.

Reinforcing bars are produced to standards established by the American Society for Testing Materials. Table 3.2 lists the standard reinforcing-bar grades that are used today and a summary of the important physical property requirements. The ASTM specifications A615, A616, and A617 prescribe requirements for certain mechanical properties. These specifications do not require any particular chemical analysis or composition of steel. (Phosphorus content is limited to 0.06% by the A615 specification.) Each manufacturer is free to use a wide range of chemical composition, so long as the specified mechanical properties are met. The critical mechanical properties are the yield strength, tensile strength, elongation, and bendability. The yield strength, the maximum elastic stress that the bar can withstand, is probably the most important mechanical property to the designer. The tensile strength, the maximum stress that the bar can withstand in tension without failing, is generally of lesser importance. Elongation is the amount of stretch that a bar undergoes when loaded. The specifications require that total stretch to failure measured over an original 8-in length of bar be not less than a certain minimum percentage that varies with bar size and grade. Bendability is a measure of the ability of the bar to be bent to a minimum radius bend without cracking. In the ASTM specifications, it is the required diameter of the pin around which the specimen must be bent without cracking. The bend-test requirements differ with grade and bar size.

The ASTM A706 specification for low-alloy steel bars has limits or controls on chemical composition because the bars are intended for welding. There are actually two limits on chemical composition. One limit is on individual chemical elements; the other limit is on carbon equivalents. The specification also has more restrictive requirements for mechanical properties than the other reinforcing bar standards. Tensile properties are controlled by a limit on yield strength, and the tensile strength cannot be less than 1.25 times the actual yield strength. The A706 specification requires larger values of minimum percentage of elongation. Bend test requirements are more restrictive (smaller bend test pin diameters) than those in the other reinforcing bar specifications. Table 3.2 includes the requirements for mechanical properties in the A706 specification. Before specifying A706 bars, local availability should be investigated.

Using the strength design method with Grades 60 and 75 reinforcing bars can often permit a substantial saving in size of members and quantity of materials required for a structure. There is, however, a practical limit on how strong the reinforcing steel should be in standard reinforced concrete construction. All strengths of steel have approximately the same amount of elongation for the same applied tensile stress. If one steel has twice the yield strength of another, we can apply twice as much stress but we will get twice as much elongation. Under fairly moderate loads, the steel reinforcing will stretch about as much as the concrete surrounding it can stretch without severe cracking. If more load is applied, the steel may safely carry the load, but the concrete cover will crack. This is not only unsightly, but generally will permit corrosion of the reinforcing. Steels with yield strengths greater than 60,000 psi generally cannot use their greater strength as standard tensile reinforcing without causing cracking of the concrete, unless special provisions are made in the design of the member.

The present trend in reinforced-concrete construction is toward the use of higher-strength-grade reinforcing bars. Use of these high-strength bars results in a significant reduction of steel tonnage and size of concrete structural members, with resulting economy in labor and other material. With smaller columns, beams, and girders, the

TABLE 3.2 ASTM Mechanical Requirements for Reinforcing Bars

Type of steel and ASTM designation	Bar numbers range	Grade	Minimum yield, psi*	Minimum tensile strength, psi	Minimum percentage elongation in 8 in		Bend test† (*d* = diameter of specimen)	
Billet, A615	3–6	40	40,000	70,000	#3	11	#3, #4, #5	3½*d*
					#4, #5, #6	12	#6	5*d*
	3–11,	60	60,000	90,000	#3, #4,		#3, #4, #5	3½*d*
	14, 18				#5, #6	9	#6, #7, #8	5*d*
					#7, #8	8	#9, #10, #11	7*d*
					#9, #10, #11,		#14, #18 (90°)	9*d*
					#14, #18	7		
	11, 14,	75	75,000	100,000	#11, #14, #18	6	#11	7*d*
	18						#14, #18 (90°)	9*d*
Rail, A616	3–11	50	50,000	80,000	#3, #7	6	For Grades 50 and 60	
					#4, #5, #6	7	#3, #4, #5	6*d*
					#8, #9		Per SI‡	3½*d*
					#10, #11	5	#6, #7, #8	6*d*
							Per SI‡	5*d*
							#9, #10	8*d*
	3–11	60	60,000	90,000	#3, #4		Per SI‡	7*d*
					#5, #6	6	#9, #10	8*d*
					#7	5	per SI‡	7*d*
					#8, #9		#11 (90°)	8*d*
					#10, #11	4½	per SI‡	7*d*
Axle, A617	3–11	40	40,000	70,000	#3, #7	11	#3, #4, #5	3½*d*
					#4, #5, #6	12	#6 through #11	5*d*
					#8	10		
					#9	9		
					#10	8		
					#11	7		
	3–11	60	60,000	90,000	#3, #4, #5		#3, #4, #5	3½*d*
					#6, #7	8	#6, #7, #8	5*d*
					#8, #9		#9, #10, #11	7*d*
					#10, #11	7		
Low alloy, A706	3–11,	60	60,000	80,000§	#3, #4		#3, #4, #5	3*d*
	14, 18		minimum;		#5, #6	14	#6, #7, #8	4*d*
			78,000		#7, #8, #9,		#9, #10, #11	6*d*
			maximum		#10, #11,	12	#14, #18	8*d*
					#14, #18	10		

*Yield point or yield strength; see ASTM specifications.

†Test bends 180° unless noted otherwise.

‡Under supplementary requirements of ASTM A616 only. ACI 318 requires rail-steel bars (ASTM A616) to meet Supplementary Requirement S1.

§Tensile strenghth shall not be less than 1.25 times the actual yield strength.

dead load is reduced, resulting in an accumulated saving for tall buildings. In addition, there are other advantages such as a gain in floor space and a reduction in story height.

3.3 Identification

With the wide range of strength available in otherwise identical reinforcing bars and the possible danger that lower-strength bars might find their way into locations where the design calls for high-strength bars, the ASTM has established a standard marking system for deformed reinforcing bars. There are two general systems of bar marking. Both systems serve the same basic purpose of identifying the maker, size, type of steel, and grade of each bar. In both systems an identity mark denoting the producer of the bar, the size number of the bar, and a symbol denoting the type of steel used are branded on every bar by engraving the final roll used to produce the bars so as to leave raised symbols between the deformations. Figure 3.3 shows the standard marking system for deformed bars. The producer identity mark which signifies the mill that rolled the bar is usually a single letter, or in some cases a symbol. The bar size follows the maker mark and is followed by a symbol indicating new billet steel (S), rail steel (I), axle steel A, or low-alloy steel (W).

The ASTM specifications for billet-steel, rail-steel, axle-steel and low-alloy reinforcing bars (A615, A616, A617, and A706, respectively) require identification marks to be rolled into the surface of one side of the bar to denote the producer's mill designation, bar size, type of steel, and minimum yield designation. Grade 60 bars show these marks in the following order.

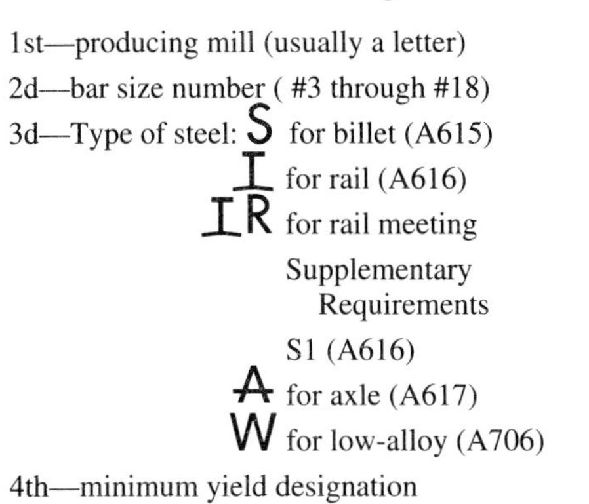

1st—producing mill (usually a letter)

2d—bar size number (#3 through #18)

3d—Type of steel: S for billet (A615)
I for rail (A616)
IR for rail meeting Supplementary Requirements S1 (A616)
A for axle (A617)
W for low-alloy (A706)

4th—minimum yield designation

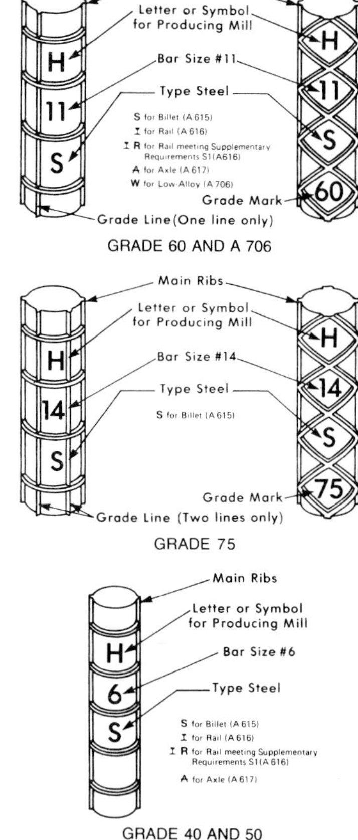

FIGURE 3.3 The identification marking system for deformed reinforcing bars. (*Courtesy of Concrete Reinforcing Steel Institute.*)

Minimum yield designation is used for Grade 60 and Grade 75 bars only. Grade 60 bars can either have one (1) single longitudinal line (grade line) or the number 60 (grade mark). Grade 75 bars can either have two (2) grade lines or the grade mark 75. A grade line is smaller and is located between the two main ribs which are on opposite sides of all bars made in the United States. A grade line must be continued through at least five deformation spaces, and it may be placed on the side of the bar opposite the bar marks. A grade mark is the fourth mark on the bar. Grades 40 and 50 bars are required to have only the first three identification marks (no minimum yield designation). *Variations:* Bar identification marks may also be oriented to read horizontally (at 90° to those illustrated). Grade mark numbers may be placed within separate consecutive deformation spaces to read vertically or horizontally.

3.4 Fabrication

The typical concrete reinforcing bar is a hot-rolled mill product. It begins as a steel billet or, in some cases, as a billet-shaped section cut from used railroad steel. The billet is heated to red heat and passed through successive stands of rolling-mill rolls. Each pass reduces the cross-sectional area and lengthens the resulting bar. The billet is reduced to the required bar size. The rolls of the last pass are generally deeply engraved to produce the standard deformations of the deformed reinforcing bar. The bars are then cropped to the standard mill length, usually about 60 ft long. After cooling, the bars are bundled in lifts of about 5 tons, tagged to permit identification of the heat of steel from which they were made, and shipped to the reinforcing-bar fabricator. Sufficient tests are conducted at the mill to ensure that each heat meets ASTM specification requirements. In the original mill length, the bars are not readily usable in very many structures. The basic job of the fabricator is to convert the mill lengths to the usable reinforcing steel desired by the concrete structural designer.

An architect or engineer designing a concrete structure prepares drawings showing the complete design of the structure. From these design drawings a set of detailed placing drawings is prepared showing the number, size, length, and bending dimensions of each piece of reinforcing steel and its location in the structure. The detail drawings may be prepared by the designer or by the reinforcing-bar fabricator in accordance with the design drawings. In either case the fabricator will generally prepare the bar lists showing the full information of each mark of bent or straight bar that is shop-fabricated. In some areas, union regulations prohibit the shop fabrication of bent reinforcing bars, requiring that this be done on the jobsite. Where possible, it is preferred that all reinforcing steel be shop-fabricated for the fabrication can be performed with greater accuracy and less expense using the special machinery in the fabricating shop.

The first basic shop operation is cutting the reinforcing bars to length. Standard fabricating practice is to shear the bars with a tolerance of plus or minus 1 in. from the detailed length. Where the cut end of the bars must be exceptionally square, or where bevels or V grooves are required the bars must be saw-cut. Bars that are to be provided as straight bars are then bundled, each size and length making a separate bundle. The bundle is then tagged to show the bar mark and the bundle is ready to ship. Much of the reinforcing will, however, have to be bent before it is bundled for shipping. Shop bending is generally divided into three classes—light, heavy, and special bending. Light bending is more expensive per hundred pounds of bars than heavy and special bending. Light bending includes the small 3 bars, all stirrups and ties, and all the larger bars (4 through 11) which are bent at more than 6 points, bent in more than one plane, radius-bent with more than one radius (three maximum) in any one bar in one plane, or a combination of radius and other bending in one plane. Heavy bending is the bending of bar sizes 4 through 11, which are bent at not more than six places in one plane, or the radius bending to one radius of these heavy bars. Special bending is bending to special tolerances, all radius bending in more than one plane, all multiple plane bending containing one or more radius bends, and all bending for precast units. Figures 3.4 through 3.6 show typical fabricating operations.

Bar mats are fabricated from individual bars. Two layers of bars are assembled at right angles to each other by clipping or welding at the intersections. ASTM specification A184 covers mats fabricated from deformed bars.

3.5 Handling and Storage of Reinforcing Bars

Reinforcing bars should be handled and stored in such a manner that they will not be bent out of the planned shape. They should not be stored directly on the ground. Open storage of reinforcing steel will in most cases result in rusting. The suitability of rusting reinforcing has

FIGURE 3.4 Reinforcing bars being sheared to length in a fabricating shop. (*Courtesy of Ceco Corporation.*)

been the subject of some concern in past years. Studies going back to 1920 have demonstrated that a rust film or tight mill scale, rather than harming the bond between the steel and concrete, actually causes an improvement in the bonding characteristics of the steel. Since the end of World War II, two major studies have confirmed the results of the earlier studies that rust on reinforcing that is to be placed has no harmful effect on the bond strength. The U.S.

FIGURE 3.5 Shop bending of reinforcing bars. (*Courtesy of Ceco Corporation.*)

FIGURE 3.6 Spiral reinforcing for a reinforced-concrete column being formed. (*Courtesy of Ceco Corporation.*)

Bureau of Reclamation Concrete Laboratory conducted an extensive series of tests that led to the conclusion that normal handling was sufficient preparation even for extremely rusty reinforcing steel, and that sandblasting, wire brushing, or rubbing with burlap gave no better bond. Tests conducted at the West Virginia University confirm that rust on reinforcing bars has no adverse effect on bond. Where the reinforcing bars are very badly rusted, the cross-sectional area may have been reduced sufficiently so that the bars are unsuited for use. This can be checked by cleaning and weighing a length of bar to ensure that it will meet the specifications.

FIGURE 3.7 A roll of welded-wire fabric. (*Courtesy of Ceco Corporation.*)

3.6 Welded-Wire Fabric

Where lighter reinforcement is required for such items as concrete pavements, driveways, sidewalks, swimming pools, reservoirs, and thin floor slabs (such as those used in concrete-joist construction), it is usually more economical to use welded-wire fabric than to place individual reinforcing bars. Welded-wire fabric consists of longitudinal and transverse cold-drawn steel wires arranged to form a square or rectangular mesh. The wires are then electrically welded together at every intersection. The appearance of the resulting fabric is

TABLE 3.3 Cold-Drawn Wire Sizes (ASTM A82)

Size	Diameter in	Area, in^2	Weight, lb/ft
W-0.5	0.080	0.005	0.017
W-1	0.113	0.010	0.034
W-1.5	0.138	0.015	0.051
W-2	0.159	0.020	0.068
W-2.5	0.178	0.025	0.085
W-3	0.195	0.030	0.102
W-3.5	0.211	0.035	0.119
W-4	0.225	0.040	0.136
W-4.5	0.240	0.045	0.153
W-5	0.252	0.050	0.170
W-5.5	0.264	0.055	0.187
W-6	0.276	0.060	0.204
W-7	0.298	0.070	0.238
W-8	0.319	0.080	0.272
W-10	0.356	0.100	0.340
W-12	0.390	0.120	0.408
W-14	0.422	0.140	0.476
W-16	0.451	0.160	0.544
W-18	0.478	0.180	0.612
W-20	0.504	0.200	0.680
W-22	0.529	0.220	0.748
W-24	0.553	0.240	0.816
W-26	0.575	0.260	0.884
W-28	0.597	0.280	0.952
W-30	0.618	0.300	1.020
W-31	0.628	0.310	1.054

shown in Fig. 3.7. Using welded-wire fabric permits lightly reinforcing large areas with a minimum need for supervision and inspection. The wire used in making the fabric is generally produced in accordance with ASTM Specification A82. The plain cold-drawn wire sizes are designated by a size number consisting of a W followed by a number indicating the nominal cross-sectional area in hundredths of a square inch. The wires are provided in sizes W-0.5 to W-31, as shown in Table 3.3. The fabric is designated by a "style" code which specifies the wire spacing and size. For example, 4 × 6-W10 × W6 indicates that the longitudinal wires are W-10 on 4-in centers and that the transverse wires are W-6 on 6-in centers. The fabric is produced in accordance with ASTM Specification A185.

It is still common to find an earlier standard welded-wire fabric that uses the American Steel and Wire Gauge numbers in place of the W-numbers. Thus 6 × 6 − 10 × 10 would refer to the equivalent of 6 × 6 − W1.5 × W1.5. These previously used wire sizes are shown in Table 3.4.

Those styles of fabric in which the transverse wire provides the minimum steel area necessary for fabricating and handling are termed "one-way fabrics." Where significant reinforcement is provided in both transverse and longitudinal directions, the fabric is called "two-way." Common stock styles of welded wire fabric are shown in Table 3.5.

TABLE 3.4 ASW Wire Sizes

Gauge	Diameter, in	Area, in	Weight, lb/ft
0000000	0.4900	0.189	0.641
000000	0.4615	0.167	0.569
00000	0.4305	0.146	0.495
0000	0.3938	0.122	0.414
000	0.3625	0.103	0.351
00	0.3310	0.086	0.292
0	0.3065	0.074	0.251
1	0.2830	0.063	0.214
2	0.2625	0.054	0.184
3	0.2437	0.047	0.158
4	0.2253	0.040	0.136
5	0.2070	0.034	0.115
6	0.1920	0.029	0.099
7	0.1770	0.025	0.084
8	0.1620	0.021	0.070
9	0.1483	0.017	0.059
10	0.1350	0.014	0.049
11	0.1205	0.011	0.039
12	0.1055	0.009	0.030
13	0.0915	0.007	0.022
14	0.0800	0.005	0.017

Note: ASW = American Steel and Wire Gauge.

TABLE 3.5 Common Stock Styles of Welded-Wire Fabric

Style designation		Longitudinal or transverse steel area, in^2/ft	Approximate total weight lb/100 ft^2
Current designation (by W-number)	Previous designation (by steel wire gauge)		
	Rolls		
6 × 6 − W1.4 × W1.4	6 × 6 − 10 × 10	0.028	19
6 × 6 − W2.0 × W2.0	6 × 6 − 8 × 8*	0.040	27
6 × 6 − W2.9 × W2.9	6 × 6 − 6 × 6	0.058	39
6 × 6 − W4.0 × W4.0	6 × 6 − 4 × 4	0.080	54
4 × 4 − W1.4 × W1.4	4 × 4 − 10 × 10	0.042	29
4 × 4 − W2.0 × W2.0	4 × 4 − 8 × 8*	0.060	41
4 × 4 − W2.9 × W2.9	4 × 4 − 6 × 6	0.087	59
4 × 4 − W4.0 × W4.0	4 × 4 − 4 × 4	0.120	82
	Sheets		
6 × 6 − W2.9 × W2.9	6 × 6 − 6 × 6	0.058	39
6 × 6 − W4.0 × W4.0	6 × 6 − 4 × 4	0.080	54
6 × 6 − W5.5 × W5.5	6 × 6 − 2 × 2†	0.110	75
4 × 6 − W4.0 × W4.0	4 × 4 − 4 × 4	0.120	82

*Exact W-number size for 8-gauge is W2.1.
†Exact W-number size for 2-gauge is W5.4.

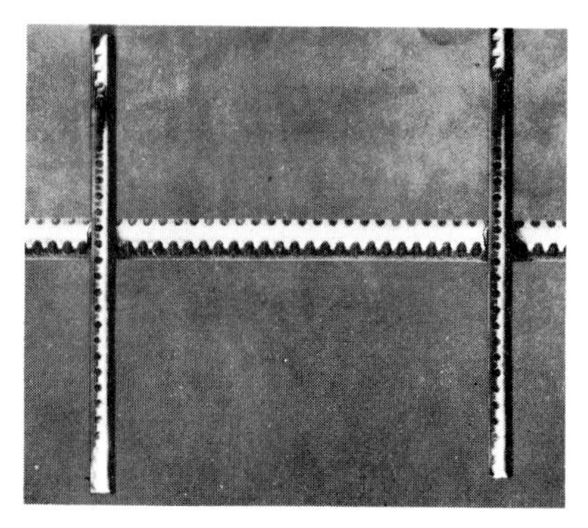

FIGURE 3.8 Typical deformed wires used for welded-wire mesh. (*Courtesy of U.S. Steel Corp.*)

Deformed-Wire Fabric. Deformed wires are also used in producing welded-wire fabric. A cold-worked, deformed steel wire, as shown in Fig. 3.8, replaces the normal cold-drawn wire. The deformations along the wires are thought to improve the bonding characteristics for better concrete crack control. The deformed wires are manufactured to ASTM Specification A496. The deformed-wire sizes are designated by a size number consisting of a D followed by a number indicating the nominal cross-sectional area of the wire in hundredths of a square inch. The wires are provided in sizes D-1 to D-31, as shown in Table 3.6. Individual deformed wires can also be used for reinforcing concrete structures in the same manner as deformed reinforcing bars.

The deformed-wire welded-wire fabric is produced to ASTM Specification A497 and is similar to standard fabric except for the use of the deformed wire for either the longitudinal or both the longitudinal and transverse wires.

3.7 Coated Reinforcement

Reinforced concrete is inherently a durable and near maintenance-free construction material under normal conditions. But structures exposed to deicing salts, or located near seacoasts, for example, can experience damage as a result of corrosion of the reinforcement. Coated reinforcement can be used as a corrosion-protection system. ASTM has issued specifications for zinc-coated (galvanized) and epoxy-coated reinforcement. These specifications are as follows: ASTM A767 for galvanized reinforcing bars, A775 for epoxy-coated bars, and A884 for epoxy-coated wire and welded-wire fabric. Similar to the style of the other ASTM standards for uncoated reinforcement, the specifications for coated reinforcement are product standards. The specifications A767, A775, and A884 prescribe the requirements for coated reinforcement to the point of shipment from the manufacturer's facility. Project specifications should include requirements for all construction operations and procedures affecting coated reinforcement as well as uncoated reinforcement on the jobsite.

3.8 Accessories

Accessories, or bar supports, are used in placing the reinforcing bars to ensure that the bars are correctly positioned in relation to the forms. The varied accessories all serve the basic function of supporting the reinforcing steel, in proper vertical position, enabling the bar setters to space and position the bars accurately. The bar supports should be heavy enough and numerous enough to support the bars properly. If too few supports are used, the bars will sag between supports and will not be correctly positioned.

Bar supports can be made of steel wire, concrete, fiber-reinforced concrete, plastic, or other materials. The CRSI "Manual of Standard Practice"* presents industry practices for

*"Manual of Standard Practice," 25th ed., 1990, Concrete Reinforcing Steel Institute, Schaumburg, Ill.

TABLE 3.6 Deformed-steel Wire Sizes (ASTM A496)

Size	Diameter, in	Area, in^2	Weight, lb/ft
D-1	0.113	0.01	0.034
D-2	0.159	0.02	0.068
D-3	0.195	0.03	0.102
D-4	0.225	0.04	0.136
D-5	0.252	0.05	0.170
D-6	0.276	0.06	0.204
D-7	0.298	0.07	0.238
D-8	0.319	0.08	0.272
D-9	0.338	0.09	0.306
D-10	0.356	0.10	0.340
D-11	0.374	0.11	0.374
D-12	0.390	0.12	0.408
D-13	0.406	0.13	0.442
D-14	0.422	0.14	0.476
D-15	0.437	0.15	0.510
D-16	0.451	0.16	0.544
D-17	0.465	0.17	0.578
D-18	0.478	0.18	0.612
D-19	0.491	0.19	0.646
D-20	0.504	0.20	0.680
D-21	0.517	0.21	0.714
D-22	0.529	0.22	0.748
D-23	0.541	0.23	0.782
D-24	0.553	0.24	0.816
D-25	0.564	0.25	0.850
D-26	0.575	0.26	0.884
D-27	0.586	0.27	0.918
D-28	0.597	0.28	0.952
D-29	0.608	0.29	0.986
D-30	0.618	0.30	1.020
D-31	0.628	0.31	1.054

all types of bar supports, including wire bar supports, precast concrete blocks, cementitious fiber-reinforced bar supports, and all-plastic bar supports and side form spacers. Typical types and sizes of wire bar supports are shown in Fig. 3.9.

Bar supports for supporting galvanized or epoxy-coated reinforcing bars should be compatible with the coated reinforcement. When reinforcing bars are used as support bars, the bars should be coated with coating material that is compatible with the bars being supported. The project specifications should include requirements for the bar supports and support bars. Such requirements are given in the ACI 301 Specifications, Specifications for Structural Concrete for Buildings.*

*Specifications for Structural Concrete for Buildings (ACI 301–89).

SYMBOL	BAR SUPPORT ILLUSTRATION	BAR SUPPORT ILLUSTRATION PLASTIC CAPPED OR DIPPED	TYPE OF SUPPORT	TYPICAL SIZES
SB	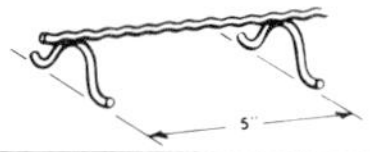 5"	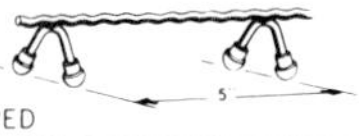 CAPPED 5"	Slab Bolster	¾, 1, 1½, and 2 inch heights in 5 ft. and 10 ft. lengths
SBU*	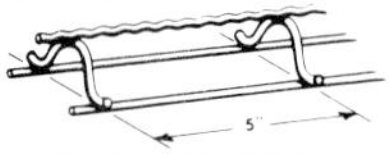 5"		Slab Bolster Upper	Same as SB
BB	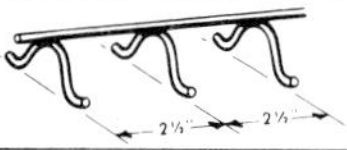 2½" 2½"	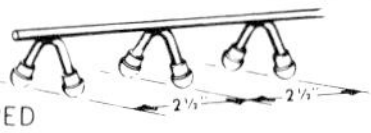 CAPPED 2½" 2½"	Beam Bolster	1, 1½, 2, over 2" to 5" heights in increments of ¼" in lengths of 5 ft.
BBU*	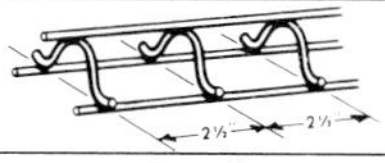 2½" 2½"		Beam Bolster Upper	Same as BB
BC	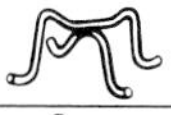	DIPPED 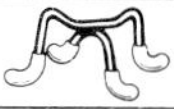	Individual Bar Chair	¾, 1, 1½, and 1¾" heights
JC	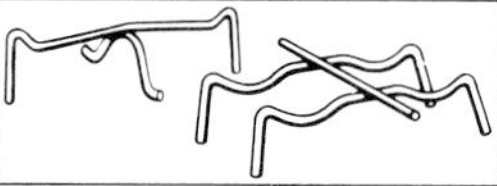	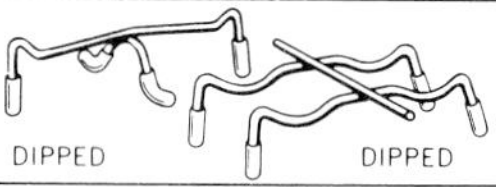 DIPPED DIPPED	Joist Chair	4, 5, and 6 inch widths and ¾, 1 and 1½ inch heights
HC		CAPPED 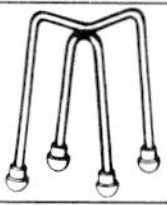	Individual High Chair	2 to 15 inch heights in increments of ¼ inch
HCM*			High Chair for Metal Deck	2 to 15 inch heights in increments of ¼ in.
CHC	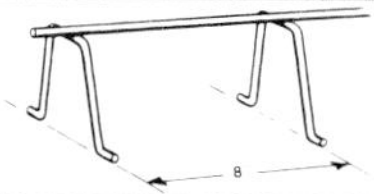 8"	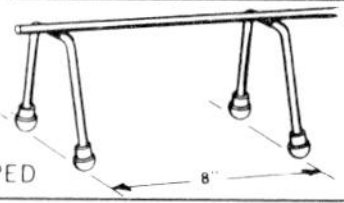 CAPPED 8"	Continuous High Chair	Same as HC in 5 foot and 10 foot lengths
CHCU*	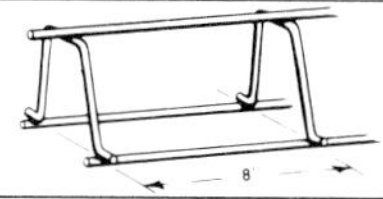 8"		Continuous High Chair Upper	Same as CHC
CHCM*	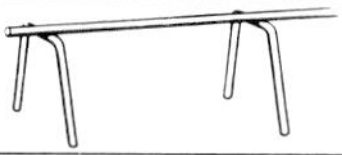		Continuous High Chair for Metal Deck	Up to 5 inch heights in increments of ¼ in.
JCU**	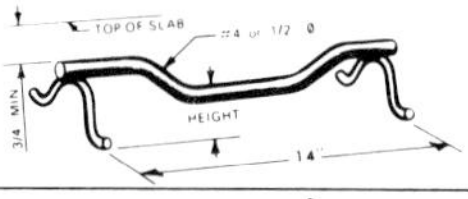 TOP OF SLAB #4 or 1/2 Ø HEIGHT 3/4 MIN. 14"	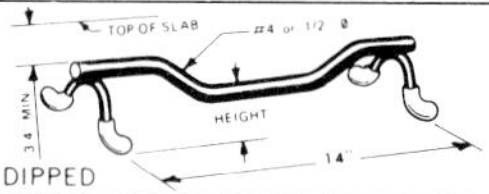 TOP OF SLAB #4 or 1/2 Ø HEIGHT 3/4 MIN. 14" DIPPED	Joist Chair Upper	14" Span Heights −1" thru +3½" vary in ¼" increments
CS	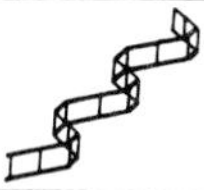		Continuous Support	1½" to 12" in increments of ¼" in lengths of 6′-8″

*Usually available in Class 3 only, except on special order.
**Usually available in Class 3 only, with upturned or end bearing legs.

FIGURE 3.9 Typical types and sizes of wire bar supports (*Courtesy of Concrete Reinforcing Steel Institute.*)

3.9 Placing

Accurate positioning of the reinforcing steel is of utmost importance, as discussed in Chap. 25. The reinforcing steel must be securely held in its proper position as shown on the design drawings and adequately tied and supported before concrete is placed, and secured against displacement within the placement tolerances specified in the project specifications.

Historically bars have been most commonly held together with tie wire. Usually No. 16½ or 16 gauge or heavier black, soft-annealed wire is used. Ten to fifteen pounds of wire are usually required per ton of reinforcing steel. Sufficient intersections of the reinforcing are tied to prevent shifting of the reinforcing. It is not necessary to tie every intersection. The tying adds nothing to the strength of the finished structure other than holding the bars in their proper position until the concrete has been placed. Although lapped splices are tied, the surrounding concrete forms the actual splice. A typical tie is shown in Fig. 3.10.

The project specifications should include requirements for handling, storage, and placing of coated reinforcing bars to minimize damage to the coating on the bars. There should also be specified limits on permissible coating damage due to shipment, handling and field placing operations, and, when required, the repair of damaged coating. The ACI 301 Specifications include requirements for these items.*

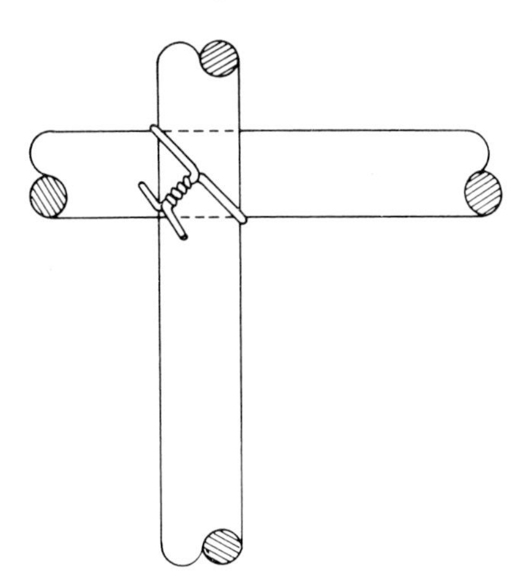

FIGURE 3.10 A typical "tie" holding crossing-reinforcing bars in place.

3.10 Splices

Splices of reinforcement are unavoidable in any sizeable structure. Properly designed splices are a key element in any well-executed design. The design drawings or project specifications must clearly show or describe all splice locations and the performance required. Three methods are used to splice reinforcing bars: lap splices, mechanical connections, and welded splices.

Lap Splices. The lap splice, when conditions permit and when it will satisfy all requirements, is generally the most common method for splicing reinforcing bars. The bars may be spaced or in contact. For bar-to-bar splices, contact splices are preferred for the practical reason that, wired together, they are more easily secured against displacement during concrete casting. In noncontact lap splices, the bars must not be spaced too widely apart, permitting a zigzag crack between bars. Design codes and specifications limit the spacing of bars in noncontact lap splices; for example, in the current ACI 318 Code, the spacing should not exceed one-fifth the lap length nor 6 in.

Design codes do not permit #14 and #18 bars to be lap-spliced, except in compression only to #11 and smaller bars.

*Specifications for Structural Concrete for Buildings (ACI 301–89).

Mechanical Connections. When the use of lap splices becomes uneconomical or impractical (long lap lengths, location of construction joints, provision for future construction), causes congestion or field placing problems, or when their use is not permitted by design codes or specifications, then mechanical connections, welded splices, or proprietary lap-splice connector systems* may be used, as appropriate. Lesser value welded splices may be used under code-prescribed conditions.

Design codes and specifications specify a minimum connection strength. For example, the ACI 318 Code requires a full mechanical connection to develop in tension or compression, as required, at least 125 percent of the specified yield strength of the bar. The use of mechanical connections of less strength than 125 percent of the yield strength of the bars, and end-bearing connections are permitted by the Code under certain conditions.

There are essentially three basic types of mechanical connections:

- Tension-compression (can resist both tensile and compressive forces)
- Compression only (also known as *end-bearing* mechanical connection)
- Tension only

A variety of proprietary mechanical connection devices are currently available. Except for the tension-only connection, most mechanical connection devices align and secure the joined ends of the reinforcing bars in an in-line connection. The methods used in commercially available proprietary devices to connect bars are as follows:

- Threading
- Cold-swaging, cold-extruding, or hot-forging
- Steel-filled coupling sleeves
- Grout-filled coupling sleeves
- Clamping or friction

Lack of space here precludes a comprehensive treatment of the commercially available proprietary mechanical connection devices. Readers are referred to a report by ACI Committee 439.† Comprehensive descriptions of the physical features, mechanical characteristics, and installation procedures of the various available mechanical connection devices are presented in the committee report.

Welded Splices. Design codes and specifications specify the required strength of welded splices. For example, the ACI 318 Code states: "a full welded splice shall have the bars butted and welded to develop in tension at least 125 percent of the specified yield strength of the bar." The use of welded splices of less strength than 125 percent of the yield strength of the bars is permitted by the Code under certain conditions.

There are many different welding methods and procedures, for steels of different chemical composition are quite differently affected by the heat of welding. The welding procedure necessary to produce sound crack-free welds depends upon the chemical composition of the reinforcing bars. A procedure that is suitable for one chemical composition can be totally unsuited for another composition of the same strength grade. It is essential that the composition of the steel to be welded be determined before a welding procedure is established. A basic rule that has often been stated is: "Know the composi-

---

*See discussion of proprietary lap-splice connector systems in "Reinforcement: Anchorages, Lap Splices and Connections," 3d ed., Concrete Reinforcing Steel Institute, Schaumburg, Ill, 1990.

†Mechanical Connections of Reinforcing Bars, in "Manual of Concrete Practice," Pt. 3, American Concrete Institute, P.O. Box 19150, Detroit, MI 48219.

tion of the material that you are trying to weld. If you don't know, find out, and then adopt the most convenient and economical procedure that will give sound crack-free welds in steel of that composition."

Sound economical welds require that the actual steel composition be determined before establishing a welding procedure. All welding should conform to "Structural Welding Code—Reinforcing Steel" (ANSI/AWS D1.4–79) of the American Welding Society.*

ASTM A706 reinforcing bars are intended for welding. Chemical composition and carbon equivalent of the steel are controlled. The carbon equivalent (CE) formula in the A706 specification is

$$CE = \%C + \%Mn/6 + \%Cu/40 + \%Ni/20 + \%Cr/10 - \%Mo/50 - \%V/10$$

The current edition of ANSI/AWS D1.4 contains the same formula for carbon equivalent.

For A706 low-alloy steel bars, mill test reports showing the results of chemical analyses of each heat of steel must be furnished by the producer. The chemical analysis (available on request) for A615 billet-steel bars is incomplete for determining welding requirements under ANSI/AWS D1.4–79. Special complete analyses may be secured, usually at extra cost. Chemical analyses are not ordinarily meaningful for rail-steel (A616) and axle-steel (A617) bars.

The most commonly used manual welding process in the field is electric arc welding. The welding heat is provided by an electric arc drawn between the reinforcing bars and an electrode. The *shielded-metal-arc method* uses a consumable metallic rod electrode coated with a material that gives off an inert gas to prevent contamination of the molten weld metal by the atmosphere. The electrodes that are to be used should be clearly specified. The coating on low-hydrogen electrodes must be thoroughly dry when used. It is vitally important that the manufacturer's recommendations be followed to the letter, and that no coated electrode that has been wet be used under any conditions.

Resistance welding is used only in shop fabrication of reinforcing steel, particularly welded-wire fabric and reinforcing bar mats. The welding is accomplished by a combination of heat and pressure. The welding heat is provided by the flow of a low-voltage current between two electrodes. Because of the equipment required, resistance welding is never used in field welding of reinforcing steel.

Types of Welded Splices. Probably the least desirable welded splice is the *lap welded splice* shown in Fig. 3.11*a*. When a lap welded splice is loaded the eccentricity of the bars causes a bending distortion, as shown in Fig. 3.11*b*. This distortion tends to split the concrete cover, producing a very unsatisfactory splice. Lap welded splices are particularly unsatisfactory for splicing the larger sizes of reinforcing bars. Where smaller bars are to be spliced and ties and stirrups are supplied to prevent splitting, satisfactory splices can be made. ANSI/AWS D1.4 limits the bar sizes in a lap welded splice to No. 6 and smaller.

The *butt-welded splices* shown in Figs. 3.12–3.14 are the preferred reinforcing-bar splice welds. The stress is transferred directly and concentrically across the joint, producing a splice that is compact and efficient. The actual detail of the butt-welded splice will vary somewhat, particularly with changes in bar sizes. For bar sizes 9 and larger, the single-V-groove weld of Fig. 3.12 is generally used. For bars smaller than No. 9, ANSI/AWS D1.4 recommends a 60° single V-groove weld with split-pipe backing.

Direct butt-welded splices for bars placed in a vertical position are shown in Fig. 3.14.

Connection of crossing bars by small arc welds, known as "tack welds," is not recom-

*Available from American Welding Society, P.O. Box 351040, 2501 N.W. 7th St., Miami, FL 33125. At press time, the American Welding Society was in the final stages of adopting a 1990 edition of the D1.4 Code.

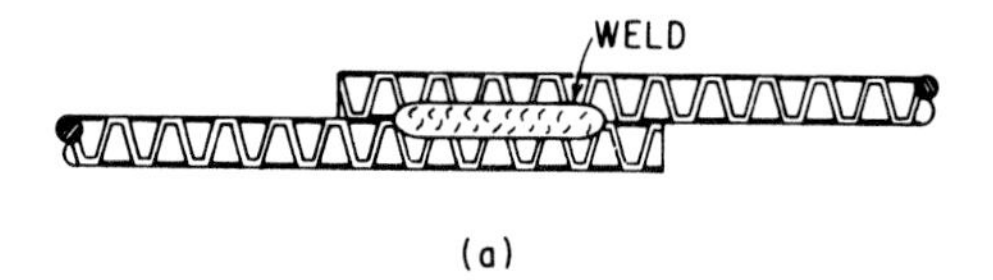

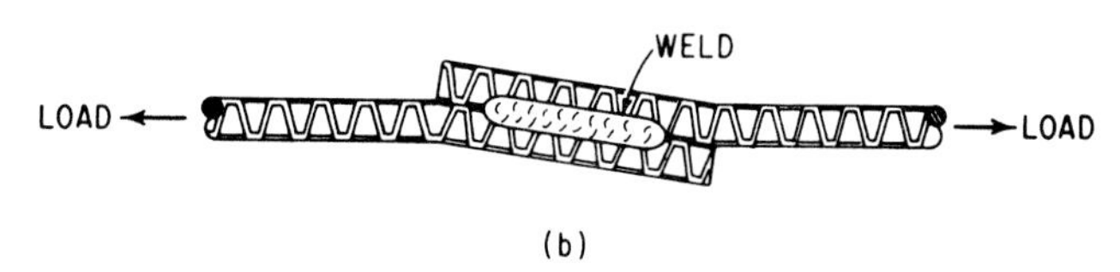

FIGURE 3.11 A welded single-lap splice.

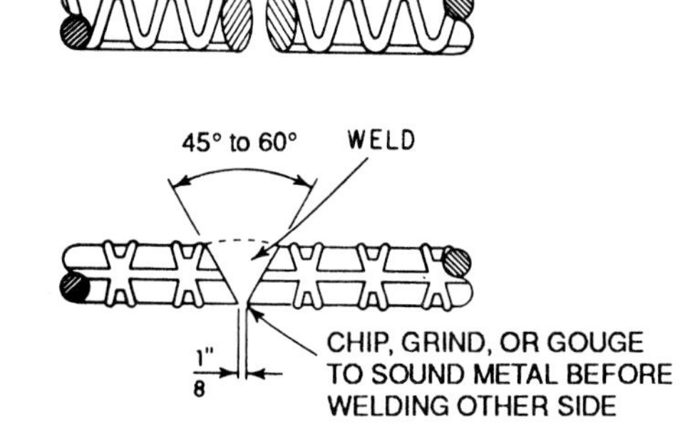

FIGURE 3.12 A single-V-groove welded splice (for bars placed in horizontal position).

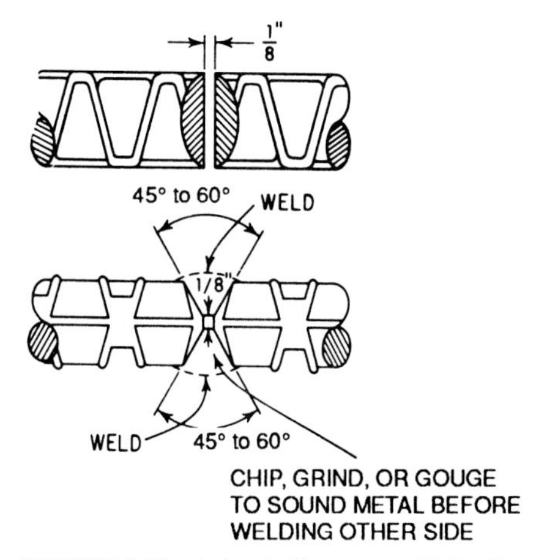

FIGURE 3.13 A double-V-groove welded splice (for bars placed in horizontal position).

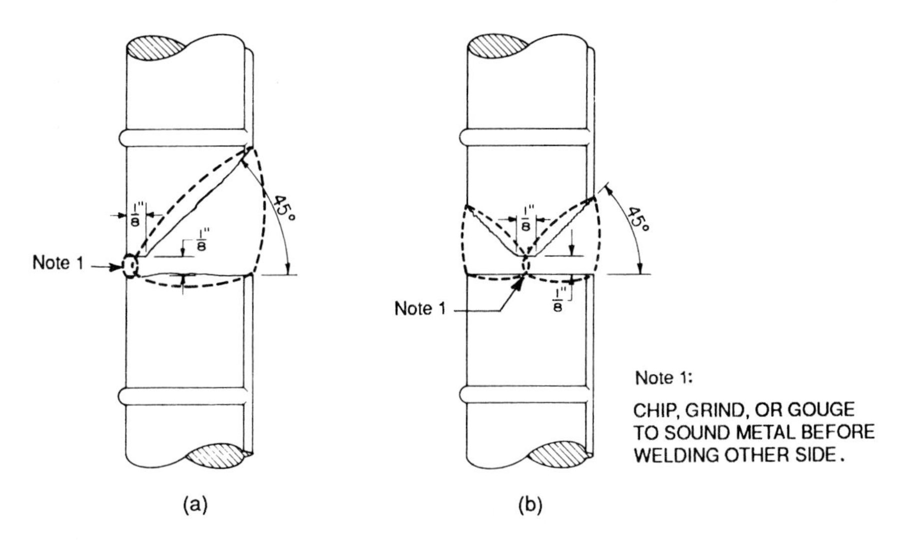

FIGURE 3.14 Direct butt splices for bars placed in vertical position. *(a)* Single-bevel groove weld; *(b)* double-bevel groove weld.

mended. Building codes and ANSI/AWS D1.4 prohibit tack welding for assembly of reinforcement unless authorized by the engineer. Unless these welds are made in conformance with all requirements of ANSI/AWS D1.4, they tend to cause a metallurgical "notch" effect and may affect the strength of the bars. In the same manner, indiscriminate striking of arcs can also weaken reinforcing bars and should also be avoided.

A dangerous situation can arise where welding is being performed in the vicinity of the high-strength tendons for concrete prestressing. No welding should be allowed near these tendons, since even minor weld splatter can cause failure of a tendon during the stressing operation. The prestressing tendon should never be used as a welding ground. Best practice does not permit any cutting or welding of reinforcing steel around the prestressing tendons, once the tendons are in place.

PRESTRESSED-CONCRETE REINFORCING

As discussed in the section on conventional reinforcing steel, concrete is a material that is weak in tension, and reinforcing steel must be supplied to carry the tensile forces which develop in normal structural behavior of the members. The concrete can, however, be prestressed in compression, so that the application of loads that would cause tension acts only to relieve the compression preload. An oversimplication of this is the simple case of lifting a stack of books as a beam. If sufficient pressure is supplied by pressing the ends of a stack of books together, the compression forces between the books counteract the tension forces set up by the bending action and the books will indeed carry their own weight as a beam. If standard reinforcing bars are preloaded by applying tension loads to the bars and the loads are maintained while the concrete that is placed around the bars sets, the prestressed-concrete member that results will be capable of carrying sizable "tensile" loads in the same manner as the stack of books.

To accomplish this in actual practice, an extremely-high-strength steel is required.

Concrete is subject to shrinkage and creep, so that over a period of time in service the length of a concrete structure may shorten as much as 0.0006 to 0.0008 in per inch of length. With steel reinforcing this would amount to a loss in preload of 18,000 to 24,000 psi. If the prestressing steel had originally been loaded to only 18,000 to 24,000 psi, this would amount to a complete loss of preload. Even with preloads of 30,000 to 40,000 psi, the irregular nature and high percentage of the preload loss would require a high factor of safety. Prestressing thus would not be economically competitive with standard reinforced concrete. The solution to the problem of the loss of preload by shrinkage and creep lies in the use of extremely high strength steels. If steel preload stresses greater than 100,000 psi can be obtained, the loss of preload from shrinkage and creep will be a small percentage of the total preload. Prestressing is the only method in which the extremely high-strength steels can be effectively used to reinforce a concrete structure. The high-strength steels must be elongated a great deal before their full tensile strength can be utilized. If such a high-strength steel is simply embedded in the concrete, the concrete surrounding the bar will have to crack very seriously before the steel can develop its full strength. By pre-stretching the high-strength reinforcing steel and anchoring it to the concrete, the most desirable stresses and strains in both materials are produced. The steel is pre-elongated, avoiding excessive lengthening under applied loads. The concrete is precompressed, avoiding cracking of the concrete under applied loads.

The methods of constructing a prestressed-concrete member are widely varied and are discussed in detail in Chap. 41. Many prestressed members are *precast.* The members are cast at a permanent plant or temporary plant at the site of the structure and then later erected. Precasting permits better control of the resulting member and is widely used for mass-producing various types of sections. *Cast-in-place* concrete is, as the name implies, cast at its final location. This requires more forming and false-work than precasting and generally is not subject to the finer control that is possible at a precasting plant. There can, however, be a great savings in the cost of transporting and erecting extremely large and heavy members. In some cases a form of composite construction is used, with part of the structure precast and erected and the remainder of the structure then cast in place. This often permits some savings in forms and falsework.

Tensioning the prestressing steel strands or bars, which individually are called tendons, can be done before or after the concrete is placed. In *pretensioning,* the tendons are tensioned before the concrete is placed. This is done by stretching the tendons between abutments before casting the member. After the member is cast and the concrete is set, the prestress is transferred from the abutment to the member by cutting the tendons between the abutments. The prestress is transferred to the concrete through bond between the prestressing steel and the concrete at the ends of the beam. Pretensioning is generally performed only at permanent precasting plants, although some pretensioned members have been field-cast where the abutments could be economically provided.

In *posttensioning,* the tendons are always tensioned after the concrete has been placed and has set. The tendons are prestressed against the hardened concrete and are then anchored to it. There is no need for heavy abutments for the tensioning operation, and posttensioning can be used on either precast or cast-in-place members. When a prestressed member is to be posttensioned, it is necessary to provide a means of preventing the prestressing steel from bonding to the concrete before tensioning. The type of prevention used depends on whether the tendons are to be bonded or unbonded after they have been stressed. Bonded tendons are bonded throughout their length to the surrounding concrete, generally by grouting. For bonded tendons, a duct or conduit is typically used to form a void in the concrete through which the tendon is placed after the concrete has set. These ducts can be made of metal or plastic. Sometimes a solid rubber core is used instead, which must then be withdrawn before the tendon can be placed. In some cases, an inflated rubber tube is used instead of a solid rubber core, to facilitate removal.

Cement grout is almost universally used where the tendons are to be bonded to the concrete after posttensioning. The cement grout also serves to protect the steel against corrosion. The grout is packed into the ducts or voids through holes provided at each end of the member. The grout is forced into one end of the member until it is discharged from the other end. For very long members it is generally applied at both ends until it is discharged from a center vent. The best practice is to wash the cables with water before grouting, forcing the excess water out of the ducts or voids with compressed air. The grout mix is determined by the space limitations around the prestressing tendons.

Where the tendons are to remain unbonded after posttensioning, the tendons are well lubricated to prevent corrosion and facilitate posttensioning and are then sheathed in plastic. The plastic sheathing should be properly lapped and sealed at all seams to prevent any leakage of mortar. Where the tendons are unbonded, the tendons must be protected from corrosion. The usual methods of providing corrosion protection for the tendons are by galvanizing or greasing.

3.11 Prestressing Steel

The steel used for prestressing concrete can be divided into three classes: strands, wires, and bars. In every case, the prestressing steel will be considerably higher in strength than standard concrete-reinforcing bars.

The ACI 318 Building Code requires prestressing tendons to conform to Standard Specification for Uncoated Seven-Wire Stress Relieved Strand for Prestressed Concrete (ASTM A416), the Standard Specification for Uncoated Stress-Relieved Wire for Prestressed Concrete (ASTM A421), or the Standard Specification for Uncoated High-Strength Steel Bar for Prestressing Concrete (ASTM A722). Other steel strands or wire or bars are permitted by the Code providing they conform to the minimum requirements of A416, A421, or A722 and have no properties that make them less satisfactory.

Prestressing Strand. The time and labor required to place and tension many small prestressing wires can be extensive as the number of wires increases. To reduce the placing and tensioning expense, a strand of cable composed of several prestressing wires twisted together can be used. Prestressing strands are the most popular form of prestressing steel. The first strands were used for pretensioned concrete beams. The bond between the concrete and steel strands is generally all that is necessary to maintain the prestress of the beam. The strands used for pretensioning are almost universally seven-wire strands consisting of six wires helically twisted around a seventh straight wire. The straight central wire is made slightly larger than the other six wires to assure that the helical wires tightly grip the center wire.

ASTM Specification A416 prescribes the requirements for two types and two grades of seven-wire strand. Low-relaxation strand is regarded as the standard type, and it is the most widely used. The other type of strand is stress-relieved (normal relaxation). The two grades are 250 and 270, which have minimum ultimate tensile strengths (breaking strengths) of 250,000 and 270,000 psi, respectively, based on the nominal area of the strand. The physical properties of seven-wire strand are shown in Table 3.7.

ASTM Specification A779 covers stress-relieved, compacted seven-wire strand. The strand is compacted by drawing through a die and subsequently stress-relieved. Compacted strand has a larger area of steel in any nominal diameter as compared to regular seven-wire strand. The larger area results in a larger breaking strength than a regular strand. Specification A779 a Supplement I, which prescribes the requirements for low-relaxation compacted strand. The physical properties of compacted strand are shown in Table 3.8.

TABLE 3.7 Properties of Seven-Wire Prestressing Strand (ASTM A416)

Nominal diameter of strand, in	Nominal weight of strands, lb/1000 ft	Nominal steel area of strand, in^2	Breaking-strength minimum, lb	Yield strength	
				Normal-relaxation minimum, lb	Low-relaxation minimum, lb
			Grade 250		
¼	122	0.036	9,000	7,650	8,100
5⁄16	197	0.058	14,500	12,300	13,050
⅜	272	0.080	20,000	17,000	18,000
7⁄16	367	0.108	27,000	23,000	24,300
½	490	0.144	36,000	30,600	32,400
0.6	737	0.216	54,000	45,900	48,600
			Grade 270		
⅜	290	0.085	23,000	19,550	20,700
7⁄16	390	0.115	31,000	26,350	27,900
½	520	0.153	41,300	35,100	37,170
0.6	740	0.217	58,600	49,800	52,740

In 1988, ASTM issued Specification A886 for indented, stress-relieved, seven-wire strand. As implied by the title of the specification, the outer wires of the strand are indented to enhance the bond of the strand to concrete. This type of strand is intended for use in pretensioned, prestressed concrete members which require shorter transfer and development lengths.

At press time, ASTM was in the process of developing new specifications for two- and three-wire strand, and epoxy-coated seven-wire strand.

Other types of strands for special applications are also available. These include galvanized strand, strand with a minimum ultimate tensile strength of 300,000 psi, strands with diameters larger than those listed in the ASTM A416 specification, and stainless-steel strand. A report* by ACI Committee 439 discusses these other types of strands. It is also suggested that manufacturers be contacted directly regarding these other types of strands.

*Steel Reinforcement—Physical Properties and U.S. Availability (ACI 439.4R), "ACI Manual of Concrete Practice," Pt. 3.

TABLE 3.8 Properties of Compacted Seven-Wire Strand (ASTM A779)

Nominal diameter of strand, in	Nominal weight of strands, lb/1000 ft	Nominal steel area of strand, in^2	Type	Breaking-strength minimum, lb	Yield strength	
					Stress-relieved minimum, lb	Low-relaxation minimum, lb
0.5	600	0.174	270	47,000	40,900	42,300
0.6	873	0.256	260	67,400	58,700	60,690
0.7	1176	0.346	245	85,430	74,300	76,800

Prestressing Wire. Stress-relieved prestressing wire is manufactured by a cold-drawing process. The drawing process reduces the diameter and increases the tensile strength of the wire. The strength of the wire is increased by cold work of each drawing; consequently, the smaller the diameter of the final wire, the higher its ultimate strength. The as-drawn wire is very stiff and has very little ductility or elongation before failure. The as-drawn wire is very difficult to handle and takes a marked coil set, with the wire tending to retain the diameter of its shipping coil. The drawing process also results in nonuniform locked-in stresses. To relieve these locked-in drawing stresses and to improve the physical properties of the wire, the as-drawn wire is subjected to a closely controlled continuous heat treatment.

When the wire is to be used in unbonded prestressed concrete where there is a danger of corrosion, the wires are often furnished in galvanized form. Where the wire is to be galvanized, the as-drawn wire is passed through a bath of molten zinc.

Since the cold-drawing process tends to increase the strength of the wire with each additional drawing, the wire strength will vary with diameter. The smaller-diameter wires which have been drawn more often will show a higher ultimate tensile strength than the larger-diameter wires. Low-relaxation wire is covered in Supplement I of Specification A421. Table 3.9 shows the physical properties of prestressing wire.

High-Strength Bars. For use in posttensioned prestressed concrete, high-tensile-strength, alloy-steel bars are available as plain (Type I) or deformed (Type II) in diameters ranging from ⅝ in (Type I) or ¾ in (Type II) to 1⅜ in. The bars are manufactured from high-alloy, hot-rolled-steel rounds that are heat-treated and then cold-stretched by loading the bar to not less than 80 percent of its minimum ultimate tensile strength. The cold stretching coldworks the bar, producing a high yield strength. The cold stretching also acts as a proof load, ensuring that any defective bars will be eliminated in the manufacturing process. The rigidity of the bars permits quick and easy placement, and they require less frequent securing in the form than other prestressing steel. The frictional losses due to tendon wobble and curvature are lower than cable-stressing systems.

The ultimate tensile strength of the high-tensile bars is considerably less than that obtained with prestressing strand or wire. Commercial high-tensile prestressing bars are produced with a minimum ultimate tensile strength of 150,000 psi.

Specification A722 has supplementary requirements which apply only when specified by the purchaser. These additional requirements cover bend tests, reduction of area for Type I plain bars, and reporting of the chemical composition of the bars.

TABLE 3.9 Properties of Stress-Relieved Wire (ASTM A421)

Nominal diameter, in	Nonimal weight, lb/ft	Nominal area, in^2	Tensile-strength minimum, psi		Yield-strength minimum, psi			
					Stress-relieved		Low-relaxation	
			Type BA	Type WA	Type BA	Type WA	Type BA	Type WA
0.192	0.0983	0.02895	—*	250,000	—*	212,500	—*	225,000
0.196	0.1025	0.03017	240,000	250,000	204,000	212,500	216,000	225,000
0.250	0.1667	0.04909	240,000	240,000	204,000	204,000	216,000	216,000
0.276	0.2032	0.05983	235,000	235,000	199,750	199,750	211,500	211,500

*Not commonly furnished in Type BA wire.

3.12 Prestressing Systems

One of the unique features of prestressed concrete is that it is perhaps the only major structural field that is young enough that patents and patents pending encompass almost the entire field. Although the basic principle of prestressing cannot be patented, the actual details of accomplishing the prestressing can. The various systems involve either the details of end anchorage or the method of applying a prestress or both. There are hundreds of patents and dozens of workable prestressing systems. The profusion of patents and systems that exist appears to be quite confusing. There is, however, a great similarity between many of the systems. It is unnecessary and perhaps impossible to know the intimate details of every system in use in the United States today. The vast majority of prestressing work is being done with very few systems or variations of these systems.

Pretensioning. The simplest systems of prestressing concrete members are the various pretensioning methods. The Hoyer system is perhaps the most basic. The tendons are stretched between buttresses or bulkheads located at the ends of a casting bed. The concrete is placed and permitted to harden. The tendons are cut loose and the prestress transfers to the concrete by bond. The casting bed and buttresses must be strong enough to withstand the preload. By placing the buttresses several hundred feet apart several members can be cast at one time by providing shutters to separate the members along the casting bed. Quick-release grips (chucks) are generally used to grip the ends of the tendons at the loading buttresses.

Posttensioning Anchorages. There a wide variety of anchorages and systems available for use in posttensioning. The ACI 318 Building Code establishes strength requirements for anchorages and couplers. For bonded and unbonded tendons, anchorages and couplers have to develop at least 95 percent of the breaking strength of the tendons when tested in an unbonded condition without exceeding anticipated set.

No attempt will be made in this chapter to describe the various posttensioning systems. The reader is referred to the Post-Tensioning Institute's Manual* for descriptions of the commercially available systems.

---

*"Post-Tensioning Manual," 4th ed., 1985, Post-Tensioning Institute, 1717 West Northern Avenue, Phoenix, Arizona 85021.

CHAPTER 4
ADMIXTURES, CURING MEDIUMS, AND OTHER MATERIALS

Raymond J. Schutz

ADMIXTURES

Admixtures are materials other than water, aggregates, and portland cement used as an ingredient of concrete and added to the batch immediately before or during mixing. Admixtures are used to modify the properties of concrete, such as to improve workability, increase strength, retard or accelerate strength development, and increase frost resistance. Generally, an admixture will affect more than one property of concrete, and its effect on all the properties of the concrete must therefore be considered.

Admixtures may increase or decrease the cost of concrete by lowering cement requirements for a given strength, changing the volume of the mixture, or lowering the cost of concrete placing and handling operations. Control of the setting time of concrete may result in economies such as decreasing waiting time for floor finishing or extending the time in which the concrete is plastic and therefore eliminating bulkheads and construction joints.

Admixtures have been used almost since the inception of the art of concreting. It is reported that the Roman builders used oxblood as an admixture in their concrete and masonry structures. Research has shown that oxblood is an excellent air-entraining agent. During the early part of this century it was common practice to add Gold Dust soap to concrete as a waterproofing agent. This soap was rich in stearates and acted as a combination air-entraining and dampproofing agent. It is doubtful whether these early uses of admixtures were carried out on a scientific basis; however, the users did realize their benefits.

4.1 Testing

Since the efficiency of an admixture may be affected by the type and amount of cement used, or modifications of the aggregate grading or mixture proportion, it is desirable to pretest all admixtures with the concrete to be used on the job. Since many admixtures affect more than one property of concrete, sometimes adversely, the admixture should be tested for more than one of the properties of concrete. To evaluate an admixture fully, tests should be carried out on the following properties and with the following test methods:

Slump. Method for test of slump of portland cement concrete; ASTM Designation C143.

Air content. Method of test for air content of freshly mixed concrete by the pressure method; ASTM Designation C231.

Time of setting. Method of test for time of set of concrete mixtures by penetration resistance; ASTM Designation C403.

Compressive strength. Compressive strength of molded concrete cylinders; ASTM Designation C39. Concrete compression and flexure test specimens, making and curing in laboratory; ASTM Designation C192.

Flexural strength. Flexural strength of concrete; ASTM Designation C78.

Resistance to freezing and thawing. Either of the following test methods would be applicable: ASTM Designations C290-61T, C291-61T.

Volume change. Method of test for volume change of cement mortar and concrete; ASTM Designation C157.

To test concrete containing admixtures only for compressive strength may lead to the choice of an admixture undesirable for the purpose intended. As an example, concrete with high compressive strength and poor durability to freezing and thawing would be undesirable for highway construction. A concrete with high compressive strength and high drying shrinkage would be undesirable for use in most structures. Evaluation should be made on the total effect of the admixture on all the characteristics necessary for the concrete in relation to its ultimate use.

4.2 Addition

Admixtures are generally used in relatively small quantities, as little as ⅓ fluid oz per sack of cement. It is therefore important that suitably accurate dispensing equipment be used in most cases. Liquid admixtures, including job mix solutions, may be added with the mixing water or added on nonabsorptive or saturated aggregates. Powdered admixtures should preferably be added onto the fine aggregate. All admixtures can be added after the concrete has been partially mixed. Under no circumstances should admixtures be added to portland cement prior to the addition of mixing water.

Liquid-admixture dispensers are available and have proved accurate and durable. These dispensers work on either a time-flow or a positive-displacement principle. A time-flow type of dispenser should be equipped with a transparent measuring tube, and each addition should be dispensed through this tube since any inaccuracy in flow rate with this type of dispensing equipment will change the volume delivered. A positive-displacement-type dispenser does not depend on flow rate for accuracy but is generally adaptable only to dispensing fixed quantities of the admixture.

It may be necessary to add two or more admixtures of different types to the concrete mixture to obtain desired characteristics in the concrete. As an example, a water-reducing admixture and an air-entraining admixture may be required. Most admixtures are compatible when mixed in concrete, but under no condition should two admixtures of different types be allowed to mix together prior to addition to the mixer. In most instances the admixtures will react, causing precipitation and loss of effectiveness. The incompatibility of two admixtures intermixed alone, or in water, does not indicate that such admixtures will be affected when combined in concrete.

4.3 Storage

Powdered admixtures generally have an indefinite shelf life if stored dry. Liquid admixtures may freeze or precipitate at low temperatures. Freezing may permanently damage some liquid admixtures; other liquid admixtures may be frozen and thawed without damage. The manufacturers' storage directions should be followed. Certain admixtures are shipped in powder form to be dissolved in water before addition to the concrete. In such cases only agitated storage tanks should be used to ensure that all the constituents of the admixture are added with each dose.

4.4 Types and Uses

Air-Entraining Agents. The purpose of any air-entraining admixture is to entrain small discrete air bubbles of proper size and spacing in the concrete mixture. Field and laboratory experience has demonstrated conclusively that proper air entrainment increases the resistance of concrete to disintegration by freezing and thawing by a very large factor.

As water freezes, it undergoes a volume change of approximately 9%. This increases in volume results in hydraulic pressure which can be sufficient to disintegrate the concrete. The evaporation of water and subsequent crystallization of deicing salts may also cause a similar phenomenon. Purposefully entrained air will provide voids spaced close enough to reduce pressures that tend to build up. It is evident that the efficiency of any air-entraining agent will depend on size and spacing of the voids induced by that agent. Numerous investigations have demonstrated that an air-void size of between 0.003 and 0.05 in and a spacing factor of 0.004 to 0.008 in will result in optimum durability. Expressed as a percentage of the volume of the concrete this range will be 4 to 7% (see Fig. 4.1).

Air-entraining agents expand the volume of the cement matrix. Where a mixture is deficient in paste volume (usually four bags of cement cubic yard or less) concrete strength will be increased. With richer mixes where sufficient paste volume is present, the dilution of paste with entrained air will weaken the mixture. However, this strength penalty is more than offset by the increase in durability imparted by entrained air.

In the plastic concrete, the entrained air voids tend to block the capillaries, which are the natural paths for escaping bleed water; therefore, air-entrained concrete will tend to bleed less than non-air-entrained concrete. In the hardened concrete these air bubbles will also tend to interrupt the capillaries, resulting in lowered absorption. Entrainment of air

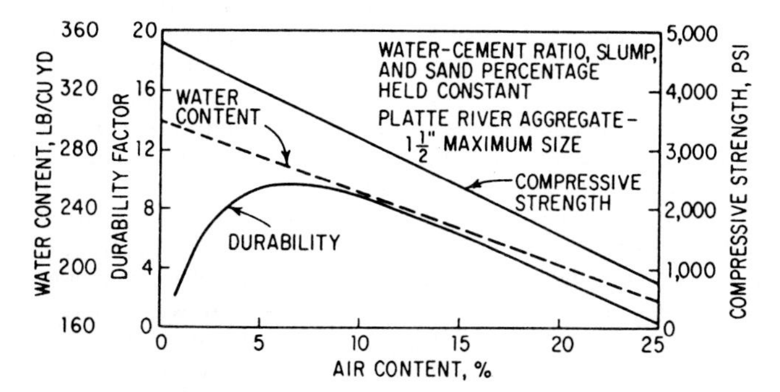

FIGURE 4.1 Effects of air content on durability, compressive strength, and required water content of concrete. (*From Ref. 1.*)

greatly improves the workability of concrete and permits the use of aggregates deficient in fines or poorly graded. Since the workability of the concrete is improved by the addition of entrained air, for equal workability mixing water can be reduced 2 to 4% per percent of entrained air. Entrained air increases the volume of the concrete mixture, and therefore to maintain proportions it is necessary to reduce the sand content of the mix in an amount equal to the volume of the entrained air. Reduction in bleeding generally permits finishing of the concrete surfaces earlier.

Even though entrained air will increase the paste volume of the concrete, drying shrinkage is not significantly increased. Entrained air occurs only in the mortar fraction, and as the mortar is replaced by aggregate with increasing maximum aggregate size, the air content for equivalent durability should be decreased, ranging from 8% for ⅜-in maximum-size aggregate to 3% for aggregate graded up to 6-in. maximum (see Fig. 4.2). Air entrainment should be used in all curbs, flat slab, or other concrete work which will be exposed to frost action or deicing salts. Air entrainment is not necessary in structural concrete made with well-graded sound and not harsh aggregates, not exposed to freezing and thawing.

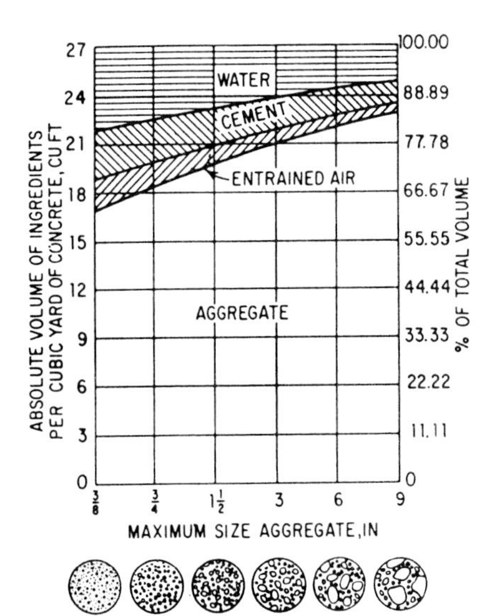

FIGURE 4.2 Effect of maxiumum aggregate size on percentage of entrained air, cement, and water. Chart based on natural aggregates of average grading in mixes having a water/cement ratio of 0.54 by weight, 3-in slump, and recommended air contents. *(From Ref. 1.)*

Examples of air-entraining agents with proved satisfactory performance are those based on neutralized Vinsol resin and the sulfonated hydrocarbon salt of triethanolamine.

Control of Air Content. The efficiency of any air-entraining agent depends on temperature, slump, cement content and fineness, aggregate gradation, mixing time, and placing techniques. For a given quantity of air-entraining agent added, air content increases with greater slump and decreases with higher temperature, longer mixing time, greater cement content, and greater proportion of fines.

It can be seen that the quantity of air-entraining agent will have to be adjusted at the mixer to control air content under varying conditions.

Air content can be measured conveniently by the use of two types of air meters available on the market. These meters are based on either the volumetric method or the pressure method.

In the volumetric-type meter (Rollameter), a given quantity of concrete is added to the meter, generally $\frac{1}{20}$ ft,^{3} the air meter is closed, and a given quantity of water is shaken with the concrete displacing the air. The volume of water used to displace the air is read directly as entrained air. This type of meter may be used with lightweight or porous aggregates.

In the pressure-type meter (Acme), a given quantity of concrete is added to the meter, usually $\frac{1}{10}$ ft,^{3} the meter is closed, and air pressure is pumped into a small chamber. This air pressure is then released on the plastic concrete, compressing any entrained air in the plastic concrete. The difference in pressure is then read directly as entrained air, the principle of Boyle's gas law applying; if there were no air in the concrete, no air could be released from the pressure chamber and the reading would be zero.

A small pocket air indicator is available (Chace air indicator) for indicating the percentage of air in mortar taken from the concrete. This indicator is handy to determine whether the concrete has entrained air or not but is subject to manipulation of the operator and should be considered a qualitative and not a quantitative instrument.

Accelerators. There are many chemicals that will accelerate the hardening of portland cement. Among these are the soluble chlorides, bromides, fluorides, carbonates, thiocyanates, nitrates, nitrites, formates, silicates, and alkali hydroxides.

The nitrates, nitrites, formates, and thiocyanates are used commercially to a minor extent, but calcium chloride is the most widely used accelerator since it is the most cost-effective. (Accelerators should conform to ASTM C494 type C or E. Calcium chloride should conform to ASTM D98. Many of these compounds may contain other admixtures as well. Triethanolamine in combination with other admixtures is also used to a very limited extent.

Since calcium chloride is the most commonly used accelerator the following discussion refers only to this chemical or admixtures containing this chemical.

The addition of calcium chloride changes the complex reactions of portland cement and water. These changes are not fully understood, but calcium chloride can be considered a catalyst which triggers the hydration of portland cement. Calcium chloride is partially consumed during hydration, probably reacting with the tricalcium aluminate to form calcium chloroaluminate.

Calcium chloride is available in flake form 77% pure, or in granular form 94% pure. Calcium chloride is a hygroscopic salt; it should be stored in a dry place. Prior to use, calcium chloride should always be dissolved in water. A convenient solution can be prepared by dissolving 4 lb of calcium chloride in sufficient water to make 1 gal. Addition of this solution can be accomplished manually or by use of automatic dispensing equipment, 1 qt of solution being equivalent to 1 lb of calcium chloride.

Calcium chloride can be used safely in amounts up to 2% (2 lb per sack) by weight of the cement. Larger amounts may be detrimental and will provide little additional advantages. The addition of up to 2% calcium chloride by weight of the cement will increase compressive and flexural strengths at early ages of all types of portland cement. This increase will be in the area of 400 to 1000 psi at both 70 and 40°F curing temperatures. This increase in strength reaches a maximum in 1 to 3 days and thereafter will generally decrease. At 1 year some increase may still be evident. Calcium chloride will not increase the flexural strength of concrete to the same degree as compressive strength, and decreases in flexural strength are generally obtained at or after 28 days.

Calcium chloride will also affect the following characteristics of concrete.

Drying Shrinkage. Calcium chloride will generally increase the drying shrinkage of concrete, this in spite of the fact that calcium chloride concrete will lose less water on storage at low humidity.

Durability. Calcium chloride will lower the resistance of concrete to freezing and thawing and attack by sulfates and other injurious solutions.

Rate of Temperature Rise. Calcium chloride will increase the rate of temperature rise due to heat of hydration and in large sections will therefore increase stresses caused by thermal contraction.

Effect on Embedded Metals: Prestressing Steel. Data are available indicating that calcium chloride may cause stress corrosion of prestressing steel.

Effect on Embedded Metals: Reinforcing Steel. Calcium chloride has a tendency to support corrosion of reinforcing steel because of the presence of chloride ions, moisture, and oxygen. ACI 318-89 recommends the following maximum chloride ion concentrations: prestressed concrete, 0.06; reinforced concrete exposed to chlorides in service, 0.15; reinforced concrete which will be dry in service, 1.00; other reinforced concrete, 0.30 (by weight of cement).

Galvanized Metal. Galvanized metal embedded in concrete containing calcium chloride may be expected to corrode at an accelerated rate.

Combinations of Metals. Combinations of materials such as aluminum alloy conduit and steel reinforcing should not be used in concrete containing calcium chloride as electrolytic corrosion may take place.

Water-reducing Admixtures and Set-Controlling Admixtures. Certain organic compounds are used as admixtures to reduce water requirements and to retard the set of concrete. In combination with other organic or inorganic compounds the characteristic of retarding the set of concrete can be offset and water reduction can be obtained without retardation. Water-reducing and set-controlling admixtures are classified by ASTM C494 as follows:

Type A Water reducing
Type B Retarding
Type C Accelerating
Type D Water Reducing and Retarding
Type E Water Reducing and Accelerating
Type F Water-Reducing High-Range
Type G Water-Reducing High-Range and Retarding

Types F and G are also covered by ASTM C1017. Types 1 and 2 are recommended for use with flowing concrete.

Raw Materials. The materials used to formulate these admixtures are generally as follows:

For the normal range water reducers:
- Salts and modifications of hydroxylated carboxylic acids (commonly called HC)
- Salts and modifications of lignosulfonic acids (commonly called lignins)
- Carbohydrates, polysaccharides, and sugar acid (commonly called polymers)

For the high-range water reducers:
- Salts of high-molecular-weight formaldehyde condensates of naphthalene sulfonic acid
- Salts of melamine formaldahyde condensates of sulfonic acid
- Salts and modifications of ligins sulfonic acid

Combinations of the above raw materials and other minor constituents are often employed in commercial formulations.

Water Reducing Admixtures, Normal Range: Characteristics. These admixtures will change the characteristics of concrete in the general manner described in the following paragraphs. Where a combination of chemicals are used, the results obtained will be additive. For simplicity the following code is used: hydroxylated carboxylic acid type, HC; the polysaccharide type, PS; the lignosulfonic acid type, lignins.

Water Reduction. When used at the manufacturers recommended normal dosage the HC and PS types will reduce water content by 5 to 8 percent. The lignin types will reduce water content by 7 to 10 percent. All types will exhibit higher water reductions at higher dosages; however, excess retardation may be incountered.

Setting Time. All types are inherently retarders, normal setting (ASTM C494, Type A), or accelerating types (ASTM C494, Type E) and may contain other chemicals such as calcium chloride and/or triethanolamine to offset retardation. At normal dosages all Type D admixtures will retard initial and final setting time by 1 to 3.5 h.

Strength. The HC, PS, and lignin types will increase the strength of concrete above that indicated by their water-reduction values[2] (Fig. 4.3).

Air Entrainment. The HC and PS admixtures do not entrain air in themselves, but since they render the concrete more workable, to obtain a given air content, less air-entraining agent is generally required. The lignin-type admixtures may add 1 to 2% entrained air to the concrete.

Bleeding. Water-reducing admixtures change the bleeding characteristics of the concrete as shown in Fig. 4.4. Reduced bleeding is generally desirable, but for placing concrete in areas of low humidity, high temperatures and high wind velocities, increased bleeding may be desirable.

Economy. Since strength can be increased at a given cement content and workability by use of water-reducing admixtures, a given strength can be obtained at a lower cement content. Since the cement saved generally costs less than the cost of the admixture, concrete of a given strength can be produced more economically (Fig. 4.5). Reducing the cement content will also reduce the unit heat of hydration, which can be very desirable when placing mass concrete.

Slump Loss. All water-reduced concrete will lose slump at a faster rate than equivalent nonadmixtured concrete; however, this is generally no problem with the HC, PS, and lignin types. Concrete containing HRW admixtures will lose slump at a much faster rate than control, very often limiting their use to job addition.

Water-Reducing Admixtures, High-Range (Superplasticizers, HRW): Characteristics. These admixtures are powerful dispersing agents and their action in portland cement pastes, mortars, and concretes can be predicted by this action. High-range water-reducing admixtures may be formulated by combining the dispersing agents with lignin-, HC-, or

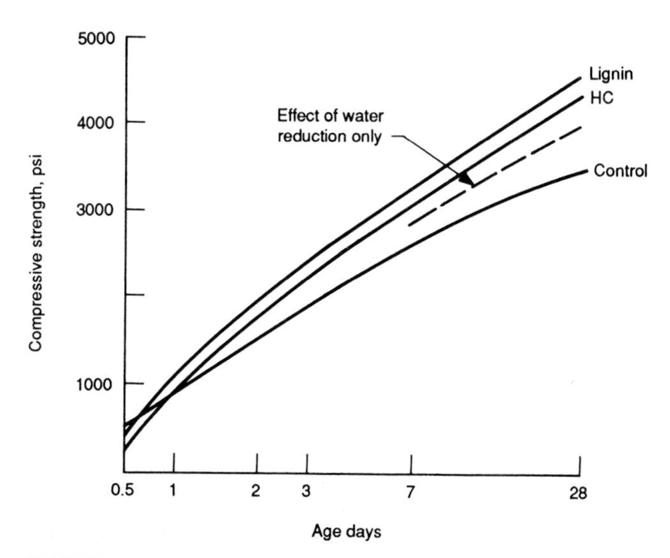

FIGURE 4.3 Strength of concrete as affected by two water reducers (5½ sacks cement; 3½-inch slump).

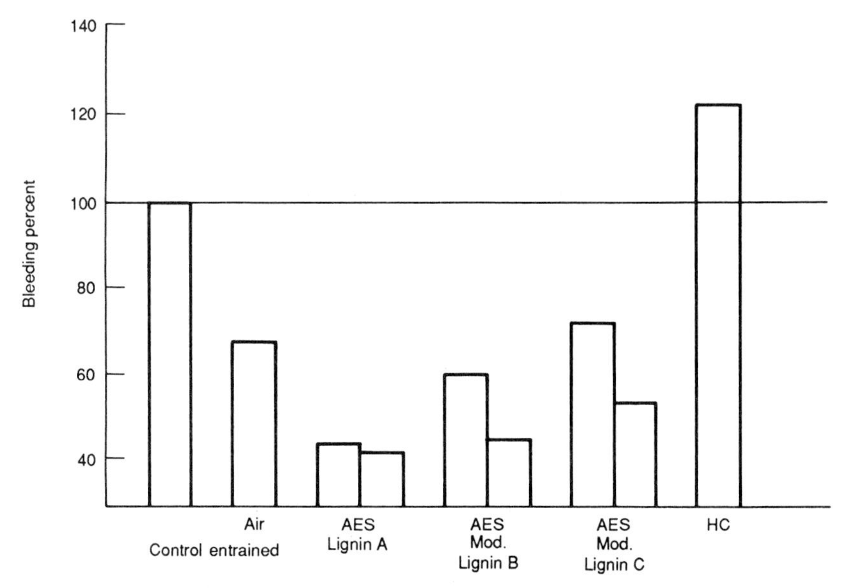

FIGURE 4.4 Relative effect of admixtures on bleeding (517 lb Type I cement; crushed limestone aggregate; AES (air-entraining solution) concretes 5 to 5.7 percent air; 3- to 4-in slump).

PS-type admixtures. These formulations may impart retardation or slightly longer slump life.

Water Reduction. At low dosages (0.05 to 0.1% by weight of the cement) the HRW admixtures reduce water to the same extent as the conventional admixtures. However, since these admixtures are not inherently retarders, they may be used at much higher dosages, and water reductions of 20 to 30 percent are not uncommon. When used without water reduction, concrete with very high workability, flowing concrete, can be produced with conservative water cement ratios.

Setting Time. When used as water reducers the unblended naphthalene formaldehyde and melamine formaldehyde high-range water reducers will have little effect on setting time as measured by ASTM C403. When blended with conventional admixtures, concrete containing the blends may exhibit retardation of 1 to 3.5 h (ASTM C494, Type G; C1017, Type II). When used to produce flowing concrete, retardation will generally occur.

Strength. Since HRW may be used to manufacture workable concrete at very low water/cement ratios, very-high-strength concrete can be manufactured in the field, and over 14,000 psi compressive strength at 28 days has been attained. Such low water/cement-ratio concrete will have very high early strengths. Such high-early-strength concrete is utilized for rapid repairs or quick reuse of forms. For example, 3000-psi strength can be routinely attained using a HRW to produce workable concrete at 0.25 or lower water/cement ratios within 8 hs.

Air Entrainment. The addition of a HRW admixture to air-entrained concrete will produce a coarse air-void system, often out of the generally accepted limits. Despite this coarse air-void system, such concrete has exhibited excellent freeze-thaw resistance.[3,4]

Bleeding. When used at moderate dosages in concrete with sufficient fines (cement, fly ash, silica fume, sand) as a water reducer, bleeding is not significantly affected. How-

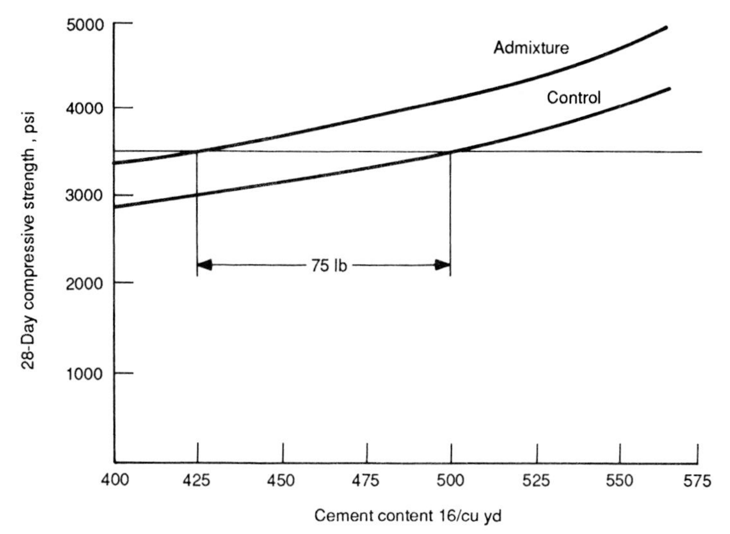

FIGURE 4.5 Savings in cement possible by use of a water-reducing admixture, 5-in slump.

ever, when used in lean concretes or concretes deficient in fines, or when used to produce flowing concrete, severe bleeding (segragation) can occur. The remedy for such an occurrence would be to increase the fines in the concrete by one or more of the following procedures, increasing the sand/aggregate ratio (Fig. 4.6) or the cement or pozzolan content, or decreasing the dosage of the HRW.

Economy. Because of their high cost, HRW admixtures can rarely be used to reduce the cement content economically. However, good economies can be realized from savings in labor required to place the concrete, with early stripping and reuse of forms or earlier use of the structure.

Slump Loss. The rate of slump loss of concretes containing HRW admixtures is greater than that of nonadmixtured concrete or concrete containing conventional admixtures. This phenomena will occur even with the blended HRW admixtures or so called second- and third-generation admixtures.[5] The short working life of HRW concrete should be taken into consideration during the planning stage.

Finely Divided Mineral Admixtures. Finely divided mineral admixtures are often added to the concrete to expand the paste volume or to offset poor gradation of the aggregates. Finely divided materials can be classified as either chemically inactive (inert), pozzolanic, or cementitious. All three classes affect plastic concrete in the same manner. The pozzolanic and cementitious materials may contribute to strength development in concrete and therefore usually require less cement to produce a given strength.

When added to concretes deficient in fines, the addition of finely divided mineral admixtures improves workability, reduces the rate and amount of bleeding, and increases strength. The deficiency in fines may be a deficiency in aggregate gradation or a mixture with insufficient cement paste to produce good workability. When mineral powders are added to concrete with sufficient fines, particularly concretes rich in portland cement, workability generally decreases for a given water content; therefore, water requirements and drying shrinkage are increased and strength decreased. These admixtures therefore

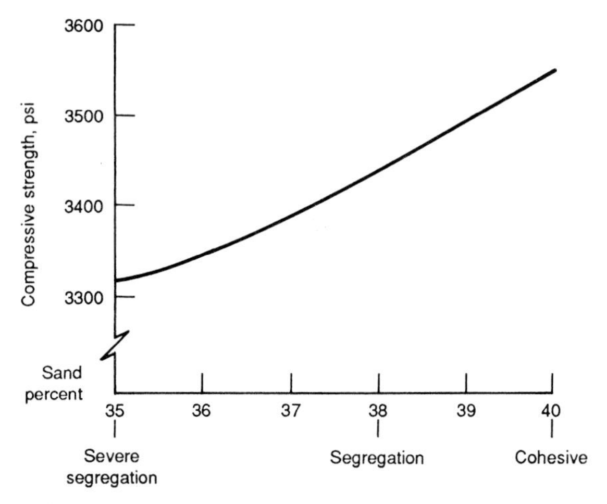

FIGURE 4.6 Effect of sand aggregate ratio on strength and segregation of superplasticized concrete (cement 517 lb; W/C 0.59; slump 7 in).

have merit only in lean concrete or concrete manufactured with aggregates deficient in material passing a 200-mesh sieve.

The addition of fine mineral powders will decrease the efficiency of air-entraining agents, and the proportion of air-entraining agent generally has to be increased when these powders are added to the mix. The additional role of pozzolanic and cementitious admixtures is covered elsewhere. Examples of finely divided mineral admixtures are as follows:

Inert Material

Ground quartz and limestone

Crushed dusts

Hydrated lime and talc

Cementitious Types

Natural cements

Hydraulic limes

Granulated iron blast-furnace slag

Pozzolanic Types

Covered under the following paragraphs

Pozzolans. A pozzolan is a siliceous or siliceous and aluminous material which in itself possesses little or no cementitious value but will in finely divided form and in the presence of moisture chemically react with calcium hydroxide at ordinary temperatures to form compounds possessing cementitious properties.

The most commonly used pozzolans are fly ash and silica fume, although natural pozzolans have been used to some extent. Pozzolans should conform to ASTM C618, Class N natural pozzolans, Class F fly ash with pozzolanic properties, and Class C fly ash with pozzolanic and cementitious properties. ASTM is presently working on a standard for silica fume.

Silica fume is a by-product of the electric arc furnace production of silicon and ferrosilicon. It is a relatively new concrete-making material. Silica fume, an extremely fine amorphous silicon dioxide, is commercially available in uncompacted, compacted, and slurry form. Silica fume will react as a conventional pozzolan, except, because of its extreme fineness, less is required. This high fineness increases water requirements. Therefore, silica fume concrete generally requires the use of a high-range water reducer. Such concrete can be produced with strengths above 15,000 psi and extremely low permeability.

In plastic concrete, pozzolans conforming to ASTM C618 will produce the same physical effects as finely divided materials; however, since pozzolans are chemically reactive, additional benefits are realized. In addition to improving the workability of concrete, pozzolans can improve quality and reduce heat generation, thermal volume change, and bleeding. They can also be used to protect concrete from the destructive expansion caused by alkali-reactive aggregates. Certain pozzolans such as calcined clay and shale may increase water requirements resulting in increased drying shrinkage and resultant cracking; therefore, care must be exercised in their selection and use. Pozzolans will affect the following characteristics of concrete:

Cement Reduction. Since pozzolans will chemically react in the presence of water with the calcium hydroxide developed during the hydration process and form compounds with cementitious properties, part of the portland cement can be replaced or substituted for equal strength (see Fig. 4-7). The strength produced by pozzolanic admixtures is developed relatively slowly, particularly at low temperatures. Moist curing must therefore be continued for longer periods in order to develop the potential strength of such concrete. Under favorable curing conditions ultimate strengths of concrete containing pozzolans as a replacement for part of the cement will generally be higher than those obtained with portland cement alone.

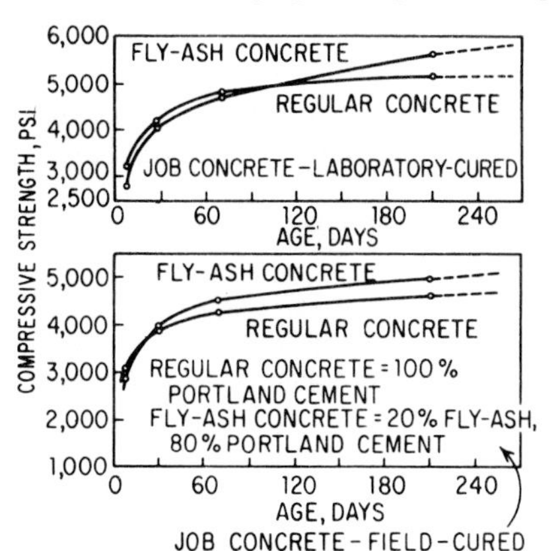

FIGURE 4.7 Strength development of concrete with and without cement replacement with fly ash. (From Ref. 6)

Cement replacement by a pozzolan will range from 5 to 30% by weight. Because of the low bulk specific gravity of pozzolanic materials, the volume used may be equal to or greater than the volume of portland cement. As an example, structural concrete mixtures containing four, five, or six bags of portland cement per cubic yard may be reproportioned to replace 94, 83, and 71 lb of portland cement with 150 to 175, 125 to 137, and 100 lb of fly ash, respectively. In this particular test series, concrete of equal strength and workability was obtained with a general reduction in water requirements of 2 gal/yd.[3]

Control of Alkali–Aggregate Reaction. The siliceous constituents of certain aggregates (alkali-reactive aggregates) will react with the alkalies of portland cement. The term alkali refers to the sodium and potassium present in small quantities and is expressed in mill reports and analyses as sodium oxide (sum of Na_2O and 0.658 K_2O expressed in percent). This reaction produces an expansive alkali-silica gel which causes excessive expansion, cracking, and general deterioration of the concrete. Laboratory tests and field experience have indicated that the use of low-alkali cements (less than 0.6% Na_2O equivalent) and/or the use of pozzolans minimize alkali–aggregate reaction, thereby allowing the use of such aggregates. Pozzolans vary in their ability to control alkali-aggregate reaction; therefore, before a choice of a pozzolan is made for this purpose, tests should be carried out to determine their efficiency. An applicable test would be Method or Test for Effectiveness of Mineral Admixtures in Preventing Excessive Expansion of Concrete Due to Alkali Aggregate Reaction, ASTM Designation C441. Pozzolans that have proved effective in reducing alkali–aggregate reaction are some opals and highly opaline rocks, clays of the kaolinite type, some fly ashes, diatomaceous earth, and calcined clays of the montmorillonite type.

Heat Development. Pozzolans are generally used as cement replacements, thereby reducing the quantity of portland cement per cubic yard of concrete. This reduction in portland cement will reduce the total heat of hydration. This is very desirable when large masses of concrete are placed, as maximum temperature is reduced with subsequent reduction in thermal stresses and cracking on cooling.

This reduction in heat generation can be undesirable when relatively thin sections are placed during cold weather.

Coloring Admixtures. Inert pigments are often added to concrete to impart color. Coloring admixtures should be fast to ultraviolet light, stable in the presence of alkalies, and have no adverse effects on the characteristics of concrete. Insofar as their physical effect on concrete is concerned, coloring admixtures can be considered finely divided mineral admixtures. Coloring admixtures or pigments are available as natural or inert colors or as synthetic materials and are used in amounts of between 2 and 10% by weight of the cement. Coloring admixtures should be blended thoroughly with the dry cement or the dry cement mixture before addition of water. The use of white portland cement in place of gray portland cement will always result in cleaner colors. White cement is much more effective than gray cement and a white pigment such as titanium dioxide.

Trial mixtures should be made to determine the type and concentration of pigment used. The dry hardened concrete will have a different color from the plastic concrete. The color of the concrete will also be affected by the water/cement ratio employed and the tools and procedures used in finishing and curing the surface.

The earth color or natural pigments available have a good service record for durability. The newer organic phthalocyanine blues and greens are satisfactory. Carbon black will impart a clear blue-gray color to concrete and mortar but has a tendency to float to the surface and requires more care in finishing than the black iron oxides.

The addition of coloring admixtures to concrete apparently does not affect durability, but a considerable increase over the normal amount of air-entraining agent may be required to produce the desired air content in concrete.

Acceptable Coloring Admixtures

Grays to black	Black iron oxide Mineral black Carbon black
Blue	Ultramarine blue Phthalocyanine blue
Bright red to deep red	Red iron oxide
Brown	Brown iron oxide Raw and burnt umber
Ivory, cream, or buff	Yellow iron oxide
Green	Chromium oxide Phthalocyanine green

Dampproofing Agents. Dampproofing agents generally are soaps based on stearates, oleates, or certain petroleum products. These admixtures may reduce penetration of the visible pores and may retard penetration of rain into porous concrete or block. Data have shown, however, that they do not resist transmission of moisture through unsaturated concrete. These admixtures cannot be considered waterproofing agents since they will not prevent passage of moisture. Concrete in itself is watertight and leaks occur only in faults such as honeycombs and cracks or other areas where a waterproofing admixture could not help. Commonly used chemicals are salts of fatty acids such as calcium, ammonium, stearate or oleate, and butyl stearate. These soaps tend to cause air entrainment during mixing, and many of the benefits obtained from their use are found in their air-entraining ability.

Other Admixtures. There are other admixtures available which are not in general use such as gas-forming, air-detraining, and grouting admixtures. These admixtures are generally used only in very specialized work, generally under the supervision of the manufacturer or a specialist in the particular field of concrete involved.

WATER

4.5 Quality

Water for mixing and curing concrete will be satisfactory if it is potable (fit for human consumption). Mixing and curing water should be reasonably clean and free from objectionable quantities of organic matter, silt, and salts. The maximum limit of turbidity should be 2000 ppm. If clear water does not taste brackish or salty it can be generally used for mixing and curing of concrete without testing. Water that is apparently hard, or tastes bitter, may contain high sulfate concentrations and should be analyzed.

Experience and tests have shown that water containing sulfate concentrations of less than 1% can be used safely; however, a strength penalty may be incurred; 0.5% sulfate has been reported as reducing strength 4 percent and 1% sulfate reduces strength 10 percent.

Ordinary salt (sodium chloride) in concentrations of 3.5% may reduce concrete strength 8 to 10 percent but may produce no other deleterious effects. Highly carbonated mineral water may produce substantial reductions in strength. Water from swamps or stagnant lakes may contain tannic, acid which may cause retardation of set and strength development. Where choice is available, the cleanest and purest source of water should be used.

4.6 Seawater

Seawater can be used successfully for mixing concrete. Seawater will contain an average salt content of 3.5%. Clean seawater has been used to manufacture concrete in many areas and the service record has been good. An 8 to 10 percent reduction in ultimate strength can be expected. Since concrete manufactured with seawater contains salts, mixing and placing techniques must be carried out with care to assure low porosity and full embedment, and cover of reinforcing steel, as the steel will rust if exposed to seawater penetrating through cracks and porosity. Many structures constructed with seawater may be exposed to freezing and thawing and alternate cycles of wetting and drying caused by waves and tides. Air entrainment should be employed and cement content should be high to ensure durability in this aggressive environment.

CONCRETE CURING MATERIALS

4.7 General

Concrete gains strength through the chemical process of hydration, a cement–water reaction. Drying of the concrete will cause this reaction to cease. Since the reaction of cement and water takes place over a prolonged period it is necessary to prevent loss of water from the concrete during this period to attain the inherent strength of the concrete mixture. As an example, concrete continuously moist-cured will attain approximately 80 percent of its 90-day strength in 28 days, 70 percent in 14 days, and 45 percent in 7 days. Drying of the concrete at these early ages will limit strength gain severely since after drying very little strength gain will take place. It can be seen that to obtain desired strength, curing of the concrete is of the utmost importance.

Concrete undergoes volume changes on alternate cycles of wetting and drying, shrinking on drying, and expanding on wetting. Flexural strength of the concrete is also developed slowly. If the concrete is alternately dried and wetted when it is young and low in flexural strength, cracking especially on the surface can occur. Therefore, curing must be continuous to prevent these detrimental volume changes.

Since any loss of moisture from the concrete will result in a strength penalty or surface cracking, curing regardless of the method should commence as soon as forms are removed or in the case of slabs as soon as the surface will not be damaged by the curing procedure.

Under certain conditions of low relative humidity, high wind, or radiant energy of the sun, or any combination, it is possible for the surface of concrete slabs to dry before hardening has taken place and while the interior is still plastic. The surface of the concrete will caseharden and shrink because of drying, causing plastic cracking to occur. Immediate application of curing, or any method that will lower wind velocity or raise relative humidity above the concrete, or will prevent the radiant energy of the sun from heating the surface, may solve this condition—fog sprays, wind breaks, and sun shades have proved effective. The application of a membrane curing compound immediately after the last finishing operation has under some conditions eliminated plastic cracking.

4.8 Types

Water. The application of water to concrete surfaces is the ideal curing medium as this cover of water prevents any loss of moisture from the concrete. Water can be supplied by continuous spraying or in the case of flat slabs by ponding with use of sand or other dikes. Water suitable for use in making concrete is suitable for curing.

Liquid Membrane-Forming Compounds. Liquid membrane-forming curing compounds are generally solvent solutions of resins and/or waxes. On evaporation of solvent these compounds leave a membrane that reduces evaporation of water from the surface of the concrete. These compounds are available pigmented in clear, white, light gray, or black. The clear compound contains a fugitive dye that helps to indicate the areas covered but fades after a day or so. Clear or gray-pigmented membrane curing compounds are preferred where discoloration must be kept at a minimum; white membrane curing compounds where ambient temperatures are high, the reflective pigment preventing absorption of solar energy; and black membrane curing compounds on structural elements not exposed to the sun, or in cool weather where the solar energy will tend to warm the concrete. ASTM Designation: C309, Liquid Membrane Forming Compounds for Curing

Concrete, is an applicable specification. A test method for determining the efficiency of these compounds is covered by ASTM Designation C156. These specifications are based on a coverage of 200 ft^2/gal and a moisture loss of 0.055 g/in^2 under test conditions. Increasing this coverage will lessen the efficiency of these curing compounds, rendering them useless for the purpose. Care should be taken that the coverage of these curing compounds have the advantage that inspection is simplified. There is no chance that the concrete will dry out because of neglect such as is possible with water curing.

Paper and Plastic Sheets. Paper and plastic sheets are available for curing concrete. These sheets are applied to the concrete as soon as possible without disfiguring the surface of the concrete. The water-retentive qualities of paper and plastic sheets are the same as those required for liquid membrane curing compounds. Waterproof paper for curing concrete is covered by ASTM Designation C171. Plastic sheeting is available either reinforced with nylon threads or backed with paper. No ASTM test method or specification is presently available for plastic sheeting; however, the water-retaining properties should be the same as those acceptable for curing paper. Both paper and plastic sheets have the advantage that the concrete is not discolored by their use; however, they do have the disadvantage that they have to be fixed in position to prevent blowing away by wind. In hot climates a white pigmented sheeting is desirable. Reinforced sheeting has proved more economical since more reuses may be obtained.

Mats and Blankets. Wet curing can be accomplished by the use of burlap or cotton curing mats. Once wetted down with water these curing mats will supply moisture to the concrete for a long period of time, eliminating the need for continuous spraying of the concrete by a worker. These mats consist of a burlap, jute, or cotton cloth covering filled with linters of raw cotton or cotton waste. A minimum of 12 oz/yd^2 of filling should be required. ASTM Designation C440 applies to this item. These mats are bulky to use and store but have proved an excellent curing medium.

For winter concreting, reusable commercial blankets are available. Ordinary commercial bat insulation may also be used. The object of these materials is to insulate the concrete from cold. To prevent loss of their insulating value these blankets must be protected from wind, rain, snow, or other wetting by use of moistureproof cover material. To be effective they must also be kept in close contact with the concrete or concrete-form surface. Table 4.1 lists the insulation requirements of concrete walls and floor slabs placed above ground as listed in the ACI Recommended Practices for Winter Concreting, Publication 604-56.[7]

JOINT SEALANTS

Joint sealants are flexible materials used to seal joints subject to movement between adjacent sections of concrete or between concrete and other construction materials. Joint sealants are available as field-molded sealants such as the common poured- or troweled-in-place sealants, and as premolded sealants such as the plastic and rubber water stops and gaskets.

4.9 Field-Molded Sealants

Sealants falling into this class are either thermosetting elastometers, thermoplastics, or mastics. The shape of these sealants in use is determined by the joint slot cast or sawed

TABLE 4.1 Insulation Requirements for Concrete Walls and Floor Slabs above Ground Concrete Placed at 50°F (10°C)

Wall thickness, ft (m)	Minimum air temperature allowable for these thicknesses of commercial blanket or bat insulation, °F (°C)			
	0.5 in (1.3 cm)	1.0 in (2.5 cm)	1.5 in (3.8 cm)	2.0 in (5.1 cm)
Cement content, 300 lb/yd³ (178 kg/m³)				
0.5 (0.15)	47 (8.3)	41 (5.0)	33 (0.56)	28 (−2.2)
1.0 (0.30)	41 (5.0)	29 (−1.7)	17 (−8.3)	5 (−15)
1.5 (0.46)	35 (1.7)	19 (−7.2)	0 (−18)	−17 (−27)
2.0 (0.61)	34 (1.1)	14 (−10)	9 (−23)	−29 (−34)
3.0 (0.91)	31 (−0.56)	8 (−13)	−15 (−26)	−35 (−37)
4.0 (1.2)	30 (−1.1)	6 (−14)	−18 (−28)	−39 (−39)
5.0 (1.5)	30 (−1.1)	5 (−15)	−21 (−29)	−43 (−42)
Cement content, 400 lb/yd³ (237 kg/m³)				
0.5 (0.15)	46 (7.8)	38 (3.3)	28 (−2.2)	21 (−6.1)
1.0 (0.30)	38 (3.3)	22 (−5.6)	6 (−14)	−11 (−24)
1.5 (0.46)	31 (−0.56)	8 (−13)	−16 (−27)	−39 (−39)
2.0 (0.61)	28 (−2.2)	2 (−17)	−26 (−32)	−53 (−47)
3.0 (0.91)	25 (−3.9)	− 6 (−21)	−36 (−38)	
4.0 (1.2)	23 (−5.0)	− 8 (−22)	−41 (−41)	
5.0 (1.5)	23 (−5.0)	−10 (−23)	−45 (−43)	
Cement content, 500lb/yd³ (296 kg/m³)				
0.5 (0.15)	45 (7.2)	35 (1.7)	22 (−5.6)	14 (−10)
1.0 (0.30)	35 (1.7)	15 (−9.4)	− 5 (−21)	−26 (−32)
1.5 (0.46)	27 (−2.8)	− 3 (−19)	−33 (−36)	−65 (−54)
2.0 (0.61)	23 (−5.0)	−10 (−23)	−50 (−46)	
3.0 (0.91)	18 (−7.8)	−20 (−29)		
4.0 (1.2)	17 (−8.3)	−23 (−31)		
5.0 (1.5)	16 (−8.9)	−25 (−32)		
Cement content, 600 lb/yd³ (356 kg/m³)				
0.5 (0.15)	44 (6.7)	32 (0)	16 (−8.9)	6 (−14)
1.0 (0.30)	32 (0)	8 (−13)	−16 (−27)	−41 (−41)
1.5 (0.46)	21 (−6.1)	−14 (−26)	−50 (−46)	−89 (−67)
2.0 (0.61)	18 (−7.8)	−22 (−30)		
3.0 (0.91)	12 (−11)	−34 (−37)		
4.0 (1.2)	11 (−12)	−38 (−39)		
5.0 (1.5)	10 (−12)	−40 (−40)		

Insulating material	Equivalent thickness, in (mm)
1 in. (31 mm) of commercial blanket or bat insulation	1.000 (25.4)
1 in. (31 mm) of loose-fill insulation of fibrous type	1.000 (25.4)
1 in. (31 mm) of insulating board	0.758 (19.2)
1 in. (31 mm) of sawdust	0.610 (15.5)
1 in. (31 mm) (nominal) of lumber	0.333 (8.45)
1 in. (31 mm) of dead-air space (vertical)	0.234 (5.94)
1 in. (31 mm) of damp sand	4.023 (102.2)

into the concrete. All sealants are solids and their volume cannot be changed by application or release of pressure; only the shape of the sealant will be changed on movement of the joint. Therefore, the shape factor (depth-to-width ratio) of the sealant is critical and will affect the performance of any field-molded sealant. As a joint slot is compressed the sealant will have to bulge up and/or down. As the joint slot is extended the sealant will have to neck down in the center as the sealants in themselves cannot expand or compress. The strain S_{max} on the outer fiber of the sealant caused by this movement will determine whether the sealant will perform successfully. The significance of S_{max} is that it is the outer fiber of the sealant that has the greatest strain, and failure of a sealant in extension or compression will occur in this outer fiber. Therefore, knowing the strain on the outer fiber that a given sealant can withstand allows one to design a joint with any depth/width ratio. The flatter the cross-sectional shape of a given sealant, the greater the extension or compression of the joint. Expressed another way, at the same extension, the flat joint will induce a smaller strain in the joint material than the narrow, deep joint. Figure 4.8 illustrates the effect of shape factor (ratio of depth to width) upon the strain on the outer fiber of a sealant. This relationship will hold true regardless of the type of sealant under consideration. The strain on the outer fiber of the sealant for a given movement and shape factor can be determined mathematically or graphically from Fig. 4.9. For example, for a joint 1 in deep and ½ in wide $D_x/W_{max} = 1/0.5 = 2$, and for a linear expansion of 100%, $S = 160\%$. To prevent bond of the sealant on the bottom of the joint slot a bond breaker must be used to allow free movement of the sealant. This bond breaker can be silicone-treated paper, polyethylene tape, or even strips of newspaper. In order to maintain the proper shape factor it is often necessary to fill the bottom of the joint slot with a compressible, nonabsorptive backup material. Materials which have proved good for backing up field-molded joint sealants are closed-cell neoprene, polyurethane, or polystyrene foam. Materials composed of wood fibers and other organic materials that will rot or saturate with water have proved unsatisfactory.

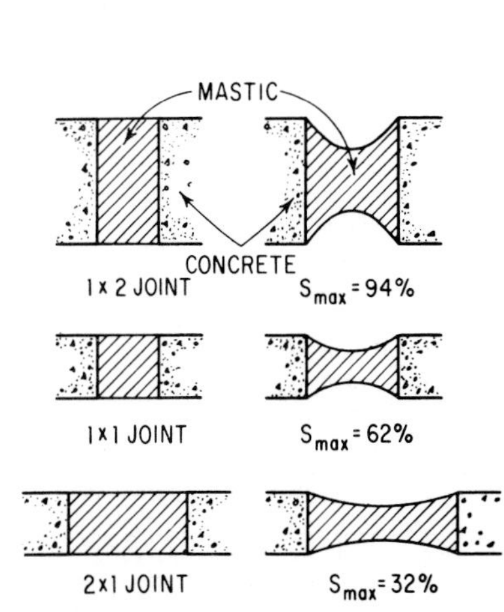

FIGURE 4.8 Strain on the extreme fiber of a sealant for a ½-in extension of different joint designs. (*From Ref.* 8.)

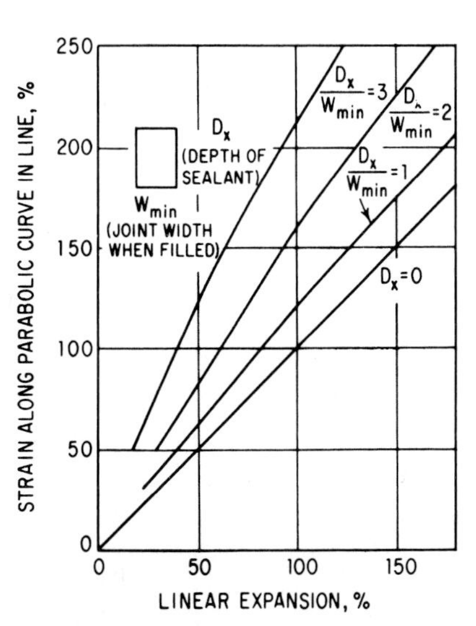

FIGURE 4.9 Maximum strain on a joint plotted against percentage of linear expansion, applicable to any joint and any sealant. (*From Ref.* 8.)

TABLE 4.2 Field-Molded Sealants

Base compound	Uses	General comments
Self-leveling and nonsag thermosetting elastomers—chemical-curing type—little or no shrinkage on curing		
Epoxy	For caulking purposes	Relatively high modulus as compared with other sealants; limited elongation
Polysulfide	Caulking, glazing, expansion and construction joints	Good weathering, and ozone resistance; high elongation; low modulus. Resistant to solvents, fuels, and other chemicals
Polyurethane	General caulking and glazing	Good weathering properties; good resistance to fuel and chemicals; high modulus; limited elongation
Silicones	Caulking and glazing	Good weathering; limited elongation; maintains properties over wide temperature range
Polysulfide coal tar	Highway, airport, and bridge joints	Low modulus; good fuel resistance Good elongation
Thermosetting sealants—solvent-release type—shrink on curing		
Neoprene	General caulking and glazing	Good weathering; moderate elongation; small cross sections
Chlorosulfonated polyethylene Hypalon	General caulking and glazing	Good weathering; nonstaining, exhibits shrinkage, for joints with small cross sections; moderate elongation
Butyl rubber	General caulking and glazing	Good weathering characteristics; limited elongation; good in all colors
Thermoplastic sealants—hot-melt type		
Lead	Joints in floors having heavy traffic	Joints subject to no movement; good impact resistance; no elongation
Rubber asphalt	Concrete highway joints	Good for use in nonfuel areas; limited elongation
Rubber coal tar	Concrete highway joints	Good for use in nonfuel areas; limited elongation
Thermoplastic sealants—solvent-release type—shrink on curing		
Acrylic	General caulking and glazing	Limited elongation
Vinyl	General caulking and glazing	Limited elongation
Rubber asphalt	Joints in canal linings	Limited elongation
Rubber asphalt emulsion	Narrow cracks	Limited elongation
Mastics—trowel or gun grade		
Oleoresinous	Caulking and glazing	Very limited elongation; short service life
Polybutene	Sealing of butt and lap joints	Good weathering properties; good adhesion; use only for compression joints and where material is concealed; limited elongation
Asphalt	Construction joints, tanks, and canals	Limited elongation

The choice of a particular sealant will depend on the movement anticipated in a given joint, physical abuse which might be encountered in service, and the physical environment of the installation (injurious solutions, temperature, water pressure, etc.).

Properties and uses of several types of field-molded sealants are summarized in Table 4.2.

4.10 Types of Field-Molded Sealants

Mastics. Recommended only for use in joints subject to little strain or no movement. The group of sealants classified as mastics are composed of viscous liquids rendered immobile by the addition of fibers and fillers. The vehicle in mastics may be a nondrying oil, a polybutene, a low-melting-point asphalt, or any combination of these materials. The filler used may be asbestos fiber, fibrous talc, or a finely divided calcareous or siliceous material.

Thermoplastics. Recommended maximum strain on the outer fiber is 5 to 10 percent depending on the formulation. Thermoplastics include such materials as asphalts, rubber asphalts, pitch, and coal tar. They will become soft on heating and will harden when cooled. For design purposes these are considered elastomers with limited extensibility.

Thermosetting Elastomers. Recommended maximum strain on the outer fiber 10 to 25 percent depending on the formulation. The thermosetting plastic sealants can be two-component liquid systems, which react after installation to form elastomers, or thermosetting plastics in solvent solution which set or cure by evaporation of solvent. These materials will not soften appreciably in applications up to 225°F.

Preparation of Joint Slot. The surfaces of the joint slot must be clean and free of oil or dust, and dry in order that the sealant may bond properly. Generally speaking, even joints cast in new concrete will be dirty; form oil, loose mortar, and dust may be present. Sawed joints will be dusty and should be blown out with oil-free air. Cast joints can best be cleaned by mechanical abrading. This can be accomplished by sandblasting, power or hand wire brushing, or power-driven mechanical routers. If any honeycombs or faults exist on the side of the joint slot these should be patched prior to application of the sealant.

4.11 Premolded Joint Sealants

Premolded joint sealants are either embedded in the concrete, termed *water stops,* or installed by compressing an extruded shape into a joint slot, commonly termed *compression gaskets.*

Water Stops. Water stops are either sheets of metal, Z- or N-shaped, or extruded natural or synthetic rubber or plastic installed in the plastic concrete by use of split forms.

Metal water stops may be of steel, stainless steel, copper, or lead. Copper is the most expensive metal water stop; it has high resistance to corrosion but is easily damaged in construction. Lead water stops enjoyed popularity in the past but are seldom used today. Steel is the stiffest of the metal water stops; if plain carbon steel is used it is subject to corrosion.

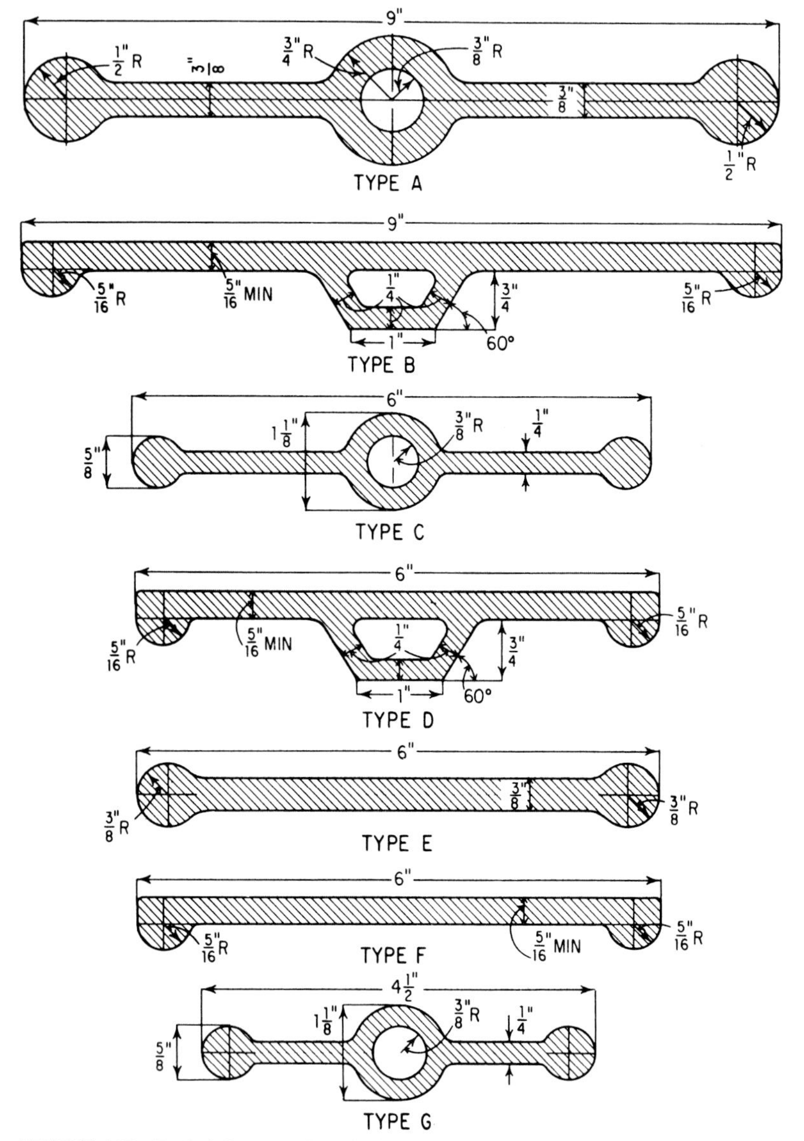

FIGURE 4.10 Typical shapes available in flexible water stops. (*From Ref.* 8.)

Splicing of steel water stops can be accomplished by welding. Copper water stops can be spliced by brazing or soldering.

Flexible water stops are made of natural or synthetic rubber, polyvinyl chloride, or other plastics. They are quite often extruded with a hollow bulb in the center so that movement can be accommodated with minimum strain on the sealant. Complex shapes are available to assure better mechanical grip to the concrete. Although natural- and synthetic-rubber water stops are available the most commonly used flexible water stops are manufactured of polyvinyl chloride.

Splicing of rubber water stops can be accomplished by use of special cements. Polyvinyl chloride water stops can be spliced by heating the ends and pressing them together when they are in the molten state. Many grades of polyvinyl chloride are available, and a grade should be chosen that will be flexible at the lowest anticipated service temperature. Figure 4.10 illustrates some of the shapes available in flexible water stops.

Installation. Water stops, both metal and flexible, are installed by the use of split forms as shown in Fig. 4.11. Care must be exercised when placing the concrete so that the water stop is not bent or misplaced. Compaction and vibration of the concrete adjacent to the water stop must be thorough so that no honeycombs or voids exist on the underside of the water stop. Since the water stop will be inaccessible after the concrete has hardened, the success for the installation will be dependent upon the position of the water stop and the thoroughness of the compacting of the concrete.

Compression Gaskets. Compression gaskets are premolded shapes of either neoprene foam, extruded cellular neoprene shapes, or bitumen-impregnated polyurethane foam. They are installed by compressing the gasket into the joint slot. The joint must be so designed that the gasket will always be under compression even at the widest opening anticipated. Sealing is accomplished only if the gasket is under compression. Compression gaskets are useful where little or no water pressure is anticipated such as in slabs on grade or above grade joints. Neoprene is the most suitable elastomer for compression gaskets as it has the greatest resistance to compression set. Quite often adhesives are used to hold the compression gaskets in place. The use of adhesives is always desirable when using compression gaskets.

OTHER MATERIALS

4.12 Adhesives

Organic adhesives are available to bond fresh concrete to hardened concrete, or hardened concrete or other materials to concrete. Adhesives used for structural bonding are based on epoxy resins. For bonding patches, overlays, etc., latex-based adhesives can be used.

Structural Adhesives. Epoxy resins used for this purpose should conform to ASTM C881. These materials will bond to wet or damp surfaces. ACI Specifications 503.1, 503.2, 503.3, and 503.4 cover the use of these materials for a variety of adherends. Wood, copper, steel, stone, etc. may be bonded to concrete.

FIGURE 4.11 Typical installation of water stops by use of split forms.

ASTM C881 covers seven types of epoxy resin systems. For general bonding, Type I hardened materials to hardened concrete; Type II, fresh concrete to hardened concrete; and Type III, binder for epoxy mortar and concrete. For structural bonding: Type IV hardened materials to hardened concrete; Type V, fresh concrete to hardened concrete; and Types VI and VII, for segmental construction. All types are available in low and medium viscosities and a paste or gel.

Nonstructural Adhesives. For bonding patches and overlays, organic latexes have a good service record. These materials should conform to ASTM C1059, Type I for installations not subject to water or high humidity and, Type II for installations subject to high humidity or immersion in service.

Surface Preparation. The strength of any adhesive system will be only as strong as the weakest link. Surface preparation cannot be overemphasized; laitance, dust, dirt, or other foreign matter will reduce the strength of the bond of these adhesives drastically. The concrete surface should be cleaned, preferably by mechanical abrading such as sandblasting, chipping, or other means. Steel-troweled or formed surfaces are usually weak, and if structural strength is required the surface should be removed. As an example, a steel-troweled surface might have a strength of only 75 psi in direct tension. Mechanically abrading the surface will raise this value to as high as 275 psi. A layer of oil can reduce this bond value to zero.

Application—Structural Adhesives. For bonding plastic concrete to hardened concrete the two components of the adhesive must be thoroughly mixed and applied as soon as possible after mixing. Concrete or mortar can then be placed at any time during which the adhesive remains tacky. This may vary anywhere from 30 min to 2 h depending on the particular system. During hot weather, the curing time of these resins is faster and the contact time will be reduced. If thin portland-cement concrete or mortar overlays are being applied, great care should be exercised that the thin overlay is kept moist, as drying will result in shrinkage, curling, and loss of bond of the overlay. The epoxy-polysulfide adhesives are vapor barriers before cure and the overlays will dry only on the top surface. This drying will tend to curl the thin overlay away from the adhesive; wet curing or the application of a curing compound should be carried out as soon as possible.

When bonding new concrete to old concrete which has high absorption, the old concrete should be wet with water to prevent the adhesive from being absorbed into the old concrete. The concrete should be wet but have no free puddles standing on the surface.

Application of the bonding compound can best be carried out by the use of brushes or brooms, although automatic spraying and mixing equipment is available for this purpose.

For bonding hardened concrete to hardened concrete or other objects to concrete, if the surfaces to be bonded do not mate, the epoxy adhesive can be mixed with fine sand to form a mortar, although for this purpose the high-viscosity or gelled epoxy-resin adhesives are easier to use.

Application—Latexes

Type I. Apply the latex as supplied by the manufacturer by brush, roller, or spray. Place concrete on the applied latex at any time, even after the latex has dried.

Type II. Mix the latex as supplied with portland cement, or portland cement and 2 to 3 parts sand to form a slurry. Scrub this slurry into the surface to be bonded. Place the new mortar or concrete on this slurry before the slurry dries or sets. For best results with both Type I and Type II latexes, use concrete or mortar containing the appropriate latex. A minimum of 15% latex polymer solids, based on the cement, should be used.

4.13 Floor Hardeners

Floor hardeners currently employed in the industry are based on either fluosilicates or organic resins. Fluosilicates combine chemically with the free lime in the concrete and reduce dusting and increase hardness slightly. The fluosilicates may be salts of zinc or magnesium or a combination thereof. Some proprietary compounds contain wetting agents and/or organic acids to improve their efficiency. Fluosilicate hardeners should be applied as aqueous solutions at the rate of approximately 60 to 150 ft^2/gal. The fluosilicate content should be between 1 and 1½ lb/gal.

Combination curing and hardening compounds are marketed. These compounds are usually chlorinated-rubber- or butadiene-styrene-based coatings. They do not actually harden the floor but apply a protective lacquer to the concrete. Since these materials are classified and used as curing compounds, they should be applied at a rate not exceeding 200 ft^2/gal as covered by ASTM Specification C309. Higher coverages will result in little or no curing of the concrete. Neither the fluosilicate nor the resinous-type floor hardeners will overcome basic faults in the concrete such as the results of overfinishing, high water content, or carbonation.

REFERENCES

1. "Concrete Manual," 7th ed., U.S. Bureau of Reclamation, 1963.
2. Tuthill, L. H., R. F. Adams, and J. M. Hemme, Observations in Testing and Use of Water Reducing Retarders, ASTM Spec. Tech. Publ. 266, Oct. 1959.
3. Schutz, R. J., Durability of Superplasticized Concrete, "Proceedings of International Symposium on Superplasticizers in Concrete," Ottawa, May 1978.
4. Whiting, D., Effects of High Range Water Reducers on Some Properties of Fresh and Hardened Concrete, *Res. Devel. Bull.* No. RD–61–01T, Portland Cement Association, Skokie, Il.
5. Whiting, D., and W. Dziedzic, Behavior of Cement Reduced and "Flowing" Fresh Concretes Containing Conventional Water-Reducing Admixtures. *Cement, Concrete and Aggregates,* Vol II, No. 1, 1989.
6. Pearson, A. S., Fly Ash Improves Concrete and Lowers Its Cost, Civil Eng., Sept. 1953.
7. ACI Committee 306 Report, Recommended Practice for Cold Weather Concreting, (ACI 306–66).
8. Schutz, R. J., Shape Factor in Joint Design, Civil Eng., Oct. 1962.

P A R T • 2

PROPERTIES OF CONCRETE

CHAPTER 5
WORKABILITY

George W. Washa

ELEMENTS OF WORKABILITY AND CONSISTENCY

5.1 General

All concrete mixtures must be properly proportioned to have satisfactory economy, workability, required strength, and durability properties. These different objectives require differences in proportioning, and consequently most concrete mixtures are compromises rather than mixtures having the best workability, or the highest compressive strength at a given age or the greatest economy. Excellent workability, for example, normally requires high cement and fine-aggregate contents with a low coarse-aggregate content and a relatively high water content. Such a mixture would certainly not be economical, and its properties would not be optimum. Consequently as the proportions of a given mixture are changed to provide greater workability the effects on the other properties must be considered, and the desired improvement in workability should be obtained by means of such changes as have the least harmful effect on the other desired properties. In other words, the concrete mixture must have sufficient workability to enable satisfactory placement under job conditions, sufficient strength to carry design loads, sufficient durability to allow satisfactory service under expected exposure conditions, and necessary economy not only in first cost but in ultimate service.[1,2]

Workability of a concrete mixture may be defined as the ease with which it may be mixed, handled, transported, and placed into its final position with a minimum loss of homogeneity. Workability is dependent on the proportions and physical characteristics of the ingredients, but there is no general agreement on the properties of a concrete mixture that may be used as measures of workability. Workability also depends on the mixing, transporting, and compacting equipment used, on the size and shape of the mass of concrete to be formed and on the spacing and size of reinforcement. Further, workability is relative because a concrete may be considered workable under some conditions and unworkable under others. For example, concrete having satisfactory workability for a pavement slab would be difficult to place in a heavily reinforced concrete column. Special considerations are usually involved for concrete to be transported by pump or placed under water.

Consistency or fluidity of concrete is an important component of workability which relates to the degree of wetness of concrete. It is measured by ball penetration, slump, and flow tests. Normally wet concretes are more workable than dry concretes, but concretes of the same consistency may vary greatly in workability. Other components of workability may be determined from a trowel test that will allow evaluation of such properties as harshness, stickiness, cohesiveness, and ease of manipulation. Another measure of workability may be obtained from the bleeding test, which determines the tendency of the water to separate from the other constituents of the concrete and to rise to the top of the concrete mass.

Since all the properties needed to determine the workability of a concrete mixture correctly are either not known or are impossible to measure, systematic visual inspection must be used along with the results of the consistency, trowel, and bleeding tests to ensure the use of concrete with satisfactory workability. Constant vigilance on the part of the inspector is necessary to avoid the undesirable effects resulting from the use of a concrete with improper workability.

MEASUREMENT OF WORKABILITY

5.2 Slump Test

Despite its limitations, the slump test is widely used to measure the consistency of concrete. It is commonly the first test made on a sample of freshly mixed concrete and frequently determines whether the batch will be accepted or rejected. Details of the slump test are given in ASTM Standard Test Method for Slump of Portland Cement Concrete (C143). The test is conducted with a standard slump mold 12 in high having a base diameter of 8 in and a top diameter of 4 in. The dampened mold is placed on a smooth, horizontal, nonabsorbent surface and is filled in three layers, each approximately one-third of the volume of the mold, while the mold is held firmly in place. Each layer of concrete is tamped 25 times with a round, straight steel rod ⅝ in in diameter and 24 in long, having one end rounded to a hemispherical tip. The bottom layer is rodded throughout its depth and each remaining layer is rodded with strokes that just penetrate the underlying layer. Strokes should be uniformly distributed over the cross section of the mold. As the third layer is rodded additional concrete, if needed, should be added to keep the level of concrete above the top of the mold at all times. After completion of rodding, any excess concrete should be removed by means of a rolling and screeding motion of the tamping rod. The slump cone is then lifted vertically in about 5 s with a steady upward lift, with care being taken to avoid any lateral or twisting motion. The slump mold is then placed gently near the settled concrete mass and the slump is measured by placing the tamping rod across the top of the mold and reading the vertical distance from the center of the settled concrete mass to the bottom of the tamping rod with ruler as shown in Fig. 12.1 (Chap. 12). Slump is usually determined to the nearest ¼ in.

The slump test may be quickly performed and provides some answers that are helpful in evaluating the workability of the concrete. However, at best, only partial answers are obtained and in some situations they may be of doubtful value. The concrete cone after removal of the metal cone will subside into one of the four forms shown in Fig. 5.1. The near-zero slump shown in Fig. 5.1a may be the result of a concrete that has all the requirements of good workability except that the water content is too low, or it may be the result of a harsh concrete mixture that is free-draining and allows the water to run out of the concrete mass without causing any significant change in subsidence. Special care must be used when lightweight aggregates are used because many are angular and interlocking and produce a harsh concrete, unless special precautions are taken to increase

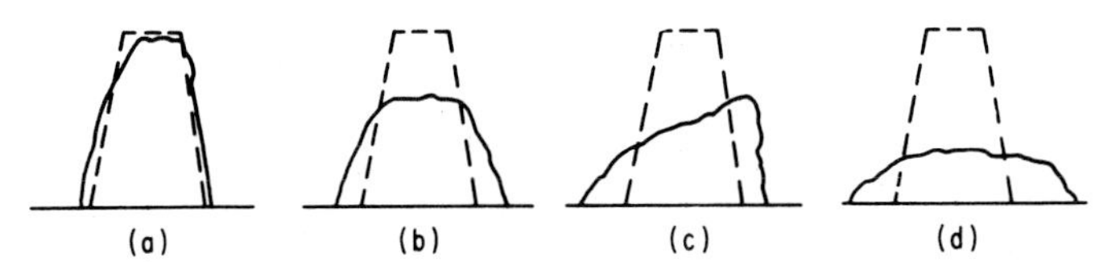

FIGURE 5.1 Types of slump: (a) near-zero slump; (b) normal slump; (c) shear slump; (d) collapse slump.

workability by additions of fine sand, air-entraining agents, or special admixtures. When the normal slump cone, as shown in Fig. 5.1b, is obtained, the concrete should usually have good to excellent workability. The slump used for placing most structural concrete ranges between 2 and 7 in and up to 8½ in with a super plasticizer. The shear slump shown in Fig. 5.1c indicates that the concrete lacks plasticity and cohesiveness and that the results of the slump test are of questionable value. The collapse slump shown in Fig. 5.1d is obtained with lean, harsh or extremely wet mixes. Concrete of this type normally has slump values ranging between 7 and 10 in, and in the slump test the grout or mortar tends to run out of the concrete, leaving the coarser material in the center unless a high-range water reducer is used. Generally, concrete exhibiting shear or collapse slumps has poor or unsatisfactory placeability.

The slump test is not considered applicable to nonplastic and noncohesive concrete, and is further limited to concrete containing coarse aggregate not over 2 in in size.

5.3 Ball-Penetration Test

A measure of consistency may also be determined by means of the ASTM Standard Test Method for Ball Penetration in Fresh Portland Cement Concrete C (360). Essentially, this test consists of placing a 30-lb metal cylindreal weight, 6 in in diameter and 4⅝ in in height, having a hemispherically shaped bottom, on the smooth level surface of the concrete and determining the depth to which it will sink when released slowly. During penetration the handle attached to the weight slides freely through a hole in the center of the stirrup which rests on large bearing areas set far enough away from the ball to avoid disturbance when penetration occurs. The depth of penetration is obtained from the scale reading on the handle attached to the weight using the top edge of the independent stirrup as the line of reference. Penetrations are measured to the nearest ¼ in, and each reported value should be the average of at least three penetration tests. The apparatus is shown in Fig. 5.2.

This test can be quickly made, is less subject to personal errors, and does not require molding a specimen as does the slump test. The test may be performed on concrete as placed in the forms prior to any manipulation, or in suitable containers such as pans, hoppers, or wheelbarrows. The depth of the concrete should not be less than 8 in or three times the maximum size of aggregate depending on which of the two values is larger. The centerline of the handle should be more than 9 in horizontally from the nearest edge of the level surface.

The ratio of slump to penetration is usually reported between 1.3 and 2.0. The variations are apparently due to differences in mix proportions, size and type of aggregate, and test conditions. In the absence of comparable field data, the ratio of slump to penetration may be approximated as 1.67. However, where possible, the ratio should be determined under actual job conditions (see Chap. 12).

FIGURE 5.2 Measuring ball penetration of concrete.

5.4 Trowel Test

This is not a standard test, but it may be used to provide valuable information about the workability characteristics of any concrete mixture. The test consists of subjecting the concrete mixture to troweling action and observing the behavior of the mixture under this action. The cohesiveness may be judged by noting if concrete sticks together and also if it tends to stick to the trowel under the troweling action. The magnitude of bleeding, if any, may be visually observed around the perimeter of the concrete batch. The appearance of the surface after troweling may suggest changes in the proportions of the concrete mixture. Figure 5.3a to c shows surface conditions of three different concrete mixtures. In Fig. 5.3a the mixture contains insufficient mortar to fill the voids in the coarse aggregate. This mixture would be difficult to place and would result in a porous concrete with honeycombed surfaces. The concrete mixture in Fig. 5.3b has an excess of mortar and is plastic and workable. It would produce smooth surfaces, but the concrete may be porous. In addition, this concrete would be uneconomical because of a low yield. In Fig. 5.3c the concrete mixture contains the correct amount of mortar. The voids between the coarse-aggregate particles have been filled with mortar under light troweling, but no excess mortar is evident. This workable mixture will provide maximum yield of concrete for a given amount of cement.

5.5 Bleeding Test

The tendency for water to rise to the surface of freshly placed concrete, known as bleeding, may be determined from the ASTM Standard Test Methods for Bleeding of Concrete (C232). The test consists of placing concrete in a ½ ft^3 container to a height of 10 ± ⅛ in using standard procedures. The surface is troweled smooth and level under minimum action, and the concrete is then allowed to stand at a temperature between 65 and 75°F on a level floor free from vibration. At specific intervals the water collecting on the surface is drawn off with a pipette and measured. The rate of bleeding and the accumulated bleeding water, expressed as a percentage of the net mixing water in the test specimen, may be obtained.

Bleeding might be regarded as desirable because the decrease in water content must

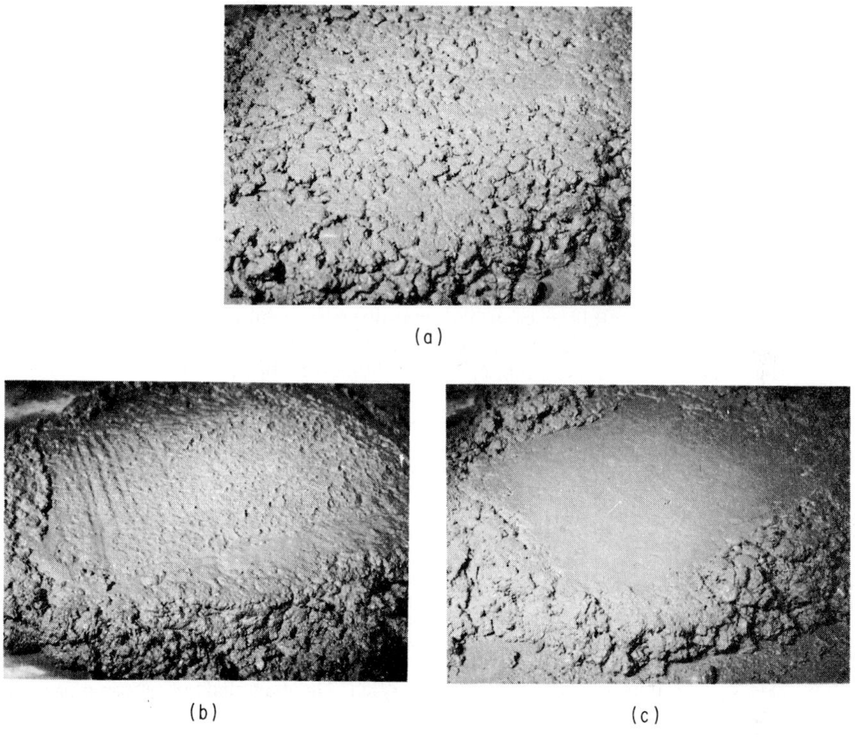

FIGURE 5.3 Trowel workability: (a) insufficient mortar; (b) excess mortar; (c) correct amount of mortar.

lead to a decrease in water/cement ration. However, bleeding disturbs the internal homogeneity of the concrete and causes other results that are not desirable. Because of the water gain in the top portion of freshly placed concrete it tends to be weak and porous and subject to disintegration by freeze-and-thaw action or by percolation of water. The water rising to the surface may carry fine inert particles of cement that weaken the top portion and form a scum called laitance which must be removed if a new layer of concrete is to be placed over the original layer. In addition, as the water rises through the concrete flow channels are formed in the concrete mass and water accumulates under the coarse-aggregate particles and under horizontal reinforcing bars. This action results in a weaker concrete structure because of the lack of bond between the paste and coarse aggregate and between the concrete and reinforcing steel. As a result, concrete with a large amount of bleeding water may be very permeable and the reinforcing steel may be subject to corrosion.

Bleeding may be largely controlled by the proper choice of constituents and the proportions of the concrete mixture. Richer mixtures made with finely ground cements having normal setting properties, minimum amounts of mixing water, smooth natural sands with an adequate percentage of fines, air-entraining admixtures, or admixtures consisting of fine particles are all helpful in decreasing bleeding of concrete mixtures. Some improvement of the properties of concrete that has had appreciable bleeding may be obtained by vibrating or tamping the concrete at the end of bleeding and at the beginning of setting.

EFFECTS OF CONSTITUENTS

5.6 Cement

The workability of a concrete mixture depends on the amount of cement, the fineness of the cement, and its chemical composition. Very lean mixtures are apt to be harsh and have poor workability because of the lack of sufficient material of cement fineness. Generally, other things being equal, workability increases as the amount and fineness of the cement increase. Extremely rich mixtures, however, may be too cohesive or sticky.

In some instances the necessary period for proper workability before stiffening may not be available. This premature stiffening or false set is believed to be due to unstable gypsum in the cement resulting from high grinding temperatures in the manufacture of the cement. It is usually possible to rework the stiffened concrete to its original workability without the addition of extra mixing water. Remixing after development of false set will prevent further stiffening until the start of true setting and hardening action. Improperly proportioned or manufactured cements may cause a flash set accompanied by an appreciable liberation of heat. This condition is permanent and cannot be improved by additional mixing or agitation.

5.7 Aggregates

The grading and shape of the fine and coarse aggregates and the maximum size of the coarse aggregate have important effects on workability. Both the fine and coarse aggregates should be uniformly graded from fine to coarse and should be free from an excessive amount of any one size fraction that would tend to cause particle interference and would result in poor workability. It appears that for any given aggregate a large number of gradings may be used which will be about equally satisfactory for workability provided that reasonable limits and uniformity of grading are maintained. Gap gradings in which one or more of the intermediate size fractions have been eliminated should be checked under job conditions before adoption.

Natural sands with rounded grains produce more workable concrete than crushed sands made of angular, flat, or elongated pieces. The latter types usually have a high percentage of voids and may cause excessive bleeding of the concrete. The coarse-aggregate particles should preferably be spherical in form. Crushed coarse aggregates cubical in shape will, if properly graded and combined with the proper amount of workable mortar, produce workable concrete. Flat, disk-shaped pieces and long, thin wedgelike particles are objectionable because they cannot be easily and closely compacted.

The maximum size of coarse aggregate must be selected for each specific construction condition. The choice will usually involve consideration of such factors as spacing of reinforcing bars, minimum width of form, and methods of placing and compacting the concrete mass.[1,2]

5.8 Admixtures

Workable concrete mixtures made with satisfactory aggregates, sufficient cement, and the correct amount of water to produce the required slump do not normally require any additions of admixtures for satisfactory workability. However, admixtures are helpful in lean, harsh concrete mixtures of poor workability and where difficult placement conditions are involved.

In order to increase workability of a lean concrete it is necessary to increase the surface area of the solids per unit volume of water. This may be accomplished by adding finely divided admixtures such as hydrated lime, diatomaceous earth, bentonite, or fly ash or by increasing the cement content. The amount of the finely divided admixture added to the concrete mixture must be carefully controlled since large quantities tend to require a higher water/cement ratio with undesirable effects on strength, durability, and drying shrinkage unless appropriate adjustments are made.

A large variety of proprietary compounds are available that affect various workability characteristics. These compounds are classified as wetting agents, dispersing agents, densifying agents, retarders, accelerators, air-entraining agents and high-range water reducers. When these compounds are properly used to overcome certain deficiencies of a concrete mixture they may be very effective. Indiscriminate use of these compounds, however, may produce harmful rather than desirable effects.[3,4]

Air-entraining agents are usually added to increase the durability properties of concrete. In addition, the billions of disconnected small air bubbles that they produce increase workability, decrease bleeding and segregation, decrease density, and may decrease strength. Air entrainment permits a reduction of the sand content in a concrete mixture approximately equal to the volume of the entrained air. It also permits a reduction of the mixing water, about 2 to 4 percent for each percent of entrained air, with no loss of slump and some gain in workability. While purposeful air entrainment improves workability of concrete mixtures it may cause problems related to finishing of horizontal surfaces. The marked reduction in bleeding of air-entrained concrete usually requires that finishing be accomplished much more quickly than for concrete containing no entrained air. This is especially true where the water evaporation from the surface is rapid as in the case of a slab exposed to hot, dry winds. Under such conditions the water that is very rapidly evaporated from the surface is not replaced by bleeding water. Concrete containing entrained air can be properly finished, but special care and understanding are required.

5.9 Temperature

Extreme temperatures produce additional problems in placing concrete and may cause undesirable changes in properties. The difficulties due to placement of concrete in cold or hot weather must be anticipated and provision must be made to avoid or to lessen undesirable effects. In general this requires that mixing temperatures be kept between 55 and 80°F and that the concrete be protected during its early life.

Concrete placed at low temperatures above freezing and kept from freezing will develop higher ultimate strength and greater durability than similar concrete placed at higher temperatures. The advantages of placing concrete under cold-weather conditions should be realized along with the complications. Generally, the temperature of freshly placed concrete in thin sections should not be less than 55°F, while the minimum temperature for mass concrete should be at least 40°F. In order to obtain these temperatures under very cold conditions it may be necessary to heat the mixing water or the aggregate. Overheating of these constituents should be avoided because of a possible increase in the water requirement, a loss in slump, and the possibility of causing flash set. The relationship between slump and temperature of concrete is shown in Fig. 5.4. Frequently air-entraining agents and accelerators may be added to concrete that is to be mixed and placed under cold temperatures. The most commonly used accelerator, calcium chloride, is added in amounts of 1 to 2% by weight of the cement. (Recommendations of ACI Comm. 201, Guide to Durable Concrete, ACI 201.2R-92 and ACI Comm. 222, Corrosion of Metals in Concrete, ACI 222R-85 should be followed.)

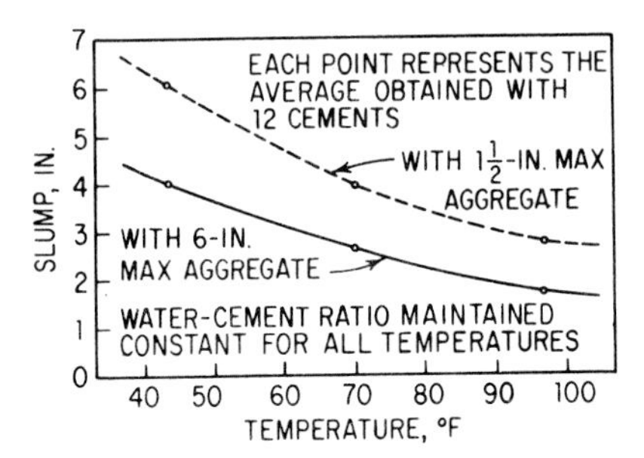

FIGURE 5.4 Relation between slump and temperature of concrete made with two maximum sizes of aggregate.[1]

Hot weather also presents special problems in manufacture and placement of concrete. High temperatures usually accelerate the setting rate, increase evaporation of mixing water, increase the required amount of mixing water, reduce ultimate strength, and increase the tendency to crack. Undesirable effects of hot weather may be minimized by cooling the ingredients, to keep the temperature of freshly placed concrete in the range 60 to 80°F, and to protect and cure the concrete promptly after placement. Water-reducing retarders are used to counteract the accelerating effects of high temperatures and to decrease the need for increased mixing water. See Sec. 5.8 for detailed discussion of hot- and cold-weather concreting.

SEGREGATION

5.10 Factors Influencing Segregation

Segregation of the various constituents of fresh concrete is detrimental, and every effort should be made to minimize it. Imperfections in hardened concrete such as rock pockets, sand streaks, weak and porous layers, crazing, pitting, and surface scaling are usually related to segregation. Repair of the damaging effects due to segregation is difficult and costly. It is much better to avoid segregation by using well-designed mixtures and by placing the concrete properly under competent supervision.

Segregation of concrete is always a problem because concrete is not a homogeneous material but an aggregation of constituent materials that vary widely in size and specific gravity. Special care to avoid defects due to segregation must be taken when the concrete mixture is very lean, or very wet, or contains rough-textured aggregate not of cubical or spherical shape or if the maximum aggregate size is large in comparison with the dimensions of the member to be formed. Segregation is not limited to the solids in a concrete mixture. The mixing water tends to rise as the heavier solid particles of aggregate and cement settle through it. This type of segregation may be apparent during placing, but it is most evident after placement.

Proper proportioning is only one of the important factors that influence segregation. Every step in the manufacturing and placing procedures must observe time-proved methods in order to minimize segregation. Accurate batching and thorough mixing are essential. After mixing, every operation involved in transporting and placing the concrete affords further opportunity for loss of uniformity. Filling and discharging concrete from hoppers, buckets, or cars and discharging from chutes, belt conveyors, or buckets, and placing concrete in the forms by hand methods or vibration provide further opportunity for segregation. Special problems are encountered when the concrete is to be pumped, when it is to be placed under water, when the reinforcement is closely spaced, or when the mold is complex and has sharp corners.

Segregation must be avoided. It will not be corrected automatically in the succeeding operations required in the construction of any concrete structure. Every operation involved in handling, placing, and consolidating the concrete mixture must be carefully planned and controlled to avoid segregation. In general, the concrete mixture should be no wetter than necessary, it should be allowed to fall vertically in a continuous flow, and it should be placed as near to its final location as possible to avoid excess lateral motion. Correct and incorrect methods of handling, placing, and consolidating concrete are discussed in detail in Chap. 26.

REFERENCES

1. "Concrete Manual," 8th ed., U.S. Bureau of Reclamation, 1975.
2. ACI Committee 211, Standard Practice for Selecting Proportions for Normal, Heavyweight and Mass Concrete, ACI 211.1–91, 1991.
3. ACI Committee 201, Guide to Durability, ACI 201.2R-92, 1992.
4. ACI Committee 222, Corrosion of Metals in Concrete, ACI 222R-85, 1985.

CHAPTER 6
STRENGTH AND ELASTIC PROPERTIES

George W. Washa*

STRENGTH TESTS

6.1 General Considerations

Since concrete is a hardened mass of heterogeneous materials its properties are influenced by a large number of variables related to differences in types and amounts of ingredients, differences in mixing, transporting, placing, and curing, and differences in specimen fabrication and test details. Because of the many variables, methods of checking the quality of the concrete must be employed. The usual procedure is to cast strength specimens at the same time that the structure is placed and to regard the specimen strength as a measure of the strength of the concrete in the structure. The reliability of this assumption should always be questioned because of different curing conditions for the specimen and the structure, because poor workmanship in placing concrete in the structure may not be reflected in tests of specimens, and because poor testing procedures may provide false results. A pattern of tests should be used rather than placing reliance on only a few tests to check uniformity and other characteristics of concrete. Statistical methods as given in ACI Standard 214 should be used where large quantities of concrete are involved.

Most concrete is proportioned for a given compressive strength at a given age, and consequently a compression test is most frequently used. A 6 × 12-in cylinder is most commonly required but larger cylinders are frequently used with mass concrete to be placed in dams. Compressive strengths may also be determined from modified cube tests made on beam specimens remaining after flexural tests and on cores of various sizes cut from hardened concrete. The details of all strength tests are given in ASTM standards.

Concrete is not normally required to resist direct tensile forces because of its poor tensile properties. However, tension is important with respect to cracking due to restraint of contraction caused by chemical activity, drying shrinkage, or decrease in temperature. Tension tests also provide an approximate measure of the compressive strength. Direct tension tests of concrete are infrequently made because of difficulties in applying the tensile forces. The splitting tensile strength of concrete is determined by applying increasing diametral compressive loads on a 6 × 12-in cylinder placed on its side until splitting fail-

*Revised by Joseph A. Dobrowolski.

ure occurs and then calculating the tensile strength from the maximum load and known dimensions (ASTM C496).

Bending tests on beams are usually required when unreinforced concrete is to be subjected to flexural loading, as in the case of highway pavements. The beams are usually end-supported and loaded with a concentrated force at the center or with concentrated forces at the third points. The stress in the fiber farthest from the neutral axis, calculated from the flexure formula $S = Mc/I$, is called *modulus of rupture*. While this is a fictitious stress value because the assumptions on which the flexure formula is based are not valid near failure, it is useful for calculation and comparison purposes. It is usually 60 to 100 percent higher than the tensile strength.

A standardized test procedure to determine the effects of variations in the properties of concrete on the bond strength between concrete and reinforced steel is provided by ASTM C234. Tests to determine bond strength are infrequently performed except for research purposes. Tests to determine fatigue strength and strength under various types of combined loading are not standardized and usually are performed only for research purposes.

FACTORS AFFECTING STRENGTH

6.2 Mix Proportions; Water/Cement Ratio

If satisfactory materials are combined into a workable concrete mixture that is allowed to age under satisfactory curing conditions, the strength of the hardened concrete at a given age will be greatly influenced by the water-cement ratio of the mixture. The exact position of the strength-vs.-water/cement ratio curve will be dependent on the properties of each of the ingredients, the proportions of the ingredients, the mixing and placing methods, and the curing methods.

Proportions of concrete mixtures are specified in many ways. Ready-mixed concrete producers are frequently asked to provide concrete with a given cement content, pounds of cement per cubic yard of concrete,and a given slump. In some instances concrete proportions may be specified by weight or volume as 1:2:4, 1 part of cement to 2 parts of fine aggregate to 4 parts of coarse aggregate. In other instances concrete proportions will be dictated by a required strength at a given age and a given slump. Regardless of the method of specifying proportions the most important consideration for good strength is the use of a workable mixture with a water/cement ratio that is as low as possible. A concrete that does not have its water/cement ratio specified may have a wide range in strength. A 1:2:4 mixture, by weight, may have relatively good or poor strength depending on the amount of water added. If a desired strength is required the amount of water (or the slump) must be given along with the proportions of dry materials.

Relations between compressive strength and water/cement ratio for air-entrained concrete made with Types I and III portland cement for ages of 1, 3, 7, and 28 days are given in Fig. 6.1. In each instance a band of values is shown rather than a single curve in order to cover variations in materials and testing procedures. On a given job with reasonably uniform materials and procedures a single curve for each age can usually be obtained. Relations between compressive strength and water/cement ratio for non-air-entrained concrete made with Types I and III portland cement for ages of 1, 3, 7, and 28 days are given in Fig. 6.2. When the compressive strength is known the flexural strength may be approximated from the equation

$$R = K\sqrt{f_c'}$$

where f_c' = compressive strength, psi
R = flexural strength (modulus of rupture), psi
K = a constant, usually between 8 and 10

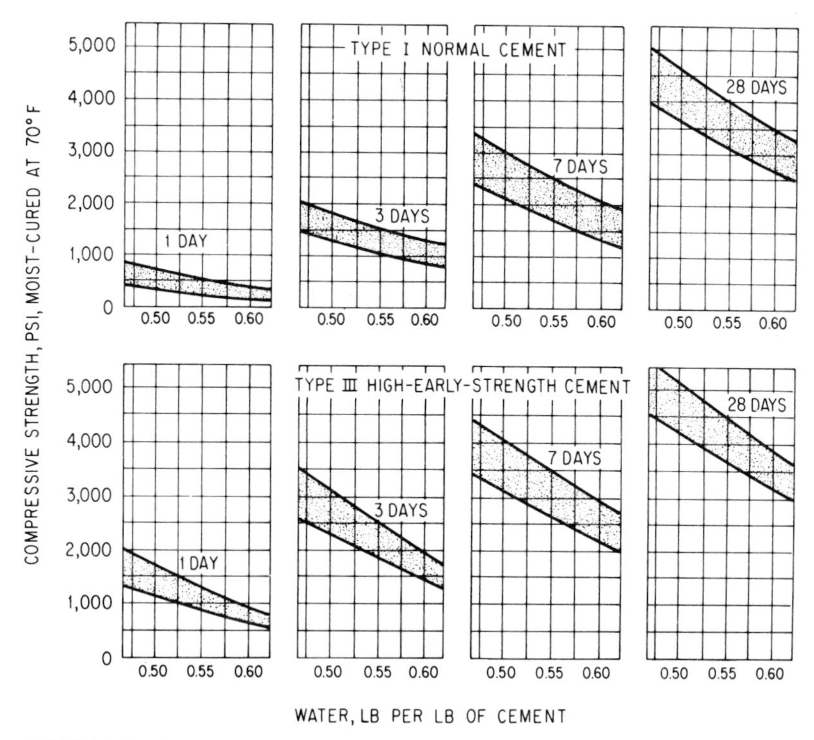

FIGURE 6.1 Age–compressive strength relations for air-entrained concrete (concrete with air content within recommended limits and maximum aggregate size of 2 in or less) made with Types I and III portland cements.[1] These relationships are approximate and should be used only as a guide in lieu of data on job materials.

6.3 Cement Content and Type

With materials, consistency, and density constant, the strength of concrete increases with the proportion of cement in the mixture until the strength of the cement or aggregate, whichever is the weaker, is reached. The data in Fig. 6.3 represent tests on moist-cured, workable concretes having the same slump, 4 to 6 in, and made from the same aggregate mixture, 38% sand and 62% gravel. For workable mixtures of drier consistencies, the corresponding curves would be higher and for those of wetter consistencies the curves would be lower. Also the use of higher-strength cements tends to move the curves higher.

The effect of cement fineness (Wagner), expressed as specific surface in square centimeters per gram of cement, on the compressive strength of concrete at four different ages is shown in Fig. 6.4. Finely ground cements are desirable in that they increase strength, especially at early ages, and also increase workability. They may be undesirable because they contribute to cracking and have lower resistance to freezing and thawing. The ASTM minimum fineness requirements for portland cement are 160 m^2/kg (Wagner) and 280 m^2/kg (Blaine).

The curves in Fig. 6.5 show the change in compressive strength of concrete with age for the five standard types of portland cement specified by ASTM. All curves cross at an approximate age of 3 months. The concrete made with high-early-strength cement (Type

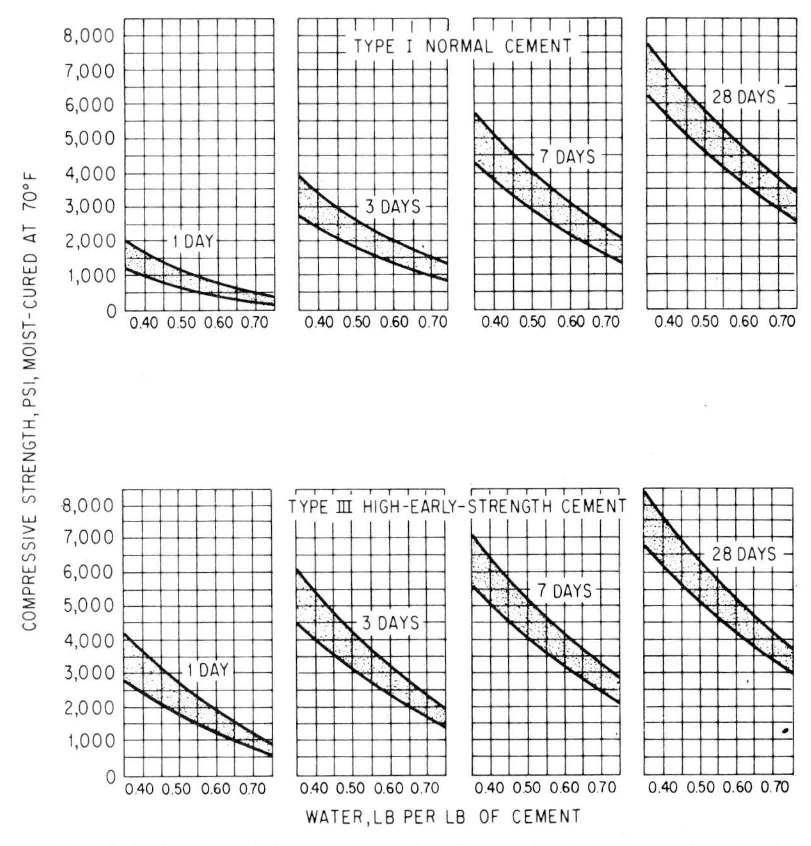

FIGURE 6.2 Age–compressive strength relations for non-air-entrained concrete made with Types I and III portland cements.[1] These relationships are approximate and should be used only as a guide in lieu of data on job materials.

III) has relatively high strength at ages up to 3 months but after that its strength is slightly lower than that of concrete made with normal cement (Type I) and considerably lower than the strengths of concretes made with modified (Type II) or low-heat (Type IV) or sulfate-resisting (Type V) cement.

6.4 Aggregate Types and Characteristics

Heavy aggregates such as magnetite, barite, and scrap iron are used for making concrete (180 to 220 pcf) for radiation shielding and concrete counterweights. Light-weight aggregates such as expanded shales and clay, expanded slag, vermiculite, perlite, pumice, and cinders are used to make insulating, masonry unit, and light-weight structural concrete weighing between 25 and 120 pcf.

The principal materials used in normal-weight concrete, usually 140 to 155 pcf, include the sands and gravels, crushed rock, and air-cooled blast-furnace slag. The more

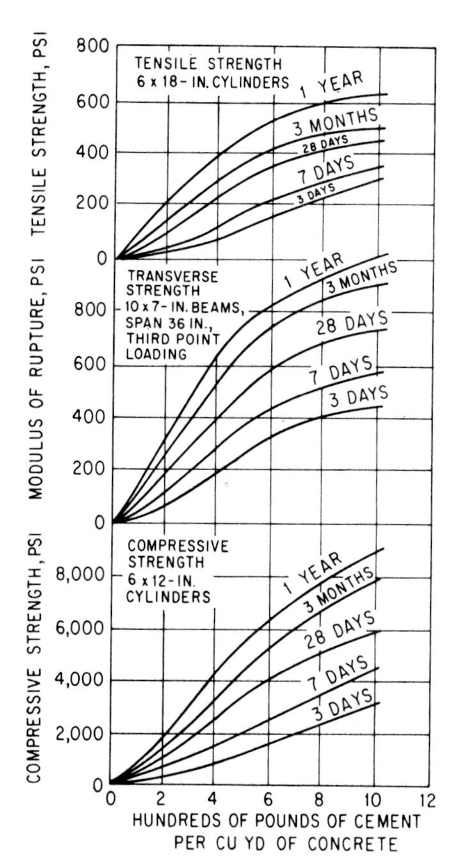

FIGURE 6.3 The effect of cement content on the strength of concrete. Materials, Elgin sand and gravel graded from 0 to 1½ in. Fineness modulus of mixed aggregate = 5.5.[2]

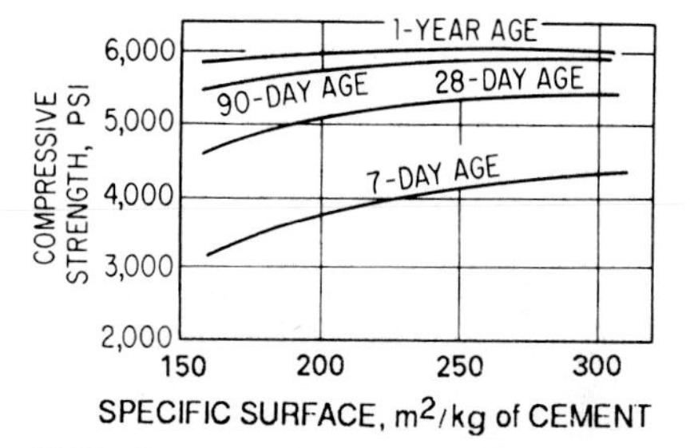

FIGURE 6.4 Relation of specific surface of cement to compressive strength of concrete.[3]

commonly used crushed rocks are granite, basalt, sandstone, limestone, and quartzite. Aggregate characteristics that affect concrete strength include particle shape, texture, maximum size, soundness, grading, and freedom from deleterious materials. Usually the effect of type of normal-weight aggregate with satisfactory properties and gradation on the strength of concrete is small because the aggregates are stronger than the cement paste.

As the maximum size of aggregate in a concrete mixture of a given slump is increased the water and cement contents in pounds per cubic yard of concrete are decreased. The influence of maximum size of aggregate on compressive strength of concrete is shown in Fig. 6.6. The curves show that the compressive strength varies inversely with the maximum size of aggregate for a minimum cement, that in the lower-strength ranges maximum size is less important, and that at the higher-strength ranges concretes containing the smaller maximum-size aggregates generally develop the greater strengths.

6.5 Admixtures

The effect of admixtures on the strength of concrete varies greatly with the properties of the admixture and with the characteristics of the concrete mixture. The admixtures with important strength effects include the accelerators, the water reducers, the air entrainers, and the finely divided minerals.

The most widely used accelerator is calcium chloride. Added in proper amounts it provides increased early strength during cold weather and also provides better protection against damage due to freezing temperatures, by shortening the time that the concrete requires protection. The amount of calcium chloride should be restricted to that necessary to produce the desired results and should not exceed 2% by weight of the cement. Frequently 1% will be sufficient to meet the requirements. The effects of calcium chloride additions, up to 3% on

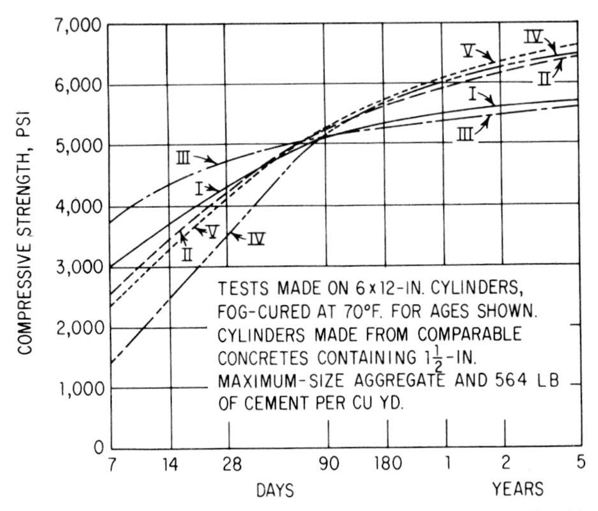

FIGURE 6.5 Rates of strength development for concrete made with various types of cement.[4]

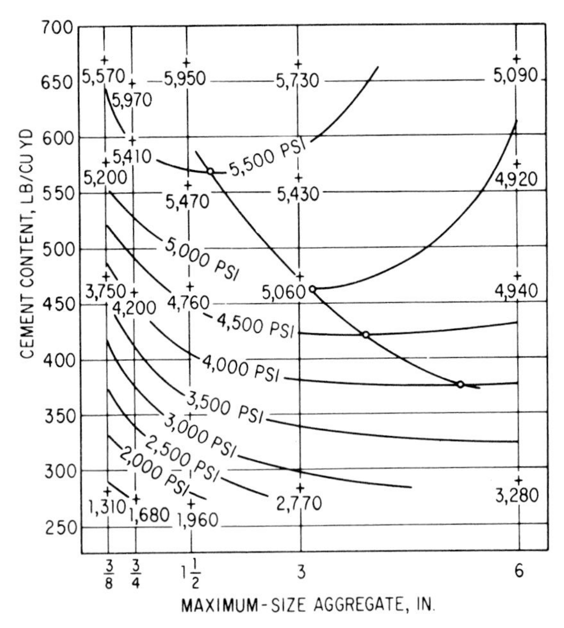

FIGURE 6.6 Variation of cement content with maximum size of aggregate for various compressive strengths. Each point represents an average of four 18 × 36-in and two 24 × 48-in concrete cylinders tested at 90 days for both Clear Creek and Grand Coulee aggregates. Mixes had a constant slump of 2 ± 1 in for each maximum-size aggregate.

the compressive strength of concrete made and cured at temperatures of 40 and 70°F, are shown in Fig. 6.7. Under some conditions the desirability of calcium chloride additions for strength purposes must be considered in terms of possible undesirable effects on other properties such as drying shrinkage, resistance to sulfate attack, resistance to alkali–aggregate reaction, and corrosion of reinforcing steel.

Certain organic compounds usually available in various proprietary forms are used to reduce the water requirement of concrete mixtures. In some cases they are also used to retard set. The usual effect of these materials is to improve compressive strength and impermeability of hardened concrete. The decrease in the water content of a concrete mixture produces a reduction in the water/cement ratio which results in an increase in strength, but the increase in strength is frequently greater than that indicated by the reduction in the water/cement ratio alone. Development of high-range water reducers have allowed low-slump concretes (1 to 2 in) to be placed at 5 to 6-in slump without reducing the characteristic strength qualities of the original 1 to 2-in slump concrete.

In order to increase workability and durability, concrete mixtures are frequently designed to contain between 3 and 6% air. The inclusion of this air may normally result in some loss of strength. When the water/cement ratio is held constant the compressive strength is reduced about 5% for each percentage of air entrained. However, when the cement content is kept constant and advantage is taken of the opportunity to reduce the water/cement ratio for a given workability the strength losses are smaller and may actualy be slight increases for the leaner mixtures. This is true because relatively large decreases in water/cement ratio in the leaner mixtures offset the strength loss due to entrained air. In the richer mixes the relatively small decreases in water/cement ratio do not offset strength loss due to entrained air. The effect of air content on compressive strength is shown in Fig. 6.8. The curves in the top portion of Fig. 6.8 were obtained for mixtures

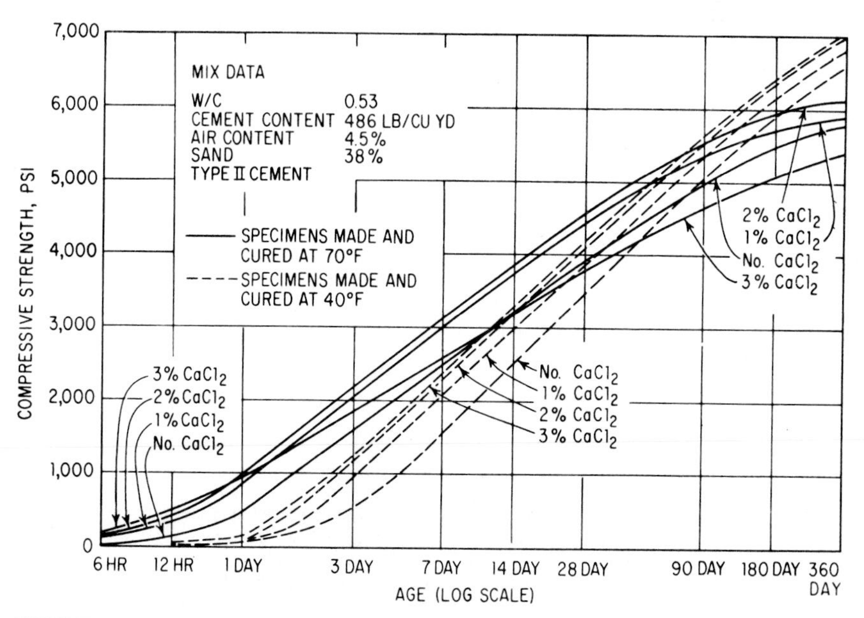

FIGURE 6.7 Effect of small additions of calcium chloride on the compressive strength of concrete.[3]

with and without entrained air when the cement content was kept constant; the curves in the bottom portion were obtained for mixtures with and without entrained air when the water/cement ratio was kept constant. Flexural-strength losses due to entrained air are usually not so great as the losses in compressive strength.

The finely divided mineral admixtures are usually classified into three types: (1) relatively chemically inert materials such as ground quartz, ground limestone, bentonite, hydrated lime, and talc; (2) cementitious materials including natural cements, hydraulic limes, slag cements, and granulated blast-furnace slag; and (3) pozzolans including materials such as fly ash, volcanic glass, diatomaceous earth, and some shales and clays. In many instances these admixtures are added primarily to improve workability but strength changes are also obtained. Generally, the strengths of

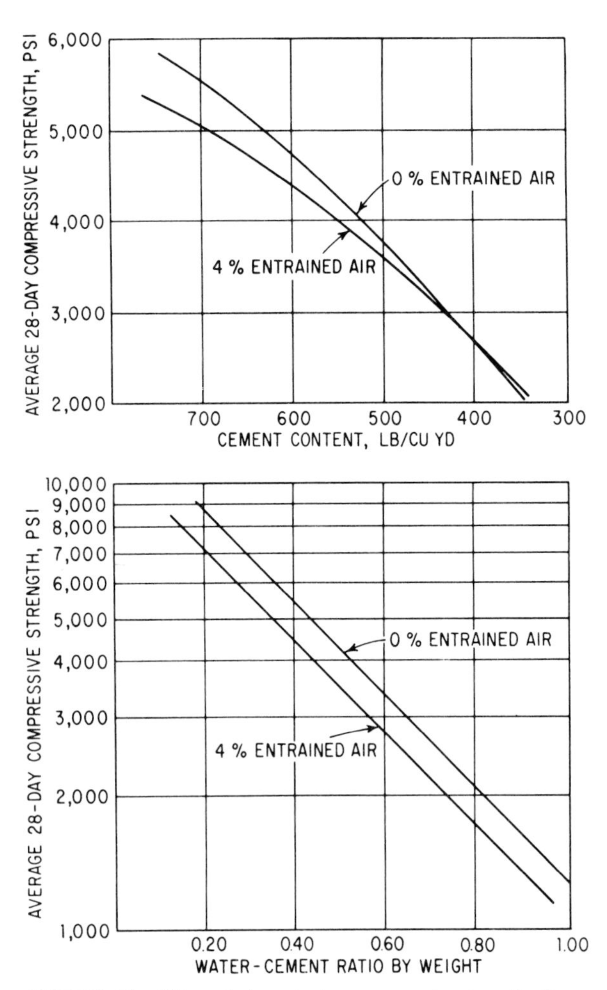

FIGURE 6.8 Effect of air content on compressive strength of concrete.[3]

lean mixtures are improved while the strengths of rich mixtures may be only slightly affected and may actually decrease.

Strength contributions caused by pozzolanic and cementitious admixtures are relatively slow, especially at low temperatures. Under continued moist curing the strengths at later ages will normally be higher than those obtained with portland cement alone. When fly ash is used as a replacement for up to 30 percent of the cement the compressive strength of concrete is usually reduced at 7 and 28 days and may be increased after 3 months. However, compressive strengths at 7 and 28 days for concrete containing fly ash may be made about equal to the strength of concrete without fly ash by proper redesign of the mixture.[4]

6.6 Curing Method

Proper curing requires a satisfactory moisture content and a favorable temperature during the early life of the concrete until satisfactory properties have been developed. When satisfactory curing is not provided serious loss of desirable potential properties occurs.

The effect of moisture during curing is shown in Figs. 6.9 and 6.10. Both figures show that compressive strength increases at a decreasing rate as the moist-curing period increases and that strength development stops at an early age if the concrete is kept in air. Figure 6.9 shows that when moist curing is discontinued the compressive strength increases for a short period but remains constant or decreases thereafter. Figure 6.10 shows that when moist curing is resumed after a period in air, strength increases are also resumed.

The effect of temperature on the compressive strength of concrete is represented in Figs. 6.11 through 6.13. Figure 6.11 shows that higher strengths are obtained at early ages with higher curing temperatures and that the strengths at 28 days for temperatures above 55°F are inversely related to the curing temperature. A curing temperature of 40°F develops low strengths at all ages, but the rate of increase at 28 days is large. The additional information on the effect of temperature on compressive strength in Fig. 6.12 shows the desirable effects of maintaining a good initial curing period for as long a period as possible. The extremely harmful effects of curing temperatures below freezing are clearly evident.

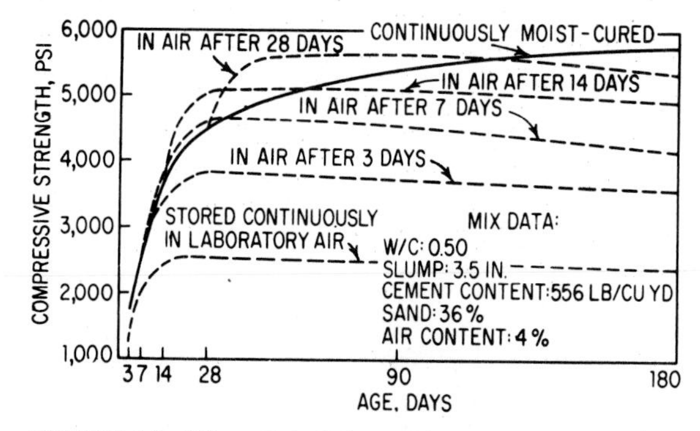

FIGURE 6.9 Effect of air drying on the compressive strength of moist-cured concrete.[4]

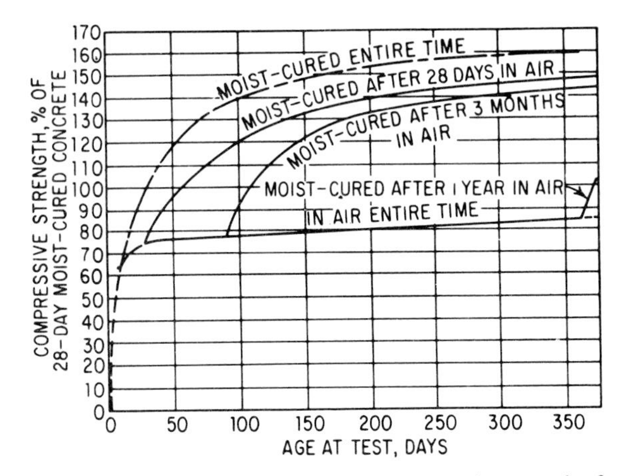

FIGURE 6.10 Effect of curing on the compressive strength of concrete.[1]

While structural concrete is usually subjected to moist curing, concrete products are frequently cured in low-pressure steam or in high-pressure steam (autoclaved). Low-pressure-steam curing is usually carried out in saturated steam at temperatures between 135 and 195°F. The effect of various steam temperatures on compressive strength is shown in Fig. 6.13. The greatest acceleration in strength gain and a minimum loss in ultimate strength are obtained at temperatures between 130 and 165°F. High-pressure-steam curing is usually performed at a maximum steam pressure of 140 to 150 psig (about 360

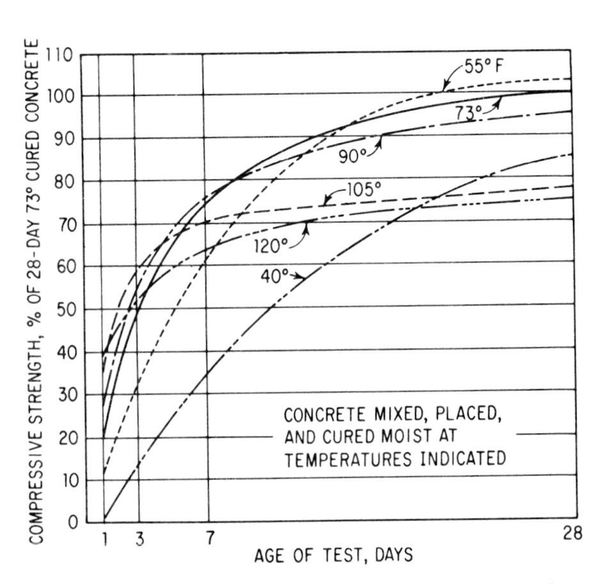

FIGURE 6.11 Effect of curing temperature on the compressive strength of concrete.[1]

to 365°F). Concrete cured in this manner attains in 24 h about the same strength that moist-cured or low-pressure-steam-cured concrete attains in 28 days.[7]

6.7 Age at Test

The compressive strength of concrete increases with age if moisture is present. In a series of tests started in 1910, Withey[8] showed that compressive strength increases with age up to 50 years and that a logarithmic equation may be used to express the relationship. The 1:2:4 and 1:3:6 mixtures used by Withey and shown in Fig. 6.14 had water/cement ratios of 0.62 and 0.90, by weight. Withey suggested that the equations $S_c = 350 + 1195 \log D$ and $S_c = 220 + 665 \log D$ for the 1:2:4 and 1:3:6 mixtures, respectively, with S_c representing compressive strength in psi and D the age in days, are conservative. On the basis of certain assumptions he further suggested that the equations $S_c = 1380 \log D$ for the 1:2:4 mixture and $S_c = 810 \log D$ for the 1:3:6 mixture appear to be satisfactory for estimating strength gain with age of similar concrete mixtures made with present Type I cements when cured under comparable conditions.

The rates at which compressive, tensile, and flexural strengths are developed under moist-curing conditions for concrete made with different types of aggregates and with a water/cement ratio of 0.532, by weight, are given in Table 6.1. The 28-day values are taken as 100 percent, and the values at all other ages are based on the 28-day values.

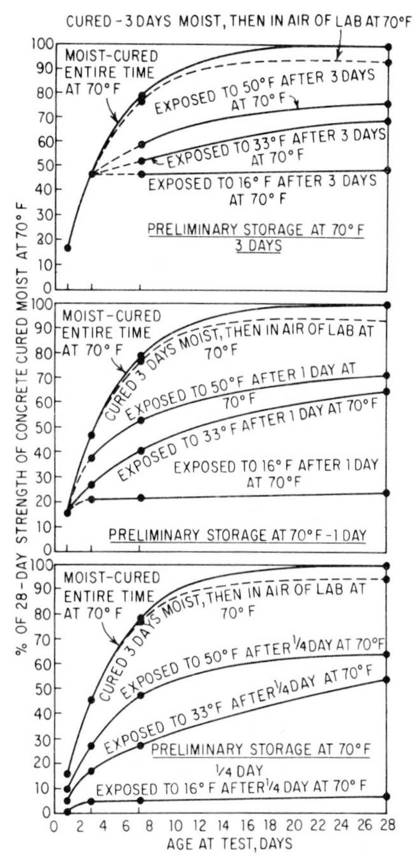

FIGURE 6.12 Relative strength of concrete as influenced by storage temperature. Water/cement ratio = 0.53 by weight, slump 3 to 5 in.[6]

TABLE 6.1 Rate of Strength Development

Type of test	Age at test				
	3 days	7 days	28 days	3 months	1 year
Compression	35	59	100	135	161
Flexure	53	71	100	126	143
Tension	46	68	100	121	150

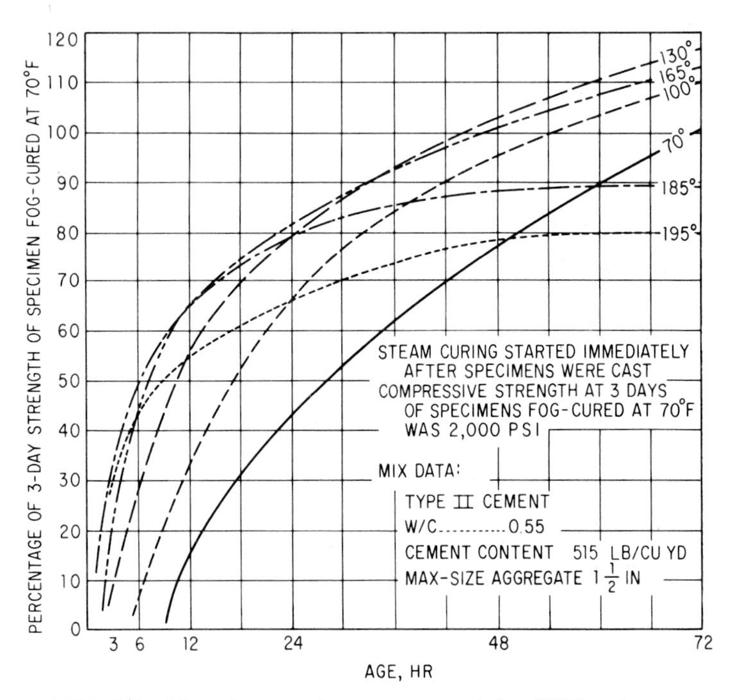

FIGURE 6.13 Effect of steam curing at temperatures below 200°F on the compressive strength of concrete at early ages.[4]

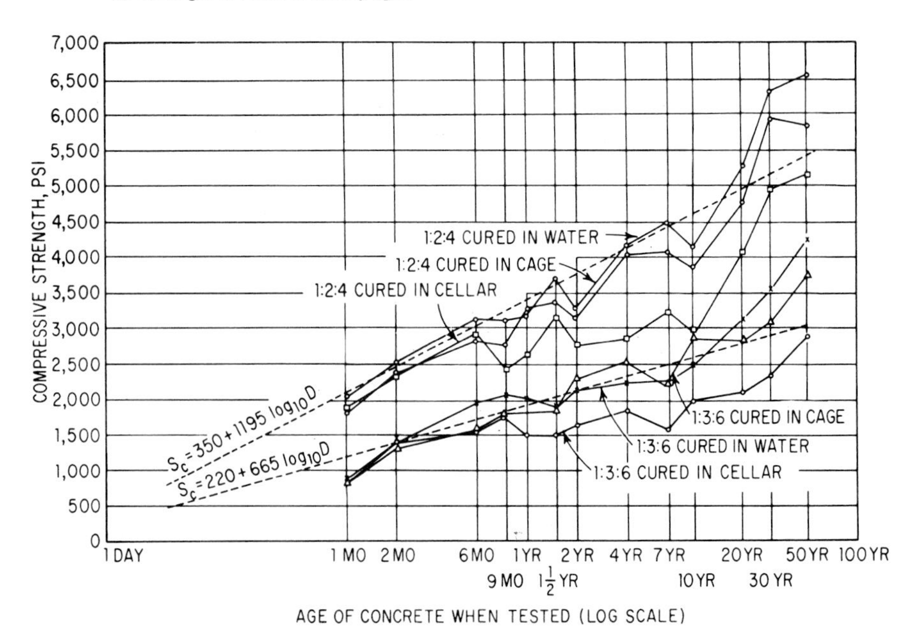

FIGURE 6.14 Relation of concrete compressive strength and age.[3]

STRENGTH RELATIONSHIPS

6.8 Relations between Various Types of Strength

Compressive strength, tensile strength, flexural strength, shearing strength, and bond strength are all related, and usually an increase or decrease in one is reflected similarly in the others, but not necessarily in the same degree. Relations between compressive strength, tensile strength, and modulus of rupture are shown in Fig. 6.15. The tensile strength usually ranges between 8 and 12 percent of the compressive strength and averages about 10 percent. The modulus of rupture usually ranges between 12 and 20 percent of the compressive strength and averages about 15 percent for concrete with a compressive strength of 3500 psi.

The relations between splitting tensile strength and flexural strength and between splitting tensile strength and compressive strength are shown in Figs. 6.16 and 6.17 for gravel concrete, crushed-stone concrete, and lightweight-aggregate concrete. The curves in Fig. 6.16 show that the ratio of splitting tensile strength to the flexural strength was 67 percent for the concrete made with crushed stone, 62 percent for the gravel concrete, and 76 percent for the lightweight-aggregate concrete. The curves in Fig. 6.17 show nonlinear relations between splitting tensile strength and compressive strength. Average ratios of splitting tensile strength to compressive strength were about 10.7 percent for the concrete made with crushed stone, 10.8 percent for the gravel concrete, and 8.0 percent for the lightweight-aggregate concrete.

Satisfactory bond between reinforcing steel and concrete is necessary if reinforced-concrete structures are to perform satisfactorily. Bond may be the result of adhesion, friction, lug action, or end anchorage. Bond-strength values determined from pullout tests are usually noted at initial slip of the free end and at a slip of 0.01 in. at the loaded end. They are significantly influenced by the type of reinforcement, the properties of the concrete, and test procedures. The curves in Fig. 6.18 show large differences between plain and

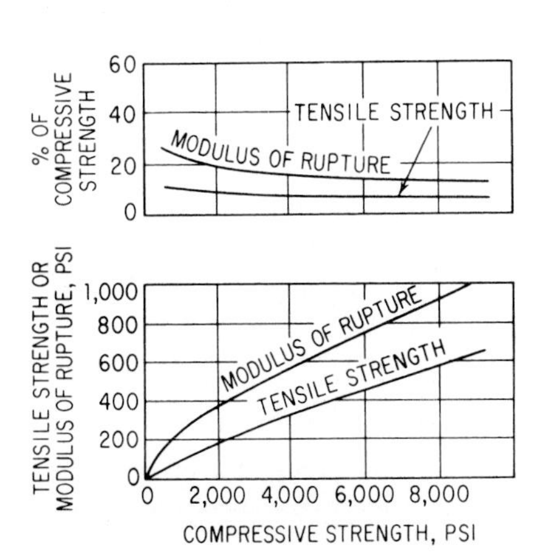

FIGURE 6.15 Relations between compressive strength, tensile strength, and modulus of rupture.[9]

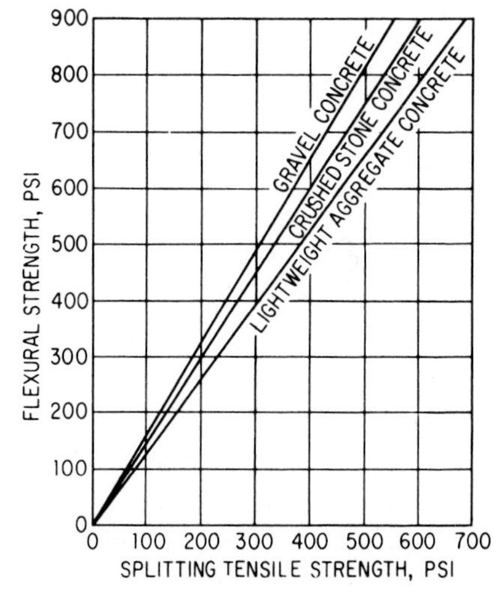

FIGURE 6.16 Relation between flexural strength and splitting tensile strength for concrete made with three different aggregates.[10]

deformed bars. It is also apparent that the bond stress at a given slip increased as the compressive strength increased up to a compressive strength of about 3000 psi, and that above 3000 psi the increase in bond stress continued, but at a decreasing rate, up to a compressive strength of about 6000 psi.

Bond strength is greatly affected by the quality of the paste that is used. It is not significantly affected by air entrainment if the air content is the normal amount of 3 to 6%. It is increased by delayed vibration that is properly timed and applied. It is less for horizontal bars than for vertical bars because of the water gain on the underside of horizontal bars. It is smaller for horizontal bars 1 ft or more from the bottom than for horizontal bars near the bottom. In general, bond strength is reduced by alternations of wetting and drying, of freezing and thawing, and of loading.

The importance of shearing strength is evident from the fact that standard concrete cylinders tested in axial compression usually fail by shearing along an inclined plane. The failure is actually due to a combination of normal and shearing stresses on the plane. Similarly, when the term "shear failure" is applied to the diagonal failure in the web of a concrete beam, it is misleading because the cause of such failure is a tensile stress resulting from a combination of tensile and shearing stresses. The resistance of concrete to pure shearing stress has not been directly determined because such a stress condition causes principal tensile and compressive stresses, equal in magnitude to the shearing stresses, acting on other planes. Since concrete is weaker in tension than in shear, failure occurs as a result of the tensile stresses. Reported shearing strengths vary widely because of difficulties and differences in test procedures. However, on the basis of available test data it appears that the strength of concrete in pure shear is about 20 percent of the compressive strength and that the shear strength may increase when compressive stresses also occur on the failure plane.

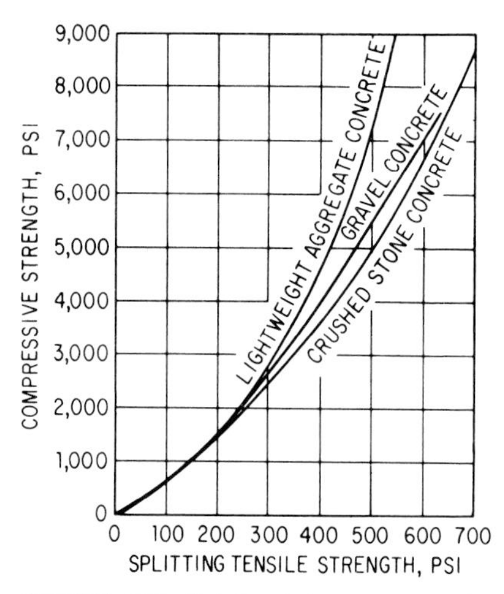

FIGURE 6.17 Relation between compressive strength and splitting tensile strength for concrete made with three different aggregates.[10]

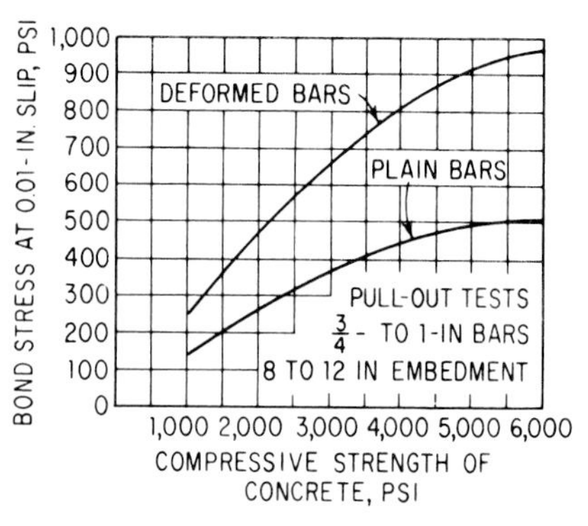

FIGURE 6.18 Variation of bond with compressive strength of concrete.[3]

CAUSES OF STRENGTH VARIATIONS

6.9 Materials

In order to maintain production of concrete having nearly constant properties, strict control of all materials and manufacturing processes is necessary. Careful control and adequate inspection will help avoid many of the difficulties.

One of the commonest reasons given for a poor concrete is the use of a cement with presumed unsatisfactory properties. While cements are variable and have a large effect on concrete properties they are only one of a large number of possible causes of poor concrete. All the other factors that contribute to the properties of the concrete must also be carefully examined before the reason for the poor concrete can be established. Lack of uniformity of portland cement is an important factor relating to strength. There are large differences in the strength-producing properties of cements of a given type from different sources, and also significant differences in the day-to-day production from a given cement mill. Obviously, also, different types of portland cement have significantly different strength properties, and variations in the amount of the cement in a given mixture will greatly influence strength properties. Consequently, while variations in portland cement must be carefully considered, they should not always be blamed for poor concrete properties. All other factors involved must also be considered.

Variations in the aggregate from batch to batch may cause significant strength changes. Aggregate from a given source should be used and the grading should be carefully controlled. Variations in grading may require changes in the amount of mixing water and result in strength changes. Changes in the moisture content of the aggregates, especially the fine aggregate, must be determined and allowed for or serious strength changes may occur. In addition, undesirable materials such as clay lumps, soft particles, organic matter, silt, mica, lignite, and easily friable pieces should be kept at a minimum to decrease harmful effects on strength.

Water for concrete mixtures should not contain any material that can have an appreciably harmful effect on strength, durability, or time of set. Normally water that is acceptable for drinking purposes has been considered satisfactory for concrete. However, relatively small amounts of sugar in water may have serious effects on time of set and strength. Substances that have harmful effects if present in sufficient quantity include silt, oils, acids, alkalies and their salts, organic material, and sewage. General limits of tolerance for the degree of contamination have not yet been developed. The possible effects of contaminated water on the properties of concrete may be evaluated by making comparative tests for time of set and soundness of the cement and for strength and durability of mortars with the contaminated water and with satisfactory water. Contaminated waters that have had no significant effect on time of set and soundness and have not produced strength decreases greater than 15 percent have been frequently used with good success. The presence of salt in mixing water and its possible effect on corrosion of the reinforcing steel must also be considered.

Satisfactory admixtures that are properly used will have beneficial effects on concrete properties. However, admixtures should not be expected to compensate automatically for improperly proportioned concrete mixtures or for poor construction practices. Great care must be taken to use the admixtures properly, and constant checks on the amounts added must be made since small changes in the very small quantities used may seriously influence properties. Overdosages must be avoided.

The use of steel and plastic fibers to influence the concrete properties is under development and initial experiences are being tabulated for design purposes.[22]

6.10 Production

Uniformity of concrete strength depends on careful attention to all the factors involved. Without such care wide variations in strength will occur. All materials should be accurately weighed, and special precautions must be taken in ensuring the correct amounts of cement, water, and admixtures. Water in the aggregates must be taken into consideration in determining the amount of water to be added. After the correct quantities of materials have been added care should be taken to prevent any losses, especially cement or water. A written record of all quantities in a given batch should be obtained as a check and for possible future reference in the event of unusual results.

The necessary mixing procedure and period for a given mixture should be kept constant. The period of mixing must be long enough to obtain a homogeneous concrete, but it should not normally be longer than necessary because of changes due to loss of water by evaporation and to grinding action resulting in an increase of fine materials. Figure 6.19 shows that when water was not added to restore slump the strength increased progressively until the concrete became too dry to allow satisfactory molding of specimens. When water was added to restore slump the strength decreased progressively with time to about one-half of the original value after a 3-h mixing period.

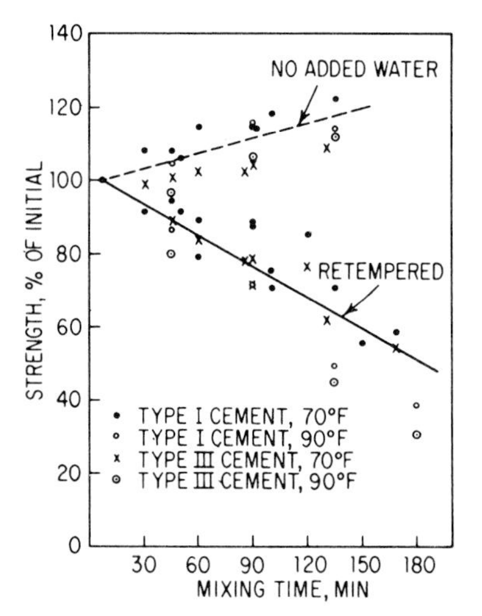

FIGURE 6.19 Effect of prolonged mixing on compressive strength of concrete.[11]

Every precaution must be taken to avoid segregation while handling and placing concrete to avoid variations in strength from one portion of a concrete structure to another. In addition, equal compaction of concrete batches from a given mixture is required if equal strengths are to be obtained.

The effect of curing on the strength of concrete has been discussed in Sec. 6.6. The importance of curing has frequently been overlooked. It is unfortunate that in many instances great effort and expense have led to proper mixture proportions and satisfactory placement procedures, but poor concrete properties have been obtained because of improper curing. In general, concrete will continue to gain strength as long as moisture is available and the temperature is satisfactory. Lack of moisture or excessively low or high temperatures will stop strength gain.

6.11 Sampling and Testing Procedures

Compression tests, usually on 6 × 12-in cylinders, are commonly made to evaluate properties of concrete. Tests may also be made on cores taken from hardened concrete. Flexural tests of beams, frequently 6 × 6 in in cross section tested over a span of 18 in, are made on concrete for highway use. Details of test procedures are given in ASTM Standards. While the tests of relatively small specimens do not necessarily provide information on the strength of concrete in a structure, they are sufficiently accurate for purposes of design and job control.

Concrete specimens may be subjected to standard curing, moist at 73.4 ± 3°F, and tested to determine potential strength, or they may be cured in the field under the conditions of moisture and temperature to which the structure is exposed. Field-cured specimens are usually tested to determine when forms may be removed and when the structure may be placed in service. Many factors affect the strength values determined from tests, and consequently great care should be used in evaluating the results.

The number of specimens that should be tested for a given structure is frequently specified. Usually the requirement is based on a given number of specimens per day, or a given number of specimens for a given volume of concrete, or a given number of specimens for a given area of paved surface. Generally a minimum of three specimens per test result are required. On large structures statistical methods are used to assess strength results and to refine design criteria. The methods used to evaluate compression-test results of field concrete are provided in detail in ACI Standard 214. Test ages most commonly specified are 7 and 28 days, although tests may also be required at other ages such as 3, 60, and 90 days and at 1 year.[2]

6.12 Bearing Conditions

Concrete strengths obtained by tests are significant only if all proper precautions are taken in conducting the tests. In all compression tests the specimen axis should be vertical, the specimen should be accurately centered in the testing machine, the bearing surfaces should be prone and should be perpendicular to the axis, and a spherically seated bearing block should be used.

All bearing surfaces that depart from a plane by more than 0.001 in should either be ground smooth or be capped with a material that has as nearly as possible the same strength and elastic properties as the concrete specimen. Materials frequently used for capping concrete cylinders include neat high-alumina cement, neat portland cement, mixtures of sulfur and granular materials, and high-strength gypsum plasters used with specimens having a compressive strength less than 5000 psi. All capping materials must be properly applied and be given sufficient time to harden before the specimen is tested.

The capping material used may have considerable influence on the compressive strength of concrete cylinders. High-strength concrete cylinders, above 5000 psi are more affected by the capping material, with strength losses as large as 40% reported for plaster of paris caps. Low-strength concrete cylinders around 2000 psi or less are not greatly affected by the capping material, with strength losses usually less than 10 percent even when poor capping materials are used.

6.13 Form and Size of Specimen

Concrete strengths are dependent on the type of specimen, cube or cylinder, and on the specimen dimensions. The effect of specimen size, for concrete specimens made and cured under the same conditions and having a fixed ratio of height to diameter of 2, is shown in Fig. 6. 20. The values are based on the average of tests at 28 and 90 days. The decrease in strength with an increase in cylinder diameter is believed to be due to a faster strength gain of the smaller-diameter cylinders and other reasons. Further, it is believed that age may equalize the difference due to difference in specimen diameter.

The effect of the ratio of length of cylinder to the diameter is shown in Fig. 6.21. It is evident that decreasing values of L/D below 2.0 result in large strength changes. The actual values of the changes are dependent on many factors such as concrete strength, type of aggregate, amount of entrained air, and method of curing. The effect of concrete strength on correction factors required for various L/D ratios is shown in Fig. 6.22*a* and *b*

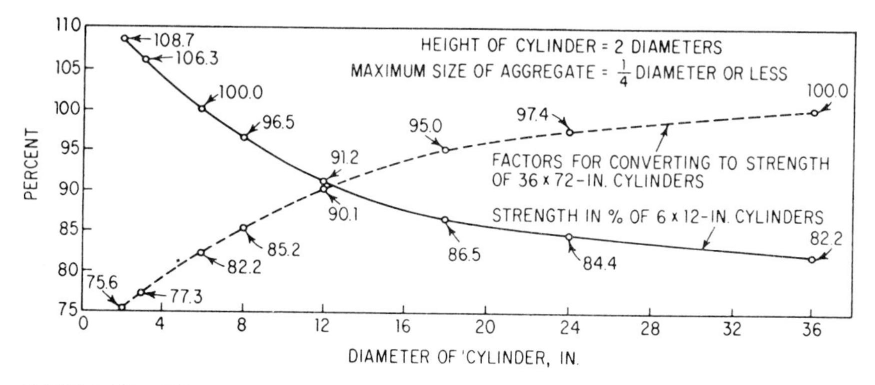

FIGURE 6.20 Effect of cylinder size on compressive strength of concrete.[3]

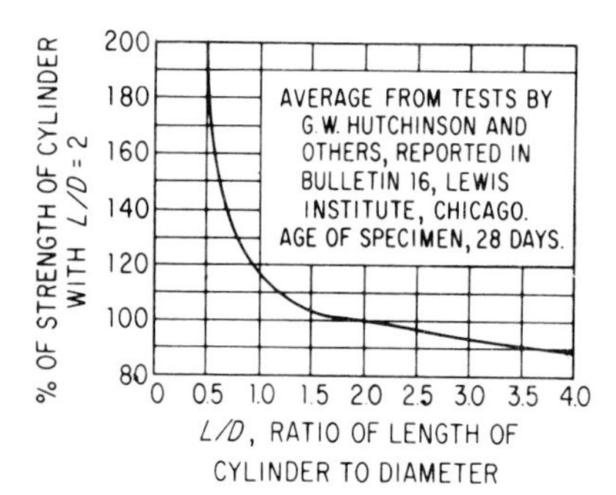

FIGURE 6.21 Relation of length and diameter of specimen to compressive strength.[3]

for normal-weight air-entrained concrete and for lightweight concrete.

The method of determining compressive strength of a concrete using portions of beams broken in flexure (modified-cube method) is given in ASTM Standard C116. The modified-cube compressive strength is usually somewhat higher than the cylinder strength, especially for low-strength concrete. The compressive strength determined from tests of 6-in cubes is normally taken as ⅘ times the compressive strength of a cylinder with an L/D of 2.0. However, the actual correction is affected by the concrete strength and other variables.

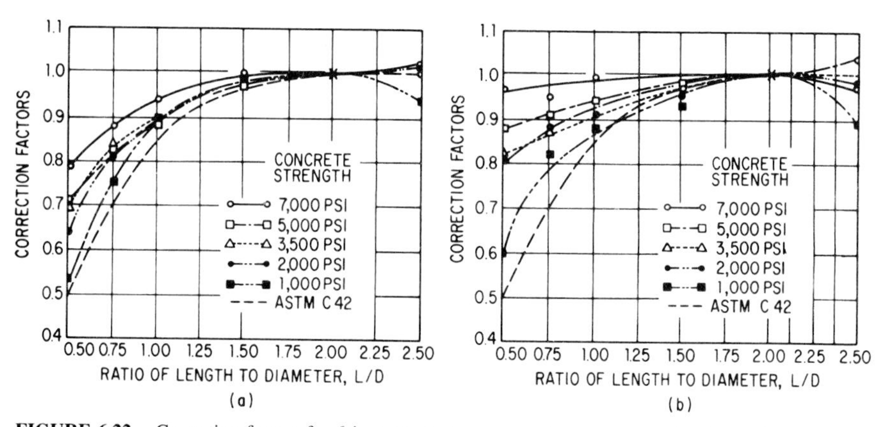

FIGURE 6.22 Correction factors for 6-in concrete specimens: (*a*) normal-weight air-entrained concrete; (*b*) lightweight concrete.[12]

6.14 Moisture Content at Test

The moisture content of a concrete specimen at the time of test has an appreciable influence on the strength. Compression specimens tested in an air-dry condition usually have strengths about 20 to 40 percent higher than similar specimens tested in a saturated condition. Consequently most test specifications require that all specimens be in a saturated condition at the time of test to eliminate indefinite effects due to partial drying. Cores taken from a structure should be soaked for about 48 h prior to test. The effect of drying on compressive strength is also evident in Fig. 6.9.

The effect of moisture on tensile and flexural strength appears to be more variable, probably because of nonuniform drying. Generally, drying appears to reduce both tensile and flexural strength, with reported losses of up to 25 percent. In any event, care should be taken to avoid partial drying of specimens prior to test.

6.15 Loading Rate

An increase in the rate at which a concrete specimen is loaded causes an increase in strength. Consequently ASTM Standard C39 requires that compression cylinders be tested at a rate of 20 to 50 psi/s if a hydraulic machine is used and that the loading head operate at an idling speed of 0.05 in/min if a screw-gear machine is used. Flexural tests specified in ASTM Standard C293 should be performed at a rate such that the increase in the extreme fiber stress approximates 125 to 175 psi/min.

Tests[13] have shown that loading at 1 psi/s reduced indicated compressive strength by about 12 percent when compared with tests at a standard rate of 35 psi/s, and that tests at 1000 psi/s increased the indicated compressive strength by about 12 percent when compared with the standard. Other tests[14] have shown that the compressive strength of concrete tested at stress rates between 3.0×10^6 and 20.0×10^6 psi/s was about 85 percent greater than that obtained from tests requiring an approximate failure time of 30 min. Investigators have also shown that concrete may be expected to sustain indefinitely only about 70 percent of the ultimate load determined from tests at a standard loading rate.

6.16 Test Temperature

Tests of concrete specimens are usually made at room temperature, and the effect of temperature on strength may be overlooked. In Fig. 6.23 mixtures 1C and 2C were made with sand and gravel aggregate and had water/cement ratios of 0.48 and 0.84, by weight, respectively. Mixture 3C was made with an expanded shale aggregate and had a total water-cement ratio of 0.81. The curves in Fig. 6.23 show that compressive strength and modulus of elasticity decreased as the specimen temperature increased to approximately 200°F, and then increased before finally decreasing again. The decrease appeared to be greatest as the temperature increased from 0 to 50°F. At about 450°F the strengths were about equal to those at 70°F, but the moduli of elasticity were considerably lower. Sand-and-gravel concrete and expanded-shale concrete responded similarly to variations in test temperature.

6.17 Lateral Restraint

Concrete that is restrained on all sides will support a much greater load than unrestrained concrete. Figure 6.24 shows that as the lateral restraint increased the axial stress also

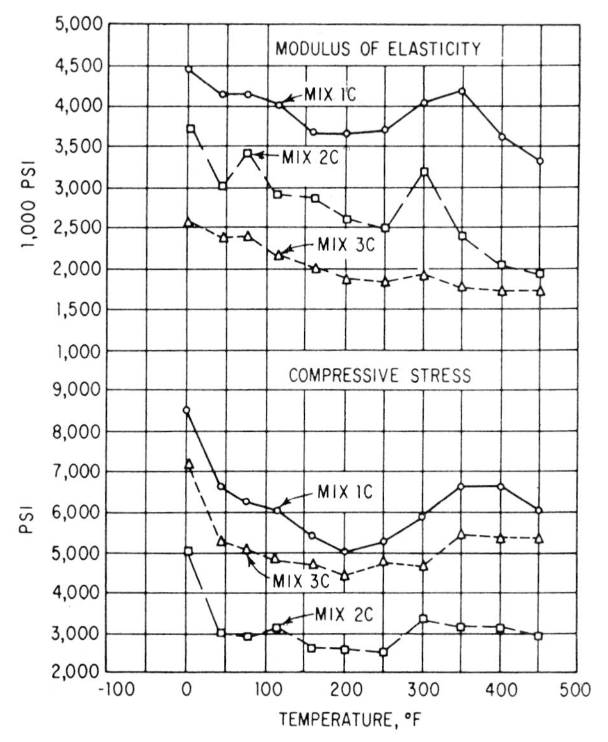

FIGURE 6.23 Variation of compressive strength and modulus of elasticity with testing temperature.[15]

increased. While concrete in normal use is not subjected to the large lateral restraints shown in Fig. 6.24, the effect of lateral restraint in concrete is frequently encountered in bearing pressures. Test results have shown that when the ratio of total concrete area to the loaded area exceeds 40 the maximum bearing pressure may be about five times the cylinder compressive strength.

STRENGTH GAIN

6.18 Retarded Strength Gain

Seriously retarded strength gains of concrete may be due to a wide variety of causes usually associated with poor concrete materials or poor concrete practices. In order to prevent serious retardation of strength it is necessary to pay particular attention to the following:

1. Aggregates and mixing water must be clean and free from detrimental amounts of silt, oil, acids, organic matter, alkalies, sewage, and alkali salts. Very small amounts of some impurities such as sugar may greatly delay the rate of setting and the rate of strength gain.

2. The effects of retarding and accelerating admixtures should be determined for the concrete used. The correct amount of admixture needed to obtain the desired result should be ascertained and used in all batches. Suitable and accurately adjusted dispensing equipment is necessary for this purpose. The effect of temperature on the required amount of the admixture must be taken into account.
3. The correct design proportions in each batch must be used. Special care must be taken to make certain that the specified type of cement is consistently used and that the correct amounts of all materials are always used. Special care must be taken to avoid low cement contents and/or high water contents.
4. The air content of each batch should be determined to avoid low strengths due to very high air contents.
5. The concrete must be properly cured. This requires proper temperature and moisture conditions for the required length of time to obtain the desired properties. The concrete should not be allowed to freeze until it has developed the required strength. Under winter concreting conditions care must be observed to vent space heaters properly to avoid surface carbonation effects.
6. Compressive-strength tests should be made in accord with ASTM standards.

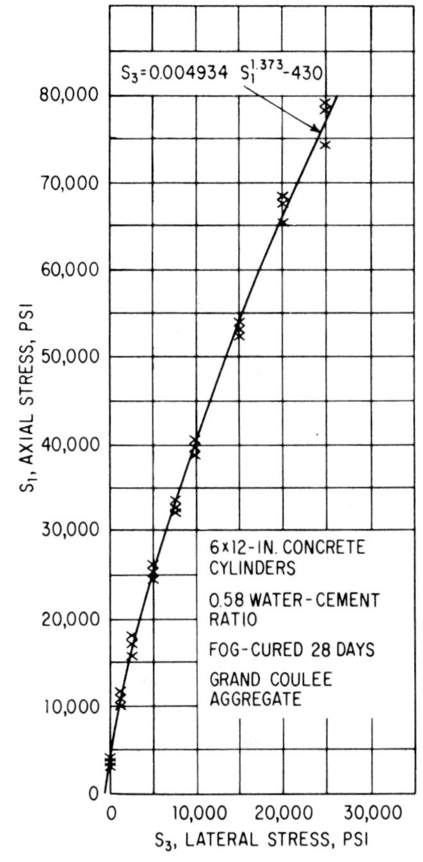

FIGURE 6.24 Relation of axial stress to lateral stress at failure in triaxial compression tests of concrete.[3]

6.19 Accelerated Strength Gain

Under some circumstances it is highly desirable that concrete develop high strength at early ages. In some instances economical reuse of forms may be required or it may be necessary to use the concrete within a few days after manufacture or placement, or placement in cold weather may require fast initial strength gains to resist freeze-and-thaw damage. High early strength may be obtained by using richer concrete mixtures made with Type I or IA portland cement; high-early-strength cement Type III; accelerating admixtures such as calcium chloride, water-reducing admixtures, high-alumina cement, high steam-curing temperatures at normal pressures, or high curing temperatures at elevated pressures (autoclaving). The method or combination of methods to be used depends on the strength requirements and economical considerations.

If special cements or admixtures are used extra care is required to make certain that they are correctly used. High-alumina cement for example, requires special precautions in placement and curing, and the instructions provided by the manufacturer should be carefully followed. In cold weather, additions of calcium chloride, up to 2% by weight of the cement, will help in increasing the rate of strength gain. The effects of admixtures and special cements on other properties of the concrete should be taken into consideration.

High temperatures during moist curing or low-pressure-steam curing accelerate the early rate of strength gain, but if the temperatures are too high they may have an adverse effect on strengths at later ages. Optimum temperatures are dependent on the materials used, the mix proportions, and the details of the curing cycle. Low-pressure steam curing widely used for precast structural elements usually includes a presteam period at room temperature in excess of 3 h, followed by a steaming period of 12 to 18 h, during which time the temperature is increased at a rate not exceeding 40°F/h up to a maximum between 140° and 160°F, and finally return to room temperature.[16]

High-pressure-steam curing of precast-concrete products greatly accelerates early strength and is generally considered to give 28-day strengths in 24 h. In addition this type of curing is generally considered to increase chemical resistance, decrease efflorescence, and increase volume stability. When this curing method is used additions of 10 to 40% of reactive silica, by weight of the cement, are made to allow reaction between the calcium hydroxide, liberated during hydration of the cement, and the reactive silica to form new fairly insoluble compounds. The curing cycle usually consists of a 1- to 4-h presteam period at room temperature followed by a buildup of steam pressure in about 3 h to 150 psig (350°F), then a period of 5 to 8 h at full pressure, and finally a pressure release in a period of 20 to 30 min. The rapid pressure release allows the final product to emerge in a relatively dry condition.

6.20 Predicting 28-day Strength

Concrete specifications are usually based on required compressive strength at an age of 28 days. Improved methods of scheduling concreting operations require methods of anticipating the age at which concrete will be strong enough for all subsequent operations. Methods of predicting strengths at later ages from values obtained at early ages have been used with fair success.

Several investigators have attempted to provide equations relating 3- and 7-day compressive strengths with the 28-day strength and have concluded that predictions based only on 3-day strengths were not reliable. On the basis of a study made on a large amount of test data for moist-cured concrete, W. A. Slater (*Proc. ASCE,* Jan. 1925) proposed the following equation relating the 7-day strength S_c to the 28-day compressive strength S_c':

$$S_c' = S_c + 30\sqrt{S_c}$$

This equation generally provides satisfactory predictions.

Rough approximations of the 28-day compressive strength of moist-cured concrete may be made by assuming that the strength at 3 days in one-third of the 28-day strength and that the strength at 7 days is two-thirds of the 28-day strength. However, it should be noted that large errors are possible because of variations in materials, proportions, manufacture, and curing and testing methods.

ASTM Standard C684 covers four procedures (warm water, boiling water, high temperature and pressure and autogenous curing) for making accelerated compression tests of concrete specimens at ages between 1 and 2 days.[17]

Another method for developing early-age compression-test values and make projections of subsequent strengths is to measure the maturity of a concrete at suitable time intervals in the age of a concrete and then determine the degree-hours (maturity), which is the age of a concrete multiplied by the weighted average ambient temperature during that interval. A prediction equation is developed by determining the compressive strength at various ages and the corresponding maturities and then plotting the compressive strength as a function of the logarithm of each maturity. The form of the equation for predicting future strengths is

$$S_M = S_m + b(\log M - \log m)$$

where S_M = predicted potential strength of maturity, M
S_m = measured compressive strength at maturity, M
b = slope of the prediction line
M = degree-hours of maturity under standard conditions
m = degree-hours of maturity of the specimen at time of early test

The full procedure may be found in ASTM C918, Developing Early-Age Compression Test Values and Projecting Later-Age Strengths.

MODULUS OF ELASTICITY

6.21 General

Concrete is not a truly elastic material, but concrete that has hardened thoroughly and has been moderately preloaded has essentially a straight-line compressive stress–strain curve within the range of usual working stresses. The modulus of elasticity, defined by the equation E = unit stress/unit strain, is a measure of stiffness or of the resistance of the concrete to deformation. The modulus of elasticity of structural concrete normally varies between 2 and 6 million psi.

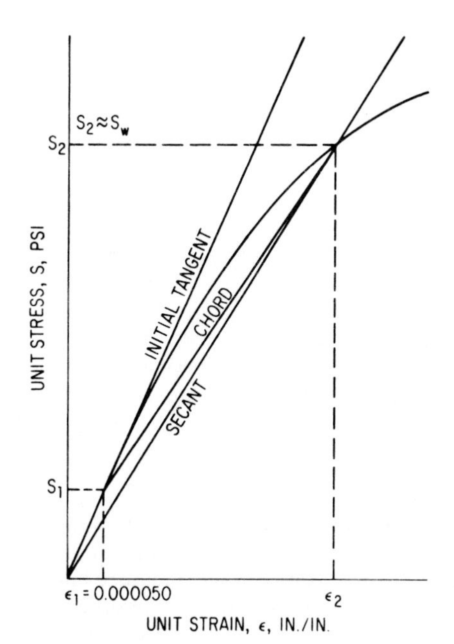

FIGURE 6.25 Stress–strain curve for concrete.

6.22 Method of Determining Modulus of Elasticity

Generally moduli of elasticity are determined from compression tests of concrete cylinders. Different values that may be determined from one test include initial tangent modulus, secant modulus, and chord modulus. Each of these may be represented by the slope of the appropriate line shown in Fig. 6.25. The secant or chord modulus is usually determined and used in calculations. Details for the determination of the chord modulus of elasticity are given in ASTM Standard C469. It is apparent that if the curve approaches a straight line up to the working stress S_w the different values of modulus of elasticity E tend to become equal.

Values of the dynamic moduli of elasticity may be obtained by measuring the natural frequency of vibration of specimens of known dimensions, ASTM Standard C215, or by measuring the velocity of sound waves as they travel through concrete. These E values are larger than the values obtained from static tests.

Moduli of elasticity values in flexure are calculated from load-deflection information obtained from flexural tests. If a simple end-supported beam is loaded at its center and if only deflections due to bending are considered, the modulus of elasticity may be obtained from the equation

$$E = \frac{PL^3}{48ID}$$

where E = modulus of elasticity, psi
P = applied central load (in the working range), lb
D = deflection at midspan due to P, in
L = distance between supports, in
I = moment of inertia of the section with respect to the centroidal axis, in^4

Appreciable effects on the flexural modulus of elasticity may result from the neglect of shear deflections, especially for beams that are relatively deep and have a short span.

Modulus of elasticity in shear, also called modulus of rigidity, may be determined from torsion tests by means of the equation[2]

$$G = \frac{L}{\phi I}$$

where G = modulus of elasticity in shear
L = torque
ϕ = angle of twist per unit length
I = polar moment of inertia of the cross section

6.23 Effects of Variables on Modulus of Elasticity

The same factors that cause strength variations in concrete also cause variations in modulus of elasticity, but there does not appear to be any direct relationship between the two. In general, factors that cause strength increases also cause increases in modulus of elasticity. Thus lower water/cement ratios, richer mixtures, lower air contents, longer curing periods, and greater ages at time of test result in improved strength and modulus of elasticity. Test variables are also important. The modulus of elasticity decreases as the rate of testing is decreased because creep effects increase with time; as the stress S_2 at which E is determined is increased because creep effects increase at higher stresses; and as the moisture content at test is decreased. The amount, type, and grading of the aggregate have important effects on E. Under comparable conditions concrete made with limestone (E = 4,000,000 psi) has a much lower modulus of elasticity than concrete made with traprock (E = 13,300,000 psi).

6.24 Modulus of Elasticity Relationships

Under comparable conditions and for loads less than 50 percent of the ultimate the static moduli of elasticity for tension, compression, and flexure are approximately equal. The dynamic moduli of elasticity are greater than the static values. The modulus of elasticity in shear appears to vary between 0.4 and 0.6 of the modulus of elasticity in compression.

A widely used equation for calculating modulus of elasticity, given in the ACI Building Code (ACI 318), relates the modulus of elasticity to the ultimate compressive strength f_c' and the unit weight of the concrete w. This equation

$$E = 33w^{1.5}\sqrt{f_e'}$$

is satisfactory for values of w between 90 and 155 pcf.

OTHER PROPERTIES

6.25 Poisson's Ratio

The ratio of lateral strain to longitudinal strain, within the elastic range, for axially loaded specimens is called *Poisson's ratio.* Values of Poisson's ratio are required for structural analysis and for design of many types of structures. The method of determining Poisson's ratio is detailed in ASTM Standard C469. Consistent relations between the values of Poisson's ratio and the usual mixture variables have not been demonstrated. There is some evidence that suggests that Poisson's ratio increases with age up to about 1 year. Most reported values for Poisson's ratio, up to an age of 50 years, fall within the range of 0.15 to 0.25. In the absence of experimental data Poisson's ratio is frequently assumed as 0.20 to 0.21.[1]

6.26 Fatigue Strength

Information on fatigue properties of plain and reinforced concrete is still meager and effects of various factors on fatigue strength have not been reliably established. The following facts appear to be generally accepted.[18]

1. The fatigue strength of air-dry plain concrete subjected to several million cycles of repeated compressive loading, for a stress range between zero and a maximum, is about 55 percent of the static ultimate compressive strength.
2. The fatigue strength of plain concrete in tension under similar loading is about 55 percent of the modulus of rupture.
3. The fatigue strength in flexure of plain concrete under several million cycles of repeated loading is about 55 percent of the static ultimate flexural strength for a range in load from zero to a maximum.
4. Plain concrete does not appear to have an endurance limit at least within 10 million load repetitions.
5. The age of the test specimens, conditions of curing, moisture content and range of stress may significantly affect the results.
6. Most fatigue failures of reinforced-concrete beams appear to be due to failure of the reinforcing steel, associated with severe cracking and possible stress concentration and abrasion effects. Beams with critical longitudinal reinforcement appear to have a fatigue strength of 60 to 70 percent of the static ultimate strength for about 1 million cycles.
7. Most fatigue failures of prestressed-concrete beams are due to fatigue failure of the stressing wires or strands and are related to the extent and severity of cracking. There

is some evidence to suggest that prestressed beams are superior to conventional beams for resisting fatigue loading.

6.27 Toughness

Plain concrete has poor toughness properties, with the energy of rupture in compression generally under 10 in-lb/in^3 and the energy of rupture in bending generally under 0.1 in-lb/in.^{3} The energy of rupture of steel in tension, for comparative purposes, is about 15,000 in-lb/in.^{3} The toughness of concrete may be slightly increased by increasing the concrete strength. A significant increase in toughness (shock resistance) may be obtained by the addition of proper amounts of randomly placed reinforcement consisting of appropriately sized nylon fibers and/or steel wires.[19] The randomly placed steel wires are also effective in increasing the tensile and bending strengths of mortar and concrete.[20]

Concrete used for structural purposes, as in bridges and buildings, is reinforced with tensile and shear steel and may also have compressive and temperature steel. The design of these reinforced-concrete structures must take into account shock loading resulting from earthquakes and other types of dynamic loadings. Properly designed reinforced-concrete structures are very resistant to blasts produced by bombing and to dynamic loadings caused by earthquakes.

Toughness, strength, and durability properties of fresh and hardened concrete have been significantly improved by impregnating the concrete with a monomer and polymerizing by either radiation or thermal-catalytic techniques. Radiation polymerization of dry concrete specimens containing from 4.6 to 6.7%, by weight, of methyl-methacrylate increased the strength almost four times, drastically reduced absorption and permeability, and greatly increased freeze-thaw durability and resistance to chemical attack.[21]

NONDESTRUCTIVE TESTS

6.28 Types and Significance of Nondestructive Tests

Nondestructive tests are used on specimens and concrete structures to determine information on certain properties of concrete. These tests may be subdivided into the following groups:

1. Indentation tests in which the rebound of a spring-driven hammer is measured and related to ultimate strength.
2. Sonic tests usually involving determination of the resonant frequency of longitudinal, transverse, or torsional vibration of small concrete specimens. These tests provide information on dynamic modulus of elasticity, damping constant and logarithmic decrement.
3. Pulse-transmission tests at sonic and ultrasonic frequencies. These tests measure the velocity of a compressional pulse traveling through the concrete and provide information on the presence or absence of cracking in monolithic concrete. They also may be used to measure the thickness of slabs one face of which is inaccessible.
4. Radioactive tests involving absorption of gamma and x rays. These tests provide information on density or quality of concrete and also on the presence or absence of reinforcing steel.

5. Penetration probe tests require measurement of the depth that a carefully made steel pin is driven into the concrete by a special gun fired with a precisely measured powder charge. In addition the Moh's hardness of the aggregate must be determined. With these known values, an estimate of the compressive strength of the concrete may be obtained from given tables or charts.

The most widely used indentation device is the Schmidt concrete test hammer. It is a valuable tool when properly used, but careless and indiscriminate use may lead to incorrect conclusions. The results of the hammer tests are affected by many variables including position of hammer (vertical, horizontal, or intermediate angle), smoothness of test surface, internal and surface moisture content, size, shape and rigidity of specimen, aggregate type, aggregate size and concentration near the surface, air pockets, age, temperature, and previous curing conditions. Standardization of test conditions will reduce or eliminate some of the variations, but the others require special calibration curves. Since the manufacturer's calibration curves provided with the hammer apply only to specific conditions, it is apparent that strength values obtained from test-hammer readings and the given calibration curve cannot be accurate or reliable except by chance. More reliable and accurate strength results may be expected when the hammer is calibrated for the specific conditions relating to the given job. The principal advantages of the indentation tests are their rapidity, simplicity, and economy. However, the results must be carefully interpreted and should be considered qualitative rather than quantitative unless a job calibration curve has been developed. The indentation tests should not be considered as a replacement for standard compression tests but as a complement to provide additional information.

Sustained sonic frequencies are primarily used for laboratory testing of beams and cylinders to follow the course of deterioration as the specimens are subjected to weathering or exposure tests. Pulse-velocity tests are usually made on concrete structures to establish information on degree of uniformity, especially with regard to the amount and position of cracking. On the basis of available test results it appears that concrete with a longitudinal wave velocity in excess of 15,000 fps (feet per second) is in excellent condition and that concrete with a longitudinal wave velocity less than 7000 fps is in very poor condition.

The penetration probe test is a valuable addition to the available test procedures used to determine concrete compressive strength. However, the test must be carefully made and the significance of the final numerical results must be carefully evaluated. Correlation of penetration probe tests with cylinder or core tests may be very helpful in evaluating final results. The penetration probe test is not completely nondestructive since it may leave craters similar to popouts in the concrete surface.

REFERENCES

1. "Design and Control of Concrete Mixtures," 13 ed., Portland Cement Association, 1988.
2. ASTM STP 169B, Significance of Tests and Properties of Concrete and Concrete-Making Materials, Dec. 1978.
3. Price, W. H.: Factors Influencing Concrete Strength, *J. ACI*, Feb. 1951, pp. 417–43[illegible].
4. "Concrete Manual," 8th ed., U.S. Bureau of Reclamation, 1975.
5. Lovewell, C. E., and G. W. Washa: Proportioning Concrete Mixtures Using Fly Ash, *J. ACI*, June, 1958, pp. 1093–1101.

6. Timms, A. G., and N. H. Withey: Temperature Effects on Compressive Strength of Concrete, *J. ACI,* January, February 1934, pp. 159–180.
7. "Manual for Quality Control for Plants and Production of Prestressed Concrete Products," PCI, 1977.
8. Withey, M. O.: Fifty Year Compression Test of Concrete, *J. ACI,* December 1961, pp. 695–712.
9. Gonnerman, H. F., and E. C. Shuman: Compression, Flexure, and Tension Tests of Plain Concrete, *ASTM Proc.,* Vol. 28, Pt. II, pp. 527–573, 1928.
10. Grieb, W. E., and G. Werner: Comparison of Splitting Tensile Strength of Concrete with Flexural and Compressive Strengths, *ASTM Proc.,* Vol. 62, pp. 972–995, 1962.
11. Gaynor, R. D.: Effects of Prolonged Mixing on the Properties of Concrete, *NRMCA Publ.* 111, June 1963.
12. Kesler, Clyde E.: Effect of Length to Diameter Ratio on Compressive Strength—An ASTM Cooperative Investigation, *ASTM Proc.*, Vol. 59, pp. 1216–1228, 1959.
13. Jones, J. P., and F. E. Richart: The Effect of Testing Speed on Strength and Elastic Properties of Concrete, *ASTM Proc.,* Vol. 36, Pt. II, pp. 380–391, 1936.
14. Watstein, David: Effect of Straining Rate of the Compressive Strength and Elastic Properties of Concrete, *J. ACI,* pp. 729–744, April 1953.
15. Saemann, J. C., and G. W. Washa: Variation of Mortar and Concrete Properties with Temperature, *J. ACI,* pp. 385–395, November 1957.
16. ACI 517.2R–80, Accelerated Curing of Concrete at Atmospheric Pressure—State of the Art, 1980.
17. Smith, P., and B. Chojnacki: Accelerated Strength Testing of Concrete Cylinders, *ASTM Proc.,* Vol. 63, pp. 1079–1104, 1963.
18. Nordby, G. M.: Fatigue of Concrete—A Review of Research, *J. ACI,* pp. 191–220, August 1958.
19. Fibrous Reinforcements for Portland Cement Concrete, U.S. Army Engineer Division, Ohio River Corps of Engineers, Cincinnati, OH, 45227, *Tech. Rept.* 2–40, May 1965.
20. Romualdi, J. P., and J. A. Mandel: Tensile Strength of Concrete Affected by Uniformly Distributed and Closely Spaced Short Lengths of Wire Reinforcement, *Journal ACI,* pp. 657–672, June 1964.
21. Dikeou, J. T., L. E. Kukacka, J. E. Backstrom, and M. Steinberg: Polymerization Makes Tougher Concrete, *Journal ACI,* pp. 829–839, October 1969.
22. ACI COMM.544: Measurement of Properties of Fiber Reinforced Concrete, *ACI Mtls. J.,* Nov.-Dec., 1990.

CHAPTER 7
DURABILITY

George W. Washa*

GENERAL CONSIDERATIONS

7.1 Weathering

Good concrete is a relatively durable material under a wide variety of exposures. Ordinary weathering conditions, however, may have harmful effects and may cause disintegration of poor concrete. Weathering effects are due to the disruptive action of freezing and thawing, to alternate wetting and drying, to undesirable chemical activity, and to temperature variations in the concrete mass. Many laboratory tests have been proposed and used for determining the durability of concrete under various types of exposure conditions, but correlation between laboratory tests and field service records is difficult, if not impossible.

Major factors that may influence concrete durability include the physical properties of the hardened concrete, the constituent materials of which the concrete is composed, the manufacturing and construction methods used, and the nature of the deteriorating influences. In order to have good durability, concrete should have a low water/cement ratio, it should be made with properly selected sound materials, it should be dense and well made, it should be properly cured, and where freezing and thawing are involved it should contain between 4 and 6% entrained air. Evaluation of the durability of concrete may also involve a study of the elastic, plastic, and thermal properties of the constituents and possible incompatibilities.

FREEZING AND THAWING

7.2 Effect of Entrained Air

All concrete contains some entrapped air, usually between 0.5 and 1.5%, which is relatively ineffective in increasing resistance to freezing and thawing. However, additions of 4 to 6% purposefully entrained air in concrete, with a 1½-in maximum-size aggregate, greatly increase resistance to freezing and thawing. As the maximum coarse-aggregate size is decreased the amount of entrained air should increase, and as an example concrete

*Revised by Joseph A. Dobrowolski

with a maximum aggregate size of ¼ in should have an entrained-air content between 8 and 10%. While the total air content is important, other factors such as size and spacing of bubbles are equally significant. Purposefully entrained air bubbles are usually very small and are closely spaced throughout the mass, while entrapped air generally consists of a smaller number of larger voids spaced at greater distances. Methods of determining air content and bubble characteristics are given in ASTM Standard C457. Entrained air bubbles usually have diameters varying between 0.003 and 0.05 in and should have a maximum spacing factor (index related to the maximum distance of any point in the cement paste from the periphery of an air void, in inches) of about 0.008 in. Estimates have placed the number of entrapped air bubbles in a cubic yard of non-air-entrained concrete, 1.5% entrapped air, at about 5 billion, and the number of bubbles in a cubic yard of air-entrained concrete (658 lb of cement per cubic yard, 7% air, ⅜-in maximum-size aggregate) at about 130 billion.

Air entrainment affects the properties of freshly mixed and hardened concrete. Air-entrained concrete is more plastic and workable, it has less tendency to bleed, and it may be placed with less segregation than non-air-entrained concrete. Hardened concrete containing entrained air is more uniform, has less absorption and permeability, and is much more resistant to the action of freezing and thawing. Additions of air to a given concrete mix will result in lower strength of the hardened concrete. The compressive strength of concrete will decrease by about 5 percent for each percent of entrained air when concretes of a given water/cement ratio are compared. However, if comparison is made on the basis of a fixed cement content and if the mixture containing entrained air is redesigned by reducing the water and sand contents to maintain the slump and volume of mortar, the strength decreases will be less and may in fact be strength increases. The latter may be secured with lean concrete mixtures containing up to about 470 lb of cement per cubic yard of concrete. Richer mixes that have been redesigned may suffer strength losses between 10 and 15 percent.

During freezing and thawing either the cement paste or the aggregate or both may be damaged by dilation. In this process stresses beyond the proportional limit may be produced which may cause permanent enlargement or actual disintegration. Dilation and associated stresses are believed to be due to (1) hydraulic pressure generated when growing ice crystals displace unfrozen water and cause it to flow against resistance to unfrozen portions of the mass, (2) growth of capillary ice crystals, (3) osmotic pressures brought about by local increases of alkali concentration caused by the separation of pure ice from the solution. The large number of closely spaced small air bubbles in air-entrained concrete are believed to serve as reservoirs for the relief of pressure developed within the freezing concrete and thus to relieve or avoid the high tensile stresses that lead to failure.

Some of the important factors involved in freezing and thawing concrete are (1) the size, homogeneity, and characteristics of the aggregates, especially the porosity and related absorption and permeability characteristics; (2) the proportions of the concrete mixture with special reference to the water/cement ratio and the amount and characteristics of the entrained air; (3) the amount and distribution of the moisture available to the cement paste and to the aggregate; (4) the degree of saturation of the paste and the aggregate in relation to the critical saturation coefficient of 0.917 (in unsaturated concrete the constituent with the finest texture tends to become more nearly saturated and consequently cement paste tends to be more saturated than the aggregate); (5) the rate of freezing and thawing; (6) the details of the freezing and thawing cycles including the medium in which freezing and thawing take place; (7) the age of the concrete when freezing and thawing start; (8) the total number of freezing and thawing cycles; and (9) the presence or absence of calcium chloride, sodium chloride, or other salts during freezing and thawing.

In summary, good durable concrete will be obtained if sound aggregates with proven records are used, if a properly proportioned dense concrete with a relatively low water-cement ratio is used, and if the proper amount and type of entrained air are present in the concrete.[11]

7.3 Freeze-and-Thaw Tests

Freezing and thawing tests are costly and are difficult to standardize and to evaluate. ASTM Standard C666 provides two procedures for conducting tests to determine the resistance of concrete specimens to rapid freezing and thawing in water, and to rapid freezing in air and thawing in water. These procedures require that the test be continued until the specimens have sustained 300 cycles of freezing and thawing or until the dynamic modulus of elasticity has reached 60 percent of the initial modulus. A measure of the durability, the durability factor *DF*, may then be calculated from the equation

$$DF = \frac{PN}{M}$$

where P = relative dynamic modulus of elasticity at N cycles, %

N = number of cycles at which P reaches the specified minimum value for discontinuing the test or the specified number of cycles at which the exposure is to be terminated, whichever is less

M = specified number of cycles at which the exposure is to be terminated

The standard states that the methods are not intended to provide a quantitative measure of the length of service that may be expected from a specific type of concrete under field conditions.

Results of freezing and thawing tests have usually been evaluated by one or more of the following measures: (1) reduction in dynamic modulus of elasticity, (2) loss in compressive or flexural strength, (3) loss in weight, (4) change in visual appearance, and (5) expansion of specimen. The reduction in dynamic modulus is widely used and provides a good means of evaluating freeze-and-thaw tests. Characteristic behaviors of a good and poor concrete are shown in Fig. 7.1. Reasonably good correlation has also been obtained between changes in dynamic modulus and changes in flexural strength. Flexural-strength loss is a good indication of deterioration but compressive strength is not. In addition, a large number of specimens are required for method 2. Methods 3 and 4 may provide little information during the early stages of deterioration, and information obtained during the later stages may be largely dependent on the interpretation of a single individual. Expansion tests satisfactorily indicate deterioration but are largely limited to research programs because of the experimental details involved.

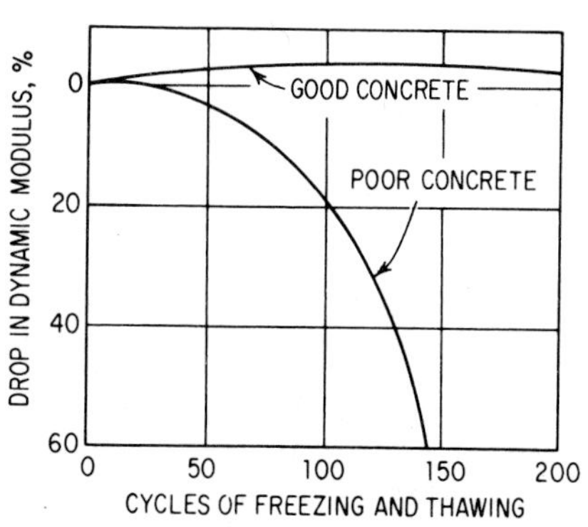

FIGURE 7.1 Relation between number of cycles of freezing and thawing and drop in dynamic modulus of elasticity.

Ultrasonic tests are used to determine deterioration of concrete structures such as dams, piles, and bridge piers in the field. The method requires the determination of the longitudinal wave velocity in the concrete structure. On the basis of previous tests it appears that if the velocity through the concrete is between 13,500 and 16,000 fps the concrete is in good to excellent condition, and that if the velocity is in the range between 10,000 to 13,500 fps the concrete is in fair or questionable condition, and velocities below 10,000 fps generally indicate a questionable to poor condition.[10]

7.4 Deicing Agents

In areas where climatic conditions are severe concrete pavements are frequently subjected to direct application of flake calcium chloride or rock salt to remove ice, or to repeated applications of granular materials impregnated with these salts. Under these conditions surface scaling has developed, usually in those locations where heavy or frequent applications of salts have been used.

Concrete that is to be subjected to freezing and thawing and to the action of deicing salts should normally contain between 4 and 7% entrained air. In addition, since concrete is more vulnerable to surface scaling damage during its early life, surface coatings are frequently applied to the concrete prior to the first winter season. Applications of two coats of either boiled or raw linseed oil to the concrete surface have produced superior resistance to scaling.

Properly proportioned concrete containing the necessary amount and type of entrained air will resist satisfactorily scaling due to the use of calcium chloride or sodium chloride for ice removal, but it will not resist the action of all salts. Deicing agents consisting primarily of either ammonium sulfate or ammonium nitrate should not be used because of their chemical attack on concrete. Solutions of these salts will disintegrate concrete under freezing or normal room-temperature conditions. The chemical composition of a deicing agent and its possible effects on concrete properties should be considered before the agent is used.

CHEMICAL ATTACK

7.5 Effect of Sewage

Except in the presence of certain industrial wastes there is little direct attack of the sewage since the usual sewage and waste water is nearly neutral, varying only slightly above or below a pH of 7.0. Whether concrete or mortar can be used in sewer constructions depends consequently on the character of the sewage and operating characteristics. Many concrete sewers have been in successful operation for long periods, but there are also published accounts of others that have failed after short periods in service. If the sewage is of such nature that a strong odor of hydrogen sulfide is evolved, sulfuric acid of sufficient strength will be formed to attack the lime compounds of the concrete and produce disintegration above the water level. The hydrogen sulfide, formed by bacterial decomposition of the sulfur compounds in the sewage, rises and combines with oxygen and moisture to form sulfuric acid on the upper portions of the sewer.

In some instances changes of the operation of the sewer may be helpful. Since the flow of sewage is not normally itself aggressive, deterioration of the crown may be prevented by keeping the sewage at lower temperatures, running the sewers full, ventilating at high velocities the space between the sewage and the crown of the sewer, or providing protective coverings or coatings. The protective coating frequently used is polyvinyl chloride, applied as a liner by placing the polyvinyl in the form when the pipe is cast, or subsequently as a painted-on coating. If proper operating conditions are not possible or if protection cannot be provided, portland cement concrete should not be used.

Some improved resistance to the action of sewage may be obtained with the use of high-alumina cement. Improved resistance may also be obtained by the use of a patented process called Ocrate concrete. In this process hardened and dried concrete in a vacuum is subjected to the action of silicon tetrafluoride gas in accord with the equation

$$2Ca(OH)_2 + SiF_4 = 2CaF_2 + Si(OH)_4$$

In a given time the gas penetrates the concrete to some depth and provides a shell resistant to acid attack.

7.6 Effect of Other Acids

Basically, no portland cement concrete is resistant to acids. When the most resistant cements are used in sound impermeable concrete that is properly placed in well-designed structures, resistance to mild acids may be satisfactory. Where strong acids are involved other materials should be used or a protective surface treatment or covering must be provided. Common acid attacks of concrete are due to lactic and acetic acids in food-processing plants. These attacks are comparatively mild but persistent, result in softening of the working floors, and may require repair or replacement at periodic intervals. Acetic acid attack of concrete silos due to silage is also fairly common.

Bacteria and fungi find good conditions for growth on the floors and walls of food-processing plants. These agencies cause damage by mechanical action and by secretion of organic acids. Antibacterial cements are available for decreasing these actions by permanently sublimating the microbiological metabolism.[1] The active component in these cements usually is a toxic material such as arsenic or copper. Antibacterial cements are effective in decreasing odors, development of slimes, and the rate of deterioration of concrete and mortar surfaces. They have been satisfactorily used in dairies, kitchens, food-processing plants, food warehouses, pharmaceutical plants, breweries, and chemical plants. They have also been used in "wet-foot traffic areas" in locker rooms, shower rooms, bathhouses, and around swimming pools to reduce the possibility of contact contagion.

7.7 Effect of Leaching

In the process of cement hydration soluble calcium hydroxide is formed. This material is easily dissolved by water that is lime-free and that contains dissolved carbon dioxide. Snow water in mountain streams is particularly aggressive because it is relatively pure, cold, and contains carbon dioxide which produces a mild carbonic acid solution with a greater capacity for dissolving calcium hydroxide than pure water. As a result of this action surfaces of water-carrying structures develop a rough sandy appearance and may suffer reduction in capacity.

In some hydraulic structures leakage of water through cracks or joints or porous concrete may carry the calcium hydroxide in solution through the concrete structure. At the surface, reaction between the calcium hydroxide and carbon dioxide will cause precipitation of a white deposit of calcium carbonate. Generally this type of leaching does not result in any serious problem, but it may over a long period of time cause serious disintegration.

The problems relating to leaching may be minimized by the use of properly proportioned dense concrete mixtures, careful attention to the proper design and placement of contraction and construction joints, provision for satisfactory drainage, and provision of effective and durable coatings where necessary. High-alumina cements, portland blast-furnace slag cements, and pozzolan cements are effective in minimizing leaching.

7.8 Effect of Sulfates

In regions where alkali is present in soil and groundwaters, deterioration of concrete structures may take place. The harmful effects are primarily due to the sulfates of magnesium and sodium. These salts react with the hydrated calcium aluminate to form crystals of calcium sulfoaluminates accompanied by considerable expansion that may result in eventual disintegration. An example of serious disintegration due to sulfate attack that took place in a few years is shown in Fig. 7.2. The rate and severity of the sulfate attacks increase as the concentration of the sulfates in the groundwater increases and as the temperature increases. Dry concrete in dry sulfate-bearing soils is not attacked. Continuous saturation in strongly sulfate-bearing water will produce rapid and severe effects. Frequently alternated saturation and drying conditions appear to produce the most harmful effects.

The deposition and growth of sulfate crystals in the surface pores of concrete may also cause disintegration. During periods of drying the alkali waters may evaporate and deposit salt crystals which will grow with periods of alternate wetting and drying until they fill the pores and ultimately develop pressures great enough to cause pitting and scaling.

The resistance of concrete to the action of sulfates may be greatly improved in several ways.

1. Sulfate-resisting cement Type V (C_3A less than 5%) should be used where concentrations greater than 0.2% water-soluble sulfates in the soil or 1500 ppm sulfates in the water are present. When the concentrations are 0.1 to 0.2% water-soluble sulfates in the soil or 150- to 1500 ppm sulfates in the water Type II cement (C_3A less than 8%), IP(MS) or IS(MS), should be used.
2. The water/cement ratio should not exceed 0.50 where Type II is used, or 0.45 where Type V is required, and the concrete should have an ample cement content. The effect

FIGURE 7.2 Concrete slab (20 × 100 ft × 4 in thick) disintegrated by sodium-sulfate-bearing soil.[2]

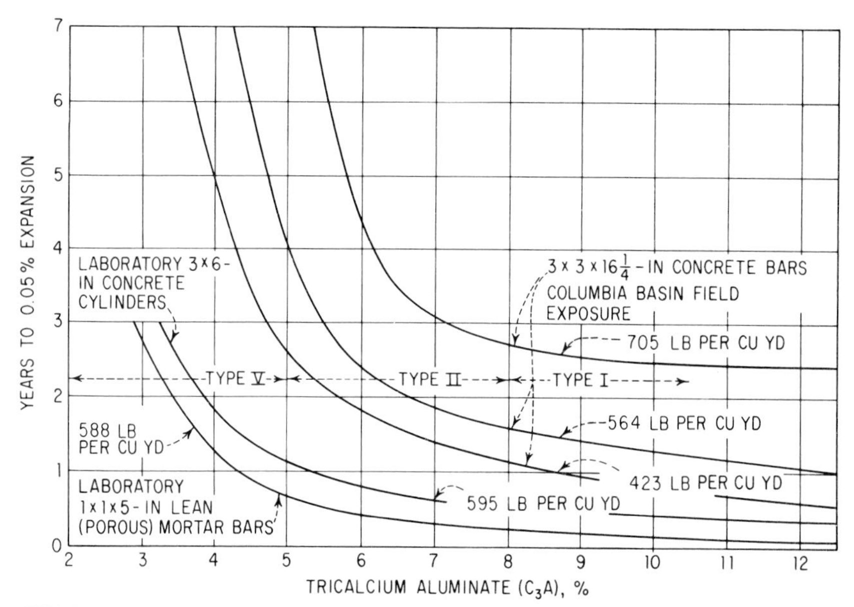

FIGURE 7.3 Effect of cement type and amount on the rate or expansion of concrete exposed to sulfate waters.[2]

of quantity and type of cement content on the rate of expansion of concrete exposed to sulfate waters is shown in Fig. 7.3.

3. Substitution of 15 to 30% of the cement, by weight, of an active pozzolanic material may be very effective. Addition of pozzulans is required for over 2% sulfates in soil and over 10,000 ppm sulfates in water. Any flyash used must meet ASTM C618 Class F.
4. Autoclaving concrete products at 350°F or higher greatly improves sulfate resistance.

7.9 Effect of Seawater

Concrete in seawater is subjected to many different effects but if it is properly proportioned, mixed, and placed it should resist these effects almost indefinitely. Wetting and drying, leaching, temperature variations, corrosion of reinforcing steel, battering of waves and tides, sulfate attack, and freezing and thawing action may all be involved.

Many of the potentially harmful effects can be largely controlled by the use of normal cement along with sound nonreactive aggregates that are properly proportioned to provide strong impermeable concrete. Reinforcing steel should be properly protected from corrosion by a minimum of 1 ½ in of concrete and maximum W/C = 0.40. In northern climates where freezing and thawing may be important the concrete mixture should contain 4 to 6% of entrained air. Since sulfate attack may be an important consideration it may be desirable to use cements low in C_3A such as Types V and II. The use of portland blast-furnace cements or high-alumina cements or additions of active pozzolans should also be considered.

7.10 Effect of Carbonation

Carbonation of concrete is a chemical combination of carbon dioxide with the hydration products of portland cement. Carbon dioxide reacts principally with calcium hydroxide but also with calcium silicate and calcium aluminate and combines with the calcium portion of these compounds to form calcium carbonate. When fresh concrete is placed in cold weather in rooms heated by improperly vented space heaters a concentration of carbon dioxide results. The carbon dioxide reacts with the fresh concrete near the surface and produces a soft crumbly surface layer between 0.10 and 0.30 in thick. The action takes place readily at a temperature between 30 and 50°F in an atmosphere that provides both carbon dioxide and moisture. If it is not possible to keep the carbon dioxide content of the air at a low level, the fresh concrete may be protected with a membrane curing compound or surface seal applied as promptly as possible to protect the concrete during the first 24 h. A carbonated surface will not respond to the action of chemical hardeners, and the only way to remove the poor surface is by surface grinding.

Carbonation also causes important changes in hardened concrete, especially lightweight porous concrete. The reaction of the atmospheric carbon dioxide with the hydration products of portland cement produces significant weight gains and irreversible shrinkages. The shrinkage values due to carbonation may be about as large as the shrinkage obtained by air-drying saturated concrete. Carbonated products may possess improved volume stability to subsequent moisture changes.

Carbonation proceeds slowly, produces little direct shrinkage at relative humidities near 100 or below 25%, and appears to be most active at about 50% relative humidity. Size of specimen, density of specimen, concentration of carbon dioxide, previous drying and carbonation history, and method of curing all appear to have significant effects on carbonation shrinkage.

7.11 Effect of Free CaO and MgO

Excessive expansion of concrete or masonry construction due to hydration of the uncombined lime and magnesia present in cement is called "unsoundness." Quantitative information on this type of expansion may be obtained from the autoclave test as given in ASTM Standard C151. If, as is usually the case, the maximum autoclave expansion for a given cement is lower than the value allowed by the appropriate ASTM standard, there should be no large expansions due to the hydration of the uncombined lime and magnesia when the cements are used. Problems due to this type of expansion are most frequently associated with the use of limes, or masonry cements containing limes, that have not been properly hydrated prior to use. High-magnesium limes, because they hydrate much more slowly, should be given special care. There are many descriptions in the literature of the serious consequences resulting from the use of materials containing incompletely hydrated lime.[3] Many relate to the use of these materials as the cementing material in brick walls.

7.12 Effect of Various Substances on Concrete and Protective Treatments, Where Required

Concrete of a suitable quality must be assumed in a discussion of the effect of various substances on concrete, and protective treatments.[4] In general, this means a properly proportioned, carefully placed, and well-cured concrete resulting in a watertight structure.

This requires the following:

1. Low water/cement ratio, not to exceed a water/cement ratio of 0.50 by weight.
2. Suitable workability, to avoid mixes so harsh and stiff that honeycomb occurs, and those so fluid that water rises to the surface.
3. Thorough mixing, at least 1 min after all materials are in mixer, or until mix is uniform.
4. Proper placing, spaded or vibrated, to fill all corners and angles of forms without segregation of materials—avoid construction joints.
5. Adequate curing, protection by leaving forms in place, covering with wet sand or burlap, and sprinkling. Concrete to be kept wet and above 50°F for at least the first week. Not to be subject to hydrostatic pressure during this period.

Many solutions, such as brines and salts, which have no chemical effect on concrete, may crystallize on loss of water. It is especially important that concrete subject to alternate wetting and drying of such solutions be impervious. When the free water in the concrete is saturated with salts, the salts crystallize in the concrete near the surface in the process of drying and this crystallization may exert sufficient pressure to cause surface scaling. Salt solutions corrode steel more rapidly than plain water. In structures which are to be subject to frequent wetting and drying by these solutions, it is essential to provide impervious concrete and sufficient coverage over the steel, and it may be advisable to provide some surface coating such as sodium silicate, linseed oil, or one of the varnishes as an added precaution.

Surface Treatments. Materials are available for almost any degree of protection required on concrete. The best material to use in a given case will depend on many factors in addition to the substance to be protected against. These include concentration of solution, temperature, taste and odor, and abrasive action. High temperatures usually accelerate any possible attack and therefore better protection is required than for normal temperatures. Bituminous materials soften at elevated temperatures, may even melt and become ineffective. Grades are available for a fairly wide temperature range and manufacturers should be consulted as to grade required for given conditions. Where taste or odor is important it should be determined whether proposed treatment will be satisfactory. As a rule, thin coatings are not so durable as heavier coatings, where there is considerable abrasion.

The more common treatments are indicated in Table 7.1, the numbers corresponding to the descriptions given below. For most substances, several treatments are suggested. These will provide sufficient protection in most cases, but any of the other treatments designated by a number higher than the highest shown would be equally suitable and often may be advisable. In making a selection, economy should be considered as well as the factors discussed above. Where continuous service over long periods is desirable it may be more economical to use the more positive means of protection rather than those of lower first cost which may be less permanent.

Protective coatings usually require dust-free, surface-dry concrete for satisfactory application.

1. *Magnesium fluosilicate or zinc fluosilicate*: The treatment consists of two or more applications. First, a solution of about 1 lb of the fluosilicate crystals per gallon of water is used. For subsequent applications about 2 lb of crystals per gallon of water is used. Large brushes are convenient for applying on vertical surfaces, and mops on horizontal areas. Each application should be allowed to dry; after the last one has

TABLE 7.1 Materials Available For Surface Treatment of Concrete

Material	Effect on concrete	Surface treatment*
	Acids	
Acetic	Disintegrates slowly	5, 6, 7
Acid waters	Natural acid waters may erode surface mortar, but usually action then stops	1, 2, 3
Carbolic	Disintegrates slowly	1, 2, 3, 5
Carbonic	Disintegrates slowly	2, 3, 4
Humic	Depends on humus material, but may cause slow disintegration	1, 2, 3
Hydrochloric	Disintegrates	8, 9, 10, 11, 12
Hydrofluoric	Disintegrates	8, 9, 11, 12
Lactic	Disintegrates slowly	3, 4, 5
Muriatic	Disintegrates	8, 9, 10, 11, 12
Nitric	Disintegrates	8, 9, 10, 11, 12
Oxalic	None	None
Phosphoric	Attacks surface slowly	1, 2, 3
Sulfuric	Disintegrates	8, 9, 10, 11, 12
Sulfurous	Disintegrates	8, 9, 10, 11, 12
Tannic	Disintegrates slowly	1, 2, 3
	Salts and alkalies (solutions)†	
Carbonates of ammonia, potassium, sodium	None	None
Chlorides of calcium, potassium, sodium strontium	None unless concrete is alternately wet and dry with the solution, when it is advisable to treat with	1, 3, 4
Chlorides of ammonia, copper, iron, magnesium, mercury, zinc	Disintegrates slowly	1, 3, 4
Fluorides	None except ammonium fluoride	3, 4, 5
Hydroxides of ammonia, calcium, potassium, sodium	None	None
Nitrates of ammonia,	Disintigrates	8, 9, 10, 11, 12
calcium, potassium, sodium	None	None
Potassium permanganate	None	None
Silicates	None	None
Sulfates of ammonia,	Disintegrates	6, 7, 8, 9
aluminum, calcium, cobalt, copper, iron, manganese, nickel, potassium, sodium, zinc	Disintegrates; however, concrete products cured in high-pressure steam are highly resistant to sulfates	1, 3, 4
	Petroleum oils	
Heavy oil below 35° Baumé‡	None	None
Light oils above 35° Baumé‡	None—require impervious concrete to prevent loss from penetration, and surface treatments are generally used	1, 2, 3, 5, 9
Benzine, gasoline, kerosene, naphtha	None—require impervious concrete to prevent loss from penetration, and surface treatments are generally used	1, 2, 3, 5, 9
High-octane gasoline		12

TABLE 7.1 Materials Available For Surface Treatment of Concrete *(Continued)*

Material	Effect on concrete	Surface treatment*
	Coal-tar distillates	
Alizarin, anthracene, benzol, cumol, paraffin, pitch, toluol, xylol	None	None
Creosote, cresol, phenol	Disintegrates slowly	1, 2, 5, 9
	Vegetable oils	
Cottonseed	No action if air is excluded	None
	Slight disintegration if exposed to air	1, 2, 5, 9
Rosin	None	None
Almond, castor, china wood,§ coconut, linseed,§ olive, peanut, poppy-seed, rapeseed, soybean,§ tung,§ walnut	Disintegrates surface slowly	1, 2, 5, 9
Turpentine	None—considerable penetration	1, 2, 5, 9
	Fats and fatty acids (animal)	
Fish oil	Most fish oils attack concrete slightly	1, 2, 3, 5, 9
Foot oil, lard and lard oil, tallow and tallow oil	Disintegrates surface slowly	1, 2, 3, 5, 9
	Miscellaneous	
Alcohol	None	None
Ammonia water (ammonium hydroxide)	None	None
Baking soda	Beer will cause no progressive disintegration of concrete, but in beer storage and fermenting tanks a special coating is used to guard against contamination of beer	Coatings made and applied by Turner Rostock Co., 420 Lexington Ave., New York, and Borsari Tank Corp., of America, 60 E. 42d St., New York
Bleaching solution	Usually no effect. Where subject to frequent wetting and drying with solution containing calcium chloride provide	1, 3, 4
Borax, boracic acid, boric acid	No effect	None
Brine (salt)	Usually no effect on impervious concrete; where subject to frequent wetting and drying of brine provide	1, 3, 4
Buttermilk	Same as milk	3, 4, 5
Charged water	Same as carbonic acid—slow attack	1, 2, 3
Caustic soda	None	None
Cider	Disintegrates (see acetic acid)	5, 6, 7
Cinders	May cause some disintegration	1, 2, 3,

TABLE 7.1 Materials Available For Surface Treatment of Concrete *(Continued)*

Material	Effect on concrete	Surface treatment*
	Miscellaneous *(Continued)*	
Coal	Great majority of structures show no deterioration; exceptional cases have been coal high in pyrites (sulfide of iron) and moisture showing some action but the rate is greatly retarded by deposit of an insoluble film; action action may be stopped by surface treatments	1, 2, 3
Corn Syrup	Disintegrates slowly	1, 2, 3
Cyanide solutions	Disintegrates slowly	7, 8, 9, 10, 12
Electrolyte	Depends on liquid; for lead and zinc refining and chrome plating use	7, 8, 9, 10, 11
	Nickel and copper plating	None
Formalin	Aqueous solution of formaldehyde disintegrates disintegrates concrete	5, 9, 10, 11, 12
Fruit juices	Most fruit juices have little if any effect as tartaric acid and citric acid do not appreciably affect concrete; floors under raisin-seeding machines have shown some effect probably due to poor concrete	1, 2, 3
Glucose	Disintegrates slowly	1, 2, 3
Glycerin	Disintegrates slowly	1, 2, 3, 4, 5, 9
Honey	None	None
Lye	None	None
Milk	Sweet milk should have no effect, but if allowed to sour the lactic acid will attack	3, 4, 5
Molasses	Does not affect impervious, thoroughly cured concrete; dark, partly refined molasses may attack concrete that is not thoroughly cured; such concrete may be protected with	2, 5, 9
Niter	None	None
Sal ammoniac	Same as ammonium chloride—cause causes slow disintegration	1, 3, 4
Sal soda	None	None
Saltpeter	None	None
Sauerkraut	Little, if any, effect; protect taste with	1, 2
Silage	Attacks concrete slowly	3, 4, 5
Sugar	Dry sugar has no effect on concrete that is thoroughly cured	None
	Sugar solutions attack concrete	1, 2, 3
Sulfite liquor	Attacks concrete slowly	1, 2, 3
Tanning liquor	Depends on liquid; most of them have no effect; tanneries using chromium report no effects; if liquor is acid, protect with	1, 2, 3
Trisodium phosphate	None	None
Vinegar	Disintegrates (see acetic acid)	5, 6, 7
Washing soda	None	None
Whey	The lactic acid will attack concrete	3, 4, 5

TABLE 7.1 Materials Available For Surface Treatment of Concrete *(Continued)*

Material	Effect on concrete	Surface treatment*
	Miscellaneous (*Continued*)	
Wine	Many wine tanks with no surface coating coating have given good results, but taste of first batch may be affected unless concrete has been given tartaric acid treatment	For fine wines the concrete has been treated with 2 or 3 applications of tartaric acid solution (1lb tartaric acid in 3 pt water); sodium silicate silicate is also effective; in a few cases tanks have been lined with glass tile
Wood pulp	None	None

* Treatments indicated provide sufficient protection in most cases, but any of the other treatments designated by a number higher than the highest shown would be equally suitable and often may be advisable. See discussion in text.

† Dry materials generally have no effect.

‡ Many lubricating and other oils contain some vegetable oils. Concrete exposed to such oils should be protected as for vegetable oils.

§ Applied in thin coats the material quickly oxidizes and has no effect. Results indicated above are for constant exposure to the material in liquid form.

dried, the surface should be brushed and washed with water to remove crystals which have formed. The treatment hardens the surface by chemical action and makes it more impervious. Fluosilicates are available from chemical dealers.

2. *Sodium silicate* (commonly called "water glass"): This is quite viscous and must be diluted with water to secure penetration, the amount of dilution depending on the quality of the silicate and permeability of the concrete. Silicate of about 42.5° Baumé gravity diluted in proportions of 1 gal with 4 gal of water makes a good solution. It may be applied in two or three or more coats, allowing each coat to dry thoroughly. On horizontal surfaces it may be poured on and then spread evenly with brooms or brushes. Scrubbing each coat with water after it has hardened provides a better condition for application of succeeding coats. For tanks and similar structures, progressively stronger solutions are often used for the succeeding coats.

3. *Drying oils:* Boiled or raw linseed oil may be used, but the boiled oil dries more rapidly. China wood oil or tung oil and soybean oil are also effective. Applied hot, better penetration is secured. The oil should be applied immediately after heating, however, as it will become more viscous if allowed to stand. Two or three coats may be applied, allowing each to dry thoroughly before the next application. Diluting the oil with turpentine, up to a mixture of equal parts, gives better penetration for the first coat. The concrete should be well cured and seasoned before the first application. The oil is sometimes applied after the magnesium fluosilicate treatment, providing a good coating over a hardened surface.

4. *Cumar:* Cumar is a synthetic resin soluble in xylol and similar hydrocarbon solvents. A solution consisting of about 6 lb of cumar per gal of xylol with ½ pt boiled linseed oil makes a good coating. Two or more coats should be applied. Concrete should be fairly dry. The cumar should be powdered to aid dissolving. It is available in grades from dark brown to colorless, and sold through paint and varnish trades.

5. *Varnishes and paints:* Any varnish can be applied to dry concrete. High-grade varnishes of the spar, china wood oil, or bakelite types and synthetic resin paints and coatings, or paints consisting largely of chlorinated rubber or synthetic rubber give good protection against many substances. Two or more coats should be applied. Some manufacturers can provide specially compounded coatings for certain conditions.

6. *Bituminous or coal-tar paints, tar, and pitches:* These are usually applied in two coats, a thin priming coat to ensure bond and a thicker finish coat. Finish coat must be carefully applied to secure continuity and avoid pinholes. Surface should be touched up where necessary.

7. *Bituminous enamel:* This is suitable protection against relatively strong acids. It does not resist abrasion at high temperatures. Two materials are used, a priming solution and the enamel proper. The priming solution is of thin brushing consistency and should be applied so as to completely cover, touching up any uncoated spots before applying the enamel. When primer has dried to slightly tacky state, it is ready for the enamel. The enamel usually consists of a bitumen with a finely powdered siliceous mineral filler. The filler increases the resistance to flowing and sagging at elevated temperatures, and to abrasion. The enamel should be melted and carefully heated until it is fluid enough to brush. The temperature should not exceed 375°F. When fluid it should be mopped on quickly, as it sets and hardens rapidly.

8. *Bituminous mastic:* This is used chiefy for floors on account of the thickness of the layer which must be applied but some mastics can be troweled on vertical surfaces. Some mastics are applied cold. Others must be heated until fluid. The cold mastic consists of two compositions—the priming solution and the body coat or mastic. The primer is first brushed on. When the primer has dried to a tacky state, a thin layer—about 1/32 in—of the mastic is troweled on. When this has dried, successive 1/32-in coats of the mastic are applied until the required thickness has been built up. The mastic is similar to the primer but is ground with finely powdered siliceous material fillers to make a very thick, pasty mass.

   The hot mastics are somewhat similar to the mixtures used in sheet asphalt pavements, but contain more asphaltic binder so that when heated to fluid condition they can be poured and troweled into place. They are satisfactory only when applied in layers 1 in or more in thickness. When ready to lay, the mixture usually consists of about 15% asphaltic binder, 20% finely powdered siliceous mineral filler, and the remainder sand graded up to ¼-in maximum size.

9. *Vitrified brick or tile:* These are special burned clay products which possess high resistance to attack by acids or alkalies. They must, of course, be laid in mortar which is also resistance against the substance to which they are to be exposed. A waterproof membrane and a bed of mortar are usually placed between the brick or tile and concrete. Some of the acid-resistant cements are melted and poured in the joints. Only materials suitable for the conditions should be used and the manufacturer's directions for installation must be followed. Silica brick and cement are not resistant to hydrofluoric acid and the hydroxides, but special brick and cement for these substances are available.

10. *Glass:* May be cemented to the concrete.

11. *Lead:* May be cemented to the concrete with an asphaltic paint.

12. *Sheets of synthetic resin, rubber, and synthetic rubber:* Thin sheets of synthetic resin, rubber, or synthetic rubber resistant to many acids, alkalies, and other substances are available. These are cemented to the concrete with special adhesives.

REACTIVE AGGREGATES

7.13 Alkali–Aggregate Reaction

Reactions between certain highly siliceous constituents of aggregates and cements having high alkali contents result in strength loss, excessive expansion, cracking, and disintegration. Cracks resulting from this action develop an irregular pattern as shown in Fig. 7.4. The expansions associated with the action may close expansion joints, cause structural members to shift with respect to others, cause machinery to be dislocated, and other undesirable effects. The reaction takes place when the alkalies Na_2O and K_2O are present in amounts exceeding 0.6% by weight, expressed as Na_2O equivalent, when aggregates contain significant amounts of reactive substances such as opal, chalcedony, tridymite, heulandite, zeolite, some phyllites, cristobalite, dacites, andesites, and some rhyolites, and when moisture is present. The expansive deterioration is believed to be caused by osmotic swelling of the alkali silica gels produced by the chemical reaction between the reactive siliceous material and the alkalies released by hydration of the cement and from other sources. All siliceous materials in concrete aggregates do not expand excessively when combined with high-alkali cement as shown in Fig. 7.5. Also it is evident from Fig. 7.6 that even when reactive aggregates are used excessive expansion occurs only when a high-alkali cement is used.

FIGURE 7.4 Typical pattern cracking on the exposed surface of concrete affected by alkali-aggregate reaction.[5]

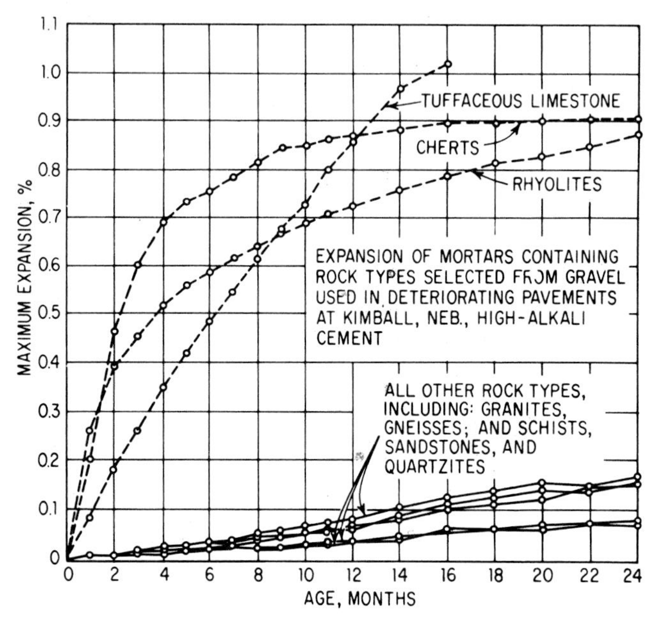

FIGURE 7.5 Relation between expansion, age, and type of aggregate when a high-alkali cement is used.[6]

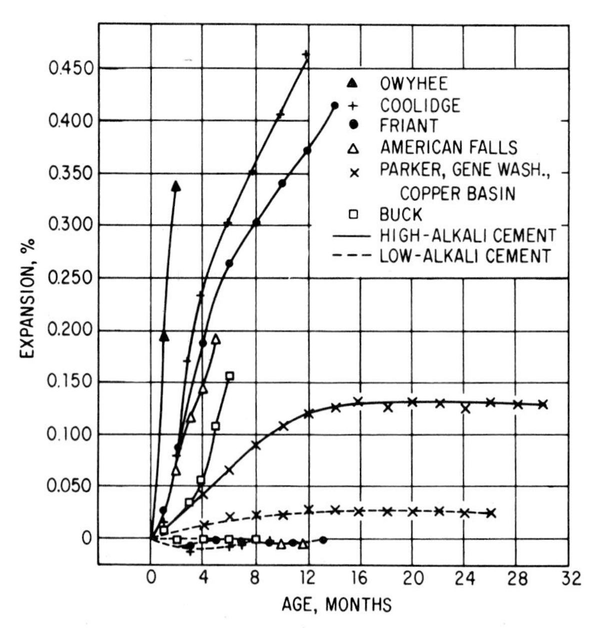

FIGURE 7.6 Relation between expansion, age, and alkali content of cement of mortars made with various aggregates.[6]

Two standard methods of test are provided by ASTM to check on the potential alkali reactivity of cement-aggregate combinations. The method given in ASTM Standard C227 is concerned with the determination of the susceptibility of cement-aggregate combinations to expansive reactions by measuring the linear expansions developed by the combination in mortar bars stored above water in covered containers kept at 100 ± 3°F. The method given in ASTM Standard C289 covers a chemical method of determining the potential reactivity of an aggregate with alkalies in portland cement concrete as indicated by the amount of the reaction during a 24-h period at 80°C between 1 *N* sodium hydroxide solution and the aggregate crushed and sieved to pass a No. 50 sieve and be retained on a No. 100 sieve. ASTM C295, Petrographic Examination, is also recommended.

Alkali–aggregate reactions may be controlled or reduced by using cements with an alkali content less than 0.60% expressed as Na_2O equivalent; using nonreactive aggregates; using certain finely ground pozzolanic materials which react chemically with the alkalies before they attack the reactive aggregates; providing exposures that cause the concrete to become and remain relatively dry; and using air-entraining agents to increase the void space within the mortar, not considered effective with large deteriorations.

7.14 Alkali–Carbonate Reaction

The study of the problems of alkali carbonate reactivity is a relatively new development in concrete research but the large amount of reported research is a good indication of its importance. Although maps of the dolomitic aggregates give excellent service in concrete, certain types of fine-grained argillaceous dolomitic aggregates may react with the cement alkalies and produce undesirable expansions.

Tests for alkali–carbonate reactivity are made by exposing concrete prisms made with the questionable aggregate and a high-alkali cement to an atmosphere at 73°F and 100% relative humidity and noting the amount of expansion for a given period (ASTM 586). Another test used to detect alkali–carbonate reactivity consists in storing prisms of rock in a 2 *M* alkali hydroxide (equal molar parts of NaOH and KOH) solution. Expansions of the questionable material are then compared with those for companion prisms of known sound limestone. Aggregates having linear expansions in excess of 0.03 percent greater than those for the control aggregate at ages up to 12 months should be regarded with concern.

Experimental results[7] indicate that the expansion of concrete containing alkali-reactive carbonate aggregates is greatly influenced by the following factors: (1) degree of rock reactivity, (2) amount of reactive component, (3) alkali content of the cement, and (4) storage conditions. The magnitude of the expansion and cracking may be greatly reduced by dilution of the reactive aggregate with a nonreactive aggregate and reduction of the alkalies in the cement. Reduction of the alkali content to 0.60% may not be adequate and reductions to 0.40% may be necessary. Pozzolans frequently used to inhibit the alkali–silica reaction are not effective in controlling the alkali–carbonate reaction.

7.15 Other Aggregate Reactions

With an increased emphasis on research of cement–aggregate reactions it becomes apparent that the actual mechanisms that produce excessive expansion are very complex and that in many instances combinations of reactions may be involved. Investigations of alkali–aggregate and alkali–carbonate reactions have provided answers to some questions, and they have also shown that in many situations the actual performance could not be predicted because of complications due to other physical and chemical properties of the reacting materials.

Under certain conditions concrete made with sand-gravel aggregates from Kansas, Nebraska, and portions of several adjacent states may suffer large expansions and possible deterioration. Apparently alkali–aggregate reaction to a varying degree, delayed hydration of free magnesia, rupture of the cement-aggregate bond, and physical and chemical phenomena that have not yet been defined may be involved. Replacement of 30% or more, by weight, of the questionable aggregate with crushed limestone may be used for partial control of the undesirable expansion.

Large expansions have also been associated with aggregates that undergo excessive drying shrinkage in air and large expansion in water. In the Union of South Africa certain sandstones containing clay minerals, and in Scotland certain basic igneous rocks of the basalt and dolerite type have given trouble. The cause of the dimensional instability of the aggregate itself under wetting and drying conditions is uncertain. The presence of clay minerals with expansive lattices such as montmorillonite offers only a partial explanation since large dimensional changes have also been observed in aggregates that do not contain such clay minerals. Further, it appears that the shrinkage and expansion of the mortar is not simply related to the dimensional change of the rock and that complex phenomena are involved.

WEAR

7.16 Abrasion

The information on the resistance of concrete to abrasion is limited, but it appears that a high degree of resistance to abrasion can be obtained with a dense concrete that has a low water/cement ratio and a minimum of fine aggregate compatible with good workability, and that has been properly placed and cured. Overworking and early troweling of the concrete surface should be avoided in order to prevent formation of an oversanded, very wet mortar at the surface. The abrasion resistance increases roughly as does the compressive strength up to about 6000 psi. Entrained air influences abrasion in a manner similar to its effect on strength and consequently the amount of entrained air should generally not exceed 4% when good resistance to abrasion is desired.

Wear of concrete surfaces has been classified as follows:[8] (1) attrition, wear on concrete floors due to normal foot traffic, light trucking, and sliding of objects; (2) attrition combined with scraping and percussion, wear on concrete road surfaces due to vehicles with and without chains; (3) attrition and scraping, wear on underwater construction due to the action of abrasive material carried by moving water; and (4) percussion, wear on hydraulic structures where a high hydraulic gradient is present—generally known as *cavitation erosion.*

Cavitation effects have caused severe damage to hydraulic structures. The action is caused by the abrupt change in direction and velocity of the water so that the pressure at some points is reduced to the vapor pressure and vapor pockets are created. These pockets collapse with great impact when they enter areas of higher pressure, producing very high impact pressures over small areas which eventually cause pits and holes in the concrete surface. It has been estimated that the impact of the collapse may produce pressures as high as 100,000 psi. Apparently cavitation damage is not common in open conduits at water velocities below 40 fps. In closed conduits velocities as low as 25 fps may cause pitting.

A large variety of abrasion and impact tests have been developed to provide information for special purposes. Tests to determine abrasion properties of concrete aggregates by means of the Los Angeles machine are given in ASTM Standards C131 and C535. Tests to determine the abrasion resistance of concrete is provided by ASTM Standards C418, C779, and C944. The abrasion resistance of concrete is determined by subjecting it to the impingement of air-driven silica sand, revolving disks, dressing wheels, and ball bearings.

The use of absorptive form linings and the vacuum process reduce wear as a result of the lowered water content. Properly applied combinations of dry cement and special aggregates decrease wear of unformed surfaces. Application of pressure through hard troweling to the concrete surface after it begins to set is also effective. Abrasion properties of hardened concrete floors may be improved by the use of surface hardeners, addition of hard sand or iron grits to surface, or painting.

CORROSION

7.17 Steel

Sufficient concrete coverage normally provided protects the reinforcement against corrosion. Special protection may be required when concrete is within tide range and exposed to water containing salt alkalies or chlorides. Some work has been reported on the use of corrosion-inhibiting admixtures, but agreement on the effectiveness of these materials over a wide variety of exposure conditions has not yet been reached. Additions of calcium chloride to accelerate rate of strength gain are considered harmful because of adverse effects on corrosion of the reinforcement. The use of calcium chloride should not be permitted in prestressed concrete, in concrete containing galvanized reinforcing steel, or in concrete that may be subjected to electrolytic action and moisture.

Disintegrations of certain reinforced-concrete structures have apparently been due to electrolytic action of stray currents from neighboring power circuits. This type of action is accelerated when the concrete is wet and when calcium chloride is present. The action results in a loss of bond between the steel and concrete due either to the oxidation of the steel and the development of high radial stress, or to the softening of the concrete around the steel.

7.18 Aluminum, Copper, Lead, and Zinc

Since some nonferrous metals are frequently used in contact with portland cement concrete the possibilities of corrosion contact must be considered. Aluminum may be used for coils, pipelines, and linings for concrete vats; copper may be used for flashings, roofing, and diaphragms in joints; lead may be used for cable sheathing and for lining tanks; zinc may be used for flashings, roofing, and leaders.

Aluminum is corroded by caustic alkalies. It is attacked by fresh and unseasoned concrete, reacting with calcium hydroxide to form calcium aluminate. The action is progressive because the corrosion products are nonadherent and do not protect the underlying metal. Aluminum should be protected with a coating of asphalt, varnish, or pitch when it is to be embedded in fresh concrete or is to come in contact with wet concrete. It has also been shown that severe corrosion of aluminum conduit may occur when the conduit is embedded in concrete containing calcium chloride and steel that is electrically connected to the aluminum. The pressures developed by the corrosion product are great enough to crack the concrete or to collapse the conduit.[9]

Copper is almost immune to the action of caustic alkalies. Copper may be safely embedded in fresh concrete and will not normally react with hardened concrete in either the wet or dry state. However, if copper is to be in contact with wet concrete, admixtures containing chlorides should not be used because of their corrosive action.

Lead in contact with fresh concrete will react with calcium hydroxide and will corrode. If it is to be in contact with fresh or green concrete it should be given a protective coating of asphalt, varnish, or pitch. Electrolytic action may also cause gradual disintegration of embedded lead under moist conditions when part of the lead is exposed to the air and the other part is embedded in concrete.

Zinc is attacked by caustic alkalies such as calcium hydroxide. Zinc in fresh concrete will react with calcium hydroxide to form calcium zincate. This corrosion product appears to form a dense, closely adherent film that protects the underlying material from further attack. If all action between the calcium hydroxide and the zinc is to be avoided the metal should be given a protective coating of asphalt, varnish, or pitch. Zinc does not corrode in contact with dry seasoned concrete.

REFERENCES

1. Levowitz, D.: *Food Eng.,* June 1952, p. 43.
2. Higginson, E. C., and D. J. Glantz: The Significance of Tests on Sulfate Resistance of Concrete, *ASTM Proc.,* Vol. 53, pp. 1002–1020, 1953.
3. McBurney, J. W.: Cracking in Masonry Caused by Expansion of Mortar, *ASTM Proc.,* Vol. 52, p. 1228, 1952.
4. "Effect of Various Substances on Concrete and Protective Treatments, Where Required," Portland Cement Association, Concrete Information ST 4.
5. "Concrete Manual," 8th ed., U.S. Bureau of Reclamation, 1978.
6. Blanks, R. F., and H. L. Kennedy: "The Technology of Cement and Concrete," Vol. 1, "Concrete Materials," Wiley, New York, 1955.
7. Newlon, Howard H., Jr., and W. Cullen Sherwood: Methods for Reducing Expansion of Concrete Caused by Alkali–Carbonate Rock Reactions, *Highway Res. Board Record,* No. 45, p. 134, 1964.
8. Kennedy, H. L., and M. E. Prior: Abrasion Resistance, *ASTM Spec. Tech. Publ.* 169, pp. 163–174, 1956.
9. Monfore, G. E., and Borje Ost: Corrosion of Aluminum Conduit in Concrete, *J. Res. Develop. Lab., PCA,* Vol. 7, no. 1, p. 10, January 1965.
10. ASTM STP 169B, Dec. 1978, Significance of Tests and Properties of Concrete and Concrete-Making Materials.
11. Portland Cement Association, Design and Control of Concrete Mixes, 1988.

CHAPTER 8
PERMEABILITY AND ABSORPTION

George W. Washa

GENERAL CONSIDERATIONS

8.1 Definitions

Water may enter a porous body as either a liquid or vapor through capillary attraction, it may be forced in under pressure, or it may be introduced by a combination of pressure and capillary attraction. Motion of water through the body may also involve osmotic effects. Absorption refers to the process by which concrete draws water into its pores and capillaries. Permeability of concrete to water or vapor is the property which permits the passage of the fluid or vapor through the concrete.

All concrete mixtures absorb some water and are permeable to a certain extent. Tests under hydrostatic heads of 100 ft have indicated that neither portland cement nor mixtures made from it are absolutely impervious. However, there is abundant evidence which shows that concrete and mortar can be made so impermeable that no leakage or dampness is visible on the surface opposite that at which the water enters. Apparently, even when the humidity is high, the frictional resistance to flow prevents the water from leaving the free surface of the concrete at a rapid enough rate to escape evaporation. In hydraulic structures where water-tightness is of great concern, permeability may be more important than strength. Permeability and absorption may also be important because of their relation to various disintegration agencies that damage concrete.

8.2 Pore Structure of Concrete

Concrete contains pores in all its constituents. Pores in the aggregates undergo only small changes with time but pores in the cement paste are subject to great changes, especially during the early life of the paste. In a freshly mixed neat cement paste the water-filled space is available for the formation of hydration products. This space, originally a function of the water/cement ratio of the paste, is continually reduced by the volume of the precipitated hydrated gel. The capillary system at any time is that part of the original water-filled space not filled with hydrated gel. It is thus apparent that hydration reduces the size and volume of capillary pores and increases the gel volume, and that the process is continuous as hydration progresses. It has been stated that if the original capillary space is low (water/cement ratio less than 0.40 by weight) the gel will eventually fill all the original water space and leave a paste with no capillary pores.[1] As the water/cement ratio is increased and as the degree of hydration is decreased the volume of capillary pores is increased. The capillary pores are of submicroscopic size, interconnected, and randomly distributed.

The pores in the gel are very numerous and are much smaller than the capillary pores. Water in the gel pores does not behave as normal free water because of the very small pore size. This is also true for water in capillary pores, but to a lesser degree, because of the larger pore size. The permeability of paste is most closely associated with the capillary pores because water in these pores is more responsive to changes in hydrostatic pressure than in the water in the gel pores.

The aggregates which constitute 70 to 75% of the total concrete volume have porosities that range from nearly 0 to about 20% by solid volume. The pores vary considerably in size, are larger than the gel pores, and may be at least as large as the largest capillary pores. Associated with the aggregates are the voids in contact with the bottom surfaces of the coarse aggregates that result from the upward flow of water through the plastic concrete. In addition, the tendency of the cement paste to settle during the plastic state results in voids between the sand particles.

Concrete usually contains entrapped air and may in addition contain purposefully added entrained air voids. The entrained air voids are generally noncoalescing and separated spheroids that reduce bleeding, tend to reduce the channel structure, and decrease permeability. The air voids usually constitute 0.5 to 6.0% of the concrete volume.

In summary, concrete contains a wide variety of pores between and within its various components. They may exist as separated spheroids or they may be interconnected and randomly distributed. Further the pore structure will change as hydration continues and causes a reduction of the capillary pores and an increase in gel pores. Since capillary pores are directly related to permeability increased hydration decreases the permeability.

ABSORPTION AND PERMEABILITY TESTS

8.3 Test Methods

Absorption tests are performed by immersing concrete in water for 48 h, weighing after surface drying, ovendrying, and again weighing. The absorption may then be calculated by dividing the loss in weight by the oven-dry weight. In some instances the concrete may be boiled for 5 hr and then dried. If rate of absorption is required the weight after a short period of immersion, as 30 min, may be compared with that after the total 24- or 48-h period. Total absorption is considered to be a criterion of concrete durability, but correlation is usually not very satisfactory. Rate of absorption appears to provide a better correlation.

Although the principal objective of permeability testing is to determine the permeability characteristics of concrete, the tests may have little direct relation to the imperviousness of the structure made with the concrete because of the presence of cracks and poor joints. However, besides providing information on the permeability properties of the concrete, the test is also useful in determining the corrosive effects of percolating waters that leach out the free lime and gradually attack the lime in the tricalcium silicate. Tests also provide information on the relative efficiencies of cements, on the use of integral and surface waterproofing agents, and provide information on the basic pore structure which is related to other properties such as absorption, capillarity, uplift, resistance to freezing and thawing, and others.

The permeability of concrete is usually obtained by determining the amount of water under pressure that is forced to flow into a specimen during a given time interval or by determining the amount that flows out of the opposite surface, exposed to air, in a given time interval. Each cylindrical specimen is usually sealed along its curved surface in metal containers and subjected to water under pressure, hydrostatic heads up to 1000 ft, on one of its flat surfaces. Means for measuring the water inflow or outflow, protected from evaporation losses, are provided. Initially outflow will be less than inflow because of absorption, but ultimately the values will be the same. The leakage, frequently given in gallons or cubic feet of water per square foot of area per hour, is usually determined at a given time after the start of the test. This is done to eliminate absorption effects during the early portion of the test and to provide a standard for comparison purposes.

The leakage of water through concrete can be calculated from Darcy's law for viscous flow.

$$\frac{Q}{A} = K_c \frac{H}{L}$$

where Q = rate of flow, Ft^3/s
A = area of cross section under pressure, Ft^2
$\frac{H}{L}$ = ratio of head of water to percolation length
K_c = permeability coefficient, unit rate of discharge in cfs due to a head of 1 ft acting on a specimen of 1-ft^2 cross section and 1 ft thick

Other factors influence the results of permeability tests. Percolating water of an aggressive nature will increase permeability due to the leaching of lime from the cement. Percolating water that contains sediment or bacteria will act to close the pores and decrease permeability. The direction of water flow through the concrete in relation to the direction of placement may be important. End effects are important because of the concentration of mortar and paste near the surface. Consequently, short specimens are less permeable per unit length than long specimens.

Tests made on Boulder Dam mass concrete containing 9-in gravel and four sacks of low-heat cement per cubic yard of concrete showed that it was relatively impermeable and that flow through its pore structure was negligible.[2] The report on these tests also included information on permeability coefficients and classes of permeability for various materials as shown in Table 8.1. An increase in the class number of unity indicates an increase of impermeability of 10 times.

FACTORS AFFECTING WATERTIGHTNESS

8.4 Materials

Test results[3] indicate that permeabilities of cement pastes made with portland cements differing in chemical composition are similar if the initial water/cement ratios, corrected for bleeding, are equal and equal fractions of the cements have hydrated. However, at a given age and given water-cement ratio, pastes containing slow-hydrating cements are more permeable than those made with fast-hydrating cements. Cement fineness is also important, with watertightness decreasing as fineness increases.

TABLE 8.1 Tabulation of Typical Permeability Coefficients* for Varying Materials

Material	$K \times 10^{12}$	Class
Granite specimen	2–10	12
Slate specimen	3–7	12
Concrete and mortar, W/C 0.5–0.6	1–300	12–10
Breccia specimen	20–	11
Concrete and mortar, W/C 0.6–0.7	10–650	11–10
Calcite specimen	20–400	11–10
Concrete and mortar, W/C 0.7–0.8	30–1,400	11–9
Limestone specimen	30–50,000	11–8
Concrete and mortar, W/C 0.8–1.0	150–2,500	10–9
Dolomite specimen	200–500	10
Concrete and mortar, W/C 1.2–2.0	1,000–70,000	9–8
Biotite gneiss in place, field test	1,000–100,000	9–7
Sandstone specimen	7,000–500,000	9–7
Cores for earth dams	1,000–1,000,000	9–6
Slate in place, field test	10,000–1,000,000	8–6
Face brick	100,000–1,000,000	7–6
Concrete, unreinforced canal linings, field test	100,000–2,000,000	7–6
Steel sheetpiling—junction open 1/1000 in with ½ in of contact—18-in sections†	500,000–	7
Concrete, restrained slabs with ¼ to ½% reinforcing—30° temp. change†	1,000,000–5,000,000	6
Water-bearing sands	1,000,000,000	3

*Coefficient represents quantity of water in cubic feet per second, per square foot of surface exposed to percolation, passing through 1 ft of substance with 1-ft head: $Q = \frac{K_c H}{L}$.

†Flow through 1 ft length of crack, 1 ft deep, $Q = 60{,}000 \frac{H}{L} D^3$.

Permeability as well as strength is related to the water/cement ratio as shown in Fig. 8.1. The permeability of concrete increases at a rapid rate as the water/cement ratio exceeds about 0.65, by weight. For given materials and conditions an increase in the water/cement ratio from 0.40 to 0.80 may increase permeability about 100 times.

The effect of aggregate size on permeability is also shown in Fig. 8.1. With a given water/cement ratio the permeability increases as the maximum size of the aggregate increases, probably because the water voids present on the underside of the coarse aggregate increase as the maximum aggregate size is increased. Sound dense aggregates having a low porosity along with proper grading are essential to the development of concrete with low permeability. Sufficient fine aggregate must be used to ensure placement of concrete free from honeycombing.

Purposely entrained air generally reduces permeability because of the increased workability, decreased bleeding, and effect of the separated voids in reducing water-channel structure. Substitution of fly ash for a part of the portland cement generally reduces the permeability of the concrete.

8.5 Proportions

The exact proportions of aggregates and the cement content are dependent on the grading and shape of the aggregate particles, on the conditions of placement, and on the desired

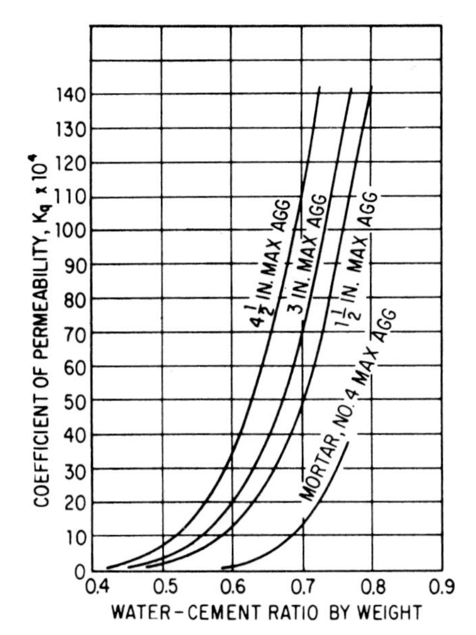

FIGURE 8.1 Relation between coefficient of permeability and water/cement ratio for mortar and for concrete with three different maximum coarse-aggregate sizes; K_q is a relative measure of the flow of water through concrete in cubic feet per year per square foot of area for a unit hydraulic gradient.[4]

permeability characteristics. The correct amount of water must be determined and used. The use of excess water reduces density and increases flow. Hand placement of a dry mix leads to pinholes and large leakages, especially with lean concrete. Generally, a slight excess of water produces less leakage than a slight deficiency. For most hand-rodded concrete the minimum slump that can be successfully used varies from 2 to 6 in, depending on the richness of the mix, the maximum size of aggregate, intricacy of reinforcing, width of forms, and amount of puddling. The relation between water/cement ratio and slump is shown in Fig. 8.2. It is apparent that the slump is especially important in mixtures with higher water/cement ratios. With vibrated concrete, slumps of ½ to 2 in. can often be used effectively to produce concrete with low permeability.

8.6 Placing and Curing

Handling, placing, and compacting properly proportioned concrete in forms is an important step in the production of watertight concrete. Every attempt must be made to avoid segregation, which may cause honeycombing or a porous structure.

The need of continued hydration of the cement to reduce the volume of pores through development of gel has been previously stated. The effect of the length of the curing period on the permeability is shown in Fig. 8.3. It is apparent that curing during the early life of the concrete is especially effective in reducing permeability. The relation between water/cement ratio and length of curing period on permeability is shown in Fig. 8.4. The

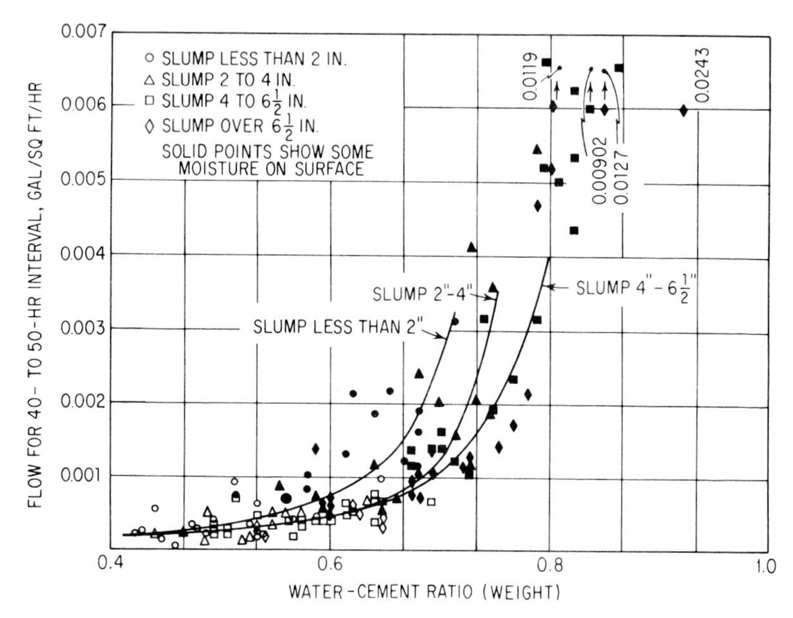

FIGURE 8.2 Effect of water-cement ratio and slump on permeability.[5]

curves show that concrete with a water/cement ratio of 0.50 became essentially impermeable after 7 days moist curing, concrete with a water/cement ratio of 0.65 required 14 days, and concrete with a water/cement ratio of 0.80 did not reach that condition even after curing 28 days.

8.7 Construction Practices

Since construction joints are a frequent source of water-leakage problems they should be avoided whenever possible. In massive concrete structures, where expansion joints are provided, concrete placement should be continuous between joints. If construction joints are necessary, precautions must be taken to ensure good bond between the old and the

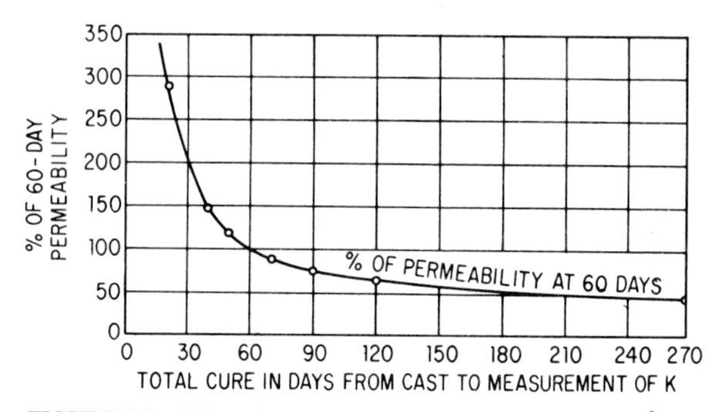

FIGURE 8.3 Effect of length of curing period on permeability.[2]

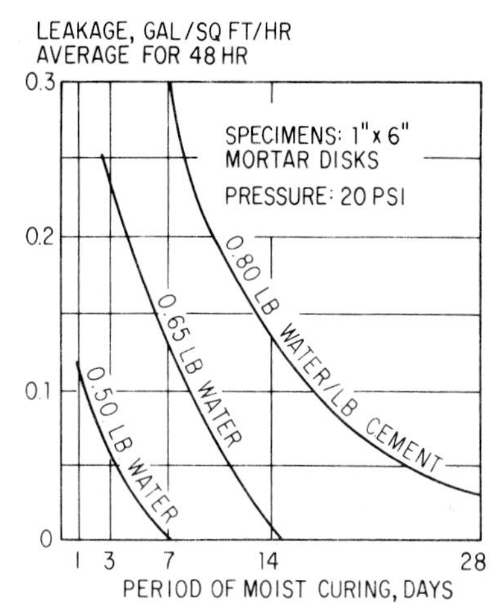

FIGURE 8.4 Effect of water/cement ratio and curing on permeabililty.[6]

new concrete. A properly proportioned mixture with a minimum of bleeding is needed to avoid formation of a weak and porous layer on the surface which will provide poor bond with the next layer. The bond between two layers of concrete may be improved by either of the following:

1. Careful cleaning of the surface of the first layer of concrete, including wire brushing to expose the coarse aggregate before the first layer has become thoroughly hard and before the second layer is placed
2. Dampening the surface of the first concrete
3. Providing a slush coat of neat portland cement grout on the surface of the first concrete layer
4. Placing, before the cement grout has attained its initial set, several inches of new concrete with less coarse aggregate to provide enough mortar to avoid coarse-aggregate pockets at the bottom of the new layer
5. Continuing placement of regularly designed concrete to complete the immediate construction (vibration of the fresh concrete is desirable)

Galvanized-iron or copper stops may be placed in construction joints of tanks. These stops are usually made of about 20-gauge sheet metal and are about 7 or 8 in wide. Half of the stop is embedded in the first layer of concrete and the other half in the second layer. Various types of rubber and plastic water stops are effectively used to make watertight joints in concrete structures.

WATERPROOFING AND DAMPPROOFING

8.8 Integral Mixtures

The terms *waterproofing* and *dampproofing* are frequently used interchangeably, but, in general, *dampproofing* refers to any method of making concrete impervious to water vapor or to liquid water under low pressure, while *waterproofing* refers to any method of making concrete impervious to liquid water that may be under high pressure.

If concrete is properly proportioned, placed, and cured, it is quite impermeable under heads as high as 100 ft, and integral mixtures or surface treatments should not be necessary. Integral mixtures cannot be expected to offset poor artinship or improper curing. They will not be effective if the concrete cracks.

A large number of proprietary materials are available as integral mixtures. They should be used in accord with the manufacturer's directions in order to be as effective as possible. The extra expense involved with the use of these materials should be considered in relation to the use of concrete made without integral mixtures but with higher cement contents. A wide variety of available integral mixtures may be classified as (1) calcium chloride; (2) water solutions of inorganic salts; (3) water suspensions of pore-filling substances; (4) soaps containing fatty acids, largely as stearates; (5) soap solutions with an evaporable solvent which react with hydrated lime; (6) finely subdivided materials; (7) combinations of solutions in successive applications which react chemically; (8) solutions of solid hydrocarbons in oil or paraffin; (9) bituminous coatings; and (10) portland cement with an added water repellent. As a result of an extensive series of tests[7] it was reported that the finely subdivided fillers generally decreased permeability, and that other types gave variable results. Some integral mixtures actually increased permeability.

8.9 Surface Applications

Materials used as surface treatments for concrete may be classified as those which penetrate the surface of the concrete and fill the pores or those which form surface films. The penetrating materials may be inert materials that simply fill the pores, or they may react with constituents in the concrete surface to form compounds of greater volume and with greater pore-filling capacity. These materials are also used as hardeners to prevent surface dusting and disintegration.[12]

A large variety of proprietary materials are available. These materials should be used in accord with the manufacturer's recommendations to secure the best results. Surface waterproofing materials are usually more effective if they are applied on the face of the concrete in contact with water, but frequently this is not economically feasible for structures in existence.

Test results[8] have shown that surface waterproofing agents are more or less effective in decreasing permeability of concrete. Properly applied and cured portland-cement mortars, asphalt emulsions, heavy petroleum distillates dissolved in volatile solvents, silane and acrylic type of sealers and transparent coatings of linseed or china wood oil were among the most effective. In general, exposure to the elements decreases the effectiveness of surface treatments. Essentially absolute imperviousness can be obtained by the use of alternate layers of fabric, such as the better grades of roof felt, and hot asphalt or coal tar. While this method of waterproofing is costly it is universally acknowledged as satisfactory.

8.10 Air and Vapor Permeability

Problems involving air and vapor permeability[9–11] of concrete are important in various types of tanks and pressure vessels. The flow of air and water vapor depend on the air or vapor pressure; the thickness of the concrete; the properties of the concrete, and the properties of the air, gas, or vapor. Concrete made with a low water/cement ratio, with well-graded aggregates, with a pozzolan addition, and properly cured for an adequate time will be relatively impermeable to the flow of air, gases, or water vapor. Severe drying increases air permeability, probably because of the formation of shrinkage cracks, but vapor permeability is not apparently increased. There appears to be no significant relation between the air, gas, and water permeabilities of concrete.

REFERENCES

1. Verbeck, George: Pore Structure, *ASTM Spec. Tech. Publ.* 169B, pp 262–270, 1978.
2. Ruettgers, A., E. N. Vidal, and S. P. Wing: An Investigation of the Permeability of Mass Concrete with Particular Reference to Boulder Dam, *J. ACI*, March-April 1935, p. 382.
3. Powers, T. C., L. E. Copeland, J. C. Hayes, and H. M. Mann: Permeability of Portland Cement Paste, *J. ACI*, Nov. 1954, p. 285.
4. "Concrete Manual," 8th ed., U. S. Bureau of Reclamation, 1975.
5. Norton, P. T., and D. H. Pletta: Permeability of Gravel Concrete, *J. ACI*, May 1931, p. 1093.
6. "Watertight Concrete," PCA Concrete Information ST-33.
7. Jumper, C. H.: Tests of Integral and Surface Waterproofings for Concrete, *J. ACI*, Dec. 1931, p. 209.
8. Washa, G. W.: The Efficiency of Surface Treatments on the Permeability of Concrete, *J. ACI*, Sept.-Oct. 1933, p.1.
9. "Air Permeability of Concrete," *Concrete and Constructional Engineering,* Vol. 60, May 1965, p. 166.
10. Henry, R. L., and G. K. Kurtz: "Water Vapor Transmission of Concrete and Aggregates," *Tech. Rept.* R244, U. S. Naval Civil Engineering Laboratory, Port Hueneme (Ventura County), Cal., June 1963.
11. Graf O.: "Die Eigenschaften des Betons," 2d ed., Springer-Verlag, Berlin-Wilmersdorff, 1960.
12. A Guide to the Use of Waterproofing, Dampproofing, Protective, and Decorative Barrier Systems for Concrete, ACI 515, 1R-79, (Rev. 1985).

CHAPTER 9
VOLUME CHANGES

George W. Washa

GENERAL

Volume changes in concrete due to variations in stress, temperature, and humidity are partly or completely reversible, but volume changes due to destructive chemical and mechanical action are usually cumulative as long as the action continues. Unrestrained volume changes due to variations in temperature, moisture, or stress are usually not important, but volume changes that are restrained by foundations, reinforcement, or connecting members may lead to stresses which may cause distress and even failure. Restrained contractions are usually more important than restrained expansions because concrete is much weaker in tension than in compression. The magnitude of volume changes is usually given in linear rather than volumetric units because linear changes can be easily measured and because they are of primary concern to engineers. A length change may be given, for example, as 900 millionths of an inch per inch or 0.09 percent or 1.08 in per 100 ft.

The causes of volume changes and the reactions of concrete to these changes are generally known, but it is not yet possible to build concrete structures such as buildings, bridges, pavements, and dams with assurance that they will not crack. However, if proper attention is given to all the factors that are involved it is possible to build these structures so that they are relatively crack-free and resistant to the action of destructive agents frequently associated with cracking.

FRESH CONCRETE

9.1 Bleeding and Setting Shrinkage

Fresh concrete in the plastic state undergoes significant volume changes. They are due to water absorption, sedimentation (bleeding), cement hydration, and thermal changes and are influenced by the temperature and humidity of the surrounding atmosphere. Absorption of water by the aggregates and bleeding of the free water to the top, where it may be lost by drainage from the forms or by evaporation, cause shrinkage. Bleeding starts shortly after the concrete has been placed and continues until maximum compaction of the solids, particle interference, or setting stops the action. Usually a large portion of the shrinkage due to absorption and bleeding takes place during the first hour after placement. Profuse bleeding along with rapid evaporation or leaky or absorbent forms will result in excessive shrinkage, which may be as high as 1 percent. Since the concrete is in a plastic or semiplastic state during bleeding, there are no appreciable stresses associated with the large shrinkage. Setting shrinkage may be kept low by the use of saturated aggregates, low cement content, properly designed concrete mixtures, moist and cool casting conditions, tight and nonabsorbent forms, and shallow lifts in placing.

9.2 Plastic Shrinkage

Plastic shrinkage and plastic cracking occur in the surface of fresh concrete soon after it has been placed and while it is still plastic. The principal cause of this type of shrinkage is rapid drying of the concrete at the surface. With the same materials, proportions, and methods of mixing, placing, finishing, and curing, cracks may develop one day but not on the next. Changing weather conditions that increase the rate of evaporation from the surface are usually involved. The highest evaporation rates are obtained when the relative humidity of the air is low, when the concrete and air temperatures are high, when the concrete temperature is higher than the air temperature, and when a strong wind is blowing over the concrete surface. This combination of circumstances, which may remove surface moisture faster than it can be replaced by normal bleeding, is most frequently obtained in summer. However, rapid drying is possible even in cold weather if the temperature of the concrete is high compared with air temperature.

Plastic shrinkage cracks are usually of shallow depth, generally 1½ to 2 in, 12 to 18 in long and usually perpendicular to the wind. They may also have a crowfoot pattern. Corrective measures to prevent plastic-shrinkage cracking are all directed toward reducing the rate of evaporation or the total time during which evaporation can take place.

HARDENED CONCRETE

9.3 Caused by Cement Hydration

Volume changes that are due to cement hydration and do not include changes due to variations in moisture, temperature, or stress are called *autogenous volume changes.* These changes may be either expansions or contractions, depending on the relative importance of two opposing factors—expansion of new gel due to the absorption of free pore water or shrinkage of the gel due to extraction of water by reaction with the remaining unhydrated cement. Generally the initial expansions obtained during the first few months do not exceed 0.003 percent while the ultimate contractions obtained after several years usually do not exceed 0.015 percent. This type of volume change is especially important in the interior of mass concrete where little or no changes in total moisture content may take place.

Test results indicate that autogenous volume changes are influenced by the composition and fineness of the cement, quantity of mixing water, mixture proportions, curing conditions, and time. The magnitude of the autogenous volume change appears to increase as the fineness of the cement and the amount of cement for a given consistency are increased. Ultimate contractions seem to be somewhat greater for low-heat cements

(Type IV) than for normal portland cement. The most significant autogenous shrinkage usually takes place within 60 to 90 days after placement of the concrete.

9.4 Caused by Thermal Changes

Unrestrained concrete expands as the temperature rises and contracts as it falls. The average value of the coefficient of thermal expansion, the rate at which thermal volume change takes place, is 5.5 millionths per degree Fahrenheit, which is fortunately close to the value for steel. While the coefficient may frequently be close to the average value, it may also vary between 2.5 and 8.0 millionths per degree Fahrenheit, depending on the richness of the mix, the moisture content, and primarily on the thermal coefficient of the aggregate.

The coefficient of thermal expansion of hardened cement paste usually varies between 5.0 and 12.5 millionths per degree Fahrenheit. Oven-dried and vacuum-saturated specimens have similar coefficients, but specimens with intermediate moisture contents may apparently have coefficients almost twice as large.

Thermal expansion of concrete is greatly influenced by the type of aggregate because of the large differences in the thermal properties of various types of aggregates and because the aggregate constitutes from 70 to over 80% of the total solid volume of the concrete. Siliceous aggregates such as chert, quartzite, and sandstone have thermal coefficients of expansion between 4.5 and 6.5 millionths per degree Fahrenheit, while the coefficients for pure limestone, basalt, granite, and gneiss may vary between 1.2 and 4.5 millionths per degree Fahrenheit. Further single-mineral crystals may have different coefficients along three different axes. As an example, feldspar has values of 9.7, 0.5, and 1.1 millionths per degree Fahrenheit along three different axes. An estimated value of the coefficient of thermal expansion for concrete may be computed from weighted averages of the coefficients of the aggregate and the hardened cement paste.

It has been suggested that significant internal stresses due to temperature variations may develop because of differences in the thermal coefficients of the cement paste and the aggregate. The relative importance of this thermal incompatibility and its effect on durability is in some doubt.

Thermal changes in mass concrete are kept as low as possible by the use of low-heat portland cements, by artificial refrigeration, and by other special procedures in order to avoid cracking as the concrete cools from the maximum to the stable temperature. Thermal changes in pavements are also important. Differential temperature gradients at night may cause the top surface to shorten in relation to the bottom surface, tending to lift the slab ends above the subgrade and decreasing the ability of the slab to support traffic loads without cracking.

9.5 Caused by Continuous Moist Storage

Moist-cured concrete starts to expand after the setting shrinkage has taken place and continues at a decreasing rate under continuous moist storage. The ultimate expansion is usually less than 0.025 percent and shows little change after 10 years of moist storage. The maximum expansion due to moist storage is usually about one-fourth to one-third of the shrinkage due to air drying. Expansions obtained with Type I cement appear to be higher than those with Types II, III, IV, and V. Replacement of cement with pozzolanic material usually increases slightly the expansion under continuous moist curing. However, the amount of the cementing material is much more important than the type, since neat cement pastes expand about twice as much as average mortars and the mortar expands about twice as much as an average concrete.

9.6 Caused by Drying

Drying shrinkage and carbonation shrinkage (due to the reaction between carbon dioxide and the constituents of cement) take place concurrently. Normally the two types of shrinkage are not separated in reported data and the total is usually designated as drying shrinkage. Continuously wet or dry structural concrete made with sand and gravel aggregates will have only a small amount of carbonation shrinkage, but the carbonation shrinkage of porous concrete dried to equilibrium in air at 50% relative humidity may approach the drying shrinkage. Drying shrinkage of hardened concrete is usually caused by the drying and shrinking of the cement gel which is formed by hydration of portland cement. It is affected primarily by the unit water content of the concrete. Other factors affecting drying shrinkage include cement composition, cement content, quantity and quality of paste, characteristics and amounts of admixtures used, mineral composition and maximum size of aggregate, mixture proportions, size and shape of the concrete mass; amount and distribution of reinforcing steel, curing conditions, humidity of surrounding air during the drying period, and the length of the drying period.

Materials. The type of cement used influences drying shrinkage. In general, finer cements exhibit slightly greater shrinkages. The tricalcium aluminate contributes most and the tricalcium silicate least to drying shrinkage. Gypsum has a large effect on drying shrinkage, and for a given cement there appears to be an optimum amount that produces the smallest shrinkage.

Aggregate particles embedded in cement paste restrain drying shrinkage. Well-graded aggregates with a large maximum size reduce shrinkage because they allow low water contents, require small quantities of paste, and encourage cracking between particles. Concrete of the same cement content and slump with ⅜-in maximum-size aggregate usually develops from 10 to 20 percent greater drying shrinkage than concrete with ¾-in maximum-size aggregate, and from 20 to 35 percent greater drying shrinkage than concrete containing 1½-in maximum-size aggregate. The actual amounts are dependent on many variables. The effect of aggregate size is also evident in drying shrinkages for neat cement, mortar, and concrete of 0.25 to 0.30, 0.06 to 0.12, and 0.03 to 0.08 percent, respectively. Aggregates having a high modulus of elasticity and those having rough surfaces offer greater restraint to shrinkage. Concrete made with sandstone, slate, hornblende, and pyroxene may shrink up to two times as much as concrete made with granite, quartz, feldspar, dolomite, and limestone. Dirty sands and unwashed coarse aggregates containing detrimental clay cause increased shrinkage.

Drying shrinkage of lightweight concrete can be relatively high, but careful selection of materials and proper attention to proportioning may provide lightweight concrete with about the same drying shrinkage obtained with normal-weight concrete. Drying-shrinkage values of structural lightweight concrete normally vary between 0.04 and 0.15 percent and are likely to be more pronounced for concrete containing aggregates that have high absorptions and that require high cement contents for strength. Moist-cured cellular products made with neat cement weighing between 10 and 20 pcf may have drying-shrinkage values between 0.30 and 0.60 percent. Autoclaved cellular products that contain fine siliceous material may weigh about 40 pcf and have drying-shrinkage values in the range 0.02 to 0.10 percent.

Admixtures appear to have a variable effect on drying shrinkage. When drying shrinkage is important admixtures should be used with care, preferably after evaluation under job conditions. Admixtures that increase the unit water content of concrete will usually increase drying shrinkage, but many admixtures that reduce the unit water content do not reduce drying shrinkage. Shrinkage is usually increased by replacing some of the cement with pozzolanic materials such as pumicite or raw diatomaceous earth. However, replacement of cement with a low-carbon, high-fineness fly ash results in about the same or slightly lower shrinkage. Compounds frequently used to increase the rate of strength development, calcium chloride, and triethanolamine usually significantly increase drying shrinkage. Within normal limits, entrained air has little effect on drying shrinkage.

Reinforcing steel restricts but does not prevent shrinkage. Shrinkage of reinforced unrestrained structures produces tension in the concrete and compression in the steel. Increasing amounts of reinforcing steel reduce the contraction but increase the tensile stress in the concrete. If sufficient reinforcement is used, the restraint may be great enough to cause cracking of the concrete. Reinforced-concrete structures with normal amounts of reinforcement may have drying shrinkages in the range of 0.02 to 0.03 percent.

Proportions. The importance of the quality and quantity of cement paste on the drying shrinkage of concrete is evident from the fact that cement paste in concrete, if not restrained by the aggregate, shrinks 5 to 15 times as much as the concrete. However, the single most important factor affecting drying shrinkage of concrete is the amount of water per unit volume of concrete. Aggregate size and grading, mix proportions, cement content, slump, temperature of fresh concrete, and other factors affect drying shrinkage principally as they influence the total amount of water in a given volume of concrete. Figure 9.1 gives the interrelation of shrinkage, cement content, and water content and shows clearly that drying shrinkage is primarily governed by the unit water content.

Size and Shape. In large concrete members, differential volume changes take place, with the largest drying shrinkages at and near the surface. Because of the large moisture variations from the center to the surface, tensile stresses are present at and near the sur-

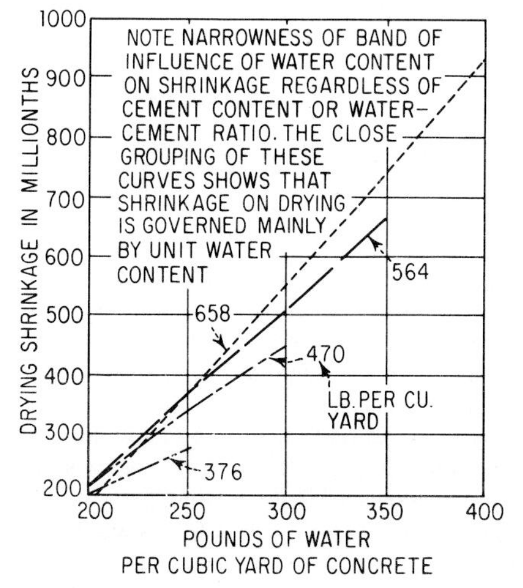

FIGURE 9.1 Relation between drying shrinkage, cement content, and water content.[2]

face while compressive stresses are present in the interior. If the tensile stresses are very high, surface cracks may appear. However, creep may act to prevent cracking and may cause permanent elongation of the fibers in tension and shortening of the fibers in compression. The rate and ultimate shrinkage of a large mass of concrete are smaller than the values for small concrete specimens, although the action continues for a longer period for the large mass.

Structural concrete slabs reinforced only in tension tend to warp, because the concrete near the top compressive surface shrinks more because of the water accumulation at that surface during placement and also because the steel on the tensile side acts to resist shrinkage.

Exposure Conditions. Moist curing of concrete beyond that required to develop the required strength has little effect on drying shrinkage. High-pressure steam curing at about 350°F is effectively used in block and precast-concrete products plants to reduce greatly subsequent drying shrinkage.

The length of the drying period and the humidity of the surrounding air have an important effect on drying shrinkage. Small specimens of neat cement paste have shown that drying shrinkage increases as the logarithm of the age, at least out to 20 years. The shrinkage of mortars and concretes is usually small after drying for 3 years. The rate and magnitude of drying shrinkage increase as the relative humidity of the surrounding air decreases and the rate increases as air movement past the member increases. Pastes, mortars, and concretes that have reached equilibrium under given drying conditions will shrink more if the relative humidity is then decreased, or will expand if it is increased. Shrinkage of concrete in dry air will be greatly retarded by coating the concrete with an impervious material which tends to prevent escape of the contained moisture.

Combined Effect of Unfavorable Factors. It has been experimentally shown[3] that the cumulative effect of the individual factors that increase drying shrinkage can be very large, and that the combined effect is the product rather than the sum of the individual effects. As shown in Table 9.1, the use of less favorable construction practices—concrete discharge temperature of 80°F rather than 60°F, a 6- to 7-in rather than a 3- to 4-in slump, a ¾-in maximum-aggregate size rather than 1½-in, and too long a mixing and waiting period—could be expected to increase shrinkage 64 percent. If, in addition, a cement with high shrinkage characteristics, dirty aggregates of poor inherent shrinkage quality, and admixtures that increase shrinkage are used the final shrinkage could be about five times as large as the shrinkage that would be obtained with the best choice of variables.

9.7 Caused by Alternate Wetting and Drying

Storage of concrete in moist air or in water produces expansion, while storage in dry air produces shrinkage. If the storage consists of alternate wetting and drying at room temperature, alternate swelling and shrinkage will result in a residual shrinkage. This shrinkage usually increases during the initial cycles and then tends to become constant. Consequently, the shrinkage after the initial cycles may be considered as completely reversible. When wetting and drying cycles are combined with alternations of high and low temperatures, residual expansions that increase with the number of cycles are caused. These expansions may be quite large. In a series of tests,[4] 120 cycles consisting of 9 h of oven drying at 180°F followed by 48-h immersion in water at 70°F and then by 15 h of air

TABLE 9.1 Cumulative Effect of Adverse Factors on Shrinkage*

Factor	Equivalent increase in shrinkage, %	Cumulative effect
Temperature of concrete at discharge allowed to reach 80°F, whereas with reasonable precautions, temperature of 60°F could have been maintained.	8	1.00 × 1.08 = 1.08
Used 6-to 7-in slump where 3- to 4-in slump could have been used.	10	1.08 × 1.10 = 1.19
Use of ¾-in maximum size of aggregate under conditions where 1½-in size could have been used.	25	1.19 × 1.25 = 1.49
Excessive haul in transit mixer, too long a waiting period at jobsite, or too many revolutions at mixing speed.	10	1.49 × 1.10 = 1.64
Use of cement having relatively high shrinkage characteristics.	25	1.64 × 1.25 = 2.05
Excessive "dirt" in aggregate due to insufficient washing or contamination during handling.	25	2.05 × 1.25 = 2.56
Use of aggregates of poor inherent quality with respect to shrinkage.	50	2.56 × 1.50 = 3.84
Use of admixture that produces high shrinkage.	30	3.84 × 1.30 = 5.00

*Based on effect of departing from use of best materials and artisanship. From Ref. 3

storage at 70°F caused expansions of 0.10 to 0.25 percent for concrete mixtures made with different cements, water cement ratios, and methods of placement.

9.8 Caused by Carbonation

When concrete is exposed to air containing carbon dioxide, it increases in weight and undergoes irreversible carbonation shrinkage that may be about as large as the shrinkage due to air drying at 70°F and 50 percent relative humidity from a saturated condition. Apparently under ideal conditions all the constituents of cement are subject to carbonation. The rate and extent of carbonation are dependent on variables such as moisture content and density of concrete, temperature, concentration of carbon dioxide, time, size of member, method of curing, and the sequence of drying and carbonation. Carbonation proceeds slowly and usually produces little shrinkage at relative humidities below 25 percent or near saturation.[5] The maximum effect appears to take place when the relative humidity is about 50 percent. Less dense concrete products, such as building block made with lightweight aggregate, are more susceptible to carbonation than dense concrete products.

Since carbonation-shrinkage values may be large and since they are irreversible, efforts have been made to carbonate concrete building blocks before they are to be placed in a wall. Some success with precarbonation treatments involving exposure of block to hot flue gases after curing has been obtained.[6] It also appears that precarbonation may improve volume stability to subsequent moisture changes.

9.9 Caused by Undesirable Chemical and Mechanical Attack

Many types of chemical and mechanical attack act to shorten the useful life of concrete. While the mechanism of destruction varies and may be quite complicated, signs of its action are usually first evident as expansions. As the action continues the expansions increase and ultimately may lead to disintegration. Some of the commoner destructive agents or actions are sewage of high acid or sulfide content, sulfate waters, electrolysis, seawater, fire, freezing and thawing, hydration of uncombined lime or magnesia or both, alkali–aggregate reaction, and reaction of alkalies in the cement with other types of aggregates. These undesirable volume changes should be avoided as much as possible by careful control of the materials used, proper proportioning, desirable construction practices, and control of exposure conditions. Prevention of undesirable reactions is much more effective than questionable remedies applied after expansions have started.

9.10 Cracking

Cracks in concrete due to restrained volume changes are largely dependent on the degree to which shrinkage is resisted by internal and external forces. The ability of concrete to resist cracking is dependent on the degree of restraint; the magnitude of the shrinkage due to carbonation, drying, and thermal effects; the stress produced in the concrete; the amount of stress relief due to creep; and the tensile strength of the concrete. In order to have good resistance to cracking, concrete should have values of sustained modulus of elasticity, thermal coefficient of expansion, drying shrinkage, and carbonation shrinkage as low as possible and a tensile strength as high as possible. Properly placed expansion joints in long concrete walls help in decreasing the amount and severity of cracking.

NONSHRINK OR EXPANSIVE CONCRETE

9.11 Prepacked Concrete

This type of concrete is made by first filling the form with clean, coarse aggregate, usually graded from ½ in to the largest practicable maximum size, and then compacting and wetting the coarse aggregate. A mortar usually consisting of water, sand, cement, and frequently a workability agent is then pumped into the forms until it fills the voids in the coarse aggregate. Because the coarse-aggregate particles are in contact and not separated by cement paste the hardened concrete has exceptionally low shrinkage. This type of concrete is widely used in resurfacing dams, repair of tunnel linings, piers, spillways, and for miscellaneous patchwork of hardened concrete. Preplaced aggregate concrete is discussed in Chap. 39.

9.12 Expansive Materials

Shrinkage may be minimized or prevented when concrete must be placed under special or difficult conditions, such as bedments under machinery or around a congestion of reinforcing steel when vibration is not possible, by the addition of small amounts of superfine unpolished aluminum powder. The aluminum powder, usually added in the range of 0.005 to 0.02% by weight of cement, reacts with the hydroxides in fresh cement paste to produce small bubbles of hydrogen gas. This gas if properly controlled causes a small expansion of the fresh concrete or mortar and reduces or eliminates settlement.

Another method of producing expanding or shrinkage-compensating mortars and concretes involves the use of finely divided or granulated iron in combination with an oxidizing material. The expansion produced is due to the increase, in solid volume of the iron as it is converted to iron oxide. Careful control of the proportion of the oxidizing catalyst is needed to obtain the desired expansion. The possibility of staining the concrete when moisture is present should be considered.

Expansive cements that compensate for shrinkage by controlled expansion of the cement are also used. In most of these cements the expansions are due to the controlled formation of hydrated calcium sulfoaluminates. These cements are also known as self-stressing cements since they are used to induce stress in the restraining steel. The chief difficulty associated with the use of expansive materials is the careful control needed to obtain the desired expansion. The production of concrete with desired predetermined volume-change characteristics requires careful consideration and control of all factors involved in the production, with special attention to compositions, proportions, and fineness of the components of the expansive cement; proportions of the mixture; time and temperature during the curing period; and the degree of restraint.

Expansive materials are used for slabs on grade and are not used for normal placement of structural concrete because of the difficulties of proper control and possible undesirable variable effects. When expansive materials are used the instruction of the manufacturer should be carefully followed.

CREEP

9.13 Nature of Creep

Elastic deformations occur immediately when concrete is loaded. Nonelastic deformations increase with time when the concrete is subjected to a sustained load. Consequently, since concrete is frequently subjected to dead loading, it usually is subjected to both types of deformations. The nonelastic deformation, creep, increases at a decreasing rate during the period of loading. It has been shown that significant creep takes place during the first few seconds after loading and that creep may increase out to 25 years. Approximately one-fourth to one-third of the ultimate creep takes place in the first month of sustained loading, and about one-half to three-fourths of the ultimate creep occurs during the first half year of sustained loading in concrete sections of moderate size. When the sustained load is removed there is some recovery but the concrete does not usually return to its original state. Figure 9.2 gives the deformation record of a specimen loaded at an age of 1 month and unloaded 6 months later. The elastic and creep recoveries are less than the deformations under load because of the increased age of the concrete at the time of unloading.

Creep may be due partly to closure of internal voids, viscous flow of the cement-water paste, crystalline flow in aggregates, and flow of water out of the cement gel due to external load and drying. The last cause is generally believed to be the most important. The magnitude and rate of creep for most concrete structures are intimately related to the drying rate, but creep is also important in massive structures where little or no drying of the concrete takes place. In these structures, most of the creep is believed to be due to the flow of the absorbed water from the gel (seepage) caused by external pressure.

The ultimate magnitude of creep on a unit-stress basis, psi, usually ranges from 0.2 to 2.0 millionths and ordinarily is about 1 millionth per unit of length. This value is about

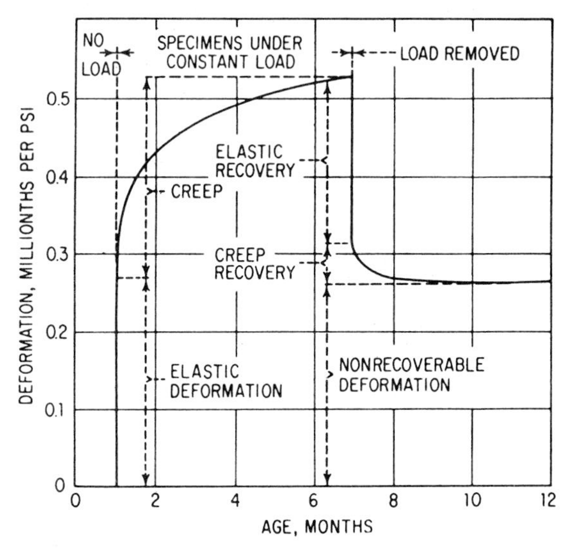

FIGURE 9.2 Elastic and creep deformations of mass concrete under constant load followed by load removal.[2]

three times the elastic deformation of concrete having a secant modulus of elasticity of 3 million psi.

In order to visualize the effects of both elastic and creep deformations at a given time, the sustained modulus of elasticity, defined as the ratio of the constant sustained stress to the sum of the elastic and creep deformations at a given time, may be used. Tests have shown that the modulus of resistance after 2 years of sustained loading varied between one-fifth and one-half of the initial secant modulus of elasticity. A reduced value of short-time secant or chord modulus of elasticity is frequently used in design to take creep into account.

Creep, unlike shrinkage, which is generally undesirable, may be either desirable or undesirable depending on the circumstances. It is desirable in that it generally promotes better distribution of stresses in reinforced-concrete structures. It is undesirable if it causes excessive deformations and deflections that may necessitate costly repairs or if it results in large losses of prestress in prestressed-concrete members.

9.14 Effects of Constituents, Proportions, and Manufacture

Concrete made with low-heat cement creeps more than concrete made with normal cement, probably because of its influence on the degree of hydration. This desirable characteristic explains, at least in part, the relative freedom from cracking of mass concrete structures as they cool to normal temperatures. The effect of cement fineness on creep appears to be variable. Pozzolanic additions to cements generally increase creep. Approved air-entraining agents added in the proper amounts appear to have no appreciable effect on creep. If creep is an important factor, proprietary compounds should not be used unless their effects have been previously evaluated.

Under comparable conditions creep and shrinkage generally decrease when well-graded aggregates with low void contents are used and when the maximum size of the coarse aggregate is increased. Hard, dense aggregates with low absorption and a high modulus of elasticity are desirable when low shrinkage and creep are wanted. The mineral composition of the aggregate is important, and generally increasing creep may be expected with aggregates in the following order: limestone, quartz, granite, basalt, and sandstone.

Creep of concrete increases as the water/cement ratio increases as shown in Fig. 9.3. In addition it appears that if two concrete mixtures have the same water/cement ratio, the mixture having the greatest volume of cement paste will creep the most. In general, lean concrete mixtures exhibit considerably greater creep than do rich concrete mixtures because the water/cement-ratio effect of increasing creep for the lean mixtures is more important than the paste effect of decreasing creep.

The tendency of concrete to creep decreases as cement hydration increases, and consequently water-cured concrete should creep less than air-cured concrete. However, the shrinkage or swelling produced during the initial during period also influences creep.

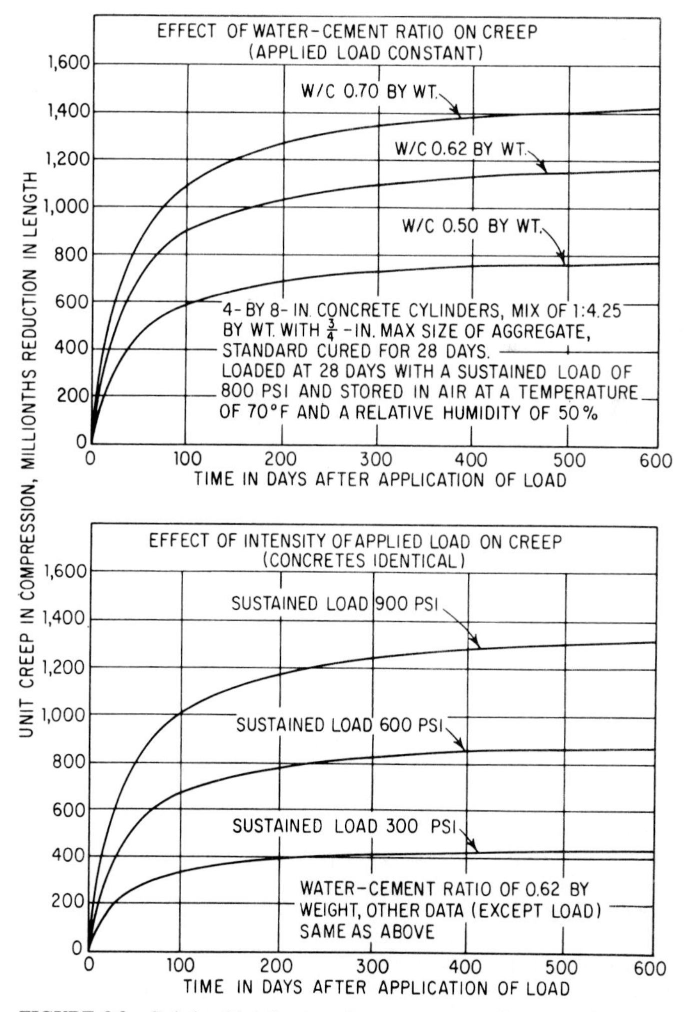

FIGURE 9.3 Relationship of creep of concrete to water/cement ratio and to intensity of applied load.[2]

Under compressive load preswelled specimens (moist-cured) creep more than preshrunk specimens (cured in dry air). Size effects are also important in curing because small specimens respond more rapidly to moisture changes than large concrete members.

Research on high-performance concrete (over 14.2 ksi) containing high-range water reducers (superplasticizers)has found that the creep after 800 days is about one-fourth to one-fifth that of normal-strength concrete.

9.15 Effects of Exposure and Loading Conditions

The rate and ultimate magnitude of creep increase as the humidity of atmosphere decreases. The relation between relative humidity and creep is not linear. Concrete under sustained load in air at 70 percent relative humidity will have an ultimate creep about twice as large as concrete in air at 100 percent relative humidity. The ultimate creep in air at 50 percent relative humidity will be about three times as large. Protection of concrete members against rapid drying is very beneficial in reducing the rate and ultimate amount of creep.

The rate and magnitude of creep generally decrease as the size of the concrete member increases. It has been estimated that the creep of mass concrete may be about one-fourth of that obtained with small specimens stored in moist air.

With a given material and sustained load the rate and magnitude of creep decrease as the age at which the sustained load is applied increases, as long as hydration continues. Creep is also approximately proportional to the sustained stress within the range of usual working stresses as shown in Fig. 9.3. Sustained stresses above the working stresses produce creep that increases at a progressively faster rate as the magnitude of the sustained stress is increased.

Most creep information for plain concrete has been obtained for sustained compressive loading. However, test information for creep under sustained tension, bending, torsion, and biaxial- and triaxial-stress conditions generally shows the same behavior pattern. The ultimate tensile creep reduced to a 1-psi stress basis is about the same as the ultimate compressive creep on the same basis.

A test method to determine the creep characteristics of molded concrete cylinders subjected to sustained longitudinal compressive load is provided in ASTM Standard C512. The method is intended to compare concrete specimens tested under controlled conditions, but it does not provide a means for calculating deflections of reinforced-concrete members in structures. The equation for total strain per psi ϵ is given by the equation

$$\epsilon = \frac{1}{E} + F(K)\log_e(t+1)$$

where ϵ = total strain per psi
E = instantaneous elastic modulus, psi
$F(K)$ = creep rate, calculated as the slope of a straight line representing the creep curve on a semilog plot (logarithmic axis represents time)
t = time after loading, days

The quantity $1/E$ is the initial elastic strain per psi, and the second term in the equation represents the creep strain per psi for any given age.

9.16 Effects of Reinforcement

Reinforced-concrete columns subjected to sustained compressive loads also exhibit creep. Under the action of shrinkage and creep there is a tendency for additional stress to be transferred to the steel, with a consequent decrease in the concrete stress. In some tests of reinforced-concrete columns in air the compressive stress in the steel increased from three to five times the design value, while the compressive stresses in the concrete decreased materially and in some instances they changed to low tensile stresses. Concrete columns in water showed much smaller stress changes with time.

Beams and slabs that are reinforced in tension only and that are not loaded will usually warp because shrinkage is resisted by the steel near the bottom and because the top concrete shrinks more than the bottom concrete because of bleeding. Consequently, beams and slabs subjected to sustained loading will be deflected and strained by warping as well as by loading. In general, the deflections and compressive strains of beams and slabs under sustained load increase at a decreasing rate while the strain at the tensile-steel level shows relatively small changes with time. Creep deflections and deformations increase rapidly as the ratio of span length to total depth increases. Inclusion of arbitrary amounts of compressive steel, not required for design strength, at the section of maximum moment in simply supported reinforced-concrete beams is effective in reducing creep. Tests have shown that inclusion of compressive steel equal in amount to the tensile steel reduces creep deflections and strains by about one-third to one-half. Consequently, it appears that when a combination of high length/depth ratio and a large sustained load (causing steel and concrete stresses approaching the limiting design value of the ACI Code) are involved, compressive reinforcement may be used to control creep. In the case of framed structures or continuous beams creep frequently, but not always, may cause a more favorable stress redistribution.

Great care is required in the manufacture of prestressed reinforced-concrete beams to minimize drying shrinkage and creep in order to avoid large reductions in the amount of prestress. Under poor conditions the loss in prestress may be as high as 50,000 psi.

REFERENCES

1. Washa, G. W.:Volume Changes, Significance of Tests and Properties of Concrete and Concrete Aggregates, *ASTM Spec. Tech. Publ.,* 1965.
2. "Concrete Manual," 8th ed., U.S. Bureau of Reclamation, 1975.
3. Tremper, B., and D. L. Spellman: Shrinkage of Concrete—Comparison of Laboratory and Field Performance, *Highway Res. Board Record 3, Publ.* 1067, 1963.
4. Washa, G. W.: Comparison of the Physical and Mechanical Properties of Hand Rodded and Vibrated Concrete Made with Different Cements, *J. ACI*, June, 1940, p. 617.
5. Verbeck, G. J.: Carbonation of Hydrated Portland Cement, *ASTM Spec. Tech. Publ.* 205, 1958.
6. Toennies, H. T., and J. J. Shideler: Plant Drying and Carbonation of Concrete Block, NCMA-PCA Cooperative Program, *J. ACI*, May, 1963, p. 617.
7. Neville, Adam M.: "Hardened Concrete Physical and Mechanical Aspects," ACI Monograph No. 6, 1971.
8. Standard Practice for the Use of Shrinkage Compensating Cement, ACI 223, 1983.
9. "Temperature and Concrete," *ACI Publ.* SP-25, 1971.
10. Prediction of Creep and Shrinkage and Temperature Effects in Concrete Structures, ACI 209R, 1982.
11. ACI SP-76, 1982, Designing for Creep and Shrinkage in Concrete Structures.
12. Rhillco, R. E., Elastic Properties and Creep, ASTM STP 169B, 1978.

CHAPTER 10
THERMAL, ACOUSTIC, AND ELECTRICAL PROPERTIES

George W. Washa*

THERMAL PROPERTIES

10.1 General

The thermal properties, as well as the strength properties, of concrete may be varied considerably by changes in the materials, proportions, and manufacturing methods. Knowledge of the thermal properties of concrete is necessary to design and to predict the performance of a large variety of concrete structures. Although concrete is generally superior to metals and natural stones in its insulating ability, it is surpassed at room temperature by such materials as asbestos, powdered magnesia, mineral wool, and pulverized cork. At high temperatures, materials such as powdered magnesia and infusorial earth are much better insulating materials. The protective value of concrete at high temperatures, proved in many large conflagrations, is due to its high resistance to fire in conjunction with relatively low conductivity and high strength.

The thermal properties of hardened concrete that are important to the engineer are thermal conductivity, specific heat, thermal diffusivity, coefficient of thermal expansion, and adiabatic temperature rise. In addition, the effect of temperature on the strength properties must be known.[13,14]

10.2 Thermal Conductivity

Thermal conductivity is the rate at which heat passes through a material of unit area and thickness when there is a unit temperature change between the two faces of the material. It is important in connection with temperature variations in mass concrete, and also with insulation and condensation properties of walls and slabs. Definitions and numerical values for the various coefficients are given in the "American Society of Heating, Refrigerating and Air-conditioning Guide and Data Book."[1]

The various coefficients used to calculate heat losses are given below:

k = thermal conductivity of a homogeneous material between the surface of the warmer side and the surface of the cooler side, Btu per hour per square foot of area per degree difference in temperature per inch of thickness

*Revised by Joseph A. Dobrowolski

C = thermal conductance of an obstruction (wall) between the surface of the warmer side and the surface of the cooler side, Btu per hour per square foot per degree difference in temperature for a stated thickness (frequently given, for example, for 4-, 8-, and 12-in concrete masonry units)

j = film surface conductance, time rate of heat flow between a unit area of a surface and the surrounding air; f_i designates the inside surface and f_o the outside surface, Btu per hour per square foot per degree difference in temperature

a = thermal conductance of an air space, Btu per hr per sq ft per degree difference in temperature

R = thermal resistance, the reciprocal of conductance, as $1/k$, $1/C$, $1/U$, etc.; the overall transmission coefficient of a composite wall may be obtained by determining the total resistance by adding the reciprocals of the various conductivity coefficients for the individual parts of the composite wall:

$$R = \frac{1}{f_i} + \frac{x_i}{k_i} + \frac{1}{a} + \frac{x_2}{k_2} + \frac{1}{C} + \frac{1}{f_o}$$

where x_1, x_2 = thicknesses of various materials

U = overall coefficient of heat transmission, Btu per hr per sq ft per degree difference in temperature between the air on the warmer side of an obstruction and the air on the cooler side:

$$U = \frac{1}{R}$$

Methods of calculating the loss through a given wall construction and also the temperature variation between the cool and warm sides of a wall are given in Figs. 10.1 and 10.2.

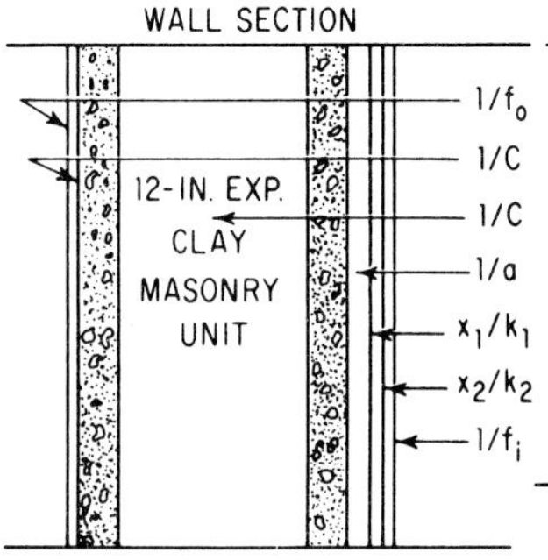

	PART	RESISTANCE TO HEAT TRANSMISSION
$1/f_o$	OUTSIDE SURFACE	0.17
$1/C$	PORTLAND CEMENT PAINT	0.19
$1/C$	12-IN. EXP. CLAY MASONRY UNIT	2.27
$1/a$	AIR SPACE (FURRING)	0.91
x_1/k_1	1/2-IN. RIGID INSULATION	1.51
x_2/k_2	1/2-IN. PLASTER	0.15
$1/f_i$	INSIDE SURFACE	0.68
	TOTAL RESISTANCE R_T =	5.88

$U = \frac{1}{R_T} = \frac{1}{5.88} = 0.17$ BTU/HR/SQ FT/°F

BTU LOSS PER 1,000 SQ FT OF WALL PER 24 HR PER 90°F TEMPERATURE DIFFERENTIAL = 0.17 x 1,000 x 24 x 90 = 367,000

FIGURE 10.1 Calculation of heat loss through wall.[2]

10.2. The various numerical coefficients have been obtained from the "Heating, Refrigerating and Air-conditioning Guide"

The mineralogic composition of the aggregate has a large effect on conductivity. Basalt and trachyte have low conductivities, quartz has high conductivity, and dolomite and limestone have relatively high conductivity. Lightweight aggregates have conductivities roughly proportional to their densities. The approximate relation between thermal conductivity and ovendry density is shown in Fig. 10.3.[3] The air content of concrete has a pronounced effect in reducing thermal conductivity. Tests[4] have shown that the thermal conductivity for both sand and gravel and lightweight-aggregate concrete increases as the moisture content of the hardened concrete increases. These tests also showed that as the temperature of the hardened concrete was increased from −250 to 75°F the thermal conductivity of sand and gravel concrete decreased, and that a similar increase in temperature caused only little change in the conductivity of lightweight aggregate concrete.

10.3 Thermal Conductivity and Condensation

The elimination of condensation on or within walls and floors may be as important as the reduction of heat loss through them. Condensation of moisture on the inside surface of an exterior wall of a building may be prevented if the overall coefficient of heat transmission is low enough to maintain the inside surface at a temperature above the dew-point tem-

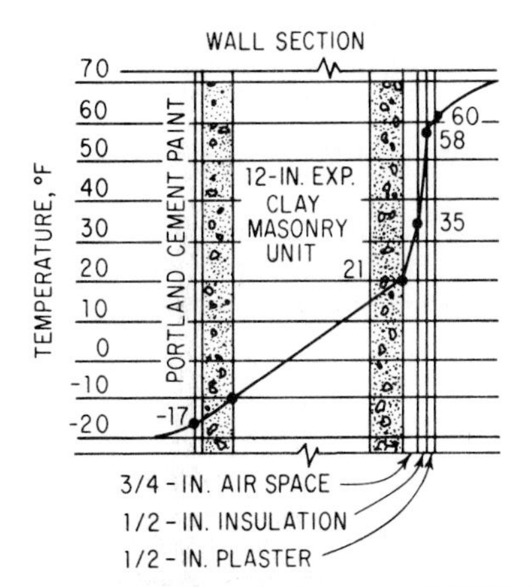

Part	Resistance	Temp change, °F
Outside-surface resistance	0.17	(0.17 ÷ 5.88)90 = 3
12 in. expanded clay masonry unit and 2 coats portland-cement paint	2.46	(2.32 ÷ 5.88)90 = 38
Air space, furring	0.91	(0.91 ÷ 5.88)90 = 14
½ in. rigid insulation	1.51	(1.51 ÷ 5.88)90 = 23
½ in. plaster	0.15	(0.15 ÷ 5.88)90 = 2
Inside-surface resistance	0.68	(0.61 ÷ 5.88)90 = 10
		90

FIGURE 10.2 Temperature variation in wall.[2]

perature t_d. The maximum heat-transmission coefficient which will prevent condensation may be computed from the equation.

$$U = f_i \frac{t_i - t_d}{t_i - t_o}$$

where U = coefficient of heat transmission
f_i = inside surface-film conductance (average value for still air is 1.46)
t_i and t_o = inside and outside temperatures, respectively
t_d = dew-point temperature (available in hygrometric tables)

Condensation on a wall may be avoided by keeping the temperature of the inside surface above the dew-point temperature (as suggested above), by reducing the relative humidity of the air inside the room, or by increasing the circulation of the air passing over the inside surface. The first of these methods is usually used. In addition, in order to reduce the possibility of condensation or dampness within the wall vapor barriers should be placed as near to the warm side of the wall as possible.

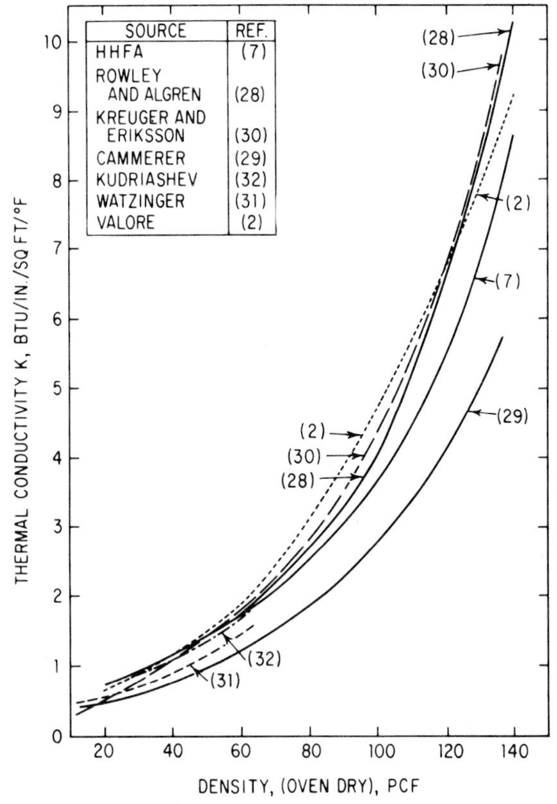

FIGURE 10.3 Relation of density and thermal conductivity. References shown here to identify the curves, are given in the original report.[3]

10.4 Specific Heat and Thermal Diffusivity

Specific heat is defined as the amount of heat required to change the temperature of 1 lb of material 1°F. The specific heat of hardened concrete usually varies between 0.20 and 0.28 Btu/lb/°F. The mineral composition of the aggregate has little effect, and for calculation purposes the specific heat of the dry materials in concrete is frequently assumed between 0.20 and 0.22 Btu/lb/°F. The specific heat of water is 1.0 Btu/lb/°F.

When calculations involving the flow of heat in concrete masses are required, thermal conductivity k, specific heat S, and density d (pcf) must be considered These three values may be used to provide the thermal diffusivity D by means of the equation

$$D = \frac{k}{Sd}$$

Thermal diffusivity is a measure of the rate at which temperature changes will take place within a mass of hardened concrete. Its value normally varies between 0.020 and 0.060 ft^2h. The value of D may also be determined experimentally.

10.5 Coefficient of Thermal Expansion

The thermal coefficient of expansion is a length change (expansion due to temperature increase and contraction due to a temperature decrease) in a unit length per degree of temperature change. It is expressed in inches per inch per degree Fahrenheit, or more commonly as millionths per degree Fahrenheit. An average value for hardened concrete is 5.5 millionths per degree Fahrenheit, although it may normally vary between 3 ½ and 7 millionths per degree Fahrenheit. The thermal coefficient of expansion of concrete varies mainly with the character and amount of the coarse aggregate. Quartz has a coefficient about 7 millionths per degree Fahrenheit while some limestones have a value as low as 3 millionths per degree Fahrenheit. Usual values of the coefficient for well-cured neat cement pastes in either a dry or saturated condition vary between 5 and 8 millionths per degree Fahrenheit, although values between 5 and 12 millionths per degree Fahrenheit have been obtained. The coefficient for concrete may be estimated as the weighted average of the coefficients of the various constituents. It appears that the coefficients of thermal expansion of oven-dry and water-saturated concretes are approximately equal but that partly dried concrete has a higher coefficient.

10.6 Temperature Rise in Mass Concrete

In concrete structures that are less than a few feet in maximum thickness the heat of hydration normally does not need to be considered because it is dissipated rapidly. The temperature rise in mass concrete is important because the heat evolved in setting is dissipated so slowly that a marked rise in temperature occurs. Eventually when the interior cools and contracts, tensile stresses are induced in the interior of the mass. If these stresses are sufficiently high, they may cause cracks, which, in turn, may unite with surface cracks and thus increase leakage and subsequent disintegration. The magnitude of the adiabatic temperature rise can be controlled by proper selection of the type of cement, proportions of the mixture, use of pozzolanic materials, rate of placement, temperature of fresh concrete, artificial cooling, and proper design and manufacturing procedures. These items are detailed in Chap. 20.

The heat generated is closely related to cement composition, with the greatest contribution per unit of compound due to C_3A and followed in decreasing order by C_3S, C_2S, and C_4AF. Low-heat portland cement, ASTM Type IV, is effective in keeping the adiabatic temperature rise at a low level. Pozzolanic materials usually are more effective in decreasing the temperature rise than the cement which they replace.

Lean mixtures made with low-heat cementing materials may produce mass concrete having the required strength and an adiabatic temperature rise less than 40°F. In massive structures where high cement factors have been used, temperature increases in excess of 100°F have been noted. The maximum temperature from which the mass concrete must cool to its final stable temperature may also be decreased by placing it at relatively low temperatures.

The rate of heat dissipation may be increased by using concrete of high conductivity, placing small rather than large units, placing units with a high ratio of exposed surface to volume, providing a long exposure period for the unit before it is covered with concrete, avoiding insulation such as backfill for as long as possible, and providing refrigeration soon after placement of the concrete.

In large mass concrete structures the thermal properties—heat of hydration, specific heat, thermal conductivity, and thermal diffusivity—are important because specifications for precooling procedures, placing temperatures, construction schedules, and design of refrigeration system are all dependent on them. In concrete structures where the thickness is less than a few feet the thermal properties other than the coefficient of expansion and the thermal conductivity are usually not considered.

10.7 Effect of Temperature on Properties

The physical properties of concrete vary with temperature.[4–6] Generally the physical properties are greater at subnormal than at room temperature and lower at high temperatures than at room temperature. Tests[5] have shown that the compressive and splitting strengths appear to reach a maximum value between −75 and −250°F. These tests indicate that the compressive strength of normal sand and gravel concrete more than tripled as the temperature decreased from 75 to a temperature around −150°F. The splitting strengths more than doubled under the same circumstances. Moduli of elasticity usually increased as the temperature decreased from 75 to −250°F. The amount of increase varied from 0 for dry concrete to about 50 percent for moist concrete. Poisson's ratio remained essentially constant as the temperature decreased from 75 to −50°F. The properties of moist concrete appeared to vary much more with temperature, between 75 and −250°F, than the properties of dry concrete.

In another group of tests[6] conducted between −75 and 450°F it was noted that at subnormal temperatures mortar and concrete properties generally increased as the temperature decreased. Above room temperature as the temperature increased the properties usually first decreased, then increased, and finally again decreased. At 450°F the strengths of the mortars and concretes were about the same as those at room temperature, but the moduli of elasticity were considerably lower. Mortar properties investigated included energy of rupture, modulus of rupture, tensile strength, compressive strength, and modulus of elasticity. Concrete properties investigated were compressive strength and modulus of elasticity.

Test results[4] between 75 and 1500°F are given for thermal expansion, density, and dynamic modulus of elasticity. These tests show that the weight loss due to loss of water was largely complete at 800°F and that weight changes at higher temperatures were related to the chemical composition of the aggregates. The coefficient of thermal expansion was significantly greater above 800°F because it is not affected by drying shrinkage of the paste at those temperatures. Moduli of elasticity at 1400°F appeared to be roughly one-third of the values at 75°F. After dehydration, for a given water/cement ratio, the modules of elasticity was essentially independent of age or curing conditions.

10.8 Fire Resistance

Concrete strength and stiffness properties decrease significantly as the temperature is increased much above 800°F. Consequently, concrete load bearing members should not be exposed continuously to temperatures above 500°F. Continuous exposure to temperatures above 900°F may result in spalling. At high temperatures, constituents of concrete may undergo important changes. Quartz, for example, changes state and expands about 0.85 percent at 1063°F, causing a severe disruptive effect. In a fire, high temperatures are initially obtained on one side or in a small portion of the structure. Under these conditions differential expansion between the hot concrete and the cold concrete will take place. The cement paste tends to shrink because of the loss of moisture and to expand because of the increase in temperature, while the aggregate expands continuously with the rise in temperature. The opposing actions lead to cracking and spalling, and in reinforced concrete to exposure of the reinforcement to the fire. The unprotected reinforcement then rapidly loses strength as the temperature rises.

The fire endurance of a concrete wall is primarily dependent on wall thickness, type of construction, type of aggregate, and quality of concrete. Tests[8] have shown that for a given aggregate type the fire-resistance period generally doubled when the thickness (weight per square foot of wall) increased 35 to 40 percent. Differences in construction such as solid concrete wall as against a hollow masonry-unit wall, full bedding of masonry units as against bedding of the face shells only, and thickness of cover for reinforcing steel all have important effects. Natural aggregates given in descending order of their fire resistance are (1) calcareous; (2) feldspathic, such as basalt; (3) granites and sandstones; and (4) siliceous, such as quartz and chert. The lightweight aggregates such as the inflated clays, shales, and slags are superior to the natural aggregates in producing fire-resistant concrete. Greater cement content is effective in increasing the fire-resistance period and increasing the load capacity of a wall, both before and after fire exposure.

The fire resistance of a wall is determined by subjecting a loaded test panel (frequently with a 9-ft minimum dimension and with an area of at least 100ft^2) on one side to temperatures rising in a prescribed manner from 1000°F at 5 min to a maximum of 2300°F at 8 h (ASTM E119). A similar procedure is followed in testing floors and roofs except that a larger test unit is usually required. During the test period the temperature of the unexposed surface is taken at a number of positions and the behavior of the test specimen is noted. Generally the specimen is considered to have passed the test for the required classification period, usually between 1 and 4h, if it has sustained the applied load without permitting passage of a flame or gases hot enough to ignite cotton waste, and if the temperature on the unexposed surface has not increased more than 250°F above the initial temperature. In some instances a hose-stream test may be required in which the exposed face is subjected at a given time to the action of water of given pressure for a given period. The cooled specimen may then be subjected to a load test. Fire tests of columns are made by subjecting columns to loads that should produce design stresses and then exposing all four sides of the columns to fire. The test is considered successful if a column sustains the applied load during the fire test for the desired classification period.

Concrete exposed to high temperatures present in a fire and subsequently cooled exhibits decreased strength and stiffness. Tests[9] of the compressive strength of walls, 6 ft high, made with 8-in hollow masonry units, showed after a 3- to 3 ½-h fire exposure that the ratio of the wall strength to the original strength of the unit averaged about 25 percent for units made with sand and gravel aggregates and about 35 percent for units made with inflated clay aggregates. The moduli of elasticity for walls prior to exposure to fire ranged between 200,000 and 750,000 psi, and after exposure the values varied between 100,000 and 300,000 psi on a gross-area basis. Other tests[8] have shown that solid concrete walls, 6 and 8 in thick and 10 ft high, carried satisfactorily uniformly distributed working loads of 400 psi during and after severe fire exposure.

ACOUSTIC PROPERTIES

10.9 Sound Absorption

The control of sound in a room must be considered with respect to the origin of the sound, whether originating within or outside the room. Consideration must also be given to whether the sound is air- or solid-borne.[7]

Control of sound originating within a room requires that the walls, ceiling, flooring, and furnishings in the room have good sound-absorption qualities. The most commonly used coefficient of sound absorption is a number that expresses the ratio of sound energy absorbed by a surface to the amount of energy incident on the surface. The *absorption coefficient* is generally given for a sound frequency of 500 Hz. The *noise-reduction coefficient* (NRC) is the average of the coefficients at 250, 500, 1000, 2000, and 4000 Hz expressed to the nearest 0.05.

Tests have shown that a porous surface with interconnected pores from the surface to the interior is the chief characteristic needed for good sound absorption. With a porous surface the energy of sound is converted into heat. If the surface pores are sealed by painting or plastering the sound absorption is markedly reduced.

The sound-absorption coefficient for plain cast concrete is about 0.02, which indicates that 98 percent of the incident sound energy is reflected by the surface. Average values of the sound-absorption coefficients for various materials are unpainted cinder block—0.45, painted cinder block—0.40, vermiculite acoustic plaster—0.55, cork—0.70, and sprayed asbestos—0.90.

10.10 Sound Transmission

Sound originating outside a given room may be transmitted as solid-borne as well as airborne sound. The solid-borne sound should be suppressed at its source. A concrete floor in a hall, for example, should be either broken at the partitions or covered by a carpet or other resilient material in order to suppress sound transmission into the room. Airborne sound transmission from outside the room may be suppressed by barriers such as masonry walls and attention to details of construction.

Transmission loss expressed in decibels (dB), a measure of sound insulation, is given by the equation

$$\text{Transmission loss (TL)} = 10 \log \frac{E_i}{E_t} = \text{STC}$$

Where E_i is the level of sound energy incident on a wall and E_t is the level of sound energy transmitted through the wall. A transmission loss of 10 dB indicates that one-tenth of the incident sound energy is transmitted, 20 dB indicates that one-hundredth is transmitted, and so on. Sound intensities vary from 0, the threshold of audibility, to about 130 dB, the threshold of painful sound or the limit of the ear's endurance. The sound intensity on an average busy street is about 60 dB.

Values of sound transmission loss (STC) frequently vary between 40 and 55 dB for walls between rooms in apartments, hospitals, and schools. Various studies[3,10] have shown that the weight per unit wall area is a very important factor influencing transmission loss of airborne sound. Figure 10.4 shows average transmission-loss data for airborne sound for various types of walls obtained from tests conducted at the National Bureau of Standards, National Physical Laboratory of Great Britain, Riverbank Laboratories, and other laboratories. The STC values are averages for frequency ranges of approximately 100 to 2000, 3000, or 4000 Hz, and the heavy line is an average STC-weight relation.[10] Porous rigid materials, such as concrete masonry, follow the STC-weight relationship only if the pores are not continuous. Transmission of airborne sound through continuous air paths in the material greatly reduces STC. Consequently painting or plastering lightweight-concrete masonry units greatly increases the STC.

The factors which provide good sound absorption tend to provide poor sound insulation. Porous concrete absorbs sound but has a poor sound-insulation value (low STC). Plastering or painting porous concrete greatly reduces sound absorption but increases sound insulation.

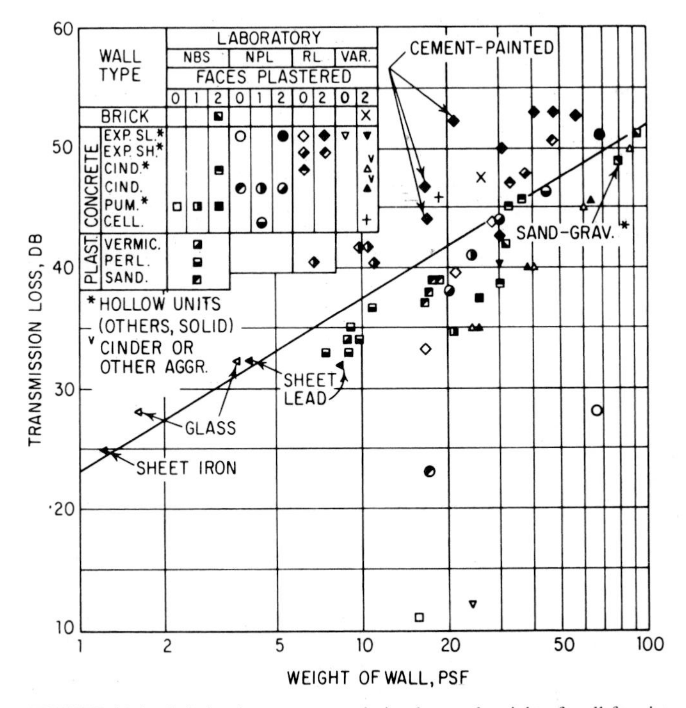

FIGURE 10.4 Relation between transmission loss and weight of wall for airborne sound.[3]

ELECTRICAL PROPERTIES

10.11 Electrolysis, Resistivity, Dielectric Strength

Disintegration of some reinforced concrete structures has apparently been due to electrolytic action of stray currents from nearby power circuits. The action is much greater for wet concrete than for dry concrete, especially if calcium or sodium chloride are present in the mixing water. It has been observed that, when the current flows from the concrete to a steel cathode, the concrete around the reinforcement softens because of a gradual concentration of sodium and potassium alkalies and the bond is destroyed. When the steel acts as the anode, splitting of the concrete surrounding the steel is caused by the oxidation and resultant volume increase of the steel.

Increasing use of concrete crossties has emphasized the need for more information on the electrical resistivity of concrete. Studies[11] including direct and alternating current have shown that moist concrete behaves essentially as an electrolyte having a resistivity of the order of 10^4 Ω-cm, which is in the range of semiconductors. When concrete has been oven-dried, its resistivity is of the order of 10^{11} Ω-cm and it is considered to be a reasonably good insulator.

The dielectric strength of an oven-dried concrete of a 1:2:4 mix made with Type I cement and with a water/cement ratio of 0.49 tested with direct current has been reported[12] as 15.9 kV/cm for the first breakdown and 12.5 kV per cm for the third breakdown. Under similar conditions, concrete made with high-alumina cement showed slightly higher values; and concrete made with Type II cement showed slightly lower values. When tests were made with 50 Hz alternating current, generally slightly lower values were obtained. The dielectric strength of air-dry concrete was approximately the same as that of the oven-dried concrete.

REFERENCES

1. "American Society of Heating, Refrigerating and Air-conditioning Guide and Data Book, Fundamentals and Equipment," American Society of Heating, Refrigerating and Air-Conditioning Engineers, Inc., New York.
2. Stone, A.: Thermal Insulation of Concrete Homes, *J. ACI,* May, 1948, pp. 849–874.
3. Valore, R. C., Jr.: Insulating Concretes *J. ACI,* Nov. 1956, pp. 509–532.
4. Philleo, R.: Some Physical Properties of Concrete at High Temperatures, *J. ACI,* April, 1958, p. 857.
5. Monfore, G. E., and A. E. Lentz: Physical Properties of Concrete at Very Low Temperatures, *J. Res. Develop. Lab., PCA,* Vol. 4, No. 2, May, 1962.
6. Saemann, J. C., and G. W. Washa: Variation of Mortar and Concrete Properties with Temperature, *J. ACI,* Nov. 1957, p. 385.
7. Randall, Frank A.: Acoustics of Concrete in Buildings, Portland Cement Association, 1974.
8. Menzel, C. A.: Tests of the Fire Resistance and Thermal Properties of Solid Concrete Slabs and Their Significance, *ASTM Proc.,* vol. 43, 1943.
9. Menzel, C. A.: The Strength of Concrete Masonry Walls after Standard Fire Exposure, *J. ACI,* November, 1932, pp. 113–142.
10. Knudsen, V. O., and C. M. Harris: "Acoustical Designing in Architecture," John Wiley & Sons, Inc., New York, 1950.

11. Monfore, G. E.: The Electrical Resistivity of Concrete, *J. Res. Develop. Lab., PCA,* vol. 10, no. 2, May, 1968.
12. Hammond, E., and T. D. Robson: The Comparison of Electrical Properties of Various Cements and Concretes, *The Engineer,* vol. 199, January 21, 1955, pp. 78–80, and January 28, 1955, pp. 114–115.
13. ACI SP–25, Temperatures and Concrete, 1971.
14. ASTM STP169B, Smith, Peter, pp 388–414, 1978.

P A R T • 3

PROPORTIONING MIXES AND TESTING

CHAPTER 11
PROPORTIONING MIXES

Joseph F. Artuso*

CRITERIA

11.1 Specifications

Project specifications differ widely from one to another in the manner in which the concrete mix or mixes are described. In general, these may be divided into two classes, "prescription specifications" and "performance specifications," with some including features of both.[6]

A straight prescription specification will spell out the exact proportions of all ingredients, one to the other. Typical of this type is the now largely outmoded but still occasionally encountered volume-proportion requirement such as 1:2:3, meaning 1 part cement, 2 parts sand, and 3 parts coarse aggregate, by volume. In these volume-proportion specifications, quantity of water is seldom limited, except indirectly by means of a maximum-slump requirement.

Prescription-type specifications more often cover only minimum cement content in pounds per cubic yard of concrete and maximum water/cement ratio usually by weight.

In either of such prescriptions, the specification writer is taking full responsibility for sufficiency of the mix design. The contractor or concrete supplier merely has the responsibility to prepare the concrete in the proportions specified, assuming no liability for the strength developed. Many people feel that this is proper. However, probably more feel that the concrete producer, be it contractor or ready-mix plant operator, should be responsible for the properties of the finished product.

Performance specifications, which tend to place the responsibility on the concrete producer, are those which spell out the strength requirement, often accompanied by certain other limiting factors. These others may include minimum cement content, or maximum water/cement ratio, and they practically always include a requirement on consistency as measured by slump, and usually include some limitation on aggregate sizes and properties. Too many limitations tend to convert the specifications back to a prescription type, and if included without full knowledge of local conditions and materials, they may make it impossible to produce concrete without violating one or more of the requirements.

If one accepts the theory that the contractor should at least share the responsibility for the final product, the specification should cover the desired strength, air entrainment, maximum-size aggregate, and slump range. Quality of aggregate, consistent with local conditions to be encountered, and with due consideration of the economics of locally available material, should be covered.

*Revised by Joseph A. Dobrowolski.

With these few limiting factors the contractor, working with his or her laboratory, can submit for consideration of the specifying agency a suitable and economical concrete mix best adapted to the local conditions.

MIX CHARACTERISTICS

11.2 Consistency

Consistency is the most easily defined method of expressing workability of plastic concrete. Consistency is measured by means of the slump test. Although the slump test does not exactly indicate other workability characteristics such as finishing qualities, bleeding, and segregation tendency, it is the best presently known test for predicting the ease of placing and consolidating concrete.

After a slump test is made, the side of the slumped specimen should be tapped lightly to investigate degree of cohesiveness. A specimen of a cohesive mix will sag and the aggregates will not fall away from it. This mix would be more easily consolidated in formed and heavily reinforced structures. A harsh mix with fewer fines but of equal slump may be satisfactory in slabs or mass concrete that can be readily consolidated by strong mechanical vibration. In order to ensure the highest quality, concrete must be placed at the lowest slump that can be handled practically, and be adequately consolidated.

Consistency can also be measured by means of the penetration apparatus (Kelly ball) described in Chap. 12.

11.3 Aggregate Size

Generally, the largest maximum size of aggregate should be used in concrete mixtures to develop the optimum properties of strength, durability, shrinkage, etc. This is primarily achieved because the larger aggregate enables the use of minimum unit water content.

For many years there has been a generally accepted rule of thumb that the largest-sized aggregate available and usable in the section to be placed would be the most desirable. It has been felt that smaller coarse aggregate would result in higher sand requirement with consequent higher water demand with all its accompanying undesirable effects. This view remains prevalent with most concrete engineers, and there is much in experience to defend it. However, data published by the National Sand and Gravel Association have indicated that, for a given cement factor and slump, the compressive strength of rich concrete mixtures increases as size aggregate decreases from 2½ to ⅜ in. This is shown in Fig. 11.1. It also indicates that from a strength standpoint, there is no significant improvement from the use of aggregate larger than ¾ in. It is noted that the range of aggregate under consideration was limited, sizes of 2½ to ⅜ in were used, and cement factors of 375, 565, and 750 lb/yd^3 were evaluated. In addition to the above considerations, the maximum size aggregate should not exceed one-fifth of the narrowest dimension between sides of forms, three-fourths of the minimum clear spacing between reinforcing bars, or one-third the depth of slabs. Also consideration must be given to the maximum size that is economically available and consistent with the above criteria.

11.4 Air Entrainment

The benefits of air entrainment in concrete mixtures have made its use very common. These advantages include improved resistance to freezing and thawing damage, reduced permeability, improved workability, and reduced bleeding. The entrainment of air in

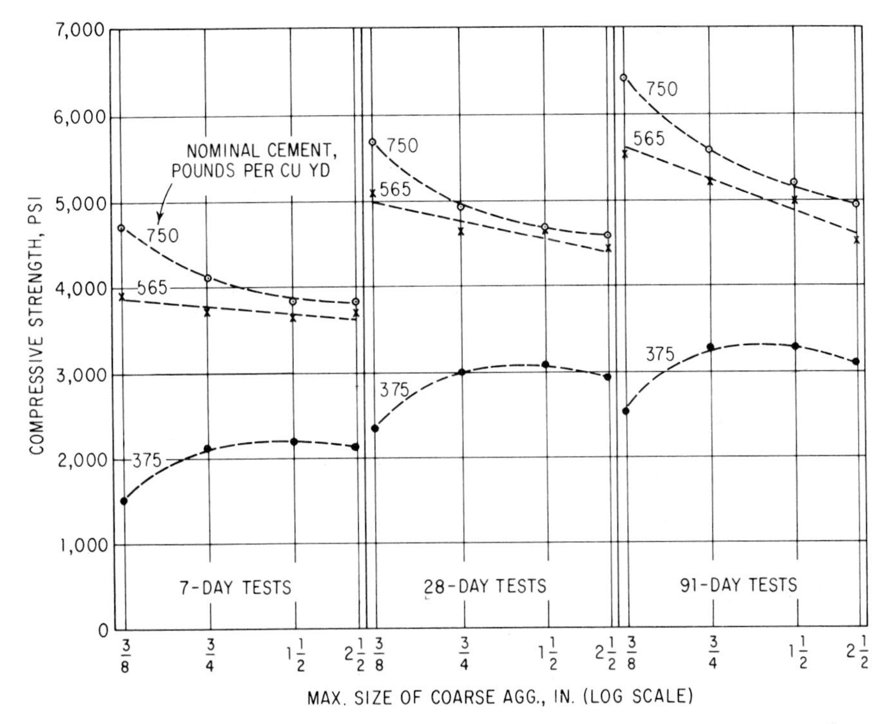

FIGURE 11.1 Effect of size of coarse aggregate on compressive strength of air-entrained concrete.[5]

concrete mixtures requires special consideration in proportioning, because of the volume aspects and the interrelationships with the other ingredients. When compared with non-air-entrained concrete mixtures, the additional volume produced by the entrained air is compensated for by both a reduction in water content and a decrease in sand quantity. Air entrainment will increase strengths of lean mixes, which may allow a slight reduction of cement. However, in rich mixes, that is, those over 600 lb of cement per cubic yard, there is usually a significant strength reduction. This requires the use of additional cement or proper water-reducing admixtures.

The amount of air entrained by a given agent or air-entraining cement is affected by many factors. Generally the air content is increased by an increase in slump, water/cement ratio, or sand content. An increase in the amount of sand size between the No. 30 and No. 50 sieve increases the air entrainment. It is decreased by an increase of fines in sand, higher cement content, higher temperature, longer mixing time, and the addition of pozzolans. The reduction in entrained air that accompanies addition of fly ash is commonly attributed to the carbon content of the fly ash and amount retained on the No. 325 sieve.

The optimum amount of air entrainment for a given concrete mixture is governed by the applicable specification. The amount will vary with the maximum-size aggregate and for concrete in buildings will generally follow the limits given in Table 11.1. Slabs on ground, such as highway pavements, are often specified to have 6 to 8 percent entrained air, regardless of maximum size of coarse aggregate.

Although the air contents are given in relation to aggregate sizes, the desirable amount is actually dependent directly on the volume of air in the cement-sand mortar part of the

TABLE 11.1 Concrete Air Content for Various Sizes of Coarse Aggregate

Nominal maximum size of coarse aggregate, in	Total air content, % by volume	
	Severe	Moderate
⅜	7.5	6.0
½	7.0	5.5
¾	6.0	5.0
1	6.0	4.5
1½	5.5	4.5
2	5.0	4.0
3	4.5	3.5

Source: *Table 4.1.1, ACI Standard 318-89.[7]

concrete. The optimum appears to be 9 percent in this mortar at all times. This is therefore reflected in the various volumes because of the reduced mortar content in the concrete with larger-sized aggregates.

The use of high-range water reducers (superplasticizers) has generated a great deal of research in the quality and quantity of air entrainment from normal dosages of air-entraining agents. Although results have been somewhat contradictory, it appears that these new admixtures do not seriously affect the benefit of air-entraining admixtures. Concretes having strengths over 5000 psi may have a 1 percent reduction in air entrainment required by ACI 318-89.[7]

11.5 Structural Requirements

The basic factors affecting proportioning are strength and durability. The concrete must be proportioned to develop sufficient strength to sustain adequately the loads to be imposed on it in service. The specific strength of the concrete mixture is established by the designer and is usually predicated on the allowable stresses used in the structural analysis. For adequate durability, the ACI Building Code[7] provides that normal-weight-aggregate concrete subject to freezing and thawing while wet shall be properly air-entrained and shall have a cement content of 520 lb/yd^3. When lightweight-aggregate concrete is used, the compressive strength f_c' shall be 3000 psi minimum. It has been established that durability is a function of both air entrainment and water/cement ratio.

Concrete used under various degrees of exposure should meet minimum requirements as given in ACI Standards on proportioning concrete mixtures. This requires the use of the proper type of cement as shown in Chap. 1, and the use of proper water/cement ratios shown in Table 11.2.

The ACI Standard 318-89 Building Code covers other specific conditions. When normal-weight-aggregate concrete is intended to be watertight, it should have a maximum water/cement ratio of 0.50 for exposure to fresh water and 0.40 for exposure to seawater. With lightweight-aggregate concrete, the specified compressive strength f_c' should be at least 3750 psi for exposure to fresh water and 4250 psi for exposure to seawater with an extra 0.5 in of concrete cover. Concrete that will be exposed to injurious concentrations of sulfate-containing solutions should conform to above-watertight concrete quality requirements and should be made with sulfate-resisting cement as noted in Table 11.3.

TABLE 11.2 Maximum Permissible Water/Cement Ratios for Concrete in Different Exposures

Exposure condition	Maximum water/cement ratio normal-weight-aggregate concrete	Minimum f_c', lightweight-aggregate concrete
Concrete intended to have low permeability when exposed to water	0.50	3750
Concrete exposed to freezing and thawing in a moist condition	0.45	4250
For corrosion protection for reinforced concrete exposed to de-icing salts, brackish water, seawater, or spray from these sources	0.40*	4750*

*If minimum allowable cover is increased by 0.5 in, the water/cement ratio may be increased to 0.45 for normal-weight concrete, or f_c' reduced to 4250 psi for lightweight concrete.

Source: Table 4.1.2, ACI Standard 318-89.[7]

TABLE 11.3 Requirements for Concrete Exposed to Sulfate-Containing Solutions

Sulfate exposure	Water-soluble sulfates (SO_4) in % by wt	Sulfate (SO_4) in water, ppm	Cement type	Normal-weight-aggregate concrete	Light weight-aggregate concrete
				Maximum water/cement ratio by wt*	Minimum compressive strength f_c', psi*
Negligible	0.00–0.10	0–150	—	—	—
Moderate†	0.10–0.20	150–1500	II, IP(MS), IS(MS) P(MS), I(PM)(MS), I(SM)(MS)	0.50	3750
Severe	0.20–2.00	1500–10,000	V	0.45	4250
Very severe	>2.00	>10,000	V plus pozzolan‡	0.45	4250

*A lower water/cement ratio or higher strength may be required for low permeability or for protection against corrosion of embedded items or freezing and thawing.

†Seawater.

‡Pozzolan that has been determined by test or service record to improve sulfate resistance when used in concrete containing Type V cement.

Source: Table 4.2.1, ACI Standard 318-89.[7]

MATERIALS

11.6 Cement

Several types of cement are commonly available to meet specific needs of various projects as described in Chap. 1. The most commonly used cement is Type I. Types II, IV, or V are specified to cover the need for special properties such as sulfate resistance and low heat development. The 28-day strength of concretes made with these cements is normally lower than that of concretes made with equal contents of Type I, but strengths at later ages will equal those of Type I. Type III, high-early-strength cement, is usually made to achieve in 7 days the comparable 28-day compressive strength of Type I cement.

Air-Entraining Cement. This cement contains an interground air-entraining agent, and is used for air-entrained concrete in lieu of a separate air-entraining agent added at the mixer. Adjustments in proportions to accommodate the air entrainment must be made, as shown later. Air-entraining cements are designated as Types IA, IIA, IIIA, IVA, VA, or ISA.

Cement Variations. It is recognized that strength-development capabilities of a given cement type may vary appreciably from plant to plant, and in some cases from shipment to shipment within the same plant. It is therefore necessary to use in trial batches for establishing design the brand of cement intended for the project, obtained from current production of the specific plant that will furnish the project.

In recent years shrinkage-compensating cement has been used to minimize the cracking caused by drying-induced shrinkage in concrete slabs, pavements, and structures. The cement is constituted so that the concrete volume will increase after setting and during hardening. The cement may be blended with the expansive additive during grinding at the mill, or the additive may be added at the jobsite under proper supervision. Recommended methods for use of this cement may be found in ACI Standard Practice for the Use of Shrinkage-Compensating Concrete (ACI 223-83).

11.7 Aggregates

Coarse Aggregates. The most important properties of coarse aggregates that affect the strength and proportions of a concrete mixture are grading, size, shape, and surface texture. These properties vary with the various conventional coarse aggregates available. The conventional aggregates are primarily gravels, crushed stone, and blast-furnace slag. (Open-hearth slags are, in general, considered to be unsatisfactory as concrete aggregates.) The gravels are available in natural rounded form, crushed, or a combination of both. The crushed stones are angular in shape and usually have significantly different surface textures from gravels. The crushed air-cooled blast-furnace slag is similar in some respects in angularity to the crushed stones. In addition it has a vesicular quality and texture that affect proportioning. The unit weight of air-cooled blast-furnace slag is usually less than that of other aggregates, and slag mixes will therefore appear quite different. In all cases, the method of determining proportions given here will automatically make allowance for the different particle shapes by use of the fixed dry-rodded volume method of ACI Standard 211.1-91[1]. The strength-development capabilities as affected by texture, structural strength, and water demand of each aggregate will appear in the trial-mix results. Aggregates consisting primarily of particles having rough texture or many flat or elongated pieces require different mix proportions from those consisting of relatively smooth, rounded, or cubical particles. These qualities will require greater proportions of sand, which requirements will develop in the calculations of mix design by the dry-rodded volume method. Grading of coarse aggregates will also be reflected in the fixed dry-rodded volume method by the void content. The variations in grading of coarse aggregates within different maximum-size limits are not reflected; however, this omission is considered of little practical importance when limits fall within conventional grading specifications.

Fine Aggregates. Fineness modulus is a convenient method of expressing the overall grading of the sand. It is a single figure determined by adding together the cumulative percentages retained on sieves Nos. 4, 8, 16, 30, 50, and 100, and dividing the total by 100. This fineness modulus of the sand is used in determination of dry-rodded volume for mix designs in ACI Standard 211.1-91. Permissible limits of fineness are given in ASTM C33, limiting the fineness modulus between 2.3 and 3.1. Generally, the ratio of sand to coarse aggregate in the mix is lower when the sand is fine than when it is coarse. A grading of sand in which one or two particle sizes predominate should be avoided. This condition would result in a large void content and would therefore require a greater amount of cement/water paste to produce a workable mixture. The greatest workability is achieved when the individual sizes of a given sand form a smooth curve within the grading limits. Compliance with the numerical limits of some specifications does not always result in a smooth curve.

Manufactured Sand. Sands produced by crushing rock are often made up of particles having rough and angular surfaces. When this quality is coupled with flat and elongated shapes, it will produce concrete mixtures that are harsh and not so workable as concrete made with rounded sands, in which case it is necessary to increase the proportion of sand and a greater content of cement paste and water content then are generally necessary. Some manufactured sands, because of their angularity and toughness, produce greater concrete compressive strength for a given cement factor, even with higher water content, than natural sand. Not all sands manufactured by crushing rock are "harsh." Some, produced with modern equipment in modern plants, behave almost the same as natural sands.

Lightweight Aggregates. There are many varieties and types of lightweight aggregates, each of which may behave differently in concrete from the others. Many are angular and have rough surface textures producing harsh mixes. (Air entrainment helps tremendously to overcome this tendency.) Natural lightweight aggregates include diatomite, pumice, scoria, volcanic cinders, and tuff; however, these are not widely used except in a few areas where they are easily available. Pumice, the most widely used of these, is a froth-like volcanic glass with a bulk density of 30 to 55 pcf.

The manufactured aggregates are marketed under a variety of names. The most common and structurally superior are expanded clay, shale, slate, and specific types of sintered fly ash. These produce concrete of excellent strength characteristics. Some are manufactured by a process that forms spherical particles with an impervious surface, in effect lowering absorption and improving workability. Another type commonly used is expanded blast-furnace slag and industrial cinders. These are more vesicular and usually are softer and more absorptive and thus result in lower compressive strength and higher shrinkage than the manufactured type. Perlite and vermiculite are much lighter in weight and consequently produce low strength and high shrinkage. They are seldom used for purposes other than insulation.

Heavy Aggregates. These include magnetite and limonite iron ores, barites, ferrophosphorus, steel punchings, steel shot, and lead. Specifications covering quality of these materials is contained in ASTM C637, High Density Aggregate. (See also Sec. 11.30, below.) Their use in concrete is primarily for nuclear or xray shielding, or in high-density concrete for such usage as movable-bridge counterweights.

The high density and angularity of the natural ores usually produce harsh mixes. These ores tend to dust, have high absorption, and may break down during handling and mixing. This complicates mix design, but satisfactory concrete can be made with densities of about 200 to 300 pcf. The artificial heavy aggregates such as steel punchings, shot, and lead produce concrete with a density as high as 350 pcf, but with a tendency to segregation during handling and placing. This is due to the considerably higher density of the metals as compared with that of the fine aggregates and cement used in the mixtures. The steel must be free of oil to develop proper bond. It is advisable to corrode the surfaces prior to use or to acid-etch the steel for improved bond.

Although boron frit is not a heavy aggregate, mention is made of it here because of its use in shielding concrete. It is efficient in the absorption of thermal neutrons when the boron content is about ½ percent by weight of the concrete. Care must be exercised in the type of boron frits that are used. A type of low solubility is necessary to prevent retarding effects on the cement paste.

11.8 Water

Water used in manufacturing concrete, including that free water on the aggregates, should be clean and free from injurious amounts of oils, acids, alkalis, salts, organic materials, or other substances that may be deleterious to concrete or steel. Also, mixing water for prestressed concrete or for concrete which will contain aluminum embedments should not contain deleterious amounts of chloride ion. See also Chap. 4.

11.9 Admixtures

Air-Entraining Admixtures. These admixtures can be interground in portland cement to achieve air-entraining cement or can be added independently to concrete mixtures. The effects for specific limits are the same. Air entrainment is one of the most important single developments in concrete technology since the invention of portland cement. Its primary purpose is to increase durability, which it does manyfold. It also improves workability. There are also side effects on strength, water requirements, and sand requirements, all of which are taken into consideration in the design of the concrete mix.

Accelerators. Accelerating admixtures are used, as is indicated by the name, to speed hardening and strength gain of concrete. They are used primarily in winter concreting to permit early finishing and to hasten the progress of cement hydration to the point that frost damage will not occur. They *cannot* be considered to be antifreeze mixtures. In permissible quantities they cannot lower the freezing point of the concrete more than 1 or 2°. They *can* and *do* hasten the set and strength gain to the point that the fresh concrete is susceptible to frost damage for a shorter length of time.

The commonest accelerator is calcium chloride, and this is the active constituent of most of the accelerators sold under many trade names. Some of the accelerators sold under various trade names contain alkali silicates and carbonates, fluosilicates, and triethanolamine.

It is recommended that calcium chloride content not exceed 2 percent by weight of cement. It is most effective in rich mixes in accelerating setting time and strength development. Calcium chloride generally does not affect unit water content; however, it may reduce air entrainment and thus require adjustment of the amount of air-entraining agent used. Some proprietary water-reducing admixtures contain an accelerator to counteract the retarding action of the admixture.

Field experience has indicated the possibility of corrosion of reinforcement in concrete whenever moisture is present. Limits for chloride amounts are listed in Table 11.4.

Retarders and Water Reducers. The common retarders available are generally a form of lignosulfonic acid or hydroxylated carboxylic acid. Both types retard the setting time about 1 to 4 h over normal setting time but do not affect normal strength gain after set has occurred. The lignosulfonic acid types tend to entrain some air and must be modified with air-entraining agents to meet levels of air entrainment for durability. Exclusive of the air entrained, normal amounts of these retarders will reduce the water requirement about 8 to 25 lb/yd^3 and increase compressive strength about 10 to 15 percent. Adjustments must be made in mix proportions to accommodate these effects.

Pozzolans. The pozzolans, such as fly ash, diamataceous earth, and silica fume, develop cementitious properties in concrete mixtures by their reaction with calcium hydroxide and water. Some pozzolans also inhibit expansion resulting from alkali–aggregate reaction. In general, they increase workability of lean or harsh mixtures.

Pozzolans are sometimes used as a replacement of a part of the portland cement. Such replacement seldom exceeds 25 percent of the cement content. In cases of replacement, strength gain of the concrete will usually be delayed somewhat, with 90-day strengths of the cement-pozzolan mixes probably being about equal to the 28-day strengths of straight portland cement mixes. When pozzolans are used as an addition to the portland cement, they frequently increase the 28-day concrete strength and will almost always increase the strengths at later ages. Some pozzolans will also affect air entrainment and will require adjustment in quantity of agents. The specific gravity of most common pozzolans is usually less than that of conventional aggregates and cement, often falling between 2.35 and 2.50. Specific gravity must be determined for proper proportioning of concrete mixtures.

TABLE 11.4 Maximum Chloride Content for Corrosion Protection

Type of member	Maximum water-soluble chloride ion (Cl^-) in concrete, percent by weight of cement*
Prestressed concrete	0.06
Reinforced concrete exposed to chloride in service	0.15
Reinforced concrete that will be dry or protected from moisture in service	1.00
Other reinforced-concrete construction	0.30

*Measured in hardened concrete at ages from 28 to 42 days when tested in conformance with AASHTO T260.

Source: Table 4.3.1, ACI 318-89.[7]

Integral Waterproofers and Dampproofers. Many of the admixtures sold specifically for this purpose are stearates or oleates, although some probably contain other chemicals. The effect or value of any should be corroborated by trial concrete mixes to determine effect on compressive strength, durability, shrinkage, and other properties of the concrete, as well as waterproofing.

Mix proportioning to adjust for a change in absolute volume of concrete is generally not necessary unless the admixture contains an air-entraining agent. Air entrainment in itself will generally reduce absorption and permeability. Most of the proprietary types which do not include an air-entraining agent do not have a significant effect on water content or compressive strength when used in normal quantities.

Other Admixtures. Other admixtures are used for a wide variety of purposes and have widely different effects on concrete mixtures. These include materials designated as air-detraining, gas-forming, expansion producers, bonding, chemicals to reduce cement-aggregate expansion, corrosion-inhibiting, fungicidal, germicidal, insecticidal, flocculating, and coloring admixtures. The individual effect on concrete properties should be verified by concrete mixes and tests. Predicted effect on proportioning can be made prior to mixing and refinements made on subsequent trial mixes.

SELECTING MIX PROPORTIONS

Numerous tables have been prepared and presented in various publications of the American Concrete Institute which are of valuable assistance in the planning of proportions and quantities of materials for the production of concrete to meet certain requirements. If no advance laboratory checks of designs are planned, the recommendations of these tables may be combined to produce a calculated design which will be on the side of safety in both strength and durability. In many cases preliminary laboratory designs, based fundamentally on the data of these tables, but with the laboratory's knowledge of local materials, may result in appreciable economies with even greater margins of safety.

11.10 Water/Cement Ratio Determinations

The water/cement ratio may be established numerically by definite project specifications. An alternative found in some specifications is specific reference to ACI Standard 211.1 [Table 6.3.4(b)]. In this table the water/cement ratio is considered primarily from the standpoint of durability under various exposure conditions. In addition to this standard, ACI Building Code Requirements for Reinforced Concrete, ACI 318-89, states that the water/cement ratio shall in no case exceed 0.45 for concrete exposed to freezing while wet. In addition to the establishing of water/cement ratios for exposure conditions, the other principal criterion used to establish a water/cement ratio is the compressive strength at 28 days f_c'. The ACI Building Code gives allowable water/cement ratios for specific strengths of concrete mixtures when preliminary tests are not made, as shown in Table 11.5.

It must be noted that the ACI Building Code does not cover, in Table 11.5, strengths in excess of 4500 psi air-entrained or 5000 psi non-air-entrained, or lightweight concrete. Under these conditions the proposed concrete mixture must be established by previous reliable tests or by a water/cement ratio curve using at least three different water/cement ratios (or cement contents in the case of lightweight concrete).

Where the concrete production facility has a record based on at least 30 consecutive strength tests[8] representing materials similar to those expected, the proportions should be

TABLE 11.5 Maximum Permissible Water/Cement Ratios for Concrete*

Specified compressive strength f_c', psi*	Maximum permissible water/cement ratio			
	Non-air-entrained concrete		Air-entrained concrete	
	Absolute ratio by weight	U.S. gal per 94-lb bag of cement	Absolute ratio by weight	U.S. gal per 94-lb bag of cement
2500	0.67	7.3	0.54	6.1
3000	0.58	6.6	0.46	5.2
3500	0.51	5.8	0.40	4.5
4000	0.44	5.0	0.35	4.0
4500	0.38	4.3	—‡	—‡
5000	—‡	—‡	—‡	—‡

*When strength data from trial batches or field experience are not available.

†28-day strengths. With most materials, the water/cement ratios shown will provide average strengths greater than indicated in Table 11.6 as being required.

‡For strengths above 4500 psi with air-entrained concrete, proportions should be selected by the methods of paragraph 10.

Source: Table 5.4, ACI Standard 318-89.

TABLE 11.6 Relationships between Water/Cement Ratio and Compressive Strength of Concrete

Compressive strength at 28 days, psi*	Water/cement ratio, by weight	
	Non-air-entrained concrete	Air-entrained concrete
6000	0.41	—
5000	0.48	0.40
4000	0.57	0.48
3000	0.68	0.59
2000	0.82	0.74

*Values are estimated average strengths for concrete containing not more than 2 percent air for nonentrained concrete and 6 percent total air content for air-entrained concrete. For a constant water/cement ratio, the strength of concrete is reduced as the air content is increased. Strength is based on 6 × 12-in cylinders moist-cured 28 days at 73.4 ± 3°F (23 ± 1.7°C) in accordance with Sec. 9 of ASTM C31. Relationship assumes nominal maximum size of aggregate about ¾ to 1 in, for a given source of aggregate, strength produced at a given water/cement ratio will increase as nominal maximum size of aggregate decreases.

Source: Table 6.3.4(a), ACI Standard 211.1-91.[1]

selected to produce an average strength at the designated test age which is the larger of the two calculated by the following:

$$f'_{cr} = f'_c + 1.34s \quad \text{or} \quad f'_{cr} = f'_c + 2.33s - 500$$

where $s = \left[\dfrac{(X_i - X)^2}{(n-1)} \right]^{1/2}$

where s = standard deviation, psi
X_i = individual strength tests which are an average of two cylinders tested at the specified test age
X = average of n strength test results
n = number of consecutive strength tests
f'_c = specified compressive strength of concrete, psi
f'_{cr} = average compressive strength for determining proportions

TABLE 11.7 Adjustments to Water Content for Changed Conditions

Changes in conditions	Effect on unit water content, %
Each 1-in increase in slump.	±3
Each 1% increase or decrease in air content	±3
Each 1% increase or decrease in sand content	±1

Source: Adapted from "Concrete Manual," 8th ed., Table 14, U.S. Bureau of Reclamation, 1975.[5]

11.11 Estimate of Total Water

The quantity of water required for a cubic yard of concrete to provide a certain slump within the fine-aggregate–coarse-aggregate proportions given here will depend primarily on the size, shape, and grading of the aggregates and amount of entrained air. The quantity of cement will generally not affect water demand for normal structural-concrete mixes. Table 11.5 will provide an adequately accurate unit water content for preliminary designs.

The selection of slump should be based on the recommendations given by ACI Committee 301. Unless otherwise permitted or specified, the concrete shall be proportioned and produced to have a slump of 4 in or less if consolidation is to be by vibration, and 5 in or less if consolidation is to be by methods other than vibration. A tolerance of up to 1 in above the indicated maximum shall be allowed for individual batches, provided the average for all batches or the most recent 10 batches tested, whichever is fewer, does not exceed the maximum limit. Concrete of lower than usual slump may be used provided it is properly placed and consolidated. The slump shall be determined by the Standard Test Method for Slump of Hydraulic Cement Concrete (ASTM C143). Recommendations to cover special situations and varying from those of the table include a suggestion that the slump of concrete made with lightweight aggregate never exceed 3 in and that concrete for steep ramps, folded plates, or similar sloping surface may, when necessary, have a slump of less than 1 in. Other conditions which will affect total water requirements are given in Table 11.7.

11.12 Cement-Factor Determination

The cement content is determined by utilizing the water/cement ratio specified, or it is taken from Table 11.2 for exposure conditions or Table 11.3 for compressive-strength requirements, whichever is lower.

TABLE 11.8 Approximate Mixing Water and Air Content Requirements for Different Slumps and Maximum Sizes of Aggregate

	Water, lb/yd^3 of concrete for indicated nominal maximum sizes of aggregate*					
Slump, in	⅜ in	½ in	¾ in	1 in	1½ in	2 in
			Nonentrained concrete			
1–2	350	335	315	300	275	260
3–4	385	365	340	325	300	285
6–7	410	385	360	340	300	270
Approx. amount of entrapped air in non-entrained concrete, %	3	2.5	2	1.5	1	0.5
			Air-entrained concrete			
1–2	305	295	280	270	250	240
3–4	340	325	305	295	275	265
6–7	365	345	325	310	290	280
Recommended average total air content, % for level of exposure						
Mild exposure	4.5	4.0	3.5	3.0	2.5	2.0
Moderate exposure	6.0	5.5	5.0	4.5	4.5	4.0
Severe exposure†	7.5	7.0	6.0	6.0	5.5	5.0

*The quantities of mixing water given for air-entrained concretes are based on typical total air content requirements as shown for "moderate exposure in this table. These quantities of mixing water are for use in computing cement contents for trial batches. They are maximum for angular aggregates having acceptable gradings within ASTM specifications. Well-rounded aggregates will require approximately 30 lb less water for non-air-entrained concrete and approximately 25 lb less water for air-entrained concrete. Water-reducing admixtures will reduce water requirements by about 5 percent or more and the admixture volume should be included in the total volume of water.

†These values are based on the premise that 9 percent air is required in the concrete mortar.

Source: Table 6.3.3, ACI Standard 211.1-91.

The unit water content given in Table 11.8 selected for the given conditions is divided by the water/cement ratio to produce the cement factor in pounds per cubic yard.

11.13 Aggregate-Quantity Determination

The quantity of coarse aggregate is determined by the use of Table 11.9. The maximum-size aggregate contemplated and the fineness modulus of the sand must be first determined. The volumes from this table are taken from an empirical relationship and will automatically provide for different types of aggregates by making allowances for differences in mortar contents as reflected by the void content. These values are so designed to provide for variations in gradations within acceptable standards. This frequently results in greater than necessary quantities of sand. Therefore, prior knowledge of the character of aggregate may warrant an increase in coarse aggregate, usually about 10 percent higher than the tabular value.

TABLE 11.9 Volume, of Coarse Aggregate per Unit of Volume of Concrete

Maximum of aggregate, in	Volume of dry-rodded coarse aggregate* per unit volume of concrete for different fineness moduli of sand			
	2.40	2.60	2.80	3.00
⅜	0.50	0.48	0.46	0.44
½	0.59	0.57	0.55	0.53
¾	0.66	0.64	0.62	0.60
1	0.71	0.69	0.67	0.65
1½	0.75	0.73	0.71	0.69
2	0.78	0.76	0.74	0.72
3	0.82	0.80	0.78	0.76
6	0.87	0.85	0.83	0.81

*Volumes are based on aggregates in dry-rodded condition as described in ASTM C29. These volumes are selected from empirical relationships to produce concrete with a degree of workability suitable for usual reinforced construction. For less workable concrete such as required for concrete pavement construction they may be increased by about 10 percent. For more workable concrete, such as may sometimes be required when placement is to be by pumping or when concrete must be worked around congested reinforcing steel, they may be reduced by up to 10 percent.

Source: Table 6.3.6, ACI Standard 211.1-91.

11.14 Absolute-volume Computations

The final determination of the proportions of each material is made by the absolute-volume method. This is shown in the subsequent examples. In order to utilize ACI Standard 211.1–91 the following data must be established: (1) gradation, (2) specific gravity, and (3) dry-rodded unit weight of the coarse aggregate.

The specific gravity of the cement can be taken as 3.15 with reasonable accuracy. The fineness modulus of the sand is computed from the gradation. The absorption and total moisture content of each aggregate must be established in order to compute the batch weights for trial mixes and field use.

11.15 Estimated Unit Weight Basis

The ACI 211 Standard Practice for Selecting Proportions for Normal Weight, Heavy-weight and Mass Concrete describes an alternate method to that of the absolute-volume method. It is based on an estimated weight of the concrete per unit volume. The selection of water/cement ratio, estimation of mixing water and air content, calculation of cement content, and estimation of coarse-aggregate content are the same as in the absolute-volume basis. The fine-aggregate content may be estimated by the weight method. The unit weight of concrete is obtained from Table 11.10 or can be estimated from experience. The required weight of fine aggregate is the difference between the weight of fresh concrete and the total weight of the other ingredients.

TABLE 11.10 First Estimate of Weight of Fresh Concrete

Maximum size of aggregate, in	First estimate of concrete weight, lb/yd³*	
	Non-air-entrained concrete	Air-entrained concrete
⅜	3840	3710
½	3890	3760
¾	3960	3840
1	4010	3850
1½	4070	3910
2	4120	3950
3	4160	4040
6	4260	4110

*Values calculated for concrete of medium richness (550 lb of cement/yd³) and medium slump with aggregate specific gravity of 2.7. Water requirements based on values for 3- to 4-in. slump in Table 11.8. If desired, the estimated weight may be refined as follows if necessary information is available: for each 10-lb difference in mixing water from the Table 11.8 values for 3- to 4-in slump, correct the weight per cubic yard 15 lb in the opposite direction; for each 100-lb difference in cement content from 550 lb, correct the weight per cubic yard 15 lb in the same direction; for each 0.1 by which aggregate specific gravity deviates from 2.7, correct the concrete weight 100 lb in the same direction; For air-entrained concrete the air content for severe exposure from Table 11.8 was used. The weight can be increased 1 percent for each percent reduction in air content from that amount.

Source: Table 6.3.7.1, ACI 211.1-91.

11.16 Examples

The use of typical examples will best explain the calculation of proportions by the design criteria of ACI Standard 211.1-91.

Non-Air-Entraining. Required: Mix proportions of concrete of the following characteristics:

1. Not exposed to weather or sulfate water.
2. Reinforced walls 6 to 11 in thick with minimum clear reinforcing spacing of 2 in.
3. Concrete design strength f'_c is 4000 psi.

Laboratory data given:

1. Dry-rodded unit weight of the angular coarse aggregate is 100 pcf.
2. Fineness modulus of sand is 2.60.
3. Saturated surface-dry bulk specific gravity of both fine and coarse aggregate is 2.65.
4. Absorption of fine aggregate is 1.0 percent, and that of coarse aggregate is 0.5 percent.

Computation of proportions:

1. Since exposure is not critical and non-air-entrained concrete is specified, the water/cement ratio will be established for compressive strength only.

2. The compressive-strength requirement will be 4000 psi plus 1200 psi because standard deviation is not known. This is 5200 psi. From Table 11.6, the water/cement ratio required to produce 5200 psi in non-air-entrained concrete is 0.47.
3. From Sec. 11.11, the slump shall be 4 in. Therefore, 4 in is used in design.
4. From the spacing of reinforcement, maximum-size aggregate is 1½ in.
5. From Table 11.8, the approximate mixing water for 4-in slump under test conditions is 300 lb/yd^3.
6. From the information in items 2 and 5, the required cement content is 300/0.47 = 638 lb/yd^3.
7. From Table 11.9, the volume of coarse aggregate per unit volume of concrete for 1½-in aggregate and sand with a fineness modulus of 2.60 is 0.73. For a cubic yard the quantity of coarse aggregate will be 0.73 × 27 = 19.7 ft^3. For aggregate of 100 pcf, the dry weight is 1970 lb/yd^3. The saturated surface-dry (SSD) weight is the dry weight multiplied by 1 plus absorption of the aggregate: 1970 × 1.005 = 1980 lb/yd^3.
8. The sand content per cubic yard is determined by calculating the solid volumes of all other ingredients and subtracting this quantity from the total volume of 1 yd^3. The calculations are usually conducted in terms of cubic feet as in the following:

Solid volume of cement:

$$\frac{\text{wt. of cement}}{\text{sp gr of cement} \times 62.4} = \frac{638}{3.15 \times 62.4} = 3.26 \text{ ft}^3$$

Volume of water:

$$\frac{\text{lb used}}{\text{lb/cu ft}} = \frac{300}{62.4} = 4.80 \text{ ft}^3$$

Solid volume of coarse aggregate:

$$\frac{\text{wt of coarse aggregate}}{\text{sp gr} \times 62.4} = \frac{1980}{2.65 \times 62.4} = 11.95 \text{ ft}^3$$

Volume of air (estimated 1% entrapped):

$$0.01 \times 27 \text{ ft}^3 = 0.01 \times 27 = 0.27 \text{ ft}^3$$

Total solid volume, all ingredients except sand = 20.28 ft^3

Solid volume of sand = 27 − 20.28 = 6.72 ft^3

Required weight of sand:

$$\text{Volume} \times (\text{sp gr} \times 62.4) = 6.72 \times (2.65 \times 62.4) = 1110 \text{ lb}$$

9. From the above trial batch, quantities per cubic yard of concrete are

Cement	= 638 lb
Water	= 300 lb
Sand (SSD basis)	= 1110 lb
Coarse aggregate (SSD basis)	= 1980 lb
Total	= 4028 lb/yd^3

$$\text{Calculated unit weight of plastic concrete} = \frac{4028 \text{ lb}}{27 \text{ ft/yd}} = 149.0 \text{ pcf}$$

Air-Entraining. Required: Mix proportions of concrete of the following characteristics:

1. Exposed moderate sulfate water and severe freezing and thawing.
2. Reinforced walls 12 to 24 in thick with minimum clear reinforcing-bar spacing of 3 in.
3. Concrete design strength f_c' is 3000 psi.

Laboratory data given:

1. Dry-rodded unit weight of the angular coarse aggregate is 100 pcf.
2. Fineness modulus of sand is 2.60.
3. Saturated surface-dry bulk specific gravity of both fine and coarse aggregate is 2.65.
4. Absorption of fine aggregate is 1.0 percent, and that of coarse aggregate is 0.5 percent.

Computation of proportions:

1. From Table 11.2, moderate section in severe freezing and thawing requires a water/cement ratio of 0.45. Type II moderate sulfate-resisting cement will be used. Air-entrained concrete will be required.
2. The compressive-strength requirement will be 4200 psi (3000 psi plus 1200 psi because standard deviation is not known). From Table 11.6, the water-cement ratio required to produce 4200-psi air-entrained concrete is 0.46. However, the water/cement ratio based on exposure (from Table 11.2), being lower, must govern. Therefore, the trial batch will be calculated at 0.45.
3. From Sec. 11.11, the maximum slump shall be 4 in, and therefore 4 in is used in the design.
4. Based on clear spacing, the maximum-size aggregate shall be 2¼ in. Since only 1½- and 3-in sizes are available, 1½-in shall be used.
5. From Table 11.8, the approximate mixing water for 4-in slump under these conditions is 275 lb/yd^3. In this table the desired air content is indicated as 5.5 percent. Therefore 5.5 percent shall be used in the design.
6. From the information in items 1 and 5, the required cement content is 275/0.45 = 612 lb/yd^3.
7. From Table 11.9, the volume of coarse aggregate per unit volume of concrete for 1½-in aggregate and sand with a fineness modulus of 2.60 is 0.73. For 1 yd^3 the quantity of coarse aggregate will be $0.73 \times 27 = 19.7$ ft^3. For coarse aggregate of 100 pcf, the dry weight is 1970 lb/yd^3. The saturated surface-dry weight is the dry weight multiplied by 1 plus absorption of the aggregate: $1970 \times 1.005 = 1980$ lb/yd.^{3}
8. The sand content is determined by the following absolute-volume method:

   Solid volume of cement:

$$\frac{\text{wt of cement}}{\text{sp gr of cement} \times 62.4} = \frac{612}{3.15 \times 62.4} = 3.12 \text{ ft}^3$$

   Volume of water:

$$\frac{\text{lb used}}{\text{lb/cu ft}} = \frac{275}{62.4} = 4.40 \text{ ft}^3$$

Solid volume of coarse aggregate:

$$\frac{\text{wt of coarse aggregate}}{\text{sp gr} \times 62.4} = \frac{1980}{2.65 \times 62.4} = 11.95$$

Volume of air (estimated 5.5% entrained):

$$0.055 \times 27 = 1.48$$

$$\text{Total solid volume, all ingredients except sand} = 20.95 \text{ ft}^3$$

$$\text{Solid volume of sand } 27 - 20.95 = 6.05 \text{ ft}^3$$

Required weight of sand:

$$\text{Volume} \times (\text{sp gr} \times 62.4) = 6.05 \times (2.65 \times 62.4) = 1000 \text{ lb}$$

9. The estimated batch quantities per cubic yard of concrete are

Type II cement = 612 lb
Water (31 gal) = 275 lb
Sand (SSD basis) = 1000 lb
Coarse aggregate (SSD basis) = 1980 lb
Total = 3867 lb

Air entrainment is obtained by use of air-entraining agent added in accordance with recommendations of the manufacturer. Adjustment should be made as found necessary by the trial mix.

$$\text{Calculated unit weight of plastic concrete} = \frac{3867}{27} = 143.2 \text{ pcf}$$

MAKING TRIAL MIX

11.17 Procedure

After the mix proportions are determined by the designs as shown in the preceding examples, a trial mix is made. This is to check the assumptions and establish the effects of the variables on water requirement and air entrainment.

	Wt/yd^3, lb	Wt per 1/10 batch, lb	% moisture	Wt of water in aggregate	Corrected batch wt, lb
Type II cement	612	61.2			61.2
Water	275	27.5		(5.0)	22.5
Fine aggregate	1000	100.0	5.0	5.0	105.0
Coarse aggregate	1980	198.0	SSD	0	198.0
Total	3867	386.7		0	386.7
Approved air-entraining agent	5.6 fluid oz	0.56 oz or 17 cm^3			17 cm^3

Calculated volume of batch = 2.7 ft^3 at 142.0 pcf.

1. For our purposes we select a trial mix of sufficient capacity to establish a 7-day and 28-day compressive strength, each with four specimens. This will require a 1⁄10 cu yd batch as shown in the following tabulation for the second example ("Air-Entraining") in Sec. 11.16 (above).
2. In order to compensate for moisture in the aggregates, the amount of free moisture is determined. Free moisture is that water over the absorbed water in the aggregate which is available as mixing water. The method of test is given in Chap. 12. The coarse aggregate was found to be saturated surface dry and therefore contained only the absorbed water. The fine aggregate contained 5 percent free moisture. Therefore, the batch weight of the sand was increased to allow for this water and the mixing water was reduced accordingly. The calculations are

$$100 \times 0.05 = 5.0 \text{ lb}$$

This weight of water must be added to the design weight of saturated surface-dry fine aggregate.

$$100 + 5.0 = 105.0 \text{ lb}$$

3. The materials at room temperature (68 to 77°F) are batched for the trial mix as follows:
   a. Weigh aggregates cumulatively beginning with the smallest size on a scale accurate to 0.03 lb or ½ oz. A sufficient size of container to hold either a full batch or a half batch may be used with tare corrections made. The fine aggregate should be moist to avoid segregation when it is weighed.
   b. Weigh cement separately.
   c. Weigh a greater quantity of water than required. Add air-entraining agent to a portion of the water. Use for this portion about two-thirds of the total amount that will be required in the trial batch.
4. The mixing procedure is as follows:
   a. Capacity of the mixer should be equal to or slightly greater than the batch size.
   b. The mixer must be "buttered" by initially mixing a small amount of mortar of the same composition as the mortar of the batch. The interior of the mixer is thus coated. Any excess should be discharged.
   c. Place the coarse and fine aggregate in the mixer with about two-thirds of the expected water which includes all the air-entraining agent. Add the cement, and mix for about ½ min. Then add water from the measured amount until the concrete in the mixer appears to have attained the desired slump.
   d. Mix the batch for 3 min, rest for 3 min, and then remix for 2 min.
   e. Dump the concrete into a watertight and nonabsorbent receptacle large enough to hold the entire batch and to permit remixing with a shovel. (A clean damp floor may be used, but a large pan is preferred.) Remix to eliminate any segregation that may have occurred during discharge of the mixer.
   f. Determine the amount of water actually used by subtracting the weight of that remaining from the weight of the amount originally weighed.
5. The following tests must be made after the above remixing:
   a. Slump by ASTM Method C143.

b. Air content by ASTM Method C231 (Pressure Method) or Method ASTM C173 (Volumetric Method). Use Volumetric Method only on concrete made with porous aggregate.

c. Unit weight, yield by ASTM Method C138.

d. Strength specimens following procedures of ASTM Method C192.

e. Concrete used for air-content tests must be discarded. Concrete from other tests can be recombined and remixed with balance of concrete batch for making the strength specimens.

6. Concrete from the trial mix will, in practically all cases, vary from the anticipated conditions slightly because of the variables of water demand and air content and possible slight differences in specific gravities and test deviations. As an illustration, the following tabulation represents the actual results of the trial mix of the second ("Air-Entraining") example in Sec. 11.16:

Type II cement	61.2 lb
Water	27.9 lb
Fine aggregate	100.0 lb
Coarse aggregate	198.0 lb saturated surface-dry
Total	387.1 lb
Slump	4.0 in
Air content	5.5%
Unit weight	143.4 pcf
Yield	$38.1/143.4 = 2.70\ \text{ft}^3$

$$W/C = 27.9/61.2 = 0.46$$

However, this calculation and the higher water content required by the mix leave a water/cement ratio of 0.46, while the specification permits a maximum of only 0.45. The water cannot be reduced without adversely affecting the desired workability; therefore the cement must be increased by the factor of 0.46/0.45, which is 1.02. Thus the cement for the trial batch should be 1.02×61.2 lb, which is 62.4 lb. To verify the calculations, we check

$$W/C = 27.9/62.4 = 0.45$$

The increase of 1.2 lb of cement to the trial-batch weights in order to adjust the water/cement ratio has added $1.2/(3.15 \times 62.4) = 0.006\ \text{ft}^3$ of absolute volume to our trial-batch volume, giving a calculated volume of $2.706\ \text{ft}^3$. To correct for this, reduce the fine-aggregate content by $0.006\ \text{ft}^3$ absolute volume, which is $(2.65 \times 62.4) \times 0.006 = 1.0$ lb. Thus the final corrected trial batch weights would be

Type II cement	62.4 lb
Water	27.9 lb
Fine aggregate (SSD)	99.1 lb
Coarse aggregate (SSD)	200.2 lb
Total	389.6 lb

Theoretically a new trial batch should now be prepared using these weights, but inherent variations in concrete materials and test procedures are such that to do so would again give experimental results indicating that new corrections should be calculated, and still another trial batch be prepared, and so on, indefinitely.

It is normally considered sufficient to make the calculated corrections as shown and to use these corrected figures as the basis for the recommended field mix. The concrete test specimens made from this trial mix are almost certainly representative of the recommended field mix within the normal concrete test variations.

Thus the batch weights per cubic yard recommended to the field in this particular example would be

Ingredient	Quantity/cu yd
Type II cement	624 lb
Water	279 lb
Fine aggregate (SSD)	991 lb*
Coarse aggregate (SSD)	2,000 lb*
Air-entraining admixture	5.8 oz

11.18 Strength Curves for Trial Mixes

The water/cement ratio law can be utilized to determine mixes for specific compressive strengths by means of a series of trial mixes. This is required in the ACI Building Code for reinforced concrete when reliable test data is not available or when proportioning new materials of unknown properties.

The usual procedure is as follows. At least three and preferably four different trial mixtures are made with different cement contents and with constant slump. The total water per cubic yard will be practically the same for all mixes. This will result in three or four water/cement ratios spaced such that a range of compressive strengths will be achieved that encompasses those required for the project. The data obtained from each mix must include water/cement ratio, percent air, slump, and workability characteristics. The yield determinations will enable calculations of water, cement, fine aggregate, and coarse aggregate per cubic yard for each mix. This calculation will be the same as the foregoing trial-mix example. Of course, it is not necessary to calculate corrections, since the results are to be plotted at their actual points on the curve. The cylinders (and beams if required) must be cured in accordance with ASTM C192. Three cylinders should be tested at 7 days and three at 28 days. Beams should be tested as required by the specifications or in a manner similar to the cylinders.

A typical compressive-strength curve developed from four trial mixes is shown in Fig. 11.2. A water/cement ratio and corresponding mix design can be taken from this curve for a given set of materials. As an example of a method of preparing and using, see the curve and notes of Fig. 11.2, which gives the following results:

If other considerations of durability do not require a lower water/cement ratio, use 0.55. Use this water/cement value to interpolate quantities of materials from the values of the two nearest mixes plotted.

*These figures have been rounded off to normal batch-plant-scale graduations.

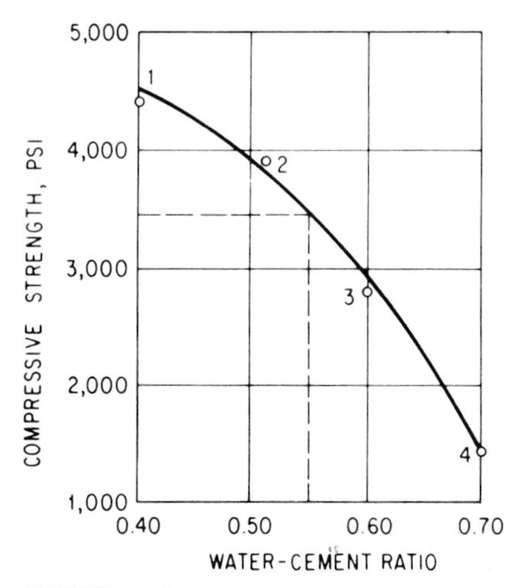

FIGURE 11.2 Trial-mix curve. For a described compressive strength of 3450 psi, enter the curve at 3450 psi, intersect the curve, then drop down and read 0.55.

	Mix 2	Mix 3	Design mix 3,450 psi
W/C	0.5	0.6	0.55
Water, lb	306	300	303
Cement, lb	612	500	556
Fine aggregate (SSD), lb	1120	1240	1180
Coarse aggregate (SSD), lb	1950	1950	1950

Compressive strengths are related to flexural strengths; however, the relationship is not exactly proportional. Characteristics of various materials affect each to different degrees. Therefore, trial mixes with beam tests are necessary to establish proportions for a given flexural-strength requirement. No attempt should ever be made to predict flexural strength from compressive-strength results.

11.19 Adjustments in Field

Laboratory-designed mixes should always be refined in the field. This is necessary because of the many variables that affect mix proportions. Temperature will affect slump and therefore will cause a change in the water/cement ratio for equal slump. Figure 5.4 shows that as the temperature increases the slump decreases. It should be noted that the use of retarding admixtures will retard setting time but generally will not affect the loss of slump resulting from temperature rise. The use of a water-reducing retarder as a means of water reduction can offset increased water when higher temperatures are encountered.

Variations in aggregate gradings and shape will affect water requirement. A change of ±0.20 in the fineness modulus of the sand will usually require a change in aggregate proportions. Generally an increase of 0.20 in the fineness modulus will require a decrease of 50 lb of coarse aggregate per cubic yard of concrete, with a corresponding increase of an equal solid volume of fine aggregate. This might in turn cause an increase in the water content of as much as 4 percent. The above would be conversely true for a decrease of 0.20 in the fineness modulus.

Changes in grading of coarse aggregates may affect the void content, which would change the quantity of sand required. This would be reflected by a change in dry-rodded unit weight. For an aggregate of given specifications a decrease in unit weight means an increase in the void content, and consequently the fine-aggregate requirement increases. A change in particle shape consisting of a larger proportion of flat and elongated pieces would require more sand and water for equal slump. The converse is equally true. It has been shown that each 1 percent increase in air content will decrease the unit water requirement by about 3 percent. This is conversely true for each 1 percent decrease in air content. In order to maintain yield, the fine aggregate should be decreased ½ percent for each 1 percent increase in air content. This is conversely true for each 1 percent decrease in air content. Generally the mix proportions need not be changed for ±1 percent air-content variation, since this much fluctuation is normal over a period of operations. However, a change should be made for a consistently high or low air content by a change in either air-entraining agent or proportions. Each 1-in slump increase will cause an increase of 3 percent in water required.

It is generally acceptable to have variations of about 0.02 in water/cement ratios without changing proportions. However, it is prudent to adjust proportions for a consistent change in one direction of the water/cement ratio. Any adjustments of proportions in the field should be made under the guidelines of the trial-mix water/cement ratio unless otherwise permitted by consistent compressive-strength development.

PREPARED MIXES

11.20 Charts

The use of previously established mixes may be desirable on small projects when the user can use judgment in selecting an overdesign for the extra factor of safety. These mixes may also be used as a guide in establishing trial designs. Tables 11.11 and 11.12 can be used for this purpose. These mixes are designed for natural sand and well-graded stone or gravel aggregates using a specific gravity of 2.65 for both aggregates. The respective compressive strengths may be approximated by the use of Table 11.6. If the use of slag is desired, compute the solid volume of the coarse aggregate and replace it with an equal solid volume of slag. Increase or decrease water per cubic yard by 3 percent for each respective increase or decrease of 1 in. slump; then increase or decrease the fine aggregate by an equal change in the solid volume of water. If manufactured sand is used, increase the sand by 3 percent and increase the water by 15 lb/yd^3, and make a respective change in the solid volume of coarse aggregate. When a less workable concrete for use in pavements is desired, decrease the sand by 3 percent and decrease the water by 10 lb/yd^3 with a corresponding change in solid volume. The solid volume of water can be taken from Table 11.13, and the solid volume of aggregates can be computed from Table 11.14. In all cases, the method of trial-mix adjustments given should be used for field adjustments.

11.21 Useful Tables

Several tables and curves, taken from Ref. 2, 3, 7, are reproduced to facilitate computations in mix designs. Table 11.14 is used for absolute-solid-volume determinations of aggregates

TABLE 11.11 Suggested Trial Mixes for Air-Entrained Concrete of Medium Consistency. (3- to 4-in slump*)

Water/cement ratio, lb/lb	Maximum size of aggregate, in	Air content, %	Water, lb/yd^3 of concrete	Cement, lb/yd^3 of concrete	With fine sand (fineness modulus = 2.50)			With coarse sand (fineness modulus = 2.90)		
					Fine aggregate, % of total aggregate	Fine aggregate, lb/yd^3 of concrete	Coarse aggregate, lb/yd^3 of concrete	Fine aggregate, % of total aggregate	Fine aggregate, lb/yd^3 of concrete	Coarse aggregate, lb/yd^3 of concrete
0.40	⅜	7.5	340	850	50	1250	1260	54	1360	1150
	½	7.5	325	815	41	1060	1520	46	1180	1400
	¾	6	300	750	35	970	1800	39	1090	1680
	1	6	285	715	32	900	1940	36	1010	1830
	1½	5	265	665	29	870	2110	33	990	1990
0.45	⅜	7.5	340	755	51	1330	1260	56	1440	1150
	½	7.5	325	720	43	1140	1520	47	1260	1400
	¾	6	300	665	37	1040	1800	41	1160	1680
	1	6	285	635	33	970	1940	37	1080	1830
	1½	5	265	590	31	930	2110	35	1050	1990
0.50	⅜	7.5	340	680	53	1400	1260	57	1510	1150
	½	7.5	325	650	44	1200	1520	49	1320	1400
	¾	6	300	600	38	1100	1800	42	1220	1680
	1	6	285	570	34	1020	1940	38	1130	1830
	1½	5	265	530	32	980	2110	36	1100	1990
0.55	⅜	7.5	340	620	54	1450	1260	58	1560	1150
	½	7.5	325	590	45	1250	1520	49	1370	1400
	¾	6	300	545	39	1140	1800	43	1260	1680
	1	6	285	520	35	1060	1940	39	1170	1830
	1½	5	265	480	33	1030	2110	37	1150	1990

0.60	⅜	7.5	340	565	54	1490	1260	58	1600	1150
	½	7.5	325	540	46	1290	1520	50	1410	1400
	¾	6	300	500	40	1180	1800	44	1300	1680
	1	6	285	475	36	1100	1940	40	1210	1830
	1½	5	265	440	33	1060	2110	37	1180	1990
0.65	⅜	7.5	340	525	55	1530	1260	59	1640	1150
	½	7.5	325	500	47	1330	1520	51	1450	1400
	¾	6	300	460	40	1210	1800	44	1330	1680
	1	6	285	440	37	1130	1940	40	1240	1830
	1½	5	265	410	34	1090	2110	38	1210	1990
0.70	⅜	7.5	340	485	55	1560	1260	59	1670	1150
	½	7.5	325	465	47	1360	1520	51	1480	1400
	¾	6	300	430	41	1240	1800	45	1360	1680
	1	6	285	405	37	1160	1940	41	1270	1830
	1½	5	265	380	34	1110	2110	38	1230	1990

*Increase or decrease water per cubic yard by 3 percent for each increase or decrease of 1-in slump, then calculate quantities by absolute volume method. For manufactured fine aggregate, increase percentage of fine aggregate by 3 and water by 17 lb/yd^3 of concrete. For less workable concrete, as in pavements, decrease percentage of fine aggregate by 3 and water by 8 lb/yd^3 of concrete.

Source: "Design and Control of Concrete Mixtures," 13th ed., Portland Cement Association, 1988, Chap. 7.[9]

TABLE 11.12 Suggested Trial Mixes for Non-Air-Entrained Concrete of Medium Consistency* (3- to 4-in slump)

Water/cement ratio, lb/lb	Maximum size of aggregate, in	Air (entrapped air) %	Water, lb/yd^3 of concrete	Cement, lb/yd^3 of concrete	With fine sand (fineness modulus = 2.50)			With coarse sand (fineness modulus = 2.90)		
					Fine aggregate, % of total aggregate	Fine aggregate, lb/yd^3 of concrete	Coarse aggregate, lb/yd^3 of concrete	Fine aggregate, % of total aggregate	Fine aggregate, lb/yd^3 of concrete	Coarse aggregate, lb/yd^3 of concrete
0.40	3/8	3	385	965	50	1240	1260	54	1350	1150
	1/2	2.5	365	915	42	1100	1520	47	1220	1400
	3/4	2	340	850	35	960	1800	39	1080	1680
	1	1.5	325	815	32	910	1940	36	1020	1830
	1½	1	300	750	29	880	2110	33	1000	1990
0.45	3/8	3	385	855	51	1330	1260	56	1440	1150
	1/2	2.5	365	810	44	1180	1520	48	1300	1400
	3/4	2	340	755	37	1040	1800	41	1160	1680
	1	1.5	325	720	34	990	1940	38	1100	1830
	1½	1	300	665	31	960	2110	35	1080	1990
0.50	3/8	3	385	770	53	1400	1260	57	1510	1150
	1/2	2.5	365	730	45	1250	1520	49	1370	1400
	3/4	2	340	680	38	1100	1800	42	1220	1680
	1	1.5	325	650	35	1050	1940	39	1160	1830
	1½	1	300	600	32	1010	2110	36	1130	1990
0.55	3/8	3	385	700	54	1460	1260	58	1570	1150
	1/2	2.5	365	665	46	1310	1520	51	1430	1400
	3/4	2	340	620	39	1150	1800	43	1270	1680
	1	1.5	325	590	36	1100	1940	40	1210	1830
	1½	1	300	545	33	1060	2110	37	1180	1990

0.60	⅜	3	385	640	55	1510	1260	58	1620	1150
	½	2.5	365	610	47	1350	1520	51	1470	1400
	¾	2	340	565	40	1200	1800	44	1320	1680
	1	1.5	325	540	37	1140	1940	41	1250	1830
	1½	1	300	500	34	1090	2110	38	1210	1990
0.65	⅜	3	385	590	55	1550	1260	59	1660	1150
	½	2.5	365	560	48	1390	1520	52	1510	1400
	¾	2	340	525	41	1230	1800	45	1350	1680
	1	1.5	325	500	38	1180	1940	41	1290	1830
	1½	1	300	460	35	1130	2110	39	1250	1990
0.70	⅜	3	385	550	56	1590	1260	60	1700	1150
	½	2.5	365	520	48	1430	1520	53	1550	1400
	¾	2	340	485	41	1270	1800	45	1390	1680
	1	1.5	325	465	38	1210	1940	42	1320	1830
	1½	1	300	430	35	1150	2110	39	1270	1990

Source: "Design and Control of Concrete Mixtures," 13th ed., Portland Cement Association, 1988, Chap. 7.

TABLE 11.13. Water Conversion Factors

Gallons	Pounds	Cubic feet
0.12	1.0	0.01607
1.0	8.33	0.1338
7.48	62.3	1.0
26.0	216.67	3.48
26.5	220.83	3.54
27.0	225.00	3.61
27.5	229.17	3.68
28.0	233.33	3.75
28.5	237.50	3.81
29.0	241.67	3.88
29.5	245.83	3.95
30.0	250.00	4.01
30.5	254.17	4.08
31.0	258.33	4.15
31.5	262.50	4.21
32.0	266.67	4.28
32.5	270.83	4.35
33.0	275.00	4.41
33.5	279.17	4.48
34.0	283.33	4.55
34.5	287.50	4.61
35.0	291.67	4.68
35.5	295.83	4.75
36.0	300.00	4.82
36.5	304.17	4.88
37.0	308.33	4.95
37.5	312.50	5.02
38.0	316.67	5.08
38.5	320.83	5.15
39.0	325.00	5.22
39.5	329.17	5.28
40.0	333.33	5.35

Source: Ref. 2,10, Table 16.8.

and cement, Table 11.15 to convert bags of cement to both weight and absolute solid volume, and Table 11.13 for conversion of gallons of water to pounds and cubic feet. It is sometimes convenient to interchange water/cement ratio by weight per bag with Table 11.16.

LIGHTWEIGHT CONCRETE

11.22 General

This discussion is limited to structural lightweight concrete incorporating lightweight aggregates meeting ASTM C330, designed with a 28-day compressive strength exceed-

TABLE 11.14. Density and Volume Relationships

Table for unit weight		
Specific gravity S	Density D	$\frac{1}{\text{Density}}$
2.45	152.64	0.006552
2.46	153.26	0.006525
2.47	153.88	0.006498
2.48	154.50	0.006472
2.49	155.13	0.006446
2.50	155.75	0.006420
2.51	156.37	0.006395
2.52	157.00	0.006370
2.53	157.62	0.006344
2.54	158.24	0.006319
2.55	158.86	0.006295
2.56	159.49	0.006270
2.57	160.11	0.006246
2.58	160.73	0.006222
2.59	161.36	0.006197
2.60	161.98	0.006174
2.61	162.60	0.006150
2.62	163.23	0.006126
2.63	163.85	0.006103
2.64	164.47	0.006080
2.65	165.10	0.006057
2.66	165.72	0.006034
2.67	166.34	0.006012
2.68	166.96	0.005989
2.69	167.57	0.005967
2.70	168.21	0.005945
2.71	168.83	0.005923
2.72	169.46	0.005901
2.73	170.08	0.005880
2.74	170.70	0.005858
2.75	171.33	0.005837
3.10	193.13	0.005178
3.15	196.24	0.005096
3.20	199.36	0.005016

Notation:

S = specific gravity;
D = density, pcf;
W = weight, lb;
V = volume, ft^3.

Basic formulas:

Weight of water is 62.3 pcf;
$D = 62.3S$;
$W = DV$;
$W = 62.3SV$;

$$V = \frac{W}{62.3S} = W \times \frac{1}{\text{density}}$$

Source: Ref. 2, 10, Table 12.6.

TABLE 11.15 Cement Conversion Factors*

Sacks		Weight and solid volume, ft^3											
		0	0.1	0.2	0.25	0.3	0.4	0.5	0.6	0.7	0.75	0.8	0.9
0	Wt	0	9.4	18.8	23.5	28.2	37.6	47.0	56.4	65.8	70.5	75.2	84.6
	Vol	0	0.048	0.096	0.120	0.144	0.192	0.240	0.287	0.335	0.359	0.383	0.431
1	Wt	94	103.4	112.8	117.5	122.2	131.6	141.0	150.4	159.8	164.5	169.2	178.6
	Vol	0.479	0.527	0.575	0.599	0.623	0.671	0.719	0.766	0.814	0.838	0.862	0.910
2	Wt	188	197.4	206.8	211.5	216.2	225.6	235.0	244.4	253.8	258.5	263.2	272.6
	Vol	0.958	1.006	1.054	1.078	1.102	1.150	1.198	1.245	1.293	1.317	1.341	1.389
3	Wt	282	291.4	300.8	305.5	310.2	319.6	329.0	338.4	347.8	352.5	357.2	366.6
	Vol	1.437	1.485	1.533	1.557	1.581	1.629	1.677	1.724	1.772	1.796	1.820	1.868
4	Wt	376	385.4	394.8	399.5	404.2	413.6	423.0	432.4	441.8	446.5	451.2	460.6
	Vol	1.916	1.964	2.012	2.036	2.060	2.108	2.156	2.203	2.251	2.275	2.299	2.347
5	Wt	470	479.4	488.8	493.5	498.2	507.6	517.0	526.4	535.8	540.5	545.2	554.6
	Vol	2.395	2.443	2.491	2.515	2.539	2.587	2.635	2.682	2.730	2.755	2.778	2.826
6	Wt	564	573.4	582.8	587.5	592.2	601.6	611.0	620.4	629.8	634.5	639.2	648.6
	Vol	2.874	2.922	2.970	2.994	3.018	3.066	3.114	3.161	3.209	3.234	3.257	3.305
7	Wt	658	667.4	676.8	681.5	686.2	695.6	705.0	714.4	723.8	728.5	733.2	742.6
	Vol	3.353	3.401	3.449	3.473	3.497	3.545	3.593	3.640	3.688	3.713	3.736	3.784
8	Wt	752	761.4	770.8	775.5	780.2	789.6	799.0	808.4	817.8	822.5	827.2	836.6
	Vol	3.832	3.880	3.928	3.952	3.976	4.024	4.072	4.119	4.167	4.192	4.215	4.263
9	Wt	846	855.4	864.8	869.5	874.2	883.6	893.0	902.4	911.8	916.5	921.2	930.6
	Vol	4.311	4.359	4.407	4.431	4.455	4.503	4.551	4.598	4.646	4.671	4.694	4.742
10	Wt	940	949.4	958.8	963.5	968.2	977.6	987.0	996.4	1005.8	1010.5	1015.2	1024.6
	Vol	4.790	4.838	4.886	4.910	4.934	4.982	5.030	5.077	5.125	5.150	5.173	5.221

*Weight and solid-volume equivalents per sack from 0.1 to 10.9 sacks: weight = 94 × sacks; volume = weight × 0.00509567, based on specific gravity of 3.15; 1 bbl = 4 sacks = 376 lb = 1.916 ft^3.

Source: Ref. 2, 10, Table 16-7.

TABLE 11.16 Conversion of Water/Cement Ratio by Weight to Gallons per Sack (Gallons per Sack = W/C by Weight × 11.28)*

Lb water per lb cement	Lb water per lb cement									
	0	0.01	0.02	0.03	0.04	0.05	0.06	0.07	0.08	0.09
0.20	2.2560	2.3688	2.4816	2.5944	2.7072	2.8200	2.9328	3.0456	3.1584	3.2712
0.30	3.3840	3.4968	3.6096	3.7224	3.8352	3.9480	4.0608	4.1736	4.2864	4.3992
0.40	4.5120	4.6248	4.7376	4.8504	4.9632	5.0760	5.1888	5.3016	5.4144	5.5272
0.50	5.6400	5.7528	5.8656	5.9784	6.0912	6.2040	6.3168	6.4296	6.5424	6.6552
0.60	6.7680	6.8808	6.9936	7.1064	7.2192	7.3320	7.4448	7.5576	7.6704	7.7832
0.70	7.8960	8.0088	8.1216	8.2344	8.3472	8.4600	8.5728	8.6856	8.7984	8.9112
0.80	9.0240	9.1368	9.2496	9.3624	9.4752	9.5880	9.7008	9.8136	9.9264	10.0392
0.90	10.1520	10.2648	10.3776	10.4904	10.6032	10.7160	10.8288	10.9416	11.0544	11.1672

*Water/cement ratio by weight equals pounds of water per pound of cement. *Example:* A W/C of 0.55 by weight equals 6.2 gal per sack; 6 gal per sack equals 0.532 by weight.

Source: Ref. 2, 10, Table 16-10.

ing 2500 psi (tested in accordance with ASTM C330) and not exceeding a unit weight of 115 pcf, as determined by ASTM C567. A portion or all of the lightweight fines may be replaced by normal-weight concrete sand meeting the requirements of ASTM C33.

The greater absorptions and higher rates of absorption of lightweight aggregates have a great effect on the proportioning and control procedures for lightweight aggregate concrete. Damp aggregates are preferable as their lower absorption during mixing has less effect on slump loss during mixing, transporting, and placing. Concrete made with saturated lightweight aggregates should be allowed to lose its excess moisture prior to exposure to freezing and thawing.

Gradation of the lightweight should be well graded to provide a minimum void content which will require the least amount of cement and provide a minimum volume change on drying. Replacement of a portion or all of lightweight fine aggregate with a well-graded concrete sand will customarily allow a greater proportion of coarse aggregate. When design strength becomes difficult to achieve, use of a smaller maximum-sized coarse aggregate should be considered.

11.23 Selecting Mix Proportions

A good start for a project is to obtain proportioning information available on previous concrete mixtures supplied to projects or from lightweight-aggregate suppliers, which can then be adjusted to meet your project requirements. If this information is not available, the Standard Practice for Selecting Proportions for Structural Lightweight Concrete (ACI 211.2-91) provides two methods for proportioning a first-trial mixture. They are Method 1 (weight method, specific-gravity pycnometer)—lightweight coarse aggregate and normal-weight fine aggregates or Method 2 (volumetric method)—all lightweight and combinations of lightweight and normal-weight aggregates.

Method 1: Weight Method (Specific-Gravity Pycnometer). Estimating the required batch weights for the lightweight concrete requires determination of the specific-gravity factor of the lightweight coarse aggregate, which can be done in accordance with ASTM C128 or ASTM C127, depending on the size of the sample to be tested. If ASTM C127 is used, the lightweight coarse material must be kept submerged by a screen. The specific-gravity factor which should be equivalent to that found by ASTM C128 is calculated by the following formula:

$$\text{sp gr } S = \frac{C}{C - E}$$

where C = weight of aggregate tested, moist or dry, in air, g
E = weight of coarse aggregate under water, g
S = specific-gravity factor

Method 2: Volumetric Method (Damp, Loose Volume). Total volume of aggregates will, generally, range from 28 to 34 ft^3/yd^3. Loose volume of fine aggregate may vary from 40 to 60 percent of the total loose volume. Both the total loose volume of aggregate and the proportions of fine and coarse are dependent on many variables such as the nature of the aggregates and the properties of the lightweight to be produced. As the required batch weights for the lightweight concrete depend on the cement content, consultation with the local lightweight-aggregate producer will be needed for recommendations on cement contents required to obtain desired strengths and unit weights. If this experience is not available, an estimate can be made by using Fig. 11.3.

11.24 Estimating First Trial-Mix Proportions

Job specifications may dictate minimum cement or cementitious materials, air content, slump, nominal maximum size of aggregate, strength unit weight, type of placement (pump, bucket, or belt conveyor), and others such as overdesign, admixtures, and special types of cement and aggregates. In any event, the following sequence can be used for selecting the proportions:

1. Choose the slump from Table 11.17, if not specified.

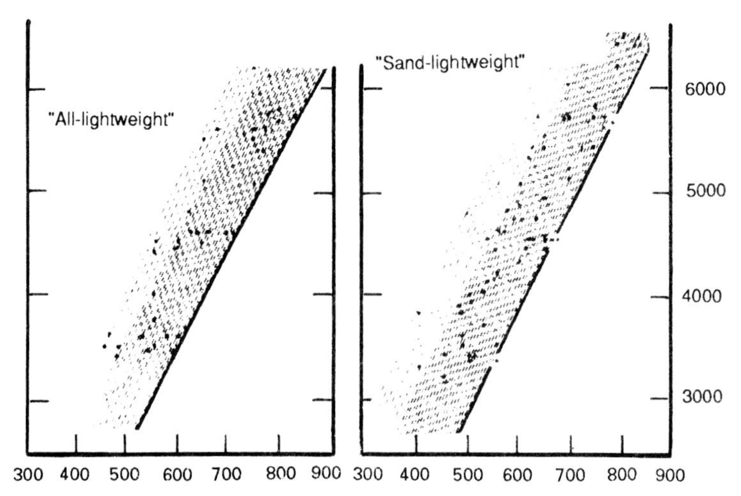

FIGURE 11.3 Relationship of compressive strength and cement content for all-lightweight or sand-lightweight concrete. *(From Fig. 3.3.2, ACI 211.2-91. Data based on actual project results using a number of sources for cement and aggregates.)*

TABLE 11.17 Recommended Slumps for Various Types of Construction*

Types of construction	Slump, in*	
	Maximum†	Minimum
Beams and reinforced walls	4	1
Building columns	4	1
Floor slabs	3	1

*Slump may be increased when chemical admixtures are used, provided that the admixture-treated concrete has the same or lower water/cement or water/cementious materials ratio and does not exhibit segregation potential or excessive bleeding

†May be increased 1 in for methods of consolidation other than vibration.

Source: Table 3.2.2.1, ACI 211.2-91.

2. Choose nominal maximum size of lightweight aggregate which would be the largest that is economically available and consistent with the dimensions of the structure.
3. Estimate amount of mixing water and air content from Table 11.8.
4. Select approximate water/cement ratio from Tables 11.3, 11.5, and 11.6.
5. Calculate cement content from estimated water content and required water/cement ratio.
6. Estimate lightweight coarse aggregate content by using the values found in Table 11.18. The volume of aggregate, in cubic feet, is found by multiplying the value from Table 11.18, which will be determined by the fineness modulus of the fine aggregate, by 27 for a cubic yard. This volume is converted to dry weight of coarse aggregate required in a unit volume of concrete by multiplying it by the oven-dry loose weight per cubic foot of the lightweight coarse aggregate.
7. If the unit weight of the lightweight concrete is known from previous experience, an estimate of the fine-aggregate content can usually be made as the remaining differ-

TABLE 11.18 Volume of Coarse Aggregate per Unit Volume of Concrete

Maximum size of aggregate, in	Volume of oven-dry-rodded coarse aggregates* per unit volume of concrete for different fineness moduli of sand			
	2.40	2.60	2.80	3.00
⅜	0.58	0.56	0.54	0.52
½	0.67	0.65	0.63	0.61
¾	0.74	0.72	0.70	0.68

*Volumes based on oven-dried condition per ASTM C29. Volumes selected empirically for usual reinforced concrete construction. They may be reduced by 10 percent for more workability when placement is to be by pumping.

Source: Table 3.2.2.4, ACI 211.2-91.

ence from the weight of the rest of the ingredients. Otherwise an approximation can be made by using the estimated weights, based on the specific-gravity factor, from Table 11.19.

8. Additional mix adjustments should be made after a trial batch. Final batch adjustments may be required during field operations.

11.25 Weight Method

Lightweight-aggregate concrete may also be proportioned by the weight method similar to that for normal-weight concrete. The sum of the weights of all materials in a mix is equal to the total weight of the mix. If the unit weight is specified or assumed, the weight of the cement and water is deducted from the total weight of the mix and the remainder becomes the total SSD weight of the lightweight aggregate. Using the specific gravity factor and the percentage of coarse aggregate from Table 11.18 provides a method for proportioning the concrete mixture.

11.26 Example of Lightweight-Concrete Proportioning

Required: Lightweight concrete with compressive strength of 3500 psi, a 4-in slump, 4.5 percent entrained air, a maximum air-dry weight of 105 lb pcf, and mild exposure.

Given: Dry loose unit weight of available fine lightweight aggregate is 60 pcf, lightweight coarse aggregate of ¾ in nominal size is 45 pcf. Fineness modulus of fine aggregate is 3.00, the specific-gravity factor of the coarse aggregate is 1.36, and the fine aggregate is 1.88, an average of 1.62. Present moisture content of the coarse aggregate is 2% with a total absorption rate of 11 percent and the fine aggregate containing 4 percent moisture has an absorption of 2 percent.

TABLE 11.19 First estimate of fresh concrete weight when lightweight coarse and normal weight fines are used

Specific-gravity factor	First estimate of lightweight-concrete weight, lb/yd^3* (air-entrained concrete)		
	4%	6%	8%
1.00	2690	2630	2560
1.20	2830	2770	2710
1.40	2980	2910	2850
1.60	3120	3050	2990
1.80	3260	3200	3130
2.00	3410	3340	3270

*Values for concrete containing 550 lb of cement and amounts of water for 3- to 4-in slump based on Table 11.17. For each 10 lb of mixing water difference, correct weight 15 lb in opposite direction. For each 100 lb cement difference from 550 lb, correct weight 15 lb in same direction.

Source: Table 3.2.2.5, ACI 211.2-91.

Solution:

1. Without local experience, the cement content can be estimated from Fig. 11.3 as 625 lb for 3500 psi.

2. From Tables 11.5 and 11.6, the maximum water/cement ratio is found to be 0.40.

3. Estimated water requirement for air-entrained concrete having a ¾-in maximum size aggregate from Table 11.8 is 305 lb/yd^3 and cement requirement is calculated to be 760 lbs.

4. Volume of the coarse aggregate per unit volume of concrete found in Table 11.18 is 0.68. Multiplying 0.68 by 27 will give the volume of 18.36 ft^3/yd. The oven-dry loose weight of coarse lightweight per cubic yard can be found by multiplying 18.36 by 45, which equals 826.2 lb. Allowing for 11 percent absorption requires 826.2 × 1.11 or 917.08 SSD lb/yd^3.

5. Without past history, a first estimate of the total weight per cubic yard may be found in Table 11.19. This is found to be 3088 lbs for each yard of concrete. As the table is based on 550 lbs of cement, the 120-lb difference would require adding 15 lb for each additional hundred pounds of cement or 15 × 1.2 = 18 lbs and the total would be 3106.

6. The proportion of fine aggregate can be found as a remainder.

Water (net for mixing)	305 lb/yd^3
Cement	760 lb/yd^3
Coarse aggregate	917.08 lb/yd^3 SSD
Total (without sand)	1982.08 lb/yd^3

SSD weight of fine aggregate = 3106 − 1982.08 = 1123.92 lb. The oven-dry weight would be 1123.92/1.02 = 1101.88 lb.

7. A trial batch of ⅒ yd^3 would be as follows:

Cement	760 × 0.1 =	76 lb
Fine aggregate (SSD)	1123.92 × 0.1 =	112.39 lb
Coarse aggregate (SSD)	917.08 × 0.1 =	91.71 lb
Mixing water	305 × 0.1 =	30.5 lb
Total		310.6 lb

As the fine aggregate contains 4 percent moisture, the water added must be reduced by 0.02 × 110.19 = 2.2 lb and the fine-aggregate batch weights to be increased by the same amount per ⅒ yd^3. The coarse aggregate has been found to have a 2 percent moisture content and an absorption rate of 11 percent. The producer has been asked to use sprinklers and presoak the coarse aggregate. The new moisture content measured is now 15 percent, which now indicates 4 percent free water which must be accounted for by decreasing the mix water by 0.04 × 82.62 = 3.3 lb and increasing the coarse aggregate batch weight by the same amount. Note that all adjustments for water are made on the oven-dry basis. The new trial batch weights for ⅒ of a cubic yard are as follows:

Cement	76 lb
Fine aggregate	114.59 lb
Coarse aggregate	95.01 lb
Mixing water	25 lb
Total	310.6 lb

8. Adjusting mix proportions of structural lightweight concrete. The specific gravity factor of the aggregate method is used to provide means for mix adjustments and proportioning. The test method is described in Sec. 12.17, Chap. 12. The specific-gravity factor is a function of the moisture content and therefore must be computed for various moisture contents and each aggregate type. An example of a typical relationship is shown in Fig. 11.4.

9. A trial batch is made and the procedure shown from items 5 through 7 is performed with the necessary adjustments for yield.

10. Additional mix designs and trial mixes should be made to provide at least three and preferably four points on a cement-strength curve so that the mix for the desired strength can be obtained. (An experienced laboratory technician can frequently calculate a trial mix of conventional-weight concrete so as to produce concrete with a given cement factor or water/cement ratio or both, to sufficient accuracy as to require only minor and insignificant corrections. In such case a second trial batch with the minor correction is usually unnecessary. With lightweight concrete there is much less probability of being able to calculate a single trial batch and have it produce a concrete so close to the desired requirements that additional mixes are unnecessary. Almost always a series of mixes will be required before a project mix for lightweight concrete can be recommended.)

11.27 Adjustment

Even more than in regular concrete, the many variables encountered during production and construction may require adjustment of the mixes in the field. It is essential that when the parameters are established in the trial mixes, these should be maintained in the field. Generally the tolerances used for regular concrete may not be completely applicable to lightweight concrete. For example, lower slumps can be used in lightweight concrete for equal placeability. The mix may appear stiff but if properly designed will flow and con-

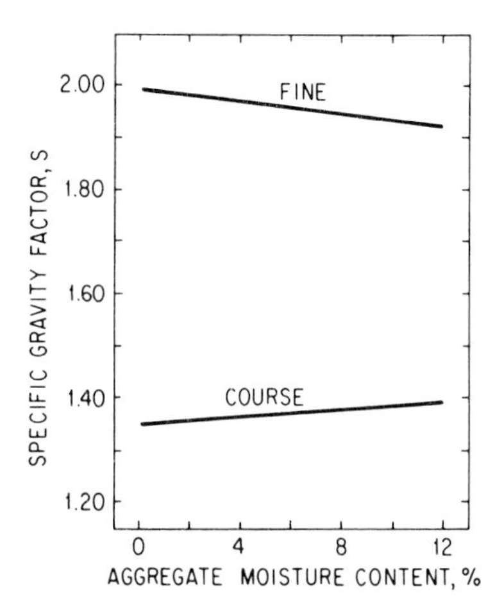

FIGURE 11.4 Example of the relationship between pycnometer specific gravity and moisture content for a lightweight aggregate. *(From Fig. A App. A, ACI 211.2–91.)*

solidate readily under mechanical vibration, and slumps of 1 to 2 in can be readily used. However, when it is found necessary to increase the water content, the cement content must be proportionally increased to maintain the design strength level. The water/cement-ratio law applies. When the moisture content of the aggregate changes, care must be taken to maintain the weight of dry aggregate constant by making the moisture correction. If the density of the aggregate has changed, the volume of dry aggregate must be kept constant to maintain yield. A completely new mix may be required if the unit weight of the concrete varies appreciably from the design mix weight. A unit weight change due to air-content variation may be corrected by changing the quantity of air-entraining agent. A unit weight change due to a change in unit weight of aggregate may possibly be corrected if the aggregates involve lightweight coarse and natural sand by changing proportions; but if the aggregate is all lightweight, a change in the aggregate being furnished will almost certainly be required. Generally the most important factors in field adjustment of a mix made of high- and consistent-quality materials are to maintain constant cement content, slump, air content, and volume of dry aggregate per cubic yard. A constant volume of dry aggregate requires continuous moisture determinations and necessary corrections.

The following rules of thumb will serve as a guide for approximate mix adjustments of structural lightweight concrete:

1. A 1 percent increase in fine aggregate increases water by 3 lb/yd^3.
2. An increase of 3 lb/yd^3 in water will require an increase of about 1 percent in cement content.
3. An increase of 1 percent in air content will decrease water content by 5 lb/yd^3 (may also decrease strength).
4. A 1-in increase in slump will require about 10 lb/yd^3 water increase.
5. An increase of 10 lb/yd^3 in water will require increased cement content of approximately 3 percent.

Mix adjustments for cement, air, and water content, using the absolute-volume method, are calculated in the same manner as normal-weight concrete proportioning. An example is shown below to illustrate the use of the specific-gravity factor and effective-displaced-volume method, which can be used for adjustments in proportions. In this case, the method is used to increase the fine-aggregate proportion of the above mix, assuming it was necessary for an increase in workability.

11.28 Field Testing and Production Control

The purpose of routine tests of lightweight concrete is to maintain uniform quality, thus assuring proper values for unit weight, cement content, and water content. It is usually quite difficult to produce a uniform mixture consistently because of the greater variations in absorption, specific gravity, moisture content, and grading of lightweight aggregate. Therefore, lightweight concrete must be thoroughly tested in the field during production and use. Unit weight, slump, and air-content tests must be made frequently. ACI Standard 301 suggests unit-weight tests of every load. Variations in unit weight will signal air-entrainment variations or improper aggregate proportioning, grading, or specific gravity. The Volumetric Method, ASTM C173, should be used for air-entraining tests. Although the design is usually based on dry weights of aggregates, it is most advantageous to use wet aggregate during production. This avoids segregation and avoids the subsequent absorption of water from the mix that causes rapid stiffening and segregation.

It also permits better control of total water content. Periodic adjustments may be necessary as predicated by the field tests that deviate from established relationships. It should always be the aim in adjustments to keep the volume of the dry aggregate constant (which is the basis of the design) unless a check reveals a change in density.

The placeability of lightweight concrete can be significantly improved by the use of air entrainment, which will also assist in improved bleeding characteristics, reduced segregation, and lowered unit weight. The amount of air-entraining agent will vary widely for a specific amount of air with the many varieties of lightweight aggregates that are used. This characteristic demands that frequent air-entrainment tests be made.

Concrete made with lightweight aggregate that has been shown to absorb less than 2 percent by weight during the first hour after inundation, based on test of a sample from the field-conditioned supply, should be batched and mixed in the same manner as normal weight concrete.

Concrete made with lightweight aggregates not conforming with this absorption limit of 2 percent maximum should be batched and mixed by adding 80 percent of the mixing water to the aggregate in the mixer, then mixed for a minimum of 1½ min (15 revolutions in a truck mixer). The balance of the materials can then be added in the normal manner.

Acceptance of lightweight concrete in the field should be based on fresh unit weight measured in accordance with ASTM C567 as described in Chap. 12. When the normal fresh unit weight varies more than 2 lb/ift^3 from the required weight, the mixture must be adjusted as promptly as possible.

HEAVY CONCRETE

11.29 Application

The greatest use of heavy concrete is in nuclear shielding. Gamma rays and xrays can be shielded against by a mass material. Concrete has proved effective and economical. The use of high-density or heavy concrete enables a reduction in thickness for the same degree of shielding. In order to shield against neutrons, it is necessary to have a mass material such as concrete containing elements whose atoms can both thermalize and capture neutrons. The hydrous atom in the form of water in the concrete thermalizes the fast neutrons, which may then be captured by atoms of elements such as cadmium or boron, which have high neutron-capture cross sections. Care must be exercised in the form in which boron is introduced into the concrete because of the high retardation effect on concrete of some forms of boron. The ACI Symposium of Heavy Concrete[4] is a good reference for information on design. The heavy concrete usually is designed to have a unit weight of 180 to 350 pcf. This is predicated upon the degree of shielding necessary and thickness of structure.

11.30 Aggregate Characteristics

High-density concrete is achieved by the use of heavy aggregates, including iron ores such as magnetite and limonite; quarry rock such as barite, steel punchings, and iron shot; or synthetic materials such as ferrophosphorus. The aggregates can be made to conventional gradings for both coarse and fine sizes. Conventional sand may be incorporated in the mixture when permitted by the allowable unit weight. All the many

TABLE 11.20 Typical Heavyweight Aggregates*

Material	Description	Specific gravity	Concrete, unit weight, pcf
Limonite Goethite	Hydrous iron ores	3.4–3.8	180–195
Barite	Barium sulfate	4.0–4.4	205–225
Ilmenite Hematite Magnetite	Iron ores	4.2–5.0	215–240
Steel/iron	Shot, pellets punchings etc.	6.5–7.5	310–350

*Ferrophosphorous and ferrosilicon (heavyweight slags) materials should only be used after thorough investigation. Hydrogen gas evolution in heavyweight concrete containing these aggregates has been known to result from a reaction with the cement.

Source: Table A4.1.1, ACI 211.1–91.

aggregates tend to produce a harsh mix because of their shape and weight. Air entrainment should be avoided because of its detrimental effect on density, and it therefore cannot be used for improved workability. Compressive strength attainable with these materials can range between 3000 and 6000 psi at 28 days. Some of the heavy aggregates are not suitable for exposure to weathering, and they should therefore be protected from the elements.

ASTM Standard C637, Aggregates for Radiation Shielding Concrete, covers important properties of heavy aggregates and should be used to evaluate this type of material in conjunction with Table 11.20.

11.31 Mix Proportioning

The procedure used for regular-weight concrete is also used to proportion mixes for heavy concrete using the solid-volume method. The cement contents will usually range from 470 to 750 lb/yd^3, with strength and other qualities similar to those of corresponding water/cement ratios of normal-weight concrete. The method of proportioning is best shown by example in which the method of establishing a required density is incorporated.

Required: Heavy concrete with a minimum compressive strength of 3500 psi and a minimum plastic unit weight of 220 pcf.

Given: Magnetite iron ore with a specific gravity of 4.2 in both fine and coarse gradation. Dry-rodded unit weight of 1½ in to No. 4 gradation is 153 pcf.

The fineness modulus of the fine aggregate is 2.70.

Iron punchings are available with a specific gravity of 6.5.

Previous experience indicates that 333 lb of water/yd^3 is required for a 4-in slump and 564 lb of cement per cubic yard is needed for the desired strength.

1. From Table 11.9, the volume of dry-rodded coarse aggregate per unit volume of concrete is 0.72 for 1½ aggregate and a 2.70 fineness modulus of sand.
2. Weight of 1½-in size magnetite, $0.72 \times 27 \times 153 = 2974$ lb/yd^3.

3. Solid volumes:

	Weight, lb	Volume, ft^3
Cement	564	2.87
Water	333	5.34
Coarse aggregate $\frac{2974}{4.2 \times 62.4}$	2974	11.35
Volume of entrapped air, 1%	—	0.27
Total, except fine aggregate	3871	19.83

Solid volume of fine aggregate $27 - 19.83 = 7.17$ ft^3. Required dry-weight fine aggregate $7.17 \times 4.2 \times 62.4 = 1879$ lb. Unit weight of plastic concrete:

$$\frac{3871 + 1879}{27} = \frac{5750}{27} = 213.0 \text{ pcf}$$

4. To achieve the desired weight of 220 pcf, the total weight per cubic yard must be $27 \times 220 = 5940$. The weight of the coarse aggregate must be increased, $5940 - 5750 = 190$ lb, without changing the solid volume of 11.35.

5. Let M = weight of magnetite and P = weight of iron punchings

Then

$$11.35 = \frac{M}{4.2 \times 62.4} + \frac{P}{6.5 \times 62.4}$$

or

$$405.6M + 262.1P = 1{,}206{,}593$$

since

$$M + P = 2974 + 190 = 3164$$

Substitute and simplify: $P = 535$ lb and $M = 2{,}629$ lb.

$$\text{Volume of magnetite} = \frac{2629}{4.2 \times 62.4} = 10.03 \text{ ft}^3$$

$$\text{Volume of punchings} = \frac{535}{6.5 \times 62.4} = 1.32 \text{ ft}^3$$

The revised mix per cubic yard is

Material	Weight, lb	Volume, ft^3
Cement	564	2.87
Water	333	5.34
Coarse aggregate:		
Magnetite	2629	10.03
Punchings	535	1.32
Fine aggregate	1879	7.17
Entrapped air, 1%	—	0.27
Total	5940	27.00

Unit weight = 5940/27 = 220 pcf.

6. A trial mix should be made similar to normal-weight concrete and the same tests. Mix adjustments are similar.

11.32 Adjustments

Variables found in the field with heavyweight concrete are similar to normal weight, and water-content adjustments may be necessary to compensate for variations in the aggregate. Additional adjustments may be necessary in the quantities of fine aggregate for workability, unit weight, and yield. Prolonged mixing should be avoided as it tends to break down many of the ore aggregates. All precautions of normal-weight concrete apply to heavyweight concrete. Slump is not a good indicator of workability, and with high slumps, segregation is more likely. Any water adjustments may require cement revision to maintain strengths.

REFERENCES

1. ACI Standard Practice for Selecting Proportions for Normal Weight, Heavyweight and Mass Concrete (ACI 211.1-91).
2. Waddell, J. J.: "Practical Quality Control For Concrete," McGraw-Hill, New York, 1962.
3. ACI Standard Practice for Selecting Proportions for Structural Lightweight Concrete (ACI 211.2-91).
4. "Concrete for Radiation Shielding," ACI Compilation 1, American Concrete Institute, 1962.
5. "Concrete Manual," 8th ed., U.S. Bureau of Reclamation, 1978.
6. ACI Specifications for Structural Concrete for Buildings (ACI 301-84) (rev. 1988).
7. ACI Standard 318-89, Building Code Requirements for Reinforced Concrete, American Concrete Institute, 1989.
8. ASTM C31-88, Standard Practice for Making and Curing Concrete Test Specimens in the Field.
9. "Design and Control of Concrete Mixtures," 13th ed., Portland Cement Association, 1988.
10. Waddell, J. J.: "Concrete Inspection Manual," International Conference of Building Officials, 1976.

CHAPTER 12
TESTING CONCRETE

Joseph F. Artuso*

ASTM TEST METHODS

12.1 Significance

The American Society for Testing and Materials[1] provides the most widely accepted standards for determining the properties of concrete materials. This is particularly true for the test methods of concrete and its components. In correlation with the determination of properties, ASTM has developed specifications denoting the minimum properties that the material must possess for a certain use. The development of both specifications and methods is not a static situation, as there is continual progress in these areas through research and application. Therefore, it is necessary to utilize the latest revision of the ASTM standards. This can be determined by noting the number following the dash in the designation listed in the latest ASTM Index. This number indicates the year of the latest revision. The test method may often be divided into tests usually performed in the laboratory and tests usually performed in the field, particularly in the case of concrete. Some of the tests are performed under both circumstances. In addition to ASTM test methods, recognition is given to other outstanding agencies such as the Bureau of Reclamation, the Corps of Engineers, the American Association of State Highway Officials, and organizations such as the American Concrete Institute. In general, most are adopting ASTM Test Methods.

LABORATORY TESTS

12.2 Cement

The specifications for all kinds and types of cement contain detailed limits on chemical compounds and physical properties. Conformance to these limits is generally checked by tests in the laboratory as detailed in applicable ASTM test methods. This can be accomplished by sampling an entire bin or by testing samples from each shipment for a given project. In many cases mill test reports are accepted; in others, independent-laboratory verification is required. In addition to the routine chemical and physical tests, cement

*Revised by Joseph A. Dobrowolski.

may be tested for special properties such as false or unusual setting, excessive bleeding or pack set.

12.3 Water

Most specifications covering water used in the manufacture and curing of concrete state that it must be clean and free of injurious impurities. In general, it is felt that water that is potable will be acceptable for use in concrete. This may not be indisputable since water contaminated with sugars could be potable but would significantly reduce the quality of concrete. When water is obtained from a treated source, it usually contains a total dissolved solids content of 200 to 1000 ppm, with only a rare source as high as 2000 ppm. With the exception of carbonates and bicarbonates, individual impurities of 1000 ppm will not affect concrete quality. It is advisable to determine the total solids content, and when concentrations exceed 1000 ppm, particularly of carbonates and bicarbonates, strength tests are in order. Excessive solids content may also warrant concrete tests for durability, volume change, setting time, flexural strength, and bond to steel. The usual physical tests performed for a check of water quality are setting time, soundness, and compressive strength as given in AASHO Standard T26. Seawater may reduce strength only slightly but is not recommended for reinforced concrete because of the corrosive effect on the steel. Nevertheless it is sometimes used in some areas because of a scarcity of fresh water. Water used for prestressed concrete should not contain more that 500 ppm chloride content.

12.4 Aggregate Tests

ASTM Specification C33 indicates the acceptable level of physical properties for general construction. The test methods given there cover each property. Routine field tests usually include the following:

	ASTM Designation
Sampling	D75
Grading	C136
Amount of material finer than No. 200 sieve	C117
Organic impurities	C40
Clay lumps and friable particles	C142
Unit weight	C29
Fineness modulus	C136
Coal and lignite (lightweight pieces)	C123

The detailed tests required for qualification prior to use in the field are more extensive and usually require laboratory facilities. These include the following:

Mortar-making properties	C87
Soundness	C88
Abrasion of coarse aggregate	C131
Reactive aggregates	C227, C289, C295, C342, C586
Freezing and thawing	C666

12.5 Concrete

Routine laboratory tests of concrete are usually confined to trial mixes to determine economical proportions necessary for strength, durability, and workability, followed by strength tests of concrete from the project. More exhaustive examinations are sometimes required for specific projects including, for example, volume change and creep. After the proportions are established, they must be checked in the field, and when changes occur in materials or conditions, additional laboratory trial mixes for new designs may be necessary. Strength tests of concrete produced on the project are the most frequent and common of all laboratory tests of concrete.

FIELD TESTS

12.6 Sampling

The procedure used in taking a sample of concrete depends on the equipment from which it is sampled. The following procedures have been summarized from ASTM Designation C172. Conformance to the proper sampling method is essential for reliable results and is considered the most important single factor in concrete testing. The test can be only as accurate as the sample.

Transit-Mix Trucks. The procedure for transit-mix trucks is as follows:

1. Take a portion of the sample from two or more regularly spaced intervals during discharge of the middle portion of the batch.
2. Obtain each portion either by passing a receptacle through the entire discharge stream or by diverting the stream completely into a container.
3. Take the composite samples to the place where the test specimens are to be molded.
4. Composite the samples into one sample to ensure uniformity, and protect it from the sun and wind while testing.
5. The time required from taking the sample to using it must not exceed 15 min.
6. Take care not to restrict flow of concrete from the mixer container or transportation unit so as to cause segregation.

Paving Mixers. From paving mixers, the entire batch should be discharged and the sample collected from at least five different portions of the pile. Compositing, protection, and time limit are as above.

Stationary Mixers. All the conditions given under transit-mix trucks apply. If discharge of the concrete is too rapid to divert the complete discharge steam, discharge the concrete into a container or transportation unit sufficiently large to accommodate the entire batch and then accomplish the sampling in the same manner as for transit-mix trucks.

12.7 Consistency

Slump. The procedure used in making the slump test is given in ASTM Designation C143 and is summarized as follows:

1. Moisten inside of slump cone and place on a flat, moist, nonabsorbent surface at least 1 × 2 ft (plank, piece of heavy plywood, concrete slab, etc.). Surface must be firm and level. Hold slump cone in place by standing on foot pieces.
2. Fill cone one-third full and rod the concrete exactly 25 times with tamping rod, distributing the rodding evenly over the area. Caution: Use standard steel tamping rod, ⅝ in diameter by 24 in long with one end rounded to a hemispherical tip. Do not use a piece of reinforcing steel.
3. Fill cone with second layer until two-thirds full; rod this layer 25 times as before with rod penetrating into but not through first layer. *Note*: One-third or two-thirds full means one-third or two-thirds of the volume, not depth of cone.
4. Overfill cone slightly and rod this layer 25 times, making sure the rod penetrates into but not through second layer. Distribute the rodding evenly over entire area.
5. Use the tamping rod to scrape off excess concrete from top of cone, and clean spilled concrete from around bottom of cone.
6. Lift the cone vertically and slowly in 5 to 10 s. Avoid torsional movement, jarring, or bumping the concrete.
7. Set slump cone on surface next to but not touching slumped concrete—lay tamping rod across top of cone. Measure amount of slump from bottom of tamping rod to top of slumped specimen over the original center of the base of the specimen (see Fig. 12.1). Discard this concrete after slump has been measured. Do not use it for making test cylinders.
8. The entire operation from the start of filling should be performed within 2½ min. If there is a decided falling away from one side or a shearing off of a portion of the sample in two consecutive tests, the concrete probably lacks the necessary plasticity and cohesiveness for the slump test to be applicable.

Ball Penetration. The consistency of concrete may be determined by the use of the ball-penetrating (Kelly ball) method in accordance with ASTM Designation C360, summarized as follows:

1. The concrete sample used shall be in suitable containers or forms and the test shall be made without preliminary manipulation. Suitable containers may be wheelbarrows, Georgia buggies, or the forms themselves.
2. The depth of the sample shall be at least three times the maximum size aggregate, but not less than 8 in. The minimum horizontal distance from the center of the ball to the free edge of the concrete shall be 9 in.
3. Strike off the concrete to a level surface and place the Kelly ball on the surface.
4. Release handle without twisting or jolting the ball.
5. Read the penetration directly on the handle.
6. The control nut should be adjusted before the test starts so that the guide will be on zero.
7. Three or more tests should be taken for a representative reading.

FIGURE 12.1 Measurement of slump of concrete.

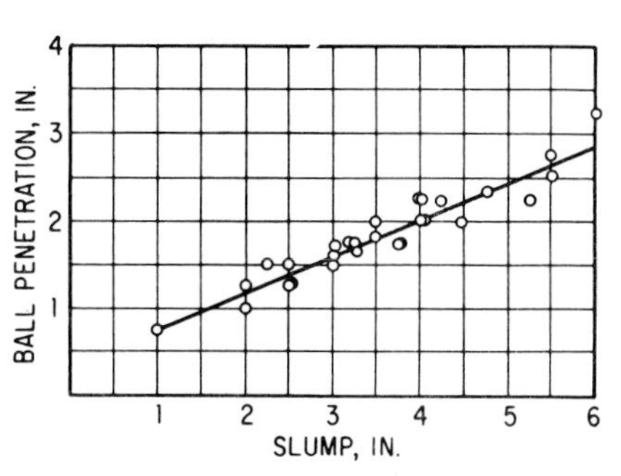

FIGURE 12.2 Typical relationship of slump to ball penetration must be determined individually for variations in ingredients.

8. The ball-penetration method should be calibrated for use in controlling consistency. This is done making both the slump test and ball test on mixes of various consistencies and each class of concrete. The relationship will vary with different aggregates, cement contents, air contents, and other variables.

A curve may be made similar to Fig. 12.2, with at least 10 tests made. As an alternate, a conversion factor can be compiled to be used within restricted limits of slump. The correlation should be checked periodically.

12.8 Air Content

The measurement of the amount of air entrained in fresh concrete during placement is essential to maintain the quality desired. It should be checked at regular intervals and at least at any change in conditions such as weather or consistency. The part used for the air test must be discarded and must not be used for any other test. There are three procedures covered by ASTM, namely, Designation C138, Gravimetric Method; C173, Volumetric Method; and C231, Pressure Method.

The gravimetric method (C138) can be made in conjunction with the yield test described here. It is calculated from the knowledge of mix proportions and specific gravities of all materials. The theoretical air-free weight can be used as a constant when the mix consistency remains the same. It is computed by the formula

$$\%\ \text{air} = \frac{\text{theoretical unit weight} - \text{measured unit weight}}{\text{theoretical unit weight}} \times 100$$

It is generally not recommended for field control because of likely inaccuracies. An error of 2 percent in moisture content of aggregate can cause an error of 1 percent in the indicated air, and an error of 0.02 in specific gravity of the aggregate can cause an error of ½ percent in computed air content.

The volumetric method (C173), commonly called the Roll-A-Meter method, is necessary for concrete made of lightweight aggregates, slag, and any other vesicular aggregates but can also be used for any type of aggregate concrete. It utilizes the principle of direct determination of air by displacement in water. The procedure is as follows:

1. Place concrete sample in the bowl in three equal layers and rod each layer in a manner similar to that described in the slump test.
2. Tap the exterior surface of the measure 10 or 15 times after each rodding, or until no large bubbles of air appear.
3. Strike off the surface flush using sawing motion with the straightedge and wipe the flange clean.
4. Clamp the top section, add the water to the zero mark, and attach screw cap.
5. Invert the unit and agitate until the concrete is free of the base, and rock and roll the unit until the air is displaced. Roll and rock the unit with neck elevated. Repeat this until there is no further drop in the water column.
6. Add isopropyl alcohol to dispel foam, using the measuring cup which is furnished with the equipment.
7. Make a direct reading to the bottom of the meniscus. The air content of the concrete is the reading plus 1 percent for each measureful of alcohol added.

The pressure method (C231) is the commonest and most accurate for all concrete except lightweight, which requires the volumetric method described above. It utilizes the principle of Boyle's law to determine air content by the relationship of pressure to volume. The manufacturer of each meter provides detailed instructions for operation and calibration for variations in atmospheric pressure. ASTM Designation C231 provides a detailed procedure. Briefly the procedure, for concrete with aggregate of less than 2-in nominal size, is (see Fig. 12.3):

1. Place the sample in the bowl in three layers, tamp, tap, and strike off the top in similar manner to that given in the volumetric method.
2. Thoroughly clean the flanges, clamp on the cover, and add water to the zero mark. Tap the sides after adding part of the water to remove entrapped air.
3. Operate the valves in accordance with the manufacturer's instructions and apply desired test pressure to read the air content directly on the graduated-gauge glass. The air content is the difference in height at specified pressure and reading at zero pressure after release of pressure.
4. Release pressure and read level at zero pressure.
5. The aggregate correction factor may be significant and should be made in accordance with ASTM C231 when the characteristics are not known. This is applied to the reading made on the concrete.

The "Washington meter," originally developed by the Washington State Highway Department, is a pressure meter that is commercially available. It determines air content by measuring the change in pressure of a known volume of air under an established pressure when released in contact with the concrete in a sealed container.

The Chace air meter is not covered by ASTM but actually utilizes the same principle as the volumetric method. Only the mortar portion of the concrete is used. The meter consists of a small graduated-glass tube and a small brass base. The mortar fraction of the concrete is placed in the base, and the air content is determined by displacement with isopropyl alcohol. Detailed instructions and mortar factors are given by the manufacturer. The Chace meter should be used only as a guide and in conjunction with a standard

FIGURE 12.3 Determination of entrained air by the pressure method.

meter. It requires less time than any of the standard methods and is generally considered to be accurate to about ½ percent of air.

12.9 Strength Specimens

The ACI Building Code (ACI 318) recommends that a test consist of two specimens made for test at 28-day age. Most specifications require one test for each 150yd,^{3} or fraction thereof, of each class of concrete placed in any one day. The samples of concrete should be taken in accordance with ASTM Designation C172 described earlier, and the specimens made and laboratory-cured in accordance with ASTM Designation C31 when the strength is determined as a basis for acceptance of the concrete. Additional test specimens cured entirely under field conditions may be required to check adequacy of curing and protection of the concrete for stripping strength, or for prestressing strength. A modified method of casting cores in place with special methods of extraction for the compression test has been used for the latter purpose.

Compression Specimens. The method for casting cylinders is described in ASTM Designation C31, and the following procedure is a summary of the procedure when applied to the standard 6-in diameter × 12-in-high cylinders.

1. The sample should be taken as described in Sec. 12.6 and should preferably be taken at the point of casting.
2. The molds may be reusable steel or single-use coated cardboard or sheet metal, as covered by ASTM Designation C470.
3. Place the molds on a smooth, firm, and level surface.
4. Fill the molds in three equal layers and rod each layer uniformly 25 times with a ⅝-in hemispherical-tipped rod. Rod the upper layers so that the rod just extends into the layer underneath about ½ in. When making two or more cylinders for one test, place

and rod the bottom layer of all cylinders, then the second layer of all cylinders, and finally the third layer. Tap the sides of the molds after rodding each layer if necessary to close any voids.

5. The last layer should contain a slight excess of concrete. After rodding and tapping, strike off the excess concrete with a straightedge.
6. In the case of low-slump concrete (less than 3 in), internal or external vibration may be used for the consolidation of concrete in the cylinder molds. The mold is filled in two layers with sufficient vibration to consolidate each layer without segregation. When internal vibration is used, the vibrating element, which should be 1½ in or less in diameter, should be inserted at three different points for each layer and must not rest on the bottom or sides of the mold and, when vibrating the second layer, should extend about 1 in into the first layer. When external vibration is used, the mold must be securely held against the vibrating element or surface. The excess concrete should be struck off in the same manner as in rodded specimens.

During the first 24 h the specimens must be maintained at temperatures between 60 and 80°F and covered to prevent evaporation. Specimens formed in cardboard molds should not be stored for the first 24 h in contact with wet sand or wet burlap or by other methods that would cause the cardboard mold to absorb water. They can be covered with polyethylene sheeting and then with wet burlap. Cylinders must not be moved or disturbed during the first 24 h. Insulated storage boxes should be used to maintain the desired temperature, using a light bulb in cool weather and damp sand or wet burlap in hot weather. A minimum-maximum thermometer can be used to check the temperature inside the box. When strength tests are made for acceptance of concrete, the specimens must be removed from their molds at the age of 16 to 24 h and stored in a moist room at a temperature of 73.4 ± 3.0°F, with free water on the surface at all times, but not exposed to running water, as shown in Fig. 12.4. If a fog room is not available, the cylinders can be immersed at the above temperature, in which case the water should be saturated with lime. Transportation to a quality laboratory will ensure that these curing conditions are maintained. Specimens should be adequately protected from breakage during transportation to the laboratory.

FIGURE 12.4 Moist-curing room. Cabinets on wall at right are for curing cement test cubes (normal fog condition has been dissipated for photographic purposes).

The cylinders should be properly tagged as they are made, and all the data concerning the concrete recorded to correlate with the cylinder marking. It is advisable to use metal tags attached with light wire to the specimens with the tags numbered consecutively. This numbering system provides for a check of all cylinders on a given project. Pertinent data recorded separately and correlated with the cylinder numbers should include project, mix proportions, date, slump, air content, temperature, location in the structure, ages to be tested, and identity of person making the cylinders.

The compressive-test specimens must have bottom and top surfaces that are plane within 0.002 in and with each surface perpendicular to the axis within 0.5° (approximately ⅛ to 12 in). Capping of both surfaces of single-use molds is usually necessary to achieve this, as shown in Fig. 12.5. The caps should conform to the requirements covered in detail in ASTM C617:

1. They should be as thin as practical, ⅛ to 5⁄16 in, and made of material with a compressive strength greater than that of the concrete. Cap should not be off center on specimen by more than 1⁄16 in.
2. Capping materials can be neat portland cement, high-alumina cement, high-strength gypsum plaster such as Hydrocal or Hydrostone, or sulfur compounds. There are a number of good proprietary compounds available. Ordinary plaster of paris should not be used.
3. The top surfaces of specimens cast in metal molds may be capped with a thin layer of neat portland cement paste before the concrete has hardened after settlement has ceased.

FIGURE 12.5 Capping cylinders with sulfur compound showing capping stands and heating ovens.

4. Hardened specimens should be capped with any of the above materials, against plane surfaces to assure the tolerance requirements. In all cases the cap must be allowed to gain strength of 5000 psi when tested. Sulfur compounds usually require 2 h.

The method of testing specimens for compressive strength, as shown in Fig. 12.6, is given in ASTM Designation C39. Some pertinent points included are as follows:

1. Cylinders must be tested as soon as practical after removal from curing and while in a damp condition.
2. The diameter of the cylinders must be measured to the nearest 0.01 in by averaging two diameters taken at right angles near the center of the specimen.
3. The specimen must be aligned in the center of the top bearing block of the testing machine. The bearing faces of the blocks must be plane to within 0.001 in. and the spherically seated bearing block must be on top and must have a face diameter at least slightly larger than the cylinder diameter. The bottom block must be at least that large but may be larger. The top bearing block must be able to rotate freely and to tilt slightly in any direction. Loading must be continuous at specified rates (34,000 to 85,000 lb/min for 6 × 12-in cylinders on hydraulic machines, or at 0.05 in/min on screw-type machines). Hand-operated machines do not comply with ASTM C39.
4. The report of compression-test results must give the maximum load and the unit strength calculated to the nearest 10 psi. It should also include information mentioned above, including age at test, specified strength, and any abnormalities observed. Some

FIGURE 12.6 Compression test of concrete cylinder.

specifications require the report to include type of break, such as "cone" or "shear." There is very little evidence that this information has any appreciable significance.

Splitting Tensile Strength. The splitting tensile strength of concrete cylinders is considered to be an accurate determination of the true tensile strength of concrete. The American Concrete Institute Standard 318-89, Building Code Requirements for Reinforced Concrete, permits use of actual splitting tensile strength in the computation of allowable stresses when structural lightweight-aggregate concrete is used.

The method for determination of this splitting tensile strength is given in ASTM Designation C496. Some of the pertinent points include:

1. The test specimens conform to the size, molding, and curing requirements given here for the compressive-strength specimens. When used for evaluation of lightweight concrete in accordance with American Concrete Institute Standard 318-89, eight specimens tested at 28 days shall be in an air-dry condition after 7 days of moist curing followed by 21 days of drying at 73.4 ± 3°F and 50 ± 5% relative humidity, as given in ASTM C330.
2. The testing machine shall conform to the requirements given here for the compressive-strength test of concrete cylinders. A supplementary bearing bar is required when the lower bearing block or upper bearing face is less than the length of the cylinder to be tested. The bar shall be machined to within ±0.001 in. of planeness. The width shall be at least 2 in, and the thickness not less than the distance from the edge of the spherical or rectangular bearing block to the end of the cylinder. The bar shall be used in such a manner that the load will be applied over the entire length of the cylinder.
3. Two bearing strips of nominal ⅛ in thick plywood, approximately 1 in wide and of length equal to or slightly larger than that of the cylinder, shall be provided adjacent to the top and bottom of the specimen during testing. Bearing strips shall not be reused.
4. Prior to positioning in the testing machine diametral lines shall be drawn on each end of the specimen using a suitable apparatus that will ensure that they are in the same axial plane.
5. Measurements shall be made of the diameter of the test specimen to the nearest 0.01 in by properly averaging three diameters, and measurements shall be made of the length of the specimen to the nearest 0.1 in by proper averaging of two lengths.
6. Positioning of the specimen, bearing strips, and bearing bar shall be made so as to ensure that the projection of the plane of the two diametral lines on the cylinder intersects the center of the upper bearing bar and bearing strips, and is directly beneath the center of thrust of the spherical bearing blocks.
7. The rate of loading shall be continuous without shock at a constant rate within the range of 11,300 to 22,600 lb/min for a 6 × 12 in cylinder.
8. The calculation of splitting tensile strength of the specimen shall be as follows:

$$T = \frac{2P}{\pi l d}$$

where T = splitting tensile strength, psi
P = maximum applied load indicated by the testing machine, lb
l = length, in
d = diameter, in

9. The report of splitting-tensile-strength test must give the strength calculated to the nearest 5 psi. It should also include identification number; diameter and length, in inches; maximum load, in pounds; estimated proportion of coarse aggregate fractured during test; age of specimen; curing history; defects in specimen; and type of fracture.

Flexural Specimens. The same ASTM Designation (C31) covers making and curing flexure-test specimens. The procedures are briefly as follows:

1. Beam molds must be rigid, watertight, and nonabsorptive. The size of beam for concrete containing aggregates up to a nominal size of 2 in is 6 × 6 × 20 in.
2. The beam molds should be filled in two equal layers and consolidated by rodding, for slumps of 3 in or more. They may be consolidated by vibration or rodding for slump between 1 and 3 in Vibration must be used for slump less than 1 in.
3. Rodding frequency for each layer is one stroke for each 2 sq in of surface. After rodding, the sides and ends are spaded with a suitable tool such as a mason's trowel. Tapping the sides may be used to close voids if necessary.
4. When vibration is used, the vibrator is applied uniformly to ensure consolidation without segregation. Internal vibrator is spaced at intervals not exceeding 6 in.
5. After consolidation, the top surface is struck off with a straightedge and finished with a wood or magnesium float. Overfinishing must be avoided and proper identification made.
6. Loss of moisture must be prevented by covering the beams with a plate, plastic, or double layer of wet burlap. All other conditions as described for making compressive-strength specimens must be followed.
7. For beams for checking a laboratory mix or for basis of acceptance, remove specimens from mold between 20 and 48 h after molding and cure the same as cylinders except that a storage for a minimum of 20 h in saturated lime water at 73.4° ± 3 prior to testing is required.

ASTM Designation C78 covers the procedure for making flexural tests. The following general requirements are necessary:

1. Loading is on the third points of the beam at a span of three times the depth of the beam. Hence, for 6 × 6-in beams, this would be 18 in span with the points of load application 3 in each side of the center.
2. The loading device must be able to apply force to the beam vertically without eccentricity, and with the load and reactions remaining parallel.
3. Specimens must be soaked in water for at least 40 h before testing. It is absolutely essential that the specimens be kept moist while testing. Uniformity of moisture conditions throughout is essential. One dry side may affect results drastically.
4. Beams must be loaded on their sides as cast. If the load points are not in full contact, capping or leather shims must be used. Loading rate for a 6 × 6 in beam is 1800 lb/min. Calculations are based on measurements of the dimensions at place of failure to the nearest 0.1 in.
5. Flexural strength is calculated as modulus of rupture, when fracture occurs within the middle third of the span length, by the formula

$$R = \frac{Pl}{bd^2}$$

where P = maximum applied load
l = span length, in
b = average width of specimen, in
d = average depth of specimen, in

Breaks outside the middle third of span length should be disregarded.

12.10 Unit Weight and Yield

To ensure conformance to specifications, field tests are made to determine yield and cement factor of the concrete being placed. ASTM Designation C138 covers this method by utilizing the unit weight of the plastic concrete. The method, generally summarized, is as follows:

1. Equipment required is a scale accurate to 0.3 percent of the test load, a standard tamping rod, a strike-off plate, and a measure of proper capacity. The size depends upon the size of aggregate in the concrete. Use the 1 ft^3 measure when the concrete contains aggregate larger than 2 in. A ½ ft^3 measure is usually used for aggregates up to 2 in nominal size. Figure 12.7 shows the necessary equipment.
2. Fill the measure in three equal layers and rod 25 or 50 strokes per layer for the ½ and 1 ft^3 measures, respectively. The method of rodding is similar to that described for the air-content test and for casting concrete cylinders. For measures of 0.4 ft^3 or greater, consolidation may be performed by internal vibration in the same manner as for cylinders.
3. After consolidation, strike off the top carefully with a rigid flat plate large enough to cover the measure, by a sawing action, while maintaining contact with the edges of the measure at all times. Striking off with a straightedge is not satisfactory as it usually yields falsely high results.
4. Weigh the measure to the scale accuracy and calculate the weight per cubic foot.
5. Total the weight of all the ingredients in the batch and divide it by the unit weight as determined above. This results in the calculated cubic feet of concrete which, when divided by 27, gives the actual number of cubic yards per batch.
6. Determine the cement factor by dividing the total quantity of cement in the batch by the number of cubic yards.

FIGURE 12.7 Equipment for unit-weight test of concrete or aggregates.

7. Determine yield in cubic feet of concrete per bag of cement by dividing 27 ft^3/yd^3 by cement factor determined above.
8. A practical consideration is the yield in actual cubic feet per cubic yard ordered. This is determined by dividing the number of cubic feet of concrete by the number of cubic yards ordered. Any amount under 27 ft^3 would indicate a low yield; anything over 27 ft^3, a high yield. A reasonable tolerance would be $\pm 1\%$.

Example. Given: The 6-yd-batch weights of the first example in Chap. 11.

$$\text{Cement} = 3{,}828 \text{ lb}$$
$$\text{Sand, including 5\% moisture} = 6{,}993 \text{ lb}$$
$$\text{Coarse aggregate} = 11{,}880 \text{ lb}$$
$$\text{Mixing water added} = 1{,}467 \text{ lb}$$
$$\text{Total} = 24{,}168 \text{ lb}$$

Weight per cubic foot from above test is 145.9 pcf.

$$\text{Volume of batch, ft}^3 = \frac{24{,}168}{145.9} = 165.65$$

$$\text{Volume of batch, yd}^3 = \frac{165.65}{27} = 6.14$$

$$\text{Cement factor, lb / yd}^3 = \frac{3828}{6.14} = 623$$

$$\text{Cement factor, bags per cu yd} = \frac{623}{94} = 6.63$$

$$\text{Yield, ft}^3 \text{ per bag} = \frac{27}{6.63} = 4.07$$

$$\text{Relative yield, ft}^3 / \text{yd}^3 \text{ designed} = \frac{165.65}{6} = 27.61$$

$$\text{Variations, \%} = \frac{+0.61}{27} \times 100 = 2.26 \text{ overyield}$$

12.11 Unit Weight of Structural Lightweight Concrete

Density of the lightweight concrete at the age of 28 days for design control is determined by ASTM Method C567, Unit Weight of Structural Lightweight Concrete. The unit weight of the freshly mixed concrete is determined in accordance with ASTM C138, Unit Weight of Concrete, except that vibration of specimens as described in ASTM C192, Making and Curing Concrete Test Specimens in the Laboratory, shall be permitted. This is used for placement control.

The air-dry unit weight is determined on 6 × 12 in cylinders, generally in the following manner. The cylinders are made in accordance with ASTM C192.

The cylinders are covered and cured at 60 to 80°F in a wrapped or sealed condition for 7 days. On the seventh day, the cylinders are removed from cylinder molds or plastic covering and dried for 21 days in 50 ± 2% relative humidity. Determine the weight of the cylinders at 28-day age as dried; then completely immerse the clinders in water for 24 h. Determine the immersed weight and the saturated surface-dry weight. Calculate the weight per cubic foot of concrete in accordance with the following equation:

$$\text{Wt/ft}^3 = \frac{62.3A}{B - C}$$

where A = 28-day weight of concrete cylinder, as dried, lb
B = saturated, surface, dry weight of cylinder, lb
C = immersed weight of clyliner, lb

12.12 Uniformity Tests of Truck Mixers

ASTM Designation C94, Standard Specification for Ready Mixed Concrete, stipulates that when the concrete is mixed completely in a truck mixer, 70 to 100 revolutions at manufacturer's specified mixing speed, it is necessary to produce the uniformity of concrete indicated in Table 12.1. Concrete uniformity tests may be made to determine if the uniformity is satisfactory and at least five of the six tests shown in Table 12.1 must be within the specified limits. Slump tests of individual samples taken after discharge of approximately 15 and 85 percent of the load may be made for a quick check of the probable degree of uniformity. These two samples shall be obtained with an elapsed time of not

TABLE 12.1 Requirements for Uniformity of Concrete

Test	Requirement, expressed as maximum permissible difference in results of tests of samples taken from two locations in the concrete batch
Weight per cubic foot (weight per cubic meter) calculated to an air-free basis	1.0 lb (16 kg/m^3)
Air content, % by volume of concrete	1.0
Slump:	
If average slump is 4 in (100 mm) or less	1.0 in (25 mm)
If average slump is 4 to 6 in (100 to 150 mm)	1.5 in (38 mm)
Coarse aggregate content, portion by weight of each sample retained on No. 4 (4.76 mm) sieve, %	6.0
Unit weight of air-free mortar* based on average for all comparative samples tested, %	1.6
Average compressive strength at 7 days for each sample,† based on average strength of all comparative test specimens, %	7.5‡

*Test for Variability of Constituents in Concrete, Designation 26, p. 558 U.S. Bureau of Reclamation, 1975. "Concrete Manual," 8th ed.[2] Available from Superintendent of Documents, U.S. Government.
†Not less than 3 cylinders will be molded and tested from each of the samples.
‡Tentative approval of the mixer may be granted pending results of the 7-day compressive-strength tests.
Source: ASTM C94-89b, Table A1.1.

more than 15 min. If these two slumps differ more than that specified in Table 12.1, the mixer shall not be used unless the condition is corrected or when operation with a longer mixing time, a smaller load, or more efficient charging sequence will permit the requirements of Table 12.1 to be met.

12.13 Uniformity Tests of Stationary Mixers

ASTM Designation C94 specifies mixing time of 1 min for 1 yd^3 mixers and an additional 15 for each cubic yard, or fraction thereof, of additional capacity.

It is considered that these minimum time requirements will produce uniformity, but they are known to be unnecessarily time-consuming for many of the large-capacity mixers. Therefore, provisions have been incorporated in the specification to permit reduction in this mixing time, provided tests are conducted to show that uniformity can be achieved in less time.

A 6 yd^3 mixer under the time restriction alone would be required to mix each full batch of concrete 2¼ min. If the owner has reason to believe that uniformity can be achieved in less time, he or she may have tests made after any predecided time of mixing, and if samples taken from near the two ends of the batch meet specified uniformity requirements, the mixing time can be reduced accordingly. The uniformity requirements are shown in Table 12.1.

12.14 Method of Test for Coarse-Aggregate Content

One measure of uniformity of a concrete batch is the percentage of coarse aggregate in two different portions of the batch (see Table 12.1). The procedure for this determination is as follows:

1. Weigh a sample of plastic concrete. In order to combine this determination partially with that for the air-free unit weight of mortar, the base of an air meter (¼ ft^3 capacity for nominal-size aggregates up to 1½ in or ½ ft^3 capacity for aggregate up to 3 in is used for unit-weight determination. The sample is then tested for air content.
2. After the air test, wash and sieve over a No. 4 screen.
3. Weigh the aggregate while immersed in water.
4. Compute the saturated surface-dry weight using the known bulk saturated surface-dry specific gravity G of the aggregate by use of the formula given in ASTM Designation C127, Specific Gravity and Absorption of Coarse Aggregate.

$$\text{SSD weight of coarse aggregate} = \frac{\text{immersed weight} \times G}{G - 1}$$

$$\text{W} = \text{coarse - aggregate content (\% by weight of sample)}$$

$$= \frac{\text{SSD weight of coarse aggregate}}{\text{weight of sample of concrete}} \times 100$$

5. In lieu of the immersed-weight method, the washed coarse aggregate may be towel-dried and weighed. However, the immersed-weight method is recommended.

12.15 Method of Test for Air-free Unit Weight of Mortar

The method for air-free unit weight of mortar is described under Variability of Constituents in Concrete, Designation 26, p. 558, in the "Concrete Manual."[2] The procedure is summarized as follows:

1. Perform the weight tests as described in Art. 13 Sec. 12.13, steps 1 through 4.
2. Determine the weight of the mortar in the sample by subtracting the SSD weight of the coarse aggregate from the weight of the concrete sample.
3. The volume of the air-free mortar is obtained by subtracting the volumes of the coarse aggregate and volumes of air.
4. The air-free unit weight of mortar is its weight divided by volume and calculated by the following formula:

$$M = \frac{b - c}{V - \left(A + \dfrac{c}{G \times 62.3}\right)}$$

where M = unit weight of air-free mortar, pcf
b = weight of concrete sample in air meter, lb
c = SSD weight of aggregate retained on No. 4 sieve, lb
V = volume of sample, ft^3
A = volume of air computed by multiplying the volume of container V by % of air divided by 100
G = SSD specific gravity of coarse aggregate

5. The following example is given to illustrate the method of unit weight of air-free mortar determinations and coarse-aggregate content test as described above.

Given: Concrete sample containing 1½ in. maximum nominal size aggregate.

G = bulk specific gravity of coarse aggregate SSD = 2.65
b = weight of concrete sample in ¼ cu ft V measure of air meter = 35.0 lb
Air content on this sample = 5.0 %
Submerged weight of aggregate retained on No. 4 sieve = 12.0 lb

Calculation:

$$c = \text{SSD weight of coarse aggregate} = 12.0 \times \frac{2.65}{2.65 - 1} = 19.27$$

$$W = \text{coarse - aggregate content (\% by weight of sample)}, \frac{19.27}{35} \times 100 = 55.0\,\%$$

$$A = \text{volume of air} = \frac{0.25\ \text{ft}^3 \times 5.0}{100} = 0.0125\ \text{ft}^3$$

$$M = \text{unit weight of air - free mortar} = \frac{35 - 19.27}{0.25 - \left(0.0125 + \dfrac{19.27}{2.65 \times 62.3}\right)} = 130.22\ \text{pcf}$$

In order to determine compliance of a truck or stationary mixer, one must first make identical determinations on each of two samples, one from near each end of the batch. Following the determination and calculations of the second sample, conducted as above, let us assume that the results of this second sample were 60 percent coarse-aggregate content and 132 pcf unit weight of air-free mortar. Then we compare results as follows:

Test	Front	Back	Variation	Specification requirements expressed as maximum permissible difference between samples. ASTM C94
Coarse-aggregate content, % of each sample retained on No. 4 sieve	55	60	5.0	6.0
Unit weight of air-free mortar	130.2	132.0	0.7 %, 1.8 pcf	1.6%*

*Percent variation based on average of comparative samples calculated as follows:

$$\text{Average unit weight of air - free mortar} = \frac{130.2 + 132.0}{2} = 131.1 \text{ pcf}$$

$$\text{Variation in unit weight from average} = \frac{131.1 - 130.2}{131.1} \times 100 = 0.7\,\%$$

Methods of comparison of other tests as given in Table 12.1 are self-evident. Test results conform to uniformity requirements of the specifications.

The above example illustrates the method of uniformity determinations by use of the given formulas. A tabular method adapted from the U.S. Bureau of Reclamation "Concrete Manual" Designation 26 may be preferred. An example of this determination of unit weight of air-free mortar is shown in Table 12.2.

12.16 Determination of Modulus of Elasticity

The modulus of elasticity is the ratio of stress to strain. Unlike most metals, concrete has no true "straight-line" portion of a stress-strain curve, although for most modern concretes the lower portion of the curve is very nearly straight. Since there is no true straight-line portion, there can be no true proportional limit and the modulus will be different at any selected stress point.

Three methods of expressing and calculating the modulus of elasticity have been proposed. These are known as "chord modulus," "secant modulus," and "tangent modulus." "Chord modulus" is probably the one most used.

ASTM Designation C469 covers the procedure for this determination. The method generally consists of measuring the strain at various loads to establish data for a stress-strain curve. This is accomplished by clamping two metal rings on a 6 × 12 in cylinder, one near the top and one near the bottom, not over 8 in apart, as shown in Fig. 12.8.

The movement of the rings is measured by a dial gauge and the strain is calculated by dividing the total measured movement by the gauge length. Electric strain gauges of at least 6 in lengths are sometimes used in lieu of rings and a dial gauge for the purpose of measuring the strain.

TABLE 12.2 Example of Computation for Unit Weight Air-Free Mortar
Plant BJA Project FBA Mixer JEA Date 8-12-65
Inspector RJA Time of test 1:45 p.m. Slump 2 in
Mixing time 1 min Batch No. 77 Mix No. 7
Maximum size of aggregate 1½ in Specific gravity of coarse aggregate 2.65

	Sample from front of mixer or first portion of batch as discharged from mixer		Sample from back of mixer or last portion of batch as discharged from mixer	
	Weight, lb	Volume, ft^3	Weight, lb	Volume, ft^3
Weight and volume of sample in air meter	35.00	0.2500	35.16	0.2500
Air content by air meter	5.00 %		4.75 %	
Volume of air		0.0125		0.0119
Weight and air-free volume of sample	35.00	0.2375	35.16	0.2381
Submerged weight of sample retained on No. 4 screen	12.00		11.50	
Computed SSD weight and solid volume of plus 4 material*	19.27	0.1170	18.49	0.1120
Weight and volume representing mortar in sample	15.73	0.1205	16.67	0.1261
Computed unit weight of air-free mortar, pcf	130.22		132.00	

Note: Comparison of unit weight of air-free mortar for compliance with the given specification is the same as shown previously when computed by the formula.

$$\text{*SSD weight of coarse aggregate} = \frac{\text{immersed weight} \times G}{G - 1}.$$

The chord modulus is the slope of a straight line originating at the point on the curve corresponding to 50 μin/in strain and extending to the stress point selected. ASTM C469 recommends that the chord modulus be calculated at 40 percent of the ultimate strength. Figure 12.9 illustrates determination of chord modulus at 1600 psi:

$$E = \frac{S_2 - S_1}{\epsilon_L - 0.000050}$$

where E = chord modulus of elasticity, psi
S_2 = stress corresponding to maximum applied load, psi
S_1 = stress corresponding to a longitudinal strain of 50 μin/in, psi
ϵ_L = longitudinal strain produced by stress S_2

$$E = \frac{1600 - 300}{0.00042 - 0.00005}$$

$$E = 3{,}513{,}500$$

The secant modulus and the tangent modulus are sometimes denoted in engineering texts. The secant modulus is similar to the chord modulus except that it is the slope of a line

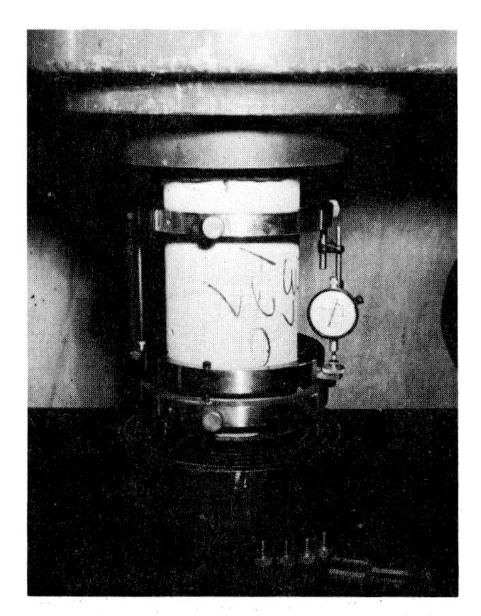

FIGURE 12.8 Modulus of elasticity test setup for determination of static Young's modulus.

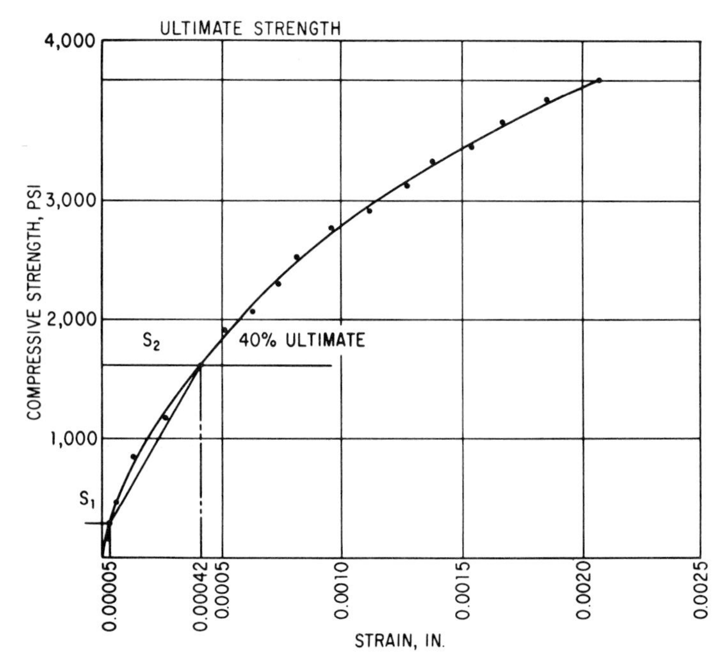

FIGURE 12.9 Typical stress-strain curve for calculation of modulus of elasticiy.

from the origin to any selected point on the curve. The tangent modulus is the slope of a line tangent to any selected point on the curve.

The modulus of elasticity of concrete will usually increase with age similar to strength development. Although modulus of elasticity is not directly proportional to strength, concretes of higher strengths have higher moduli of elasticity. This is recognized by the ACI Building Code. The determination of modulus of elasticity is not necessary for normal concrete testing but is sometimes required for special study.

The modulus of elasticity can also be calculated at any desired stress, although it is usually at 40 percent of the ultimate strength, using the conventional formula

$$E = \frac{\text{unit stress}}{\text{unit strain}}$$

12.17 Determination of Specific Gravity Factor of Structural Lightweight Aggregate

This specific gravity factor, as given in ACI 211.2-69, is the relationship of weight to displaced volume for lightweight aggregates and is determined either dry or moist for use in making adjustments in proportions of lightweight concrete. The basic apparatus is a pycnometer consisting of a narrow-mouth 2 qt mason jar with a spun brass pycnometer top. Two representative samples are obtained for each size of lightweight aggregate to be tested. The first is weighed, and then dried in an oven at 105 to 110°C or on a hot plate to constant weight.

The dry aggregate weight is recorded, and the aggregate moisture content (percent of aggregate dry weight) is calculated. This value is used to compute the specific gravity factor and moisture-content curve shown in Fig. 11.4 (Chap. 11). The second aggregate sample is weighed (weight C in grams). The sample is then placed in the empty pycnometer; it should occupy one-half to two-thirds of the pycnometer volume.

Water is added until the jar is three-fourths full, and the entrapped air is removed by rolling and shaking the jar. After all the air is dispelled, the jar is filled to full capacity and weighed. The pycnometer specific gravity factor is calculated by the formula

$$S = \frac{C}{C + B - A}$$

where A = weight of pycnometer charged with aggregate and then filled with water, grams
B = weight of pycnometer filled with water, grams
C = weight of aggregate tested, moist or dry, grams

The larger test samples of coarse aggregate that cannot be evaluated in the pycnometer can be tested for specific gravity factor by the equivalent weight in air and water procedure of ASTM C127. The top of the container used for weighing the aggregates under water must be closed with a screen to prevent light particles from floating.

Specific-gravity factors by this method are calculated by the formula

$$S = \frac{C}{C - E}$$

where C = weight in air at particular moisture content, grams
E = weight of coarse aggregate sample under water, grams
S = specific gravity factor at particular moisture content

The specific-gravity determinations are made at various periods of immersion dependent upon conditions contemplated in the field.

12.18 Durability

The usually accepted method of evaluating durability is by freezing and thawing in accordance with ASTM Designation C666, Standard Test for Resistance of Concrete to Freezing and Thawing, Procedure A, Rapid Freezing and Thawing in Water, and Procedure B, Rapid Freezing in Air and Thawing in Water, using equipment similar to that depicted in Fig. 12.10. These methods will not yield an accurate correlation with months or years of natural exposure, but are valuable for comparison purposes. Concrete specimens of specific size are cast and cured for a specified time and method. These are then placed in freezing and thawing apparatus equipped as necessary with refrigerating units, heating units, fans, and water pumps to produce reproducible cycles with specified temperature and time requirements. This is usually attained by use of automated equipment utilizing electronically programmed cycling apparatus. Approximately 8 cycles per day can be achieved in many of the units. Fundamental transverse frequency determinations are made periodically on each specimen, and from these the relative dynamic modulus of elasticity is calculated after 300 cycles, or until the relative dynamic modulus of elasticity of the specimen drops to 60 percent of the initial modulus. The durability factor is then computed for each specimen.

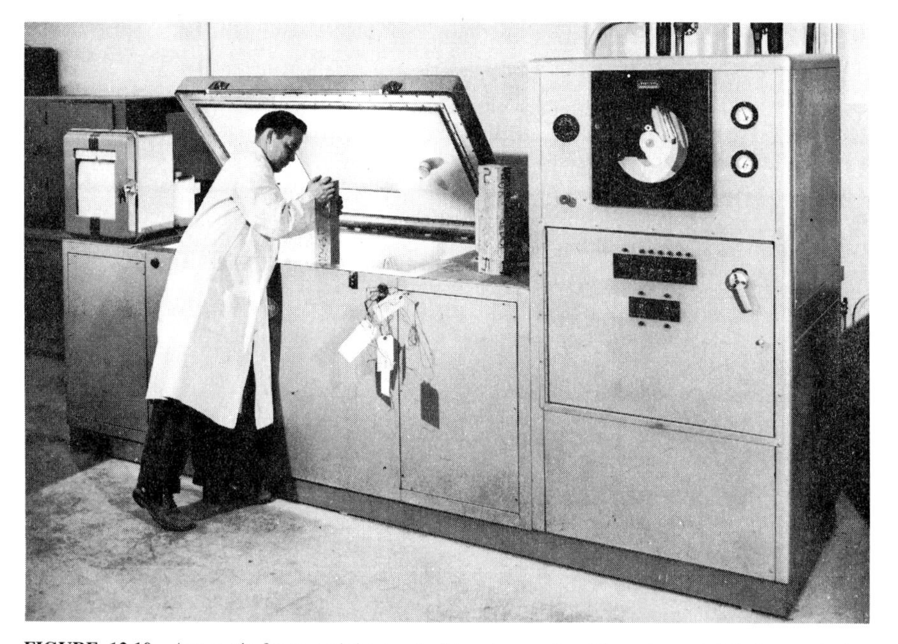

FIGURE 12.10 Automatic freeze-and-thaw unit, 12 cycles per day. (Recorder on left is for monitoring temperatures in bar with thermocouple wires shown in front. Recorder in cabinet is part of control mechanism and records cycles and cabinet temperatures.)

12.19 Cores from Hardened Concrete

The compressive strength of concrete is sometimes determined from cores extracted from a structure. This is usually performed when the compressive strength of a given concrete is not known or because of questionable concrete cylinder test results. The cores are secured by use of a core drill normally utilizing diamond drill bits of the type pictured in Fig. 12.11. Care must be exercised not to cause overheating and to avoid obvious defects such as rock pockets and joints. When the core is used for compressive-strength determinations, the length should preferably be as near as possible to twice the diameter, and the diameter should be at least three times the maximum size of aggregate in the concrete. Cores having lengths less than the diameter are unsuitable for compression testing. The ends should be sawed if necessary to produce an even surface, and the core should be ground or capped to enable proper compression test. Moisture condition at time of test will affect the results. Concrete tested while dry will have a higher indicated compressive strength than concrete tested in the saturated condition. For many years the only recognized moisture condition was that obtained by immersion in lime-saturated water for at least 40 h immediately prior to testing. However, a note has now been added to the standard method of testing cores to the effect that other moisture conditions may be used at the option of the agency for which the test is being conducted. The details of this test are covered in ASTM C42. The correlation between indicated core strengths and cylinder

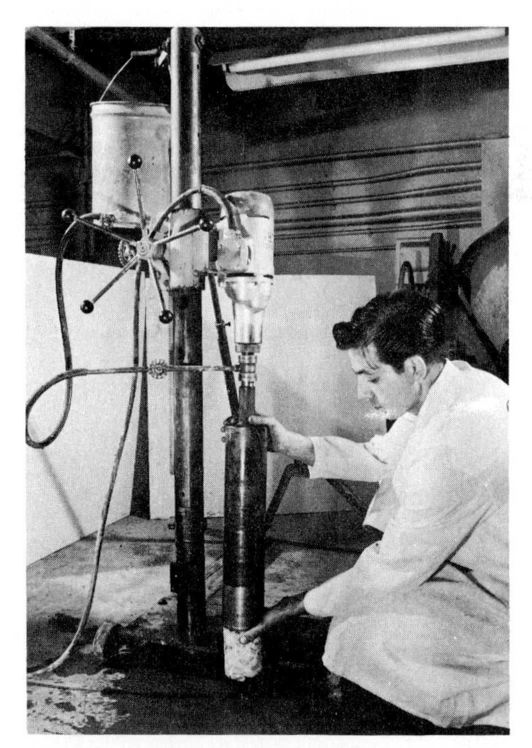

FIGURE 12.11 Core drilling inside building with portable equipment and using diamond bit.

strengths of the same concrete is the subject of much controversy. Some excellent investigators have found excellent agreement; others have found core strengths to be higher; others, lower.[3–5]

12.20 Impact Hammer

An approximate or relative compressive-strength determination can be made by the impact hammer, sometimes called a "Swiss" hammer. This hammer utilizes a spring-driven plunger, and the rebound of the plunger after impact with the concrete surface is recorded on a built-in scale. The instrument carries a chart which correlates the rebound number to compressive strength. This correlation should be accepted with caution and skepticism. It is recommended that determinations made with this instrument be used exclusively for comparative purposes on concrete of the same composition, age, and moisture content, and that judgment be used in the interpretation. This is critical because the rebound is affected by many factors such as moisture content of the concrete, type of surface finish, type of aggregate, and age of the concrete. Although there are many data purporting to show the accuracy of the instrument, data shown in Table 12.3 taken from several projects, comparing impact-hammer results with immediately adjacent cores, show the potential fallibility of accepting the correlation given on the instrument.

It is not intended to suggest that the extreme lack of correlation shown in Table 12.3 is typical. These are unusual cases, but they do show the danger of accepting impact-hammer test results with full confidence.

Probably in the majority of cases, results taken from the curve which is attached to the hammer are reasonably accurate, but there are enough wide variations from this relation to introduce doubt into any specific investigation.

If the hammer is calibrated against actually determined strengths of a given mix of concrete for a given project, under approximately comparable moisture conditions and approximately uniform age, it can be an excellent tool. It is of great value—when used with proper understanding and discretion—for checking a specific project to investigate and isolate an area of suspected low strength.

TABLE 12.3 Comparison of Core Strengths per ASTM C42 and Indicated Strengths as Determined by Impact Hammer

Core No.	Strengths as determined on cores per ASTM C42, psi	Indicated strength* of immediately adjacent concrete based on impact hammer tests, psi	Core No.	Strengths as determined on cores per ASTM C42, psi	Indicated strength of immediately adjacent concrete based on impact-hammer tests, psi
W 1	3400	4000	1	3890	6500
W 2	3150	4200	2	3840	6800
W 3	2700	3750	3	5040	7400
S 4	2480	4400	4	4650	8000
S 5	3450	3250	5	4630	7400
E 6	2190	3000	6	4490	7600
E 7	2280	4200			
E 8	2200	3250			
E 9	2340	3250			
N 10	3130	3250			

*Each impact-hammer "indicated strength" is based on the average of 10 most uniform of 15 readings and on the curve that accompanies the hammer.

12.21 Penetration Probe Gauge

A proprietary method for the nondestructive determination of compressive strength of concrete in the field is known as the *Windsor probe* and utilizes the principle that the penetration of the probe gauge is inversely proportional to the compressive strength of the concrete being tested. Generally, three probe gauges are driven into the concrete using a powder-actuated device of a specific concrete muzzle energy-producing force (Fig. 12.12).

The compressive strength is determined by correlation of the kinetic energy to compressive strength. An approximate relationship of probe penetration to compressive strength is given in a table furnished by the manufacturer. The moh hardness rating (scratch test) of the coarse aggregate is required for use of the table. The average of three probe values (exposed height of probe gauge) is used with the table (ASTM C803-82, Penetration Resistance of Hardened Concrete).

Intensive study of this method was made in the laboratories of the National Ready Mix Concrete Association. The conclusions of this study are that this probe method is comparable to the concrete test hammer in accuracy. Therefore this test should not be used as an alternate to ASTM C39, Test of Compressive Strength of Molded Cylinders, but is suitable for in-place, nondestructive approximation of strength.

12.22 Pullout Test

The pullout test was developed to determine in-place strength of concrete in order to commence posttensioning, remove forms and shoring, and terminate winter protection and curing. The method is used to determine the force necessary to pull out an embedded metal insert and its attached concrete fragment from a concrete specimen or structure. The metal insert is embedded in fresh concrete and pulled by a jack reacting against a bearing ring.

FIGURE 12.12 The Windsor probe is based on the principle that the penetration of a powder-driven probe into the concrete is inversely proportional to the strength of the concrete.

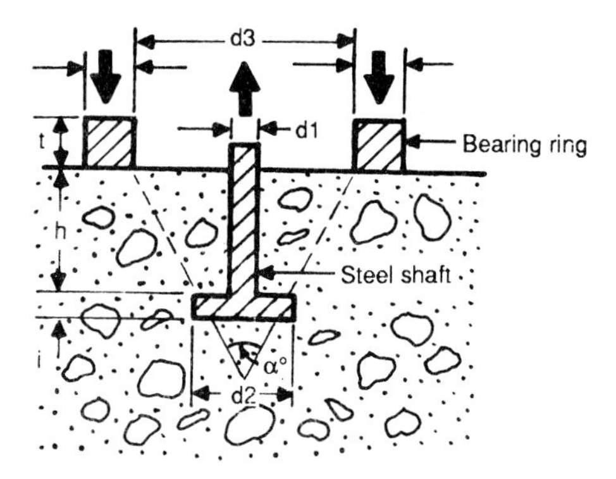

FIGURE 12.13 Apparatus for pullout test.

A relationship can be developed to other strength results when both pullout test specimens and other test specimens are of consistent size and cured under similar conditions. A schematic diagram of the apparatus can be seen in Fig. 12.13. The test method is outlined in ASTM C900, Standard Test Method for Pullout Strength of Hardened Concrete.

12.23 Other Tests

Other vital tests on concrete include volume change, bcnd to steel, bleeding, and setting time. These are usually performed in the laboratory under applicable ASTM designations and primarily as qualification tests. Structural load tests are usually performed in accordance with the ACI Building Code Specifications and are generally performed when there is a question of the concrete strength in a specific structure.

Pulse-velocity tests can be used on structures to detect progressive deterioration, concealed cracking, honeycombing, and other properties. Some investigators have attempted to determine compressive strengths by pulse-velocity determinations. Few, if any, authentic data are now available indicating that this can be done with any degree of accuracy. This test is excellent when questionable concrete can be compared against known concrete having satisfactory strengths and of the same proportions and materials.

ANALYSIS OF STRENGTH RESULTS

12.24 Terminology

The discussion of analysis of strength results involves the mathematics of statistical analysis. Some of the terms and their customary symbols used may be somewhat unfamiliar to many engineers, scientists, and technicians. We are therefore providing a series of definitions of these terms with the symbols customarily used in connection with them:

$\overline{X}$ average strength—the numerical average of all strengths under consideration deviation—the numerical difference between any given strength and the average strength

σ standard deviation—the rms (root-mean-square) deviation of the strengths from their average. The standard deviation is found by extracting the square root of the average of the square of deviation of individual strengths from their average; thus

$$\sigma = \sqrt{\frac{(X_1 - \overline{X})^2 + (X_2 - \overline{X})^2 + \ldots + (X_n - \overline{X})^2}{n}}$$

n total number of tests

V coefficient of variation—the standard deviation expressed as a percentage of the average strength

$$V = \frac{\sigma}{\overline{X}} \times 100$$

R range of strengths—the numerical difference between the highest and lowest strengths of tests under consideration

f_c' specified strength—self-explanatory

f_{cr} required average strength—the average strength, determined mathematically, that will be required to produce not more than a specified number of test results below the specified strength of f_c'

$$f_{cr} = \frac{f_c'}{1 - tV}$$

where t = a constant dependent on proportion of tests permitted below f_c' and on a number of tests used to establish V (see Table 12.4)

σ_1 within-test standard deviation—standard deviation between individual specimens of a given test, preferably calculated on at least 10 sets of specimens

$$\sigma_1 = \frac{1}{d_2}\overline{R}$$

where $\frac{1}{d_2}$ = a constant depending on number of samples (see Table 12.5)

$\overline{R}$ average range—the average range R of the sets of specimens used to determine σ_1

V_1 within-test coefficient of variation—the variation between results of cylinders in a given batch expressed as the coefficient (see V above)

$$V_1 = \frac{\sigma_1}{\overline{X}} \times 100$$

Test: A group of specimens (usually two or three) all made from the same sample and all tested at the same age.

TABLE 12.4 Values of t

Percentage of tests falling within the limits $\overline{X} \pm t\,\sigma$	Chances of falling below lower limit	t
40	3 in 10	0.52
50	2.5 in 10	0.67
60	2 in 10	0.84
68.27	1 in 6.3	1.00
70	1.5 in 10	1.04
80	1 in 10	1.28
90	1 in 20	1.65
95	1 in 40	1.96
95.45	1 in 44	2.00
98	1 in 100	2.33
99	1 in 200	2.58
99.73	1 in 741	3.00

Source: ACI Standard 214-77 (reapproved 1988), Table 4.1.

12.25 Quality-Control Requirements

The control of concrete quality encompasses the wide spectrum of conformance of each component of concrete to a specification, as well as to the mixing, placing, and curing techniques. Generally, the efficiency of concrete manufacture, testing, and material control will determine the degree of uniformity and quality of concrete. In order to determine the degree of concrete uniformity and quality, it is necessary to apply statistical methods of analysis. The American Concrete Institute Recommended Practice for Evaluation of Compression Test Results of Field Concrete[4] was developed for this purpose. In order to develop this type of study, a sufficient number of field compression tests must be made in accordance with the acceptable standards. Concrete strength-test results will fall in the normal frequency-distribution curve as shown in Fig. 12.14. The primary conclusion

TABLE 12.5 Factors for Computing Within-test Standard Deviation

No. of specimens	d_2	$1/d_2$
2	1.128	0.8865
3	1.693	0.5907
4	2.059	0.4857
5	2.326	0.4299
6	2.534	0.3946
7	2.704	0.3698
8	2.847	0.3512
9	2.970	0.3367
10	3.078	0.3249

Source: ACI Standard 214-77 (reapproved 1988), Table 4.1.

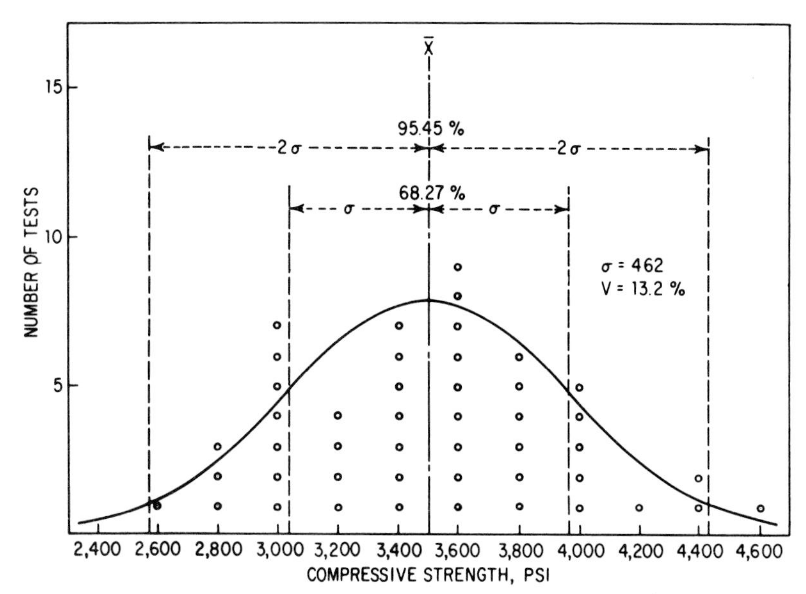

FIGURE 12.14 Normal frequency distribution of strength data from 46 tests.[4]

from this analysis is the fact that variations will occur and an absolute minimum compressive-strength specification is unrealistic.

It can be noted that the law of probability indicates that one strength out of every six tests will be more than standard deviation σ below the average; one out of every 44 tests will be more than twice the standard deviation below average and one out of every 741 tests will be more than three times the standard deviation below average.

12.26 American Concrete Institute Code Strength Requirements

The American Concrete Institute Building Code criteria for evaluation of concrete are based on the condition that the average strength of concrete produced must always exceed the specified value of f_c' that was used in the structural design phase. The probability concept is used to provide for single test values that are lower than f_c' It specifies, "The concrete strength is considered to be satisfactory as long as averages of any three consecutive tests remain above the specified f_c' and no individual test falls below the specified f_c' by more than 500 psi."

The ACI Building Code requires that samples for strength tests of each class of concrete shall be taken not less than once a day nor less than once for each 150 yd^3 of concrete or for each 5000 ft^2 of surface area placed. The samples for strength tests shall be taken in accordance with ASTM C172, Method of Sampling Fresh Concrete. Cylinders for acceptance tests shall be molded and laboratory cured in accordance with ASTM C31, Method of Making and Curing Concrete Compressive and Flexural Strength Test Specimens in the Field and tested in accordance with ASTM C39, Method of Test for Compressive Strength of Molded Concrete Cylinders. Each strength test result shall be the average of two cylinders from the same sample tested at 28 days or the specified earlier age.

If individual tests of laboratory-cured specimens produce strengths more than 500 psi below f_c' and computations indicate that the load-carrying capacity may have been significantly reduced, tests of cores drilled from the area in question may be required in accordance with ASTM C42, Method of Obtaining and Testing Drilled Cores and Sawed Beams of Concrete. Three cores shall be taken for each case of a cylinder test more than 500 psi below f_c'. Concrete in the area represented by the core tests will be considered structurally adequate if the average of the three cores is equal to at least 85 percent of f_c' and if no single core is less than 75 percent of f_c'. To check testing accuracy, locations represented by erratic core strengths may be retested. If these strength acceptance criteria are not met, the responsible authority may order load tests conducted in accordance with the ACI Building Code.

12.27 Quality-Control Charts

A convenient and reliable method of graphically analyzing the trend of concrete strengths is by the use of a control chart. Figure 12.15 is an example of a control chart for structural concrete showing individual 28-day compressive-strength tests. A curve of this type shows how many cylinders fall below the specified strength and thus indicates whether there is compliance with the specification. The daily plotting of this curve conveniently indicates whether a detrimental trend has occurred. If so, prompt remedial action such as a change in mix design, materials, mixing, etc., should be taken. The specified strength f_c' (design data) and the required average strength f_{cr} are included in this curve. The f_{cr} is calculated from the following formula:

$$f_{cr} = \frac{f_c'}{1 - tV}$$

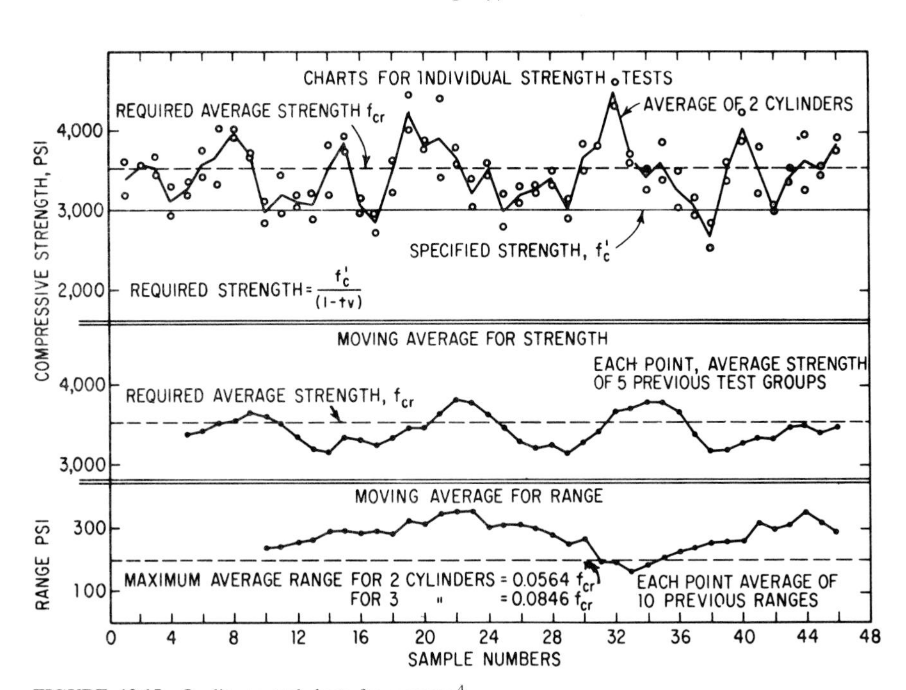

FIGURE 12.15 Quality-control charts for concrete.[4]

where t is taken from Table 12.4 and V is the coefficient of variation.

Another curve which is helpful in this type of analysis is the moving-average curve, also shown in Fig. 12.15. The moving average for strength is plotted for the previous five sets of companion cylinders, and by specification the lower limit is the specified strength f_c' The moving average for range is plotted for the previous 10 groups of companion cylinders.

It is considered that the within-test variation V_1 should be a maximum 5 percent for good control. Therefore, the maximum for average range in control becomes

$$\overline{R}m = (0.05 \times 1.128) f_{cr} = 0.0564 f_{cr} \quad \text{(for two cylinders in a group)}$$
$$\overline{R}m = (0.05 \times 1.693) f_{cr} = 0.08465 f_{cr} \quad \text{(for three cylinders in a group)}$$

12.28 Statistical Evaluation

Statistical data are applied to concrete strength of each class or type of concrete separately. In addition to the use of statistical formulas for establishing control charts, prescribed statistical methods are used for computing standard deviation and coefficient of variation. The procedure used is as follows:

1. Compute $\overline{X}$, the average 28 day strength of all test specimens (preferably after at least 30 tests are available), by the following formula:

$$\overline{X} = \frac{X_1 + X_2 + X_3 + \cdots + X_n}{n}$$

where $X_1, X_2, X_3 \ldots, X_n$ are strengths of individual tests and n is the total number of tests.

2. Compute the standard deviation σ (see the normal-frequency curve for graphic illustration) by the following formula:

$$\sigma = \sqrt{\frac{(X_l - \overline{X})^2 + (X_2 - \overline{X})^2 + \cdots + (X_n - \overline{X})^2}{n}}$$

3. Compute the coefficient of variation V by the formula

$$V = \frac{\sigma}{\overline{X}} \times 100$$

Table 12.6 gives ratings that are commonly assigned to various coefficients of variation. As can be noted from Fig. 12.16, the higher the coefficient, the higher the average strength must be and, therefore, the more costly the design mix will be. This emphasizes the value of good quality-control operations.

4. As an adjunct to the above statistical evaluation of compressive strengths, it is desirable to determine the uniformity of companion cylinders. This will, as previously discussed, indicate the efficiency of fabrication, curing, and testing. The within-test standard deviation σ_1 and within-test coefficient of variation V_1 are computed as follows:

$$\sigma_1 = \frac{1}{d_2} \overline{R}$$

$$V_1 = \frac{\sigma_1}{\overline{X}} \times 100$$

TABLE 12.6 Standards of Concrete Control

	Standard deviation for different control standards, psi				
Class of operation	Excellent	Very good	Good	Fair	Poor
		Overall variation			
General construction testing	< 400	400–500	500–600	600–700	> 700
Laboratory trial	< 200	200–250	250–300	300–350	> 350
Coefficient of variation for different control standards, percent					
Class of operation	Excellent	Very good	Good	Fair	Poor
		Within-test variation			
Field control testing	< 3.0	3.0–4.0	4.0–5.0	5.0–6.0	> 6.0
Laboratory trial batches	< 2.0	2.0–3.0	3.0–4.0	4.0–5.0	> 5.0

Source: ACI Standard 214–77 (reapproved 1988).

in which $\bar{X}$ = average strength
$\bar{R}$ = average range of groups of companion cylinders
$\frac{1}{d_2}$ = a constant depending on number of cylinders in each group, taken from Table 12.5

In the case of within-test computations, a coefficient V_1 of 5 percent is considered satisfactory for field control (see Table 12.6).

An example of the calculations of Sec. 12.24, items 1 through 4 is given in Table 12.7.

12.29 Frequency-Distribution Curve

It may be desirable in the statistical evaluation of a given project to determine the skewness of data. This is achieved by plotting a histogram from the data. If the resultant curve does not follow the normal-distribution curve, some influencing factor has developed requiring further study. The histogram is constructed by arranging all the strength results into groups or cells. (A cell as used in this sense is a strength bracket of a given maximum and minimum. Each test between these figures is assigned to that cell.) There should be at least 10 cells. The cell boundary selected must be such that any single value belongs definitely in a specific cell. Since compressive strengths are computed to the nearest 10 psi, this can easily be accomplished by having the cell boundaries end in 5, such as 2405 – 2605 – 2805, etc.; or by having one boundary end in 0 and the other in 9, such as 2400 – 2599; 2600 – 2799. Table 12.8 shows a convenient method of tabulation.

The normal curve is then plotted by computing the ordinate at $\bar{X}$, at $\bar{X} \pm \sigma$, and at $\bar{X} \pm 2\sigma$, and then drawing a smooth curve through these points with the points of inflection at $\bar{X} \pm \sigma$ as shown in Fig. 12.17. The equations for computing these points are

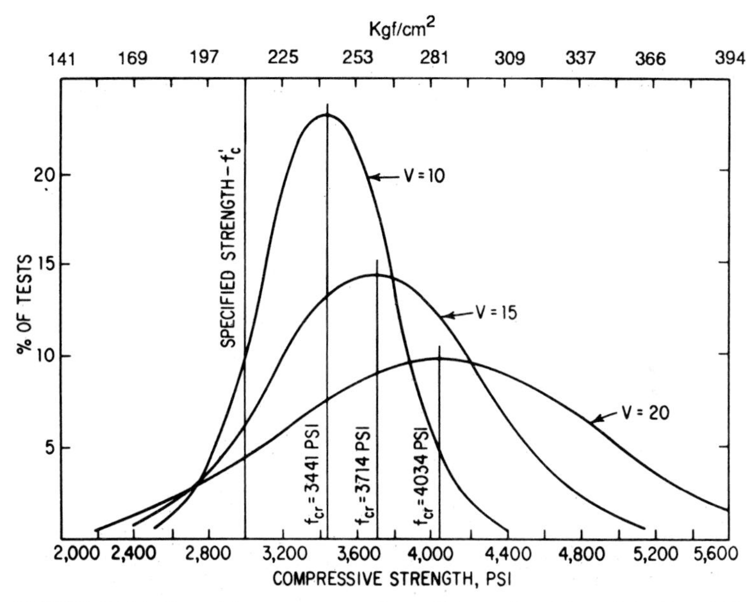

FIGURE 12.16 Normal-frequency curves for coefficients of variation of 10, 15, and 20 percent. Required average strength f_{cr} based on a probability of 1 in 10 that a test will fall below a specified strength f_{cr} of 3000 psi.[4]

At $\bar{X}$,

$$y = 0.3989 \frac{nc}{\sigma}$$

At $\bar{X} \pm \sigma$,

$$y = 0.242 \frac{nc}{\sigma}$$

At $\bar{X} \pm 2\sigma$,

$$y = 0.054 \frac{nc}{\sigma}$$

in which $\bar{X}$ = average strength
n = total number of specimens
σ = standard deviation
c = cell size or range in strength of each cell

Figure 12.17 is a graphical comparison of the data from Table 12.8 and the corresponding normal-distribution curve.

12.30 Significance of Statistical Analysis

The ACI Building Code utilizes the statistical concept of classification by attributes. This is based on minimizing the frequency of strength tests below the specified strength f_c'. If this limiting number is not exceeded, the concrete is acceptable. In order to satisfy this requirement, the average strength f_{cr} must obviously be in excess of f_c'.

The amounts by which the average strength f_{cr} should exceed the specified strength f_c'

TABLE 12.7 Sample Calculations

Test No.	Cylinder 1			Cylinder 2			Average X 1 and 2	Range R	Moving average	
	X	$X-\bar{X}$	$(X-\bar{X})^2$	X	$X-\bar{X}$	$(X-\bar{X})^2$			Strength	Range
1	3,300	170	28,900	3,100	370	136,900	3,200	200		
2	3,700	230	52,900	3,800	330	108,900	3,750	100		
3	3,200	270	72,900	3,300	170	28,900	3,250	100		
4	3,600	130	16,900	3,400	70	4,900	3,500	200		
5	3,200	270	72,900	3,000	470	220,900	3,100	200	3,360	
6	3,800	330	108,900	3,500	30	900	3,650	300	3,450	
7	4,000	530	280,900	3,800	330	108,900	3,900	200	3,480	
8	3,300	170	28,900	3,600	130	16,900	3,450	300	3,520	
9	3,200	270	72,900	3,000	470	220,900	3,100	200	3,440	
10	3,000	470	220,900	3,200	270	72,900	3,100	200	3,440	200
11	3,000	470	220,900	3,100	370	136,900	3,050	100	3,320	190
12	3,200	270	72,900	3,300	170	28,900	3,250	100	3,190	190
.	.	.	.	.	.	.	.	.	.	.
.	.	.	.	.	.	.	.	.	.	.
.	.	.	.	.	.	.	.	.	.	.
30	3,400	70	4,900	3,500	30	900	3,450	100	3,280	150
	103,500		3,581,600	104,700		5,360,400		5,400		

$n = 60$ (30 tests, 2 cylinders each)

$$X = \frac{103,500 + 104,700}{60} = 3470$$

$$\sigma = \sqrt{\frac{3,581,600 + 5,360,400}{60}} = 386 \text{ psi}$$

$$V = \frac{386}{3470} \times 100 = 11.1\,\%$$

Within-test variation:

$$\bar{R} = \frac{5400}{30} = 180$$

$$\sigma_1 = 180 \times 0.8865^* = 160$$

$$V_1 = \frac{160}{3470} \times 100 = 4.6\,\%$$

$$\bar{R}m = (0.05 \times 1.128) \times 3470 = 196$$

*Factor from Table 12.5

have been calculated by procedures outlined in the report of ACI Committee 214, Recommended Practice for Evaluation of Strength Test Results of Concrete. The listed values are required to meet all three of the following criteria, using the maximum standard deviation from the range shown in each case:

1. A probability of less than 1 in 10 that a random individual strength test will be below the specified strength f_c'
2. A probability of 1 in 100 that an average of three consecutive strength tests will be below the specified strength f_c'
3. A probability of 1 in 100 that an individual strength test will be more than 500 psi below the specified strength f_c'

Using values of t from Table 12.4, formulas for calculating the required average strengths reduce to the following for the respective criteria above:

1. $f_{cr} = f_c' + 1.28\sigma$
2. $f_{cr} = f_c' + \dfrac{2.33\sigma}{\sqrt{3}} = f_c' + 1.34\sigma$

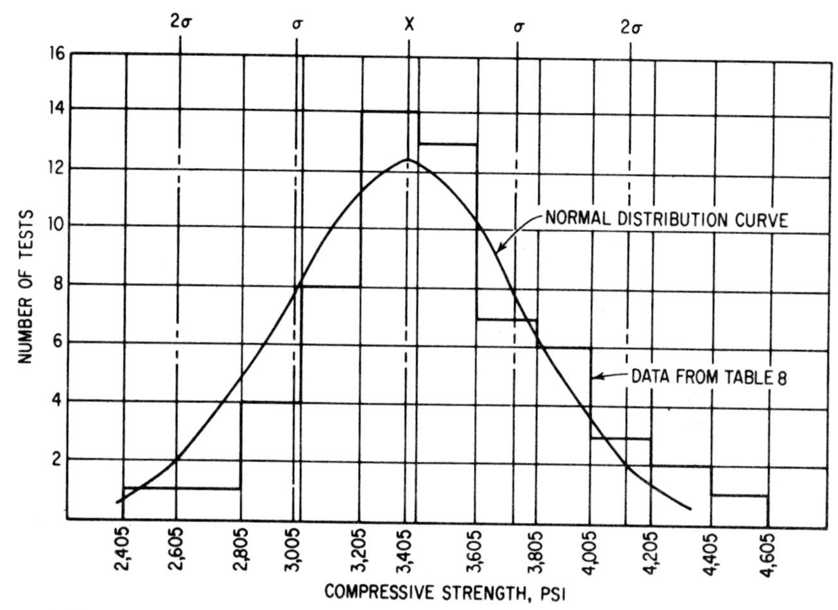

FIGURE 12.17 Normal-frequency-distribution curve with example of job data (Table 12.8) superimposed.

3. $f_{cr} = f_c' - 500 + 2.33\,\sigma$

where f_{cr} = average strength to be used as the basis for selecting concrete proportions, psi

f_c' = strength level used in the design of the structure, psi (specified f_c')

σ = standard deviation of individual strength tests, psi

If probabilities other than those given in the ACI requirements are desirable, replacement of the values of t given in the above values can be obtained from Table 12.4.

TABLE 12.8 Frequency of Tests Tabulation

Cell boundary	No. of test results in cell	Total No.
2405		
2605	/	1
2805	/	1
3005	////	4
3205	~~////~~ ///	8
3405	~~////~~ ~~////~~ ////	14
3605	~~////~~ ~~////~~ ///	13
3805	~~////~~ //	7
4005	~~////~~ /	6
4205	///	3
4405	//	2
4605	/	1
	Total	60

It can be seen that criterion 2 always produces a required average strength higher than criterion 1. Criterion 2 will produce a higher required average strength than will criterion 3 for low to moderate standard deviations; however, criterion 3 governs, i.e., limiting the expected frequency of tests more than 500 psi below the specified f_c' to 1 in 100.

The use of these criteria permits a change in mix design whenever sufficient data (normally 30 or more successive tests of a given class of concrete) are developed on a specific project. As an example, assume that the original 5000 psi design was over designed to 6200 psi for a given project. Development of proper strength data indicated that the standard deviation of this mix was 500 psi. In accordance with the above stated probability requirements, the mix may be adjusted from the water-cement ratio curve to produce the following strength level:

1. Criterion 1: $f_{cr} = f_c' + 1.28\sigma$
$f_{cr} = 5000 + (1.28 \times 500)$
$f_{cr} = 5640$

2. Criterion 2: $f_{cr} = f_c' \dfrac{2.33\sigma}{\sqrt{3}}$
$f_{cr} = f_c' + (1.34 \times 500)$
$f_{cr} = 5670$

3. Criterion 3: $f_{cr} = f_c' - 500 + 2.33\sigma$
$f_{cr} = 5000 - 500 + (2.33 \times 500)$
$f_{cr} = 5660$

Therefore, $f_{cr} = 5670$ governs. This strength is used to enter the original water cement ratio curve to develop the water cement ratio requirements for a 5670 psi strength instead of 6200 psi and a more economical mix for the balance of the project.

12.31 Mix Design Requirements

The ACI Building Code emphasizes a statistical basis for establishing the average strength required to assure attainment of the design strength level f_c'. If an applicable standard deviation for the project concrete is known, the required average strength can be established. If this standard deviation is unknown, the proportions must be selected to produce an excess of 1200 psi over the required f_c'. These average strength or overdesign formulas are given in Sec. 11.10 (of Chap. 11).

The ACI Building Code also stipulates maximum water cement ratios for various exposures which can be found in Sec. 11.10.

REFERENCES

1. ASTM Standards, Parts 04.01 and 04.02, American Society for Testing and Materials, 1991.
2. "Concrete Manual," 8th ed., U.S. Bureau of Reclamation, 1975.
3. ASTM STP169B, Significance of Tests and Properties and Concrete-Making Materials, 1978.
4. ACI Recommended Practice for Evaluation of Strength Test Results of Concrete, ACI 217-77 (reapproved 1988).
5. ACI Building Code Requirements for Reinforced Concrete, ACI 318-89, American Concrete Institute.

CHAPTER 13
INSPECTION AND LABORATORY SERVICE

Joseph F. Artuso*

INSPECTION REQUIREMENTS

13.1 ACI Recommendations

A recommended practice has been developed by the ACI to encourage and enable more effective inspection of concrete construction.[1,2] Some of the pertinent recommendations include the following:

1. The responsibility for inspection must remain with the architect-engineer.
2. Inspectors must be well qualified and properly supervised.
3. The types of inspection that must be provided are given as
  a. Inspection and approval of batching and mixing facilities, as well as batching inspection when the size of job or type of concrete warrants it.
  b. Inspection, testing, and approval of materials.
  c. Inspection of forms, reinforcing steel, shoring, bracing, embedded items, joints, and so on.
  d. Inspection of concrete handling and placing equipment such as buckets, chutes, buggies, hoppers, vibrators, and pumps.
  e. Inspection of concrete handling, placing, consolidation, finishing, curing, protection, and repair or patching.
  f. Inspection at the plant of precast items including prestressed work for strength, dimensions, and special properties.
  g. Inspection of stripping forms and removal of shoring.
  h. Preparation and testing of concrete specimens, testing of consistency, air entrainment, unit weight.
  i. Daily reporting of all these items.

13.2 Batch-Plant Inspection

A vital requirement for concrete control is the inspection of the batching operation, including tests of aggregates for grading and moisture content.

*Revised by Joseph A. Dobrowolski.

Briefly, the usual routine duties of batch-plant inspectors include witnessing the weighing and measurement of all constituent materials, including cement, aggregates, water, and admixtures, if any, and the checking of aggregates for size and moisture content. Adjustment of aggregate weights and amount of added water to compensate for the free moisture in the aggregates is an important part of the control at this point. Periodically the scales and other measuring devices should be checked for accuracy by calibration with standards furnished by the National Bureau of Standards.

The batch plant should be so equipped that cement can be weighed on a scale entirely separate from the aggregate scales. Various-sized aggregates may all be weighed in a single scale hopper on the cumulative basis. Details of batch plants are covered in Part 5 of this handbook.

The accuracy to which measurement of various ingredients should be made is shown in Table 13.1.

13.3 Aggregate-Moisture Determination

An important function of aggregate inspection at the batch plant is to ensure that correct moisture corrections are made. Several methods are available.[3]

1. Sample dried in an oven or hot plate. The sample is completely dried and a correction is made for absorption to determine amount of free moisture. An approximate formula of sufficient accuracy is given by

$$\%\ \text{moisture} = 100\left(\frac{\text{wet weight} - \text{dry weight}}{\text{dry weight}}\right) - \%\ \text{absorption}$$

   Figure 13.1 may be used for convenience. A 400 g sample is dried and weighed. The percent moisture is taken from the intersection with the curve of the appropriate absorption.

2. A convenient method for free moisture in sand is that based on U.S. Bureau of Reclamation Designation 11-B for use with Table 13.2. The specific gravity must be known.

Apparatus. Chapman-type volumetric flask of 500 mL capacity accurate to 0.15 mL balance of 2 kg capacity accurate to 0.1 g, and pipette, $\frac{1}{4}$ in diameter glass tube, of sufficient length to enable water-level adjustment.

Procedure. Fill the flask with water at room temperature to the calibrated 200 mL mark. Use the pipette to adjust the water level (lower part of the meniscus) to the 200 mL

TABLE 13.1 Accuracy to Which Concrete Ingredients Should Be Measured in a Batch Plant

Ingredient	Accuracy tolerance, %
Cement	±1
Aggregates	±2
Water	±1
Admixtures	±3

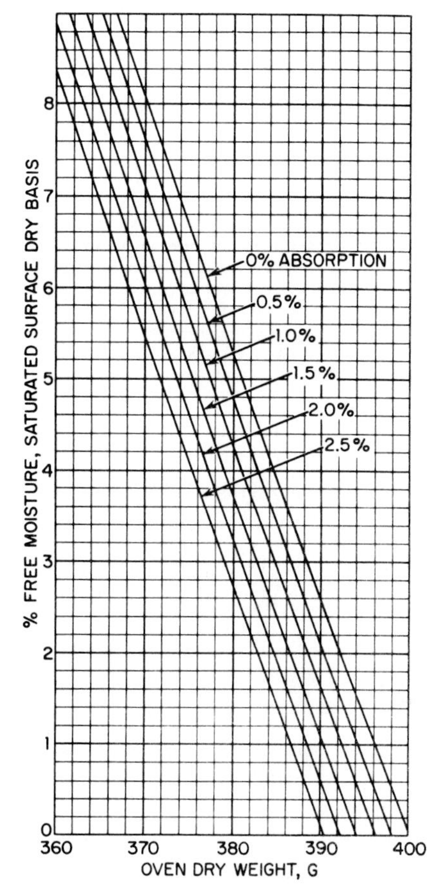

FIGURE 13.1 Chart for determining moisture content of aggregate. (1) Use 400-g wet sample. (2) Dry to oven dryness. (3) Enter chart at dry weight, intersect curve for correct absorption, and read percent moisture on scale at left.

$$\%M = 100\,\frac{400 - A(1 + X)}{A(1 + X)}$$

where A = oven-dry weight
X = absorption expressed as a decimal
(*From Ref. 3.*)

mark. Introduce 500 g of representative sand into the flask. Roll to dispel all entrapped air. Break foam. A drop or two of ether, alcohol, or commercial wetting agent will accomplish this readily. Stirring with a fine wire is tedious and time-consuming, although usually a successful method. Read the volume of the combined water and fine aggregate directly on the graduated scale. The percentage of free moisture can be taken directly from Table 13.2.

TABLE 13.2 Percent Moisture in Sand

V	Bulk specific gravity, SSD basis																				
	2.55	2.56	2.57	2.58	2.59	2.60	2.61	2.62	2.63	2.64	2.65	2.66	2.67	2.68	2.69	2.70	2.71	2.72	2.73	2.74	2.75
382																					0.1
383																				0.2	0.4
384																		0.1	0.3	0.5	0.7
385																	0.2	0.4	0.6	0.8	1.0
386															0	0.3	0.5	0.7	0.9	1.1	1.3
387														0.1	0.4	0.6	0.8	1.0	1.3	1.5	1.7
388												0	0.2	0.4	0.7	0.9	1.1	1.4	1.6	1.8	2.0
389											0.1	0.3	0.5	0.8	1.0	1.2	1.5	1.7	1.9	2.1	2.3
390									0	0.2	0.4	0.6	0.9	1.1	1.3	1.6	1.8	2.0	2.2	2.4	2.6
391							0	0	0.3	0.5	0.7	1.0	1.2	1.4	1.7	1.9	2.1	2.3	2.5	2.7	3.0
392						0	0.1	0.4	0.6	0.8	1.1	1.3	1.5	1.8	2.0	2.2	2.5	2.6	2.8	3.1	3.3
393				0	0	0.2	0.5	0.7	0.9	1.2	1.4	1.6	1.9	2.1	2.3	2.6	2.8	3.0	3.2	3.4	3.7
394			0	0.1	0.3	0.6	0.8	1.0	1.3	1.5	1.7	2.0	2.2	2.4	2.7	2.9	3.1	3.3	3.5	3.8	4.0
395		0	0.1	0.4	0.6	0.9	1.1	1.4	1.6	1.8	2.1	2.3	2.5	2.8	3.0	3.3	3.4	3.7	3.9	4.1	4.3
396	0	0.2	0.5	0.7	1.0	1.2	1.5	1.7	1.9	2.2	2.4	2.6	2.9	3.1	3.4	3.6	3.8	4.0	4.2	4.4	4.7
397	0.3	0.6	0.8	1.1	1.3	1.5	1.8	2.0	2.3	2.5	2.7	3.0	3.2	3.5	3.7	4.0	4.1	4.3	4.6	4.8	5.0
398	0.7	0.9	1.1	1.4	1.6	1.9	2.1	2.4	2.6	2.8	3.1	3.3	3.6	3.8	4.1	4.3	4.5	4.7	4.9	5.1	5.4
399	1.0	1.2	1.5	1.7	2.0	2.2	2.5	2.7	2.9	3.2	3.4	3.7	3.9	4.2	4.4	4.6	4.8	5.0	5.3	5.5	5.7
400	1.3	1.6	1.8	2.1	2.3	2.6	2.8	3.0	3.3	3.6	3.8	4.0	4.3	4.5	4.7	4.9	5.2	5.4	5.6	5.8	6.1
401	1.7	1.9	2.2	2.4	2.6	2.9	3.1	3.4	3.7	3.9	4.2	4.4	4.6	4.8	5.1	5.3	5.5	5.8	6.0	6.2	6.4
402	2.0	2.2	2.5	2.7	3.0	3.2	3.5	3.8	4.0	4.3	4.5	4.7	5.0	5.2	5.5	5.7	5.9	6.1	6.3	6.5	6.8
403	2.4	2.6	2.8	3.1	3.3	3.6	3.9	4.1	4.4	4.6	4.8	5.1	5.3	5.6	5.8	6.0	6.2	6.5	6.7	6.9	7.1
404	2.7	2.9	3.2	3.4	3.7	4.0	4.2	4.5	4.7	4.9	5.2	5.4	5.7	5.9	6.2	6.4	6.6	6.8	7.0	7.3	7.5
405	3.0	3.3	3.5	3.8	4.0	4.3	4.6	4.8	5.1	5.3	5.5	5.8	6.0	6.3	6.5	6.7	7.0	7.2	7.4	7.7	7.9

406	3.4	3.6	3.9	4.2	4.4	4.7	4.9	5.2	5.4	5.7	5.9	6.1	6.4	6.6	6.9	7.1	7.3	7.5	7.8	8.0	8.2
407	3.8	4.0	4.2	4.5	4.8	5.0	5.3	5.5	5.8	6.0	6.3	6.6	6.7	7.0	7.2	7.5	7.7	7.9	8.1	8.4	8.6
408	4.1	4.3	4.6	4.9	5.1	5.4	5.6	5.9	6.1	6.4	6.7	6.9	7.1	7.3	7.6	7.8	8.1	8.3	8.5	8.7	9.0
409	4.4	4.7	5.0	5.2	5.5	5.7	6.0	6.3	6.5	6.8	7.0	7.3	7.5	7.7	8.0	8.2	8.4	8.6	8.9	9.1	
410	4.8	5.1	5.3	5.6	5.8	6.1	6.4	6.6	6.9	7.1	7.4	7.7	7.8	8.1	8.3	8.6	8.8	9.0	9.3		
411	5.2	5.4	5.7	6.0	6.2	6.5	6.7	7.0	7.2	7.5	7.8	8.0	8.2	8.5	8.7	8.9	9.2				
412	5.6	5.8	6.1	6.3	6.6	6.8	7.1	7.4	7.6	7.9	8.1	8.3	8.6	8.8	9.1	9.3					
413	5.9	6.2	6.4	6.7	7.0	7.2	7.5	7.7	8.0	8.3	8.5	8.7	9.0	9.2							
414	6.3	6.5	6.8	7.1	7.3	7.6	7.8	8.1	8.4	8.7	8.8	9.1									
415	6.7	6.9	7.2	7.4	7.7	8.0	8.2	8.5	8.8	9.0	9.2	9.5									
416	7.0	7.3	7.6	7.8	8.1	8.4	8.6	8.9	9.1	9.4	9.6										
417	7.4	7.7	7.9	8.2	8.5	8.7	9.0	9.2													
418	7.8	8.0	8.3	8.6	8.8	9.1	9.4														
419	8.2	8.4	8.7	9.0	9.2	9.5															
420	8.6	8.8	9.1	9.4																	
421	8.9	9.2																			
422	9.3																				

$$\% \text{ moisture} = \frac{V - (500/\text{sp gr}) - 200}{200 + 500 - V}$$

where V = flask reading
sp gr = bulk specific gravity, SSD basis
See "Concrete Manual," 8th ed., p. 524, U.S. Bureau of Reclamation, 1975.
Source: Ref. 3.

LABORATORY FACILITIES

13.4 Permanent Laboratories

There are two conventional types of testing facilities. One is a federal, state, county, or city governmental agency designed to serve the needs of government-sponsored construction. The other type is the commercial independent laboratory serving all types of private and commercial and, frequently, of governmental construction. The permanent laboratory of either type is normally equipped to perform detailed investigations and tests for evaluation of concrete materials which include physical and chemical testing. Concrete is routinely tested for strength, often for other properties such as durability, volume change, modulus of elasticity, and permeability.

ASTM E329-77 (reapproved 1983) describes the requirements and functions of the testing agency in achieving reliability in the inspection and testing of materials used in construction so as to ensure compliance to the contract documents. The relationship between the agency and other parties concerned is also indicated.[4] Recently adopted ASTM C1077-91A, Standard Practice for Laboratories Testing Concrete and Concrete Aggregates for Use in Construction and Criteria for Laboratory Evaluation, will provide future guidance for the Cement and Concrete Reference Laboratory and practicing laboratories as to procedures and requirements for acceptance and evaluation.[5] Some states require a professional engineer to be in responsible charge of a testing agency and inspectors to be certified by the American Concrete Institute.

13.5 Jobsite Laboratories

Governmental agencies and commercial laboratories often establish a field laboratory on a major construction project. The size will depend on the magnitude of construction and may range from one equipped to check gradations of aggregates and to make compressive-strength tests of concrete only, to elaborate facilities such as those found in permanent laboratories. Many large private and commercial projects will require field laboratories because of distance from permanent laboratory installations. Generally these laboratories are equipped only for routine compression testing of specimens and for checking gradations and deleterious substances of aggregates. The primary equipment required for this scope of testing includes the following:

Cylinder molds

Identification tags and data cards

Slump cone

Tamping rod

Platform scales of 100 or 200 lb capacity accurate to 0.1 lb

Laboratory scales of 100 and 2000 g capacity accurate to 0.2 g

Air meters (pressure type for stone and gravel concrete—volumetric types for slag or lightweight-aggregate concrete)

Thermometers (steel-dial type and maximum-minimum type)

Pocket rule

Calibrated measures ($\frac{1}{4}$, $\frac{1}{2}$, or 1 ft^3 capacity dependent on aggregate size)

Buckets, scoops, shovels, pans, marking pens, sample bags

Curing tanks and lime

Motor-driven adjustable-speed campression-testing machine

Project specifications and reference methods of test

CEMENT AND CONCRETE REFERENCE LABORATORY

13.6 Functions and Scope

The primary function of the Cement and Concrete Reference Laboratory, an agency jointly sponsored by the American Society for Testing and Materials and the National Bureau of Standards, is the inspection of cement and concrete testing laboratories. This inspection is available to any interested testing laboratory and is performed only by request. A nominal fee is charged. Although there is no direct regulatory action, many agencies, architects, and engineers require that a laboratory engaged in their work be regularly inspected by the Cement and Concrete Reference Laboratory.

The inspection includes examination of the apparatus used in testing concrete, aggregates, and cement to ASTM procedures. Procedures are also evaluated and instructions are given regarding any deviations from standards that are detected. A detailed report is submitted to the inspected laboratory which contains an evaluation of every piece of apparatus and of every method inspected. This can be used as a guide for any corrections that should be made. The report may also be shown as evidence of compliance with ASTM methods and use of approved apparatus to any party using the facilities of the laboratory. This is one of the most reliable methods of determining the capability of a specific laboratory.

The present published scope of the work of the Cement and Concrete Reference Laboratory is "To promote uniformity and improvement in the testing of cement and concrete through field inspection of testing laboratories; the study of special problems evolving from such inspections; instructions in methods of testing; sponsorship of comparative test programs, and contributions to the work of ASTM Technical Committees."

13.7 Sources

The engineer-architect has the privilege and should exercise the right to approve or disapprove all materials and sources proposed for use in the concrete for the specific structure. It is desirable to have the contractor submit the proposed source of each material well in advance of start of concreting. In the case of established manufacturers of cement, reinforcing steel, and prestressing items, approval can be granted on the basis of experience and knowledge of the source. This can also apply to aggregates. However, the many variables encountered in the production of aggregates and sometimes cement warrant current evidence in the form of recent test data. This is particularly important, usually essential, for the product of new deposits. The approval of the source does not obviate the need for continuing evidence of conformance to the applicable specifications. During construction, mill test reports of cement, supplemented by tests by an independent testing laboratory, provide a high level of quality assurance. Aggregates should be continuously tested for grading and cleanliness and periodically for inherent properties of durability and abrasion.

13.8 Off-Site Testing and Inspection

It is often desirable to perform inspection and testing of specific materials at the point of production. This expedites their use in construction with the assurance of compliance with the applicable contract specifications and drawings. This is accomplished by use of a testing agency that can have inspectors at the plant at the time of production. The engineer should arrange for this inspection, and the contractor should indicate on the purchase order that inspection will be made by the specified testing agency in sufficient time to have this inspection and testing performed prior to shipment. Individual items must be properly tagged, stamped, or identified in some manner for shipment and acceptance at the project. In the case of cement, the inspecting agency can sample the cement as a bin is loaded for the exclusive use of a given project. The bin is sealed and the tests are performed by the testing agency to verify compliance with the applicable specifications. When conformance is established, shipments can be made to the project. The testing-agency inspector seals and appropriately identifies each shipment with tags and supporting documents for proper acceptance at the project. Aggregate stockpiles can be tested in the same manner. It is essential that a reliable method of identification be maintained for acceptance of the material at the project.

13.9 Acceptance on Basis of Manufacturer's Certifications

If agreeable to the engineer, materials of standard manufacture may be accepted for a project on the basis of certification by the manufacturer. This is accomplished by the submission of certified test results which show complete compliance to the applicable specifications on each lot of consignment. In addition, a statement certifying that the process used in the manufacture has remained constant should be included. Proper identification must be shown on all items to correlate respective lots. It is prudent to supplement acceptance by this method with check tests by an independent testing agency. It has been found that such check has disclosed errors in classifications of types of steel, such as between intermediate-grade and high-strength grade. It has also revealed variations in strengths and air contents of cement which have explained in part fluctuations in concrete strength. Variations in admixtures may cause appreciable difficulties, but the standard laboratory tests of these products are too time-consuming to be of much practical value in controlling uniformity of shipments.

13.10 General Sampling Methods

The first step in testing is often one of the most important, and this is certainly true in the case of sampling. It is essential that every effort be made to secure a sample which is representative of the material. The test, obviously, can only be as accurate as the sample. Sampling is often based on the mathematics of probability and a statistical analysis is applied. In sampling of concrete and concrete materials, methods are covered by appropriate standards of the American Society for Testing and Materials. The method for sampling aggregates is covered by ASTM D75 and has been described previously here. The method for sampling portland cement is given in ASTM C183. Liquids such as air-entraining agents, chemical admixtures, and curing compounds must usually be sampled from large drums. In this event, appropriate measures must be taken to ensure that the contents are thoroughly mixed by mechanical agitation, preferably using motorized agitators, before the sample is taken. When individual units are sampled, the number of samples should equal the cube root of the number of units in the lot that is inspected. As an example, 10 units would represent 1000 units in a lot.

The American Association of State Highway Officials and various other organizations also publish many applicable standards, many of which are very similar to and often identical with those of ASTM.

13.11 Shipping and Handling Samples

When samples are shipped to a laboratory, complete information should be included in such a manner that it is securely attached or identified with the sample. The information should include project, sample number, complete material identification as to type and grade, source, date of sampling, quantity represented, location to be used in the structure, tests required, and shipper's identification. The samples must be adequately packed to prevent damage in transit.

13.12 Inspector Certification

The American Concrete Institute has developed a training and examination program through its local chapters for Concrete Construction Inspectors Grade II and In-Training, Concrete Field Testing Technician Grades I and II, and Concrete Laboratory Testing Technician Grades I and II. Completion of each program leads to certification, which is being increasingly required by municipalities and states.

REFERENCES

1. "Manual of Concrete Inspection," American Concrete Institute Special Publication SP-2, 1981.
2. ACI Guide for Concrete Inspection (ACI 311-88), 1988.
3. Waddell, J. J.: "Practical Quality Control for Concrete," McGraw-Hill, New York, 1962.
4. ASTM E329-77 (reapproved 1983), Standard Recommended Practice for Inspection and Testing Agencies for Concrete, Steel and Bituminous Materials as Used in Construction.
5. ASTM C1077-91A, Standard Practice for Laboratories Testing Concrete and Concrete Aggregates for use in Construction and Criteria for Laboratory Evaluation, 1991.

PART • 4

FORMWORK AND SHORING

CHAPTER 14
FORMWORK OBJECTIVES AND TYPES

Jerome H. Ford

INTRODUCTION

There have been more significant changes in concrete formwork in the United States and Canada in the last 15 years than probably any other period in history. The impetus for these changes is the de facto existence and influences of the design-build contractor. For hundreds of years, the concrete construction industry, and concrete formwork, in particular, was a three-sided industry including the owner, the designer, and the builder. In almost all cases the designer was an agent of the owner, and input into design changes to increase productivity was "unacceptable." The design-build industry started off under the same parameters, but as money became more dear and as the designer and the builder were essentially under the same roof, the builder started influencing the designer. This influence was not instantaneous, nor was it founded on poor engineering practices. The significance of these changes was to add two objectives to concrete formwork, which were logistics and productivity.

Gang forming, which had existed for many years, was a matter of the contractor putting large sections of formwork into a gang, assembled on site, on the ground, or on a platform. The logistics of these early gangs was at best a nightmare as the reuses of the formwork were at best similar rather than identical in configuration. What the design-build contractor did was to organize the formwork and emphasize the extreme costs caused by variations in offsets, haunches, corbels, pilasters, and opposite-hand details. The organization of these complications into a design, where they were produced identically, symmetrically, or eliminated, changed formwork. The financial success of these contractors did more to influence the construction industry and concrete formwork than any other.

The effects of these changes and the material and equipment developed have changed and continue to change the concrete-forming industry. The computer and computer-activated drafting also helped to stimulate the changes by their abilities to organize and change data and assumptions in seconds rather than months. Concrete delivery and placing equipment also changed dramatically, all greatly improving logistics, productivity, and the resultant bottom line. The development and improvement of the extruded aluminum beam and the laminated wood beam directly changed the flexibility of form design and the weight/strength ratios for gang form design.

The laminated-beam "I" section was developed in the 1950s in Europe and has not yet influenced the U.S. and Canadian markets as in Europe. Use of the laminated beam as a rectangular section occurred initially extensively in Europe during the 1960s and slowly developed in the United States and Canada during the 1970s. The I section is extensively used today in both vertical and horizontal formwork throughout the United States and Canada. The aluminum beam was developed in Canada during the early 1970s and is now extensively used in the United States and Canada.

With the intense impact on both logistics and productivity resulting from the design-build contractors' production implementations, along with flexibility of design and weight/strength ratios of the beams, changes in both delivery and placement equipment, lumber forming, and modular forming applications have also changed.

As the world is continuing to build, the soil conditions where buildings are built require more durable concrete. This has changed the concrete industry from a predominantly Type I cement market to a Type II cement market and additions of fly ash. This change has had dramatic if not somewhat devastating effects on the concrete forming business because of the slower setting and strength gain time of Type II cements or cements with fly-ash additions. This has increased form pressures and duration. In some areas, cement manufacturers have increased the fineness of cement to offset this slower gain.

In 1989 over three-quarters of all concrete shipped within the United States contained some form and/or degree of water-reducing admixture. This is often beneficial for the concrete but may create serious form pressure problems unless properly used. Mineral admixtures such as fly ash will also increase the time and the amount of form pressures during low ambient temperatures.

The size of ready-mix concrete trucks has increased to 10 yd or more. Most concretes are now placed by pump because of its flexibility horizontally and vertically and its distance capability. Both have produced a tendency to use formwork to its capacity and in some cases beyond. Developments in OSHA (Occupational Safety and Health Administration) requirements have thrown the full responsibility for rate of placement on the contractor. A supplier of lumber, beams, or fabricated formwork is only responsible for notifying the contractor of the safe working loads at a 2:1 and a 1½:1 safety factor.

Another piece of equipment changing forming industry is the Dowel Bar Substitute. This a male-and-female threaded two-piece unit in which one section is fastened to the form, and on stripping, the male end is threaded into the female end to complete the connection so that the reinforcing splice does not have to pass through the form.

Types of plywood usage has also changed in the last 20 years. Twenty years ago only 20 percent of the plywood used in forming was overlaid with other materials. In the 1990s use of the overlaid type was estimated at 80 percent of the plywood forming. Overlaid plywood is manufactured in a wide variety of woods, number of laminations, and densities of the coating.

OBJECTIVES

The objectives to be considered in selection of formwork are concrete finish, cost, reuses, logistics, and productivity, which are discussed in detail below.

14.1 Finish

Cast-in-place architectural concrete, to be left as cast, is not easy to accomplish as the desired resulting appearance varies from one region of the country to another, from one designer to another, and from one contractor to another. True cast-in-place architectural concrete requires a desire of all concerned such as the designer and all construction personnel involved to have the same objective. Cooperation is mandatory but is difficult to obtain when needed. Tolerances need to be determined well in advance. Additional information can be obtained from Chaps. 43 and 44.

14.2 Cost

The cost of formwork is the total cost of forms, accessories, labor, and supervision to accomplish the job. Often the labor utilized in a more expensive forming system can cost considerably less than that in a cheaper system, but this may require greater reuse, better logistics, or increases in productivity. All these factors should be assessed in advance of beginning work. If not done, a costly form can be *very* costly in short order.

14.3 Reuse

"Reuse" is a simple term but is horribly abused or totally unused by too many people. If concrete formwork costs $20 per square foot or rents for $2 per square foot per month, the number of times the form is used has a significant impact on both costs and logistics. An ideal approach is to strive for 20 reuses by careful planning, cooperation of personnel, and execution to achieve a good profit margin.

14.4 Logistics

Logistics is the movement of personnel and materials in an organized fashion within a certain time frame. Many books have been written on the subject, and the computer has provided valuable assistance, but concrete formwork and its labor are still one of the highest *profit risk* areas for the contractor. The controlling of this risk factor by cooperation between contractor and designer in the design-build firms has produced a positive long-term effect in concrete formwork based on logistics and reuse.

14.5 Productivity

Simply defined, *productivity* is the units of formwork accomplished in a given period of time, usually expressed in square feet per worker-hour. Profits increase as units increase. Productivity requires mutual cooperation between designer, contractor, and suppliers well in advance of construction.

The objectives of the people involved vary greatly. The designer's priorities are finish first, cost second, and productivity last. The contractor rates them as logistics first, reuse second, cost third, productivity fourth, and finish last. The supplier of the formwork rates reuse first, cost second, productivity third, logistics fourth, and finish last.

VERTICAL FORMWORK

All vertical formwork has four basic components, which include a face sheet that comes directly into contact with the concrete so as to bear the load, a vertical or horizontal member called a *stud* which backs up the face sheet and gathers its load, a vertical or horizontal member called a *wale* that is always perpendicular to the stud and which backs up the stud and gathers its load, and a member that holds the forms together or apart and gathers the load from the wales. It is generally a tensile member, but when a compression member is seperate from the tie, it is called a *spreader.*

TYPES OF VERTICAL HANDSET FORMWORK

14.6 Job-Built Handset Forms

There are four types of vertical handset formwork, known as *twisted-wire, single-stud/single-wale; single-stud/double-wale;* and *single-wale with load-gathering hardware.*

Twisted wire with a built-in compression member or using a wood spreader is probably the oldest handset forming method employed today. Flexibility is its long suit as it can use either a board or a plywood face sheet with dimensional or nondimensional lumber, or just about anything else as studs or wales. The limiting component of the system is the low tensile strength of the ties, usually with 1000 lb or less working capacity.

The single-stud, single-wale system usually incorporates a proprietary piece of hardware and tie. This system needs 75 percent less effort than the twisted-wire system and 48 percent less than the stud and wale and single-wale systems. Most of this saving is a decrease of labor cost over other job-built systems. It is limited by a scarcity of straight lumbers in long lengths.

The single-stud, double-wale system (see Fig. 14.1) is associated with snap ties and hairpins. It consists of a plywood face sheet, usually ⅝ or ¾ in thick, used with a 2 × 4 or 2 × 6 stud and a double wale of 2 × 4 or 2 × 6, a long-end snap tie with either a cone or a washer as part of the tie to act as spreader, and a wedge-shaped piece of hardware to tighten the assembly and take the tie load. Flexibility of design and good potential for both architectural and exposed structural concrete are its strong suit, but it is labor-intensive.

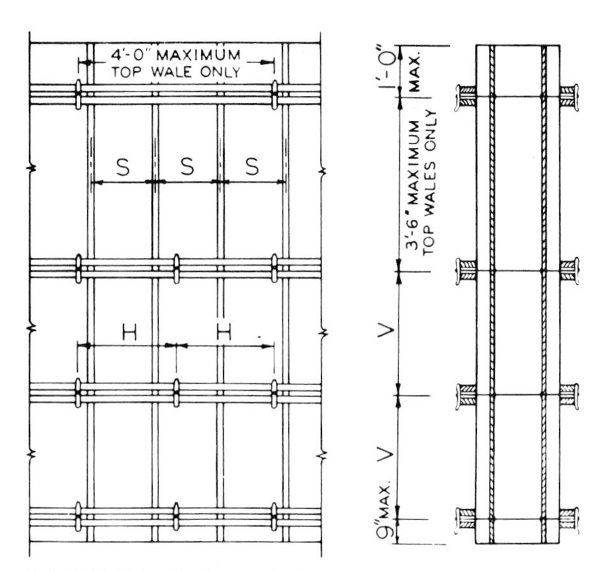

FIGURE 14.1 Single-stud, double-wale system.

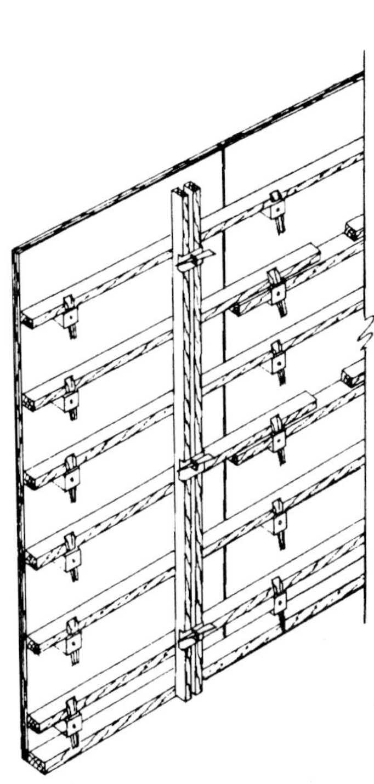

FIGURE 14.2 Single-wale with load-gathering hardware system.

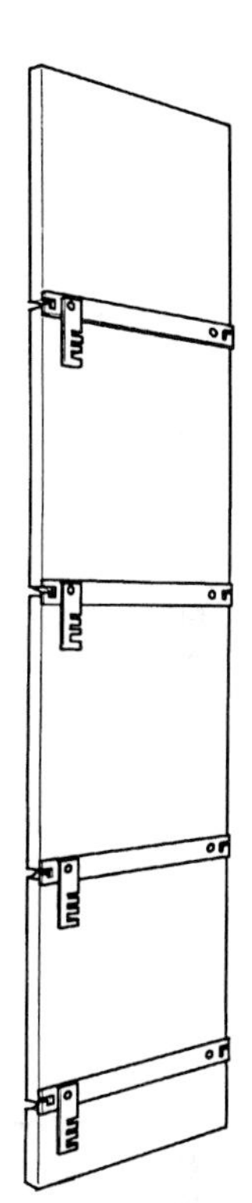

FIGURE 14.3 The 1⅛ in with attached hardware system.

Single-wale with load-gathering hardware (Fig. 14.2) is the common name for the system defined as single-stud with load-gathering hardware. It incorporates a plywood face sheet having a thickness of ⅝ or ¾ in and 2 × 4 studs and a proprietary bracket to gather the load from the studs and act as tie hardware. The system is widely used and has good potential for both cast-in-place architectural or exposed structural concrete, but is limited by both the strength of the plywood and the 2 × 4's.

These systems are available in the United States and Canada through the local telephone directory and the World of Concrete Source Book. Strengths of plywood or lumber can be found in publications of the American Concrete Institute, The Douglas Fir Plywood Association, and the National Forest Products Association.

14.7 Factory-Built Handset Forms

Factory-built handset or modular forms consist of five types: (1) 1⅛ in with attached hardware, (2) all-aluminum, (3) steel frame with plywood face sheet, (4) aluminum frame with plywood face sheet, and (5) all-steel. Type 1, the 1⅛ in with attached hardware system (Fig. 14.3) consists of a face sheet with steel backing and attached hardware acting as wales and squared round rod stock simply stamped for a tie. This system is extremely labor-efficient when used in a single lift, where the height of the concrete is equal to or

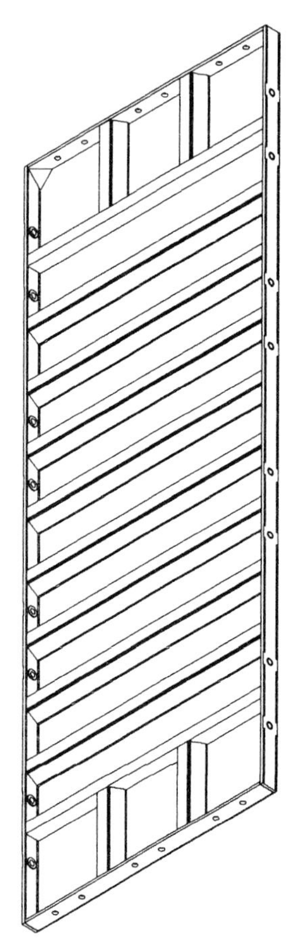

FIGURE 14.4 An aluminum factory-built forming system.

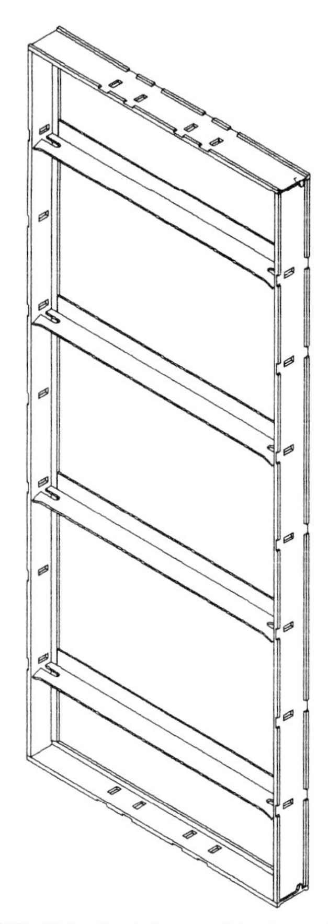

FIGURE 14.5 Steel frame with plywood face sheet.

less than the height of the form and high reuse is expected. Equipment, when new, can produce acceptable exposed structural concrete. It is not well adapted to multiple-lift placements, or for structures that require tie break back greater than 1½ in. This system is used extensively in residential, light commercial, and light industrial construction.

All-aluminum forms (type 2 in preceding list; see Fig. 14.4) have an aluminum face sheet usually ⅛ in thick, a cross member to act as a stud, siderails that gather the load from the cross members, or wales made from aluminum and steel strap ties. The system is usually in a 3 × 8 ft module and is more efficient than systems having a module of 2 × 8 ft. The major drawbacks are the cost of aluminum, a high potential for theft, and an unfavorable chemistry between aluminum and concrete which can cause gases at the interface and bonding of the concrete to the forming. Strongly acidic release agents used to prevent this problem can cause dusting and increased absorption of the surface.

The steel frame with plywood face sheet (type 3 in preceding list; see Fig. 14.5) has a face sheet of ½ or ⅝ in thick overlaid plywood, steel cross members as studs, and steel siderails as wales. The plywood is recessed in the steel frame, which protects the edges of the plywood. The system is extremely efficient for multiple lifts, where the concrete is greater in height than the height of the form panel. It is highly used in commercial, industrial, nonbuilding, heavy, and residential construction applications. The drawbacks are an inflexibility in tie spacing and a required liner when used for architectural concrete.

The aluminum frame–plywood face sheet system (type 4 in preceding list; see Fig. 14.6) incorporates a ½ or ⅝ in thick overlaid plywood face sheet, with aluminum siderails that act as wales. These panels are widely used in commercial, industrial, nonbuilding, heavy, and residential construction. The major drawbacks are an inflexibility of tie spacing, a design which is unsuitable for architectural concrete use, the high price of aluminum and potential for theft, and the potential problem of reaction with aluminum and concrete which would cause concrete buildup on the cross and side members, which is difficult to remove.

An all-steel system (type 5 in preceding list) has a steel face usually ⅛ in thick and steel studs and wales. Because of the weight of these components, this system is generally in a 2 × 4 ft module which is less efficient then the 2 × 8 ft or 3 × 8 ft modules. Its strong suit is durability if the steel face does not get stretched from overloading or warped by poor storage practices. Its drawbacks are inflexible tie spacing and design unsuitable for architectural surfaces.

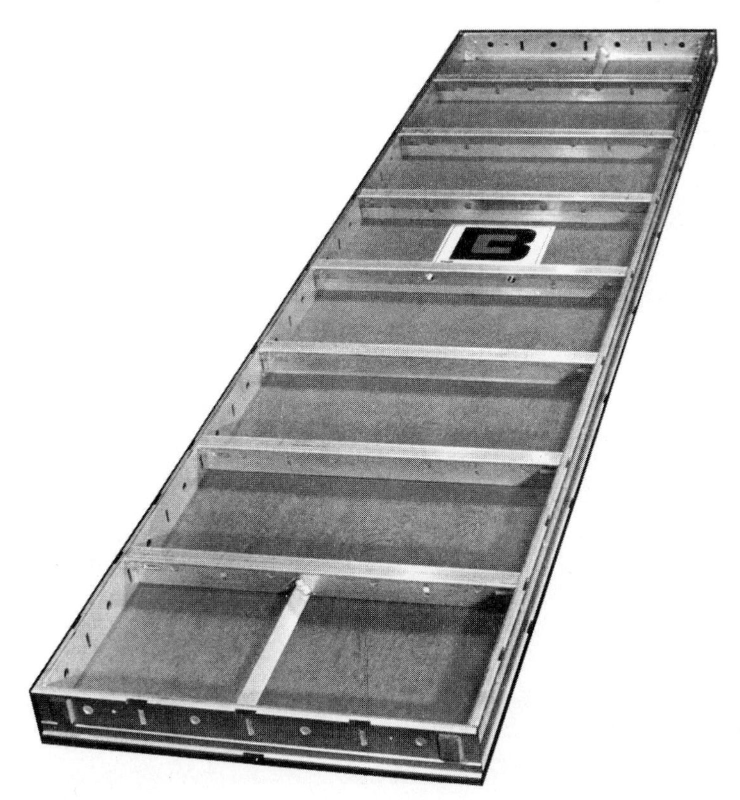

FIGURE 14.6 Aluminum frame with plywood face sheet. (*Courtesy of the Burke Co.*)

There are many different brand names of each of these systems. Information can be obtained from the local phone directory or the World of Concrete Source Book. It is essential to determine the ultimate strength of the ties and use either a 1½:1 or a 2:1 safety factor. Handset formwork can be one of the most risky from a profit standpoint and in terms of parts of concrete construction and requires an in-depth study with respect to reaching the particular objectives.

COLUMN FORMS

All the factory-built or modular systems can be used for either square or rectangular columns, with or without ties, depending on column size. Strict attention should be paid to the strength of the form and its ability to resist the concrete pressures.

There are three factory-built circular column forms which are paper tubes, wrapped plastic sheathing, and molded fiberglass.

Paper tubes (see Fig. 14.7) are ridged cylinders of wrapped paper that is glued and sealed with wax. These are single-use forms capable of withstanding high-capacity loads and are used extensively throughout the industry. Major drawbacks are lack of reuse, poor durability in bad weather conditions, and design unsuitable for use with architectural concrete.

Wrapped plastic sheathing is a sheet of plastic that has a series of slots for installing a strapping. One piece of sheathing can accomplish a large number of size changes. They require a good deal of labor for setting and stripping, but the resulting finish is better than that with the paper tube.

FIGURE 14.7 Paper-tube forming.

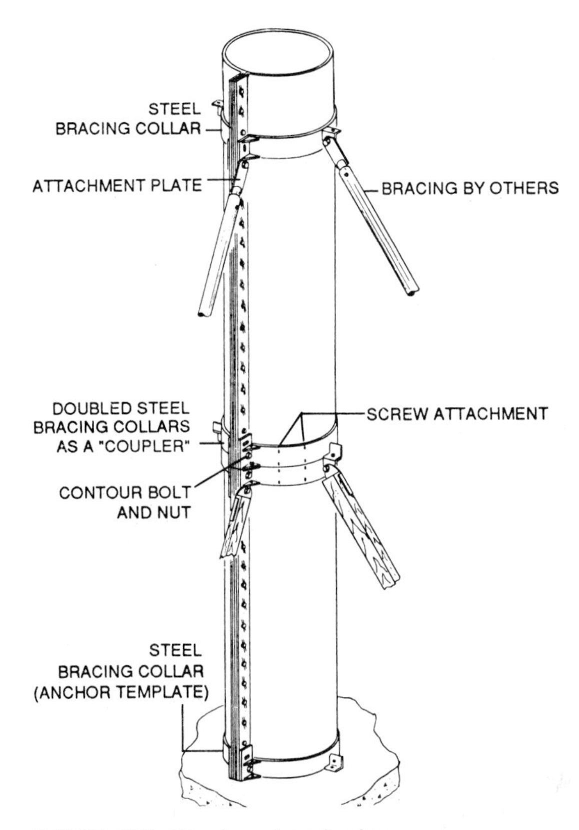

FIGURE 14.8 Fiberglass column forming.

Fiberglass forms for round columns (see Fig. 14.8) are very efficient and produce architectural concrete if the concrete mixture is properly proportioned and the vibrator operator is sufficiently skilled. Disadvantages are that they form only one size per unit and are brittle and easily damaged without strict care in their use.

GANG FORMING FOR WALLS

Assembling formwork into large sections is not a new innovation as transferring and erection of large sections of forms with a stiff legged derrick was used successfully in the 1880s. Making it cost-effective and logistically practical, has only recently been accomplished on a regular basis. This is possible when the designer, contractor, and suppliers cooperate or have their objectives aligned.

The following eight different types of gang forms are available: (1) job-built wood gangs, (2) job-built wood gangs with prefabricated wood studs and wales, (3) job-built wood gangs with prefabricated wood studs and steel wales, (4) job-built wood gangs with prefabricated aluminum beams and prefabricated aluminum wales, (5) job-assembled prefabricated forms with light-duty ties (tensile capacity of ≤10,000 lb), (6) job-assembled prefabricated forms with heavy-duty ties, (7) heavy duty prefabricated forms with steel wales and heavy-duty ties, and (8) all steel prefabricated forms. For engineering design, see Chap. 15.

Job-built wood gangs are composed of either wood or plywood face sheets, dimensional lumber as studding, and usually dimensional lumber as wales, and use proprietary tying systems with proprietary tie hardware. They can be designed to incorporate heavy-duty ties and hardware. No attempt will be made to cover all systems here as they are too numerous.

The size and shape of the gang is dictated by job conditions, reuse, logistics, cost, and finish. When these criteria are established, safety should be the next consideration, which includes form pressures, tie capacities, support systems, scaffolding, bracing, movement, and storage. The primary advantage of this system is availability as lumber can be obtained in all parts of the United States and Canada. The main drawback is strength and integrity of dimensional lumber, especially when used in a design that generates maximum loading of all four components of the form.

Job-built wood gangs (see Fig. 14.9) with a wood or plywood face sheet and studs and wales incorporating prefabricated wood beams or joists in either the "I" or box shape are used in the United States, Canada, and Europe. Caution should be exercised as different countries have different design and safety considerations. The hardware and ties are proprietary. Advantages are a known design criterion and an almost infinite design flexibility. Disadvantages are limited availability, which is diminishing as a problem.

Job-built wood gangs with a wood or plywood face sheet and proprietary fabricated wood studs and steel or aluminum wales plus proprietary ties and hardware (see Figs. 14.10 and 14.11) require the same attention to detail and have same advantages and disadvantages as the system described in the last paragraph. Their flexibility of design and weight/strength ratio is probably the best of these types.

Job-built wood gangs having a wood or plywood face sheet, extruded-aluminum beams or joists as studs, extruded-aluminum double channels as wales, and proprietary ties and hardware (see Fig. 14.12) are available from a variety of suppliers. Their primary

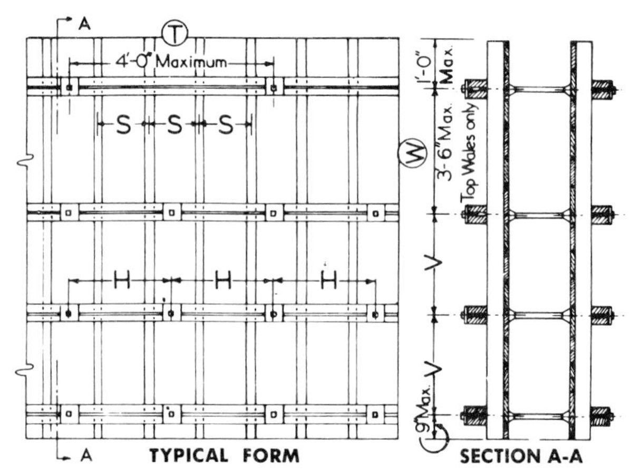

FIGURE 14.9 Job-built wood gang form.

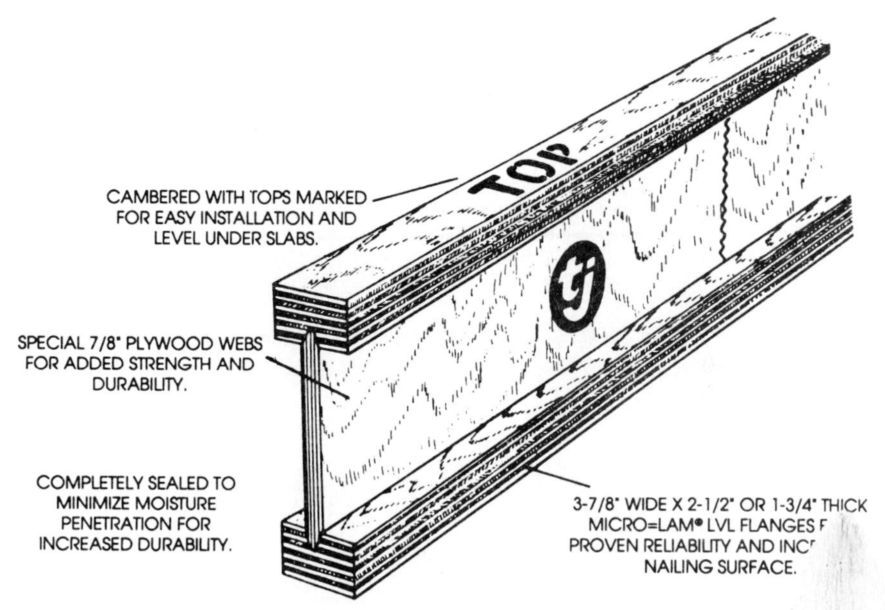

FIGURE 14.10 Example of job-built gang form with prefabricated wood stud and wale. (*C Trus Joist Corporation.*)

FIGURE 14.11 Job-built wood gang form with prefabricated wood studs or aluminum or steel wales proprietary ties and hardware.

FIGURE 14.12 Wood gang form with extruded-aluminum studs, aluminum double-channel wales, and proprietary ties and hardware.

benefit is a very advantageous weight/strength ratio and a nearly infinite flexibility of design. As this system developed in Canada, the design criteria should be checked to ensure compatibility to U.S. standards. In addition to the problem of chemical reaction between any spilled concrete coming in contact with the aluminum, as a result of the low modulus of elasticity of extruded aluminum, form overloads may cause a beam or joist to deflect and crack or break without warning.

Job-assembled prefabricated forms with light-duty ties have a prefabricated form panel as the face, its cross members as the studs, and the siderail as the wale with lumber backing to maintain alignment only (see Fig. 14.13). Ties and hardware are usually specially designed and proprietary for the system.

The system is tie-intensive as it requires more ties than the handset applications. The system is inflexible as to tie location, has a weight/strength ratio identical to that of the handset forming, and is completely rentable, allowing rental charges to be expensed to the job and not amortized over a multiyear period. It is efficient when high wall forming is required with limited reuse and design requires a gang form on one side and handset forming on the other.

Job-assembled prefabricated forms, with load gathering wales and heavy-duty ties and hardware, incorporate the prefabricated form as the face sheet, a cross member and siderail serving as the stud, double-steel channels as wales, and heavy-duty ties and hardware (see Fig. 14.14). There is no flexibility of design, and there is limited tie flexibility and a less efficient strength/weight ratio. The system is rentable and can be expensed to each project, often selected for an application requiring gang and handset forming not done concurrently, and results in little waste.

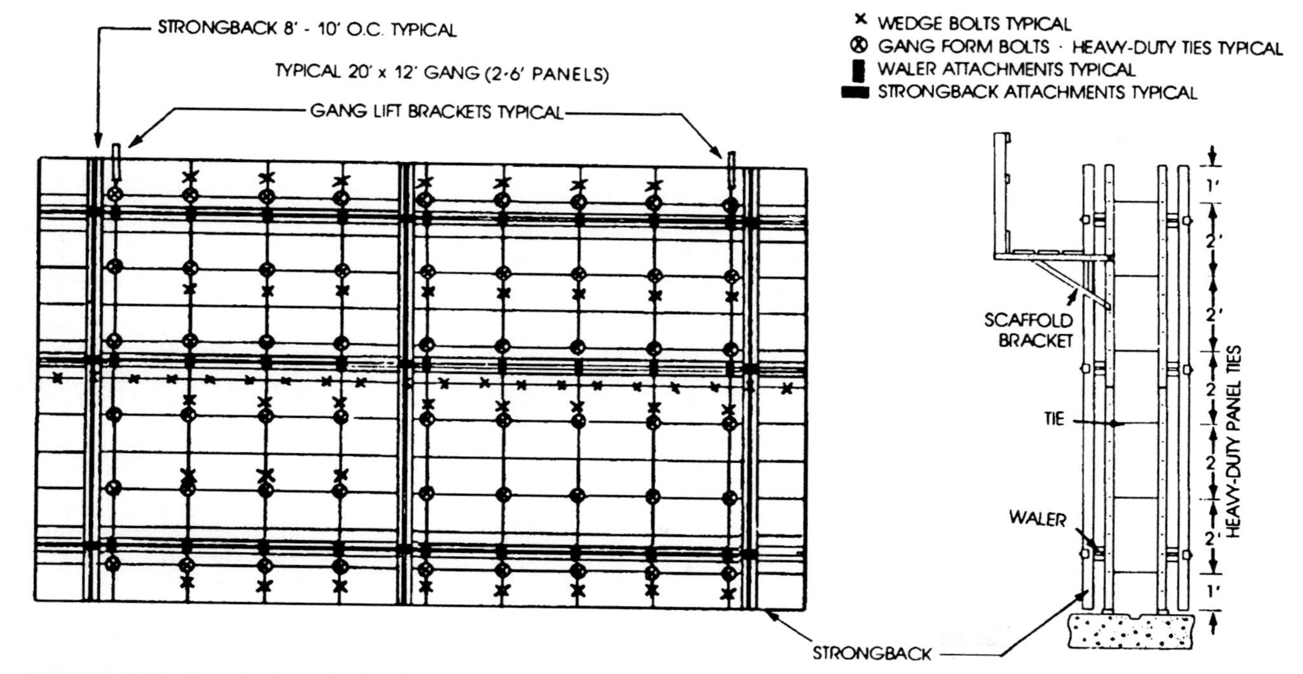

FIGURE 14.13 Example of assembled prefabricated forms having the cross member as stud and siderail as wale and proprietary ties and hardware.

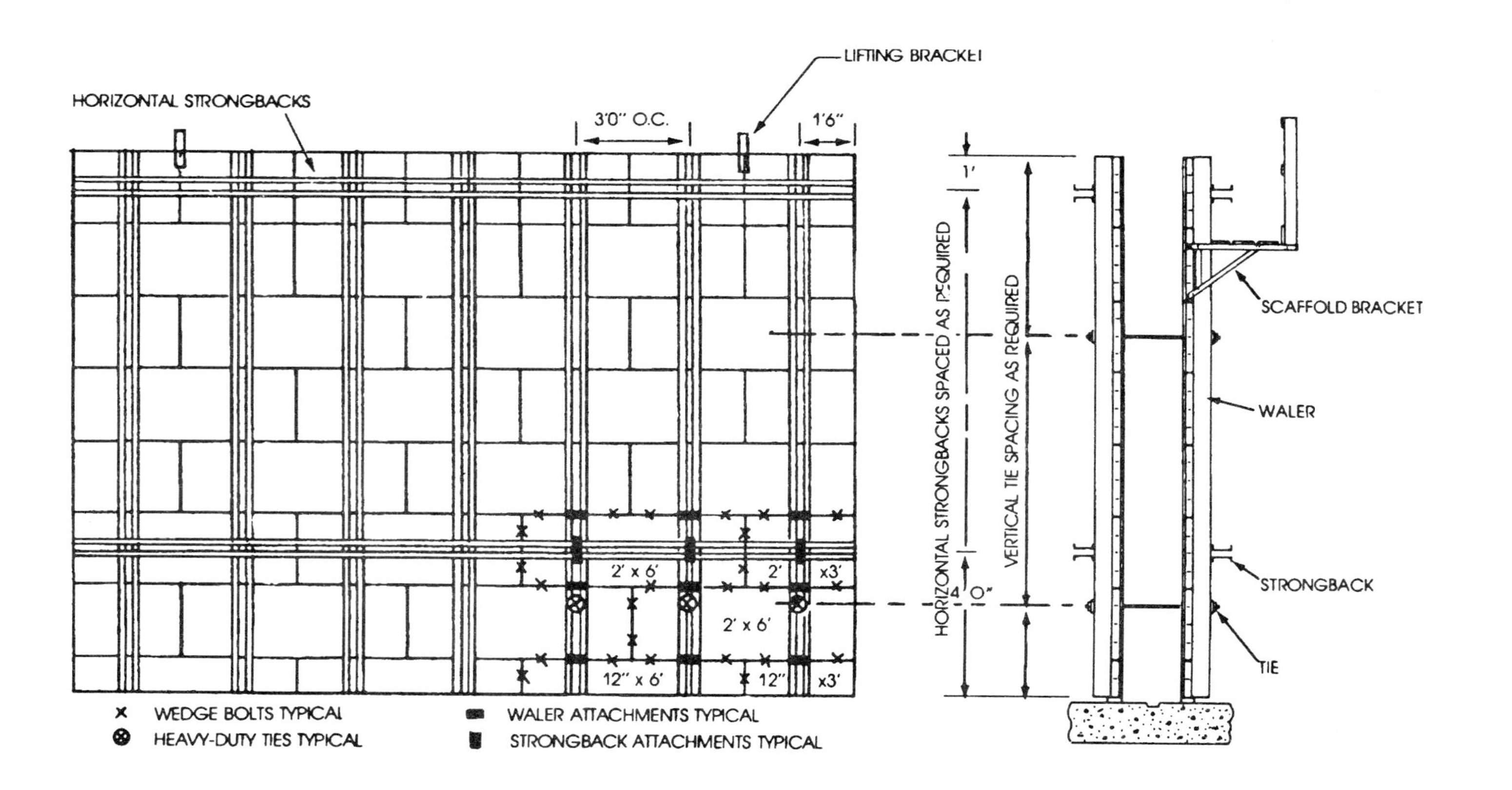

FIGURE 14.14 Job-assembled prefabricated form with cross member and siderail serving as stud and double-steel channels as wales.

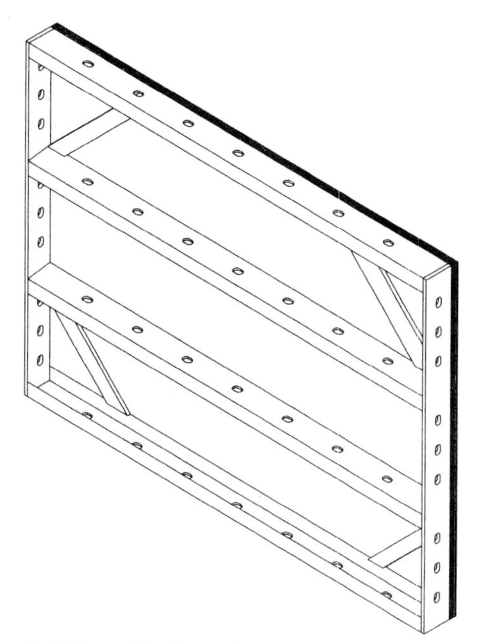

FIGURE 14.15 Heavy-duty prefabricated forms with steel wales and heavy-duty ties and hardware.

Heavy-duty prefabricated forms with steel wales and heavy-duty ties and hardware (see Fig. 14.15) contain ¾ in thick overlaid plywood as the face sheet attached to a steel channel used as studs and bordered with nonstructural end members. The system is completed by double-channel wales and heavy-duty ties and hardware. This is an old system used before the advent of laminated wood beams and joist or the aluminum beams and joist. They have lost popularity as a result of design inflexibility and an inferior strength/weight ratio. Although a fully rentable system, the apparent cost-saving benefit cannot be realized as this system is usually used on long-term projects. Its great advantage is the bolted connections and related safety factor.

All-steel prefabricated forms consist of a 3⁄16 to ¼ in thick steel face sheet, bent Z-shaped ribs as studs, and a top and bottom rail serving as an integral wale welded together to form a heavy-duty all-steel welded panel (see Fig. 14.16). The system is available from a variety of suppliers in a number of sizes and variations which can include a plywood face sheet. Advantages are a capability of use in a self-spanning system which requires little shoring and excellent durability. Disadvantages are inflexibility of design and tie spacing and a less efficient weight/strength ratio.

GANG-FORMED COLUMNS

Gang column forms are wood forms, with or without laminated wood or aluminum beams, prefabricated and ganged into large units which require handling and moving by crane. Such systems have special proprietary hinging and latching devices.

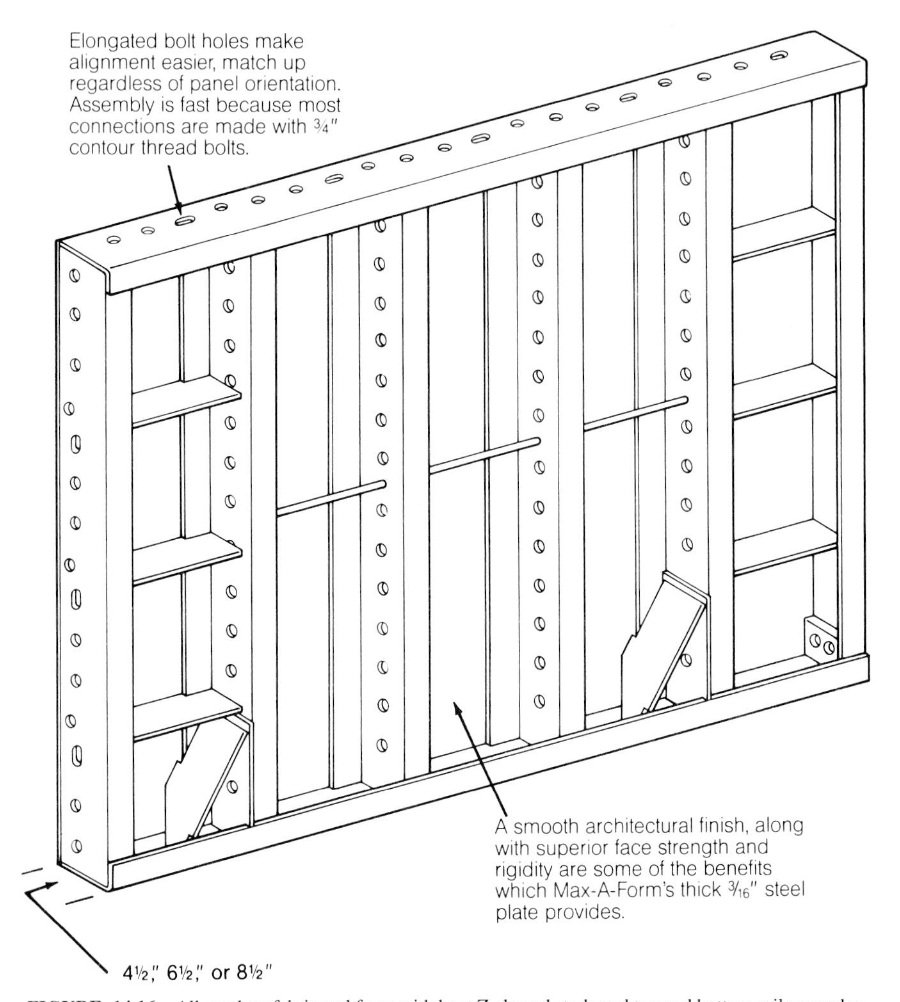

FIGURE 14.16 All-steel prefabricated form with bent Z-shaped studs and top and bottom rails as wales.

FORMWORK FOR HORIZONTALLY PLACED CONCRETE

Formwork for slabs on grade less than 24 in in thickness can be divided into five categories such as wood, wood and proprietary hardware, factory-built wood, factory-built prefabricated with special hardware, and all-steel (see Figs. 14.17 through 14.19).

For slabs less than 24 in in thickness, the face sheet is usually dimensional lumber with wood or steel stakes for studs, and no wales or ties. Slabs over 24 in in thickness are formed similar to walls.

Wood forms with proprietary hardware are used primarily for slabs under 24 in in depth and hardware consisting of a steel stake and pocket.

FIGURE 14.17 Factory-built wood forms with steel hardware.

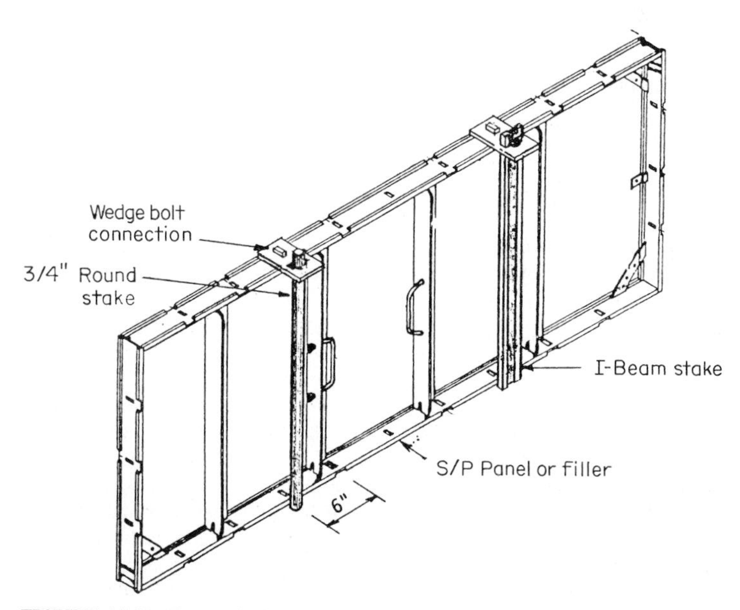

FIGURE 14.18 Factory-built prefabricated panels with special hardware and stakes.

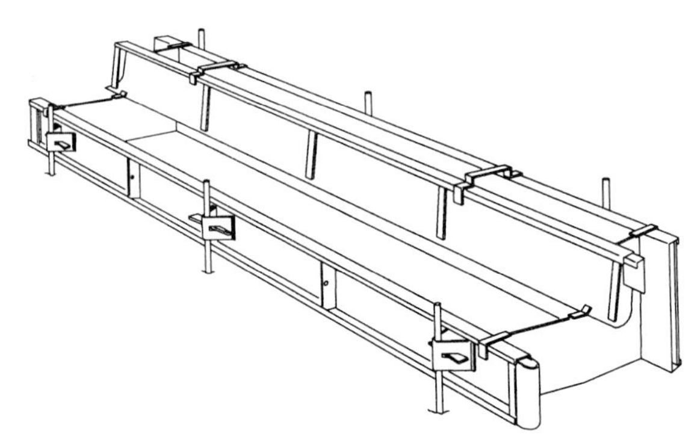

FIGURE 14.19 All-steel form with steel stakes.

FORMWORK FOR ELEVATED SLABS

Formwork for elevated slabs has improved greatly within the last 15 years for the same reasons as gang forms. New and better equipment have become more practical as designs recognized the need for more productive high-reuse formwork. Elevated-slab forms use a face sheet similar to that in vertical forms. The stud becomes a beam or a joist, the waler is now called a "stringer," and the tie becomes the support post or shoring frame. Elevated slab forms can be categorized by all-wood forms, wood forms with proprietary hardware, prefabricated forms which are rarely used today, wood forms with proprietary shoring frames, wood forms with adjustable horizontal beams, wood forms with aluminum beams or joist and prefabricated shoring frames, wood forms with laminated wood beams, or joist with prefabricated shoring frames, trusses, and steel decking on a frame.

All-wood forms have wood or plywood face sheets, dimensional lumber beams, stringers, and posts. The availability advantage of these forms is becoming less important as other systems become more available throughout the country. The disadvantages include variation of the wood, limited spans, and lack of adjustability. Design criteria may be found in ACI "Formwork for Concrete," "Form Design" by the Douglas Fir Plywood Association, and "National Design Specifications for Wood Construction" published by the National Forest Products Association.

Wood forms with proprietary hardware incorporate a wood or plywood face sheet with dimensional lumber beams or joist, dimensional lumber stringers with proprietary clamps, and "U" heads (see Fig. 14.20), which presents adjustability lacking in the all-wood system. Advantages are adjustability and availability. Disadvantages are lack of consistency in strength of lumber, limited spans, and the amount of labor required to disassemble and assemble the forms.

Wood forming with proprietary shoring is widely used and incorporates a wood or plywood face sheet or decking, dimensional lumber as joist, and a standard steel section as stringer with modular shoring frames having adjustable screwjacks as supports (see Fig. 14.21). Advantages are flexibility of design, higher allowable loading for the shoring frames and stringers than wood members, and availability for rental. The disadvantages are a weakness of the wood beams, which are presently being supplanted by the laminated or aluminum beams.

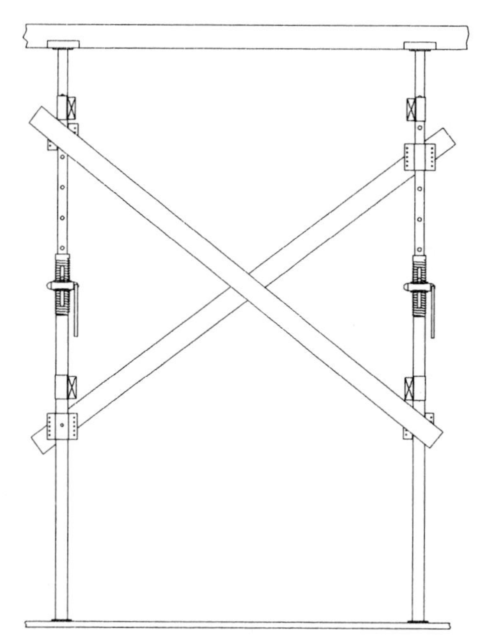

FIGURE 14.20 Wood forms with proprietary hardware, dimensional lumber stringers, and wood posts and proprietary wood clamps and U heads.

Wood forms with adjustable horizontal beams (see Fig. 14.22) have wood or plywood decking with horizontal beams and steel stringers on modular steel shoring or steel post shores. Advantages are ability to function at peak efficiency and allowance for adjustments in the joist beam area. One disadvantage is the beam built-in camber, which may not always be required.

Wood forms with aluminum beams or joists having either steel or aluminum structural members as stringers are supported by either steel or aluminum shoring frames (see Fig. 14.23). This system uses each member to its optimum working capacity. Disadvantages include chemical reaction between aluminum and spilled concrete, the tendency of the aluminum to crack rather than bend, and the possibility of pilferage due to the high scrap value of aluminum.

Wood forms with laminated wood beams or joists, with either steel or aluminum stringers supported by steel or aluminum shoring frames (see Fig. 14.24), use each member to its ultimate capacity and do not have the aluminum chemical problem with concrete or the cracking tendency under overloads.

Trusses are the most widely used ganging method in shoring applications (see Fig. 14.25). They have a lumber or plywood deck with either wood laminated or aluminum beams, and steel or aluminum trusses as both the stringer and shoring.

Steel decking on steel stringers, where the decking and supports are left in place, is popular in certain parts of the country. This method is associated with a structural steel frame building where the steel decking is welded to the steel frame. It can be easily assembled but is lost in the building and must be reassembled for each floor.

FIGURE 14.21 Wood forming with dimensional lumber joists and standard section steel stringers on modular shoring frames supported by adjustable jacks.

CONCRETE PROBLEMS ASSOCIATED WITH CURRENT FORMWORK PRACTICES

The ACI Building 318 Building Code specifies that concrete mix designs for projects have a history of successful use on 10 projects or be a specially designed concrete mixture. Concrete mixtures designed specifically for vertical formwork, where strength is more important than workability, are not readily available. Readily available mixtures are those proportioned for flatwork workability containing high quantities of fines, and mini-

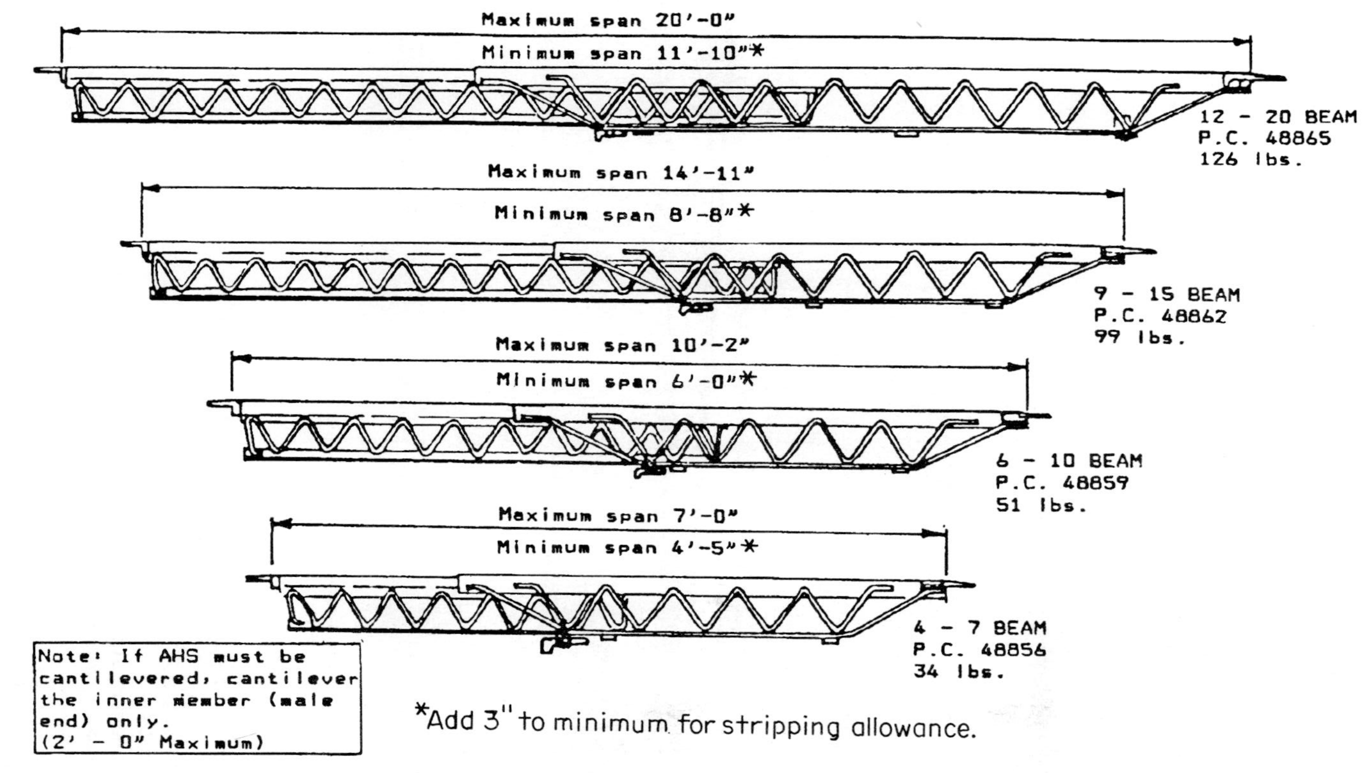

FIGURE 14.22 Wood forms with adjustable horizontal beams.

FIGURE 14.23 Wood form with steel or aluminum stringers and beams or joists supported by steel or aluminum shoring frames.

mal coarse aggregate. Their use leads to excessive blemishes such as form and amount of bugholes, scouring, and discoloration.

The changes in cement type usage and admixtures have led to higher form pressures which must be closely controlled to prevent form failure.

The improvements in concrete formwork occurring in the last decade are leading to more economical construction. Cooperation between all elements has been the key. We can look forward to more improvements as additional present-day research results are applied.

FIGURE 14.24 Wood forms with laminated wood beams or joist, with steel or aluminum stringers supported by aluminum or steel shoring frames.

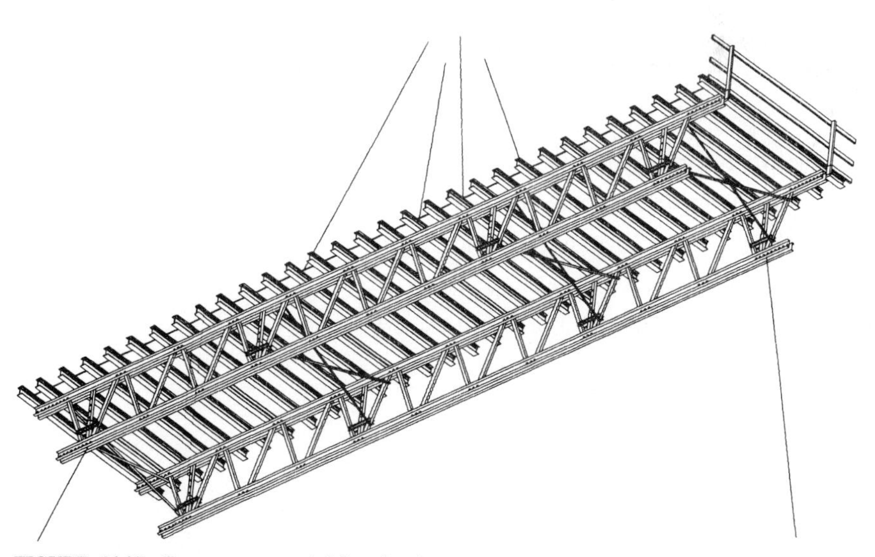

FIGURE 14.25 Truss-type suspended floor forming.

CHAPTER 15

FORM DESIGN, SAFETY, AND FORM RELEASE AGENTS

Jerome H. Ford

FORM DESIGN

15.1 Introduction

Concrete form design begins with the selection of a forming system, covered in Chap. 14, which points up the objectives to follow, the different types of forming available, and the many suppliers that can assist in the selection. These objectives are used by most large- and medium-sized contractors but may be neglected by the smaller or less experienced larger contractor, a job superintendent, or a project manager. Unfortunately, a successful superintendent may be made a project manager on another project. Often this new project is not as involved or complicated but they are locked into a system or technique which is impractical for the project. Many jobs that should have been handset are gang-formed because a decision was made without analysis of the job requirements but rather on the basis of a subjectively preferred system. Such decisions are no longer practical in the present and foreseeable competitive market.

15.2 Form Layout

Form layout must be simple, clear, and consistent. It is always less costly to work out solutions to forming problems on paper than through field trials. Mistakes that can be easily corrected on paper can be very expensive when corrections are required in the field. It is important to make shop drawings of formwork layouts that can be used in the office for estimating purposes and for construction in the field or plant. If these shop drawings are made by a formwork supplier, they should be reviewed by the contractor for any necessary corrections or revisions and redrawn to prevent future errors in construction. Shop drawings made by the contractor should have a scale easily read by field personnel with as few notes as possible, and contain fabrication instructions, list of required materials, and a logistical layout. Embedded items, boxouts, and appropriate safety instructions need to be clearly shown on the shop drawings. The rate of concrete placement which will not exceed the safety factors is passed on to the field personnel, and its maintenance is now the responsibility of the contractor.

15.3 Form Design by the Contractor

A contractor who decides to do the form design should review proposed form pressures, basic formulas, and loads. The following tables and charts will help in this regard. If additional information is required, consult ACI 347, Formwork for Concrete, *DFPA Bulletin,* and the *National Forest Products Bulletin.*

FORM PRESSURES

The load or forces imposed on a form that must be considered by the designer are classified as live load and dead load. Live load is the weight of workmen and equipment such as pipelines and buggies, and other temporary loads that are supported by the form during concrete placing and finishing. Dead load consists of the weight of the form itself plus the concrete and reinforcing steel. However, in most cases the form weight is negligible, so the concrete weight is of prime importance to the designer. The weight of reinforced concrete varies from 50 to over 400 pcf depending on the type of aggregate used. However, these very light or very heavy concretes are designed for special applications, and the form designer is usually concerned with normal-weight structural concrete at 150 pcf or lightweight concrete at 110 pcf. The same principles of form design apply in any case, regardless of the concrete density.

When concrete is placed in or on forms it is in a quasi-liquid state, exerting hydrostatic pressure on its confining surfaces in the same manner that any liquid does. Because of this, the pressure in all directions at any given point is calculated the same as for other liquids. That is, the pressure, in pounds per square foot, at any point equals the depth, measured in feet, below the fluid head (top of liquid) times the density (150 pcf for standard concrete or 110 pcf for lightweight concrete). Two important factors that influence the hydrostatic pressure of fresh concrete are the rate of placement, measured in feet of depth per hour, and the ambient temperature, measured in degrees Fahrenheit, unless the concrete itself is heated or cooled to a controlled temperature, in which case the controlled temperature is used.

As concrete hardens it changes from a liquid to a solid state, eliminating the lateral pressure. Temperature of the concrete influences this rate of hardening; thus temperature must be taken into consideration when calculating the hydrostatic pressure of concrete. Concrete hardens faster in warm weather than it does in cold (assuming there is no retarder in the concrete). The rate of placement (feet per hour) controls the height of the fluid head of concrete, making it necessary to use this rate when calculating hydrostatic pressure.

Many other factors can affect the pressure of concrete on the forms, including the type of vibration used to consolidate the concrete (external vibration, internal vibration), impact loading caused by free fall of concrete when it is discharged into or on forms, and slump of the concrete. The influence that each of these variables has on the lateral pressures that concrete exerts on forms, and the limit of hydrostatic pressure to be expected, is apt to be assigned differing degrees of importance by different persons. However, ACI Committee 347 has developed workable formulas for calculating the lateral pressures exerted by fluid concrete on wall and column forms.[1] These formulas take into consideration the many variable factors that are involved. For wall forms with rate of placement not exceeding 7 ft/h,

$$p = 150 + \frac{9000R}{T} \quad \text{(max 2000 psf or } 150h, \text{ whichever is least)} \qquad (15.1)$$

and in walls, with rate of placement greater than 7 ft/h,

$$p = 150 + \frac{43400}{T} + \frac{2800R}{T} \quad \text{(max 2,000 psf or } 150h, \text{ whichever is least)} \tag{15.2}$$

For columns,

$$p = 150 + \frac{9000R}{T} \quad \text{(max 3000 psf or } 150h, \text{ whichever is least)} \tag{15.3}$$

in which p = maximum lateral pressure, psf
R = rate of concrete placement, ft/h
T = temperature of concrete in the forms, °F
h = maximum height of fresh concrete in form, ft

The lateral pressure of fluid concrete (dead load) acts as a force against the walls of its mold. This force must be held static by the mold to ensure that the required shape can be obtained. To do this the concrete form must be constructed of form members, selected by the form designer, with physical properties adequate to withstand the stresses of compression, bending, and shearing that are exerted on them.

15.4 Basic Formulas

For many years form members have been selected and spaced on the basis of the designer's experience, by adapting materials that worked satisfactorily in the past to similar applications for the job at hand, often called "design by assumption." This practice fortunately has given way to design by application. Tables are readily available to the form designer giving the physical properties of plywood, lumber, and steel, these being the materials most commonly used for form members. Information regarding plywood properties is available from the American Plywood Association.[2] There are tables in the *Wood Handbook* that give directly the required beam sizes, loadings, shear, fiber stresses, and other data for almost any loading and construction condition. The "Manual of Steel Construction" gives the physical properties of structural shapes for design and detailing.

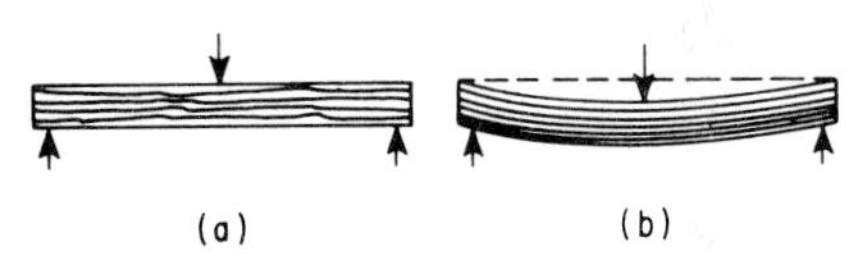

FIGURE 15.1 *(a)* Bending strength is a measure of the load-carrying capacity of a beam. *(b)* Fiber stress in bending results when the beam is subjected to a load that causes deflection.

Deflection When a beam is subjected to an external moment the fibers on one side elongate, while the fibers on the other side shorten (Fig. 15.1), causing the beam to deflect. Design of beams may be based on strength requirements (load-carrying capacity) or on stiffness, when only a small deformation, or deflection, is acceptable. The modulus of elasticity, which controls stiffness, varies with different types of lumber and, in any one kind of lumber, varies as the moisture content varies.

Form members, including walers, studs, and sheathing, all act as beams under load when supporting fresh concrete vertically, horizontally, or in some intermediate direction.

Maximum deflection for a simple beam uniformly loaded is given by the equation

$$D_{\max} = \frac{5wl^4}{384EI} \tag{15.4}$$

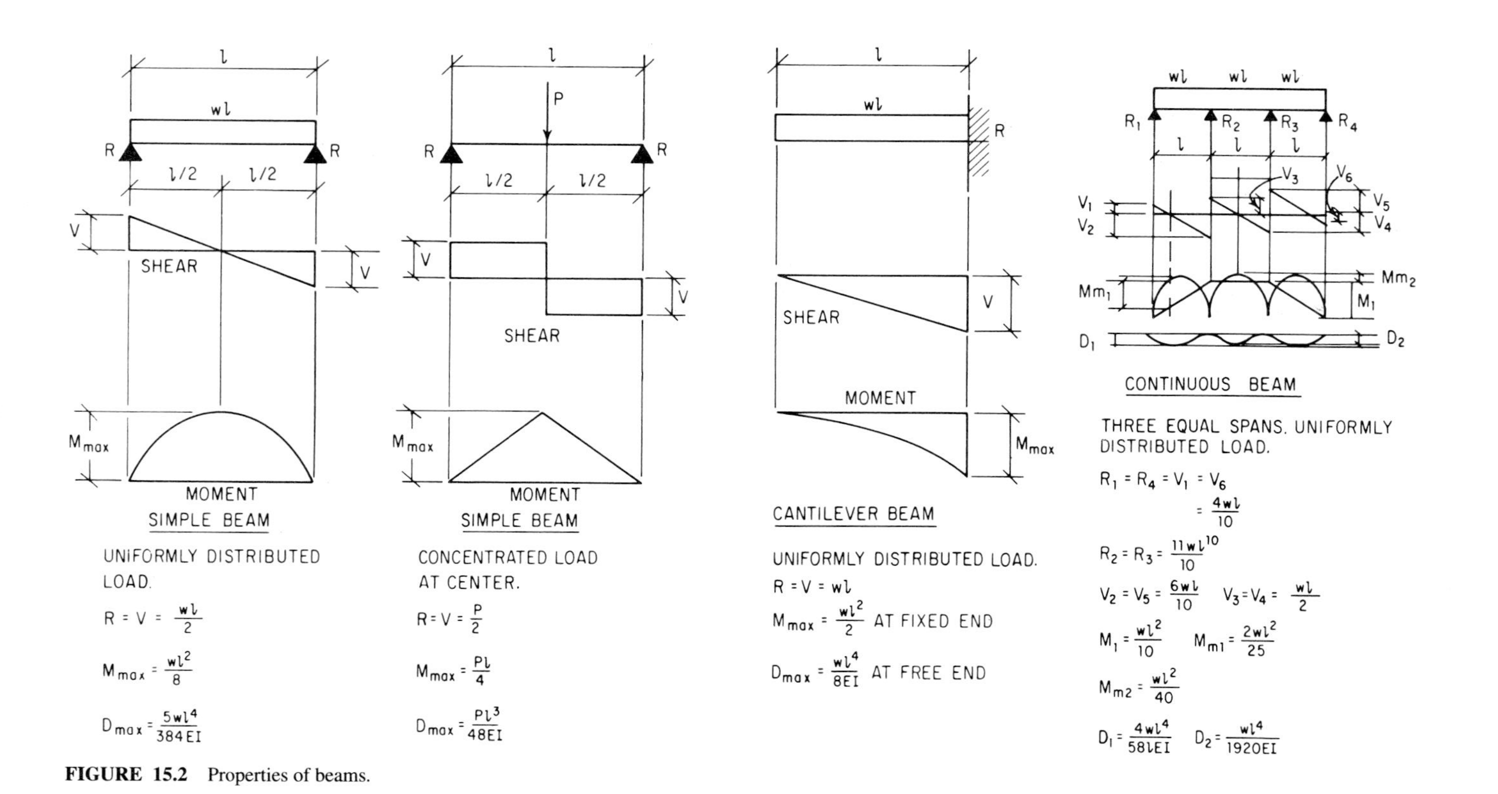

FIGURE 15.2 Properties of beams.

in which D_{max} = maximum deflection, in
w = uniformly distributed load, lb/lin ft
l = length of beam between supports, ft
E = modulus of elasticity of beam material, psi
I = rectangular moment of inertia, in^4

Other beam formulas used in design of formwork are shown in Fig. 15.2.

When loads acting perpendicular to the longitudinal axis of a form member cause deflection, the member is subjected to fiber stress in bending. The stress is calculated by the equation

$$f = \frac{M}{S} \tag{15.5}$$

in which f = extreme fiber stress, psi
M = induced bending moment, in-lb
S = section modulus, in^3

For a simple uniformly loaded rectangular beam, this becomes

$$f = \frac{3wl^2}{4bh^2} \tag{15.6}$$

in which b = width of beam, in
h = depth of beam, in

Computed values of f should not exceed those in Ref. 3, which gives allowable values of f for various types and grades of wood.

If a form is designed so the deflection of its members is maintained within reasonably close limits, poor-appearing concrete, as well as the problems that often arise when it is time for form removal, can be prevented. Deflected forms can increase the cost of form stripping as well as add the expense of chipping off unwanted hardened concrete.

The hydrostatic pressure that fluid concrete will exert on form-facing material is determined from Tables 15.1 and 15.2, using the anticipated rate of placement and expected

TABLE 15.1 Maximum Lateral Pressure for Design of Wall Forms*

Rate of placement R, ft/h	p, maximum lateral pressure, psf, for temperature indicated					
	90°F	80°F	70°F	60°F	50°F	40°F
1	250	262	278	300	330	375
2	350	375	407	450	510	600
3	450	488	536	600	690	825
4	550	600	664	750	870	1050
5	650	712	793	900	1050	1275
6	750	825	921	1050	1230	1500
7	850	938	1050	1200	1410	1725
8	881	973	1090	1246	1466	1795
9	912	1008	1130	1293	1522	1865
10	943	1043	1170	1340	1578	1935

**Note:* Do not use design pressures in excess of 2000 psf or 150 × height of fresh concrete in forms, whichever is less.

TABLE 15.2 Maximum Lateral Pressure for Design of Column Forms*

Rate of placement R, ft/h	p, maximum lateral pressure, psf, for temperature indicated					
	90°F	80°F	70°F	60°F	50°F	40°F
1	250	262	278	300	330	375
2	350	375	407	450	510	600
3	450	488	536	600	690	825
4	550	600	664	750	870	1050
5	650	712	793	900	1050	1275
6	750	832	921	1050	1230	1500
7	850	938	1050	1200	1410	1725
8	950	1050	1178	1350	1590	1950
9	1050	1163	1307	1500	1770	2175
10	1150	1275	1435	1650	1950	2400
11	1250	1388	1564	1800	2130	2625
12	1350	1500	1693	1950	2310	2850
13	1450	1613	1822	2100	2490	3000*
14	1550	1725	1950	2250	2670	
16	1750	1950	2207	2550	3000*	
18	1950	2175	2464	2850		
20	2150	2400	2721	3000*		
22	2350	2625	2979			
24	2550	2850	3000*			
26	2750	3000†				
28	2950					
30	3000†					

**Note:* Do not use design pressures in excess of 3000 psf or 150 × height of fresh concrete in forms, whichever is less.

†3000 psf maximum governs.

temperature. With this information, calculations can be made to determine size and spacing of support members, whose deflection limit should be the same as that for the form-facing material. For ordinary walls not subject to close scrutiny, and the underside of most slabs, a deflection of 1⁄270 of the span is acceptable. Spacing of support members can be computed with the aid of Fig. 15.3. Vertical-facing support members, commonly called studs, should be modularly spaced to fit the size of the facing material. For example, if computations indicate that a stud spacing of 18 in on centers (OC) is adequate to support an 8 ft length of plywood, we can see that 16 in on center requires no more studs (Fig. 15.4) but does give uniformity to the form. Therefore, the latter spacing should be used.

Compression Wood form members are apt to be subjected to compression in either of two directions, compression parallel to the grain of the member, and compression perpendicular to the grain (Fig. 15.5). Compression parallel to the grain is most commonly encountered when lumber is used as shoring, in which case the member is acting as a column and the load capacity is calculated by using the slenderness ratio, which is defined as the relationship of the unsupported length of the shore (acting as a column) to the cross-sectional dimension of the face under consideration, usually the narrowest of the two faces. It is expressed as l/d, in which l = unsupported length, in inches, and d = net

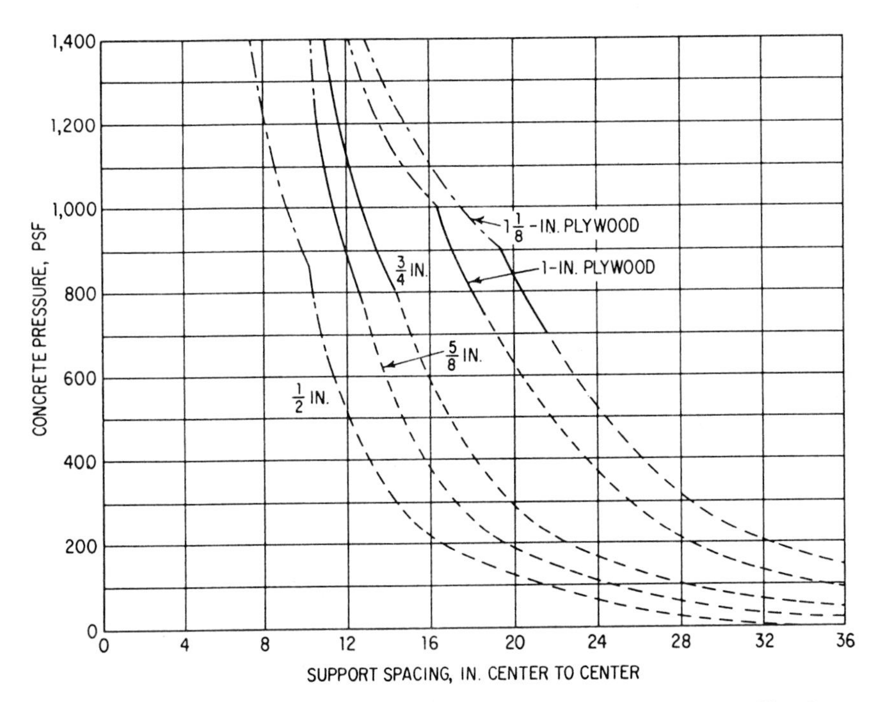

FIGURE 15.3 Spacing of joists or studs to limit deflection of plywood sheathing to 1/270 of span between studs or joists, based on flexural stress of 2000 psi and shear in plane of panel of 94 psi for Douglas fir interior plyform or exterior all-sanded grades. Panels are continuous across two or more spans. When panel is used on a single span, stiffness is only half that shown on chart. For face grain of plywood across supports. When face grain is parallel with supports, use the following percentages of the load given in the chart for each thickness: ½ in, 40 percent; ⅝ in, 51 percent; ¾ in, 73 percent; 1 in, 90 percent; 1⅛ in, 90 percent.

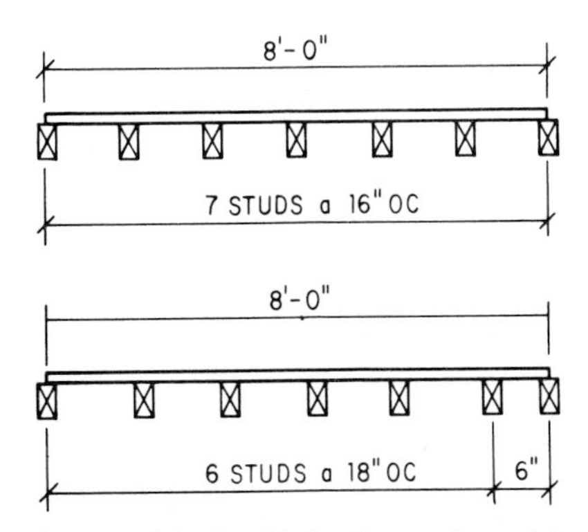

FIGURE 15.4 Careful planning permits modular spacing of studs, making a better-looking form that is more economical.

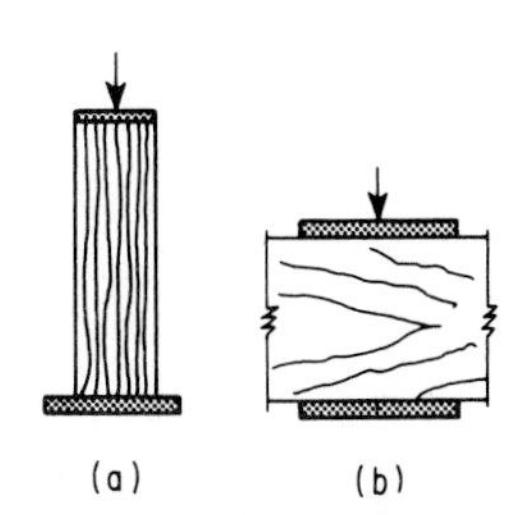

FIGURE 15.5 (*a*) Compression parallel to the grain is a measure of the capacity of a short column to withstand loads acting in a direction parallel to the long axis. (*b*) Compression perpendicular to the grain is a measure of the bearing strength of wood across the grain.

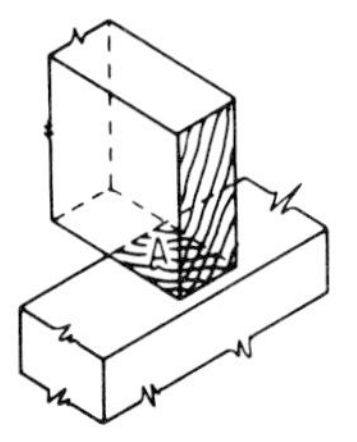

FIGURE 15.6 The end-bearing area A should be such that the reaction V divided by A will not exceed the allowable unit stress in compression perpendicular to the grain. For bearings shorter than 6 in located 3 in or more from the end of the member, the stress can be increased as follows:

Length of bearing, in	½	1	1½	2	3	4	6+
Factor	1.75	1.38	1.25	1.19	1.13	1.10	1.00

dimension in inches of face under consideration. The slenderness ratio for simple solid columns should not exceed 50.

Allowable unit stresses for compression perpendicular to the grain are of concern to the form designer when form members are subjected to loads that would cause the material to fail by being crushed by its support. An area of form members where this situation can occur is shown in Fig. 15.6.

Values for compression perpendicular to grain are adjusted for duration of loading the same as other strength properties. No allowance need be made for the fact that as members deflect the pressure on the inside edge of the bearing area is greater than it is on the open side. When the bearing is not more than 6 in in length and more than 3 in from the end of the member, an increase in the allowable unit stress for compression perpendicular to the grain is permitted, as shown in Fig. 15.6. For round bearing areas the diameter is used for the length.

Shear Vertical and horizontal form members must be selected so they are capable of resisting failure in shear (Fig. 15.7). Vertical shear is the tendency of the load forces to cut the member into two pieces perpendicular to its longitudinal axis, while in horizontal shear the load forces tend to split the member into two pieces parallel with the longitudinal axis (Fig. 15.8). Vertical and horizontal shear are of equal magnitude at any point on the member acting perpendicular to each other. These terms, horizontal and vertical, were originated for viewing beams in the horizontal position. However, if their relationship with the beam axis is remembered, they can be applied to beams at any angle, such as studs being used to form either vertical or inclined surfaces. Horizontal shear stress is common to fibrous material such as lumber.

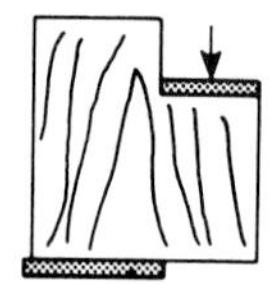

FIGURE 15.7 Shearing strength parallel to the grain is a measure of the ability of the wood to resist slipping of one part on another along the grain.

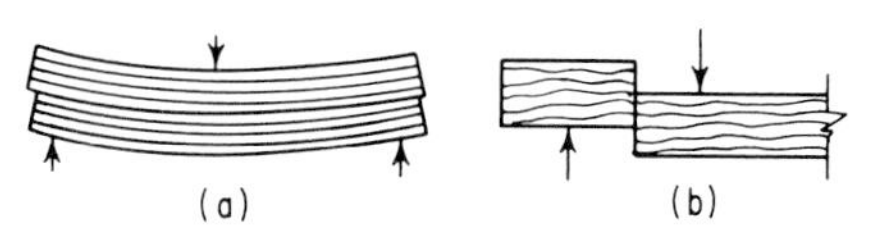

FIGURE 15.8 (*a*) Horizontal shear is the resistance to the tendency of the upper half of the beam to slide on the lower half when the beam is loaded. (*b*) Loads and reactions at the supports are vertical forces tending to shear or cut the beam across. This is vertical shear.

Horizontal shear is a maximum at the neutral axis and is given by the equation

$$H = \frac{3V}{2bd} \tag{15.7}$$

and vertical shear by

$$V = \frac{wl}{2}$$

in which b = width of beam, in
d = depth of beam, in
l = length of beam, ft
w = uniformly distributed load, lb
V = total vertical shear at supports

The top of a round column is often flared the last 2 to 3 ft below the deck the column is supporting. This section of the column, commonly called the column capital, can be formed by using the Denform system, consisting of a series of small curved modular steel panels that can be assembled into a variety of capital sizes. Denform capitals have a seat for fitting on top of round fiber-tube forms. Column capitals are also formed by sheet-metal forms, particularly applicable where beams run into the capital for support. When this is the case a section of the sheet-metal capital is cut out and fitted to the beam soffit and sides.

When it is necessary to design a form to fit unusual configuration and there is a good reuse factor, the form designer is wise to consider steel- or plastic-reinforced fiber glass. The form should be fabricated strong enough to be tied with the minimum of labor for the rate of concrete placement already established. A good example is illustrated in Fig. 15.11.

15.5 Suspended Slabs and Beams

The principal concern of every form designer when designing forms for suspended structural concrete is to be sure that the form will carry the load until the concrete has attained sufficient strength to support itself. The type of form support depends upon how the suspended slab itself is to be supported.

When the permanent horizontal support for the suspended slab (commonly called a deck) is furnished by either structural steel or a precast-concrete member, some type of hanging device is usually used to support deck forms as shown in Fig. 14.25. However, if the suspended member is to be self-supporting or to be supported by integral beams, shoring will be required until the concrete is self-supporting. Various types of shoring methods and equipment are described in Chap. 14, and installations are pictured in Fig. 14.21, 14.22, 14.23, 14.24. Elements of scaffold shoring are detailed in Figs. 15.9, 15.10 and 15.12.

Loads (dead and live) imposed upon deck forms by concrete being placed on them must be correctly calculated to ensure against failure. The weight of concrete (pounds per cubic foot) times the concrete depth in feet (in decimals) equals the normal dead load (Table 15.3). An additional amount, allowed for the live load imposed by men and equipment used to place and finish the concrete, varies from 40 to 100 psf depending on the particular job requirements. Most form designers allow 50 psf as the basic live-load allowance. It can be adjusted to compensate for impact loading of concrete being dropped or for placing concrete with mechanized equipment such as power buggies.

FIGURE 15.9 Heavy-duty scaffold shoring is combined with Jr. beam stringers and lightweight horizontal shores for the large deck form. (*Courtesy of American Pecco Corporation.*)

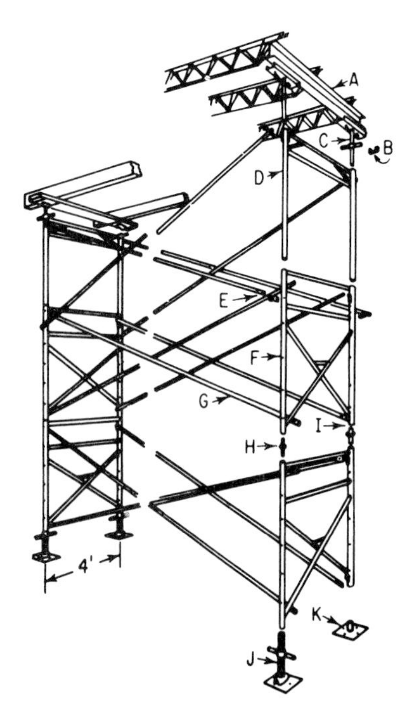

FIGURE 15.10 Elements of a scaffold shore. (*A*) 4 × 8-in steel Jr. beams serve as stringers (*B*) Beam clamps secure the stringers to the jack head in such a way as to prevent rotation of the beam. (*C*) Screwjacks provide 18 in of vertical adjustment. (*D*) Extension scaffold frames telescope into base frames to give height adjustments of 1, 2, 3, 4, or 5 ft. (*E*) Adapter pins fit into holes on legs of base frames to support the extension frames at the required height. (*F*) Base frame with X bracing (*G*) Tubular-steel cross bracing. (*H*) Coupling insert pins. (*I*) Special locking devices attach cross bracing to frame. (*J*) Swivel-base screwjacks compensate for uneven ground conditions and give 12 in of height adjustment, fitting between legs of frame and base plate (*K*). (*Courtesy of Superior Scaffold Company.*)

FIGURE 15.11 A good example of pier and cap configuration which permits multi-use of forms. (*Symons Manufacturing Co.*)

The normal steps taken to design deck forms after the dead and live loads have been determined are

1. Select the thickness of decking material; minimum ⅝ in plywood.
2. Calculate the size and maximum spacing of joists. Joists should be loaded to the maximum safe capacity for economy.
3. Calculate the size and maximum ledger or stringer spacing.
4. Calculate the best spacing for shores, and select a shore that will safely support the loads anticipated.
5. Work out the best method for reshoring subsequent deck loads, where required.

The above method of deck forming design sequences applies equally well to flat slabs, flat slabs with dropheads, and beam-and-slab structures.

Designing deck forms for the waffle or grid-slab system, in which dome pans or fiberboard boxes are used to form a two-way joist system, is similar to the method for flat slabs. Table 15.4 will assist the form designer in calculating dead load for this deck system.

Pan joists use a joist spacing that has been determined by the structural engineer for the project at hand, the commonest center-to-center spacing being 30 in; however, 20 in spacing is also used. Standard metal pans for either spacing are available for rent or to be installed in place. The pan form creates a void in the slab which cuts down on the

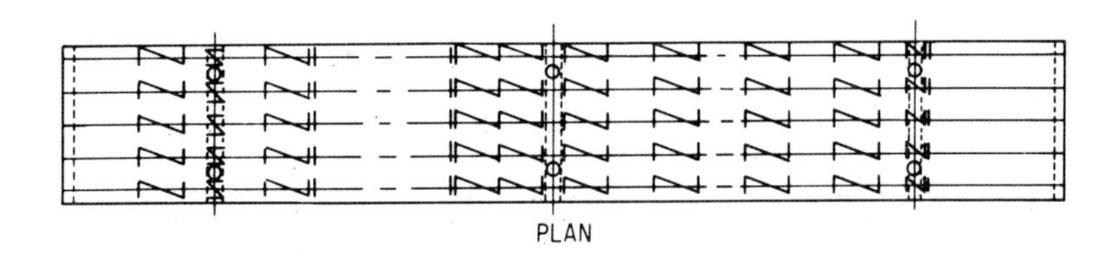

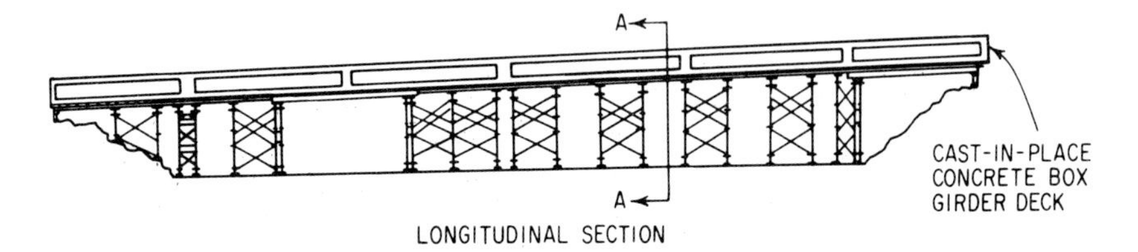

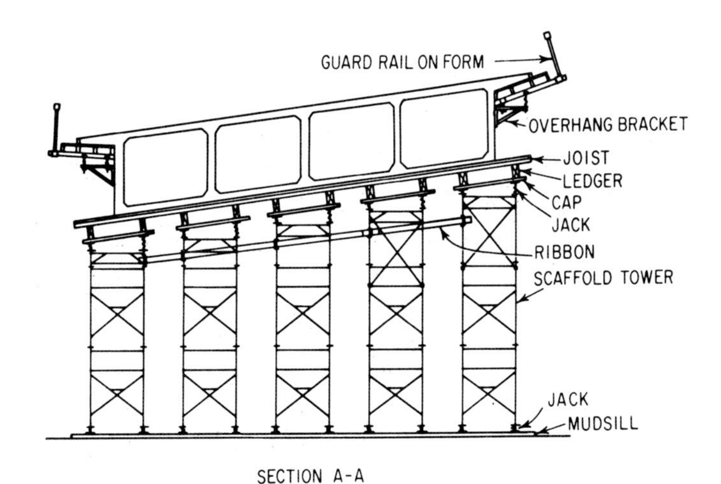

FIGURE 15.12 Scaffold framing and forming for a cast-in-place box-girder bridge on a curve. (*Superior Scaffold Company.*)

TABLE 15.3 Weight of Concrete, psf

Slab thickness, in	Slab thickness, ft	Wt/ft², 150 pcf concrete	Wt/ft², 110 pcf concrete
3	0.25	37.5	27.5
4	0.33	49.5	36.3
5	0.42	63.0	46.2
6	0.50	75.0	55.0
7	0.58	87.0	63.8
8	0.67	100.5	73.7
9	0.75	112.5	82.5
10	0.83	124.5	91.3
11	0.92	138.0	101.2
12	1.00	150.0	110.0

TABLE 15.4 Equivalent Flat-Slab Thickness for Dome Pans*

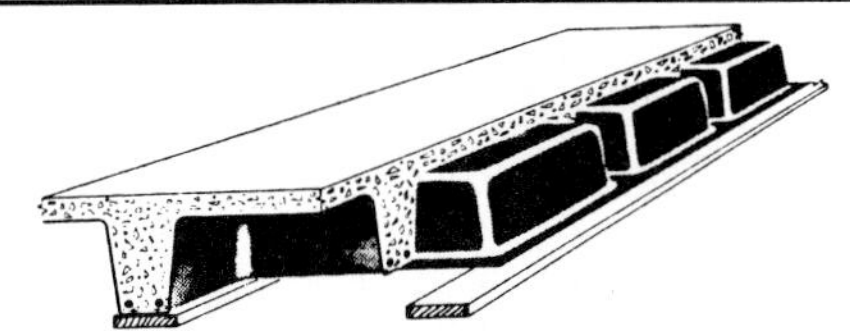

Depth of dome form, in	Equivalent slab thickness, in		
	Regular pan, 30 × 30 in	Filler pan, 20 × 30 in	Filler pan, 20 × 20 in
8	3	3½	4
10	3½	4½	5
12	4½	5½	6
14	5½	6½	7

**Notes:* (1) Add topping to above equivalents; (2) all joists in the above system are 6 in wide.

required concrete volume, at the same time creating a joist that acts as a small beam. Table 15.5 includes information for form design with this type of pan. The two types of joist beam are manufactured for both center spacings of the joist and for five different joist depths. The flange type has a flange that supports the pan by resting on the wood-form joist. Pans of different depths are required for the different depths of joist. The second type of pan is an adjustable type that slips in between the joists and can be adjusted up or down to the required joist height and nailed to the edge of the joist.

FORMWORK SAFETY

The "Guide to Safety Procedures For Vertical Concrete Formwork" is reproduced here in entirety with permission from Scaffolding, Shoring & Forming Institute, Inc. See pages 15.15–15.23.

FORM RELEASE AGENTS

Form release agents have changed as much as formwork has in the last 15 years because of changes in the oil market and not productivity requirements. The standards for petroleum oils have decreased twice in the last 15 years to the detriment of form release agents because of an increased amount of impurities and a lowering of most physical and chemical standards.

The two most dominant changes in form release agents noticeable to contractors was a lack of color consistency and a more pronounced odor. The color of petroleum oil is a function of the ash content, which darkens the oil. The oil gets darker with age after opening until it becomes black. Such oil must be discarded, which is becoming more difficult because of environmental controls. The changes in smell were due to an increase of impurities such as kerosene, unsaturated hydrocarbons, and sulfur. The odor due to these impurities becomes nauseating as the amount of impurities increases. Sulfur is a strong bonding agent, and as it increases in concentration above 1 percent, sticking of concrete to the form can be expected.

TABLE 15.5 Equivalent Concrete Thickness for Pan Joists*

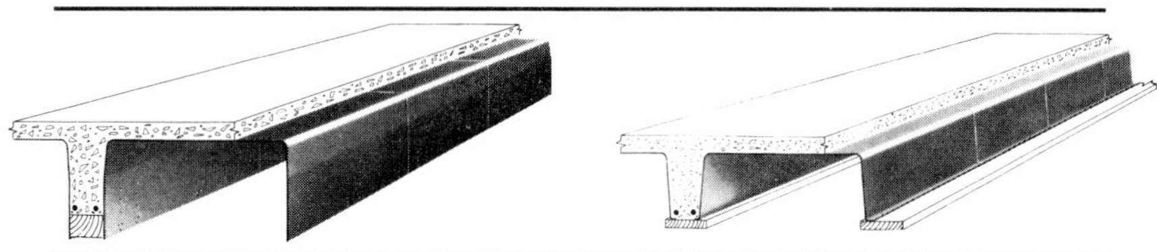

Depth of pan, in	20-in widths		30-in widths	
	Width of joist, in	Equivalent inches of concrete	Width of joist, in	Equivalent inches of concrete
6	3½	1	5	1
6	4	1	5½	1
6	4½	1½	6	1½
6	5	1½	6½	1½
6	5½	1½	7	1½
6	6	1½	7½	1½
8	3½	1½	5	1½
8	4	1½	5½	1½
8	4½	2	6	1½
8	5	2	6½	2
8	5½	2	7	2
8	6	2½	7½	2
10	3½	2	5	2
10	4	2	5½	2
10	4½	2½	6	2
10	5	2½	6½	2
10	5½	2½	7	2½
10	6	3	7½	2½
12	3½	2	5	2½
12	4	2½	5½	2
12	4½	2½	6	2½
12	5	3	6½	2½
12	5½	3	7	3
12	6	3½	7½	3
12	7	3½		
14	3½	3	5	3
14	4	3	5½	2½
14	4½	3	6	3
14	5	3½	6½	3
14	5½	3½	7	3½
14	6	4	7½	3
14	7	4		

**Notes:* (1) Add topping to slab equivalent shown above for overall thickness; (2) if tapered end spans are used, add ½ in to equivalent inch of concrete.

SCAFFOLDING, SHORING & FORMING INSTITUTE, INC.

SSFI

Guide to Safety Procedures for Vertical Concrete Formwork

SCAFFOLDING, SHORING & FORMING INSTITUTE, INC.
1230 KEITH BUILDING, CLEVELAND, OHIO 44115 • (216) 241-7333

FOREWORD

The "Guide to Safety Procedures for Vertical Concrete Formwork" has been prepared by the Forming Section Engineering Committee of the Scaffolding, Shoring & Forming Institute, Inc., 1230 Keith Building, Cleveland, Ohio 44115. It is suggested that the reader also refer to other related publications available from the Scaffolding, Shoring & Forming Institute as follows:

Guide to Horizontal Shoring Beam Erection Procedure

Guide to Scaffolding Erection Procedure

Guide to Steel Frame Shoring Erection Procedure

Guide to Safety Requirements for Shoring Concrete Formwork

Safety Requirements for Suspended Powered Scaffolds

Horizontal Shoring Beam Safety Rules

Suspended Powered Scaffolding Safety Rules

Steel Frame Shoring Safety Rules

Single Post Shore Safety Rules

Flying Deck Form Safety Rules

Rolling Shore Bracket Safety Rules

Scaffolding Safety Rules

Scaffolding Safety Do's and Don'ts Slide Presentation

Shoring Safety Do's and Don'ts Slide Presentation

Suspended Powered Scaffolding Safety Do's and Don'ts Slide Presentation

Forming Safety Do's and Don'ts Slide Presentation

TABLE OF CONTENTS

	Page
INTRODUCTION	1
SECTION 1—GENERAL	2
SECTION 2—ERECTION OF FORMWORK	2
SECTION 3—BRACING	3
SECTION 4—WALKWAYS SCAFFOLD BRACKETS	3
SECTION 5—SPECIAL APPLICATIONS	4
SECTION 6—INSPECTION	4
SECTION 7—CONCRETE PLACING	5
SECTION 8—STRIPPING FORMWORK	5
SECTION 9—GLOSSARY OF FORMING TERMINOLOGY	5
APPENDIX: SOME COMMON CAUSES OF FORM FAILURES	14

INTRODUCTION

Because of the widespread and constantly growing use of forming in construction today, it is vital that it be used properly and safely.

At present, the subject of forming is not covered in most standard reference works on safe construction practices; therefore, two objectives were established before preparing this guide.

1. **To fill the need for information on forming and the safe and proper use thereof and —**
2. To provide a guide to various federal, state and local authorities having jurisdiction over construction work in developing their own codes.

However, the information and recommendations contained herein do not supersede any applicable federal, state or local code, ordinance or regulation.

Safety precautions prescribed by OSHA and other government agencies should be followed at all times and persons working with forming equipment should be equipped with requisite safety devices.

The procedures outlined in this Guide describe conventional procedures for erecting and dismantling forming systems. However, equipment and forming systems differ, and accordingly, reference must always be made to the instructions and procedures of the particular manufacturer or supplier whose equipment is being used. Since field conditions vary and are beyond the control of the Institute and its members, safe and proper use of this equipment is the responsibility of the user and not the Institute or its members.

SECTION 1
GENERAL

1.1 All form components and/or hardware must be kept clean, and if appropriate, lubricated to insure proper performance and to allow for proper inspection.

1.2 All form components must be inspected regularly for damage or excessive wear. Equipment found to be in these conditions must be replaced immediately and **not** re-used.

1.3 Field repair of modular formwork components (other than plywood repairs) must not be undertaken without consulting the manufacturer's qualified representative.

1.4 The forming layout shall be prepared or approved by a person qualified to analyze the loadings and stresses which are induced during the construction process.

1.5 Forming installation and pouring procedures must comply with safe practices and with the requirements of the law and governmental regulations, codes and ordinances.

SECTION 2
ERECTION OF FORMWORK

2.1 Do not deviate from layout drawings when erecting formwork without the approval of a qualified designer.

2.2 Be certain that all wall ties are in place and secured as per manufacturer's recommendations. Do not weld, bend or otherwise alter wall ties as it may seriously reduce their strength.

2.3 Adequate temporary bracing must be in place while initially setting formwork. Assure that formwork is properly braced and stabilized against wind and other external forces.

2.4 Safe working platforms must be installed as per applicable safety standards and as stated in Section 5 herein.

2.5 When gang forming, lifting devices must be properly spaced, and securely attached as per manufacturer's recommendations. Rigging must be arranged so that any one lifting bracket is not overloaded and that lifting cables are not at excessive angles, which will reduce allowable loads. Spreader beams are recommended for all but simple two point lifts. Follow manufacturer's recommended procedures concerning capacity and use of lifting hardware.

2.6 A minimum of two tag lines must be used to control movement of crane handled formwork. Do not allow personnel on or directly under any gang form while it is being moved or suspended in air.

2.7 Do not erect gang forms when jobsite wind conditions prevent safe maneuvering of gangs. Assure that all rigging connections are properly made in accordance with safe practices and procedures.

2.8 Formwork should be adequately braced, re-anchored, or otherwise secured prior to releasing lifting mechanism.

2.9 Wall forms must not be erected so as to support deck concrete loading unless the wall forms are a designed part of the deck support system.

SECTION 3
BRACING

3.1 Braces are considered only as alignment devices with no provisions for withstanding concrete pressure or any portion thereof. Maintain forms plumb during pour to ensure that braces are not supporting or stabilizing concrete pressures.

3.2 If bracing is to withstand concrete pressure, a qualified formwork designer must be consulted.

3.3 Unless specified, wind loading is not considered in the manufacturer's bracing recommendations. Contractor must verify adequacy of bracing to withstand anticipated wind conditions for each project.

3.4 The adequacy of stakes, dead-men, sills, anchor-bolts, etc., must be determined to assure safe support of the imposed brace loads. The responsibility for adequate anchorage of braces should be assigned only to those personnel with sufficient experience to assure sound judgement.

3.5 Assure that the concrete has attained sufficient strength to safely support the imposed load at support locations.

3.6 Do not exceed allowable brace loads given by manufacturer. Inspect installed braces immediately after installation for correctness of spacing and proper attachment device.

SECTION 4
WALKWAYS/SCAFFOLD BRACKETS

4.1 All walkways must be properly positioned, spaced and fastened as per manufacturer's specifications and all applicable safety regulations.

4.2 Walkways must be in place along the upper level of formwork. Workmen must never attempt to walk or stand on top of forms.

4.3 Scaffold or walkway brackets must be attached with the manufacturer's recommended connectors. Never use substitutes or make-shift devices. Never hang brackets from wall ties after removal of forms.

4.4 All walkway platforms must utilize at least two (2) planks laid side by side, and must overlap their support ledger by not less than 6 inches. Unsupported ends of scaffold planks must not project more than 12 inches past their support ledger.

4.5 Scaffold planks must be minimum 2 inch x 10 inch nominal lumber and must be scaffold grade as recognized by approved grading rules for the species of lumber used, or must be of materials having equivalent or greater strength. Scaffold planks must safely support a minimum of 25 pounds per square foot over a maximum span of 8 feet.

4.6 Scaffold planks shall be either of such length that they overlap the brackets at each end by at least 6 inches, or must be nailed and clinched, bolted or otherwise positively secured against dislodgement from effects of wind, weather, gang form lifting operations or the like. Bolt heads and nails must be driven flush with tops of planks to prevent tripping hazards.

4.7 All scaffold bracket platform must be equipped with guardrails, midrails and toeboards along all open sides and ends and be maintained secure and in good condition at all times. Guardrails must be of at least 2 x 4 nominal sized lumber, with minimum 1 x 6 or 2 x 4 nominal midrails, with toeboards at least 4 inches high, supported by 2 x 4 nominal lumber uprights spaced not more than 8 feet apart, or must be of other materials providing equivalent or greater strength and protection.

4.8 Maximum spacing between walkway backets is eight feet. Never exceed this distance. Follow manufacturer's recommendations as to loading of walkway brackets. Walkway brackets are designed to support a maximum load of 25 lbs. sq. ft., when spaced on 8 foot centers and are not designed for the additional loads imposed from stacking rebar or placing other equipment on walkways.

4.9 Always brace and or otherwise secure forms and scaffold from overturning due to attachment and use of walkway brackets.

4.10 Never allow persons to work on one level of walkways if others are working directly below or overhead unless proper protection is provided, such as safety nets.

4.11 It is unsafe and unlawful for persons to occupy any form walkway while the form is being moved.

4.12 Access ladders or other suitable safe methods must be used to obtain access to walkway platforms. Do not position ladders so that their weight while being used can effect the strength or stability of the scaffold and formwork.

4.13 Do not use form panels as a ladder.

4.14 If using scaffold bracket walkways is not practical, personnel must be protected against falls by means of safety belts and lanyards attached to components having adequate strength to safely support the imposed loads, or by safety nets or other equivalent protection. Personnel protected by safety belts must not be required to handle excessively heavy or unwieldly formwork components.

SECTION 5
SPECIAL APPLICATIONS

5.1 The size (weight) of ganged formwork must be limited to the following: (1) the safe handling capacity of the crane; (2) prevailing wind conditions to maintain safe tagline control.

5.2 All support brackets, friction collars and other friction devices must be torqued to the manufacturer's specification during erection and re-checked immediately prior to placing concrete.

5.3 Assure that anchor brackets are attached properly and with required thread engagement per manufacturer's recommendations. The proper anchor bolts and inserts must be used and be of adequate strength for combined shear and tension loadings. Assure that the concrete has attained sufficient strength to properly allow the designed use of these devices with the required safety factors.

5.4 Assure that anchor brackets are installed so that they are equally loaded and are installed correctly prior to setting the formwork. Anchor bolts must be tight and brackets level.

5.5 When erecting forms for battered walls, allowance must be provided for proper anchorage to offset resulting uplift forces.

SECTION 6
INSPECTION

6.1 Inspect completed formwork prior to placing concrete to assure proper placement and secure connections of ties and associated hardware. All threaded connectors, such as ties, inserts, anchor bolts, etc., must also be checked for proper thread engagement.

6.2 Inspect erected scaffolds before each use. Assure that bolts, nuts, and other connections are maintained securely.

6.3 Inspect bracing attachments and form alignment after each form cycle. Inspect installed forms and braces immediately prior to pour and during pour.

SECTION 7
CONCRETE PLACING

7.1 The contractor must verify prior to and during concrete placing that the method of placement and rate of pour is consistent with formwork design. DO NOT OVERLOAD FORMWORK.

7.2 Concrete must not be placed in any manner which imposes impact loads that exceed the safe working capacity of the form.

7.3 Verify concrete mix to assure that the maximum formwork design pressure is not exceeded.

7.4 Instruct personnel on proper vibration. Do not use vibrator to move concrete. Do not vibrate further than one-foot into the previous lift. Avoid vibrator contact with wall ties. External vibrators must not be attached to formwork unless it was designed for their use.

SECTION 8
STRIPPING FORMWORK

8.1 Follow manufacturer's recommended field procedures—generally, reverse the order of procedures used in erection of formwork. Be certain that concrete has sufficiently set to carry its own weight and any imposed loads prior to stripping formwork.

8.2 When gang-forming, secure the lifting mechanism prior to removal of ties, anchors and or bracing.

8.3 Use extreme caution for all formwork to assure that no panel, walkway bracket, brace or any other form component is unfastened prematurely.

8.4 Assure that all disconnects have been made and the bond of the formwork to concrete has been broken prior to lifting of gang form.

SECTION 9
GLOSSARY OF VERTICAL FORMING TERMINOLOGY

ALIGNER Lumber or metal members used to align vertical formwork.

ANCHOR BRACKET A projecting member designed in combination with a specified anchor to attach to a previous concrete pour so as to support the dead weight of the subsequent formwork and live loads specified.

ANCHORS Devices used to secure formwork, braces or accessories to previously placed concrete, either embedded during placement or set in holes drilled in hardened concrete. There are two basic parts: the embedded anchor device and the external fastener which is removed after use.

BATTER WALL Wall with one or both faces slanting from the vertical, usually creating a wall thicker at its base than at its top.

BEAM A structural load bearing member which usually supports a slab.

BEAM FORM The entire formwork to form the bottom and both sides of a beam.

BEAM POCKET Opening left in a vertical member in which a beam is to rest; also an opening in a column or girder form where forms for intersecting beams will be framed.

BEAM SIDE Vertical side panels or parts of a beam form.

BEARING WALL Wall supporting a vertical load other than its own weight.

BENCH MARK An elevation mark on a permanent object used as reference.

BOX-GIRDER Usually a long span hollow structural beam.

BOX-OUT An opening or pocket formed in concrete positioning a box-like form within the wall forms.

BRACE Any external structural member used to hold forms in alignment and/or to resist horizontal forces exerted on the forms such as wind loads.

BRACKET Projecting member from a structure to support weight beyond its face.

BREAK-BACK The distance from the face of concrete to the end of the remaining imbedded portion of a tie (snapped off wire-tie, or the face of concrete clearance of a three-piece tie inner unit) (also referred to as Cut-Back).

BRICK LEDGE (Brick Seat) Ledge on wall or footing to support a course of masonry.

BUCK Framing to void an opening in a wall, such as a door buck, which forms the opening for a door.

BUG HOLE Void on the surface of formed concrete caused by an adhering air or water bubble not displaced during consolidation.

BULKHEAD A partition in the forms blocking fresh concrete from a section of the forms or closing the end of a form, such as at the construction joint.

CAMBER An inward curvature of a wall or an upward curvature of an elevated slab or beam form to improve appearance or to compensate for anticipated load deflection.

CANTILEVER The portion of a structural member, which projects beyond its support.

CAPITAL The tapered upper section of a column under the drop head. Conical shaped with round columns, pyramidal shaped with square columns.

CAULK To use a putty-type material to seal form joints from grout leakage.

CHAMFER A beveled external corner. It is usually formed in the concrete work by use of a chamfer strip placed in the form at the outside corner to provide a rounded or beveled corner.

CHASE An elongated void or opening formed into a concrete surface.

CLEANOUT An opening in the forms for removal of refuse, closed before the concrete is placed.

CLEAT Small board used to connect two or more pieces of formwork lumber together.

CLIMBING FORM A form which is raised vertically for succeeding lifts of concrete in a given structure, usually supported on anchor bolts or rods embedded in the top of the previous lift. The form is moved only after an entire lift is placed and (partially) hardened; this should not be confused with a slip form which moves during placement of the concrete.

COIL BOLT The hex-head outer unit of a three-piece wall tie with external contoured threads to engage the helical threads of a coil tie inner unit.

COIL TIE The non-reusable inner unit or center part of a three-piece wall form tie. Ties are made with two or more straight wire struts with helix coils welded at each end foming female threads.

COLUMN CLAMP Any of the various types of stiffening or fastening units to hold a column form sides together.

CONSTRUCTION JOINT The surface where two adjacent placements of concrete meet, frequently with a keyway or reinforcement across the joint.

CONTROL JOINT Formed, saw cut, or tooled groove in a concrete surface to regulate the location of shrinkage cracks.

CORBEL The projection from the face of a concrete wall which is used to support a beam or elevated slab.

CROSSMEMBER Intermediate stiffening member of a form panel connected at both ends of the perimeter frame.

CRUSH PLATE An expendable strip of wood used as a pad to protect either the form or concrete surface from damage during prying action to strip forms.

DADO Rectangular groove in the perimeter frame of a form which allows for the passage of ties without leaving a gap between forms.

DEAD LOAD The load of forms, stringers, joints, reinforcing rods, and the actual concrete to be placed.

DEADMAN A steel beam, block of concrete or other heavy item used to provide anchorage for a guy line or form brace.

DECK The form upon which concrete for a slab is placed, also the floor or roof slab itself.

DECKING Sheathing material for a deck or slab form.

DESIGN PRESSURE The predetermined load per square foot at form face predicated by pressure, temperature, rate of concrete placement and height of concrete above point considered.

DIAPHRAGM Cross walls positioned between long span, deep beams to provide lateral stability to the beams.

DOME The square prefabricated pan form used in two-way (waffle) concrete joist floor construction.

DOUGHNUT A large washer of any shape to increase bearing area of bolts and form ties, also to act as a shim.

DRAFT The slight taper difference between opposite sides of a form so that it will readily strip out the concrete.

DROPHEAD or DROP PANEL The thickened structural portion of a flat slab usually square in shape, that is in an area surrounding the top of columns.

DUTCHMAN Usually a solid lumber thickness utilized to fill in under one side of equal height wall forms such as on a side slope footing, also to compensate for lineal dimension variation between opposing forms due to a slight angle corner or curved wall.

ELEVATION A drawing showing a specific area projection of a structure on a vertical plane.

EMBEDMENT An insert, anchor bolt or other device attached at the form face so as to be encapsulated by the concrete for future attachments or structural performance.

END-BARS Perimeter frame members similar to end-rails but are usually perpendicular to crossmembers.

END-RAILS Perimeter frame members of prefab form panel which are perpendicular to side-rails.

EXPANSION JOINT A thickness of flexible material between consecutive placements of concrete to absorb linear expansion of concrete.

FACTOR OF SAFETY Ratio of ultimate load to allowable load. Also, in design, the ratio of the ultimate load strength (or yield point) of the material to the working stress assumed in the design.

FALSEWORK The temporary structure erected to support work in the process of construction, such as shoring or vertical post to support an elevated wall or spandrel beam.

FASCIA An exposed flat member or band on the elevation of a structure, such as the edge beam on a bridge or exposed roof slab edge on a building.

FILLER A non-standard width form panel used to take up odd dimensions.

FILLER STRIP Piece of wood, metal or other material placed between large ganged slab form areas and vertical surfaces to permit easy stripping.

FILLET A beveled or rounded inside corner.

FLANGES The parallel flat portions that describe the cross-sectional shape of structural members such as I-beams, channels, or T-beams.

FLAT PLATE A flat slab without column capitals or drop panels.

FLAT SLAB Reinforced concrete floor construction of uniform thickness; eliminate the underdrop of beams or drop heads at columns.

FLYING FORMS Any large form or built-up gang, either deck or wall, capable of being moved by crane.

FORM COATING Anti-bonding material applied to form face surface to induce easy stripping.

FULL LIQUID HEAD Concrete pressure where the entire pour is still in a liquid state.

GANG FORM A large area of wall form with independent structural integrity. May also be a grouping of panels to be used as a unit for convenience in erecting, stripping and reusing.

GIRDER Rectangular-shaped member, larger than the beams, which is supported by the columns and supports the beams.

8

GIRDER FORM Self-supporting form system where the load is carried in bending by the side panels.

GRADE STRIP A temporary wood strip secured to form face prior to concrete placement to denote finished grade elevation.

GUYS (Guy Wire) Cable anchor from ground to top of wall form to brace in one dirdirection through tension.

HAIRPIN The wedge used to tighten some type of form ties, also a hairpin-shaped anchor set in place while concrete is plastic.

HANDSET FORM A modular form erected and stripped by hand rather than a crane.

HANGER A device used for suspending one object from another such as the hardware attached to a building frame to support deck forms. Refer also to beam hanger.

HAUNCH A projection built on a wall or column used to support a load outside the wall or column.

H-BEAM Steel beam with wider flanges than an I-beam.

HE-BOLT The outer unit of a three-piece wall tie, of which the external threads of the outer units engage the internal threads of an inner unit such as a coil tie.

HEAD (Liquid Head) The vertical height measurement of liquid concrete in wall form.

HONEYCOMB Undesirable voids left in the formed concrete surface revealing unbonded coarse aggregates.

INITIAL SET An early state of the concrete curing process at transformation from a liquid to a solid.

INNER UNIT (Inner Tie) The non-reusable center part of a three-piece she-bolt tie.

INSERT A female threaded connector embedded in concrete to which a male anchor device can be connected.

INVERT The lowest visible surface; the floor of a drain, sewer, tunnel, culvert or channel.

JOIST A horizontal structural member supporting deck form sheathing, usually rests on trusses, stringers or ledgers.

JUMBO Traveling support for forms, commonly used in gang-formed tunnel work.

KERF To make a series of cuts or notches in order to curve a wood member.

KEYWAY A recess or groove created in an earlier pour of concrete which is filled with concrete of the next pour giving shear strength to the joint.

KICKER A piece of wood (block or board) or metal attached to a formwork member to take the thrust of another member.

KIP Kilopound, a unit weight equal to 1,000 pounds or 454 kilograms.

KNEE BRACE A brace between horizontal and vertical members in a building frame or formwork to make the structure more stable.

9

LEDGER A horizontal structural member supporting shoring joists secured to the top of vertical shoring members or secured to a concrete wall supporting forms.

LIFT BEAM (Spreader Bar or Spreader Beam) A beam utilized to distribute the weight of a ganged form through two or more equalized vertical pick-up points.

LIFT BRACKET Special brackets attached to top of ganged forms to facilitate fast, safe attachment of crane sling lines.

LIFTOR A mechanical lifting device used to vertically elevate ganged forms to subsequent vertical reuses.

LINER Any sheet or layer of material attached directly to the inside face of the forms to improve surface quality, alter the texture, or to imprint specific architectural patterns on the finished concrete.

LIVE LOAD The total weight of workmen, equipment, buggies, vibrators and other loads that will exist and move about due to the method of placement, leveling and screeding of the concrete pour.

MONOLITHIC Concrete placement technique in which the slab, the beams, the columns, and the walls or any combination of the above elements are poured at the same time.

MUDSILL A plank, or concrete slab, on the ground, to provide a level surface and support to concrete forms or shoring.

MUD SLAB A 2" to 6" layer of concrete below structural concrete floor or footing over soft, wet soil.

MULTI-LIFT The vertical stacking or forms in tiers for any height wall. A wall requiring more than one row of forms is generally referred to as multi-lift.

NAILER Strip of wood or other material attached to or set in concrete or attached to steel to facilitate making nailed connections.

OFFSET A displacement or abrupt change in line or the distance between two parallel lines; such as a change in wall thickness which will create a vertical offset.

ONE-WAY SLAB Concrete slab with reinforcing steel rods provided principally in one direction.

OPEN CENTERING A shoring system often used for pan-joist construction where the only sheathing used is soffit boards directly under the ribs of the slab.

PALLET A portable platform of wood or other material used for forming. A flying deck form without vertical support members attached is referred to as a pallet.

PAN Prefabricated rectangular form unit used in one way concrete joist floor construction.

PANEL A section of form sheathing constructed from boards, plywood, metal sheets, etc., that can be erected and stripped as a unit. Panels can be built on jobsite or prefabricated factory built.

PAN-JOIST A light slab with ribs normally 24 to 36 inches on center acting as beams. The joists or ribs run at right angles to primary beams or girders.

PARAPET That part of a wall that extends above the roof level.

PENCIL ROD Metal rod (wire), usually about 1/4" diameter, used in conjunction with special set-screw bearing clamps to perform as a wall form tie.

PENETRATION Any concrete embedment device that must pass through the form face (such as anchor bolts, rebar, or dowel rods).

PERMANENT FORM Any form that remains in place after the concrete has developed its design strength. The form may or may not become an integral part of the structure.

PILASTER Column built with a wall, usually projecting beyond the wall face.

PLATE A flat horizontal member such as a 2 x 4 placed on the footing for leveling and upon which the forms are set, sometimes referred to as a "shoe."

PLUMB Vertical or the act of making vertical.

POST-TENSIONED CONCRETE Re-inforced concrete in which, after the concrete has set and sufficiently hardened, the desirable distribution of stress is achieved by post tensioning steel tendons, bars or wires.

PRECAST CONCRETE Concrete units (such as beams, joists, deck panels, or wall panels) cast elsewhere than its final position and then set in place.

PRESTRESSED CONCRETE A system for utilizing the compressive strength of concrete by producing required compressive stresses with highly stressed tension rods, tendons or wires.

REBAR Abbreviation for "Reinforcing Bar."

RETAINING WALL A wall, which is designed to resist horizontal loads such as those imposed by soil or water.

RIBS Parallel structural members backing sheathing in a prefabricated form. Same as crossmembers.

RUSTIFICATION A groove formed in the concrete formed by securing a strip to the face of the formwork. Also referred to as a "feature strip."

SCAB A small piece of wood fastened to two formwork members to secure a butt joint.

SCAFFOLDING An elevated platform supporting workmen, tools, and materials, either attached to wall forms or free standing.

SCREED The tool used to control the top surface elevation of freshly placed concrete.

SCREED BAR Pipes or lumber set with their top edge at a controlled elevation to provide a reference elevation for the screed and are supported by screed chairs. They are spaced according to the length and type of strike-off and run parallel to the direction of the pour.

SCREED CHAIR A support which holds the screed bar at a definite distance above the sheathing. Chairs are of either fixed height or adjustable by means of a bolt and lock nut.

SCREEDING	Is the first step in finishing a concrete slab through the operation of pulling a straightedge over screed bars.
SHEAR HEAD	That portion of the slab, directly over the column, which is reinforced with additional steel rods to withstand the shear load of the slab. Takes the place of the drop head.
SHEATHING	The material forming the contact face of forms, also called lagging or sheeting.
SHE-BOLT	The outer unit of a three-piece wall tie that contains female threads to engage the external threaded inner unit (rod).
SHIM	Thin pieces of material used to bring abutting members to an even, level bearing.
SIDERAIL	Perimeter frame member of prefab form panel which is perpendicular to crossmembers.
SILL	The lowest part of an opening in a wall such as a door sill or window sill, also horizontal bearing member as a plate.
SKIN PLATE	The steel form face of an all-steel form.
SLAB	The thinner portion of the floor, usually of uniform depth, that is between the drop heads or beams.
SLING	A length of cable with a loop at each end, usually the cable line from the crane hook to ganged form.
SLIP FORM	A form which moves, usually continuously, during placing of the concrete. Movement may be either horizontal or vertical. Slip forming is like an extrusion process with the forms acting as moving dies to shape the concrete.
SNAP TIE	A wire-type tie with or without spreader washer or cones. After forms are released, the protruding tie ends are snapped off by twisting at a predetermined break-off crimp usually about 1" in the concrete.
SOFFIT	The underside of subordinate part or member of a building such as a beam, stairway, arch, etc.
SOLDIERS	Vertical wales used for strengthening or alignment.
SPANDREL BEAM	A beam in the perimeter of a building, spanning between columns and usually supporting floors or roof. An up-turned spandrel depth dimension extends above the floor, and a down-turned spandrel extends below the floor.
SPREADER	A brace, usually of wood, inserted in forms to keep the form faces the proper distance apart until the concrete is placed.
SPUD	Adjustable bolt-like strut extending between the skin of tunnel forms and bored rock tunnel walls to provide position support of the formwork.
STAKE	A pointed wood or metal object driven in ground to attach brace or to support form sides in footing forming.
STIFFENER	A structural member for the support of the plywood face or skin plate on panel forms sometimes called ribs.

STRIP	To remove formwork from concrete.
STRIPPING BAR	A solid bar positioned in between form panels or adjoining ganged forms which is the first unit stripped thereby providing relief to readily strip the large form panels; also referred to as "wrecking strips."
STRONGBACK	A load gathering member attached to the back of the formwork on the outside of the walers for added strength, to hold proper alignment (sometimes referred to as "stiffbacks").
STUD	Supporting member to which sheathing is attached.
TAGLINE	Line connected to gang form or flying form to control free swing movement during crane lifting.
TAPER TIE	A one-piece reusable steel form tie with a slight taper to facilitate removal.
TELLTALE	Any device designed to indicate movement of the formwork.
TEMPLATE	Thin plate or board frame used as a guide in positioning or spacing of form parts, reinforcements, anchors, etc.
TIE	A concrete form tie is a tensile unit adapted to holding concrete forms secure against the lateral pressure of unhardened concrete, with or without provision for spacing the forms a definite distance apart, and with or without provision for removal of metal to a specified distance back from the finished concrete surface.
TIE DADO	Half-slot thickness dado's at the siderails of adjoining forms provide the tie location slot common to many prefabricated form systems.
TOENAIL	To drive a nail at an angle.
TRAVELER	Traveling support and bracing for ganged tunnel and culvert formwork.
WALER	Load gathering members used to hold studs or panel forms in position.
WATERSTOP	Rubber, plastic, or other material inserted in a construction joint to prevent the seepage of water through the joint.
WEDGE	A piece of wood or metal tapered to a thin edge, used to adjust elevation, tighten formwork, etc.
WEDGE BOLT	A two-way action designed wedge which contains a slot to facilitate its function as a connecting bolt also.

APPENDIX

SOME COMMON CAUSES OF FORM FAILURES

Exceeding design (working) pressure

(1) Excessive rate of pour

(2) Mix design not taken into account

(3) Improper vibration

(4) Temperature not taken into account

Ties not placed properly.

Ties in place, but improperly fastened (insufficient tread engagement, etc.)

Improper design — particularly when layouts were not provided by manufacturer.

Job-built fillers, corners and bulkheads which are not adequately designed by field personnel and become the weak-point of the system.

Failure to brace at least one side of formwork.

Connecting hardware not installed.

Lack of inspection by qualified personnel to see that form layout has been interpreted properly.

The chemical release agents have been developed within the last 25 years to become highly effective in the interface between concrete and formwork. They function through a chemical reaction between the alkali in concrete and the fatty acids contained in the chemical release agents. This reaction produces a low-grade soap at the interface which retards the set of the concrete for approximately 24 h. The result is excellent within this time frame, but sticking may result if form stripping is delayed further. Formwork removal must fit into this cycle for the new productivity requirements.

This timing of formwork removal requires use of a membrane curing compound for continuation of curing. Experience has shown this to be a better method of moisture retention than keeping the forms in place. Economically speaking, a form can cost $5 to $25 per square foot, while a curing compound can cost only a few cents per square foot. Therefore, it is not productive to leave expensive forms in place when a curing membrane does a better job, especially with absorbent forms. Subsequent surface coatings or toppings may dictate or prohibit the use of certain types of membrane.

Problems may occur with overapplication or rain washoff. Overapplication problems are less common because present-day release agents have less reactive material and the additional experience gained over the years by contractors. An overapplication will tend to increase the concrete water/cement ratio at the interface, which can produce a more desirable lightening of the concrete but may also produce a striping effect caused by the more intense reaction next to a vibrator. The striping effect is more enhanced when an overapplication occurs.

Rain, which is generally alkaline, has the same chemical reaction with the release agent as with the concrete and may lower the effectiveness of the release agent, since it is difficult to reapply the release agent after the reinforcement or prestressed tendons are in place.

Certain neutralized chemical release agents and buffered compounds are more effective than the chemically reactive types in keeping formwork clean. These compounds reduce stripping efficiency slightly because of less product generated by the agent than the chemically reactive types, but still are sufficiently effective for the job.

As soap is a better lubricant, its potential for removal of surface air voids is greater and the cleaner forms lead to more production. Like other facets of formwork, the pros and cons of form release agents should be reviewed by trial prior to the production concrete. Recent research has shown that a fan-type spreader and 100 percent coverage with as thin a layer as possible produces overall successful surfaces.

REFERENCES

1. ACI 347R-88, *Guide to Formwork for Concrete,* American Concrete Institute, 1988.
2. *Plywood for Concrete Forming,* American Plywood Association, Tacoma, Washington, V345 1177, 1973.
3. *Wood Handbook,* USDA, Agricultural Handbook No. 72, 1974.

PART • 5

BATCHING, MIXING, AND TRANSPORTING

CHAPTER 16
PLANT LAYOUT

Robert W. Strehlow

COMPONENTS

A concrete plant consists of equipment and facilities for receiving, storing, handling, and proportioning the materials required to make concrete, and for delivering the proportioned materials to the transport equipment, either before or after mixing the ingredients. For convenience, plants are clarified as *mass concrete, paving, ready mix,* or *concrete products* plants; by function as *central-mix,* or *transit-mix* plants; and by ease of movement as *permanent, portable,* or *mobile* plants. Plants where materials flow continuously downward as they are stored, proportioned, mixed, and discharged into the delivery unit are called *gravity* or *tower* plants; plants that elevate the aggregates (and cement) after proportioning are called *low-profile* plants. Thus a plant may be described as a *low-profile, mobile,* or *central-mix paving* plant. Standards for concrete plants are published by the Concrete Plant Manufacturers Bureau (CPMB). Their Plant Mixers Manufacturers Division (PMMD) publishes standards for central mixers and their Control Systems Manufacturers Division (CSMD) publishes the standards for batching controls. The CPMD is located at 900 Spring Street, Silver Spring, MD 20910.

The profile plant in Fig. 16.1 shows the basic elements of all batch plants, i.e., aggregate storage, cement storage, batchers, feed system, and mixer charging.

With very few exceptions, all concrete plants consist of the same basic components: a compartmented bin for storage of the different grades of aggregate, storage bins for each type of cement, weigh batchers for proportioning the cement and aggregate, a means of batching water (and/or ice), and some means for controlling the batching devices. The arrangement and sizing of these components is determined by the mission of the plant.

16.1 Aggregate Storage

The volume of aggregate storage need not be any greater than that necessary to hold the volume needed for one batch, provided the means for filling the bin never lets it run out of any material, and act essentially as a funnel to the batcher. If the bin is also to provide a retention period for conditioning the aggregates, such as heating, cooling, or draining, the size will depend on the production rates and the desired condition of the materials when they reach the batch gates. In cold climates aggregate storage bins are enclosed and often sized to allow for overnight heating of sufficient materials for the first round of

FIGURE 16.1 Basic plant components. This low-profile plant picture shows the basic elements of all batch plants: aggregate storage, cement storage, batchers, feed system, and mixer charging. *(Courtesy of McNeilus Companies.)*

ready-mix trucks. Where the rate of concrete production varies during the day, a larger aggregate storage acts as a buffer to a bin filling system that has been sized on the average production rate. In that the volume of each aggregate used per batch is not the same, and because major bin compartments are usually the same size, the total bin size may be determined by the desired storage for the most-used material multiplied by the number of materials to be stored at one time.

For aggregate storage bins that carry CPMB rating plates, the rated volume is based on bin water level plus a heaping formula. The practicable obtainable volume is usually less since to fill a bin to the rated volume would require that the bin be filled with a clamshell bucket until it overflowed at all points around the rim. The rating plate is a useful guide, and the water-level volume is usually available. The water-level rating is a close approximation to what the bin will actually hold.

The "mobile" plant in Fig. 16.2 is arranged with added aggregate, cement, and fly-ash storage, plus dust collection, for a permanent installation.

16.2 Cement (and Pozzolan) Storage

Each cement or pozzolanic material must have its own storage compartment. The absolute minimum volume of any compartment is equal to the volume of the delivery unit plus a "fluff" factor. Almost all cement is being delivered by special bulk tankers that transfer the cement to the bin or silo pneumatically. The volume that they transport is governed by the local highway weight laws. When sizing a fly-ash storage compartment one must remember that fly ash weighs about 50 pcf and the volume that is hauled may be considerably more than the volume of cement hauled per truck. One should check with the hauling agent regarding the volumes being hauled per truckload. Almost all pneumatically transferred powder materials entrap air, which leads to the fluff factor. Freshly transferred cement has been observed to weigh 63 pcf. This means that for every cubic foot of cement hauled there must 1½ ft^3 of storage provided. The cement will slowly lose some of this fluff, but the safe rule of thumb for sizing storage is to use a 1.5 fluff factor. To determine the volume above the minimum that a bin should hold, you must first determine the *maximum rate* of cement consumption. Then ask the cement supplier how long it takes to receive a load after ordering. You must have enough cement in the silo to cover this period. Now multiply this volume by a safety factor to cover flat tires, weather, and broken promises. Cement bins and silos that carry CPMB rating plates are rated in cubic feet of gross air volume. This is a safe volume to use if you apply a 1.5 fluff factor to the delivered cement.

16.3 Water and Admixture Storage

Water storage is important only if it is necessary to provide a place to heat or cool the water. The sizing should be done in conjunction with the supplier of the heating or cooling equipment. Where admixtures are delivered in bulk by tank trucks the size of storage will be determined by the production rate as it relates to the admixture consumption and the suppliers delivery arrangement.

16.4 Weigh Batchers

For specification work cement is weighed in a seperate weigh batcher. Most specifications permit the cumulative weighing of cement and fly ash, but require the cement to be weighed first. Cement and flyash are batched directly from overhead storage using "jog" type gates and vane feeders. Fluidized gravity conveyors, more commonly referred to as

FIGURE 16.2 Adapting a mobile plant. Shown here is a "Mobile" plant arranged with added aggregate, cement, and fly-ash storage, plus dust collection, for a permanent installation. (*Courtesy of the Concrete Equipment Co.*)

FIGURE 16.3 Belt-charged aggregate batchers. Here the starting and stopping of belt feeders is used to batch aggregates on a low-profile, high-production, central-mix paving plant. Note the use of the individual batchers to speed up the batch cycle. The cement batcher is behind the three aggregate batchers. *(Courtesy of the Heltzel Company.)*

"airslides" (Airslide is a trademark of the Fuller Company), and screw conveyors are also used to charge cement batchers.

The starting and stopping of belt feeders (Fig. 16.3) is used to batch aggregates in a low-profile, high-production, central-mix paving plant. Note the use of individual batchers to speed up the batch cycle. The cement batcher is behind the three aggregate batchers.

Aggregates may be weighed in cumulative or individual batchers. Individual weigh batchers have the advantage of providing the maximum rate of aggregate batching for high-production plants. There is very little difference in the accuracy of batching between cumulative and individual batching when automatic tare-compensated batching controls are used. Cumulative aggregate batching is the most popular method. Where many mixture designs are used, this method is simpler, more compact, and far less costly. Aggregate batchers are charged directly from overhead bins or belt conveyors. Some aggregate batchers are charged by front-end loader and employ decumulative weighing as a means of measuring.

Water batchers are the most accurate to batch water. They also offer a higher production capacity in some higher-production arrangements. However, the water meter is the most popular means for batching water. When 100 gal or more water is batched, the water meter meets the same delivery tolerance requirements and improves its accuracy with larger batches.

In Fig. 16.4 an inclined screw conveyor is used to feed crushed ice to an ice-and-water batcher located over the central mixer on a low-profile mass concrete plant.

The required delivery tolerances commonly specified can be easily met by either method. Ice is always batched by weight. It may be weighed in a batcher that is charged by a screw conveyor handling crushed ice, or it may be weighed in block form on a platform scale, chipped, and blown into the mixer. The choice of ice system depends on the volume of ice needed and the availability and cost of commercial block ice. While any of the admixtures could be weighed, they are usually handled by volumetric methods, such as meters, sight tubes, or both.

16.5 Scales and Controls

The scale and means for controlling the batching devices can be considered as a necessary component of any weigh batcher. The two most popular types of scales are the springless dial and the load-cell type. The springless dial presents a continuous reading from zero to full scale. While it is a mechanical device connected to the batcher by levers with knife edges, a linear potentiometer can be attached to its dial shaft to send an analog weight signal that is used by the batching control, and this also provides a digital display of the weight in the batcher.

Note in Fig. 16.5 the slump meter for central mix control, CRT screen for process monitoring, temperature meter, admixture dispenser, and printer for documentation.

Both systems provide equal results, and both are required to meet the same degree of accuracy. The advantages attributed to the dial scale are that it is easier to read and, because it is mechanical, is available for manual batching if there is an automatic batching control failure. The advantages attributed to the load-cell system are the ability to be read to a finer degree, absence of moving parts and hence less maintenance requirement, and lower cost. While it is possible to control the batching devices with manually operated levers or push buttons, the explosion of electronic technology in the 1980s and early

FIGURE 16.4 Ice batching. Here an inclined screw conveyor is used to feed crushed ice to an ice-and-water batcher located over the central mixer on a low-profile mass concrete plant. (*Courtesy of the Erie Strayer Co.*)

FIGURE 16.5 Automatic control station. Note the slump meter for central-mix control, CRT screen for process monitoring, temperature meter, admixture dispenser, and printer for documentation. (*Courtesy of Auto Control, Inc.*)

1990s has resulted in the lowering of automatic control prices to the point where virtually all concrete plants use some form of automatic batching control. Often these controls do more than batch, such as controlling the charging rate of the mixer, printing load tickets, recording batch information or providing input to truck dispatch systems, and accounting.

Low-profile means that the aggregates must be elevated to get to the mixer. Figure 16.6 shows large modular storage compartments combined with an elevated control room and attached garage.

16.6 Dry Plants (Truck Mix Plants)

A "dry" plant is one that only batches the ingredients and does not include any means for mixing them. The "truck or transit mix" plant is the most prevalent dry plant. Dry plants are available in all shapes and sizes. The size is determined by the logistics related to the desired production rate. The shape or configuration requires consideration of the required size of the storage components, the location including environmental requirements, the mobility required, and the erected-ready-to-operate cost. The two basic styles are *gravity* and *low-profile.* Gravity plants usually are assembled wholly or in part using modular sections. One advantage of a gravity plant is that gravity charging of a mixer results in "choke charging" where the batcher charging gates are fully opened and the mixer then draws the material at its maximum rate. No unusual operator control attention is required. They are usually easier to enclose, require less ground space, require a more expensive filling conveyor, and in general are more difficult to maintain than a low-profile plant.

In Fig. 16.7 both the cement and aggregates are batched directly from a combination bin.

FIGURE 16.6 Low-profile truck mix plant. Low-profile means that the aggregates must be elevated to get to the mixer. Here large modular storage components are combined with an elevated control room and attached garage. *(Courtesy of Rexwerks, Inc.)*

FIGURE 16.7 Gravity-style truck mix plant. Here both the cement and aggregates are batched directly from a combination bin. *(Courtesy of the Heltzel Co.)*

FIGURE 16.8 Modular sections. Most plants are built in some degree of factory-assembled modules. Here a large truck mix plant is being erected by stacking up the preassembled units. (*Courtesy of Rexwerks, Inc.*)

Low-profile plants are further categorized as *permanent, portable,* or *mobile.* Mobile plants have wheels (or provisions for wheels) for the purpose of moving them from site to site. Portable plants are assembled from modular sections that can be moved by truck. Most so-called permanent plants are assembled from factory-built (portable) modules.

Most plants are built with factory-assembled modules. Figure 16.8 shows a large transit mix plant being erected by stacking preassembled units.

There is no clear distinction between a "portable" class plant (one that can be trucked in modules) and a permanent plant made of factory-assembled modules, unless local codes have different requirements. Portable plants are very seldom moved after the initial setup. Probably 85 percent of all mobile plants are never moved from the original setup.

In Fig. 16.9, the relatively large mobile truck mix plant, shown in its erected position, transforms into a single portable towed unit. Note the inclusion of the dust collection bag house between the aggregate and cement bins.

The appeal of using a mobile plant in a fixed location is its cost-effectiveness. The delivery and erection costs are the minimum of those for all types. The plants are complete, and added costs for additional needed items are small. They are not difficult to enclose where required by local environmental ordinance and custom. A check should be made to determine the need for a building permit prior to erection.

The large central mix concrete plant in Fig. 16.10 is an excellent example of the many concrete production plants that have been enclosed for environmental acceptance. Most plants are designed to be enclosed for pollution control and safety.

The low-profile-style plant is the most prevalent type of ready-mix plant. Batchers are generally sized to match truck mixer size. A cumulative aggregate batcher is used. Fly ash is used when available and is batched accumulatively following the cement. Water is batched volumetrically using a water meter controlled automatically.

FIGURE 16.9 Mobile truck mix plant. This relatively large mobile truck mix plant, shown in its erected position, transforms into a single transportable towed unit. Note the inclusion of the dust collecting bag house between the aggregate and cement bins. *(Courtesy of the Vince Hagan Co.)*

High-volume mobile central-mix plants (Fig. 16.11) are the standard means of producing all mass concrete, highway, and airport paving. Note the use of the standard dump truck for transporting the mixed concrete to the placement.

16.7 Wet (Central-Mix) Plants

Today there is little difference between a mass concrete plant, a central-mix paving plant, and a central-mix ready-mix plant except that a paving plant will have more mobility, and a mass concrete plant will have a cooling system for the materials. Many paving plants were modified slightly and used for the mass concrete of atomic energy electrical generating plants. This involved the use of ice and fly ash, being equipped with batch recorders and the same automatic batching controls found in central-mix and transit-mix plants. They produced long continuous pours for the casting of reactor bases. At the completion of a project, they became commercial ready-mix plants. Parallel to this example are the many central-mix plants that became ready-mix plants when construction of the federal highway Interstate System ended. Most of the plants had cooling systems installed for thick airport runway slabs.

Figure 16.12 illustrates the use of the same basic high-production mass concrete plant's portable components in a central mix ready mix operation. The general configuration of the portable plant is similar to that of a low-profile dry ready-mix plant except for the addition of a tilt drum central mixer. As a result of extensive research by the then Bureau of Public Roads, mixers in present use are sized to match the size of the allowable haul load. This research demonstrated that properly charged central mixers could produce uniform mixtures in 60 s or less. All of these mixers are charged with a high-capacity aggregate charging belt. This belt not only provided a means of blending the different aggregates (and sometimes cement) on the belt prior to entering the mixer but also speeded up the charge cycle by virtually throwing the material into the drum. As weighing of

FIGURE 16.10*A* Dust collection. Shown here are bag-house dust collectors on the top of the cement bin and vacuum type dust pickups at the mixer and dry loading chute (rear). *(Courtesy of the Erie Strayer Co.)*

FIGURE 16.10*B* Good neighbors. This large central-mix concrete plant is an excellent example of one of the many concrete production facilities, both large and small, that have been enclosed for environmental acceptance. Most plants are designed with the thought of enclosing, pollution control, and safety as prime considerations. *(Courtesy of the Ross Co.)*

the next batch occurs during the mixing cycle, cumulative aggregate batching is generally used for plants producing up to 65 batches per hour.

Cement is gravity-batched from elevated storage. Water is usually metered into a holding tank from which it can be rapidly charged under sequential control into the mixer. The same arrangement is used for admixture batching and charging. Controls are always automatic for both batching and mixing control.

Figure 16.13, taken during the erection of a gravity products plant, shows the arrangement of the components. Material flows from the overhead storage to the batchers, then to the mixers, and finally to the casting machinery. Note the modular sections.

16.8 Concrete Products Plant

Concrete products plants with mixers are a breed of their own. They are always plants with either spiral- or turbine-type mixers. As the concrete produced is on site, space for the plant is a major consideration. The aggregate sizes used are usually smaller and the mixtures often dry and stiff. Products produced include structural and architectural units,

FIGURE 16.11 High-volume central mix. High-volume mobile central-mix plants are the standard means of producing almost all mass concrete, highway, and airport paving. Note the use of the standard dump truck for transporting the mixed concrete to the pour. (*Courtesy of Rexwerks, Inc.*)

pipe, block, burial vaults, and paving bricks. Each product presents its own set of plant considerations and methods of handling the mixed concrete. Except for the plants casting large structural units, operations are usually under cover.

Figure 16.14 shows two front-end-loader-charged hoppers, supported on load cells, which retrogressively weigh onto a transfer conveyor belt that charges the spiral blade mixer. Cement and water are weighed in a seperate batcher over the mixer.

To maintain a uniform moisture content, aggregates are stored under cover and not exposed to the weather. Selecting components to meet production requirements is not difficult as the same components can be used for high- or low-production facilities. In operations such as block or pipe production where molding equipment is close to the mixer, tower plants are still popular because they occupy minimum floor space. Where the casting operation calls for the use of an overhead crane to handle the cast product, the plant must be located at the end of the side of the building, which is ideal for a low-profile plant.

Plants producing architectural units often require both white and regular cement and many different aggregates. These requirements have led to a configuration using retrogressive aggregate batching from front-end-loader-charged hoppers suspended on load cells that batch directly onto a mixer charging belt. Since the accuracy of any scale is based on full-load capacity, there is a practical limit to a minimum size of batch. With retrogressive weighing, special attention must be paid to ensure that the storage hoppers do not run empty or over tolerance batches which are difficult to correct once the batch is in the mixer. The low-profile products plant in Fig. 16.15 demonstrates the use of a weigh belt (a belt conveyor hung on load cells) being used to weigh the aggregates from a row of front-end-loader-charged hoppers. These belts can be made up to 120 ft long and used under direct truck-charged hoppers.

FIGURE 16.12 Portable central-mix plant. This picture illustrates the use of the same basic high-production mass concrete plant's portable components in a central-mix ready-mix operation. *(Courtesy of Rexwerks, Inc.)*

A slightly different configuration that allows hoppers of any size is equipped with deep skirts and suspended on multiple load cells. The weighing sequence is the same as weighing into any suspended batcher. When the weighing is complete, the belt is started and transfers the batch to the mixer. Correction of an overtolerance condition can be done normally.

The alternate use of white and gray cements and color additives may require compartmented cement batchers and seperate mixers to produce products with uniform color. In selecting size of plant productive capacity, the fact that concrete is needed only when forms or molds are available for filling must be considered. Time is required for stripping, cleaning, and preparing the forms or molds. In many plants, concrete production is required for only 25 percent of the workday. The use of automatic controls is prevalent and recording of batch weights is used where required for structural members. Such automatically printed records provide a means of duplicating batches and architectural units to meet customer requests.

FIGURE 16.13 Gravity-style products plant. This picture, taken during erection of the gravity products plant, shows the arrangement of the plant's components. Material flows from the overhead storage to the batchers to the mixers to the casting machinery. Note the modular sections. (*Courtesy of Erie Strayer Co.*)

FIGURE 16.14 Load-cell retrogressive batching. Pictured are two front-end-loader-charged hoppers which are supported on load cells and retrogressively weigh onto the transfer conveyor belt which charges the spiral blade mixer. Cement and water are weighed in a separate batcher over the mixer. *(Courtesy of Mixer Systems, Inc.)*

FIGURE 16.15 Load-cell weigh belts. This picture of a low-profile products plant demonstrates the use of a weigh belt (a belt conveyor hung on load cells) being used to weigh the aggregates from a row of front-end-loader-charged hoppers. These belts can be made up to 120 ft long and used under direct truck-charged hoppers. *(Courtesy of Mixer Systems, Inc.)*

CHAPTER 17
BATCHING EQUIPMENT

John W. Strehlow

The accepted way of producing concrete is in measured batches. The process is called *batching*. The ingredients requirement for a particular concrete mix is determined by the absolute volumes required to produce a homogeneous mass without voids. There is no practical way to measure most of the materials by volume in a commercial production operation. Cement volume varies with entrapped air content; aggregates "bulk" or change volume depending on the method of handling and moisture content. As the specific gravity of the materials is relatively constant, proportioning by weight with appropriate corrections is the almost universal method of batching. Water and liquid admixtures are commonly batched by volume using liquid meters.

BATCHING TOLERANCES

It is not clear if the batching tolerances in use are the result of scientific necessity to apply these tolerances to obtain uniform concrete and the batching devices designed to accomplish this, or if the industry in its wisdom has adopted a range of tolerances that is as tight as can be produced both consistently and economically with the available production equipment. In either event the equipment available does meet the accepted delivery tolerances with reliable and basically simple batching devices. Table 17.1 lists batching tolerances as published by the Concrete Plant Manufacturers Bureau (CPMB) in its standards as of 1986. These tolerances are consistent with the majority of other specifications published by other concrete specifying agencies or organizations.

WEIGH BATCHING

The common required items in any weigh batching system are a hopper and a scale (a weigh batcher). Materials may be weighed into or out of the batcher. A batcher that is used to weigh only one material is called an *individual batcher.* A batcher that is used to weigh more than one material is a *cumulative batcher.* Most batchers measure as weight is added to the batcher. When measuring is done by withdrawing the batch from the hopper, it is called *retrogressive batching.*

TABLE 17.1 Batching Tolerances

For individual batchers, the following tolerances shall apply based on the required scale reading:

Cement and other cementitious materials*
±1% of the required weight of material being weighed *or* ±0.3% of scale capacity, whichever is greater

Aggregates
±2% of the required weight of material being weighed *or* ±0.3% of scale capacity, whichever is greater

Water
±1% of the required weight of material being weighed *or* ±0.3% of scale capacity, whichever is greater

Admixtures
±3% of the required weight of material being weighed *or* ±0.3% of scale capacity, *or* ± the minimum dosage rate for one sack of cement, whichever is greater

For cumulative batchers with tare-compensated controls, the above tolerances shall apply based on the required weight of each material

For cumulative batchers without tare-compensated controls, the following tolerances shall apply to the required cumulative weight.

Cement and other Cementitious

Materials or Aggregates
±1% of the required cumulative weight of material being weighed *or* ±0.3% of scale capacity, whichever is greater

Admixtures
±3% of the required cumulative weight of material being weighed *or* ±0.3% of scale capacity, *or* ± the minimum dosage rate per sack as it applies to each type of admixture, whichever is greater

For volumetric batching equipment the following tolerances shall apply to the required volume of material being batched:

Water
±1% of the required volume of material being batched *or* ± 1 gal, whichever is greater

Admixtures
±3% of the required volume of material being batched *or* ± the minimum recommended dosage rate per sack of cement, whichever is greater

Range of accuracy. For ingredients batched by weight the accuracy tolerances required of the batching equipment shall be applicable for batch quantities between 10 and 100% of scale capacity

For water or admixtures batched by volume, the required accuracy tolerances shall be applicable for all batch sizes from minimum to maximum, as is determined by the associated cement or aggregate batcher rating

*Other cementitious materials are considered to include fly ash and other natural or manufactured pozzolans.
Source: Concrete Plant Manufacturers Bureau, 1986.

The scale provides the readout of the weight in the batcher. There are two types of scales in common use. Going into the 1980s the springless dial scale was the generally accepted batcher scale. Going into the 1990s the load cell is the most used device. This change is again one of improvements in technology leading to a greater degree of reliability accompanied by a reduction in cost and maintenance expense. While the mechanical scale's structural considerations limited the shape of the batcher, load cells allow long narrow live bottom batchers by summing the load on multiple cells. The advertised or warranted accuracy of either type of scale is 0.1 percent of full-scale capacity. Thus the greatest weighing accuracy is obtained when the batch weight is close to the scale capacity. Just because a digital readout can register 1 lb on a ≥10,000 lb scale does not mean that the scale is more accurate.

Besides the batcher and scale, some means of charging the batcher and controlling the charging is required. The most simple means would be a gate equipped with a manual control lever. Add an air cylinder and a push button, and the resulting accuracy is still dependent on the operator's observation of the scale. Automatic batching allows target weights to be set, and the automatic controls then regulate the charging devices.

17.1 Aggregate Batching Devices

Two basic systems of aggregate weighing are currently being used: (1) gravity weighing through a mechanical clam or slide gate and (2) starting and stopping a belt or vibratory feeder that discharges into the weigh batcher. This applies to progressive as well as retrogressive batchers. The flow rate of most fine and coarse aggregates is quite uniform, and the control accuracy is not complicated. Gates or belt feeders can be sized to match the production rate requirements. They must be sized so that the opening is at least 2½ times the size of the maximum aggregate. Batching rates through gates usually ranges from 1.5 to 6 s per cubic yard. The batching speed obtainable with a belt conveyor depends on the transfer capacity of the conveyor selected. The batching rate of the aggregates is seldom the bottleneck in any production cycle.

17.2 Cement Batching Devices

Unlike aggregate, the flow rate of cement (fly ash, or pozzolans) is not uniform or as predictable. These materials respond to aeration as a means of changing their flow-rate characteristics and their unit weights. The unit weight of cement may vary from 60 to 94 pcf and fly ash from 35 to 60 pcf depending on the amount of aeration introduced. The commonly used devices for weighing these materials are gates, valves, screw feeders, vane feeders, airslides, and pneumatic transfer systems.

Where the cement (or fly ash) storage is above the batcher, rotary plug and butterfly valves are the most common types of gates used for batching cement and fly ash. They may be used signally or in pairs for higher-production requirements. The opening size is somewhat dependent on the production requirements, but the most important consideration is the ability to close sufficiently fast to obtain the target weight within the required tolerance. Batching is often done in two stages in order to gain the required accuracy. Observed rates of batching range from 100 to 600 lb/s. Gravity weighing of cement seldom becomes a bottleneck in the production cycle. While rotary-vane feeders draw material from an overhead bin, their flow-rate capacity is based on their size and speed of rotation. Their accuracy is limited to the contents of one vane pocket.

Where storage bins cannot be located over the cement batcher, screw conveyors or air slides are used as the batching devices that transfer the material to the batcher. The capacity of a screw conveyor is dependent on its diameter, rotational speed, and flight design. The capacity of an airslide depends on the exposed width of its aeration membrane. On the basis of a 45 percent screw conveyor loading and running at the maximum recommended speed, screw conveyor capacity is compared to generally accepted airslide capacity.

Screw conveyor		Airslide	
Diameter, in	Capacity, ft^3/h	Width, in	Capacity, ft^3/h
4	100	4	400
6	900	6	1000
9	1250	8	2000
10	1700	10	3000
12	2240	12	4000
14	4000	14	6000

Screw conveyors can be inclined upward at any angle, including vertical with special flighting; however, a maximum angle of 45° is generally accepted as desirable for tubular conveyors with standard flighting. Airslide conveyors must be declined at approximately 7° to ensure uniform flow. If there is no bridging and the cement is well aerated over the inlet to the screw conveyor, a uniform rate of discharge can be expected. The low-pressure air that an airslide uses to create flow also serves to aerate the cement in the bin and in total the airslide conveyor does produce a uniform rate of flow that can operate with a simple single-stage cutoff to obtain the required batching accuracies.

17.3 Water Batching

Water is the easiest ingredient to batch because its flow rate is predictable and does not present any undue handling problems. It can be weighed in a water batcher or metered with a conventional water meter. The batching tolerances required for water is not difficult to obtain and can often be accomplished with the most simple cutoff valve. Batching water by weight is the most accurate method, but the use of a water meter is the most popular way of batching water. The speed of weigh batching water is limited only by the size of the valve used if the water is stored in an overhead reservoir. Where water is batched directly from the supply, the rate is affected by the pipe size and water pressure. Water temperature has no effect on the speed of weigh batching but is a factor in the selection of a water meter as shown in Table 17.2.

Either a nutating piston or a propeller drives an indicator showing the *volume* of water that has passed through a water meter. Nutating-piston (or disk) meters are generally accurate within 1 percent in the ranges shown in Table 17.2 and are commonly more accurate over a wider range of flows than are propeller meters. Propeller meters are accurate within 1 to 1.5 percent from about 40 to 140 percent of rated continuous capacity, and within 2 to 3 percent for the balance of their operating range. They have higher maximum capacities and lower head loss than do nutating-piston meters. Meters read low at very low flows, slightly high at low flows, and slightly low at flow rates near the maximum. In Table 17.2, three capacity figures are given: minimum, maximum continuous, and maximum intermittent flow. Batching meters are operated intermittently between the latter two values. When used with nearly constant water pressure, repeatability is better than accuracy. Cold-water meters cannot usually be used with hot water, but hot-water meters can be used with cold water at cold-water flow rates with some sacrifice in accuracy. A strainer (about 50 mesh) is used ahead

TABLE 17.2 Typical Water-Meter Capacities*

Nominal size, in	Cold water, U.S. gal/min, 32–100°F			Hot water, U.S. gal/min, 100–150°F			Hot water, U.S. gal/min, 150–200°F		
1½	5	66	100	9	36	50	9	18	36
2	8	110	160	13	52	92	12	30	52
	18	**90**	**170**	16	100	160	16	100	160
2½	10	150	225	20	80	112	19	45	80
3	16	210	330	23	100	140	23	55	100
	30	**185**	350	30	150	350	30	150	350
4	40	330	500	50	200	350	45	110	200
	50	**300**	600	50	225	600	50	225	600
6	50	660	1000	100	400	500	90	220	400
	85	**625**	1400	90	500	1400	90	500	1400

*Three capacity figures are given: minimum, maximum continuous, and maximum intermittent flow. Figures in boldface type are for propeller meters; other figures are for nutating-piston meters. Manufacturer's data should be consulted for capacities of specific meters.

of the meter to prevent damage from solid particles. Some meters can be installed vertically; most must be horizontal. Meters must be protected from freezing and from pressure surges (water hammer) in water lines. Electrical or mechanical controls to stop flow at a preset amount and automatic, remotely controlled meters are available.

Water meters that are used with automatic batching controls produce a calibrated pulse per a specified volume. This pulse is then output to the control panel where it is summed and modified to control the cutoff valve.

17.4 Admixture Dispensers

A wide variety of additives or admixtures are used in the production of concrete, and the quantities used may be from 1 oz to 2 qt per 100 lb of cement. Since many admixtures become somewhat gummy when in contact with air, it is advisable to arrange the piping so that air will not contact the solenoid valves used with admixtures because the solenoid valves would not open or close completely, would overheat, and eventually would fail completely. If the solenoid valves are exposed to the air they should be washed out thoroughly with water at the end of each day.

Some admixtures are not compatible, and if two or more are put into the water batcher or water piping together, they may react with each other and nullify the effect of all. It may be necessary to add one admixture to the water batcher, one to the sand, and one or more directly into the mixer truck.

Because of the small quantities required per batch liquid admixture are almost universally batched by volume using calibrated timers, transparent check tubes, liquid meters, or a combination of these. Although the delivery tolerances are plus or minus 3 percent, these admixtures are very powerful in their affect on the finished product. Because they represent such a small relative percentage of the total volume care must be observed to ensure that the dispensing system is preforming as required. Most specifications require a visible check tube and most admixture suppliers furnish dispensing equipment for their brand of admixture. Like the water meter, the liquid meter is most commonly used for admixture dispensing and when equipped with a pulse generator provides the necessary input to an automatic batching panel.

17.5 Compartmented and Combination Batchers

A combination batcher is used to weigh two different materials and is usually compartmented to separate these materials. This applies to cement(s) and all aggregates. Two or more compartments in a batcher are used to separate white and regular cements or to separate cement and fly ash. Multicompartmented retrogressive batchers are not considered to be combination batchers. Most specifications require a separate cement batcher. However, combination batchers are quite common in product plants or where there are low-production requirements. Generally these applications produce batches of only one size and the scale can be sized to ensure the accuracy required.

Retrogressive batchers are most often compartmented. While these batchers can provide a low-profile configuration, care must be taken to ensure that the size of the batch produced is within the minimum range of the scale. There is no easy corrective recourse if a retrogressive batch is overweight. Care must be exercised to ensure there is sufficient material in the batcher before the retrogressive batching is started. If the batch is charged directly into a mixer, no blending of materials is possible during charging of the mixer, and care must be exercised to select the most efficient batching sequence.

FIGURE 17.1 Live bottom ice batcher. Pictured is a twin-auger discharge ice batcher. Chipped ice is delivered to this batcher by screw conveyor or pneumatically. *(Courtesy of C. S. Johnson.)*

17.6 Ice Batching

Ice batching is done generally in two ways depending on the source of ice. Crushed or chunk ice can be manufactured on the site (usually a very large project), or block ice can be purchased from a commercial source, delivered, and stored in an insulated van. When the ice is manufactured on the site it is usually transferred from the ice manufacturing house to the weigh batcher by a screw conveyor. The ice batcher (Fig. 17.1) is usually located over the mixer charging belt and is equipped with a screw conveyor discharge (live bottom). This batcher may be a compartmented batcher that allows all water, liquid or frozen, to be weighed on the same scale.

When block ice is used, the blocks or partial blocks may be weighed on a platform scale and then crushed and blown directly into the mixer. The observed weight is then entered into the batching control and the balance of water is determined and added last through a separate water batch system. Block ice may also be used by chipping and blowing a set number of blocks of ice into an ice batcher with a separate water compartment and bringing the scale into balance by batching the remaining water required.

MOISTURE COMPENSATION

No discussion of batching equipment is complete without a review of moisture compensation considerations. The natural state of the fine and course aggregates used in concrete is one where there is water clinging to the surface of these materials. The amount of this excess surface water in a material is usually expressed as a percentage of its saturated surface-dry (SSD) weight. For sand or fine aggregate this percentage does not normally exceed 10 percent without having some free water running from it. The moisture content of course aggregates seldom exceeds 2 percent and generally varies so little that an average of 1 percent is commonly used. A 5 percent moisture content for fine aggregates is a common average. Most manufactured and lightweight aggregates do not follow these rules. The drying method as described in ASTM C566 is widely used to determine the moisture content of fine aggregate.

To determine the target weights for batching you must first determine the percent moisture content of the aggregate and then determine the weight of free water involved per batch.

$M\%$ = percent of aggregate moisture
DWA = design weight of aggregate
DWW = design weight of water
TWA = target batch weight of aggregate
TWW = target batch weight of water
FWW = free-water weight*

Then

TWA = DWA/(100 − $M\%$)
FWW = TWA − DWA
TWW = DWW − FWW*

The importance of moisture compensation is demonstrated in the following example. Assume that 1 yd^3 mix requires 1200 lb of sand, 2000 lb of course aggregate, and 25 gal of water.

A 1 percent variation in the moisture content of the sand would equal 12 lb or approximately 1½ gal of water. This changes the water requirement by ± 6 percent. If the old rule of thumb that "a 1 gallon per cubic yard change in the water content will cause a ¾ in change in slump" is true, the importance of moisture is evident. ASTM Specification C94 for Ready Mixed Concrete calls for a ± 3 percent tolerance on total mixing water from all sources. From the example it is obvious that to meet this, moisture variation must be kept to less than 1 percent.

Keeping sand moisture variations to a minimum is more of an art than a science and requires constant monitoring of the final product and proper response by the plant operator. This job is made easier if the plant is equipped with a moisture meter in the fine aggregate since there is no way that one can work with the current moisture and be drying samples. The moisture meter is only a tool that one must learn to use and does not alter the need for constant monitoring as the results of any sampling are related to the fact that only a very minute portion of the aggregates is sampled. It is not necessary that the meter

*The free-water weight is a sum of values for each aggregate used.

be highly accurate as it is that the meter be responsive and have a high degree of repeatability. Thus when the meter shows a change, it's time to make an adjustment.

In certain situations the moisture variations in the supply make it impossible to reach the desired slump, even with the most contentious efforts of the plant operator. In these cases the answer is to reduce the design water content and produce a mix with lower slump and then add only enough water on the job (or in the central mixer using a slump meter) to increase the slump into the desired range. For many product plants that require very low slump mixes, batch water is often controlled by reading the conductivity of the concrete in the mixer, across several probes, and then automatically introducing only enough water to reach a preset current value. No compensation for sand weight is made for various sand moisture contents.

BATCHING CONTROLS

The last link in the chain of devices required for batching is the means to control the batching devices. While some low-capacity products operate with manual levers or "push-button" controls that require visual observation of the scale display for batching accuracy, most plants use some form of automatic batching control. It is virtually impossible to describe all the automatic batching controls available. They range from simple one-formula dedicated controls to computer-controlled systems that include features that go far beyond the batching operation.

All automatic batching controls (Fig. 17.2) convert electrical signals from the dial scale, load cells, or volumetric pulse generators to weight values. These are then com-

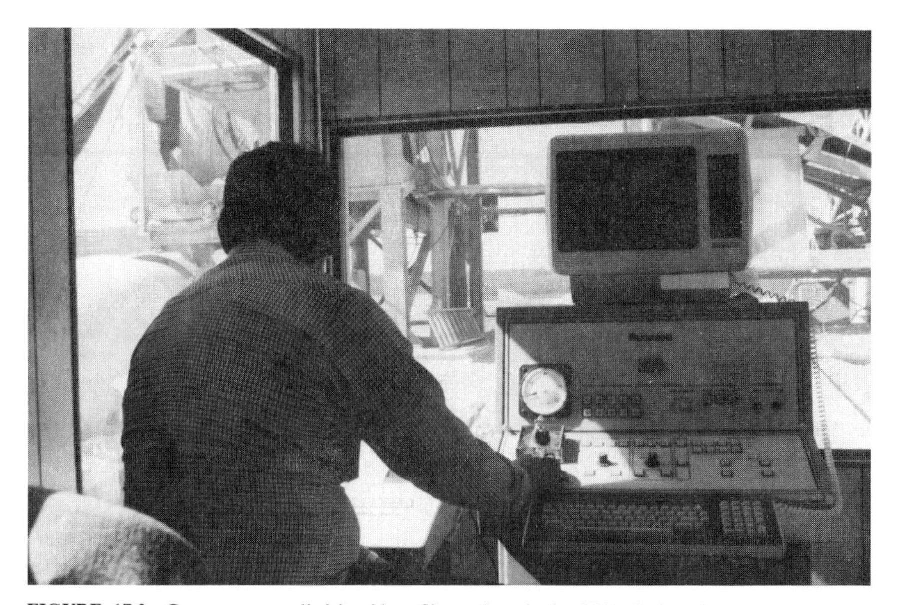

FIGURE 17.2 Computer-controlled batching. Shown here is the CRT display, the keyboard for data entry, a slump meter, and a mix timer for the central mixer. Almost all of the lights and switches on the panel are for manual override operation. (*Courtesy of Rexwerks, Inc.*)

pared to the batch target values and the associated batching devices controlled. Most batching specifications require that an automatic control meet the following conditions:

1. The scale must be at zero before starting a batch. (The last batch must have been completely discharged.)
2. The charging device cannot be actuated if the charging device is open.
3. The discharge device cannot be actuated if the charging device is open.
4. The discharge device cannot be actuated until all the required materials are batched and are within the applicable delivery tolerances.

Today almost all automatic batching controls use tare-compensated weighing where a cumulative weighing method is used. Tare-compensated weighing treats the weighing of each cumulatively weighed material as if it were being weighed in an individual batcher and then checks the delivery tolerance using the allowed values for an individual batcher. The old controversy as to the accuracy of individual batching versus cumulative batching is no longer valid since the same batching device is used in either case and the same tolerances are applied. For many controls the batching is more efficient because past experience has taught batcher designers and batching operators to automatically adjust for the amount of midair material and scale bounce, thus eliminating the need for manual control trim operations. Where these controls use a cathode-ray tube (CRT) to display batch operation, the target weights or volumes and the actual delivery values—and, in some cases, variations—are displayed. Error messages pop up when needed, and input instructions are displayed along with formula modifiers.

These controls are available from simple to complex, and the following is a list of features available. Select the ones you need, and there will be someone who builds it.

Formula storage. The number of mix formulas that can be preset in a batching control is a function of the memory provided. Some vendors offer several choices, but 100 formulas is quite common.

Batch size. Where the size of the batch varies between batches, mix formulas are stored in memory in 1 yd^3 values and the operator enters both the formula number and a batch size, plus any modifiers at the time of batching.

Dual display. Some controls with a CRT display both the information of the batch in progress and provide a setup screen for modifications to the next batch, thus increasing productive efficiency.

Moisture compensation. Moisture compensation can be applied individually to each aggregate used.

Water holdback. This feature allows holdback of some water to be used to flush down the mixer charging chute after all solid materials have been charged.

Slump control. This feature allows a quick modification of a batch formula by adding some water or deleting some water so as to change the slump.

Automatic charging. Some controls offer automatic mixer charging by reading the rate of draw-down as the batchers discharge and then control the batcher discharge gates to provide both blending of materials going into the mixer and controlled sequential entry. Some controls allow this charging method to be modified for each truck mixer.

Inventory. Some controls keep running totals of the weight or volume of each material used.

Recordation. Creating a record of the delivered weights or volumes or of the inventory are natural and simple extensions of any computer controlled batching control. Standard low cost dot matrix printers are in common use.

Load ticket printing. Load ticket printing is an extension or recording, and unless there is a interface with a dispatch control, it requires the control to provide for customer names, job locations, and possibly accounting information input. Interfacing and integrating the batching, central dispatch, and billing has been accomplished.

Industry standards for batching controls are established by the Control System Manufacturers Division of the Concrete Plant Manufacturers Bureau, 900 Spring Street, Silver Spring, MD 20910.

TYPICAL MIX DESIGNS

Tables 17.3 through 17.6 assemble a group of mix designs that were observed being used on various types of projects and for various applications. These mix designs are included for reference only and are not intended for use without proper trial-mix testing.

TABLE 17.3 Typical Composition of Highway and Airport Paving Concrete*

	Lowest observed value	Second centile	Median	Ninety-eighth centile	Highest observed value
Paving mixes, all mixes					
Cement	385	442	557	674	677
Water	157	150	231	312	316
Sand	975	955	1185	1415	1580
Total coarse aggregate	1660	1790	2030	2300	2380
Total aggregate	3020	3040	3230	3500	3560
Wet concrete	3906	3896	4018	3140	4142
Paving mixes, 2 aggregates					
Sand	1066	1043	1180	1359	1342
Coarse	1660	1775	2013	2193	2163
Paving mixes, 3 aggregates					
Sand	975	950	1190	1430	1580
Coarse, smaller quantity	360	520	930	1220	1190
Coarse, larger quantity	956	910	1160	1410	1740
Coarse, smaller aggregate	360	520	1000	1300	1274
Coarse, larger aggregate	683	750	1100	1410	1740
Paving mixes, 4 aggregates					
Sand	986	—	—	—	1311
Coarse, smallest quantity	143	—	—	—	528
Coarse, middle quantity	733	—	—	—	992
Coarse, largest quantity	850	—	—	—	1203

*Values in lb/yd^3 of resultant concrete based on saturated surface-dry aggregates.

TABLE 17.4 Typical Composition of Commercial Ready-mix Concrete*

	Lowest observed value	Second centile	Median	Ninety-eighth centile	Highest observed value
All mixes except pavement and lightweight mixes					
Cement	188†	330	552	785	877
Pozzolan	0	—	—	—	75
Cement + pozzolan	230	330	522	785	877
Water	210	209	269	372	383
Sand	985	1019	1342	1596	1670
Total coarse aggregate	1454	1524	1888	2150	2267
Total aggregate	2755	2990	3233	3410	3442
Wet-concrete weight	3809	3845	4011	4148	4180
1½ in maximum, 2 aggregates					
Cement	370	352	490	655	658
Water	210	208	246	302	292
Sand	1018	950	1310	1550	1550
Coarse aggregate	1780	1750	1950	2150	2143
Total aggregate	3017	3000	3265	3450	3442
Wet-concrete weight	3844	3820	3990	4165	4165
1½ in maximum, 3 aggregates					
Cement	380	426	533	640	750
Water					
Sand	985	960	1200	1440	1410
Fine aggregate	750	660	1010	1400	1362
Coarse aggregate	607	650	1010	1250	1155
Total coarse aggregate	1736	1690	2010	2300	2267
Total aggregate	3040	3000	3220	3440	3410
Wet-concrete weight					
1½ in maximum, 4 aggregates					
Cement	432	—	—	—	658
Water	321	—	—	—	339
Sand	1122	—	—	—	1311
Fine aggregate	201	—	—	—	202
Medium aggregate	1167	—	—	—	1170
Coarse aggregate	644	—	—	—	646
Total coarse aggregate	2012	—	—	—	2018
Total aggregate	3134	—	—	—	3329
Wet-concrete weight	4086	—	—	—	4113
¾ to 1 in maximum, 2 aggregates					
Cement	188	305	505	705	750
Water	225	225	270	345	347
Sand	1040	1080	1360	1630	1670
Coarse aggregate	1660	1670	1860	2140	2170
Total aggregate	3021	3040	3235	3430	3440
Wet-concrete weight	3809	3860	4005	4150	4131

*Values in lb/yd^3 of resultant concrete based on saturated surface-dry aggregates.
†Plus 45 lb of pozzolan.

TABLE 17.4 Typical Composition of Commercial Ready-mix Concrete (*Continued*)

	Lowest observed value	Second centile	Median	Ninety-eighth centile	Highest observed value
$^3/_4$ to $^5/_8$ in maximum, 2 aggregates					
Cement	393	360	585	910	877
Water	267	250	325	400	383
Sand	1000	1210	1430	1650	1625
Coarse aggregate	1454	1400	1680	1920	1885
Total aggregate	2755	2710	3160	3320	3300
Wet-concrete weight	3960	3970	4020	4065	4180
Lightweight insulating or fill concrete					
Cement	318	300	530	770	700
Water	401	360	460	560	509
Lightweight aggregate	214	180	460	2350	1900
Wet-concrete weight	989	850	1600	3050	2265
Lightweight structural concrete					
Cement	461	415	590	900	852
Water	250	220	365	510	473
Sand	250	100	1200	1410	1380
Lightweight aggregate	710	640	850	1680	1530
Total aggregate	1592	1450	2030	2270	2260
Wet-concrete weight	2410	2310	3000	3340	3274

TABLE 17.5 Typical Concrete Mixes (in lb/yd^3) for Specific Uses

Use	Cement or cement plus pozzolan	Water	Sand	Total coarse aggregate	Total aggregate	Wet-concrete weight
Grout	590–2360	318–730	316–2700	—	316–2700	3406–3645
Pumped concrete	500–580	280	1555	1485	3040	3820–3900
Nuclear shielding	564	300	900	3860	4760	5624
Prestressed concrete	658–752	267–287	1125–1200	1920	3045–3120	4045–4064
Lightweight prestressed concrete	846	459	180	1293	1473	2780
Regular concrete block	455	177	1425–1950	1425–1950	3375	4007
Lightweight concrete block	340	280–560	—	—	1750–2200	2490–2940
Concrete pipe	640–830	250–420	1120–2230	985–1860	2960–3330	4115–4315
Porous concrete	596	213	—	2835	2835	3644

TABLE 17.6 Typical Composition of Mass Concrete*

	Lowest observed value	Second centile	Median	Ninety-eighth centile	Highest observed value
Grout mixes					
Cement	590	—	810	—	1020
Pozzolan	0	—	—	—	260
Cement + pozzolan	590	—	—	—	1020
Water	318	—	325	—	550
W/(C + P), %	32	—	39	—	78
Sand	2065	—	2350	—	2700
Wet weight	3465	—	3600	—	3645
¾ in maximum mixes					
Cement	318	410	610	740	711
Pozzolan	0	—	—	—	105
Cement + pozzolan	423	410	610	740	711
Water	225	270	307	355	342
W/(C + P), %	45	44	51	63	59
Sand	991	970	1255	1445	1450
No. 4 to ¾ in aggregate	1355	1685	1775	1835	2080
Total aggregate	2801	2765	3025	3155	3280
Wet-concrete weight	3794	3790	3920	4090	4043
1½ in maximum mixes					
Cement	282	375	590	805	765
Pozzolan	0	—	—	—	100
Cement + pozzolan	374	375	590	885	852
Water	163	160	270	310	310
W/(C + P), %	25	20	45	64	60
Sand	825	795	990	1180	1180
No. 4 to ¾ in aggregate	830	810	1050	1210	1576
¾ to 1½ in aggregate	444	990	1110	1330	1372
Total coarse aggregate	1844	1960	2180	2330	2300
Total aggregate	2996	2990	3145	3300	3397
Wet-concrete weight	3823	3820	3990	4190	4277
3 in maximum interior mixes					
Cement	188	160	280	550	500
Pozzolan	0	—	—	—	89
Cement + pozzolan	255	240	320	550	500
Water	152	140	185	290	252
W/(C + P), %	50	49	57	79	74
Sand	748	680	875	1050	997
No. 4 to ¾ in aggregate	642	590	800	1010	938
¾ to 1½ in aggregate	745	685	830	980	927
1½ to 3 in aggregate	750	840	1010	1160	1115
Total coarse aggregate	2490	2430	2660	2870	2769
Total aggregate	3330	3250	3540	3850	3766
Wet concrete	3980	3930	4090	4250	4193

*Weights in lb/yd^3 of resultant concrete based on saturated surface-dry aggregates. W/C or W/(C + P) indicates water/cement ratio or water cement + pozzolan ratio by weight.

TABLE 17.6 Typical Composition of Mass Concrete (*Continued*)

	Lowest observed value	Second centile	Median	Ninety-eighth centile	Highest observed value
3 in maximum exterior mixes					
Cement	245	250	400	550	625
Pozzolan	0	—	—	—	82
Cement + pozzolan	286	275	425	575	625
Water	160	145	210	280	266
W/(C + P), %	42	41	49	64	61
Sand	685	650	845	1140	1061
No. 4 to ¾ in aggregate	395	300	760	980	910
¾ to 1½ in aggregate	675	650	840	980	927
1½ to 3 in aggregate	750	700	1030	1450	1370
Total coarse aggregate	2445	2420	2610	2920	2884
Total aggregate	3210	3140	3470	3840	3750
Wet-concrete weight	3980	3940	4100	4260	4235
6 to 8 in maximum interior mixes					
Cement	148	150	205	400	400
Pozzolan	0	0	56	110	118
Cement + pozzolan	198	200	255	410	400
Water	94	108	165	305	309
W/(C + P), %	36	40	64	85	86
Sand	714	705	825	1210	1170
No. 4 to ¾ in aggregate	410	415	565	725	721
¾ to 1½ in aggregate	333	380	580	770	913
1½ to 3 in aggregate	675	650	815	1140	1096
3 to 6 or 8 in aggregate	433	700	950	1150	1150
Total coarse aggregate	2560	2540	2900	3260	3211
Total aggregate	3542	3520	3750	3980	3925
Wet-concrete weight	3914	3910	4150	4390	4306
6 to 8 in maximum exterior mixes					
Cement	247	247	307	441	441
Pozzolan	0	0	50	105	106
Cement + pozzolan	247	247	368	441	441
Water	130	132	175	232	226
W/(C + P), %	41	41	49	56	56
Sand	710	680	790	1030	1022
No. 4 to ¾ in aggregate	280	260	500	650	679
¾ to 1½ in aggregate	201	460	575	670	675
1½ to 3 in aggregate	635	620	790	1060	1069
3 to 6 or 8 in aggregate	764	760	950	1290	1230
Total coarse aggregate	2500	2420	2885	3090	3055
Total aggregate	3440	3410	3630	3970	3913
Wet-concrete weight	4036	4040	4185	4310	4290

CHAPTER 18
CONCRETE MIXERS

Robert W. Strehlow

Mixers for portland cement concrete are either stationary (plant mixers) or mobile (truck or transit mixers). The industry association that sets the standards for plant mixers is the Plant Mixer Manufacturer Division (PMMD) of the Concrete Plant Manufacturers Bureau (CPMB). For truck mixers, the Truck Mixer Manufacturers Bureau (TMMB) sets the standards. Both groups are affiliated with the National Ready Mixed Concrete Association (NRMCA) and are located at 900 Spring Street, Silver Spring, MD 20910.

Mixer design is mostly empirical and based on the experience of the manufacturers. The industry standards set up standard sizes, and minimum (and maximum) drum volume formulas, plus any common performance criteria. The ASTM C94 Committee on Ready Mixed concrete sets the standards for delivered concrete. For paving concrete the performance standards are set by the state Highway Commissions. The U.S. Army Corps of Engineers, the U.S. Bureau of Reclamation, and other federal agencies also set delivery specifications. With few exceptions (such as mass concrete), the specifying agencies specify the type of mixer. It is up to the producer to determine the best mixer to meet the specifications. The PMMD has presently established standards for the following size and types of mixers:

Single-compartment two-opening nontilting type—1, 2, 3½, and 4½ yd^3

One-opening tilting type with 15° inclined axis and two-opening horizontal-axis tilting type—2, 4½, 6, 8, 9, 10, 12, and 15 yd^3

Vertical-shaft type—½, 1, 1½, 2, 3½, and 4½ yd^3

Horizontal-shaft type—1, 2, 3, 4, 6, 8, 9, 10, and 12 yd^3

ROTATING-DRUM PLANT MIXERS

With the publication of research on mixing in large central mixers by the U.S. Bureau of Public Roads, full-size, rotating-drum tilting mixers became the accepted cost-effective means of mixing concrete for paving and any mass-concrete project. Research found that mixer size had no bearing on mixing time if concrete ingredients were properly blended (ribbon-loaded) and the charging sequence controlled. Proper blending and sequential entry is essential because large, longer drum mixers lack the ability to obtain end-to-end

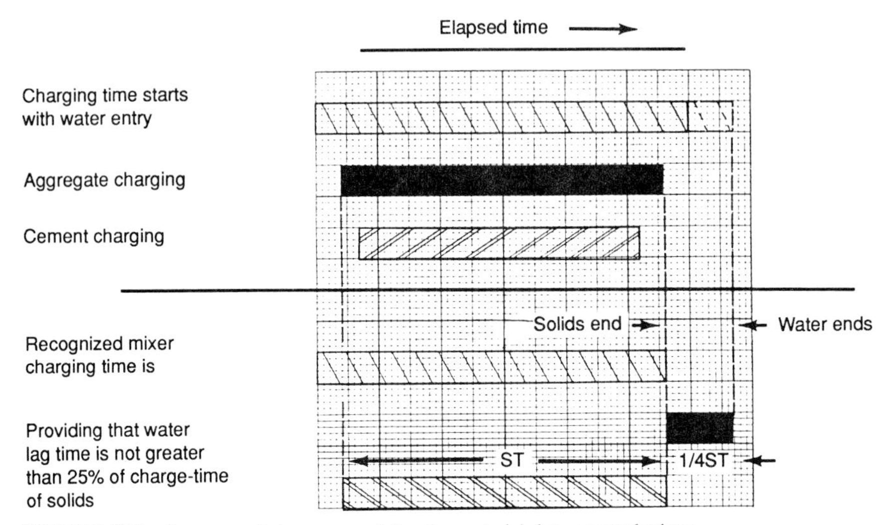

FIGURE 18.1 Recommended sequence of charging materials into a central mixer.

uniformity in a short mixing time if improperly charged. For mixers with proper blending, a mixing time of 60 seconds is the accepted standard and in most cases can be lowered if the performance can be demonstrated with uniformity tests. Figure 18.1 shows the accepted recommended charging sequence.

There are two basic tilting mixer styles. Both styles mix concrete by using a rotating drum that has internal blades designed to lift and tumble the ingredients onto itself. Since these mixers tilt to discharge the mix, the blades have only one purpose, and that is for mixing. With both styles, tilting is accomplished using hydraulic cylinders. The most popular style is the horizontal drum with a charging opening at the rear and a discharge opening at the front (Fig. 18.2). The other style has a single drum opening through which the materials are charged and discharged (Fig. 18.3). Since air is displaced when the mixer is charged, it is more difficult to accomplish dust collection on a single-opening mixer.

The main difference between brands of mixers is the tilting mechanism. Each manufacturer attempts to reduce installation height and maintain a relatively simple mechanism. All use electric motors to rotate the drum and require 10 horsepower (hp) per cubic yard. The size of the package hydraulic tilting motor varies with the speed of tilting required and the tilting arrangement. Where mixers discharge into open-top-haul units, such as for paving, a high-speed tilt package is often used. For charging ready-mix trucks the tilt speed is set to match the ability of the truck mixer to charge and some times remote-controlled adjustable-tilt speed controls are used for maximum truck charging efficiency.

While the tilting, rotating-drum mixer can mix zero-slump concrete, it is slower discharging and where "roller-compacted concrete" is used for base course paving, the use of a pugmill-type mixer has been used successfully.

The standard rating plate for a tilting mixer specifies a maximum slump of 3 in. The purpose of this limitation is that higher-slump mixes will spill out during mixing (not as much stays in suspension). Thus the ready-mix producer who ships 4- and 5-in slump concrete must go to the next larger-size drum. There is no problem in mixing the high-slump mixes if the charging protocols are followed.

FIGURE 18.2 Rear charge front-discharge mixer. Modern 12 yd^3 forward-tilt-point mixer includes integral dust hood. (*Courtesy of McNeilus Companies.*)

FIGURE 18.3 Front-charged, front-discharge tilt mixer shown here charged by a swing conveyor and discharging into open top (agitor) haul unit. (*Courtesy of the C. S. Johnson Co.*)

THE SLUMP METER

One of the accessories available on most plant mixers is the slump meter (Fig. 18.4). A slump meter is nothing more than a long-scale ampmeter that registers the amperage draw of the mixer motor(s), a factor which is directly related to the slump. The rotating-drum mixer is the most responsive to slump changes as it lifts concrete on its blades. The stiffer it is, the more concrete is lifted and the more power it takes. Some of the force mixers show little change of readings with changes in slump. However, if the same target reading is always reached, it is most likely that the slump is acceptable.

The "slump meter" should be called a "manual moisture correction" device because it is impossible to accurately determine the total free water in a large aggregate batch. That makes the slump meter a quality control tool. If all the batch tolerances and moisture meter errors occur in one direction, it is possible to have a free-water variation of 1 to 2 percent or at least a gallon (4 L) per cubic yard. The slump meter allows the operator to leave some water out of the mix and then, based on observations as the batch mixes, to add only the water needed to reach the desired mix. Certain automatic controls work just this way, but none has ever reached the efficiency of a good operator.

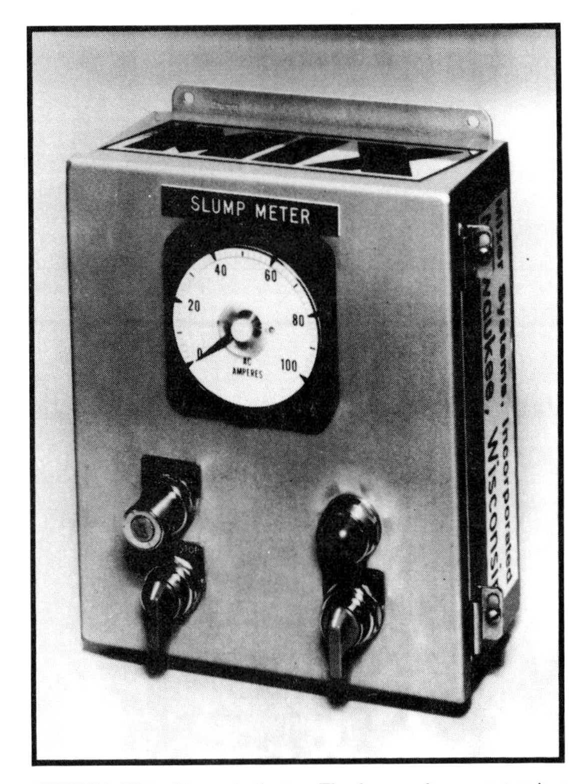

FIGURE 18.4 Slump indicator. The long-scale ampmeter is a common form of slump meter. Here the scale has been selected to be 100, which is then calibrated to the slump. *(Courtesy of Mix Systems, Inc.)*

FORCE-TYPE PLANT MIXERS

The term "force"-type mixers comes from European specifications that call for zero-slump mixes. Force mixers accomplish mixing by forcing the mixing blades through the concrete. Turbine or pan type, spiral blade type, and pug types fall into this category. Only the spiral-blade mixer is found in sizes over 4½ yd^3. These mixers are not designed to mix >2-in (>51-mm) aggregates or high-slump mixes, but they do an excellent job on those applications requiring a stiff or zero-slump mix, and they do it fast. While they mix high-slump mixes equally well, they have not gained great acceptance in ready-mix production because of their small size and a reputation for high blade and liner wear. The average force mixer requires approximately 30 hp input per rated cubic yard. With few exceptions, force mixers are the prevalent type found in the concrete products industry. Turbine mixers (Fig. 18.5) are more often used where high production is required and for mixing high-strength concrete. Spiral-blade mixers (Fig. 18.6) are found more often in lower-production applications and in dry-mix (block) type applications or in the less structural wet-cast products.

When selecting the size of a force-type mix, you should be aware that the same mixer may carry two volume ratings. Wet mixing causes the ingredients of concrete to consolidate or shrink to their in-place volume. Dry-mixed concrete does not shrink to the same extent and is consolidated by force or vibration after being placed in the form. Thus the same-sized mixing drum can mix a greater in-place volume and usually requires more power to mix the added material. The manufacturers of these mixers publish both the mixing tank volume and the maximum load weight.

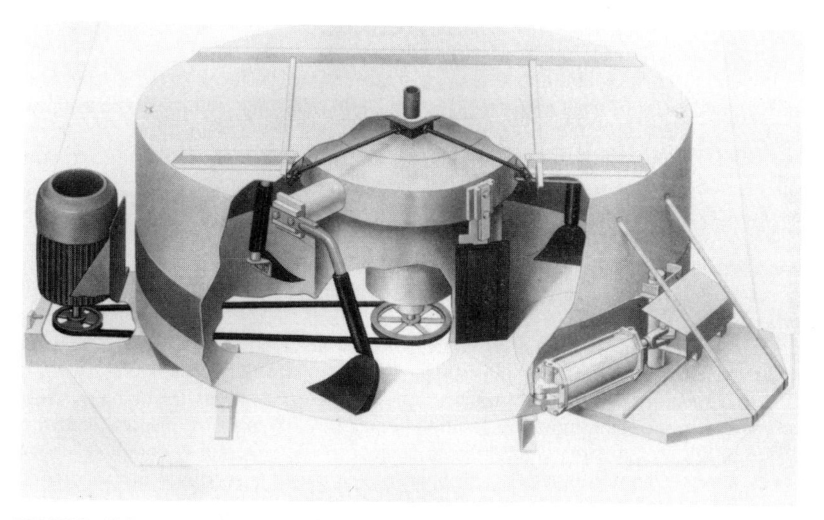

FIGURE 18.5 The vertical-shaft mixer. The turbine or vertical-shaft mixer mixes by forcing its high-speed plow-shaped paddles through the mix. The mix is discharged through one or more bottom doors. (*Courtesy of Mixer Systems, Inc.*)

FIGURE 18.6 The horizontal-shaft mixer. The rugged horizontal-shaft or spiral-blade mixers now range in size from ½ to 12 yd^3 and are no longer used only for concrete products. Here a large central-mix ready-mix plant employs a large horizontal-shaft mixer for production. (*Courtesy of Appco-Besser Co.*)

TRUCK MIXERS

Truck mixer sizes are listed in Table 18.1.

Inclined-axis truck mounted mixers are available in two styles, the rear-discharge type (Fig. 18.7) and the front-discharge type (Fig. 18.8). In the front-discharge type usually the discharge chute is controlled hydraulically from inside the cab to allow the driver to direct the placement of the concrete. The front-discharge mixer is significantly more expensive, and the vote is not yet in on which style is most cost-effective. From the concrete contractor's point of view, the front-discharge mixer seems to be preferred in that it reduces the contractor's labor.

Truck mixers are designed to mix concrete in an inclined-axis configuration that can be mounted on a truck or trailer. The rear-discharge mixer is usually mounted on a customer's choice of truck chassis. The front-discharge mixer is usually a custom chassis designed to keep the height down. Both are powered by an engine-driven, variable-speed hydraulic system. The drum rotation is reversed to discharge.

Concrete specifications set the minimum number of revolutions for complete mixing as well as the mixing and agitating rotational speeds. Mixers that transport plant-mixed concrete are allowed to carry a slightly larger load if the concrete is completely mixed. However, a very common practice with plant-mixed concrete is to mix the concrete to a point where the plant slump meter indicates that the desired slump is predictable and then finish mix on the way to the jobsite. This practice, referred to as "shrink mixing," is covered in delivery specifications.

TABLE 18.1 Standard Truck Mixer Sizes*

As a mixer	As an agitator†
6	7¾
7	9¼
7½	9¾
8	10½
8½	11¼
9	12
10	13¼
11	14¾
12	16
13	17½
14	19
15	20¼

*In yd^3; as established by the Truck Mixer Manufacturers Bureau.

†Maximum agitator capacity may be less as designated on the manufacturer's data plate.

Central-mixed concrete shows one of its greatest economic advantages by requiring fewer haul units than truck-mixed concrete operations. This is because the haul cycle time is shortened. Wet mix charges at least twice as fast as a dry charge. There is no time loss in adjusting slump on the job or in the yard, and there is no need to turn the drum for some minimum number of revolutions on short hauls of less than, say, 5 min.

FIGURE 18.7 The rear-discharge truck mixer. The rear-discharge mixer is normally mounted on the owner's choice of truck chassis. (*Courtesy of Rexwerks, Inc.*)

FIGURE 18.8 The front-discharge mixer normally includes its own integral truck chassis. (*Courtesy of the McNeilus Companies.*)

CHAPTER 19
MATERIALS HANDLING AND STORAGE

John K. Hunt*

The extreme difference in characteristics of the materials used for concrete necessitates different methods of handling and storage. Aggregate is usually elevated by clam-shell bucket, belt conveyor, or bucket elevator; cement by bucket elevator or air blower; and water and admixtures by pumps. The methods used to transport materials to the plant site are a major factor in determining the handling system used.

Aggregate conveyor belts and bucket elevators should be rated in tons per hour, cement bucket elevators and other cement-conveying equipment should be rated in barrels per hour, aggregate bin storage should be stated in terms of cubic yards of aggregate, and cement bin or silo storage should be stated in terms of barrels of cement. Capacities are defined by Concrete Plant Standards published by Concrete Plant Manufacturers Bureau, 900 Spring St., Silver Spring, MD 20910.

AGGREGATE HANDLING

Other than a few low-production concrete production plants, the in-plant aggregate storage represents only what is needed to allow the supply equipment and logistics to keep the bin filled. In the great majority of cases this means that there is some other storage for aggregates at the plant site. A plant located in a quarry may be the exception. Bulk aggregate storage is usually in the form of open stockpiles, stockpiles over a funnel conveyor (separated with dividing walls), large bunkers, or concrete stave silos (Fig. 19.1).

19.1 Unloading and Stockpiling

Delivery of aggregates to the plant may be any combination of railroad car, truck, or barge. Most rail delivery systems make use of bottom dump cars which discharge into a hopper which feeds the stockpiling or storage system (Fig. 19.2). These arrangements generally include a hopper to include the delivery of aggregates by truck. Clam-shell buckets are the most common way of unloading barge-delivered aggregates. Truck-deliv-

*Revised by Robert W. Strehlow.

FIGURE 19.1 Bulk aggregate storage. A shuttle conveyor delivers rail- or truck-delivered aggregates to stockpiles or bunkers. (*Courtesy of Atlas Conveyor Co.*)

ered aggregates are deposited directly on stockpiles, or into bunkers, or into a hopper that feeds a stockpiling system such as a stacker or shuttle system. It seems that the ideal system is one in which the delivered materials are placed in position for ready use, with the next-best alternative being one in which materials are in position for unattended recovery or transfer to the "use" location. These handling systems generally add to the investment which must be balanced against the increased efficiency. Few plants, if any, operate without the availability of a front-end loader and, this being the case, the front-end loader is the most prevalent device used for stockpiling or charging a plant. Low-profile and mobile plants are often charged directly by the front-end loader transferring material from the stockpiles to the plant bin. For the larger or elevated storage plants, the loader feeds a

FIGURE 19.2 Rail delivery of aggregate. Shown here is a materials-handling arrangement that accepts aggregates delivered by rail (or dump truck). Material moves by conveyor from a receiving hopper to a stacker, to concrete stave silos, where it is recovered by a tunnel conveyor, to an incline conveyor to a turnhead. (*Courtesy of Rexwerks, Inc.*)

hopper over an inclined conveyor utilizing a "turnhead" or rotary distributor; or feeds a stacking conveyor that is moved to each bin compartment.

19.2 Clam-Shell Buckets

The transfer of aggregates from stockpiles directly to the "use" bin by means of a crane equipped with a clam-shell bucket is quite common with mobile or portable concrete paving plants.

Rehandling buckets are most suitable for handling aggregate materials. Sand may be piled in a cone. Damp sand tends to segregate less in handling and storage than very dry sand. Coarse-aggregate stockpiles should be made by placing each bucketful in a consistent pattern and building up layer by layer instead of indiscriminately creating many small stockpiles and filling in between them. The material should be laid in place, not thrown at the pile. Throwing the material causes breakage and degradation and also causes the larger particles to run down the sides of the pile, which creates segregation.

19.3 Bucket Elevators

The aggregate bucket elevator is an efficient means of elevating material, but even before the imposition of OSHA it was on the verge of economic extinction. It is mentioned here as a means to be used only if there is no other way to arrange the plant and materials-handling system.

19.4 Belt Conveyor Systems

Whether it is for stockpiling or charging an elevated bin, the belt conveyor has provided the most economical choice. It is a reliable, simple, low-maintenance device that has a wide range of capacities that are well documented (Fig. 19.3). The capacity of a belt conveyor depends primarily on the width of the belt, the belt speed, and the degree of troughing.

Since a cubic yard of concrete requires approximately 38 ft^3 of unmixed aggregates, the required capacity of a system can easily be determined by applying an efficiency factor to the system size as determined by the plant production requirements.

The hardest determination in sizing a belt conveyor system is what to use for an efficiency factor (Fig. 19.4). A common value is 50 percent for "visually" charged plants and those using simple automatic-refill systems. Certain high-efficiency computer-controlled automatic bin-refill controls can obtain 75 to 90 percent efficiency. As a general rule, the more materials that are in regular use, the less efficient the system is. The smaller a bin is, the more often the system will have to refill a bin compartment. Each time the bin-fill system must seek another compartment to fill, a gap must be created on the belt to allow the turnhead to turn. Thus, the faster the turnhead rotates, the shorter this gap and the higher will be the system efficiency. At this time the fastest turnhead on the market makes a full rotation in 8 s, and the longest takes over a minute. The average falls between 15 and 25 s. There are other factors, but these can be neutralized by the right selection of the automation. Turnhead speed cannot.

Unlike concrete batching controls, which are standardized by an industry standards group, there is no such organization for bin filling controls. However, the "supercharger" model developed by Strehlow Engineering and now marketed by Mixer Systems is an example of the potential extent of automation. It is expected that there will be other control manufacturers adopting many of their advanced features.

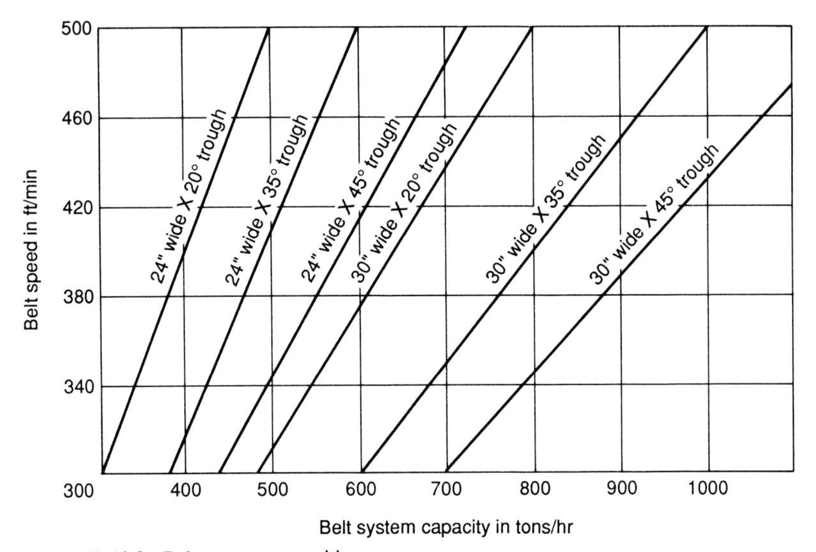

FIGURE 19.3 Belt conveyor capacities.

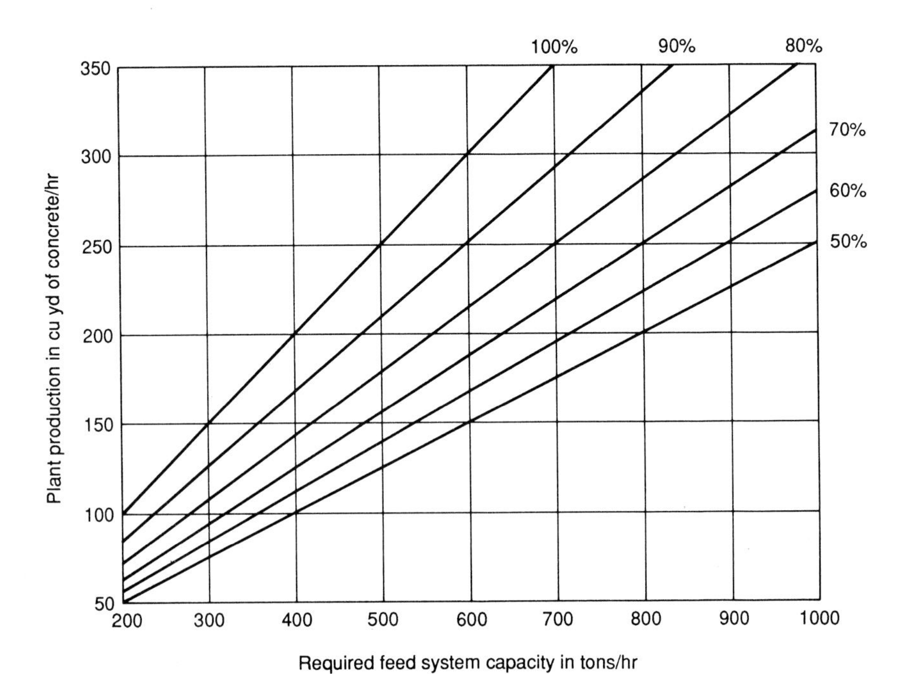

FIGURE 19.4 Plant production versus system efficiency.

19.5 Distributors

The "turnhead" or rotary distributor is the most common device used for feeding multi-compartmented bins. However, the use of two- and three-position flop gates are often used on bins with a few compartments. Flop gates are very efficient in that they change position very fast, but are more subject to sticking and are generally higher-maintenance units as they require an air supply and lubrication.

Rotary horizontal reversing belt conveyors have also been used to distribute aggregates to multicompartment bins. They possess the advantage of a lower height with a greater reach.

For multicompartment bins with up to four compartments in line, a powered stacking conveyor does the same function as a turnhead and does provide the lowest profile configuration (Fig. 19.5). It is well suited to the mobile plant configuration. Stackers are often used to charge stationary storage silos but are generally limited to a maximum of five silos. For more silos or bunkers, an overhead shuttle conveyor is a common means of bulk storage.

19.6 Aggregate Material Level Signals

No bin-fill system can operate without some means of knowing when to fill a bin compartment and when to stop filling. There are many types of bin signals on the market but the most commonly used one for aggregate level determination is the rotary-paddle signal. Its disadvantages are that its paddles must be protected from falling aggregates and it is mechanical and is subject to wear. Yet with all its faults it is simple, easy to adjust, does not cost an arm and a leg, and when properly installed is quite a reliable device. Nuclear level signals are very good but present a hazard and are quite expensive. Sonic probes and optic probes can experience failure due to dust. Capacitance probes work very

FIGURE 19.5 Individual feed system. Shown are highly portable individual material feed conveyors on a mobile central-mix paving plant. (*Courtesy of the Ross Co.*)

well as long as there is no chance of material clinging to them (as occurs with sightly dirty aggregate).

Tilt-type switches work well, but there is always the danger of their becoming tilted back upright by the shifting pile or being forced into a vertical position by a bin wall. Hanging tilt-type level signals are ideal for use as emergency shutdown signals if the "high" cutoff signal fails. In this application they cannot be trapped, are not subject to wear, and are inexpensive. No automatic bin-fill system should be operated without emergency high signals.

A single "high"-demand signal is all that is necessary in any bin-fill system. The demand to fill a compartment is created by this high signal. While some systems use a low emergency signal it can leave the compartment partially empty if the level does not drop to actuate it at the end of the day. Low-level override can be accomplished with timers or interfacing the materials-handling system with the batching system.

19.7 Stone Ladders

When the bulk aggregate storage system results in a condition in which large aggregates can have a large vertical drop that results in segregation or degradation, stone ladders should be used. These are vertical structures with a series of platforms or steps that intercept the falling material and spill over to the next-lower step or platform. They are eventually buried in the pile. The pile must be unloaded in a manner so the ladder is not toppled. The platforms or steps should not be more than 30 in apart.

RESCREENING OF AGGREGATE

After aggregate has been screened to establish the proper gradation for each size and stored in stockpiles, it is often desirable and sometimes necessary to rescreen the aggregate to reduce the effects of degradation and contamination due to handling.

19.8 Purpose and Methods

Since there is usually only one screen deck for material of each size, rescreening can only reject undersize material passing through a screen and cannot and is not intended to rearrange the gradation. Usually the material is blended in the approximate proportions to be used by feeding onto a belt conveyor simultaneously from several stockpiles with the belt conveyor feeding the screens. Sending one material at a time to the screens would overload one screen and would result in wasting the oversize and undersize material. Material passing the No. 4 screen is normally the only material wasted.

Since the specifications for gradation usually apply to the material as batched, it is more desirable to place the screens on top of the batch plant with the material going directly into the proper compartments instead of screening on the ground and then elevating the material, because screening on the batch plant eliminates one handling operation after screening and reduces the possibility of more contamination and degradation.

Rescreening structures on top of a batch plant must be carefully designed to keep vibration at a minimum. If very much vibration is transmitted to the automatic batching controls, scales, and recording, the operation may be seriously affected.

19.9 Screen-Size Selection

The most important factors affecting screen capacity are as follows:

1. Screen size
2. Amount of oversize
3. Amount of undersize
4. Density of material
5. Whether crushed or gravel material
6. Amount of moisture
7. Whether screen is top, second, or third deck on one machine
8. Shape of screen opening
9. Open-area ratio of screen
10. Desired efficiency

On a typical installation for mass concrete, the coarse aggregate is divided into four sizes: 6 to 3 in, 3 to 1½ in, 1½ to ¾ in, and ¾ in to No. 4.

For the larger installations either two double-deck screens—one above the other—or two parallel sets of two double-deck screens (four double-deck screens) are used. For small installations one triple-deck screen can be used with the bottom deck split for two materials.

Required capacity is computed on the basis of average mix design, the maximum production, or the expected hourly production, and considering use of crushed stone. For the three top sizes 90 percent efficiency is normally adequate because most gradation specifications allow 15 to 20 percent of slightly undersize material. The efficiency of the No. 4 screen could be as low as 50 percent since in general removal of one half the undersize material below No. 4 mesh will bring the material within the usual gradation specifications. This usually permits carrying a depth of material on the No. 4 screen from 6 to 10 (or more) times the screen opening.

The highest loaded screen usually determines the size of all the screens, although in some installations the upper set of screens may be one size larger than the lower set.

When the same conveyor is used for coarse aggregate and sand a diverting gate is usually used to allow the sand to bypass the coarse-aggregate screens and go directly to the sand-bin compartment. Approximately one-third of the total loading time will be needed for sand; thus only two-thirds of the total loading time is available for rescreening of the coarse aggregate. Higher-capacity plants use a separate conveyor for the sand, which increases the available screening time by 50 percent. Typical screen sizes used in plants with one conveyor handling both the coarse aggregate and the sand are shown in Table 19.1.

CEMENT AND POZZOLAN HANDLING

Since the handling problems of cement and pozzolan are similar the following comments apply generally to both. It is safer to use separate structures (Fig. 19.6) and handling facilities for cement and pozzolan to avoid contamination, but if the same handling equipment is used for both materials, the valves that determine to which compartment each is fed should be watched carefully and inspected regularly. A double wall or extrastrong partition should be placed between two adjacent compartments.

TABLE 19.1 Typical Screen Sizes Used in Plants with One Conveyor Handling Both Coarse Aggregate and Sand

Mixers, yd^3	Production, yd^3/h	Screen
2–2	80–120	One 5 × 14 or 5 × 16 triple deck
2–4	130–160	Two 5 × 12 double deck or one 5 × 16, 5 × 20, or 6 × 20 triple deck
3–4	200–240	Two 5 × 14 or 5 × 16 double deck
4–4	250–320	Two 6 × 16 or 7 × 16 double deck or two parallel sets of two 5 × 12 double deck
6–4	400–480	Two 6 × 16 or 7 × 16 double deck or two parallel sets of two 5 × 12 double deck (with a separate conveyor for sand)

Practically all cement, fly ash, and pozzolanic materials are delivered by trucks or rail cars having a pneumatic transfer system. These transfer systems fluff the cement up to an increase of 50 percent in volume. This volume shrinks as the entrapped air escapes. This has no effect on the batching system, but it is an important consideration when sizing a storage silo.

RETURNED CONCRETE

It is only fitting to end any discussion of materials handling with some comments on what to do with waste concrete. Pollution-control laws limit polluted runoff, so one must treat the washout water. For such cases, washout and reclaim units are available (Fig. 19.7). Some just reuse the wash water, some return wash water and grout to the mix, some separate the solids, and some do not. Most reclaimed solids can be reused.

FIGURE 19.6 Multiple cement storage. Shown is storage of cement and fly ash in separate silos arranged for gravity batching. Note the large low-profile aggregate bin. (*Courtesy of Appco-Besser Co.*)

FIGURE 19.7 Waste concrete reclaimer. This truck-mixer washout unit includes the separation and reclaiming of sand, stone, and grout. Cement and water slurry are discharging into a settling tank. (*Courtesy of the Vince Hagan Co.*)

CHAPTER 20
HEATING AND COOLING

John K. Hunt*

This chapter is concerned with all aspects of heating and cooling concrete in the plant. Chapter 29 covers hot- and cold-weather operations.

Concrete must often be heated in cold weather to ensure that it can be placed and protected before it freezes. Concrete is also cooled, principally in mass-concrete work, to permit lower water and cement contents for a given strength and slump, and to offset temperature rise in mass concrete caused by heat of hydration of the cement. Drying shrinkage of concrete is reduced by the lower cement content and more particularly by the lower water content of cool concrete. Plastic or drying cracking of slabs is reduced by slower moisture loss at reduced temperature. Depending on the type of cement, hydration releases 80 to 180 Btu per pound of cement in the first 3 days and about twice this amount in 28 days. In mass concrete this heat develops faster than it can be radiated by the concrete and results in temperature gains of about 25 to 30°F in 28 days for concrete with 376 lb of cement per cubic yard. To offset these effects, concrete is often placed at 40 to 50°F so that it will be substantially cooler than the average ambient temperature.

20.1 General

Heat is energy that raises the temperature of a substance as the amount of stored heat increases. Heat flows from a substance at a higher temperature to a substance at a lower temperature by *radiation,* by *conduction* where they are in contact, and by *convection* when heat is carried by a moving gas or liquid. Heat flow requires time.

Time required for heating or cooling depends on a variety of factors. Neglecting surface resistance, the minimum time T in minutes required to change the initial temperature t_0 of an aggregate particle to an average temperature t_a when its surface temperature is held at t_s is given by

$$T = \frac{CD^2}{\alpha} \qquad (20.1)$$

where C is a value from Table 20.1 depending on the percentage of heating or cooling remaining to make t_a equal to t_s. Either centigrade or Fahrenheit temperatures may be used if used consistently. Typical values of thermal diffusivity α in square feet per hour

*Revised by Robert W. Strehlow.

TABLE 20.1 Values of C for Eq. (20.1)

% of heating or cooling remaining $\dfrac{t_a - t_s}{t_0 - t_s}$	C
0.001	0.0678
0.005	0.0510
0.01	0.0436
0.02	0.0361
0.03	0.0318
0.05	0.0264
0.10	0.0191
0.20	0.0118
0.30	0.0077
0.40	0.0051
0.50	0.0032
0.60	0.0019
0.70	0.0009
0.80	0.0004
0.90	0.0001

are given in Table 20.2; and $\alpha = k/c\rho$ where k is conductivity in Btu/(h)(ft^2)(°F/ft), c is specific heat in Btu/(lb)(°F), and ρ is density in pounds per cubic foot. This equation for α is only approximately correct for porous materials; D is particle diameter in inches. Root-mean-square values of D^2 for common mixtures of particle sizes are

Diameter range, in	D^2
3–8	32.33
3–6	21.00
1½–3	5.25
¾–1½	1.31
No. 4 to ¾	0.27

and for typical blended aggregate for mass interior dam concrete are

Diameter range, in	D^2
No. 4 to 8 in	12.34
No. 4 to 6 in	8.64
No. 4 to 3 in	2.50

These values of D^2 may be used with Eq. (20.1) to give minimum times for a mixture of particle sizes to reach an average temperature t_a.

Since time is proportional to D^2, note the effect of large aggregate sizes on cooling and heating times. If oversize cobble is permitted, minimum cooling and heating times

TABLE 20.2 Thermal Characteristics of Concrete Materials*

Material	Specific heat c: Range of reported values	Specific heat c: Typical value for computations	Conductivity k, Btu/(ft^2)(°F/ft)	Density ρ, pcf	Thermal diffusivity α ft^2/h
Cement, portland	0.12–0.22	0.22	0.17	94	0.008
Pozzolan	—	0.22			
Water	—	1.00	0.32	62.4	0.0052
Ice at 32°F	0.49–0.50	—	1.28	57	0.045
At 14°F	0.49	0.50	1.35		
At −4°F	0.47	—	1.41		
Sand, saturated surface-dry	0.16–0.22	0.20	0.19	95	0.011
Sand, 4% moisture	—	0.20	0.40	100	0.020
Crushed rock, 4% moisture	—	0.20			
Granite	—	—	0.62	110	0.030
Traprock	—	—	0.50	110	
Gravel, 4% moisture	—	0.20	0.75	110	0.036
Aggregate particles, saturated surface-dry	—	0.20			
Basalt	0.18–0.24				
Cinders	0.18				
Dolomite	0.19–0.22	—	1.0	167	0.03
Gneiss	0.18–0.20				
Granite	0.18–0.20	—	1.6	168	0.050
Limestone	0.18–0.22	—	1.2	160	0.030
Magnetite or hematite	0.16–0.17				
Marble	0.21–0.22	—	1.3	168	0.037
Rhyolite	0.18				
Quartzite	0.16				
Sandstone	0.21–0.25	—	1.5	144	0.043
Serpentine	0.25–0.26				
Traprock	—	—	—	—	0.029
Concrete, hardened	0.15–0.25	0.22	0.54	144	0.019
Concrete, wet, fresh	0.22–0.28	0.25			
Grout, wet, fresh	0.27–0.38	0.32			
Steel	0.11–0.12	0.12	26	490	0.48
Steam, constant pressure, 15 psia†	—	0.48			
150 psia	—	0.61			
Air at 14.7 psia	—	0.24	0.014	0.081	0.72
Calcium chloride "standard solution" (29.12%) at 60°F	—	0.66	—	80.1	

*These are typical values. In some instances, as with aggregate, the values may vary considerably depending on the sample.

†Pounds per square inch absolute.

are substantially increased. Inclusion of 10 percent of 9 in "plums" in 3 to 6 in aggregate increases the time for this material by 29 percent, for example. Cooling cobble graded 3 to 7 in requires about 25 percent longer than cobble graded 3 to 6 in.

It is often assumed that particles of the same material of any shape other than a sphere will cool more rapidly than a sphere. This is true only if the particle has the same volume as the sphere. Aggregate, however, is screened by linear dimension, and it is probable that cooling time is not greatly different from that for spheres. For example, cooling a $6 \times 6 \times 6$ in cube requires about 20 percent more time than a 6 in sphere. Cooling a $6 \times 6 \times 12$ in "brick" requires about 65 percent more time than a 6 in sphere. Cooling a $4 \times 6 \times 9$ in "brick" requires very nearly the same time as a 6 in sphere.

In practice, heating and cooling times for concrete aggregates are always longer than the minimum times given by Eq. (20.1). The surface of the particles being heated or cooled cannot be maintained at the temperature of the surrounding medium. Often part of the aggregate surface will be in contact with the surface of other particles, exposing less than the full surface to the cooling medium. Also in most heating or cooling processes some time is required to change the temperature of the surrounding medium to t_s. In addition, the cooling medium must be able to remove heat as rapidly as it can be given up by the aggregates.

Figure 20.1 shows typical practical cooling times for various processes. Of the cooling methods shown, only inundation, vacuum, and airflow through aggregates in deep beds are presently used in large concrete plants in North America. The curves for tumbling and airflow through shallow beds are from tests by J. Poulter of the Koehring Company using high-velocity airflow. Vacuum-cooling tests are by E. Goldfarb, and tests of air cooling of deep aggregate beds are by T. B. Appel and G. L. Marks for the C. S. Johnson Division, Koehring Company. Time required to fill and empty tanks is excluded, but time to lower temperature of coolant in contact with the aggregate is included in Fig. 20.1. Minimum curves shown for 1½ to 3 and for 3 to 6 in material are for A equal to 0.03.

Temperature (or dry-bulb temperature) is a measure of the property of a substance that determines whether it will gain or lose heat in proximity with another substance. Temperature is usually determined by a thermometer based on expansion or vapor pressure of a liquid. It may also be measured by expansion or pressure of a gas or by its electrical effect. While the 0 and 212°F points can be determined accurately, intermediate points are somewhat arbitrary and may not agree exactly on different thermometers even when they are properly calibrated. Also thermometers can get out of calibration by several tenths of a degree as a result of alternate heating and cooling and must be calibrated against a standard for accurate work.

Temperature of water, flake or slush ice, sand, cement, pozzolan, and wet concrete can be taken by inserting a thermometer well into the material. In fresh concrete the thermometer bulb should not be placed against large aggregate because its temperature may be higher or lower than the other ingredients until sufficient time has passed for the temperature of the concrete to stabilize. If coarse aggregate is warmer or cooler than the mortar, temperature of the mortar may increase or decrease several degrees after concrete leaves the mixer. Temperatures of small and medium aggregates may be taken with a thermometer inserted in a sample in an insulated bag or container. Temperatures of block ice and aggregates over about 1 in are determined in an insulated container with a hole in the lid for a thermometer (called a *calorimeter*) by mixing a known weight of ice or large aggregate with a known weight of water at a known temperature. For example, if a sample of aggregate weighing m lb and with a specific heat of c (see below) is placed in a calorimeter containing w lb of water at a temperature t, and the mixture is allowed to reach a stable temperature t_2, the initial temperature of the aggregate t_1 is computed from

$$t_1 = \frac{(w + K)(t_2 - t)}{cm} + t_2 \tag{20.2}$$

where K is the *water equivalent* of the calorimeter and is the amount of water that requires the same amount of heat to raise or lower its temperature 1°F as was required to raise or lower the calorimeter temperature 1°F; K is determined as follows. Fill the calorimeter with water at a temperature near that used in making tests (say, 60°F). After about 10 min measure the temperature t_3, empty the calorimeter, and fill it with a known weight W of water at a higher temperature t_4. After an additional 10 min read the temperature t_2 of the water in the calorimeter:

$$K = \frac{W(t_4 - t_2)}{t_2 - t_3} \tag{20.3}$$

The value of K should approximate the weight of the calorimeter lining multiplied by its specific heat; K equals about 12 percent of the weight of the lining if it is steel.

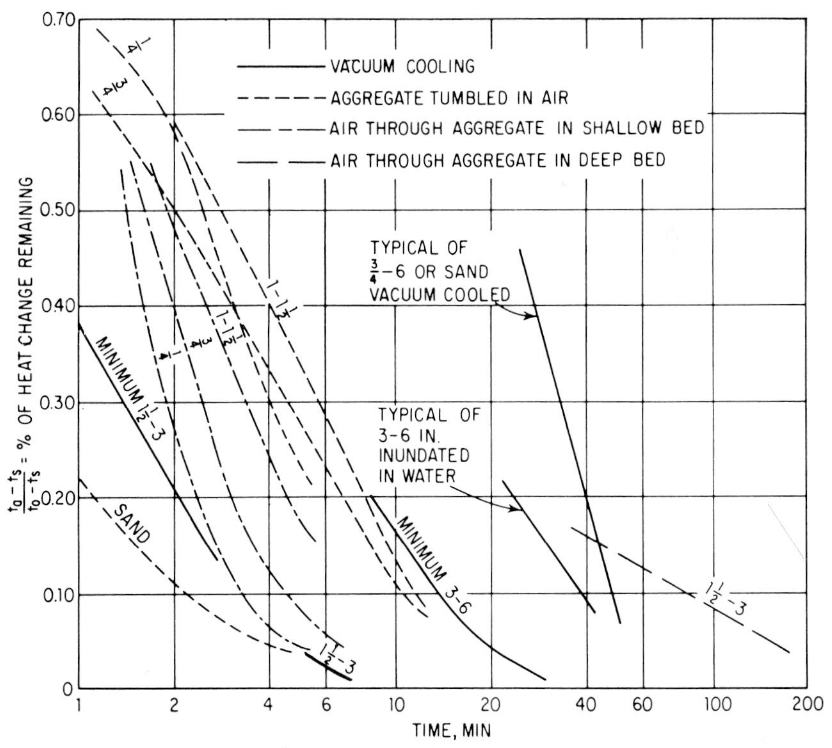

FIGURE 20.1 Typical samples for vacuum cooling, inundation, and air cooling in deep beds are obtained with initial high surrounding temperatures (surface temperatures), which are then gradually reduced (depending on size of cooling units) to the final surface temperatures t_s on which the shown percentages are based (t_a = initial uniform aggregate temperature, t_0 = average aggregate temperature, t_s = temperature of cooling medium). Minimums shown are for ideal conditions with thermal diffusivity = 0.030.

A British thermal unit (Btu) is a unit of heat. One Btu will raise 1 lb of water by 1°F, and 180 Btu are required to heat 1 lb of water from 32 to 212°F.

Specific heat is the number of Btu required to raise the temperature of 1 lb of a substance by 1°F. The specific heat of water is 1. Table 20.2 gives specific heats of various materials. Specific heat of most solids and liquids varies only slightly with temperature insofar as concrete computations are concerned. Specific heat of cement is assumed to be 0.22, and that of aggregate is 0.20. These values are generally used in computations. Specific heat of aggregate is determined in a calorimeter when the weights and temperatures of aggregate and water are known.

Heat of fusion of water is heat absorbed by ice in melting or given off by water in freezing, 144 Btu/lb.

Heat of vaporization is heat absorbed in changing a liquid to vapor or given off by vapor in condensing to liquid, for water 970.2 Btu/lb at normal atmospheric pressure (14.7 psi and 212°F). Heat of vaporization of water is higher at lower temperatures, about 1,070 Btu/lb at 38°F. Because of its high specific heat and high heats of fusion and vaporization, water can store or release large amounts of heat when it changes temperature, as shown in Fig. 20.2.

Vapor pressure of water is the pressure exerted by water in a gaseous state (Table 20.3). It is also called *elastic pressure, gaseous pressure,* and *vapor tension.* This pressure exists at all temperatures when water is present. When the

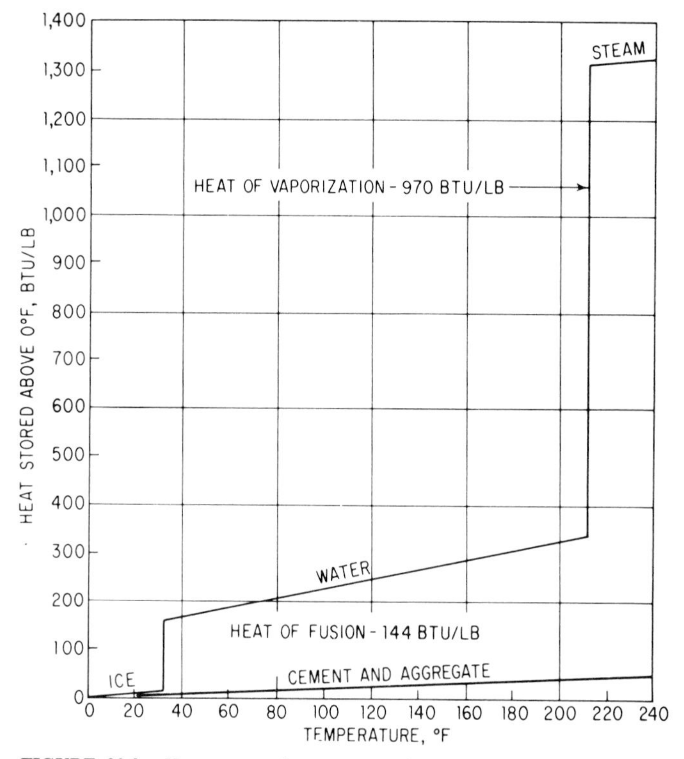

FIGURE 20.2 Heat content of concrete materials under a pressure of 14.7 psi.

TABLE 20.3 Vapor Pressure of Water

Air temp t, °F	Vapor pressure V, in Hg	t	V	t	V
30	0.164	60	0.517	90	1.408
31	0.172	61	0.536	91	1.453
32	0.180	62	0.555	92	1.499
33	0.187	63	0.575	93	1.546
34	0.195	64	0.595	94	1.595
35	0.203	65	0.616	95	1.645
36	0.211	66	0.638	96	1.696
37	0.219	67	0.661	97	1.749
38	0.228	68	0.684	98	1.803
39	0.237	69	0.707	99	1.859
40	0.247	70	0.732	100	1.916
41	0.256	71	0.757	101	1.975
42	0.266	72	0.783	102	2.035
43	0.277	73	0.810	103	2.097
44	0.287	74	0.838	104	2.160
45	0.298	75	0.866	105	2.225
46	0.310	76	0.896	106	2.292
47	0.322	77	0.926	107	2.360
48	0.334	78	0.957	108	2.431
49	0.347	79	0.989	109	2.503
50	0.360	80	1.022	110	2.576
51	0.373	81	1.056	111	2.652
52	0.387	82	1.091	112	2.730
53	0.402	83	1.127	113	2.810
54	0.417	84	1.163	114	2.891
55	0.432	85	1.201	115	2.975
56	0.448	86	1.241	116	3.061
57	0.465	87	1.281	117	3.148
58	0.482	88	1.322	118	3.239
59	0.499	89	1.364	119	3.331

Source: National Weather Service.

vapor pressure equals the pressure at the water surface, water vaporizes freely (boils).

Wet-bulb temperature is the minimum temperature to which water can be cooled by evaporation under existing conditions of pressure, dry-bulb temperature, and humidity. It is obtained with a *psychrometer,* which consists of a dry-bulb thermometer and a wet-bulb thermometer that has a wick saturated with distilled water. When fanned or whirled vigorously, the wet-bulb thermometer will show a wet-bulb temperature lower than the dry-bulb temperature. Relative humidity and actual vapor pressure can be obtained from tables such as the "Smithsonian Meteorological Tables" published by the Smithsonian Institution, Washington, D.C., if these temperatures and the barometric pressure are known.

Relative humidity is the actual vapor pressure divided by the saturated vapor pressure (Table 20.3) at the dry-bulb air temperature.

Dew point is the temperature at which the water vapor in the air is just sufficient to saturate the air (increase the relative humidity to 100 percent). Moisture condenses (changes from vapor to water) on objects with temperatures below the dew point in still air and condenses if the air temperature is reduced to the dew point, releasing the heat of vaporization, about 1000 Btu per pound of water condensed.

Boiler horsepower is a unit of steam boiler capacity equal to 33,472 Btu/h.

Steam quality is a measure of the liquid water carried in steam. The liquid water has lower heat-carrying capacity than water vapor. Quality is the dry steam (vapor) weight per pound of wet steam.

A ton of refrigeration is the cooling necessary to make 2000 lb of ice from 32°F water in 24 h. It is equivalent to 12,000 Btu/h.

20.2 Typical Material Temperatures

Cement is usually discharged from the grinding mills at 200 to 300°F and may be cooled to 120 to 180°F in cement coolers before being placed in storage. Thermal conductivity of dry cement is fairly low, about 2 Btu/(h)(ft^2)(°F/in), and it loses heat very slowly in storage. A typical loss is 3 to 10°F per month in large concrete silos, and slightly more in large steel silos. A further reduction of about 10 °F may occur as cement is handled for shipment. Shipping temperatures are usually 120 to 180°F, with the higher temperatures probable in late summer or when cement stocks are low and shipments are made from recently ground supplies. Shipping temperatures as high as 300°F are not unknown but are not common because cement is nearly always cooled after grinding. There is often little temperature loss in shipment (but it may be 20°F or more), and unloading (by other than compressed air) may reduce temperatures by a few degrees. Thus cement usually arrives at a concrete plant at about 90 to 150°F, and occasionally as high as 170°F.

Water from wells typically has a temperature of about the mean annual air temperature, but there is a large variation. For example, a study of about 700 wells by the Illinois State Water Survey shows an average temperature of 56°F at the pump discharge, but a range from 47 to 77°F. Water from surface streams commonly approximates the mean monthly air temperature but may be 5 to 15°F lower in spring and summer and 5 to 15°F higher in fall and winter.

Aggregate is usually assumed to be at approximately the mean monthly temperature of air in the shade as reported by the U.S. National Weather Service. It may be 10 to 25°F warmer than this if dry and stored or handled in the sun, and 2 to 10°F cooler if sprayed with water or stored moist in shade. Aggregate that is processed by washing and placed in large storage piles without delay tends to remain at the wash-water temperature for several days.

20.3 Heat Balance

Formulas for concrete temperature are complicated because of the heat of fusion of water and the variety of materials. They can be derived as needed by remembering that the number of Btu above the final temperature of the mixture must equal the Btu below the final temperature, or computed as in the following example:

Material	lb/yd^3	Specific heat above 32°F	Btu/°F above 32°F	Initial temp., °F	Btu/yd^3 above or below 32°F (ice melted)
Cement	516	0.22	113.5	80	113.5(80 − 32) = 5448
Sand	1340	0.20	268.0	18	268(18 − 32) = −3752
Moisture (ice) in sand, 6%	80.4	1.00	80.4	18	80.4[0.50*(18 −32) − 144*] = −12,140
Coarse	1885	0.20	377.0	20	377(20 − 32) = −4524
Moisture (ice) in coarse, 1%	18.9	1.00	18.9	20	18.9[0.50*(20 − 32) − 144*] = −2835
Added water (270 − 80.4 − 18.9)	170.7	1.00	170.7	160	170.7(160 − 32) = 21,850
	4011		4011)1028.5		1028.5)4047
Specific heat of wet concrete = 0.256					3.9
					+32.0
				Concrete temp =	35.9°F

*Specific heat of ice is 0.50; heat of fusion is 144 Btu/lb. Notice the important effect of ice in the aggregate, that hot water is an important source of heat, and that it is not easy to make warm concrete with frozen moisture in the aggregate. Computations such as this are accurate only if temperatures, specific heats, and material weights are accurate. For this example of typical ready-mix concrete:

An error of	Changes concrete temp., °F, by
1°F in water temp.	0.17
1°F in sand temp.	0.30
1°F in coarse-aggregate temp.	0.38
1°F in cement temp.	0.11
1% in sand moisture	1.83
1% in coarse-aggregate moisture	2.66
1% in water weight	0.21
1% in sand weight	0.17
1% in coarse-aggregate weight	0.09
1% in cement weight	0.05
0.01 in specific heat of sand	0.23
0.01 in specific heat of coarse aggregate	0.29
0.01 in specific heat of cement	0.22

Two quarts of standard calcium chloride solution at 40°F would lower the concrete temperature by about 0.5°F. These changes apply only to this set of computations but indicate the magnitude of changes. When the temperature of any material is near the final concrete temperature, the effect of an error in specific heat becomes less important.

HEATING CONCRETE

20.4 Heating Water

Figure 20.3 shows mix-water temperatures required to produce 60°F concrete with aggregate at various temperatures and moisture contents for a typical commercial mix. Moisture percentage for all aggregate is usually near half the moisture percentage for the sand. Unless sand is quite dry, hot water alone will not produce 60°F concrete from frozen aggregate. It is customary to use heated water if air temperatures are below 45°F.

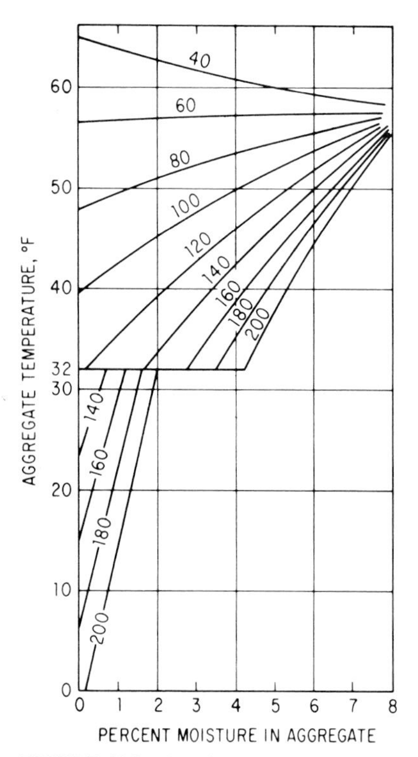

FIGURE 20.3 Required temperature of added water to produce concrete with a temperature of 60°F. Batch weights (per cubic yard) are cement 516 lb at 80°F, aggregate (SSD) 3225 lb at temperatures shown, and water 270 lb total.

Flash set results when cement contacts water that is too hot. This may occur at relatively low temperatures if cement and water are combined directly, and can be avoided by mixing aggregate and water together for about 15 to 20 s before adding cement if either the water or aggregate is hotter than 100°F. This increases mixing time, both because of the above delay and because the cement and aggregate are not blended before reaching the mixer. Water as hot as 150°F is mixed with blended cement and cold aggregate in some plants, and water as hot as 200°F can be used by mixing the hot water with cold aggregate for about 30 s before adding cement.

Heat required depends on temperatures and plant capacity. Commercial concrete is usually delivered at 55 to 75°F when air temperatures are near freezing, and 60 to 80°F when temperatures are substantially below freezing. Mass concrete is usually produced at 40 to 60°F. A plant heating 60°F water to 180°F, using 200 lb/yd^3/h needs $100 \times 200(180 - 60) = 2{,}400{,}000$ Btu/h. Allowing 15 percent for losses in piping, this requires $2{,}400{,}000 \times 1.15 \div 33{,}472 = 82.5$ bhp minimum, or about $2{,}400{,}000 \times 1.15 \div 1015.6 = 2718$ lb/h of steam at 15 psig (pounds per square inch gauge) at sea level if the steam temperature is reduced to 180°F in heating the water and steam quality is 100 percent. The value 1015.6 is obtained from pressure tables for saturated (100 percent quality) steam, and is the difference in total heat of 1 lb of steam at 29.7 psia and the total heat of water at 180°F. If steam quality is 90 percent, 2,996 lb of steam is required per hour.

Sparger nozzles in a water reservoir will heat water rapidly by dispersing steam in the water, where it condenses and remains. A 2 in pipe nozzle with 32 holes ¼ in in diameter, with deflectors to mix steam and water, will heat 24 g/min through an 80°F rise at 5 psig steam pressure at the nozzle and will heat 36 g/min at 15 psig. A steam valve controlled by a thermostat regulates water temperature. If boiler feedwater does not require treatment this is the most economical method of heating large quantities of water rapidly, but when feedwater treatment is required to prevent boiler scale or sludge, the large quantities of steam used make the water treatment too expensive. About 1 lb of steam is used for each 8 lb of batch water at 180°F. Heating with live steam cannot be done in a batcher because water weight is increased by the added steam. The reservoir where the water is heated should have a vent and overflow pipe.

Steam coils in a water reservoir heat the water without loss of boiler water because the condensed steam is returned to the boiler. Feedwater treatment is required to make up losses due to leaks only. The required coil surface area in square feet is given by

$$\frac{W}{U} \ln \frac{t_s - t_1}{t_s - t_2} \tag{20.4}$$

where W is pounds of water heated per hour, t_s is steam temperature at the pressure supplied, t_1 is initial water temperature, t_2 is final water temperature, all in degrees Fahrenheit, and U is the coefficient of heat transmission in Btu per square foot of coil surface area per hour. From condensing steam to stationary water in a tank, U is usually about 125 for thin tubes and about 100 for pipe, perhaps as low as 50 for old steel pipe. U values for hot water to water would be about half those for steam to water. In the example above for 100 yd^3/h production and 200 lb/yd^3 added water heated from 60 to 180°F using 15-psig steam, the pipe coil outside surface area required would be about

$$\frac{20,000 \times 1}{100} \ln \frac{249.75 - 60}{249.75 - 180} = 200\text{ft}^2$$

and that is about 266 ft of 2½-in pipe. Smaller pipe would cause excessive pressure loss. Pressure losses can be reduced by using a system of coils in parallel.

In-line water heaters have steam coils built into an enlarged section of water pipe. They are installed on the inlet to a water reservoir and heat the water as it flows through the pipe automatically with thermostatic control. Heat transfer is improved because both water and steam are moving, so much less heating surface is required than with tank coils. Some auxiliary heat in the reservoir may be required for startup and to make up reservoir heat losses during erratic production.

Package water heaters with water coils, either light-oil- or gas-fired, with automatic startup and thermostatic control are convenient for heating water where little other heating is required. They start fast, recover rapidly, and are lower in cost and occupy less space than boilers. Efficiency is about 70 percent of fuel input. Water treatment is less essential than with boilers or steam generators because fewer minerals are precipitated at normal heating temperatures. Typical units are in the 175,000 to 1,000,000 Btu/h heating range [5 to 30 (brake horsepower)].

Steam generators of the flash type produce steam in a few minutes from a cold start. They are usually gas-fired. They are compact and low in cost but of limited capacity and durability. Feedwater treatment consisting of at least softening is needed in most areas. Typical capacities are 5 to 30 bhp.

Package or scotch boilers are multiple-pass fire-tube units, automatic in operation and typically available from 50 to 300 bhp with insulation, burners, and controls installed. Larger units from 400 to 700 bhp are also available. Usual fuels are natural gas regulated to 1000 Btu/ft^3 or fuel oils in grades 1 to 3 (light, medium, and heavy domestic) or grades 4 to 6 (light, medium, and heavy industrial). Grades 5 and 6 are also called "Bunker B" and "Bunker C." Typical caloric value is 18,500 to 20,000 Btu per lb (140,000 to 150,000 Btu per U.S. gal). Industrial-grade oils usually require heating before use. Coal, now seldom used in concrete plants, has caloric values of 7,000 to 14,000 Btu/lb. Boiler fuel-conversion efficiency is about 60 to 75 percent for stoker-fired coal, 70 to 80 percent for oil or gas. These efficiencies are for steady load, and fuel consumption is higher in practice because of warmup and operation at part load. Large amounts of fresh air are required for combustion, about 10 ft^3 per cubic foot of gas, for example. While package boilers can be used to heat water or oil, they normally make steam. Whether high- (above 15 psig) or low-pressure boilers are used depends somewhat on local regulations, amount of pipe, and steam quality required. Operators are frequently required for boilers operating over a fixed pressure or capacity. Some steam systems operate with more efficiency slightly above or below atmospheric pressure by using vacuum pumps on condensate return lines. Feed-water treatment is nearly always required to protect the investment in and efficiency of a good boiler.

Fixed boilers are high-capacity built-in units most suitable for permanent plants where large quantities of steam are required. Other comments on package boilers apply to these as well.

Storage tanks are useful in reducing the size of heating equipment required by allowing it to operate over longer periods. Tanks of 1000 to 1500 gal capacity are used in smaller plants, tanks up to 20,000 gal in large plants. The tanks and piping should be insulated. Large tanks are sometimes buried 3 ft or more below ground, which provides effective insulation, and heated at night when heating equipment is operating at less than capacity. With small tanks temperature of mix water is regulated by the heater thermostat. Large tanks are better kept at a temperature as high as the maximum used and the batch water temperature regulated by a thermostatic mixing valve, which blends hot and cold water to produce the set temperature. Heating of a closed tank (kept full of water under pressure) is by steam coils or by circulating the water through a water heater with a pump typically 30 to 50 gpm at 80 to 120 ft head (Fig. 20.4). The pump should move about the quantity that can be heated through a 40°F rise at the pressure needed to overcome friction in the heater and piping. Buried tanks must be kept nearly full or well anchored to prevent their "floating" out of the ground. Small open tanks (not under pressure) are usually located above the batchers and filled through a water heater, by circulating water through a heater, or with steam coils.

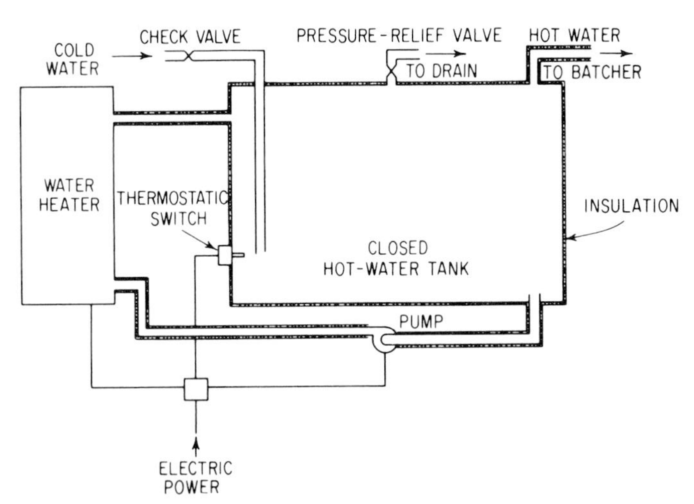

FIGURE 20.4 Connections for a closed-tank, circulating hot-water system.

Storage-tank size is determined by demand and heating capacity. For example, a ready-mix plant makes the following amounts of concrete for 10 successive hours of a peak 500 yd^3 winter day: 94, 70, 65.5, 46.5, 59, 44.5, 46, 49.5, 20, 5. Water in the tank is heated to 175°F at the start of the day and is to be kept above 165°F minimum. The water heater or steam valve will cut in at 174°F and cut out at 176°F. An average cubic yard uses 230 lb of added water, and supply water is at 60°F. A heater that will provide something over half the load for the peak hour is desirable: ½ × 94 × 230(170 − 60) = 1,189,100 Btu/h heating required. Say 1,500,000 Btu/h heater capacity is provided. This will make 165°F water faster than it is used if production is less than

$$\frac{1,5000,000}{230(165-60)} = 62.1 \text{ yd}^3/\text{h}$$

so for a 3 h period the demand will exceed the supply. During this period approximately (94 + 70 + 65.5)(230)(170 − 60) = 5,806,350 Btu will be used and 1,500,000× 3 = 4,500,000 Btu heated; so 1,306,350 Btu will have to come from storage without reducing tank temperature by more than 10°F. Thus a minimum of 1,306,350 ÷ 10 = 130,635 lb of water or 16,100 U.S. gal storage will be needed (water weighs 8.13 lb/gal at 170°F).

By limiting the amount of water fed into the reservoir to

$$1{,}500{,}000/(170 - 60) = 13{,}636 \text{ lb per h}$$

(good for 59.3 yd^3/h) and using the stored hot water when production exceeds 59.3 yd^3/h, the minimum reservoir size for the above application (using a 1,500,000-Btu/h heater) is

$$(94 + 70 + 65.5)(230) - (3 \times 13{,}636) = 11{,}877 \text{ lb}$$

or 1461 U.S. gal storage needed.

If the heating unit can be operated 24 h per day another arrangement which could possibly be the most economical would be to size the heater for the total heating required over 24 h, 500 × 230 × (170 − 60) ÷ 24 = 527,083 Btu/h, say a 600,000-Btu/h heater, and limit the water rate into the reservoir to 600,000/(170 − 60) = 5455 lb/h (good for 23.7 yd^3/h). The minimum storage required is then

$$(94 + 70 + 65.5 + 46.5 + 44.5 + 59 + 46 + 49.5)(230) - (8 \times 5445) = 65{,}610 \text{ lb or } 8070 \text{ U.S. gal}$$

In the first example given above, it is assumed that the storage tank will be kept full of water at all times at a temperature between 165 and 175°F. This would ensure availability of more than a 10-h supply of hot water if the heating system had to be shut down for any reason. In the second example the storage tank is full at the beginning of the day but would be almost empty at the end of the first 3 h. After the third hour water would be entering the tank faster than it is used and at the end of the day the tank would be full. In the third example the storage tank is full at the beginning of the day but would be almost empty at the end of 8 h. At the end of the day there would be 5160 lb or about 63.5 U.S. gal in the tank. About 11 h would be required to finish filling the tank for the next day.

Heat stored by a steel tank is negligible. If a 20,000 gal tank 12 ft in diameter by 24 ft long is used, it has a surface of 1131 ft^2. Two inches of polystyrene-foam insulation with a k value of 0.32 (see Table 20.4; k is increased by 0.05 because of higher temperature) and a 30°F air temperature will have a heat loss of about

$$0.32 \times 1{,}131 \frac{175-30}{2} = 26{,}240 \text{ Btu/h}$$

TABLE 20.4 Thermal Conductivity k* of Various Materials

Material	Approximate k, Btu/(h)(ft^2) (°F/in) thickness
Urethane foam	0.15
Polystyrene foam	0.27
Commercial insulating blanket materials of vegetable fiber, hair, glass fiber, or rock fiber	0.28
Corkboard	0.30
Board of pressed corn, cane, or wood fiber	0.33
Glass foam	0.37
Sawdust, wood shavings	0.40
Wood across grain:	
White pine, fir, redwood	0.80
Cypress	0.90
Yellow pine	1.0
Oak	1.1
Air space greater than ½ in	1.2
Sand or gravel, dry	2.3
Sand or gravel, moist	5–15
Soil, wet	5–13
Concrete, regular, dry	6–10

*For most insulating materials k increases with density and also increases about 0.05 per 100°F. Most insulating materials become less effective as temperatures increase. Values given are for about 70°F and for typically low densities for insulating materials

and 20,000 gal × 8.13 × 1 = 163,000 Btu/°F, so tank temperature will drop about 26,240 ÷ 163,000 = 0.16°F/h if no water is used. If thick insulation were used, the mean area of the insulation would be used in the above computation in place of the tank surface area.

Estimation of heat loss from buried water tanks is complicated. Burial under 3 ft of earth is roughly equivalent to 1½ to 3 in of insulation, under 8 ft of earth equivalent to 2 to 4 in of insulation, for 5000 to 30,000 gal tanks. The higher equivalent insulation thicknesses apply to larger tanks with the smaller diameter range for each size.

20.5 Heating Aggregate

Heating of frozen aggregate to a temperature above 32°F is usually required to make warm concrete. Aggregate should not be heated above about 100°F because of the danger of flash set, and if all aggregate were heated to this temperature, the concrete would be too hot. If aggregate can be kept above 32°F in storage, additional heating can be kept to a minimum. Heat required to thaw and warm frozen aggregate is given by

$$\text{Btu} = A[0.20(t_2 - t_1) + p(128 - 0.5t_1 + t_2)] \tag{20.5}$$

where A is the weight of saturated surface-dry aggregate, t_1 and t_2 are initial and final temperatures in degrees Fahrenheit, and p is surface moisture expressed as a decimal. If the aggregate weight available is total weight including moisture, substitute $A/(1 + p)$ for A. Heating 3200 lb of aggregate plus 3 percent average surface moisture at 30 to 60°F for a typical cubic yard of concrete requires about 35,800 Btu or about 36 lb of steam. Total boiler capacity for aggregate and water heating combined is in the range of 1 to 2½ bhp/yd^3 per hour of concrete depending on severity of winter temperatures. Heating aggregate in covered stockpiles may be necessary to permit handling in cold weather, stabilization of moisture content, or rescreening of the aggregate before it is placed in the bin. Stockpile heating also permits round-the-clock heating for more efficient use of heating capacity and may be essential where small bins are used in high-production plants because of the short retention time in the bin.

Live-steam jets at pressures of 10 to 25 psi in storage bins will heat aggregate rapidly because steam disperses through the material and condenses on the aggregate particles, giving up the heat of vaporization. Time required for heating is about five times that given by Eq. (20.1). The condensed steam usually raises the moisture content of the aggregate to saturation and water drains from the material. Total water in concrete is in the range of 6 to 12 percent of the aggregate weight. When moisture reaches this level, no added batch water is needed and aggregate must then be heated to an average temperature near the final concrete temperature without having the water in the aggregate exceed the requirement for total water. If live-steam heating is continued long enough, temperature of aggregate near the nozzles will approach 212°F, while aggregates farther away are cooler, resulting in uneven heating. Boiler feedwater treatment may be expensive because of the large amount of makeup water required. Despite these objections, steam jets are a useful method, but control of the heating process and control of water in the mix become difficult. Two or more ½ in pipe fittings with ½ in globe valves on each bin compartment with a 2 in steam header are usually adequate. Drain channels can be built into a bin to remove much of the water drainage before it reaches the fill gates, but some water will nearly always drip from the gates. Live-steam heating of storage piles where some remixing and drainage take place before material is delivered to an overhead bin is more easily managed.

Steam, hot-water, or hot-oil coils in the aggregate heat the material slowly because sand and gravel are relatively poor heat conductors. This method tends to dry the aggregate some and stabilize the moisture content and so has few of the objectionable features of live-steam heating. Also the aggregate must be heated only sufficiently to melt any ice because generous quantities of hot batch water can be used. Where usage is moderate and concrete mixes dry, as in products plants, coil heating may be the only practical method. Coils should be located near the center of a bin or pile to minimize heat loss and near the main material flow. When used in stockpiles to prevent freezing slowness of heating is not so objectionable, but frozen lumps may still be drawn down through the piles without thawing unless coils are located to intercept them. The amount of coil surface area in square feet is given by

$$\frac{0.20A}{U} \ln \frac{t_s - t_1}{t_s - t_2} \tag{20.6}$$

where A is pounds of aggregate heated per hour, t_s is steam, (or water or oil) temperature, t_1 is initial aggregate temperature, and t_2 is final aggregate temperature, all in degrees Fahrenheit; U is heat transmitted in Btu per square foot of coil surface area per hour, and is probably in the range of 2 to 10. Oil heated in a boiler similar to a steam boiler or heater similar to a water heater has an advantage in that corrosion, need for water treatment, and danger of coil freezing are eliminated. The specific heat of oil is about half that of water and viscosity is higher than water, requiring more pump capacity or larger pipe. Temperatures up to 600°F are possible, but temperatures over 212°F will have the undesirable effect of removing absorbed water from the aggregate. A pump is required for circulation of oil or hot water. The possibility of oil leakage into the aggregate is the main objection to the use of oil.

Electric resistance-heating elements attached to a bin bottom and covered by insulation can effectively prevent freezing in the lower part of the bin and are easily controlled by a thermostat. Each 1000 W of heating element supplies 3413 Btu/h, of which about 90 percent may reach the aggregate.

Radiant heaters or space heaters below a bin bottom provide an effective means of keeping aggregate temperature above freezing but usually do not produce enough heat to melt ice at any appreciable distance from the bin bottom. If a bin is enclosed and the enclosure is spaced out from the sides of the bin, warm air from the batcher space below the bin will envelop the bin, provide some heat, and effectively prevent freezing if aggregate can be delivered from a stockpile at temperatures above freezing.

Inundation heating by immersing aggregates in hot water is seldom used. Dewatering of the aggregate is difficult.

Vacuum heating wherein air is partially exhausted from a tank of aggregate and replaced by steam is very effective but too expensive to be used unless the same equipment is provided for cooling. If a tank containing 400,000 lb of aggregate at 20°F and 1 percent surface moisture is reduced to a pressure of 5 psia and the air removed replaced by 2900 lb of steam at 144 psia, the steam condenses on the aggregate and raises the temperature to about 50°F. Air is then introduced until the tank returns to atmospheric pressure. Moisture is increased from 1 to about 1.8 percent, and some of the increase in moisture can be removed by a barometric drain or pump at the bottom of the tank. Heating is fairly uniform because steam is drawn into the mass.

Warm-air heating is feasible for coarse aggregate but quite expensive and seldom used.

COOLING CONCRETE

Commercial concrete usually can be produced at reasonable temperatures in hot weather without use of refrigeration by using the coldest water available (from wells, for example), by shading and sprinkling aggregate piles, and by painting storage, handling, and mixing equipment white if it is exposed to the sun. While most common surfaces absorb solar radiation nearly as well as they radiate heat at lower temperatures, and some metals absorb solar heat more efficiently than they radiate, white paint radiates heat at 100°F several times as efficiently as it absorbs heat from the sun, and white-painted surfaces will remain cooler than other surfaces in sunlight.

When concrete must be cooler than about 80°F in hot weather, refrigeration is usually required. Cooling by refrigeration is more expensive than heating and requires more expensive equipment, so equipment must be accurately sized. It is even less economical to provide cooling for the maximum possible temperatures than to provide heating for the minimum possible temperatures. Some calculated risk must be taken after study of reported climatologic data for the area and the cost of providing excess cooling capacity. Even then it must be assumed that combinations of climatic conditions can occur that will make operation at the design temperature impossible.

The following heat balance for a typical 6 in maximum exterior dam will be used for illustration, but actual mix designs should be used when available. Material temperatures and moisture content of aggregates will also vary depending on the method of processing and cooling and the local climate. Concrete has a maximum placing temperature of 50°F in this example. Allowing for 2°F temperature in transporting to the forms, concrete must come out of the mixers at 48°F.

Cooling required per 100 yd^3/h of concrete is $100 \times 43{,}743/12{,}000 = 365$ tons of refrigeration. About 15 percent of this can be accomplished by sprinkling coarse aggregate; so a minimum of 310 tons of mechanical refrigeration will be required per 100 yd^3/h.

Material	Lb/yd³	Specific heat, Btu/°F	Btu/yd³/°F	Initial temp.	Btu/yd³ above 48°F
Cement	310	0.22	68.2	120	4,910
Pozzolan	58	0.22	12.8	120	922
Sand	780	0.20	156.0	90	6,552
Water in sand 8% maximum	62.4	1.00	62.4	90	2,621
Aggregate:					
3–6 in	964	0.20	192.8	90	8,097
1½–3 in	800	0.20	160.0	90	6,720
¾–1½ in	584	0.20	116.8	90	4,906
No. 4 to ¾ in	506	0.20	101.2	90	4,250
Water in aggregate, ½% ¾–6	11.7	1.00	11.7	90	491
2% No. 4 to ¾	10.1	1.00	10.1	90	424
Mix water (180 − 62.4 − 21.8)	95.8	1.00	95.8	70	2,108
Concrete weight	4,182		)987.8		42,001
Specific heat of concrete			0.236		
Heat generated in mixing 0.70 × 10 hp · min/yd³ × 42.4 Btu/hp · min) × 2.5 = (assumes 70% of rated motor power goes into concrete)					742
Heat gain in screening and handling materials					1,000
					987.8)43,743
					44.3
					48.0
Concrete temperature out of mixer without cooling					92.3°F

The effect of errors in estimating temperature and specific heat of coarse aggregate is most important. For this typical mass-concrete mix:

An error of	Changes the cooling required, %, by
1 min in mixing time	0.68
1°F in water temp.	0.22
1°F in sand temp.	0.50
1°F in coarse-aggregate temp.	1.35
1°F in cement temp.	0.16
1% in sand moisture	0.55
1% in No. 4 to 3/4 moisture	0.38
1% in water weight	0.05
1% in sand weight	0.21
1% in weight of one of coarse aggregate	0.11–0.19
1% in cement weight	0.11
0.01 in specific heat of sand	0.75
0.01 in specific heat of coarse aggregate	2.74
0.01 in specific heat of cement	0.51

At least 43,700 Btu/yd³ must be removed by cooling to make 48°F concrete. There are several ways of removing heat; in approximate order of increasing cost:

	Approximate practical concrete temp. reduction, °F
1. Sprinkle coarse-aggregate stockpiles with water (10°F reduction)	6
2. Chill mix water to 34 or 35°F	3
3. Substitute ice for 80% of chilled mix water	12
4. Cool coarse aggregate by vacuum to 35 or 38°F	31
5. Cool 3/4- to 6-in coarse aggregate by air to 40°F	25
6. Cool coarse aggregate by inundation to 40°F	30
7. Cool sand to 34–80°F	2–12
8. Cool cement to 80°F	3

Sand can be cooled to about 50–70°F by contact coolers. The higher-temperature reduction is for vacuum cooling of sand in bulk. Ice is limited to about 80 percent of the added mix water because some water must be used to dilute admixtures.

Some combination of these methods adding up to more than 92.3 − 48 = 44.3°F will be required, and the combinations cannot include various methods of cooling the same materials. For example, vacuum cooling of coarse aggregate and sand plus chilled mix water gives 31 + 12 + 3 = 46. Cooling coarse aggregate by inundation, chilled mix water, and substitution of ice for 80% of mix water gives 30 + 3 + 12 = 45. As a practical matter in this example coarse aggregate must be cooled to make 48°F concrete.

20.6 Plant Startup

The time of plant startup is important in making cool concrete regardless of the method of cooling used. It is best to schedule the start of pours for early-morning hours when air and material temperatures are lowest. After starting with proper concrete temperatures, the plant will take higher ambient temperatures up to the design temperature in stride. If a plant must be started late in a hot day, all materials must be as cold as possible and equipment in contact with the materials or concrete such as batchers, mixers, hoppers, and buckets must be washed thoroughly with cold water.

20.7 Sprinkling Coarse Aggregate

Sprinkling coarse aggregate with clean water will, under most conditions, perform a substantial part of the coarse-aggregate cooling, reducing refrigeration requirements. Sprinkling may be intermittent on a short time cycle but must be sufficient to keep the aggregate continuously wet, which requires more than enough water to replace moisture evaporated. Where the active zone (Fig. 20.5) of a large-aggregate storage pile holds more than is used in 1 day, the temperature of the pile can be approximated by balancing the heat gained and lost from material additions, radiation, convection, and evaporation. With all temperatures above 32°F, a formula for approximate pile temperature t in degrees Fahrenheit is

$$t = \frac{t_x A(s+p) + t'_x A'(s+p') + t_w W + t'_w W' + R' + R + t_a C + t'_a C' - HE - H'E'}{A(s+p) + A'(s+p') + W + W' + C + C'} \qquad (20.7)$$

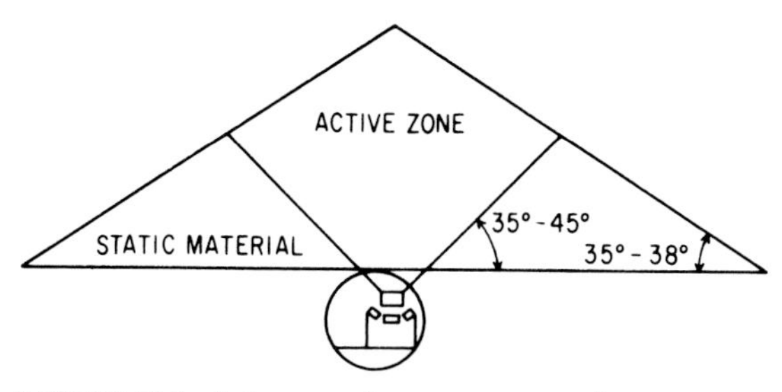

FIGURE 20.5 Active zone of aggregate storage pile.

where t_x = daytime temperature of aggregate added to pile
t_w = daytime temperature of water sprinkled on pile
t_a = daytime temperature of air
A = lb aggregate added during day
W = lb water sprinkled during day
R = Btu radiation gain (or loss) during day
C = convective-heat gain (or loss) during day, Btu/°F
s = specific heat of aggregate
p = percentage of moisture in aggregate as received at the pile during the day, expressed as a decimal
H = heat of vaporization of water at temperature t_w, Btu/lb
E = lb water evaporated during the day

Prime symbols are the same except that they are for night conditions. In some applications it is necessary to consider day and night conditions separately as above, but where $t_x = t_x'$, $A = A'$, $p = p'$, $t_w = t_w'$, $W = W'$, and where operation is continuous with day and night of equal length, this formula reduces to

$$t = \frac{2t_x A(s + p) + 2t_w W + R + R' + t_a C + t_a' C' - HE - H'E'}{2A(s + p) + 2W + C + C'} \tag{20.8}$$

with A, W, E, and E' in pounds per hour, R and R' in Btu/hour, and C and C' in Btu per hour per degree Fahrenheit.

The difficulty in applying the above formulas is in evaluating evaporation, radiation, and convection, as discussed below.

Evaporation occurs when the dew-point temperature of water vapor in the air is less than the water temperature. Approximate evaporation for large areas, adapted from the formula developed be R. E. Horton (*Engineering News-Record,* Apr. 26, 1917), is

$$E_i = 0.0167\,(WV_w - V_a) \tag{20.9}$$

where E_i is evaporation in inches per hour, V_w is vapor pressure in inches of mercury at the water temperature t_w (Table 20.3), V_a is vapor pressure in inches of mercury at the air temperature t_a, and W is a factor from Table 20.5 to correct for wind velocity and convective vapor removal at low wind velocities. Actual evaporation over small areas can reach a value of $E_i = 0.0167\,(WV_w - hV_a)$, where h is relative humidity, when the vapor is removed rapidly as on the windward side of a pile; otherwise the accumulation of vapor near the surface increases h to 1.00. Evaporation always occurs if $WV_w > V_a$. When $V_a >$

TABLE 20.5 Evaporation from Larger Areas: Values of Factor W

Wind velocity,* mi/h	0	1	2	3	5	10	15	≥30
$t_w \leq t_a$	1.00	1.18	1.34	1.45	1.64	1.86	1.95	2.00
$t_w - t_a = 1$	1.18	1.34	1.45	1.64	1.64	1.86	1.95	2.00
$t_w - t_a = 5$	1.38	1.49	1.58	1.64	1.64	1.86	1.95	2.00
$t_w - t_a = 10$	1.48	1.57	1.64	1.64	1.64	1.86	1.95	2.00
$t_w - t_a = 20$	1.64	1.64	1.64	1.64	1.64	1.86	1.95	2.00

*Where t_w = temperature of water; t_a = temperature of air.

$WV_w > hV_a$, some evaporation occurs on windward areas but cannot be relied on for cooling aggregate piles. When $hV_a > V_w$, condensation occurs in still air. Evaporation with a 10-mph wind, 70% relative humidity, 80°F air temperature, and 70°F sprinkler water is

$$E_i = 0.0167[(1.86 \times 0.732) - (0.70 \times 1.022)] = 0.11 \text{ in/h}$$

maximum over small areas, and a reasonable figure for large piles is

$$E_i = 0.0167[(1.86 \times 0.732) - 1.022] = 0.0057 \text{ in/h}$$

With an active pile area of 30,000 ft^2, this is 890 lb/h of water; and at 1053 Btu/lb, 940,000 Btu/h is removed by evaporation. Evaporation increases rapidly at higher water temperatures or lower air temperatures.

Radiation of heat from the sun and sky generally increases aggregate temperatures during the day, and aggregate piles may lose heat by radiation at night. Radiation gain varies with season, time of day, altitude, latitude, and weather. Maximum radiation at the earth's surface at altitudes of up to a few thousand feet is about 250 to occasionally 300 Btu/(ft^2)(h) and varies very little with latitude. In midlatitudes there is little variation in the maximum on a normal surface summer and winter. Average daytime radiation is about 58 percent of the peak values. Average total radiation considering weather at mid-latitudes is about 750 Btu/(ft^2)(day) in winter and about 2000 Btu/(ft^2)(day) in summer. The average exposed normal area of a typical wedge-shaped aggregate pile three times as long as its width during a day is 67 to 71 percent of the horizontal area, depending on pile orientation. Resultant radiation at the surface of the pile is about 2000 Btu/(ft^2)(day) or about 145 Btu/(ft^2)(h) average during a summer day.

The ability of a material to absorb (or emit) thermal radiation is its *emissivity* B expressed as a percent of the radiation absorbed by a perfect absorber; B values for aggregates with an effective increase because of roughness vary from about 0.40 to 0.90. Assuming a typical effective B of 0.70, about 100 Btu/h is absorbed by the aggregate per square foot of horizontal pile area.

Heat from the pile is lost by radiation, mostly to the sky but some to other surroundings, if their temperature is lower than that of the aggregate-pile surface. Net heat loss is on the order of $17.3 \times 10^{-10} B(t_1^4 - t_2^4)$, where B is emissivity of the aggregate and t_1 and t_2 are absolute Fahrenheit temperatures of the aggregate surface and of the surroundings, respectively ($t = 459 + °F$). The daytime effective sky temperature is a few degrees above air temperature. A cloudy night sky has an effective temperature a degree or two lower than air temperature; a clear night sky may have an effective temperature as low as −50°F. Heat loss on a cloudy night with aggregate temperature of 90° and air temperature of 70°F is about $17.3 \times 10^{-10} \times 0.70(549^4 - 529^4) = 15$ Btu/(ft^2)(h). This is usually neglected.

Convection heat gain or loss per degree Fahrenheit difference between aggregate and air temperature is about as follows:

Air velocity, mi/h	2	5	10	15	20	30	40
Btu/(ft^2)(h)(°F).	1.7	2.7	4.3	5.8	7.3	10.0	12.5

While this applies to the surface area of the pile, horizontal area can be used to correct approximately for reduced velocity at the material surface.

Figure 20.6 shows computed aggregate-pile temperatures assuming 913,000 lb per hour of aggregate is added day and night at 90°F to a pile 100 by 300 ft, with air temperatures of 80°F day and 60°F night and wind velocities of 5 mi/h day and 2 mi/h night. The results indicated for sprinkling with high-temperature water are not attainable in practice because the cooling of water drops and films by evaporation reduces their temperature and thus reduces evaporation. The opposite is true if cold water is used because aggregate and air warm the water. Thus aggregate temperatures shown for sprinkling with small quantities of water would be lower than shown for cold water and higher than shown for warm water, with the temperatures more nearly correct as the quantity sprinkled increases. As a practical matter, ⅛ to ½ in/h must be sprinkled to keep piles wet, and it is better to use the smallest quantity that will keep the pile wet if the water is warm. If water is cool, larger quantities are beneficial. The spray must be coarse because water evaporating from a fine spray will increase the relative humidity near the pile and decrease the water temperature, both of which decrease evaporation. Sprinkler heads of the type used on golf

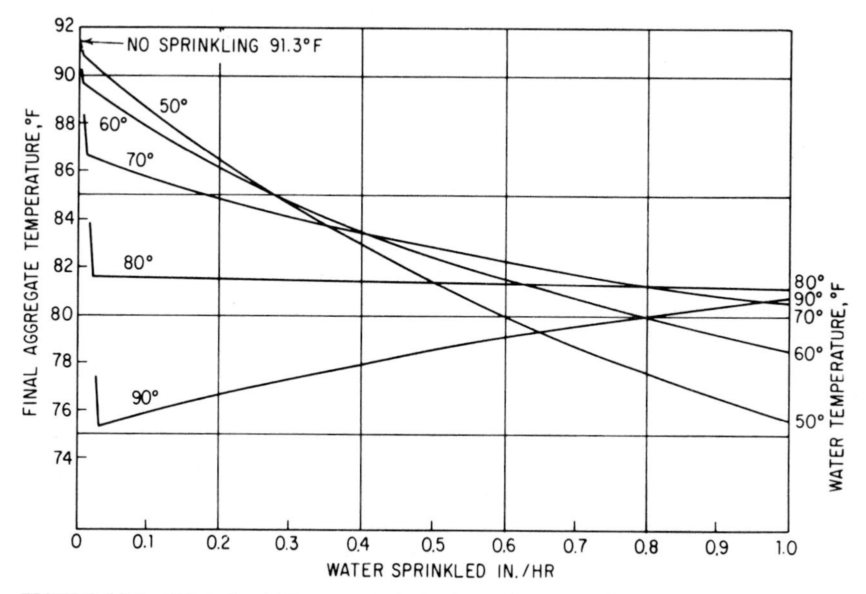

FIGURE 20.6 Effect of sprinkling water on temperature of aggregate pile.

courses can put out as much as 150 g/min at 100 psi on a 250 ft diameter, with quantities up to ½ in/h. Above 3 mi/h wind velocity, the diameter decreases by about 2 percent for each added mile per hour of wind velocity. Drains are needed to remove excess water. Sand and some fine aggregates cannot be cooled in practice by sprinkling because they accumulate too much water and are difficult to dewater.

20.8 Cooling Water

Chilled water is useful mostly as an adjunct to other cooling or for cooling other materials. Water can be chilled economically by vacuum or by vapor-compression refrigeration units and heat exchangers. Most heat-balance computations assume near maximum moisture in aggregates, resulting in minimum batch-water requirements. Excess chilled batch-water capacity is needed for times when aggregate moisture content is low.

Vacuum water chillers used in concrete plants usually have two or three cooling chambers. Water is partly chilled in the first chamber, then goes to the second and then to the third chamber, in each of which it is further chilled. From the last chamber it is pumped through insulated pipe to an insulated batcher reservoir or other point of use. Overflow or return water is fed to the intake of the chiller to reduce the cooling load. Water level is automatically controlled by a float valve and temperature automatically controlled by sensors in the unit. Safe temperature is about 34°F with temperatures of 32.5 to 33°F possible with careful adjustment and a constant rate of water use. About 0.2 connected electric horsepower and 0.75 bhp are required per ton of refrigeration, including condensed-water pumps and cooling-tower fans.

Vapor-compression water chillers are available as packaged units in capacities from 12,000 to 3,000,000 Btu/h (1 to 250 tons) with about 1.0 to 1.2 electric motor horsepower per ton of refrigeration. When a water-cooled condenser is used it is usually built into the unit, but air-cooled or evaporative condensers are often separate units. Where the chilled water is used in heat exchangers about 2 percent makeup water is required and the unit remains reasonably clean. Where water is used as mix water or for inundation of aggregates, makeup or return water may require treatment to prevent fouling the chiller. Automatic control for varying rates of water use is by compressor unloading or starting and stopping multiple compressors depending on water temperature or suction pressure. Temperature of chilled water leaving the unit cannot be less than about 35°F without danger of freezing the water. Condenser water is required at 3 to 5 g/min per ton of refrigeration for water-cooled units. Chillers will not function properly if condenser water or air temperature (for air-cooled units) is too low.

The use of liquid nitrogen (LIN) to cool water is becoming more prevalent due to economical improvements (Fig. 20.7). LIN has been used to turn the water to slush by cooling to 33°F (0.5°C) or using the cooled water to cool the aggregates. As the systems are new, further information on their use and economics in each case should be obtained from a manufacturer.

20.9 Cooling with Ice

Ice is very effective in cooling concrete, but the quantity that can be used is limited. Ice can be delivered to the plant in blocks and ground up for use in batching. It is then elevated in bucket elevators to an insulated daily use tank or bin (commonly of wood) holding several tons of ice at 60 ft^3 per ton with ice depth not more than 8 or 9 ft, and moved by a screw conveyor to an ice weigh batcher similar to an aggregate batcher. In smaller plants ice may be added to the water batcher, but it tends to jam in conventional water batchers.

FIGURE 20.7 Low-profile mass concrete. Pictured are two low-profile plants with a combined capacity of over 1000 yd^3/h in paving the Houston airport. Liquid nitrogen was used to control concrete temperature. (*Courtesy of Rexwerks, Inc.*)

Ice does not flow readily in bins and hoppers. Ice usually enters the bin below 32°F, frequently about 10°F. If the insulated bin can be kept free of water and moist air, handling is easier. Some refrigeration of the bin is required for long-time storage. Ice melts slowly in contact with cold aggregates. Flake ice usually melts during about 2 min of mixing if only coarse aggregate is cold; more time is required if all other ingredients are cold or ice particles larger.

Automatic icemakers are available as packaged, electricity-driven units with capacities to 30 U.S. tons of ice per day. They usually make ice at 0 to 10°F to keep the ice from sticking together, but this subcooling is not relied upon when computing concrete temperatures because much of it is lost in handling. A battery of ice machines is commonly required for a high-production concrete plant. Machines that make flakes instead of cubes or chunks are used to reduce melting time in the mixers. Several methods are used but basically water is frozen on a stationary or revolving cylinder or belt and mechanically broken off into a storage bin as white ice. A few machines thaw ice loose from tubes or plates and crush it. Input water temperatures are 40 to 110°F, with more efficiency at the lower temperature. Vapor-compression refrigeration is used. The condenser may be air-cooled, but water cooling is used for larger units. Power required is about 2 to 3 hp per U.S. ton of ice per day, not including condenser-water pumps or ice-transporting equipment. About 5000 gal of condenser water at 70°F is required per ton of ice.

An advantage of ice is that it can be stored, and icemakers can operate nearly continuously. A shutdown for 2 h each day is allowed for maintenance.

20.10 Cooling Coarse Aggregate

Aggregate can be cooled at ground level before being placed in a bin over the batchers or after it is placed in the bin. When cooled on the ground it may be cooled in separate size groups and conveyed to the bin as sized material or blended and separated into size groups by screens on the bin. It may also be blended before cooling and separated into sizes as it is fed into bin compartments, and this method reduces the size of cooling, conveying, and rescreening equipment but requires more careful process control. When aggregates are cooled before conveying them to the bin they pick up about 2°F from radiation, conduction, convection, and condensation of atmospheric moisture as they are moved to the bin. Handling equipment should be rinsed with cold water before startup, shielded from wind, and painted white on surfaces exposed to the sun. Coarse aggregates can also be cooled after they are placed in separate size groups in the overhead bin structure. Aggregate temperatures at the batchers depend on the method of cooling used and where the cooling is done. In order of increasing cost for large installations, cooling may be by vacuum, cold air, or inundation in cold water. In India, aggregates have been cooled at rates of 240 U.S. tons per hour by 36 to 40°F water and 42°F air while moving slowly on belts in long tunnels. Exposure time is 20 to 25 min. Cooling by spraying with large quantities of cold water and by spraying while forcing cold air through the material is also feasible but is not currently used in large concrete plants.

Cold aggregate gains about 1°F/h in insulated storage bins at 80 to 90°F air temperature. Regardless of the cooling method used, the storage bin (other than a vacuum tank) should be enclosed on the sides and bottom and covered with about 2 in of commercial insulation or 1 in of urethane foam, which may be sprayed in place. The bin should be roofed. Cooling coils may be built into the bin bottom for cold-water circulation, to reduce heat gain primarily while the plant is shut down temporarily and chilled water is not in use elsewhere.

Vacuum cooling of coarse aggregates from No. 4 to the largest size used is done by placing a batch of moist aggregate in one of a group of vacuum tanks and removing the air and vapor to maintain a pressure corresponding to the desired temperature (Table 20.3). Typical tanks hold 150 to 300 tons. Three tanks are typical of ground installations, eight of an overhead installation. The amount of water removed as vapor is about 1 percent of the aggregate weight, and this can be replaced by sprays inside the tank. Even so, for satisfactory operation the aggregate should be sprinkled in storage piles so that it is saturated and wet on the surface before being placed in the vacuum tank to ensure even moisture distribution and to provide a reserve for recooling should it be needed. Dry aggregate may absorb 1 to 3 percent moisture. Dewatering is not a problem because aggregates can be drained through a pump or barometric drain for at least an hour while in the vacuum tank. When aggregates are held in a tank under vacuum or recooled after a shutdown, water vapor continues to be extracted. Once a tank is cool it can be sealed under vacuum and the temperature held for several hours because heat gain is quite slow.

While ducts are arranged to permit reasonably free egress of vapor, all parts of the tank are not at the same pressure during the cooling process and some margin must be left to prevent freezing of material near the duct entrance. Practical average temperature is about 34°F. When aggregate is accidently frozen, it thaws very slowly; so steam jets built into the tank are used to break it up.

The process sequence is complicated by filling and cooling tanks in rotation, and by interchange of vacuum between warm and cold tanks, but basically a sealed tank is *primed* for about 5 min to remove air and reduce the pressure to about 0.7 in Hg. Then a *booster* is connected until the pressure in the tank is reduced to about 0.19 in Hg. The pressure may be held at this level for a short time for added cooling effect. About 40 to 50 min is required to cool a tank, depending on the aggregate temperature. Cooling time (Fig. 20.1) is controlled more by the time required for the process than by the time required for heat flow from the aggregate. Sensitive vacuum gauges are used to determine pressure and thus temperature. After the desired pressure is reached the main duct is shut off and the tank is held under vacuum until the material is required. Vacuum is broken by an air-inlet valve or partially broken by connecting the cold tank to a warm tank rapidly, and then by an air inlet to bring the tank to atmospheric pressure before opening the vacuum seals and discharging material. About 0.18 connected electric horsepower and 0.75 bhp are required per ton of refrigeration, including power for condenser-water pumps and cooling-tower fans.

Cold-air cooling of sized coarse aggregates from ¾ in to the largest size used to about 40°F is usually done in overhead bins by circulating about 38°F air through them. Aggregates smaller than ¾ in have such high resistance to airflow that they are not cooled by this method economically. Cold air is forced by a fan through shielded openings in the bin bottom, travels up through the aggregate, and is reclaimed from the top of the bin and rechilled. Some cooling is accomplished by evaporation, especially initially in the smaller sizes. Vapor-compression refrigeration is used, with air passing over the evaporator to be chilled. Air temperatures below about 35°F at the evaporator may result in stoppages caused by icing of the evaporator. Package air-cooling units may be mounted on the bin or large units may be on the ground. Cooling time is about 2 h because of the poor heat transfer from air to aggregate, and bins should be sized to hold material for this time and should be kept well filled. About 1.9 electric horsepower is required per ton of refrigeration.

Inundation cooling of coarse aggregates from No. 4 to the largest size has been used on several projects. It is usually done on the ground in groups of 5 to 10 tanks, each of about 120 to 180 tons of aggregate capacity. Vapor-compression refrigeration has been used to supply chilled water, but vacuum water chilling could be used economically. A typical cycle starts with a tank about one-third full of 35°F water to which the aggregate is added during about 10 min. Water at 35°F is then pumped into the bottom of the tank, flows through the aggregate and over weirs near the top, and is piped to a reclaiming (surge) tank, whence it is rechilled and reused. The water has a tendency to accumulate fines that may coat aggregate particles, necessitating filtering or changing the water. After about 40 min, the tank is drained and the material is held in the tank to drain until it is needed. It may be conveyed over vibrating dewatering screens (sometimes with air suction or blowers) before delivery to the overhead bin. Cycle time is about 2 h for each tank. About 1.2 electric horsepower is required per ton of refrigeration.

Table 20.6 shows advantages and disadvantages of the various methods of aggregate cooling in use in North America in 1967.

20.11 Cooling Sand

Sand is cooled most efficiently by the *vacuum process*. Because it impedes vapor flow, sand must be exposed to the vacuum in thin layers or small increments to obtain uniformly low temperature. One method uses a two-compartment vacuum tank. Sand from the upper chamber is fed to the lower chamber in the sealed tank, is cooled quickly as it is exposed to the vacuum, and accumulates in the lower chamber. With delicate control sand can be frozen into granules like popcorn by this method, but temperatures of 33 or 34°F are more easily controlled. Another type of sand vacuum tank has a built-in bucket elevator to circulate sand from the bottom to the top. Most cooling is performed as sand enters the elevator buckets, in the buckets, and at the elevator discharge. Sand can be frozen into chunks small enough to pass through material gates, but again 33 to 34°F sand is more easily controlled. Freezing the moisture in sand can remove about an additional 7000 Btu/yd^3 of concrete. This is useful in emergencies but should not be relied on for routine operation. About 1½ percent sand moisture is removed when cooling sand from 90 to 34°F, 1 percent when cooling from 70 to 34°F.

Sand can also be cooled in *screw-conveyor coolers* with jackets and hollow flights through which 35 to 40°F water or 20 to 30°F brine circulates. Because the sand is han-

TABLE 20.6 Methods of Cooling Aggregates

Cooling method	Typical aggregate temp. at batchers, °F	Advantages	Disadvantages
Vacuum cooling of sized aggregate overhead	35	Low operating cost; all sizes including sand can be cooled; aggregates can be recooled after shutdown; high initial aggregate temperature requires relatively small added cooling time; easily used for heating; electric-power requirement is low	Moderately high equipment cost; about 1 h may be required for plant startup; boilers require 1 h additional; requires fuel supply for boilers; fine particles accumulate in water pumps
Vacuum cooling of blended aggregate on ground	38	Low equipment and operating cost; all sizes including sand can be cooled; high initial aggregate temperature requires relatively small added cooling time; easily used for heating; electric-power requirement is low	Aggregate left in bin over long shutdown must be removed; about 1 h may be required for startup; boilers require 1 h additional; requires fuel supply for boilers; fine particles accumulate in water pumps
Cold-air cooling of sized aggregate in overhead bin	40	Low equipment cost if extreme low temperatures not required; aggregate can be held at low temperature for long periods; continuous process; can be all-electric power; can be used for heating with addition of heat supply	Requires large overhead bin capacity; No. 4 to ¾ in cannot be cooled economically; sand cannot be cooled; cooling time is extended if initial aggregate temperature high; efficiency lowered by air leakage; dust accumulates in cooling system; power requirements high because of large fans
Inundation cooling of sized aggregate on ground	40	Water chillers available in package units; can be all-electric power; can be used for heating with addition of heat supply	Aggregate left in bin over long shutdown must be removed; high equipment cost; finer aggregate difficult to dewater; sand cannot be cooled; chilled water picks up fine material and may coat aggregate if not filtered or changed frequently

dled in thin layers while exposed to air, cooling below the dew point results in condensation on the sand, and 40 to 50°F is about the lower limit for efficient cooling with brine, 50°F with chilled water.

20.12 Cooling Cement

Cement and pozzolan are cooled when necessary by contact with metal jackets chilled by 35 to 40°F water. One type of cooler is like the screw-conveyor cooler described for sand. In another type thin layers of cement move upward in contact with the inside of a vertical metal cylinder with cold water running down the outside of the cylinder. Capacities to 265 bbl/h with a 100°F temperature reduction, or higher capacity with less reduction, are available. Cement cooling is relatively expensive because cement loses heat slowly. Cement should not be cooled below the dew point of air in contact with it, or it will pick up moisture from the air. This limits cement temperature to 60 to 70°F. About 20 hp and 25 g/min of 35°F water are required per 100 bbl/h cooled.

20.13 Practical Cooling Combinations

As a practical matter certain combinations of cooling methods show more economy than randomly selected methods. In many combinations the same equipment (such as boilers or water chillers) can be used for several materials. For example, if coarse aggregate, sand, and water are cooled by a vacuum installation, additional chilled water can be made economically for cooling cement if further cooling is required, and addition of icemakers and an ice batcher is not economical. If coarse aggregate is cooled by air, ice is usually required. If inundation cooling is used for coarse aggregate, either ice or chilled water and sand cooling are usually needed, and cement cooling may be required. In any event, the combination must remove enough heat units to reduce concrete temperature to the desired level, and specialists in concrete-material cooling should be consulted for economical design of other than the most simple systems.

CHAPTER 21
PLANT OPERATION

John K. Hunt*

For maximum economy consistent with good practice it is necessary that the complete operation of the plant from the arrival of the basic materials to the final delivery of the concrete be carefully planned and checked periodically. Preventive maintenance is essential to avoid very costly shutdown time. Properly operated and maintained equipment is necessary for personnel safety.

21.1 Materials Handling

Planning and scheduling material delivery to coincide with production needs is difficult but important. Many costly errors can be prevented by checking and carefully recording material deliveries.

21.2 Batching

To produce the desired quality of concrete operators should know as much as possible about what is happening throughout the plant. They should know any specifications they must meet in the batching operations and how to use the equipment at hand needed to accomplish this task. It is not necessary to know how or why modern controls work, only how to use them. They need to know what must be done to compensate for moisture changes, possible temperature changes, and slump and air-content changes. Operators also need to know how to produce the same-quality concrete using the manual back-up controls should there be a failure of the automatic controls.

21.3 Mixing

Regardless of the type of mixer used, the operators should have a good understanding of how to control the charging and blending sequence to produce the optimum results. If there is a central mixer with a slump meter, they should understand how to use it.

*Revised by Robert W. Strehlow.

21.4 Dispatching

In a ready-mix operation the job of order taking and truck dispatching ranks as one of the most important. Not only must dispatchers be skilled at matching the demand for concrete to ability of the equipment to produce and deliver this concrete, they must also be diplomats when things go wrong. They must be knowledgeable about concrete problems and have some understanding of mix designs. Two-way radio communication between dispatchers and truck drivers is a must, and the use of a computerized truck dispatch system is a modern, cost-effective way to deliver the most possible concrete at the lowest cost (Fig. 21.1). Delivery costs are a very significant portion of concrete cost.

21.5 Plant Maintenance and Repair

Manufacturer's recommended practices should be followed for cleaning, lubricating, and general maintenance of all equipment. A maintenance check sheet and repair record for each piece of equipment is very useful in determining what spare parts are necessary and in predicting and preparing for breakdown of equipment.

Scales should be inspected frequently to ensure that there are no rubbing parts or chipped knife edges. A regular scale inspection and testing procedure should be set up and carried out to ensure that any potential scale problems are found and remedied before they become expensive problems.

Minor adjustments on a scale system can be made by the operator, but any major overhaul or repair should be done by a specialist.

FIGURE 21.1 Computer-aided dispatching. Computer-aided dispatching not only does delivery scheduling but can provide ticket writing and accounting input. (*Courtesy of Auto Controls Inc.*)

Bins and hoppers should be inspected regularly and cleaned and painted when necessary. The bottoms and sides of bins and hoppers should be checked for wear and liner plates replaced as needed.

Elevators, screw conveyors, and belt conveyors should be kept cleaned, painted, checked for wear, and lubricated as needed.

Water meters should be calibrated regularly. They should be drained in cold weather to prevent damage due to freezing.

Mixers should be kept cleaned, lubricated, and properly aligned. Blade wear should be checked periodically and worn blades and liners repaired or replaced.

Recorders should be kept clean at all times.

Controls are particularly susceptible to dust and moisture. Damaged or malfunctioning parts should be repaired or replaced.

Many of the sophisticated and highly complex control systems must be serviced by experts. Troubleshooting some of this equipment requires specialized tools or testing equipment. Any changes in the control systems should be recorded on the wiring diagrams so they will be current at all times.

Gates should be cleaned and lubricated.

Hydraulic equipment must be kept free from contamination.

Air equipment must be kept free from moisture and well lubricated.

Caution! Historically most deaths and injuries occur to personnel when the plant equipment is being cleaned, maintained, or repaired. During these times many of the safety devices are bypassed as a matter of convenience. There must be safe ways to accomplish these tasks—this is a mandate of OSHA. Always keep all required warning signs in good condition and visible. Cleanup and maintenance should be done *only* by personnel fully knowledgeable of how to do the assigned job safely.

CHAPTER 22

PLANT INSPECTION AND SAFETY

John K. Hunt*

INSPECTION AND TESTING

Inspection and testing procedures are discussed in detail in Part 3 of this handbook, and need not be repeated here. However, there are several points that need to be emphasized. Inspection and testing at the batch plant are intended to ensure close control over all phases of the operation with the objective of uniform quality of concrete. Since concrete manufacturing is not an exact science, it requires some exercise of judgment.

22.1 Materials Handling

Materials must be received, stored, and used without excess handling and in a way to reduce contamination, segregation, and degradation. Adequate storage must be provided. All materials-handling vehicles must be emptied as completely as possible and full and empty weights checked to reduce unintentional shorting of material.

Samples of aggregate should be taken at the batcher discharge for checking gradation.

22.2 Batching Equipment

Hoppers and valves should be inspected for faulty operation. Scales should be inspected for drag or jerky operation and to be certain they are operating within the applicable delivery tolerances. Scales should be given a static-load test with standard weights periodically. Graphic recorders should be checked frequently during the day to be sure the pens are marking properly and accurately. Digital recorders should be checked frequently during the day to be sure all digits are printing correctly.

Water and admixture meters should be calibrated as often as necessary. Admixture timers and the calibration of the flow controls should be checked frequently to be sure they are as they should be. If calibrated sight glasses are used this can be done by simply watching the sight glasses for a few batches.

*Revised by Robert W. Strehlow.

Moisture meters should be calibrated at least daily.
Any unusual condition should be noted and checked immediately.

22.3 Mixers

Mixers should be cleaned daily to prevent buildup of material, and blade and liner wear should be checked as needed. Mixer-drum speed should be checked against the manufacturer's recommended speed. Any changes in material-charging sequence to the mixer should be noted and the mixing action and results checked. Mixing time should be spot-checked.

22.4 Dispatching

Each batch of concrete should be identified by type and the time of dispatching recorded. A record should be kept of wasted batches and the reasons for them.

SAFETY AND ACCIDENT PREVENTION

22.5 Responsibility

The federal Occupation Safety and Health Administration has set the *minimum* standards for the manufacture, installation, and use of all types of equipment. However the matter of safety and accident prevention is primarily a responsibility of the owner and operator of the plant. However, construction is inherently a hazardous occupation, and it is the responsibility of every person employed on the job to be alert to dangerous conditions and to take the necessary precautions for personal safety as well as that of others.

Construction is a highly competitive industry in which deadlines impose the need for speedy completion regardless of delays occasioned by unfavorable weather. Much of the work is done outdoors where the workers are exposed to heat and cold, rain, wind, and all other vagaries of the weather. Because of these hazardous conditions, accident prevention requires even closer attention to be effective.

Industrial commissions and similar political bodies of the states provide rules and regulations governing nearly all phases of construction and machinery operation, and most of them employ safety inspectors to enforce these laws. The National Safety Council and insurance companies supply posters and safety literature to their clients—material that can be posted about the jobsite, especially in the vicinity of dangerous conditions. Many large contractors employ a safety engineer whose duties are to educate the workers by use of posters, literature, motion pictures, and similar aids, as well as to supervise installation and maintenance of safety equipment, first-aid stations, fire-fighting facilities, machinery guards, and similar safeguards. Smaller companies usually designate one employee as a safety supervisor in addition to other duties.

22.6 Methods

Accident prevention is primarily a matter of education and organization. Accident prevention and safety, to be successful, must be publicized and must be presented to the workers so they are constantly aware of the program and voluntarily cooperate in observing safety rules. A new employee should be introduced to the company's safety program at the time of employment by being given a letter briefly explaining the accident-prevention features.

One method of maintaining safety awareness is to establish a system of awards in which the department, section, or gang with the best safety record receives publicity for its record, and perhaps a small cash bonus.

Most persons are apt to measure an accident by the spectacular results of the accident, the injury or death of workers, or property damage. Actually, even though there was no personal injury or property damage, any unexpected or unintentional interruption of the orderly progress of the work is an accident and should be analyzed as such to determine why it happened and how it can be prevented in the future.

Accidents should not be hidden. It is wrong to "cover up" for any reason whatever. Openly investigating and reporting an accident makes it possible for the workers to benefit thereby, so they can avoid a similar situation in the future.

Specific recommendations and warnings in and about concrete construction include those relating to formwork, shoring, hoists and cranes, conveyors, bucket elevators, pneumatic and electric machinery and tools, motor trucks, railroad cars, pumps, handling and storing aggregates and cement, ladders, scaffolding, and runways. Special regulations govern the use of boilers for supplying hot water or steam for heating aggregates and curing concrete. Quarry and aggregate excavation operations have special dangers in drilling, blasting, and excavating.

Protective clothing is necessary for some occupations. Welders and sandblasters require special helmets, face guards, and goggles. Gloves should be worn only when necessary for protection. Loose-fitting clothing should be avoided. Every new employee should be issued a safety helmet ("hard hat") with instructions to wear it at all times on the job. Protective shoes ("hard toes") are recommended.

These are but a few of the precautions. The safety engineer or supervisor should become familiar with all potential hazards on the job. One way to do this is to study the literature that is available from manufacturers, state industrial safety commissions, insurance companies, the Associated General Contractors, and similar organizations.

CHAPTER 23
TRANSPORTING CONCRETE

John K. Hunt*

Concrete should be transported so that segregation of ingredients, loss of ingredients, and loss of slump are at a minimum. Because of its plastic consistency, low-slump air-entrained concrete is more easily handled without detriment than the runnier mixes used in the past. Some segregation of the larger aggregate is usually not objectionable if the aggregate is recombined with the fine materials or ends up on top of the concrete when it is placed in its final location, because vibration will consolidate the materials. Deleterious segregation is avoided by dumping, dropping, or chuting concrete only when necessary, and then so that materials are recombined as they are discharged or so that the segregated larger aggregate, if any, ends up on top of the concrete. Concrete lowered vertically in full tubes or trunks segregates very little. Concrete turned by baffles as it is dropped or chuted is usually recombined satisfactorily, but the best location for the baffles may have to be determined by trial. Stirring of the concrete while it is being moved reduces segregation and permits more time to elapse between mixing and placing. Excessive jolting or vibration without stirring tends to segregate concrete.

Loss of ingredients, usually grout, is avoided by using gates that do not leak grout, by coating mixing and handling equipment with grout at the start of a pour (often by adding cement, sand, and water to the initial batch), by preventing excessive buildup of grout on equipment in contact with the wet concrete, and by not dropping concrete through water. Belt conveyors need scrapers to keep fine materials from carrying around head pulleys.

Loss of slump is caused by increase in concrete temperature and by drying as well as loss of grout. It is best avoided by handling concrete expeditiously, protecting concrete from sun and wind, and by painting containers exposed to the sun gloss white if concrete is held in them more than a short time.

CONCRETE HOPPERS

23.1 Normal-Slump Hoppers

Mixers do not discharge vertically into concrete hoppers, and there is a tendency for larger materials to roll to the far side of the hopper. Materials are best recombined if the hopper discharges vertically on the centerline of a symmetrical hopper. Thus hopper gates are

*Revised by Robert W. Strehlow.

best located centrally with double-clam gates for vertical discharge from the hopper. Rubber gate wipes may be required to reduce loss of grout. The gates must be large for efficient handling of low-slump concrete. Operating forces required for gates are typically 3 to 4 psi of gate area. Hopper shape may be conical or nearly conical with six or more sides. Hoppers with four sides require steep side slopes because the net valley angle should be not less than 60° from the horizontal for held concrete, and the sharper corners tend to accumulate the smaller ingredients. A guide chute or hopper through which concrete moves without stopping should have net valley angles of at least 50°. Baffles may be used to reduce segregation in hoppers, but their optimum location must be found by trial. Regardless of planned capacity, concrete hoppers should be designed for a full load of concrete of the maximum weight shown in Tables 17.1 through 17.4.

23.2 Low-Slump Hoppers

Hoppers for very dry concrete, as used in base courses and concrete products, present a different problem. Shape is less critical because the material is usually a granular mixture of coated particles and does not tend to segregate as much as regular concrete. However, it is tacky and tends to build up in corners unless valley slopes are steep and the hopper is vibrated each time it is nearly empty.

CONCRETE BUCKETS

Concrete buckets are available in a variety of types and sizes for specific purposes. Typical sizes are shown in Table 23.1.

TABLE 23.1 Sizes and Typical Weights of Concrete Buckets*

	Typical bucket weights, lb			
Size, yd^3	Lightweight buckets‡	General-purpose buckets	Lay-down buckets	Low-slump buckets
⅓	300–400	330–450		
½	**330–460**	430–530	500–550	
¾	**440–590**	450–1030	500–630	
1	**500–700**	**640–1170**	**620–1580**	**970–1910**
1½	**560–940**	**730–1560**	**860–2040**	1050–2080
2	610–1080	**1100–1680**	**1080–2480**	**1700–3600**
3	—	1550–2600	1460–3500	2500–4230
4	—	1800–3800	—	**3650–5180**
5	—	—	—	4400–5200
6	—	—	—	4900–7200
7	—	—	—	5800–7000
8	—	—	—	6600–8300§
8†	—	—	—	8600–9800†
12	—	—	—	9500–12,700§

*Sizes shown in boldface type are most commonly used.
†These weights are for an 8 yd^3 bucket with two 4 yd^3 compartments.
‡Aluminum buckets weigh about three-fourths of the minimum weights shown.
§Some 8 yd^3 buckets weigh as much as 13,000 lb. Some special 12 yd^3 buckets weigh as much as 14,500 lb.

FIGURE 23.1 Concrete bucket for low-slump concrete. *(Blaw-Knox Company.)*

23.3 Low-Slump or Mass-Concrete Buckets

Some specifications limit the amount of concrete that can be placed in one pile to 4 yd^3; so larger buckets are sometimes compartmented with separate gates on each compartment. Low-slump buckets (Fig. 23.1) have bottom slopes of 60 to 70° and large gate areas for handling 1- to 3-in.-slump concrete with aggregate up to 8 in. The smaller sizes may be operated manually, but compressed-air (or sometimes hydraulic) operation is used on larger sizes because of the large gate forces required. Air-operated buckets are operated by air connections from hoses at the placing site, either by making or breaking the air connection or by a manual valve, or occasionally by air lines connected to the bucket from the crane and manual valves, or by built-in rechargeable high-pressure air receivers and manual valves. About 10 ft^3 of free air at 125 psi should be allowed for each gate cycle, and about two cycles per bucketload of concrete. Gates should close automatically if the valve is released or if the air line fails. Hydraulic buckets have devices built into them to use the bucket weight to provide energy to operate the gates several times.

23.4 Lightweight or General-Purpose Buckets

These buckets, as shown in Fig. 23.2, are used for handling 2 to 6 in-slump concrete with about 2½ to 3 in maximum aggregate. They are usually manually operated and often have rubber spout attachments to aid in placing walls and columns. Gate areas are smaller than in low-slump buckets and hopper slopes are 50 to 60°.

23.5 Lay-Down Buckets

Normally lay-down buckets (Fig. 23.3) are used in building construction where they are filled from transit mixers and low filling height is required. They usually have small manual gates of the single-clam type.

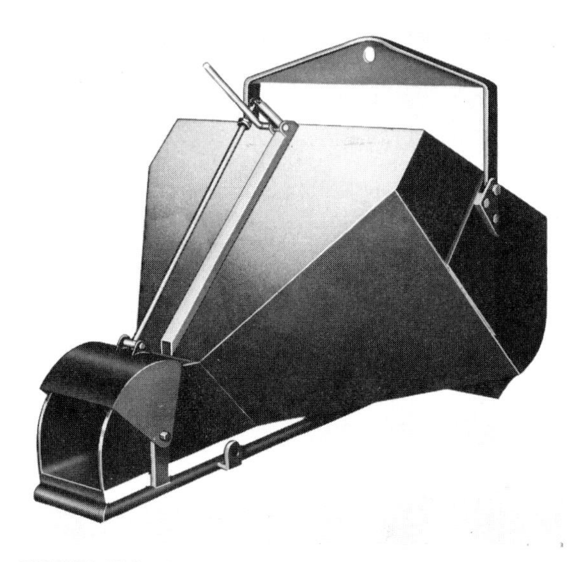

FIGURE 23.2 One type of lightweight lay-down concrete bucket for general usage. (*Gar-Bro Manufacturing Company.*)

23.6 Special Buckets

Special buckets are made for placing concrete underwater, lowering concrete in caissons, and other special applications. *Underwater* or *tremic buckets* are frequently covered to prevent loss of grout while the bucket is being lowered and operate on contact with the bottom or with a gate line from above (see Chap. 26 also).

FIGURE 23.3 A lay-down bucket frequently used in building construction. (*Gar-Bro Manufacturing Company.*)

While some buckets for special applications must compromise on desirable features, to minimize segregation, concrete buckets should have hopper slopes steep enough to discharge thoroughly and have a symmetrical vertical discharge. An asymmetrical discharge opening causes a suspended bucket to kick sideways during discharge. The minimum dimension of the discharge opening should be at least five times the nominal size of the largest aggregate. While low-slump buckets may open full and close on release of the valve, buckets for building construction require regulated gate openings for feeding concrete into forms and do not always close automatically.

In building construction the concrete bucket is often filled from a transit mixer and handled by a crane. Mass concrete is usually hauled from the mixing plant in buckets with two or more buckets on a rail car or truck and lifted and moved to final position by a trestle crane or cableway. An empty bucket is often placed on the car or truck before a full bucket is lifted; so there is usually space for one bucket more than the number of full buckets carried on the hauling unit.

AGITATING TRUCK BODIES

23.7 Truck Mixers

When used as agitating hauling units, truck-mixer capacity is shown in Table 18.2, unless otherwise limited by truck chassis capacity or legal load limits. Drums (or paddles) are revolved at 2 to 6 hr/min during agitation. Discharge times are the same as for mixers and can be quite long for low-slump concrete. Typical discharge times are 30 to 90 s/yd^3, but mixers built especially for low-slump concrete may discharge in 5 to 10 s/yd^3. Standards for agitators are published by the National Ready Mixed Concrete Association, 900 Spring St., Silver Spring, MD 20901.

23.8 Open-Top Hauling Units

Units of this type have specially shaped bodies with rotating blades (Fig. 23.4). Concrete discharges down a chute at the rear by tilting the body. Typical capacities are 4, 6, 8, and 12 yd^3. Discharge times are comparable with those of transit-mix trucks except that low-slump concrete can be discharged more readily. Typical discharge time is 30 to 40 sc.

NONAGITATING TRUCK BODIES

23.9 Side-Dump and End-Dump Units

These nonagitating hauling units are available in capacities of 4, 6, and 8 yd^3. Truck-trailer side-dump combinations increase capacity to 16 yd^3 per unit. Nonagitating units work well with plastic air-entrained concrete with slump up to about 3 in. Typical discharge time is 30 to 40 sc. In addition, ordinary dump bodies may be converted to haul low-slump concrete by welding fillets in the corners where necessary to prevent concrete buildup, and by making sure the tail gate seals well enough to prevent loss of grout. Dump bodies discharge rapidly, and the rate is difficult to control. The tail gate may be chained to provide some control of discharge rate. On hauls up to 8 or 10 mil covers are seldom used on hauling units, but they may be required if drying due to exposure is trou-

FIGURE 23.4 Agitating open-top hauling unit. Used for transport of central-mixed concrete. Provides controlled chute discharge. *(Courtesy of Maxon Industries, Inc.)*

blesome. A vibrator or frequent scraping may be required to avoid accumulation of concrete. Nonagitating units are also suitable for hauling very dry concrete like that used for base courses.

NUMBER OF HAULING UNITS REQUIRED

23.10 Estimating the Number of Haul Units Required

Hauling, in the context of concrete transport, is a batch process over time and distance. To estimate the number of hauling units that can be utilized by any batch plant two factors must be determined: (1) a constant, which is the cycle time it takes to produce a (repetitive) batch of concrete, and (2) a variable, which is the complete round trip cycle time of the haul unit.

The batch plant manufacturer will rate a plants production at its maximum output without any regard to charging the hauling equipment. ("On the ground" is an expression often heard). While this is a practical measure of comparing the productive capacities of two plants, it is not a practical means of calculating plant output. The plant cycle time we need here is the time it takes between starting to load one haul unit to starting to load the next haul unit. For our calculations this is best expressed in seconds per batch cycle. Note that this is a constant that is based on absolute maximum efficiency. Anything that delays this cycle time causes a loss in efficiency which cannot be recovered. By tracking your performance over a long period of time, you can determine the efficiency of your operation.

The haul unit cycle time must also be expressed in seconds per haul cycle. This is a variable in that haul cycle time is determined by distance, travel speed, traffic conditions, unloading time, rest stops, washout, gas-up, and waiting in line to be loaded. All the slowdown factors do not occur in every cycle, so you need a practical component "fudge factor" applied to this cycle time. You are concerned with this cycle time only during the peak production period as this is the only time you need to be concerned with a haul unit shortage.

If Hauling to Only One Project. Once you have determined these two cycle time values you can calculate the maximum number of haul units you will need as long as the trip cycle time remains the same.

Just divide the haul unit or truck cycle by the plant cycle and drop the fraction or multiply by an efficiency factor if you have determined one.

$$N = \frac{TC}{PC} \times Eff$$

where TC = the truck cycle time in seconds
PC = the plant cycle time in seconds
N = the maximum number of haul units required
Eff = the efficiency factor

If Hauling More Than One Project. Now you need to know the consumption rate of each project, in loads per hour. You need to know the truck cycle time on each project. You need to know the plant capacity in loads per hour. Note that the use of cubic yards per hour is not used because the maximum plant capacity is truly a fairly constant value when expressed in loads per hour. The total consumption rate of all projects being serviced *cannot* exceed the plant capacity. There is no simple formula for dispatching trucks to multiple jobs, but there are many dispatch aids ranging from truck tracking boards to sophisticated computer programs.

THE CONCRETE-MOBILE

The Concrete-Mobile is a concrete manufacturing plant that can be truck-chassis mounted, truck semitrailer-mounted, or in-plant stationary-mounted. It is a self-contained plant designed to handle stone and sand aggregate, water, and cement in separate bins and tanks (Fig. 23.5). Concrete materials, mechanically dispatched at a specified engine speed through calibrated aggregate bin openings and constant cement discharge, are discharged into a rapid mix-conveyor of proprietary design. No scales are required. A predetermined continuous flow of water is introduced into the mix-conveyor through a calibrated flow-control valve.

The mixing system consists of a U-trough with a tough, flexible rubber boot (bottom). Through it, longitudinally, runs a hydraulically driven shaft that rotates at a predetermined speed during the mixing operation. During mixing, the mix-conveyor system is operated at an angle from the horizontal, with the receiving end lower than the discharge end. A combination of auger flights and angled paddles is attached to the lower or receiving end of the shaft. From this point in the mix-conveyor system, auger flighting moves the mixed concrete to the discharge point, where placing or distributing chutes can be

FIGURE 23.5. The batch plant on wheels. Shown is the "concrete-mobile" which is a volumetric batch plant with a mixer designed to deliver freshly mixed concrete at the job site. It is covered by separate ASTM specifications. *(Courtesy of C. S. Johnson Co., Div. of Koehring Co.)*

attached. Mixing action is completed in about 20 s. The mix-conveyor is so constructed that the operator and inspector can look into the mixer and observe the continuous mixing and discharge operation.

Before concrete-mix design formulas can be calculated, it is necessary to know the physical characteristics of the aggregates, including gradations, specific gravities, and dry rodded unit weights, as well as the type of cement to be used. Separate mix-setting charts are designed for each machine for each set of materials and mixes, including admixtures. Any mix design, whether stated by weight or by volume, can be converted into a mix-setting chart from which the operator sets each aggregate dial and the water-control valve. The cement-meter register indicates the cement being used.

Advantages claimed are that transit time is eliminated because dry ingredients are mixed in any volume required on the job. This eliminates the danger of partial set of the concrete in transit and in waiting prior to placement. Close control of workability is possible, as the operator and inspector can see the concrete being mixed and can adjust water as required to maintain slump.

PART • 6

PLACING CONCRETE

Placing concrete consists of moving or transferring the concrete from its delivery point at the jobsite, placing it in forms, and consolidating it to provide a concrete structure having adequate structural integrity, durability, serviceability, and appearance in accordance with the design and specifications. Placing concrete to provide such a finished structure requires knowledge of a few fundamental facts about the behavior of concrete when handled under different conditions, selection of equipment suitable for the job, and close attention to details that make the difference between a satisfactory job and a mediocre job. This has resulted in a large background of "do's and don'ts" and "correct and incorrect practices and methods" which are discussed in this section.

Roller-compacted concrete is discussed in Chap. 30.

CHAPTER 24
METHODS AND EQUIPMENT

Robert F. Adams

DISTRIBUTING CONCRETE ON THE SITE

24.1 Equipment for Moving or Handling Concrete

Most commercial concrete is delivered from a batch or central-mix plant by transit-mix concrete trucks having capacities of up to 12 yd^3 used as either mixers or agitators. Concrete may also be delivered by bottom-dump trucks, dump trucks, Dumpcrete, or Agitor haul units. Concrete is moved from the delivery point to the structure or placing area by chute, crane and bucket, conveyor, auger, or pump. Very little concrete is handled by wheelbarrows, buggies (Georgia buggies), or motorized buggies except on very small jobs because it is labor-intensive and slow. Large concrete construction jobs situated where concrete is not readily available from commercial ready-mix plants, or for other reasons, may have on-site batching and mixing plants.

Equipment should be carefully selected to move concrete from its delivery point into place in the finished structure.[1] The slump, sand content, maximum size of aggregate, or concrete mix should not be governed by the equipment; rather, the equipment should be capable of expeditiously handling, moving, and discharging concrete of such slump, sand content, maximum aggregate size, or mix proportions considered otherwise suitable, and which can be placed by vibration or other suitable placing methods.[2] Description of some of the basic equipment used for moving and handling concrete after it is received at the jobsite follows. (Equipment for transporting concrete to the job is covered in Chap. 23.)

Chutes. Chutes (Fig. 24.1) are a simple and expeditious way of transferring or moving concrete to a lower elevation. The commonest example is the chute used to discharge concrete from a mixer into a bucket or other equipment, or directly into the forms. Chutes must have sufficient slope so that concrete will readily move down them by gravity. Flat chutes requiring high slump, lower-quality concrete should not be used (Fig. 24.2).

Wheelbarrows, Concrete Buggies (Georgia Buggies), or Motorized Carts. These vehicles may be used on small jobs on level ground (Fig. 24.3). They are slow and labor-intensive by present-day standards.

Buckets. Buckets are transported or handled by crane, derrick, hoist, monorail, truck, railway car, cableway, or helicopter. Two types of buckets, the conventional type, either

FIGURE 24.1 Chute delivering concrete from transit-mix truck into hopper of drop chute in bridge pier. (*California Department of Water Resources photograph.*)

cylindrical or square with bottom discharge (Fig. 24.4), and the lay-down type (Fig. 24.5), are available. The conventional cylindrical or square bucket is available in many sizes from ⅓ to 12 yd^3 and in many designs having various slopes to their sides and sizes of gate openings. Some, having steep side slopes and large gate openings, are suitable for low-slump and mass concrete. The lay-down bucket is for use where the headroom for charging the bucket is more limited. These are available in capacities of up to 5 yd^3. Discharge of the buckets may be by either hand-operated gates for smaller buckets or air-actuated gates. Air-actuated gates are operated by air supplied and connected at the placing site or by an air tank carried on the bucket. The gates for buckets should be grout-tight, particularly if transported by truck or railway car for some distance. Lightweight buckets made of magnesium are available and advantageous where weight is a factor, such as transporting them by helicopter (see Fig. 24.4).

Conveyors. Belt conveyors (Figs. 24.6 and 24.7) are used to transfer concrete horizontally and modest distances vertically. Conveyors are relatively inexpensive and may eliminate the need for other more expensive auxiliary equipment, such as cranes and pumps. Conveyors are particularly useful in areas of limited room, such as in tunnels for transferring concrete from the delivery car or truck to the hopper on a concrete pump or tunnel-invert lining machine, but are widely used on large areas such as floor slabs and bridge decks. A drop chute should be used at the discharge end of a conveyor to help control segregation. A side-discharge feature on a conveyor (Fig. 24.7) provides more flexibility in discharging concrete at the desired location on bridge decks or slabs. Conveyors are particularly useful for lower-slump, langan-maximum size aggregate concrete.[3]

Concrete Pumps. Concrete pumps (Fig. 24.8) are most commonly used to move concrete from the delivery truck to the place where it is used.[4] Modern concrete pumps have capacities of over 175 yd^3/h. Most pumps have lines 5 in in diameter or less, which limits

FIGURE 24.2 A very flat chute requires higher slump, lower-quality concrete for the concrete to flow down the chute. Such a condition sometimes results in liberties being taken with the concrete. If lower-slump concrete is used, much effort must be used to push or pull the concrete down the chute, resulting in a very slow concrete placing operation. An attempt should be made to eliminate such an undesirable condition.

FIGURE 24.3 Carts or buggies transport concrete short distances. Carts having rubber tires are best. (*California Department of Water Resources photograph.*)

FIGURE 24.4 Concrete bucket used to place concrete in bridge foundation. Bucket is transported from mixing plant on shore by helicopter. Use of lightweight bucket reduces deadweight and increases payload. *(Sikorsky Aircraft photograph.)*

the maximum size of the aggregate in the concrete to 2 in or less. Pumps can pump concrete more than 2500 ft horizontally and over 1000 ft vertically in high-rise building construction. Some pumps have articulating booms which can rotate 360° and can reach as much as 170 ft. See Chap. 36 for more information on concrete pumps.

Receiving Hoppers. Receiving hoppers (Fig. 24.9) are used in combination with other equipment to provide temporary storage or surge capacity or to facilitate transfer of concrete from one type of handling equipment to another. Hoppers with center discharge and with side slopes steep enough to facilitate discharge of the concrete are preferred.

Drop Chutes or Elephant Trunks. These devices (Fig. 24.10) are designed to deliver concrete to a lower elevation without segregation or for the purpose of correcting segregation that might otherwise occur because of concrete's hitting reinforcing steel or other obstructions. They also keep the reinforcing steel and forms from becoming coated with mortar, where this may be considered objectionable. (Where concrete placing is rapid enough that this mortar coating does not dry out, this may not be objectionable.) These are made of rubber tubing, plastic tubing, sheet metal, or short sections of steel tubing fastened together, so they are flexible and can be readily shortened. Hoppers for drop

FIGURE 24.5 Lay-down-type concrete bucket, used when headroom is limited, discharging concrete in wall form. *(California Department of Water Resources photograph.)*

chutes should be large and steep enough to readily discharge the concrete. Drop chutes which impede the concrete flow or which are difficult to move become "bottlenecks" in the concrete placing operation.

PLACING CONCRETE IN THE FORMS

24.2 Correct and Incorrect Methods of Handling and Placing Concrete

Figures 24.11 through 24.15 (from Ref. 1) show some of the correct and incorrect ways of handling and placing concrete. A study of these figures and their captions will give a better understanding of the principles involved and will greatly assist in selecting proper equipment for placing concrete, using equipment properly, and analyzing and correcting placing difficulties.

A basic rule is that concrete should be deposited as nearly as possible in its final location. The production of uniform concrete which materially assists the placing operations is discussed in Chaps. 16 through 21. Concrete consists of coarse aggregate and mortar and will separate if conditions and opportunity exist. To minimize segregation or separation, it is desirable to drop concrete vertically, rather than at an angle, whenever possible

FIGURE 24.6 The belt conveyor system provides convenient method of placing concrete on a large placement. Note drop chute at end of conveyor to prevent segregation of concrete. Note also conveyor elevating concrete from delivery truck onto truck-mounted conveyor. (*Photograph from ROTEC Industries.*)

FIGURE 24.7 Belt conveyor placing concrete on a bridge deck. Side-discharge feature is movable and permits discharge of concrete at any point across span of conveyor, facilitating coverage of area. (*Morgen Manufacturing Co. photograph.*)

FIGURE 24.8 Upper photograph: small trailer-mounted pump pumping into a small concrete placement. Lower photograph: large concrete pump placing concrete on a building deck. Boom which rotates can cover a large area. (*Photograph from Associated Concrete Pumping.*)

FIGURE 24.9 Receiving hoppers provide temporary storage or surge capacity, or facilitate transfer of concrete. (*Gar-Bro Manufacturing Co. photograph.*)

FIGURE 24.10 Drop chutes used in dropping concrete to a lower elevation. (*California Department of Water Resources photograph.*)

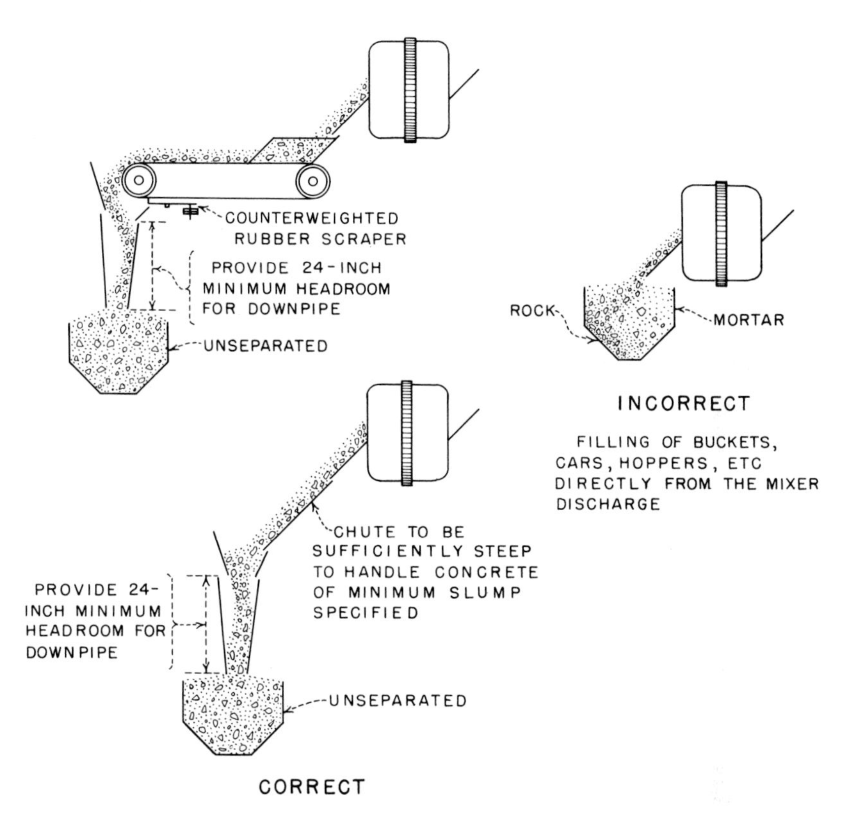

FIGURE 24.11 Unless discharge of concrete from mixers is correctly controlled, uniformity resulting from effective mixing is destroyed by separation. *(By permission of American Concrete Institute.*[1])

(Figs. 24.10 through 24.13). There should be a short vertical drop chute at the end of sloping chutes or the end of belt conveyors. Baffles should generally be avoided as they may only cause separation to occur on the opposite side. Baffles that can be moved or tilted can sometimes be used satisfactorily provided they are properly positioned to direct the stream of concrete into previously placed concrete. If baffles are used, they must be changed as the height of the concrete or other conditions change to be used effectively and intelligently. Conveyor belts should have a scraper or wiper to remove mortar adhering to the belt and return it to the concrete (Fig. 24.12).

A hopper should be filled by dropping concrete into the center and should discharge vertically from a center opening; hoppers with side discharge or sloping gates should be avoided as they may cause separation (Fig. 24.13).

Drop chutes or elephant trunks should be avoided when the concrete can be satisfactorily placed without them. A crane bucket moving along the top of and discharging directly into a wall form is entirely satisfactory provided segregation does not occur. Although some specifications may limit free fall of the concrete to 5 or 6 ft, much greater free fall

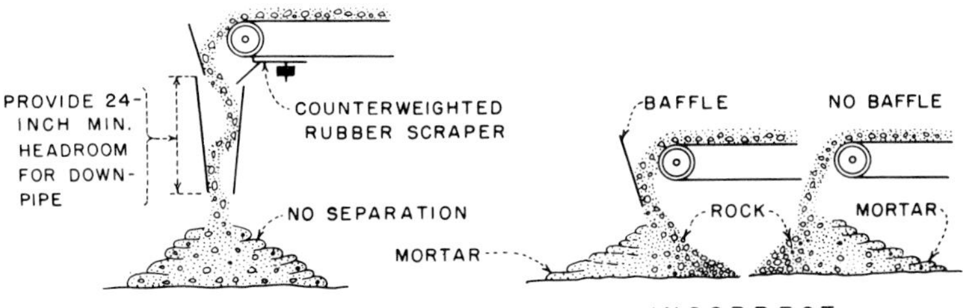

CORRECT

THE ABOVE ARRANGEMENT PREVENTS SEPARATION OF CONCRETE WHETHER IT IS BEING DISCHARGED INTO HOPPERS, BUCKETS, CARS, TRUCKS, OR FORMS.

INCORRECT

IMPROPER OR COMPLETE LACK OF CONTROL AT END OF BELT.

USUALLY A BAFFLE OR SHALLOW HOPPER MERELY CHANGES THE DIRECTION OF SEPARATION.

CONTROL OF SEPARATION OF CONCRETE AT THE END OF CONVEYOR BELT

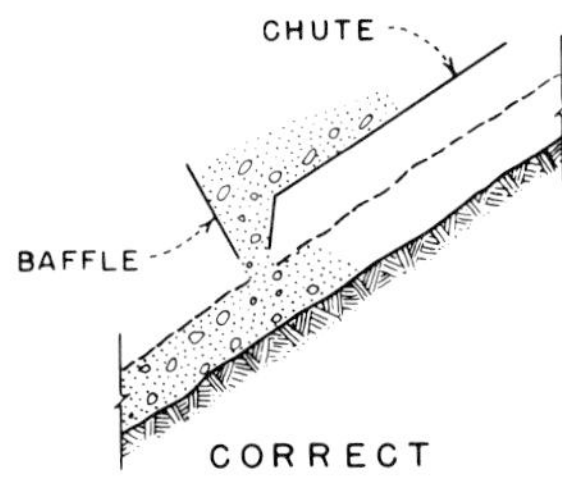

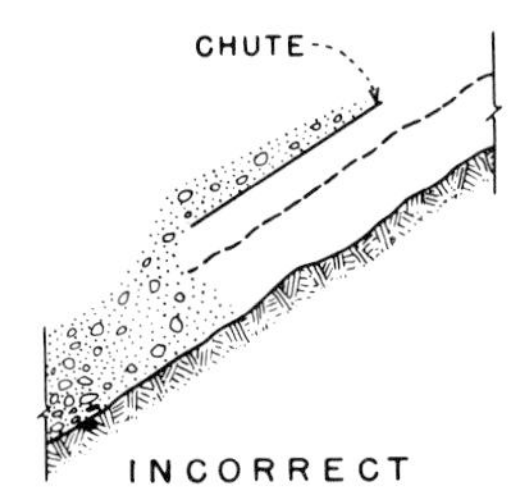

CORRECT

PLACE BAFFLE AND DROP AT END OF CHUTE SO THAT SEPARATION IS AVOIDED AND CONCRETE REMAINS ON SLOPE.

INCORRECT

TO DISCHARGE CONCRETE FROM A FREE END CHUTE ON A SLOPE TO BE PAVED. ROCK IS SEPARATED AND GOES TO BOTTOM OF SLOPE.

VELOCITY TENDS TO CARRY CONCRETE DOWN SLOPE.

PLACING CONCRETE ON A SLOPING SURFACE

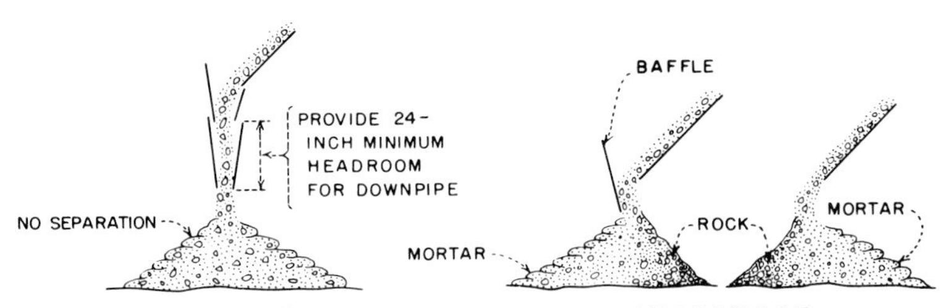

CORRECT

THE ABOVE ARRANGEMENT PREVENTS SEPARATION, NO MATTER HOW SHORT THE CHUTE, WHETHER CONCRETE IS BEING DISCHARGED INTO HOPPERS, BUCKETS, CARS, TRUCKS, OR FORMS.

INCORRECT

IMPROPER OR LACK OF CONTROL AT END OF ANY CONCRETE CHUTE, NO MATTER HOW SHORT.

USUALLY A BAFFLE MERELY CHANGES DIRECTION OF SEPARATION.

CONTROL OF SEPARATION AT THE END OF CONCRETE CHUTES

THIS APPLIES TO SLOPING DISCHARGES FROM MIXERS, TRUCK MIXERS, ETC AS WELL AS TO LONGER CHUTES, BUT NOT WHEN CONCRETE IS DISCHARGED INTO ANOTHER CHUTE OR ONTO A CONVEYOR BELT.

FIGURE 24.12 Control of separation of concrete. (*By permission of American Concrete Institute.*[1])

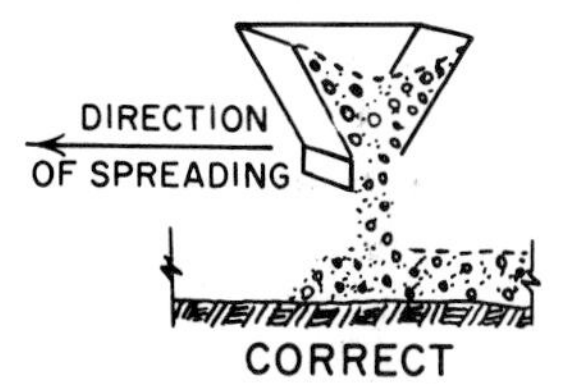

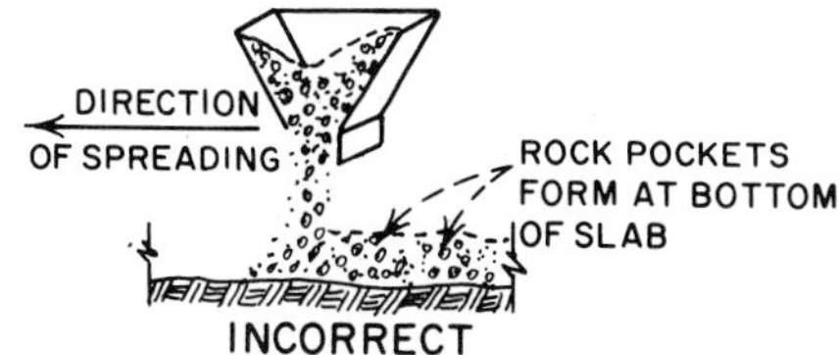

CORRECT

TURN BUCKET SO THAT SEPARATED ROCK FALLS ON CONCRETE WHERE IT MAY BE READILY WORKED INTO MASS.

INCORRECT

DUMPING SO THAT FREE ROCK ROLLS OUT ON FORMS OR SUBGRADE.

DISCHARGING CONCRETE

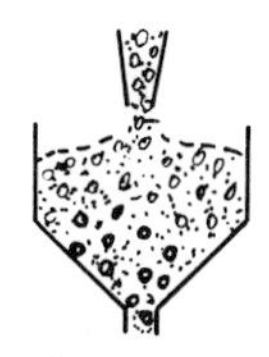

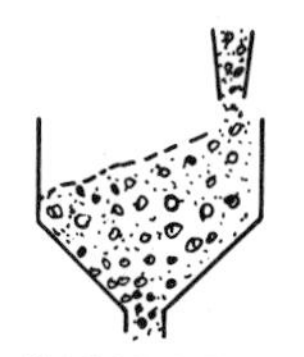

CORRECT

DROPPING CONCRETE DIRECTLY OVER GATE OPENING.

INCORRECT

DROPPING CONCRETE ON SLOPING SIDES OF HOPPER.

FILLING CONCRETE HOPPERS OR BUCKETS

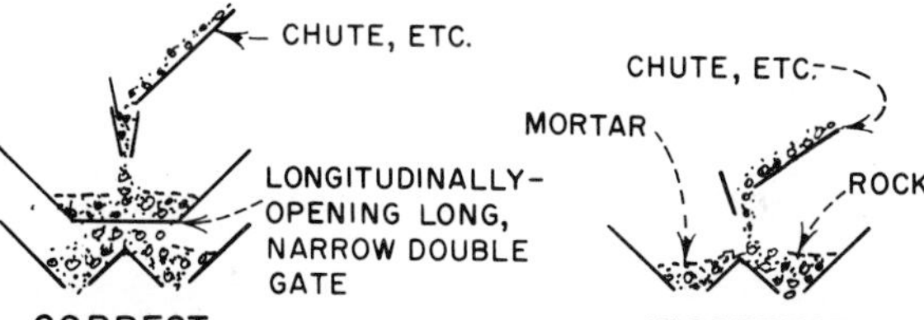

CORRECT

THE ABOVE ARRANGEMENT SHOWS A FEASIBLE METHOD IF A DIVIDED HOPPER MUST BE USED. (SINGLE DISCHARGE HOPPERS SHOULD BE USED WHENEVER POSSIBLE.)

INCORRECT

FILLING DIVIDED HOPPER AS ABOVE INVARIABLY RESULTS IN SEPARATION AND LACK OF UNIFORMITY IN CONCRETE DELIVERED FROM EITHER GATE.

DIVIDED CONCRETE HOPPERS

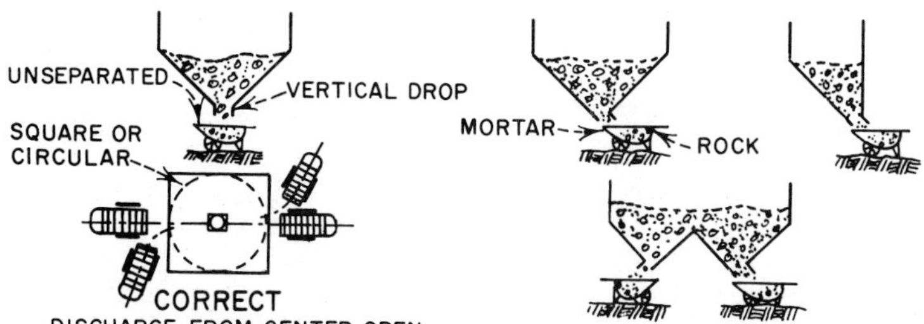

CORRECT

DISCHARGE FROM CENTER OPENING PERMITTING VERTICAL DROP INTO CENTER OF BUGGY. ALTERNATE APPROACH FROM OPPOSITE SIDES PERMITS AS RAPID LOADING AS MAY BE OBTAINED WITH OBJECTIONABLE DIVIDED HOPPERS HAVING TWO DISCHARGE GATES.

INCORRECT

SLOPING HOPPER GATES WHICH ARE IN EFFECT CHUTES WITHOUT END CONTROL CAUSING OBJECTIONABLE SEPARATION IN FILLING THE BUGGIES

DISCHARGE OF HOPPERS FOR LOADING CONCRETE BUGGIES

FIGURE 24.13 Correct and incorrect methods for loading and discharging concrete buckets, hoppers, and buggies. Correct procedure minimizes separation of coarse aggregate from mortar. (*By permission of American Concrete Institute.*[1])

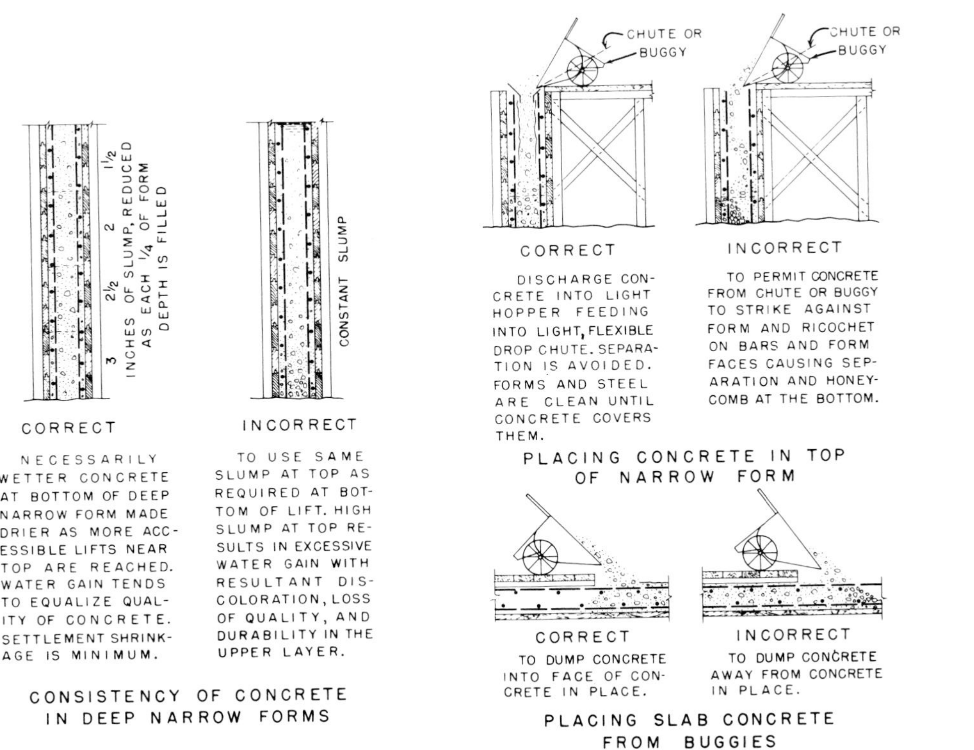

FIGURE 24.14 Placing concrete in curved and narrow form and in slabs. (*By permission of American Concrete Institute.*[1])

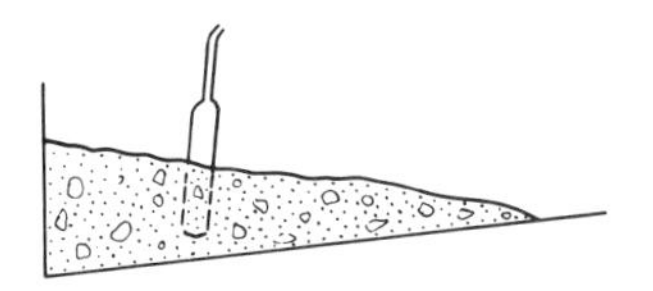

CORRECT

START PLACING AT BOTTOM OF SLOPE SO THAT COMPACTION IS INCREASED BY WEIGHT OF NEWLY ADDED CONCRETE. VIBRATION CONSOLIDATES.

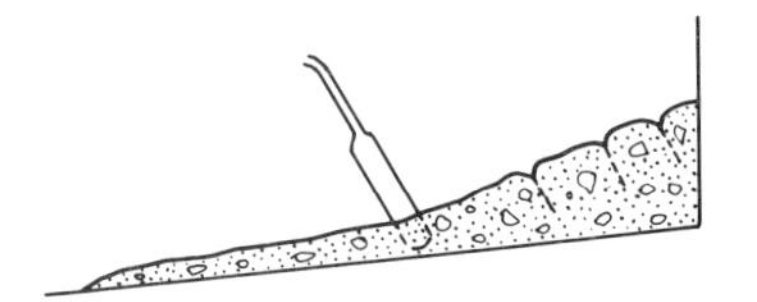

INCORRECT

TO BEGIN PLACING AT TOP OF SLOPE. UPPER CONCRETE TENDS TO PULL APART, ESPECIALLY WHEN VIBRATED BELOW, AS VIBRATION STARTS FLOW AND REMOVES SUPPORT FROM CONCRETE ABOVE.

WHEN CONCRETE MUST BE PLACED IN A SLOPING LIFT

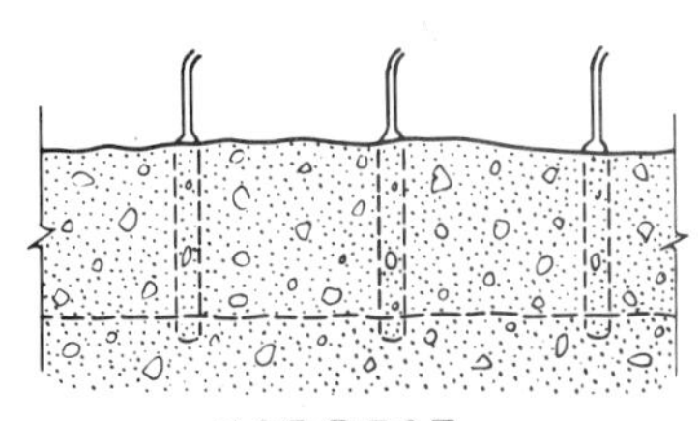

CORRECT

VERTICAL PENETRATION OF VIBRATOR A FEW INCHES INTO PREVIOUS LIFT (WHICH SHOULD NOT YET BE RIGID) AT SYSTEMATIC REGULAR INTERVALS FOUND TO GIVE ADEQUATE CONSOLIDATION.

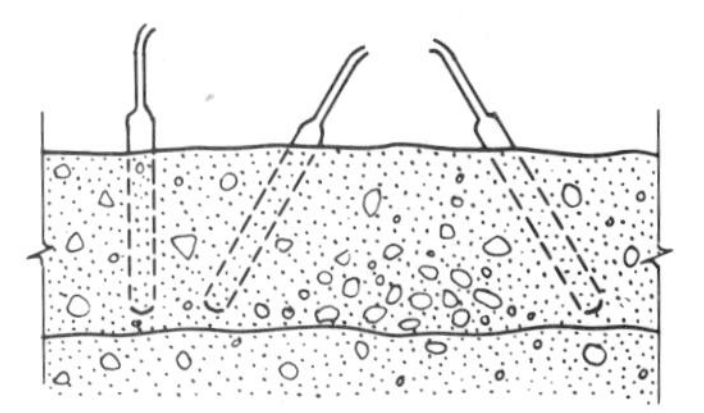

INCORRECT

HAPHAZARD RANDOM PENETRATION OF THE VIBRATOR AT ALL ANGLES AND SPACINGS WITHOUT SUFFICIENT DEPTH TO ASSURE MONOLITHIC COMBINATION OF THE TWO LAYERS.

SYSTEMATIC VIBRATION OF EACH NEW LIFT

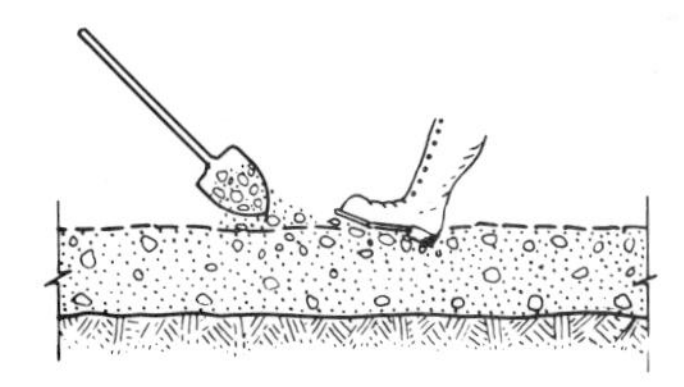

CORRECT

SHOVEL ROCKS FROM ROCK POCKET ONTO SOFTER, AMPLY SANDED AREA AND TRAMP OR VIBRATE.

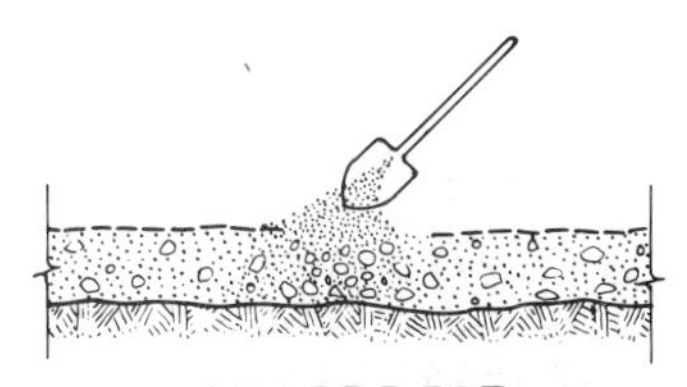

INCORRECT

ATTEMPTING TO CORRECT ROCK POCKET BY SHOVELING MORTAR AND SOFT CONCRETE ON IT.

TREATMENT OF ROCK POCKET WHEN PLACING CONCRETE

FIGURE 24.15 Proper use of vibrator and treatment of rock pocket. (*By permission of American Concrete Institute.*[1])

can give entirely satisfactory results provided the fall is vertical and segregation does not occur (see Chap. 26). When used, drop chutes should be arranged so that they can be quickly moved and shortened, or a sufficient number should be supplied to cover the placing area without moving. If the use of drop chutes or elephant trunks slows concrete placing, it only adds to the problems as concrete loses slump rapidly. The ends of elephant trunks should remain vertical; to push them to the side results in a sloping condition and causes separation. When concrete is placed in a deep or curved wall form through a port in the side of the form, the concrete should drop vertically into an outside pocket, then overflow into the form without separation (Fig. 24.14). Sometimes such a drop chute and pocket can be built into the form, between studs.

Concrete should be discharged from equipment and placed rapidly enough so that "stacking" does not occur. When concrete dribbles slowly from a bucket, chute, drop chute, or elephant trunk, it piles up and coarse aggregate rolls down the sides, separating. By opening the bucket faster or speeding up the flow of concrete it does not stack and separate and its impact force materially assists in its placing and consolidation. If concrete separates as it comes out of a drop chute, such as one at the end of a conveyor, the separation can be easily corrected by operating a vibrator in the pile of discharged concrete.

Concrete being placed in a slab should be dumped into the face of the concrete in place, not away from it (Fig. 24.14).

If concrete has separated, the coarse aggregate should be shoveled onto concrete and incorporated into the concrete. Concrete or mortar should not be shoveled onto separated coarse aggregate (Fig. 24.15).

In placing concrete on a slope, placing should start at the bottom of the slope and move up the slope (Figs. 24.12 and 24.15).

Because concrete usually bleeds, that is, the solids settle and water moves to the top, concrete in a wall placement should be placed with a lower slump (lower water/cement ratio) in the uppermost layer, thus compensating for water gain, providing concrete of uniform appearance, and ensuring more durable concrete at the top where freezing and thawing conditions are usually more severe (Fig. 24.14). It is helpful to slope the tops of walls so that "birdbaths" are eliminated and water drains off.

Concrete should be deposited as nearly as possible in its final location, as mentioned previously. It should not be moved long distances by vibration. Rather, it should be deposited as close intervals and each pile consolidated by vibrating it in its final location, particular attention being given to vibrating the junction between piles.

Considerable thought should be given to the job conditions and in selection of equipment to place the concrete.[5] For example, placing concrete in a floor slab or foundation walls directly from a ready-mixed concrete truck discharging from only a limited number of places results in having to move the concrete a considerable distance with a vibrator or gravity, resulting in demands for wet, lower-quality concrete to accomplish this movement readily, or the use of a long chute with a flat slope which would require wet concrete to flow down the chute. The wet concrete would have less strength and durability than concrete of moderate slump, and is more subject to cracking, dusting, abrasion, leakage, or dampness, and other undesirable qualities. The desirable alternative to the above is to provide more access to the placing site so that the truck can discharge the concrete at as many locations as needed to cover the area properly, or to use other additional placing equipment to move the concrete from the truck mixer to the location where it is deposited. This equipment, discussed previously, might include wheel-barrows, concrete carts, crane buckets, conveyors, or pumps.

Concrete in walls or in sections having considerable depth (more than 18 in) should be placed systematically in layers having a thickness of not more than 15 to 20 in. Each layer should be systematically vibrated, care being taken to ensure that the junction between deposits is adequately vibrated (Fig. 24.15).

Large areas having several layers such as mass concrete should be placed systematically in a stair-step manner. In placing a large area or large section, the placing should be arranged to proceed systematically and have the least area of concrete exposed, thus helping to avoid cold joints. More details on placing procedures will be covered in Chap. 26 for different kinds of construction.

CONSOLIDATION

Concrete after being deposited in forms must be consolidated into a homogeneous solid mass without such defects as rock pockets, voids, or sand streak. Until the advent of mechanical equipment that consolidates concrete by vibration, concrete was consolidated by laborers using spades, shovels, hand tampers, or their feet. It should again be pointed out that the purpose of vibration is to consolidate concrete, not move it. Almost all concrete is now compacted or consolidated with vibrators of many designs and characteristics. High-frequency vibration transforms low-slump concrete into a semiliquid mass so that gravity causes it to flow together into a compact mass. ACI Guide 309R[6] lists the different types of internal vibrators and their characteristics and gives considerably more detail on consolidation than is covered here. Table 24.1, which shows the performance of internal vibrators, is reproduced from this report.

24.3 Equipment

Vibrators used on construction jobs for mass, structural, paving, or similar concrete are usually high-frequency low-amplitude vibrators. Low-frequency high-amplitude vibrators are sometimes used, particularly in fixed installations in plants producing precast elements or in concrete-block machines. One power float for heavy-duty concrete floors imparts a tamping action to the concrete which amounts to low-frequency vibration, effectively tamping the surface of earth-moist concrete.

Most vibrators are powered by either electricity or air. Hydraulic vibrators are used in some applications. Small gasoline engines are also used, either direct-connected or driving a generator to supply electricity. For electric operation, adequate electric service is required. For air-operated vibrators, an adequate air supply is likewise required. Difficulties might occur with air vibrators slowing or stopping during cold weather because of freezing of moisture in the air supply. Sometimes the use of alcohol-type antifreeze agents or air driers may alleviate this difficulty; ethylene glycol should not be used since it "gums up" the vibrator. All vibrators, regardless of type, should be kept in good repair in order that they will operate satisfactorily. A marked loss in frequency will seriously reduce the effectiveness of a vibrator. The vibration frequency of a vibrator can be determined by a small inexpensive tachometer of the vibrating-reed type called a Vibra-Tak (Fig. 24.16). The frequency should be determined with the vibrator operating in concrete. A marked difference in the frequency when operating in and out of concrete or a frequency lower than normal or previously obtained for the same vibrator may indicate repairs are necessary, or an inadequate power or air supply, and corrective action should be taken. Pressure gages can be used to measure the air line pressure near the vibrator (Fig. 24.17). Some vibrators are variable-frequency to accommodate concrete of different consistencies.

Vibrators are likely to fail frequently and for this reason require considerable maintenance. Standby vibrators must be readily available for replacement should any in use fail

TABLE 24.1 Range of Characteristics, Performance, and Applications of Internal Vibrators

(1)	(2)	(3)*[a]	(4)	(5)[b]	(6)[c]	(7)[†,F]	(8)[d, e]	(9)
			Suggested values of			Approximate Values of		
Group	Diameter of head, in. (cm)	Recommended frequency, vibrations per minute (Hz)	Eccentric moment, in · lb (cm · kg)	Average amplitude, in (cm)	Centrifugal force, lb (kgf)	Radius of action, in (cm)	Rate of concrete placement, yd^3/h per vibrator (m^3/h)	Application
1	¾–1½ (2–4)	10,000–15,000 (170–250)	0.03–0.10 (0.035–0.12)	0.015–0.03 (0.04–0.08)	100–400 (45–180)	3–6 (8–15)	1–5 (0.8–4)	Plastic and flowing concrete in very thin members and confined places; may be used to supplement larger vibrators, especially in prestressed work where cables and ducts cause congestion in forms; also used for fabricating laboratory test specimens
2	1¼–2½ (3–6)	9000–13,500 (150–225)	0.08–0.25 (0.09–0.29)	0.02–0.04 (0.05–0.10)	300–900 (140–400)	5–10 (13–25)	3–10 (2.3–8)	Plastic concrete in thin walls, columns, beams, precast piles, thin slabs, and along construction joints; may be used to supplement larger vibrators in confined areas
3	2–3½ (5–9)	8000–12,000 (130–200)	0.20–0.70 (0.23–0.81)	0.025–0.05 (0.06–0.13)	700–2000 (320–900)	7–14 (18–36)	6–20 (4.6–15)	Stiff plastic concrete [< 3-in (8-cm) slump] in general construction such as walls, columns, beams, pre-stressed piles, and heavy slabs; auxiliary vibration adjacent to forms of mass concrete and pavements; may be gang mounted to provide full width internal vibration of pavement slabs

1	2	3	4	5	6	7	8	9
4	3–6 (8–15)	7000–10,500 (120–180)	0.70–2.5 (0.81–2.9)	0.03–0.06 (0.08–0.15)	1500–4000 (680–1800)	12–20 (30–51)	(15–40) (11–31)	Mass and structural concrete of 0–2-in (5-cm) slump deposited in quantities up to 4 yd^3 (3 m^3) in relatively open forms of heavy construction (powerhouses, heavy bridge piers and foundations); also auxiliary vibration in dam construction near forms and around embedded items and reinforcing steel
5	5–7 (13–18)	5500–8500 (90–140)	2.25–3.50 (2.6–4.0)	0.04–0.08 (0.10–0.20)	2500–6000 (1100–2700)	16–24 (40–61)	25–50 (19–38)	Mass concrete in gravity dams, large piers, massive walls, etc.; two or more vibrators will be required to operate simultaneously to melt down and consolidate quantities of concrete of 4 yd^3 (3 m^3) or more deposited at one time in the form

[a]Column 3—While vibrator is operating in concrete.
[b]Column 5—Peak amplitude (half the peak-to-peak value), operating in air.
[c]Column 6—Using frequency of vibrator while operating in concrete.
[d]Column 7—Distance over which concrete is fully consolidated.
[e]Column 8—Assumes insertion spacing is 1½ times the radius of action, and that vibrator operates two-thirds of time concrete is being placed.
[f]Columns 7 and 8—These ranges reflect not only the capability of the vibrator but also differences in workability of the mix, degree of deaeration desired, and other conditions experienced in construction.

Source: ACI Guide 309R-87 by permission of American Concrete Institute.[6]

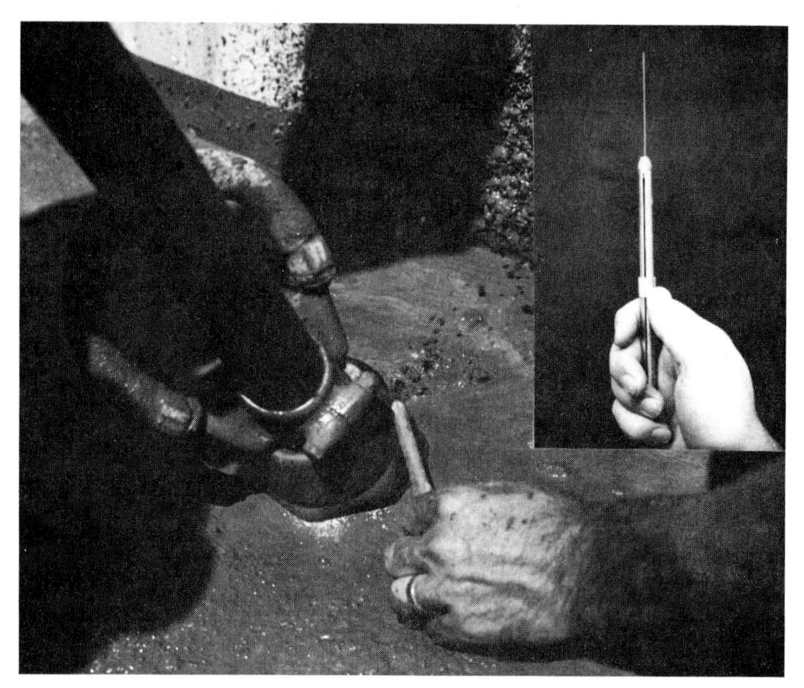

FIGURE 24.16 Checking vibration frequency of a concrete vibrator with a Vibra-Tak. Measurement is made while vibrator is operating in concrete and measures vibrations per minutes. (*California Department of Water Resources photograph.*)

to operate properly. One standby vibrator is a bare minimum where only one or two vibrators are used on a job. Larger jobs require several standby vibrators.

Vibrators may be grouped into three classes, described in the following subsections.

Internal Vibrators or Immersion Vibrators. These devices (Figs. 24.18 through 24.23), are used internally in the concrete. These vary considerably, ranging from vibrators with small vibrating heads on flexible shafts suitable for thin walls or small sections congested with reinforcing to large vibrators suitable for mass concrete. In some, the motor is in the head; in others, it is separate with the vibrating element on a flexible shaft. Figure 24.19 shows a short head or "stubby" electric vibrator. This is most suitable for consolidation of concrete in thinner slabs and floors since it permits complete vertical insertion of the vibrator head in the concrete. This eliminates the undesirable practice of dragging longer head vibrators through the concrete horizontally as is frequently done. Some have a saberlike extension on the vibrating element to use in congested areas such as around prestressing steel (Fig. 24.21). Several vibrators may be mounted together such as the multiple-tube or gang vibrators used on paving or canal-lining machines (Fig. 24.22) or gang-mounted on tractors as sometimes used for mass concrete, the latter application being more common in other countries (Fig. 24.23).

External Vibrators. These devices (Fig. 24.24) are clamped or attached to the forms and vibrate the concrete from vibration of the form. External or form vibrators are usually used for thin or congested sections, and for locations such as in small tunnels or the arch

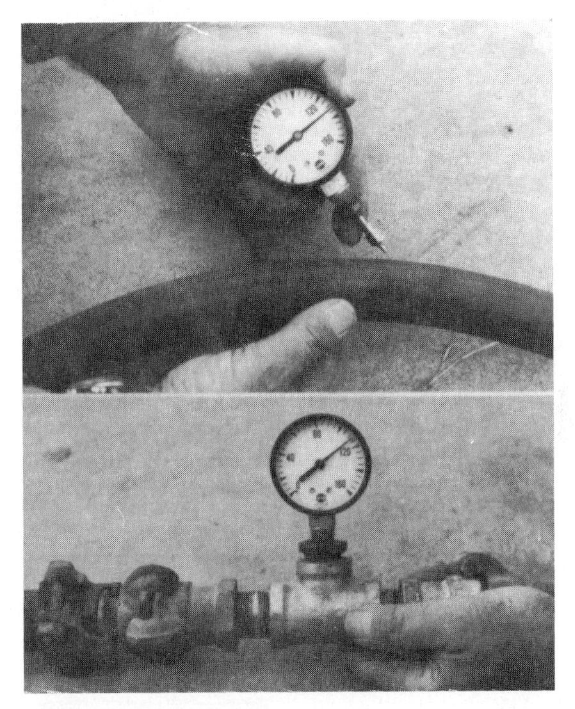

FIGURE 24.17 Pressure gauge inserted in air line near vibrator to determine adequacy of air pressure to vibrators. Insertion is by hypodermic needle or line couplings.

of larger tunnels, where internal vibration is not possible or practical. They are also particularly useful for tunnel lining, either alone or with internal vibrators, and in plants for concrete pipe and other precast-concrete products.

Surface Vibrators. These vibrators consist of vibrating-pans or vibrating screeds which vibrate the concrete from the surface—usually at the time the concrete is struck off or screeded. Surface vibrators are usually used as screeds for slabs or pavements. They are

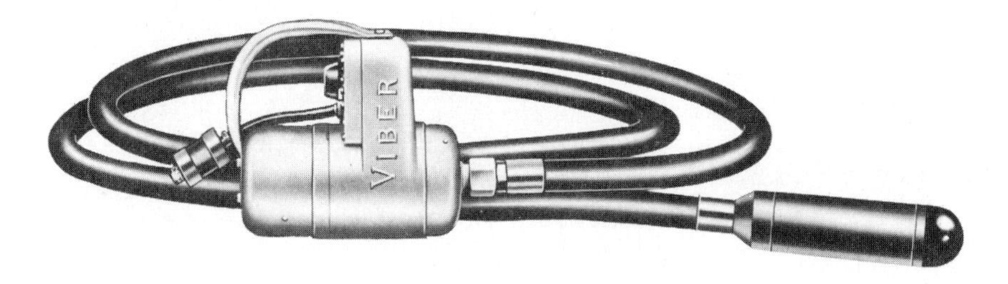

FIGURE 24.18 Electric internal vibrator. *(Viber-Kelley Co. photograph.)*

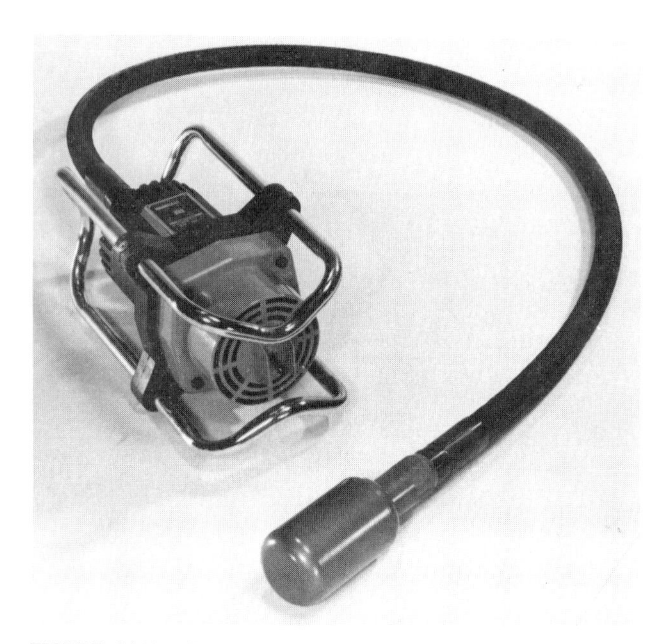

FIGURE 24.19 Short head or "stubby" electric vibrator most suitable for consolidating concrete in thinner slabs and floors. This vibrator is 2½ in in diameter and 4 in long and has a 10 in radius of action. (*Wyco Tool Co. photograph.*)

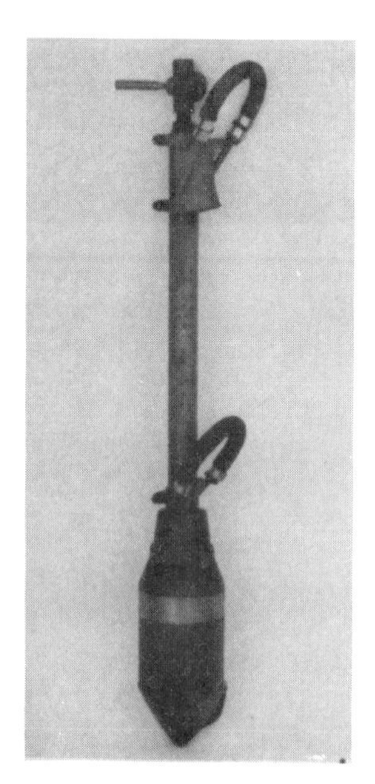

FIGURE 24.20 Malan 6 in one-operator pneumatic vibrator for mass concrete. (*Malan Vibrator Co. photograph.*)

FIGURE 24.21 Saber vibrator for vibrating thin sections and sections containing closely spaced reinforcing steel. Vibrator is homemade using a pneumatic external vibrator mounted on a piece of ¼ × 1½ in steel flat bar. (*Model made by John Gebetsberger of Lathrop Construction Co.*)

FIGURE 24.22 Gang-mounted internal vibrators vibrating 15 in airfield parking apron. (*Viber-Kelley Co. photograph.*)

effective to a limited depth. The ability to vibrate the greater depth should be determined or internal vibrators should be used in conjunction with a vibrating screed. A vibrating screed is particularly useful as a screed for horizontal surfaces (see Chap. 26).

24.4 Consolidating Concrete with Vibration

Concrete should be consolidated to its maximum practical density so that it is free of rock pockets and entrapped air and closes snugly around all form surfaces and embedded materials. Vibration should be applied systematically to cover all areas immediately after the concrete is deposited, as shown in Fig. 24.15. The vibrator should penetrate the layer of concrete vertically and into the underlying layer previously vibrated when possible. Immersion vibrators are sometimes used in slabs and pavements where they cannot be used vertically. In this usage, the vibrator should be kept immersed. The contact between adjacent piles or loads of concrete should be vibrated thoroughly. Vibrators should not be used to move concrete in the form appreciable distances.

Vibration should be continued until the concrete flattens and takes on a glistening appearance, the rise of entrapped air ceases, the coarse aggregate blends into the surface but does not completely disappear, and the vibrator, after an initial slow-down when first inserted in the concrete, resumes its speed. The vibrator should then be slowly withdrawn to ensure closing the hole resulting from insertion of the vibrator. Satisfactory mixes, properly placed, seldom require more than 5 to 15 or 20 s of vibration at any one location to accomplish the above.

Overvibration can and does occur, although it is unlikely in well-proportioned concrete with normal-weight aggregate having the correct slump. Overvibrated concrete will

FIGURE 24.23 Internal vibrators mounted on a tractor vibrating mass concrete on a dam in Japan. Note bulldozer used to spread and level concrete prior to vibration.

have excess mortar on the top and a frothy appearance. The correction is to reduce the slump and also to reduce the sand content if the mix is oversanded. Overvibration can occur in lightweight-aggregate concrete, resulting in "floating" the coarse aggregate causing finishing difficulties, particularly if the concrete is too wet. Overvibration can occur in heavy-weight concrete as there is a greater tendency for the heavy aggregate to settle in the fresh concrete, particularly if the concrete is too wet.

It is difficult to vibrate concrete on a sloping surface or where a grade must be held on a lower surface, such as a sloping wall surface or a step, unless positive means are taken to hold the concrete at lower elevations. Such situations frequently result in inadequate compaction of upper concrete as the amount of vibration is frequently controlled by the difficulties caused by bulging of concrete at lower unformed surfaces, rather than by the needs of the concrete being vibrated above. Such a condition should and usually can be eliminated by proper selection of equipment or construction practices, such as use of temporary holding forms or heavy sliding forms. Examples of these are given in Chap. 26, in discussions of placing mass, canal-lining, or slope-paving concrete. Stub walls, which frequently are not satisfactorily vibrated, can and usually should be eliminated.

Revibration of previously vibrated concrete is generally beneficial when done at a time the vibrator will still penetrate the concrete with its own weight and make it plastic again. Revibration will improve strength, improve bond to horizontal reinforcing steel by eliminating the water pocket under the steel, remove air and water pockets, and eliminate settlement cracks such as occur at the junction of floor slabs and columns in the same placement or over window or door openings or other blockouts.[7] Revibration into a layer of concrete partially hardened to the extent that it cannot be made fluid with vibration may cause a wavy line on the surface of walls, which is undesirable on exposed concrete.

Vibrators should not be allowed to strike the forms as the forms may be damaged, resulting in an unsightly blemish on the surface of the concrete, as well as damage to the forms (Fig. 24.25). See Ref. 7 for other consolidation-related surface defects in concrete.

FIGURE 24.24 External or form pneumatic vibrators on forms for concrete pipe. (*Viber-Kelley Co. photograph.*)

It is sometimes useful to vibrate the reinforcement in congested or inaccessible areas, although it is not advocated that the reinforcement system be regularly used for this purpose. This is best done with form vibrators. Fear that vibration transmitted to the reinforcing steel is detrimental is unfounded, even when part of the steel is embedded in partially hardened concrete. Contact between a vibrator and reinforcing steel will not cause damage (unless the steel is displaced—and it should be adequately tied and supported to prevent displacement).

In some instances, especially in architectural concrete, vibration may be supplemented by rodding or spading along the form, especially at corners or angles. Such spading as well as additional vibration will sometimes minimize or eliminate "bug-holes." It is virtually impossible to eliminate bugholes from inward-sloping forms. Fear that ample or extra vibration necessary to reduce or eliminate bugholes will adversely affect the durability of air-entrained concrete because of removal of part of the air is unfounded, provided the concrete originally contained the amount of entrained air recommended for dura-

FIGURE 24.25 Blemish on concrete surface caused by vibrator hitting plywood form surface and damaging it.

bility in ACI Standard 211.1[8]. Research has shown that the remaining air-void system is unimpaired in its ability to improve durability in freezing and thawing. Less than the recommended initial air content may result in not enough air remaining after vibration to provide adequate durability.

24.5 Consolidating Concrete without Vibration

Modern concrete technology and construction practice assume the use of vibration in placing concrete. Even for small jobs, vibrators are relatively low in cost or can be rented, and their use is encouraged. Where vibration is not used, concrete can be adequately placed but with more physical effort. The concrete will need to be more workable for hand-placing methods with more slump and perhaps a higher sand content, although excessive high slumps and sand contents are not necessary and are undesirable as they result in lower-quality concrete with greater potential for shrinkage, lower strength, and less durability. After the concrete is placed in the forms, it must be worked with spades, tampers, shovels, or with the feet to consolidate it, eliminating voids. Simply dumping concrete into a form from a ready-mixed-concrete truck without some means being taken to consolidate it, even with high-slump concrete, cannot achieve a satisfactory job. This is seen in observing the results of such practices, particularly on small jobs such as foundations for houses or small buildings or sidewalks, slabs, or other flat work. Slabs such as those in Fig. 24.26 had an excellent surface finish but were inadequately compacted—and full of "rock pockets."

FIGURE 24.26 Concrete slabs removed from a structure. Surface finish was excellent. Concrete below surface was inadequately compacted and full of rock pockets, which may have contributed to its failure.

REFERENCES

1. ACI Guide 304R, Recommended Practice for Measuring, Mixing, and Placing Concrete.
2. "Concrete Manual," 8th ed., Chap. VI, U.S. Bureau of Reclamation.
3. ACI Guide 304.4R, Placing Concrete with Belt Conveyors.
4. ACI Guide 304.2R, Placing Concrete by Pumping Methods.
5. Waddell, J. J.: "Practical Quality Control for Concrete," McGraw-Hill, New York, 1962.
6. ACI Guide 309R-87, Guide for Consolidation of Concrete.
7. ACI Guide 309.2R, Identification and Control of Consolidation-Related Surface Defects in Formed Concrete.
8. ACI Standard 211.1-91, Standard Practice for Selecting Proportions for Normal Heavyweight and Mass Concrete.

CHAPTER 25

PREPARATION FOR PLACING CONCRETE

Robert F. Adams

Soon after award of the contract and prior to start of concrete work, a meeting between the contractor's and owner's representatives who will be responsible for the concrete work should be held. Some job specifications require the contractor to submit the proposed materials, equipment, and procedures for approval by the owner. A meeting provides an opportunity to review the specification requirements, any interpretations of them, the standards which the owner expects, and a review of the contractors proposal for materials, equipment, and procedures.

Much preparatory work needs to be done before concrete is placed; foundations must be excavated, cleaned, and prepared; forms must be built; reinforcing steel must be placed; construction joints against previously placed concrete must be cleaned; embedded items such as bolts for anchoring machinery or equipment, pipe, conduit, castings for manholes or catch basins, water stops, blockouts for gate guides, machinery bases, doors, windows, drains, and sumps must be prepared and placed.

All the above must be supervised and inspected by the contractor's representative and by the owner's engineer during progress of the work and given a final inspection before concrete is ordered for placing. The supervision and inspection are for the purpose of building the structure required by the owner as shown in the plans and in accordance with the specifications.

FOUNDATIONS AND JOINTS

25.1 Foundation Preparation

Foundation surfaces against which concrete is to be placed should conform to the specified location, size, and shape and should have adequate bearing capacity to carry the anticipated loads, both during and after construction.

As foundation excavation and preparation proceed, construction forces should be alert to changed or unforeseen conditions that would affect design of the foundation. Such conditions should be reported to the engineer for review and investigation.

Excavation. Rock should be excavated to sound material and be completely exposed, and rock surfaces should be normal to the direction of load. (Sloping rock surfaces have caused failure due to sliding of the foundation or footing on the rock.) Blasting should be done so that the foundation rock which remains is not damaged and excessive overbreak does not result. Controlled blasting, supervised by specialists in this field, can minimize rock problems caused by careless blasting and result in much better construction. Much blasting is carelessly done and results in considerable damage to remaining rock, or overbreak that must be removed and replaced with concrete, increasing the cost for concrete and for the job. Good practices, such as closer and better spacing of drill holes, controlled powder charges, and delays, can result in closer control of rock excavation to required line and grade and less damage to the rock (Fig. 25.1).

Surfaces of rock subject to air slaking or raveling need special attention prior to placing concrete on them. A covering of shotcrete, bituminous material, or other sealer is used to protect such rock from air slaking until concrete can be placed over it (Fig. 25.2). Covering rock with shotcrete or other sealer usually improves working conditions, particularly if wet conditions are encountered. The workmen do not work in mud; tools and equipment, forms, and reinforcing steel are kept cleaner; and cleanup prior to placing concrete is minimized.

Loose or unsound rock or debris should first be removed and open fissures cleaned to a suitable depth and to firm rock on both sides. Rock should be cleaned by use of brooms, picks, sandblast, water jets, air-water jets, or other effective means followed by thorough washing and blowing out the remaining water by air jets (Fig. 25.3). The rock surface should be dry and free of any free surface water (indicated by a shiny surface) that will prevent bond of concrete to the rock.

FIGURE 25.1 Close spacing of drill holes and careful blasting result in rock excavation in accordance with the plans. (*California Department of Water Resources photograph.*)

FIGURE 25.2 Excavation in shale which slakes covered with shotcrete to eliminate slaking and mud and improve working conditions. In contrast, note muddy area on right side outside shotcrete area and structure limits. (*Delta Pumping Plant of California Water Project. California Department of Water Resources photograph.*)

FIGURE 25.3 Cleaning rock surface with air-water jet and removing loose rock preparatory to placing concrete. (*California Department of Water Resources photograph.*)

Earth foundations should be generally excavated to original undisturbed material if practical. If the original undisturbed material is unsatisfactory for the foundation, it should be removed and backfilled. Backfill should be suitably compacted to the specified density. Subgrades or bases for pavements or slabs should be filled with specified material and suitably compacted to the required density in accordance with the plans and specifications. Soft spots in the foundation are eliminated by removal of unsuitable material and replacement with suitable material adequately compacted (Fig. 25.4) or lean backfill concrete. Overexcavation may be corrected by backfill with suitable earth material adequately compacted, or lean backfill concrete. Backfill concrete has sometimes been used with cement contents as low as 1 sack per cubic yard. Earth subgrades should be damp or moist but not wet when concrete is placed on them.

Drainage. All foundations should be free of debris, frost, ice, snow, mud, or water. Water which enters a foundation area should be drained to a sump and pumped out or drained to the outside of the foundation area. Sometimes permanent concrete or clay pipe or porous gravel underdrains are placed to provide drainage for water entering the area and to relieve uplift pressures on the concrete (Fig. 25.5). Such drains are usually designed as part of the job and are provided for in the plans and specifications. Drains may be constructed by placing the concrete or clay pipe on a clean gravel or crushed-stone blanket in a trench or depression in the foundation, backfilling around it with the same clean gravel or crushed stone (¾ × ⅜ in or 1½ × ¾ in.). Usually the aggregate is covered with burlap to prevent mortar from infiltrating the gravel when concrete is placed over it, and it may be covered with a thin layer (1 to 2 in) of ¾ in. maximum-size-aggregate concrete to protect it from the hazards of construction if some time will elapse before the concrete is placed over it. The concrete or clay pipe is laid with open joints so that the water drains into the pipe. Sometimes the concrete or clay pipe is omitted, and gravel alone serves as the drain. Some drains may be temporary and may be grouted later.

FIGURE 25.4 Backfilling and compacting an overexcavated foundation for a concrete culvert. (*California Department of Water Resources photograph.*)

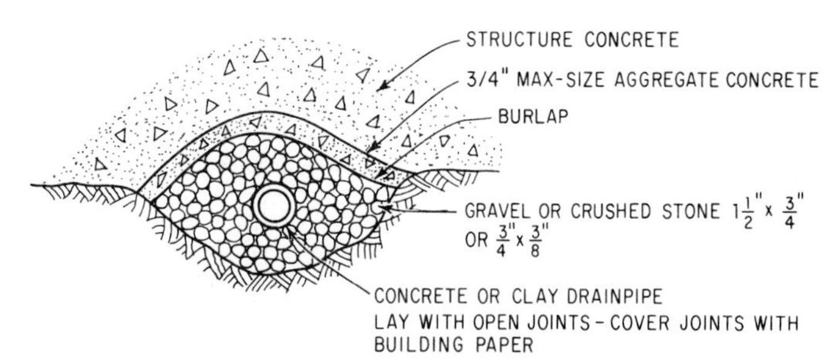

FIGURE 25.5 Underdrain for temporary or permanent use.

Sometimes the entrance of water into a foundation excavation can be prevented by installation of wells or well points and pumps outside the excavation, or it can be controlled by diverting the flow by constructing dikes of sandbags or sacked concrete.

Concrete should not be deposited in water, particularly in running water, unless approved by the plans and specifications of the engineer and suitable underwater concrete-placing procedures such as tremies or special concrete buckets are used. When concrete is placed underwater and it is impossible to dewater the excavation for cleanup, cleanup can be accomplished by suction pumps to remove mud and other debris. Underwater concrete-placing procedures are discussed in more detail in Chap. 26. In some cases, inspection of the underwater area by divers may be required.

Construction of forms and other related work is facilitated where it is difficult to control water in an excavation if a foundation seal is placed. This is a slab of concrete placed in the bottom of the excavation usually within a sheetpile cofferdam or caisson to seal off the flow of water. If such a seal is placed underwater, it should be placed by one of the underwater concrete-placing methods. Concrete for the seal is placed so the top is approximately at the elevation of the bottom of the foundation.

25.2 Construction-Joint Preparation

When fresh concrete is placed on or adjacent to previously placed concrete, whether old or new concrete, and where a bond between the two surfaces is required, it is essential that the surface of the previously placed concrete be clean and properly prepared. This is particularly important where watertight and durable joints are required (Fig. 25.6). The surface should be free of laitance, carbonation, scum, dirt, oil, grease, paint, curing compound, and loose or disintegrated concrete and should be slightly roughened. Several methods of cleanup are available, depending on the size of areas to be cleaned, age of the concrete, skill of the workers, and availability of equipment.

Whichever method of surface cleanup is used, it is necessary only to remove undersirable surface dirt, laitance, or unsound mortar. It is necessary only to expose the sand in sound surface mortar (Fig. 25.7). Removal of sound mortar or concrete to expose coarse aggregate or to create roughness is not considered essential. Deep cutting, resulting in removal or undercutting of coarse aggregate, is also considered unnecessary and is undesirable if the coarse aggregate is loosened.

Wet sandblast and high-pressure water blast are considered to be the most satisfactory methods of joint preparation (Fig. 25.8). If there is any question about the adequacy of other joint-preparation methods, their results should be compared with that obtained with

FIGURE 25.6 Evidence of leakage of water through a poorly prepared construction joint. In many cases, concrete deteriorates at this point because of freezing and thawing.

wet sandblast. A properly prepared joint will have a clean, clear, sharp appearance, similar to that of a fresh break in sound concrete.

Frequently construction-joint cleanup, particularly on horizontal surfaces, can be facilitated by foresight and care in seeing that the surface of the concrete placement is left relatively smooth and free of depressions, foot tracks, or other surface irregularities that will impede cleanup which is to be performed later. Excessive job traffic and overworking of the surface of fresh concrete should be particularly avoided. On mass concrete, finishing beyond that left by the vibration of the concrete is not necessary. On other

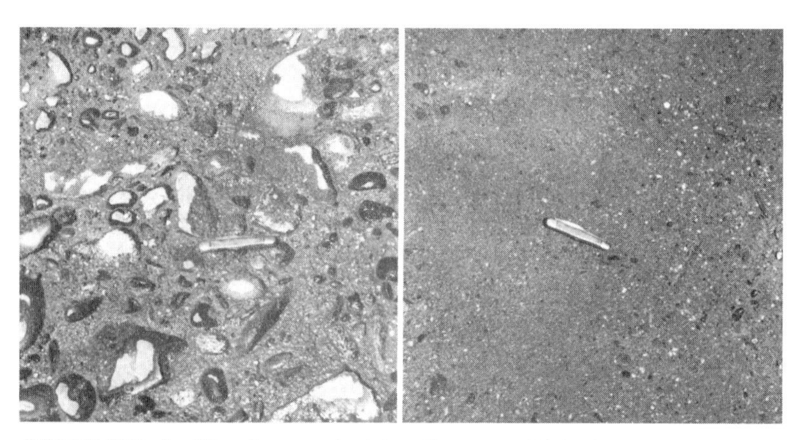

FIGURE 25.7 Sandblasted construction joints. Concrete on left has been sandblasted more than really necessary, exposing a considerable amount of coarse aggregate. Additional cutting which would have undercut aggregate would have been undesirable. Concrete on right has been sandblasted only enough to remove laitance and soft mortar to expose sound mortar. The appearance is clean and bright.

FIGURE 25.8 Left, wet sandblast of a large area of mass concrete. Smaller sandblast units for smaller jobs are available. Right, high-pressure water jet cleanup. (Pipe grid shown on concrete surface is cooling pipe for next concrete placement.)

concrete, finishing beyond a screed strike-off or float finish is not required. Workmen who find it necessary to walk on the surface of fresh mass concrete to place form anchors for future placements or for other reasons should wear "snowshoes" (oval wooden boards strapped to the feet) to reduce footprint depressions that make cleanup more difficult (Fig. 25.9). Planks can be laid on the surface for other foot traffic—or traffic should be detoured.

Hand tools such as wire brushes, wire brooms, hand picks, or bushhammers, used to remove dirt, laitance, and soft mortar, may produce satisfactory results, but their use is time-consuming and considered practical only for small areas. If bushhammers or chippers are used, particularly those powered by pneumatic or electric hammers, care must be exercised to see that their use does not result in broken or shattered aggregate at the surface, leaving an undesirable surface for bonding new concrete, and consequently reduced bond.

FIGURE 25.9 "Snowshoes" made from ¾ in plywood permit workers to work on top of concrete without leaving depressions which interfere with construction-joint cleanup.

Etching with acid is considered practical only for small areas. The concrete should be thoroughly wet with water prior to application of the dilute acid so the acid does not soak into the concrete. Commercial hydrochloric acid or muriatic acid diluted with water in the proportions of about 1 part acid to 3 or 4 parts water should be satisfactory for initial trials. The acid should be applied by spray, a bristle brush, or broom and vigorously scrubbed onto the concrete. After application of the acid and scrubbing, the surface should be thoroughly washed with water. The acid water should be disposed of where it will not create problems. Although the danger connected with using hydrochloric acid is not great, workers should be cautioned regarding the danger and should be protected with rubber gloves, rubber boots, goggles, and rubber or plastic outerwear to the extent necessary to remove such danger.

Green cutting or initial cleanup method consists of washing the surface of freshly placed concrete with high-pressure water or air-water jets at the proper time in the hardening process (when the concrete has hardened sufficiently so it will not ravel) to remove surface laitance. This is satisfactory only for horizontal surfaces and is usually used only on large jobs. Considerable skill is required to get a satisfactory job with this method because the cutting must be done at the proper time, not too early so as to avoid too much cutting, and not too late in order to obtain sufficient removal of the surface laitance. The proper time is dependent on the hardening characteristics of the concrete, which in turn is affected by many factors, such as temperature, characteristics of cement, use of admixtures, and water/cement ratio. (This technique is similar to that used to produce exposed-aggregate concrete on horizontal surfaces, although the removal of mortar to expose aggregate is more than is considered necessary for construction-joint cleanup.) Water and a wire broom may be found effective for small areas instead of air-water jets.

Use of retarder consists of treatment of the surface with a retarder and subsequent removal of the unhardened surface mortar with water and brushes or brooms, high-pressure water, or air-water jets. Retarders suitable for use on both horizontal and vertical surfaces are commercially available. The retarder is most conveniently applied to a horizontal concrete surface by a sprayer after the final finishing operation and before the concrete has set. For vertical surfaces, the retarder is applied to the forms. The manufacturer's instructions for application and coverage rate should be followed. The removal of the retarded surface layer by wire brushes or brooms and water or air-water jets may be done the following day or two. The longer it is delayed, the less retarded surface layer will be removed. This method is subject to some of the same disadvantages as the green-cutting method but to a lesser extent. It is dependent on the hardening characteristics of the surface, which are dependent on the same factors as mentioned previously plus the effect of the retarder. (This technique is also used for exposed-aggregate concrete surfaces. See Chap. 44.)

Wet sandblast is considered the most satisfactory method of joint preparation, as mentioned before. This is done with water in order to eliminate the dust problem (Fig. 25.8). Sandblasting is considered practical for both large and small construction projects and is suitable for either horizontal or vertical surfaces. Suitable equipment is readily available for both large and small jobs.

Sandblasting should be delayed as long as practical, preferably until just before the forms are erected for the next placement. If done after forms are erected, the forms must be protected from the sandblasting. Embedded items such as conduit or water stop must not be damaged by the sandblasting. As mentioned previously, only laitance and soft mortar should be removed. Overcutting resulting in exposure and undercutting of coarse aggregate is not necessary and is considered undesirable (Fig. 25.7). The advantage of wet sandblast over green cutting or the use of surface retarders (which must be done at the right time) is that sandblasting can be delayed, thereby reducing the chance that the surface will become unsatisfactory because of a scum forming from curing water, second rise of laitance, carbonation, or dirt from construction activity, and thus it can be done at a more convenient time.

Sand for sandblasting should be hard, not readily broken down, and sufficiently dry that it feeds through the equipment satisfactorily. Specially prepared dried sand from which the particles larger than No. 4 or No. 8 mesh and smaller than No. 16 or No. 30 mesh are removed is most satisfactory. Sand can be prepared on the job by using simple hand-screening methods, or on large jobs in mechanical screening equipment. Prepared sandblast sand is also commercially available in most areas.

Water blasting using a high-pressure water jet has become popular recently because of the development of suitable equipment and is considered by many to be equal to and an acceptable alternate to wet sandblast for construction joint cleanup (Fig. 25.8). This method is different from "green cutting" in that much higher water pressure is used, up to 10,000 psi, and no air. Water blasting is also done at a later time than possible in green cutting. The time at which it is done is not so critical as in green cutting; although the longer it is delayed, the less the production will be, as with sandblasting. One advantage of water blasting over wet sandblast is elimination of the sand problem (supply, scatter, and disposal). Some of the same comments above regarding wet sandblast apply equally to water blasting. Equipment for water blasting is available from several manufacturers.

Protection of construction joints is required after construction-joint cleanup by any of the above methods. All require a final thorough washing of the surface to remove the debris resulting from the cleanup operations. Once a joint is properly prepared, it should be protected to prevent its becoming unsatisfactory again. Protection includes some of the curing procedures such as damp sand, burlap, or cotton mat covers, and waterproof paper or plastic covering. Should a prepared surface become unsatisfactory because of a second rise of laitance, carbonation, or dirt, the cleanup should be redone.

The Mortar Layer. A rock surface or a construction joint in previously placed concrete should be properly prepared to receive the concrete placement, as discussed previously. Research by the Corps of Engineers[1–3] indicates that it may not make much difference whether or not a layer of mortar is broomed onto a properly prepared surface of either structural or mass concrete prior to placing concrete upon the surface. This research does indicate that a thin mortar either flowed or broomed onto the surface is superior to a thick mortar. In some cases superior joints were obtained without mortar. Also dry joints appeared to give superior results to wet joints.

Tests by the California Department of Water Resources showed that satisfactory construction joints were obtained in mass concrete including lean mass concrete (cementing materials equal to 188 lb/yd^3 in the Oroville Dam core block) without the use of a layer of mortar on rock or a previously placed concrete lift. Inspection of cores taken across and along the joints and tests confirms this (Fig. 25.10). In some cores, it has been difficult to locate the construction joint. Successful elimination of mortar in structural concrete with a starter mix using smaller maximum-size aggregate is also being satisfactorily accomplished.

The advantage of eliminating mortar is that some of the problems involved in obtaining and using the mortar are eliminated. The quantities of mortar may be rather small. Often only large equipment is available for its delivery. The use of mortar sometimes results in coating forms, reinforcing steel, and equipment with mortar which sometimes dries before concrete is placed over it and must be cleaned, or is otherwise "messy."

Regardless of whether mortar is used on rock or concrete construction joints, good construction-joint cleanup or rock cleanup is also necessary and the surface should be saturated, surface-dry, or free of water which would be indicated by shininess. More than usual care must be taken to ensure that the first layer on the construction joint is adequately vibrated to secure good bond.

Mortar, if used on rock or concrete construction joints, should be made with the same materials and the same mix as in the concrete except that the coarse aggregate is omitted. On some jobs where a small amount of mortar is needed, materials and a small mixer are kept on the job to supply the mortar as needed. As mentioned above, thin mortar flowed on the surface gives as good or better results than a thick mortar broomed in. Mortar will readily spread itself on a large area if dumped from a concrete bucket held at a height of

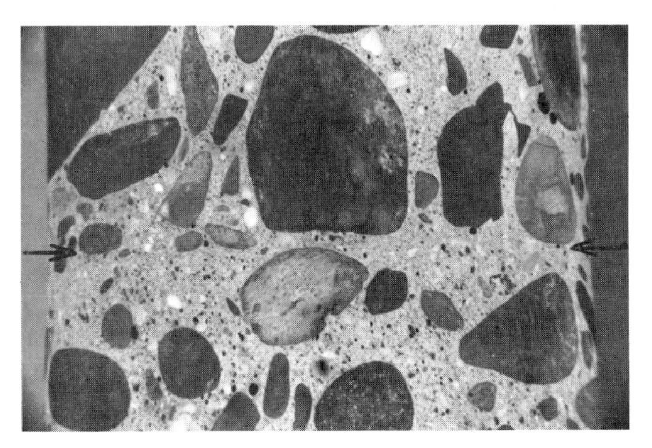

FIGURE 25.10 Polished concrete core showing excellent construction joint between two lifts of mass concrete. Joint was wet-sandblasted. Upper layer of concrete was placed on joint without use of mortar coating. (Arrows locate construction joint.) In some cores it has been difficult to locate the construction joint.

several feet, although the resulting splatter may be objectionable. The surface on which mortar is placed should be free of water puddles.

In structural concrete, such as wall sections where mortar is not used, a more workable bottom layer of concrete is sometimes used. A mix with more sand, more slump, or with part of the coarse aggregate omitted or smaller maximum size of coarse aggregate may be used. This provides not only a more suitable mix for bonding but a softer concrete in the bottom of the form for the next layer to work into.

FORMWORK

25.3 Form Preparation

The design and construction of forms are discussed in Sec. 25.4 and in Refs. 4 through 6.

Details of form construction are covered in Sec. 25.4, but certain details are important to those concerned with preparation for and placing concrete. Particular care must be taken with forms for concrete that will be permanently exposed to view or in which special architectural effects are required. Since deflections in formwork cause waviness in finished surfaces that show under certain light conditions and may be objectionable, forms for architectural concrete must be designed carefully, and deflections may govern rather than design loads. For example, waviness in panels caused by deflection of plywood or sheathing between the studs would not be objectionable in a location not subject to view, but such waviness would be objectionable where good appearance is desirable, and heavier plywood or closer stud spacing would be required. Frequently it is desirable to construct test sections or panels to determine that proposed material or techniques will achieve the desired architectural effect, and to acquaint the workers with techniques to be used. Sometimes this test section can be incorporated in a portion of the structure constructed at the start of construction and which will not be exposed.

Forms should be mortartight. Vibration will liquefy the mortar in the concrete, allowing it to leak from any openings in the forms, leaving voids, sand streaks, or rock pockets (Figs 25.11 and 25.12). Sometimes wood-board forms will need to be soaked with water before concrete placing starts, to close the joints and prevent leakage of grout or mortar.

Forms should be tied together in an approved manner and braced so that movement does not take place (Figs. 25.13 and 25.20). Formwork should be anchored and braced to adjacent sections and to shores below so that upward or lateral movement of any part of the formwork system will be prevented during concrete placing. Tie rods which act as spacers for forms are available and are preferred. Wood spreaders, if used, should not remain in the concrete, and means should be provided so that they can be easily removed. Sometimes they can be tied to wires and lifted out as concrete placing proceeds. It is the responsibility of the inspector and placing supervisor to see that they are removed. Wire form ties may be used in unimportant work if both sides of the concrete will be covered with backfill or not exposed to view. Wire ties should be cut off flush with the concrete surface. For concrete surfaces exposed to view, metal ties or anchorages should terminate not less than 1 to 1½ in from the surface of the concrete. Such ties should be made so that the projecting portion can be easily removed without causing spalling of the face of the concrete. She-bolts and tie rods or other similar form-tie devices should have sufficient threads engaged to develop full strength of the tie without thread stripping. Ties which when removed will leave an opening through the concrete should not be permitted. Directions for filling cone holes and other holes left by form ties are given in Chap. 44.

Line and grade on forms should be checked before and during placing (Fig. 25.14). Sometimes telltale devices are desirable so that any undesirable movement or excessive deflection of the forms can be detected. These can consist of lines or wires stretched between reference points with intermediate measurements, plumb bobs, plumb lines, check with a surveyor's transit or level, check with carpenter's level, or other methods.

When placing successive lifts of concrete on previously hardened concrete (construction joints), rustication strips (Fig. 25.15) or grade strips (Figs. 25.15 and 25.16) can be

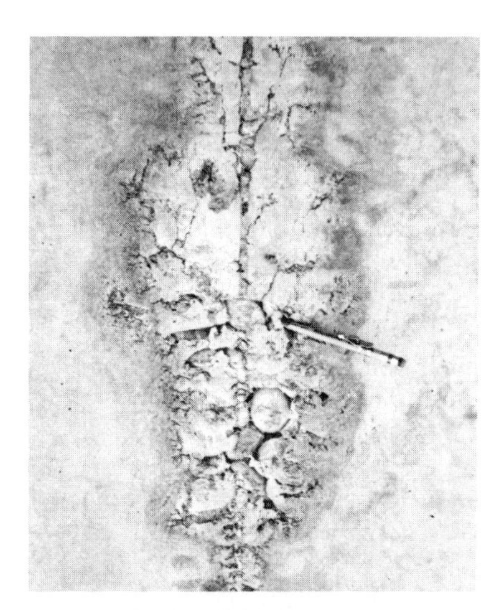

FIGURE 25.11 Sand streaks and rock pockets left in concrete because of loss of mortar at gap between form panels during vibration of concrete.

FIGURE 25.12 Rock pocket left because of leakage of mortar at opening for conduit during vibration of concrete.

FIGURE 25.13 Well-braced form for downstream-sloping face of a spillway. Upper row of ties used to tie forms for lift above this one. Struts are removed after form is partially filled with concrete. Grade strip is on form at top of oiled area.

used to obtain a straight and neat horizontal joint. The grade strip (Fig. 25.16) should be set as shown with its bottom edge about ½ in below the finished elevation for the placement. Rustication strips are frequently required at locations other than construction joints to provide a pattern for architectural effect and at control or expansion joints. The grade and particularly the rustication strip should be planned beforehand and should be straight and continuous across the structure so as to give a pleasing appearance. Form anchorage should be provided about 4 in below the top of a construction joint to anchor the forms for the placement above (Fig. 25.16). When the form is set for the succeeding lift, the sheathing should overlap the previous concrete not more than 1 in, as shown, and the forms should be drawn tightly to the existing concrete by means of the anchorage below and by ties close to the bottom of forms for the upper lift. This ensures a neat joint and will prevent the unsightly offset which many times occurs (Fig. 25.17). (More than 1-in lap prevents pulling the forms tight against the concrete.)

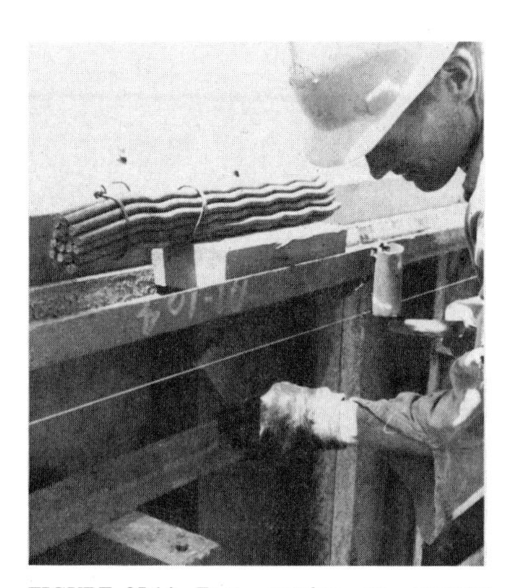

FIGURE 25.14 Form watcher or carpenter checking alignment of form during placement of concrete by checking clearance of form from string stretched between ends of form. "Pigtails" are placed in concrete to serve as anchors for forms for next lift.

All elements of form ties, such as nut washers (sometimes called cat-heads),

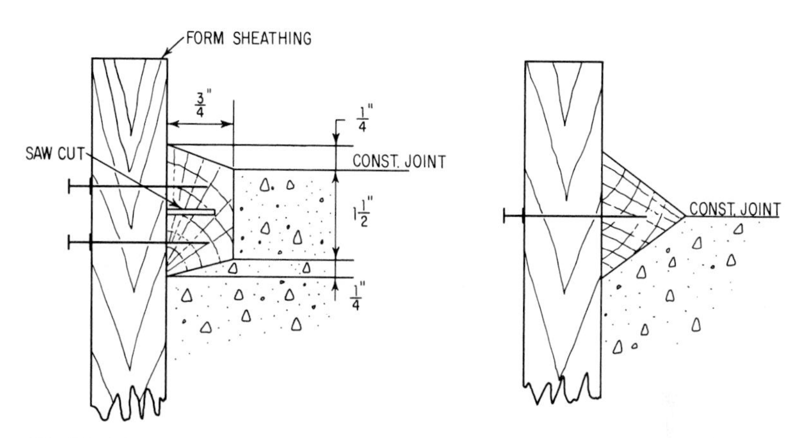

FIGURE 25.15 Rustication strips for construction joints or architectural effects. Use of double-head form nails facilitates removal of forms without removing strip. Saw cut facilitates removal of strip.

clamps, wedge clamps for snap ties, and wooden wedges used for alignment, should be secured to {the forms with nails or other means so that vibration of the concrete does not loosen them, causing loss of form alignment or more serious difficulties (Fig. 25.18). Likewise, form shoring, and bracing should be well secured to prevent similar difficulties.

Side openings should be provided and strategically located in the side of tall forms to facilitate placing of concrete, vibration, or inspection, when their use will facilitate such operations or where required. They should be made so they can be conveniently and quickly closed by the form carpenter when necessary and will not cause a blemish on the surface of the concrete where appearance is important. Runways should rest on the forms, not on the reinforcing.

Falsework, shoring, or centering should be adequately designed, supported, and braced to prevent undue sagging or movement and provide adequate support of the forms. Particular attention should be paid to bracing forms, shoring, and falsework against lateral movement. Provision should be made to leave all or at least part of falsework or shoring in place under beams, slabs, or arches after concrete has hardened until the concrete has obtained sufficient strength to permit safe removal. Forms should be designed so they can be removed and the shoring left in place when required. Vertical shores for multifloor forms must be set plumb and in alignment with lower tiers so that loads from the upper tiers are transferred directly to the lower tiers. Lateral loads must be transferred to the ground or to completed construction of adequate strength. All joints in falsework and shoring members should be butt joints when practical, not lap splices. Adequate spread footings, mudsills, or other supports should be provided. Mudsills should not rest on frozen

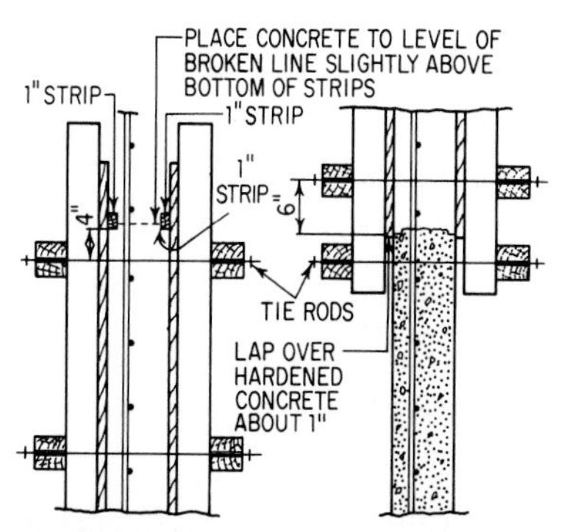

FIGURE 25.16 Grade strip at top of placement ensures neat straight line at construction joint. Minimum overlap of form permits tightening form to prevent unsightly joint (see Fig. 25.17). *(Reproduced by permission of American Concrete Institute.)*

FIGURE 25.17 Unsightly construction joint caused by form not fitting tightly against previous placement.

FIGURE 25.18 Nailed form-tie wedges and nut washers prevent their loosening when concrete is vibrated during placing.

ground—a thaw would be disastrous. Sometimes jacks or wedges can be used in falsework or shoring to adjust the forms or take up settlement before or during concrete placement. Wedges should be nailed in place to prevent their coming loose.

Tolerances for the concrete work are outlined in the job specifications normally. ACI Committee 117 Report, "Standard Tolerances for Concrete Construction and Materials," provides a referance for dimension and material tolerances. Dimension tolerances are those required in the completed concrete work, not form tolerances. Tolerances specified should be reasonable and only that required with concern for the use, serviceability, and aesthetic requirements of the job. Tighter tolerances than required may increase costs unnecessarily, if enforced.

25.4 Use of Forms

Safety Provision. Provision should be made, particularly on wall forms, for safe working conditions for the placing crew. Ladders, working platforms, hand railing, kick boards, and other safety necessities should be provided, as required by all safety codes (Figs. 25.19 and 25.20).

Form Coatings. Forms must be treated with a form oil or other coating material to prevent adhesion to the concrete. Many form oils and parting compounds, such as plastic, lacquer, or shellac, are available. Major oil companies formulate or recommend form oil for either metal or wood forms. Sometimes the same oil will work for both. The form coating should be formulated for the particular usage and material to which it is to be applied and should protect the form from water, strip readily from the concrete, not interfere with subsequent curing, painting, or other surface treatments, or stain or cause softening of the surface. Form oil or coatings should preferably be applied before the forms are erected. Care should be taken to avoid getting form oil on the reinforcing steel or other embedded items or on construction joints.

FIGURE 25.19 Safety provisions on a column for a freeway viaduct. Note ladder and safety cage, working platform and railing, and barricade around excavation.

Cleaning within Forms. All dirt, sawdust, shavings, loose nails, and other debris should be removed from within the forms before concrete placing is started. Provision should be made in wall or similar forms, particularly those for thin sections, for washing or blowing out all such debris, by leaving a panel or the bottom of the form loose or otherwise providing an opening at the bottom or end of the form for washout. Sometimes nails and wire clippings can be picked up by a magnet on a pole. The ends of reinforcing-steel tie wires should not rest against the form where they will cause unsightly rust stains.

Concrete Placing. The placing inspector and contractor's form watcher should carefully watch the forms during concrete placing for any signs of difficulties. Should any develop, placing should be immediately stopped or slowed as conditions warrant so that the conditions can be inspected and corrected.

Form Removal. Factors to be considered in determining when forms are to be removed are the effect of the form-removal operations in damaging the concrete, the structural strength or deflection of the concrete, curing and protection, finishing requirements, and requirements for reuse of the forms. Forms may be removed from a few minutes to days after the concrete is placed.

Forms should be removed using tools and equipment that will not damage either the concrete or the forms if the forms are to be reused or salvaged. Wooden wedges, not steel tools, should be used to separate the forms from the concrete. Sometimes air or water pressure can be used to remove forms such as for larger slabs or pans for waffle slab con-

FIGURE 25.20 Form insulated with 2½ in insulation placed between studs. Form is well braced by ties from top of other structures located outside picture. Note working platform and hand railing.

struction, in which case fittings are provided for attaching the air or water lines. Forms should be designed and made so that they can be easily removed without damage to either the concrete or forms, particularly those which are to be reused.

Strength of the concrete and structural requirements determine the form-stripping time for arches, beams, girders, and similar load-carrying structural members. Strength of the concrete for this purpose is best determined by strength tests of field-cured concrete cylinders subjected to the same curing conditions as the structure. Some requirements specify a safety factor of 2 or more for form removal for such structures.

Form removal at 1 day is common on walls, columns, sides of beams, girders, and slabs, and other places where strength is not a problem. Forms of the soffits of arches, beams, and girders and floor slabs are commonly left in place 7 days, 14 days, or more depending upon strength and deflection considerations and strength gain of the concrete.

In summer concreting, forms are not considered curing and forms should be removed as soon as practical and specified curing started immediately, as discussed in Chap. 29. Sometimes forms are loosened without being removed and curing water is allowed to run down between the forms and concrete. Only under this condition should forms be left on as long as practical. However, it should be noted that leaving the forms in place is better than removal if it is known the concrete will not be properly cured, a not uncommon practice despite specification requirements for curing.

In winter concreting when insulated forms are used, they should be left on the concrete as long as practical to protect the concrete from cold, thermal shock, or freezing conditions (Fig. 25.20). Reference 7 gives further information on the insulated forms.

Early form removal is sometimes desirable to permit finishing the concrete to eliminate bug holes such as in inverts, on spillway surfaces, or for other reasons, or to facilitate repairs to the concrete while it is still "green" and bond of the repairs is improved. Early form stripping to permit earlier reuse of the forms may be facilitated by acceleration of the strength gain of the concrete. This can be accomplished by such means as warmer concrete, steam curing, or heated enclosures (with care taken to prevent drying), more cement (lower water/cement ratio), lower slump (lower water/cement ratio), use of high-early-strength cement, use of accelerators (when permitted by specifications), vacuum-processed concrete, and insulation of forms. Form removal in a few hours to permit reuse of the forms on a 24 h cycle or sooner is common in tunnel lining and in precast-concrete-products plants.

Reuse of Forms. As mentioned previously, forms should be designed to permit easy removal and reuse. Forms should be carefully removed, cleaned, repaired, handled, and stored so that they are not damaged. Cleaning may be done by a wooden scraper and stiff fiber brush on wood forms or steel scrapers and wire brush on steel forms. Mechanical wire brushes are used for cleaning large numbers of form panels on large projects. Damaged places are repaired as required. The holes are patched with metal plates, corks, or plastic inserts. When coating with some coating materials or oils, a drying period is required before stacking or storing forms.

Reusable form hardware such as form ties, she-bolts, nut washers, and wedges should be sorted, cleaned if necessary, and stored for reuse.

REINFORCEMENT

Reinforcing steel, commonly called "resteel" or "rebars," is placed in concrete to reinforce the concrete structure adequately so that it will support expected loads. Concrete is weak in tension, and most reinforcing is used to provide tensile reinforcement. Steel is also used to provide compressive reinforcement, in columns and arches, for example, and to resist shrinkage or temperature stresses. Reinforcing should be accurately placed. Inaccuracies can reduce the strength of the structural unit containing the steel or, if the steel is placed in the wrong place (the wrong side of a cantilever beam, for example), might lead to failure of the structure. It is particularly important that reinforcing steel be properly placed in areas subject to earthquakes, tornadoes, and hurricanes. Investigation of structural failures following such disasters shows many violations of specifications and good practice in placing the reinforcing steel and making adequate connections between structural elements. This caused or contributed to the failure. Also indicated is a lack of competent inspection.

Design drawings and specifications give the grade of steel, its location in the structure, and other requirements. On larger jobs, specialty contractors prepare bar lists and schedules from the design drawings, fabricate, and place the reinforcing steel. Bar lists and schedules are submitted to the designer for approval.[10]

On larger and more sophisticated structural designs such as ductile frame high-rise buildings and others, considerable effort and thought has gone into the design of the reinforcing steel. In such cases, the structural designer may be the best one to prepare the details of the reinforcing steel and may be commissioned to do so. Prior to placing concrete, inspection by both the contractor and owner's representatives should be made to see that the reinforcing steel is properly placed. Particular attention should be given to grade of steel; size, number, location, and spacing of bars; correctness of bends, laps, or splices; clearance from forms or required cover; cleanness of steel, supports, and ties; and that the sections of the structure are satisfactorily tied together by the resteel.

Those concerned with reinforcing steel will find Chap. 3 and Refs. 8 through 13 helpful. The ACI Building Code, ACI Standard 318, is quite detailed.

The various grades of reinforcing steel and their properties, sizes and weights of bars available, and markings used to designate the producer, size, and grade are discussed in Chap. 3. Bars received on the job are tagged or marked for identification. Bent bars are identified with tags giving order number, number of pieces in the bundle, size, length, and the bar mark. The reinforcing steel should be checked for conditions that will limit or make concrete placing difficult or interfere with concrete placing and consolidation (see Fig. 25.21). Sometimes remedies for such situations may be found. It may be necessary to limit the maximum aggregate size or or permit a greater slump in the concrete. However, the designer should be consulted and approve any changes in the reinforcing steel or other conditions which might affect the design.

Epoxy-coated reinforcing steel is required on some jobs to provide greater protection of the reinforcing steel from corrosion. Epoxy-coated steel must be handled carefully to prevent damage to the epoxy coating. Bundles of steel should be handled with fabric slings. Repair of damage to epoxy-coated bars is usually required.

Spacing of Bars. Design and detailing should provide for proper spacing of the reinforcing steel in accord with the ACI Building Code or other requirements. This limits the closeness of the bars to each other in various situations. These conditions have been provided for by the reinforcing-steel detailer. Sometimes bars are bundled: for instance, up to four bars may be placed together. Sometimes temporary removal or moving of bars, particularly in heavily reinforced mats, may be done to facilitate placement of the concrete.

FIGURE 25.21 Reinforcing steel in a ductile frame design structure prior to erection of forms. Such conditions. make placing of concrete very difficult. On seeing such conditions, it was recommended that concrete with ⅜ in maxim size aggregate, a high-range water-reducing admixture (superplasticezer), and a 7 in slump be used. Some of the reinforcing, steel and ties in the section were moved or cut out, with permission and supervision of the designer, to permit getting the concrete pump line and vibrators in place so that concrete could be placed and consolidated.

Splices. Splices are frequently necessary because of manufacturing, fabrication, handling, or transportation considerations. Splices should be located as shown on the design or detailing drawings. Splices should occur at locations of minimum stresses in the structure.

Splices may be made by lapping the bars a distance determined by the bar size, grade of steel, and concrete strength as given in the ACI Building Code or by the designer.

Bars may also be connected by welding or by various connectors, some of which are suitable for tension splices and others, for compression splices only.[12] The American Welding Society has requirements for welding reinforcing steel.[13]

Cover. Concrete protects reinforcing steel from corrosion and serves as fire-proofing. Adequate cover or embedment of the steel in concrete to provide this protection is specified in the ACI Building Code.

The following minimum concrete cover for cast-in-place concrete is given. Cover for precast and prestressed concrete may vary from this, consult the ACI Building Code.[8]

Cast-in place concrete (nonprestressed)	Minimum cover, in*
Cast against and permanently exposed to earth	3
Exposed to earth or weather:	
No. 6 through No. 18 bars	2
No. 5 bars, W31 or D31 wire, and smaller	1½
Not exposed to weather or in contact with the ground	
Slabs, walls, joists:	
No. 14 and No. 18 bars	1½
No. 11 and smaller	¾
Beams, girders, columns:	
Principal reinforcement, ties, stirrups, or spirals	1½
Shells and folded plate members:	
No. 6 bars and larger	¾
No. 5 bars, 5/8-in wire, and smaller	½

*The cover specified is that to the outside, not the center of the bar.

In corrosive atmospheres or severe exposure conditions, the amount of concrete protection should be suitably increased, and the denseness and nonporosity of the protecting concrete should be considered, or other protection should be provided.

Exposed reinforcing bars, inserts, and plates intended for bonding with future extensions should be protected from corrosion.

When a fire-protective covering greater than the concrete protection specified in this section is required, such greater thicknesses should be used.

Corner and Wall Intersection Bars and Connection Details. Horizontal bars at corners and at all intersections are required to hook around the corner or into the intersecting wall to provide strength at the corner or wall intersection. This is usually accomplished by either putting a hook on a long bar or by using a short corner or elbow bar. It is important that good connections between various structural elements be made in order that the structure be tied together and perform satisfactorily.

25.5 Bar Supports, Spacers, and Ties

Bars should be supported, anchored, and tied to hold them in place before and during concrete placing (Fig. 25.22). Supports, anchors, or ties should not permit subsequent leakage of water into the hardened concrete, which would cause corrosion.

FIGURE 25.22 Footing showing reinforcing neatly and accurately placed and well supported. *(California Department of Water Resources photograph.)*

In footings, mats of bars can be supported by precast concrete blocks made for this purpose, or chairs. Chairs made for this purpose have a large bearing pad of sheet metal (called a "sand pad") to prevent their sinking into the soil. Mats or particularly wire mesh should not be placed on the ground with the expectation they can be pulled up into place after concrete is placed. Such a procedure is uncertain at best—it is far better to support the mat or mesh adequately beforehand or to place concrete to the mat elevation, place the mat or mesh on it, then resume placing. This latter practice is commonly used in paving reinforced with mats.

For joists, slabs, beams, and girders, various wire bar supports shown in Chap. 3, are available.

Precast mortar or concrete blocks should not be used to support or block steel where they will be exposed and detract from the appearance of the structure. (Precast blocks, sometimes referred to as "dobies," are many times made of poor-quality mortar or concrete. The quality of the blocks should be specified.)

Wire ties are commonly used to assemble and support steel in mats and walls. Wire used for this purpose is No. 16, 15, or 14 gauge, black, soft iron wire; No. 14 gauge wire should be used for heavy work. No. 18 gauge wire is too light to do a good job usually.

Wall reinforcing should be adequately supported to prevent its being pushed toward or against the form when concrete is placed. This may be done by tying it to the form tie rods, tying two curtains of steel together to prevent their spreading, by concrete blocks between the steel and form where such is permissible, or by driving nails into the form at the proper locations so that bars can rest against the nailheads, and be tied to the nails, thus properly spacing the steel from the form. (This method is used only where the nails would be unobjectionable. Nails might rust, causing unsightly staining on exposed concrete.)

25.6 Cleaning Reinforcement

Reinforcement at the time of placing concrete should be free of mud, oil, or other foreign matter that will reduce or destroy bond. Mill scale or a light coating of rust tightly bonded to the steel is not objectionable. Loose, flaky, scaly rust, which would affect the bond, should be removed by scraping, shock treatment (dropping, hammering, vibration), wire brushing, or possibly sandblasting. It is not necessary to have a bare metal surface, only to remove loose, flaky rust. Bars rusted to the extent that their size is reduced below the ASTM requirement for new bars should not be used. This can be determined by removing the rust and weighing a measured length of bar. Bars should be kept free of mud and mud should be washed off before using. Loose mortar should be removed. Mortar so tightly bonded to the steel that it is not removed by vigorous wire brushing can remain. Grease or oil should be removed by a propane torch, using care not to overheat the bars, or washed off with kerosene or gasoline with adequate safety precautions for using such materials. When oiling forms, oil may get onto the bars. If the form oil evaporates or is a very thin coating, it will do no harm. If it is a thick coating, it should be removed.

EMBEDDED ITEMS AND BLOCKOUTS

25.7 Embedded Items

Most concrete contains various embedded items such as catch basins, manholes, sumps, traps, conduit, pipe, grounding cable; bolts for anchoring machinery, equipment, or fixtures; inserts for supporting or attaching wall material or panels; water stops, drains, and form hardware, as shown in Fig. 25.23. The plans and specifications should be carefully checked to determine the presence of and location of such embedded items.

Usually embedded items are installed in the forms prior to start of placing concrete and held firmly in place by attaching to the forms or reinforcing steel, or by templates. Bolts for anchoring machinery or equipment must be accurately placed. Commercially available bolt anchors (Fig. 25.24) or job-made anchors (made from steel plate, pipe, and bolts) provide a greater length of bolt to take strain and permit some shifting of the bolts when the machinery is installed to allow for inaccuracies in the machine base or in locating the bolts. The location of embedded items that are to be installed during concrete placing should be determined and referenced on the form, and the items should be available for embedment during placing. Any sizable embedded items such as heating ducts must be securely anchored to prevent floating due to uplift of the concrete. Pipes, con-

FIGURE 25.23 Concrete placement for small pumping plant showing several embedded items (conduit, anchor bolts, sumps) and blockouts (circular blockout in foreground is damaged and will need to be replaced). (*California Department of Water Resources photograph.*)

duits, and other such items having voids should be capped or plugged to prevent concrete from getting into them.

Water stop should be placed as shown on the drawings. Water stop should be firmly held in place so that it is not displaced during concrete placing operations. Expansion-joint material should cover the entire face of the concrete when it is used. Narrow strips of expansion-joint material should not be used. Expansion-joint material should not be nailed to concrete—use an adhesive.

Aluminum conduit or other aluminum products should be avoided in concrete. The high alkalinity of concrete (pH~13) causes corrosion of aluminum and distress in the concrete with cracking or spalling (see Fig. 25.25) sometimes, some aluminum products, such as aluminum handrail, are permitted if their embedded portion is liberally coated with an epoxy, asphalt, or some other barrier to protect the aluminum from the concrete.

Some embedded items are to bond to the concrete. These should be kept free of mud, oil, grease, form oil, or other foreign material that would affect bond. Others should not bond to concrete and should be coated with asphalt, tar, grease, or similar material to prevent bond.

25.8 Blockouts

Blockouts are openings made in concrete to provide places where gate guides, handrail posts, seals, doors, or windows can be placed later (Fig. 25.23). Blockouts may be formed of wood, Styrofoam, or metal. Rigid embedments or blockouts may restrain the concrete and initiate cracks. This may be alleviated by using expansion-joint material or a resilient or compressible wrapping around the embedment or blockout. Wood blockout forms

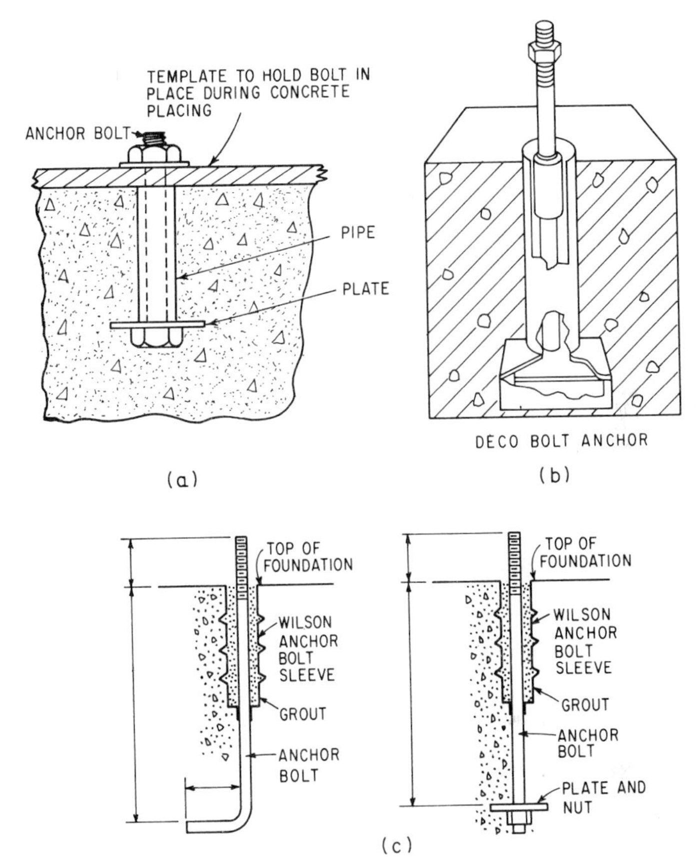

FIGURE 25.24 Anchor-bolt details. Pipe-sleeve device as shown or commercially available bolt anchors provide more flexibility for fit of bolt to machine base and longer length of bolt to take strain. *(Photographs courtesy of Wilson Anchor Sleeve, Inc., and Decatur Engineering Co.)*

should be made so as to facilitate removal later. This may be done by making them in several sections, splitting sections, making tapered pieces, and making angle cuts. Wood blockouts should be soaked in water ahead of concrete placing so that they will not expand and crack the concrete if located close to an edge or corner. Blockouts made of Styrofoam can be easily removed. Blockouts formed with steel (cans) should be removed before the final installation; otherwise the steel may rust, causing an unsightly rust stain on the concrete surface. Rusting can also cause the concrete to spall if close to an edge or surface. The use of aluminum cans can also cause similar problems if the cans are left in place.

Blockout openings are filled later, after installation of the item to be placed, using concrete, mortar, or grout as appropriate. Sometimes the use of aluminum powder in the concrete, mortar, or grout or the use of nonshrink or fast-setting cements may be considered desirable.

FIGURE 25.25 Precast bollard showing sever cracking in the concrete due to heavy corrosion of the aluminum light fixture box. Disassembly of a bollard showed the aluminum box to have as much as ⅛ inch or more of white corrosion product, resulting in severe cracking of the concrete around the box.

Shifting or cutting reinforcing steel to permit installation of embedded items or block-outs should be permitted only if shown on the plans or specifications or if approved by the engineer after consideration of the effect of such shifting or cutting. The effect of the location of items on the strength of structural members should be considered and approval of the engineer obtained if there is any question.

FINAL INSPECTION

During all phases of construction activity prior to start of placing concrete, continuing inspection of the preparations by both the contractor and the inspector representing the owner should ensure that the work is progressing satisfactorily. The progress of the work is the responsibility of the contractor. The inspector should be available for consultation and interpretation of the plans and specifications. The inspector should refrain from directing or running the job for the contractor. To do so assumes responsibility for the work.

On large jobs, it is desirable to have an inspection card such as that shown in Fig. 25.26 at the location of each placement being prepared. As each item is completed it is inspected by both the foremen and inspector assigned to inspect that particular item and

DEPARTMENT OF WATER RESOURCES

FEATHER RIVER FISH HATCHERY

CHECK OUT SHEET FOR CONCRETE PLACEMENT

Monolith No. 5-9 Lift Elev. Foundation Rock to Elev. 143.0

Placement No. 4

CHECK ITEM	APPROVED			
	Frazier-Davis		Dept. Water Resources	
	Date	Initials	Date	Initials
Foundation Adequacy	7-17-62	A.E.M.	7-18-62	G.H.
Rock Cross Sections Taken	—	—	7-18-62	R.C.B.
Formwork - Line & Grade	7-19-62	K.L.S.	7-19-62	E.R.G.
Formwork - Construction	7-19-62	K.L.S.	7-19-62	E.R.G.
Embedded Items-Waterstop	7-19-62	K.L.S.	7-19-62	E.R.G.
~~Resteel~~	—	—	—	—
~~Mechanical~~	—	—	—	—
~~Piping~~	—	—	—	—
~~Misc. Metalwork~~	—	—	—	—
~~Electrical~~	—	—	—	—
Wet Sandblast	—	—	OK w/o	G.H.
Formwork - Oiled	7-19-62	A.E.M.	7-19-62	CHE
Final Concrete Cleanup	7-19-62	A.E.M.	↓	CHE
Concrete Placing Equipment	7-19-62	A.E.M.	↓	CHE
Concrete Curing Equipment	7-19-62	A.E.M.	↓	CHE
OK for Concrete	7-19-62	A.E.M.	↓	CHE

Remarks *Close washout openings in form before placing starts. 9/19/62 RFA*

FIGURE 25.26 Checkout sheet for concrete placement. (*California Department of Water Resources form.*)

their signatures are entered on the card after completion and approval. After all applicable preparations are completed and approved, the placing inspector makes a final inspection to see that all applicable items have been previously inspected and are still satisfactory. The inspector should review details of the concrete to be used such as slump, cement content or water/cement ratio, maximum size of aggregate, air content, temperature, concrete-delivery schedule, and placing procedure. The placing inspector determines that suitable equipment for delivering and handling the concrete; vibrators including standbys; sufficient workers; preparations for protection of the concrete in either hot, cold, or rainy weather; equipment for finishing and equipment and material available for curing are all ready or arranged for. If concrete placing or finishing is to continue after dark, adequate lighting should be ready.

Quite obviously many of the details concerning the concrete placing should be discussed and agreed on in discussions or meetings with the contractor prior to the final inspection. If previous discussions, preparations, and inspections are adequately done, the final inspection takes only a few minutes.

The inspector authorizes concrete to be ordered only after confirming that everything is or will be ready.

REFERENCES

1. Investigation of Methods of Preparing Horizontal Construction Joints in Concrete, U.S. Army Engineer Waterways Experiment Station, *Tech. Mem.* 6–518, July 1959.
2. Investigation of Methods of Preparing Horizontal Construction Joints in Concrete, U.S. Army Engineer Waterways Experiment Station, *Tech. Mem.* 6–518, Report 2, Feb. 1963.
3. Wuerpel, C. E.: Tests of the Potential Durability of Horizontal Construction Joints, *ACI Proc.,* vol. 35, p. 181, June 1939.
4. ACI Standard 347–88 Guide to Formwork for Concrete.
5. Hurd, M. K.: Formwork for Concrete, *ACI Spec. Publ.* SP-4. 4th Ed., 1979.
6. "Forms for Architectural Concrete," Portland Cement Association, Skokie, Ill.
7. Tuthill, L. H., R. E. Glover, C. H. Spencer, and W. B. Bierce: Insulation for Protection of New Concrete in Winter, *ACI Proc.,* Vol. 48, p. 253, Nov. 1951.
8. ACI Standard 318–89 Building Code Requirements for Reinforced Concrete.
9. CRSI, "Reinforcing Bar Detailing Textbook."
10. ACI Detailing Manual, *ACI Publ.,* SP–66(88).
11. "CRSI Recommended Practice for Placing Reinforcing Bars," Concrete Reinforcing Steel Institute, Chicago.
12. ACI Committee Report 439.3R, Mechanical Connections of Reinforcing Bars.
13. AWS D1.4, Structural Welding Code—Reinforcing Steel.

CHAPTER 26
PLACING CONCRETE

Robert F. Adams

STRUCTURES AND BUILDINGS

Structures and buildings are made up of various elements such as foundations, floors, walls, columns, and beams (Fig. 26.1). The general principles of placing concrete in these structures are covered in Chap. 24; preparations are covered in Chap. 25.

Equipment should be carefully selected, as discussed in Chap. 24, in order to handle satisfactorily and expeditiously the concrete specified and otherwise considered satisfactory.

FIGURE 26.1 Typical high-rise concrete building frame consisting of cast-in-place and precast elements. Columns, floors, and central service core containing elevators, plumbing, and other services are cast in place. The central core may be slipformed. Floor beams are precast. Precast panels will be placed later in exterior walls. Movable crane is used for setting precast elements and for placing concrete as well as handling other materials. Note shores in place supporting floors.

Concrete mixes are usually specified or may be selected from the principles outlined in Chap. 11, based on strength, durability, workability, and other considerations.[1,3,12-15]

26.1 Foundations

Foundations for structures are important in that they support the structure. Foundations vary widely depending upon the structure, subsurface conditions, and the loads to be supported. Their design for important structures must be carefully considered by design engineers and by experts in foundation engineering. Construction personnel should be alert to foundation conditions differing from those assumed in the design or expected from foundation explorations which might affect the structural integrity of the structure.

Foundations vary from piling of various types to caissons to spread footings to mat or raft foundations. With such a wide variety of possibilities, procedures for placing concrete can vary considerably depending on the job situation, but they should follow the principles outlined in Chaps. 24 and 25 and in other parts of this book. Some foundations are massive, and the principles of placing mass concrete discussed later in this chapter apply.[2]

Problems caused by water lead to many difficulties in foundation concrete, and water should be controlled where possible. In some cases it may be necessary to place concrete underwater—which is covered later in this chapter.

Concrete in cast-in-place piles should be carefully placed to be sure that voids are not left. The shells or drilled openings should be carefully inspected immediately prior to placing the concrete to determine that they are open and free of water.

26.2 Buildings

Construction of buildings by such techniques as slipform, tilt up, precast, prestressed, or lift-slab is covered in Chaps. 33, 34 and 35.

Slabs in buildings are used for either floors or roofs. They might be cast on the ground, in which case they may or may not be reinforced; if cast on forms they are reinforced (Fig. 26.2). Construction of floors is discussed in Ref. 4 and heavy-duty floors are discussed in Chap. 27. Floors on grade may be placed in sections with control joints as discussed in Chap. 28. For a large building some of the paving equipment discussed later in this chapter can be used to place the floor slabs on grade. Also light equipment is available (Fig. 26.3). Warehouse buildings where lift trucks move at high speed and where stacks are high require "superflat" floors for which more than the usual floor finishing techniques are required. Such work must be done by those experienced in meeting the tolerances required for superflat floors.

On-grade slabs frequently experience problems with dampness and efflorescence. The use of plastic sheets under the slab remedies this. However, concrete placed on plastic sheeting may crack severely—but this is remedied by putting two or 3 in of sand over the plastic before placing the concrete. Another problem with on-grade slabs is shrinkage cracking caused by control joints being placed too far apart or other inadequate jointing in the slab. On-grade slabs must be properly jointed to control cracking.[21,41]

Walls, columns, and beams provide the framework of the building. For economy, the design should be such that maximum simplicity and reuse of forms are achieved. It is likewise important that forms be tight and have a satisfactory finish for the sake of appearance.

FIGURE 26.2 "Krete-Placer" distributing concrete from a pump line onto a floor placement in a multistory building. The flexibility of this equipment permits placement of the concrete at any place in the slab. This equipment eliminates most of the heavy labor in dragging the end of the pump line around. (*Photograph from Construction Forms, Inc.*)

Construction joints against which fresh concrete is to be placed should be cleaned by one of the methods outlined in Chap. 25. The use of drop chutes will depend upon the placement conditions. If used, they should be of a size and shape that can readily handle the concrete considered otherwise suitable. Drop chutes should not be a "bottleneck." If necessary to move drop chutes, they should be so arranged that they can be quickly and easily moved or shortened as necessary and not hold up placing concrete. If concrete leaving the drop chute "stacks" and causes segregation, this can be alleviated by keeping a vibrator working in the pile to move concrete away from it, lowering the pile and thus eliminating the segregation. It is good practice for the first few inches of concrete in a wall or column to be more workable, i.e., have more sand (or less coarse aggregate) to assist in starting. Concrete should be systematically placed in uniform layers of not more than 15 to 18 in and be followed by systematic consolidation effort, vibration preferably. It is also good practice to decrease the slump of the concrete as the concrete level rises to compensate for bleeding, providing more uniform and durable concrete. This should be done particularly in the top 2 ft of walls (see Fig. 24.14). In placing concrete around blockouts or door, window, or ventilation openings, care should be taken to determine that concrete fills the underside of the opening (Fig. 26.4). This can be assured by vibrating concrete placed on one side of the opening until it fills under and emerges at the opposite side to the level of the bottom of the opening, then placing it on the other side.

Where walls and floor slabs or columns and beams and floor slabs are placed integrally in one placement, cracking can occur at the junction of the two elements because of bleeding and settlement of the fresh concrete in the plastic state and restraint of the forms.

FIGURE 26.3 Truss screed (sometimes called "Texas screed") used for vibrating and screeding concrete slabs or floors. Equipment consists of a light truss which is pulled along the slab by means of wires anchored ahead of the screed and a hand winch. Screed is vibrated by eccentric weights on a shaft rotated by a small gasoline engine located at one end of the screed. This is effective for vibrating thinner slabs. Thicker sections are vibrated with internal vibrators with the screed being used to vibrate and screed the surface.

This can be prevented by placing the concrete to the top of the wall, or column and waiting until the bleeding of the concrete ceases and the concrete has stiffened considerably before placing the beam or floor concrete. Vibration of the concrete at the junction of the wall or column and beam or floor slab after bleeding has ceased will accomplish the same end. A similar condition exists at the top of blockouts such as door or window openings, and similar preventive measures should be taken. Joints in walls deserve attention to control cracking. The use of water stop or other joint-sealing methods are required for watertight joints.[22,42]

FIGURE 26.4 Void left below window opening due to failure to properly place and vibrate concrete.

It is also good practice to cross-slope the tops of walls which will later be permanently exposed to the elements so that water will drain off rapidly.

Tiltup buildings provide economical construction for one- or two-story commercial or industrial buildings. In this type of

FIGURE 26.5 Tiltup construction. Concrete panels cast flat on the building floor are being erected. Several panels can be seen in the foreground and are ready for erection.

construction, the floor slab is constructed first. The wall panels are cast on the floor slab using simple side forms. This results in low forming costs. The wall panels are erected very quickly by crane (Fig. 26.5).

Arches, Domes, Paraboloids, and Folded Plates. Structures of these types are characterized by thinner concrete sections, more complicated formwork with curves, slopes, intricate reinforcing, and difficult concrete placing and finishing conditions, and with considerable hand placing and finishing. Steep slopes require top forms to hold the concrete in place during vibration.[5] These should be made in small panels and be readily removable for finishing. Concrete mixes for these types of structures need to be proportioned to provide the required strength and to be readily placeable. Sometimes the maximum aggregate size is limited to ½, ¾, or 1 in for thinner sections.

Domes are sometimes built without form support by constructing the dome on the original earth which has been trimmed to the required shape or on fill trimmed to the desired shape. The earth is later excavated. In some cases, the completed dome is lifted into place. Air-inflated balloons provide forms for some structures such as culverts, particularly for shotcrete application of the concrete.

Whatever the type of structure, concrete placing should be consistent with the general principles outlined in Chaps. 24 and 25, and every effort should be made to obtain good-quality, dense concrete.

26.3 Heavy Piers and Retaining Walls

The construction of heavy piers and retaining walls varies from that of structural to mass concrete depending on the design and size of the structure. Adequate foundations are most important, as discussed earlier in this chapter and in Chap. 25. Construction of the portions above the foundations should follow the principles outlined in Chaps. 24 and 25.

26.4 Bridges and Viaducts

As with most other structures, foundations are most important and should be given the consideration discussed earlier in this chapter and in Chap. 25. The portions of the structures above the foundations should be constructed utilizing the general principles outlined in Chaps. 24 and 25 (see Figs. 26.6 through 26.8). Particular care should be taken in constructing the bridge deck, particularly bridges for high-speed highway traffic, to obtain a smooth surface for safe travel. A number of finishers are available for finishing the surface (Figs. 26.9 through 26.11). Care should be taken in supporting and aligning the forms and the surfaces the finishers operate from to assure a smooth surface. The forms should be carefully watched for excessive deflections during concrete placing.[23,24] In some cases, designers require the bridge deck to be placed in sections following a planned sequence or in alternate sections to provide uniform loading. This is true particularly of the deck in a suspension bridge and may be required in continuous-beam and arch bridges.

Bridges around water in climates subject to freezing and thawing conditions are particularly vulnerable to deterioration from these conditions. Bridge decks are most vulnerable when ice-removal salts are used. When these conditions are to be encountered, every effort should be made to use concrete having a sufficiently low water cement ratio and adequate air entrainment to ensure durability. Reinforceing steel should have adequate cover. Only competent and conscientious construction and inspection can ensure that a structure with the ability to withstand the deteriorating influences of freezing and thawing is achieved.

FIGURE 26.6 Freeway viaduct. Hollow-box girder is about 330 ft. long; columns are about 110 ft centers.

FIGURE 26.7 Shoring and forms for hollow-box-girder freeway viaduct. Forms are designed and constructed to permit frequent reuse, the shoring remaining in place until concrete has attained sufficient strength to permit removal.

FIGURE 26.8 Underside of a simple concrete bridge across a flood channel showing excellent concrete placing. Foundations are cast-in-place piles, columns cast in fiber-tube forms; deck is a simple flat slab. (*California Department of Water Resources photograph.*)

FIGURE 26.9 Placing concrete in a slab. Similar equipment is used for bridge decks. After concrete is distributed onto the slab or bridge deck by a side discharge conveyor and vibrated (not shown), concrete is screeded by an auger, rolled with a ribbed roller to embed the coarse aggregate, finished with a counterrotating smooth roller, finished with two smooth drag plates, and given a nonskid surface finish with a carpet drag.

MASS CONCRETE

Mass concrete may be defined as any large volume of concrete cast in place generally as a monolithic structure incorporating a high proportion of large coarse aggregate and a low cement content, and intended to resist loads by virtue of its mass. Mass concrete is generally thought of in connection with gravity dams but is used in arch dams, spillways, large massive foundations, and in large structures such as power and pumping plants or

FIGURE 26.10 Clary screed finishing a bridge deck. (*Clary Manufacturing Co. photograph.*)

FIGURE 26.11 Capital finisher finishes lightweight concrete in a deck of a suspension bridge. Deck was placed in sections following the sequence required by the designers to balance loading as required for such structures. (*California Department of Water Resources photograph.*)[16]

bridge piers. Considerable information on mass concrete is given elsewhere in the literature.[2,6,25-29,40]

Mass concrete is basically no different from any other concrete except that it generally uses large-sized aggregate. The larger aggregate size and grading permit a lower sand, water, and cement content. Such concrete, because of its lower cement content, undergoes a lower temperature rise which minimizes the buildup of heat and consequent temperature problems in the structure.

Mass concrete is placed in blocks, or monoliths, from 30 to 70 ft wide and 10 to 200 ft long. With modern temperature control of mass concrete there is no limit to the length of a block. In the United States, lifts are usually 5 or 7½ ft high (Figs. 26.12 and 26.13).

Roller-compacted concrete developed in recent years for mass concrete and other applications is discussed in Chap. 30.

26.5 Mixes

Concrete for mass concrete usually contains aggregate with a maximum size of 6 or 7 in but smaller maximum sizes, such as 4½, 4, or 3 in, may be used. There is no advantage to using aggregate larger than about 6 in. Concrete with aggregate larger than 6 in shows little gain in any benefits, is more difficult to use, and is harder on the equipment. Cement contents vary from about 4 to as low as 2 sacks per cubic yard. It is not uncommon to use mixes having different cement contents in the same placement. For example, a mix with 2½ or 3 sacks of cement per cubic yard might have sufficient strength and would be used for the interior concrete. For the exterior concrete exposed to freezing and thawing, sulfate, or other deteriorating conditions where durability is a problem, a mix having additional cement such as 3½ or 4 sacks might be used. Mass-concrete mixes have from

FIGURE 26.12 Mass concrete being placed in Thermalito Dam of California Water Project. Twelve-cubic-yard buckets of concrete are transferred from rail-mounted concrete-delivery car from mixing plant by cableway. Forms are steel. Concrete is placed in blocks and 7½ ft lifts. Concrete placed in five 18 in layers in each lift, in "stair-step" arrangement. Note forms for gallery (extreme right side of photograph) (*California Department of Water Resources photograph.*)

about 19 to 24 percent sand by absolute or solid volume of total aggregate. Type II, moderate heat- and moderate sulfate-resisting portland cement, is largely used for mass concrete now and has replaced Type IV, low-heat cement. Modern temperature-control methods make it practical to use Type II cement. Type V cement may be used where more severe sulfate conditions are encountered.

FIGURE 26.13 Another view of dam shown in Fig. 26.12. Note numerous blocks of concrete at various stages of construction. Note burlap shading for recently placed concrete and water curing. (*California Department of Water Resources photograph.*)

Natural sand and gravel aggregates are considered best because of the better workability of concrete made with them. However, crushed aggregates are satisfactorily used. Crushed aggregates should be prepared in equipment which produces good particle shape, with minimum "flats and slivers."

The coarse aggregate may contain as much as about 35 percent of the 6 to 3 in fraction with successively smaller percentages of smaller sizes. Gap gradings are sometimes used in other countries. A study of the aggregate sources and trial mixes should be made for large jobs to utilize the aggregate deposit and available concrete materials most efficiently. For example, in the materials investigations for Glen Canyon Dams,[29] the aggregate deposit contained 10 percent 6 × 3 in size. Computations of aggregate requirements for the various classes of concrete showed that the coarse aggregate for the mass concrete could contain 13 percent of the 6 × 3 in fraction and utilize the deposit most efficiently. Trial mixes showed that this amount of cobbles (6 × 3-in fraction) could be satisfactorily used. A larger amount of cobbles would have required overexcavation and considerably more waste. Similar studies can sometimes justify the use of smaller maximum aggregate size or sand grading different from that usually specified.

Admixtures are widely used in mass concrete. Air entrainment is particularly beneficial in improving workability as well as durability for the surface concrete. The amount of air entrained is usually in the range of 3 to 3½ percent. Pozzolans can be used as a part of the cementing material, usually about 15 to 35 percent. Pozzolans can have many effects such as improving workability, reducing the temperature rise of the concrete, and sometimes reducing cost. Water-reducing and -retarding admixtures are beneficial in that they reduce the water required by concrete, thus improving the quality of the concrete. They can also permit a small cement reduction with less temperature-rise potential with the same strength potential. Retardation may be beneficial in placing the concrete. The strength of mass concrete is usually judged on the basis of longer age tests than the standard 28 day strength for structural concrete. Common ages are 3 to 6 months or 1 year.

A mass-concrete mix looks harsh to those unfamiliar with this type of work because of the large aggregate (Fig. 26.14). The consistency of mass concrete should be judged by the consistency of the mortar; not by the looks of the total concrete. A properly designed mix with 1 to 2 in slump responds well to vibration and is easily placed (Fig. 26.15). The slump test and sometimes air-content test are made on the minus 1½ in fraction of the concrete obtained by removing the larger aggregate by either hand picking or wet

FIGURE 26.14 A 4 yd^3 yard bucket of mass concrete just dumped, ready for vibration. (*California Department of Water Resources photograph.*)

FIGURE 26.15 Vibrating mass concrete shown in Fig. 26.12 with three single-operator 6 in pneumatic vibrators. Note how readily concrete responds to vibration. (*California Department of Water Resources photograph.*)

screening. The unit weight of the full mass mix is sometimes made in a 1 ft^3 measure; however, a larger container, 3 to 5 c ft^3, is preferred.

26.6 Preparation for Placing

Forms for mass concrete can be of wood or steel. They are usually made in large sections and handled by crane or cableway or by other lifting equipment. Forms are usually anchored to the rock foundation by drilling and setting anchors, or to previously placed concrete by embedded pigtails, hairpins, or other form-anchor accessories, and are braced by struts (Figs. 26.12, 26.13, 26.20, 26.21, and 25.13). For the downstream-sloping surface of a dam forms may be hinged at about half height to permit depositing concrete in the lower half of the lift closer to the form. After concrete is placed to hinge level, the top half is folded in place and anchored.

Forms are designed for maximum reuse and economy. Forms for curved surfaces in either a horizontal or vertical direction or double curvature are designed and built with the required curvature or with adjustable curvature if curvature of the structure varies. Cantilever forms are sometimes used. These eliminate the internal bracing or ties that sometimes interfere with concrete placing. Deflection rather than strength usually governs in the design of cantilever forms. Starter forms must be specially constructed to conform to the irregular rock surface.

Cleanup and joint treatment prior to placing concrete are accomplished by the methods outlined in Chap. 25. The joint may be grouted or not as specified and as discussed in Chap. 25. All pools of water should be removed from the rock or concrete surface prior to grouting or placing concrete on them. Grout, if used, may be spread by dumping it from the bucket from a height allowing it to spread itself or by brooms, pushing it across the block as placing proceeds. Too much surface should not be grouted ahead of placing and allowed to dry out. If grout is not used, the surface of the rock or concrete should be at a saturated surface-dry condition. The surface can be kept at approximately this condition by blowing off excess water with air, and by preventing excess drying by periodic fogging with an air-water jet (Fig. 26.16).

FIGURE 26.16 Periodic fogging of construction joint with air-water fog nozzle keeps joint at saturated surface-dry condition. Surface or edge of concrete may also be fogged to prevent excessive drying during hot or windy weather.

FIGURE 26.17 Shoveling cobbles which rolled down edge of pile onto concrete to correct segregation.

26.7 Placing and Finishing

Placing mass concrete starts at one side of the form. Buckets are dumped systematically across the width of the placement. If a richer face or exterior concrete is used, the concrete is delivered to provide the number of buckets of exterior concrete required to cover the width of the placement. Buckets may be flagged to indicate the kind of concrete being delivered. Each bucket of concrete is vibrated after it is dumped (Figs 26.14 and 26.15). Usually 6 and 4 in vibrators are used. If grout is not used, particular care is taken to vibrate the bottom layer thoroughly to secure good bond to the preceding lift. Particular care is taken to vibrate thoroughly the place where adjacent piles meet. Sometimes this is done with the smaller vibrator. If large-sized coarse aggregate rolls to the edge of the pile, this is shoveled onto the concrete to eliminate the segregation which has taken place (Fig. 26.17). For 5 ft lifts, the concrete may be placed in four 15 in layers or three 20 in layers. For a 7½ ft lift, five 18 in layers can be used.

Placing proceeds across the placement, "stairstepping" the layers to keep a minimum amount of concrete exposed (Fig. 26.12). In hot weather, drying of the surface of the concrete can be prevented by frequent fogging with the air-water jet mentioned previously (Fig. 26.16), care being taken only to keep the concrete damp, replacing only the water that has evaporated. Also, the use of air-water foggers during hot weather can keep the air temperature in the vicinity of the placing several degrees cooler (Fig. 26.18 and Ref. 28).

The top layer in a lift should be vibrated sufficiently to embed all the coarse aggregate and provide a relatively smooth surface which will aid cleanup for the next lift. Further finishing of the surface is not necessary unless the lift being placed is the top lift in the block, in which case the required finish should be provided as specified and discussed in Chap. 27. Once the top layer in a lift is completed, workmen should be kept off the completed surface except as necessary to place pigtails, hairpins, or other embed-

FIGURE 26.18 Air-water foggers keep air temperature several degrees cooler during hot weather. (*California Department of Water Resources photograph.*)

FIGURE 26.19 Weighted slipform screeding mass concrete on a sloping spillway floor. Note finishing behind slipform which is supported on pipe screeds and is slowly pulled up the slope by hand-operated winches (not shown). (*California Department of Water Resources photograph.*)

ded items. Workmen doing this should wear "snowshoes" to keep from leaving footprints (Fig. 25.9).

Placing mass concrete in sections having flat slopes such as in spillway buckets, spillway crests, or spillway slopes having little slope imposes problems. It is difficult to screed concrete having large aggregate satisfactorily. In addition it is sometimes desirable to finish the concrete, particularly where water flows over it, to eliminate bug holes and provide a satisfactory surface; therefore, any forms used must be removed in time to permit finishing. There are two ways in which placing of such sloping surfaces can be accomplished. One is to use a heavy slipform similar to that used in slope paving (Fig. 26.19). The slipform is pulled up to the slope slowly as placing proceeds, the concrete ahead of the form being vibrated. Finishers follow the trailing edge of the form. The other method is to provide temporary holding forms, which can be easily and rapidly placed and removed as placing proceeds. Forms at the lower part of the placement can be removed at the proper time to permit finishing operations. Placements of this type in a spillway bucket and a spillway-crest placement are shown in Figs. 26.20 through 26.22.

FIGURE 26.20 Removable top panels used to hold concrete in place in bucket section of Thermalito Dam spillway. Ribs hold panels in place. Panels are quickly set in place and removed. After concrete has stiffened, panels are removed and concrete finished by hand. Mortar which squeezes into gaps in panels provides mortar for finishing. Ribs are removed also to permit finishing. (*California Department of Water Resources photograph*).

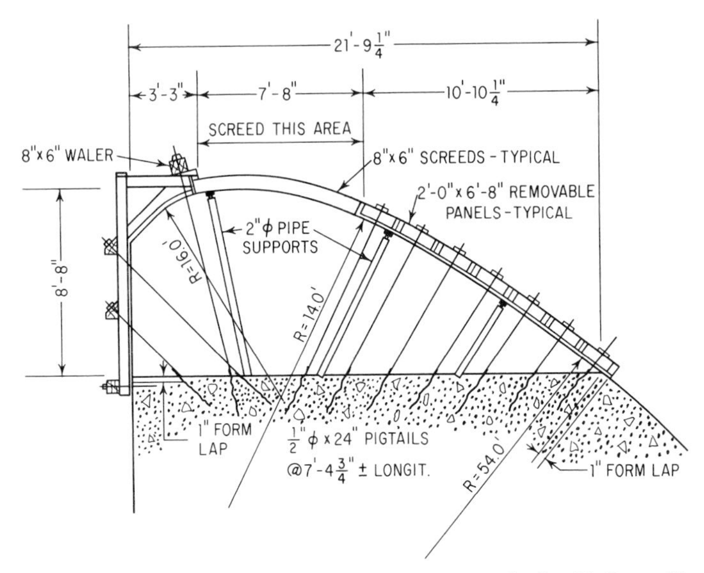

FIGURE 26.21 Sketch for removable-panel arrangement for Oroville Dam spillway crest. Note ribs, removable panels, form ties, and supports.

26.8 Temperature Control

Control of temperature in mass concrete can be achieved in various ways. Temperature control and cooling of concrete are discussed in detail in Chap. 20. For large structures, it is common practice in the design stages to make a temperature study of the structure and concrete to provide information to determine the most economical and satisfactory way to provide temperature control.[26,27]

FIGURE 26.22 Spillway crest for Oroville Dam (see Fig. 26.21).

Temperature control can be achieved by several means. Pipe cooling, that is, embedment of pipe coils in the structure, can be used (Fig. 25.8). The coils are usually placed at construction joints and either river water or refrigerated water or refrigerant is circulated through the coils to remove the heat generated by hydration of the cement. The concrete may be cooled initially to provide a lower starting temperature and consequently a lower final temperature in the structure, resulting in less temperature drop to the stable temperature range of the structure. Initial concrete temperatures as low as 40, 45, and 50°F are sometimes specified. Initial cooling can be accomplished by various methods or combinations thereof. The temperature of the concrete can be computed by means of the equations given in Chap. 20.

Aggregate can be cooled to a limited extent by shading the aggregate storage bins and piles, by water sprinklers, and by circulating refrigerated air or water through storage bins. Evaporative cooling of aggregate in storage bins resulting from a vacuum created by steam jets is sometimes used also.

Cement can be cooled by a longer storage period after manufacture or by cement coolers at the cement mill. Cooling cement is less effective than cooling aggregate or water, since the amount of cement is small in relation to the aggregate and the heat capacity of water is five times that of cement. Water can be cooled by refrigeration, or flake, tube, or crushed ice can be substituted for a portion or all of the mixing water. Cooling of water, aggregate, and concrete can also be accomplished by the use of liquid nitrogen (see Chap. 20). Limiting the minimum period for placing a lift on a previous lift (such as 72 h) and limiting the height difference between adjacent lifts (such as 20 ft) are sometimes specified for temperature-control purposes. Subsequent to placing concrete, shading the concrete can assist in keeping the surface cooler during hot weather (Figs. 26.13, 26.20, and 26.22 and Ref. 28).

Mass concrete is usually water-cured using sprinklers.

PAVEMENTS AND CANAL LININGS

26.9 Street, Highway, or Airport Paving

Paving streets, highways, airport runways, or other large areas is usually accomplished by a "paving train" consisting of several pieces of equipment designed to spread, consolidate, screed, float, and finish the concrete. Two types of equipment are now in general use, the type that utilizes and rides on side forms, and the slipform type that does not use side forms. The slipform type of paver has become popular in recent years because it eliminates the need for and considerable labor involved in placing, removing, and cleaning the side forms as well as the investment in them. Slipform pavers are capable of providing paving with excellent riding qualities. Lighter-weight paving equipment of the type which utilizes side forms is also available for small jobs such as city streets. Considerable helpful information is contained in Refs. 13, 30 and 31 and in Chap. 27.

Foundation. The subgrade, subbase, and base for the paving should be adequately compacted in accordance with the plans and specifications and should extend a foot or more beyond the form line to support properly the forms and equipment moving on it. Greater width is required where slipform paving is used, to support the slipform equipment adequately. When compacted, the final grade should be slightly higher than the finish grade to permit fine grading to the finish grade, providing firm foundation for the forms and equipment which must operate on the subgrade. The subgrade material should be of a type suitable for the purpose; if not suitable, it should be replaced.

Many types of trimmers are available to accomplish the final grading between the forms, varying from a grader to specially built trimmers which operate from the side forms or which operate from preerected grade lines set on either one or both sides of the paving. Such equipment has electrical, electronic, and hydraulic controls that follow the grade line to trim the base accurately. Accurate trimming of the base outside the paving is desirable for slipform paving because of its effect on concrete quantity, slab thickness and uniformity, and resulting paving smoothness and riding quality. Fine grading to close tolerances sometimes restricts the size of the aggregate in the base material to about ¾ in, particularly for slipform paving. When cement-treated bases are used, their construction and fine grading must be completed in a limited period of time. A longer time period can be allowed for construction of untreated bases.

Grade lines, where used, should be set accurately and supported at 50 ft intervals on tangents and 25 ft intervals on vertical and horizontal curves. The same set of level control stakes can and should be used for the subgrade, subbase, and base, for setting the side forms, or for the slipform guidelines to reduce discrepancies and inaccuracies in surveying. The same set of stakes and brackets can be used to support guidelines for different operations. The line may be removed and replaced. Guidelines are usually piano wire or nylon cord (fishing line is good).

Concrete for paving can be mixed and delivered by several methods as discussed in Chap. 23. The current trend for large jobs is to use large central-mix plants having (8 or 10 yd^3 mixers, delivering the concrete in trucks (end-dump, Agitor or Dumpcrete, or transit mixers used as agitators). Central-mix plants may be relocated several times on a paving job to provide a shorter haul of concrete to the placing site. Uniform concrete is essential for a smooth paving operation, particularly for slipform paving. Concrete can be made with 1½ or 2 in maximum-size aggregate for paving having 8 or 9 in thickness and up to 3-in aggregate for greater thickness as in airport runways. Mix proportioning should be in accordance with methods outlined in Chap. 11 and in Standard Practice for Selecting Proportions for Normal, Heavyweight, and Mass Concrete, ACI 211.1. Slumps should be in the lower range, 2 to 2½ in or less, and sand contents should be suitable for paving concrete. Cement content should be adequate to provide required strength and with water cement ratio for required durability. Air entrainment should be used for durability, and adequate inspection should ensure by test that the required amount of air is obtained and that poor construction practices such as finishing too soon or overfinishing of overwet concrete do not result in loss of entrained air in the top surface where durability is most important.

Side-Form and Slipform Paving. Forms must be set accurately to line and grade and be supported uniformly by a firm foundation. The smooth-riding quality of the paving depends on the care with which the forms are set and maintained since the finishing equipment rides on the forms. Forms should be maintained in good condition. After final form alignment with instrument, a final check should be made by sighting along the tops of the forms—any gross errors will be readily apparent.

The paving train capable of paving widths up to 36 ft may consist of either one or two spreaders (two are sometimes used if reinforcing mats are used), a finishing machine, longitudinal float, mechanical belt, and burlap or pipe drag (Fig. 26.23).

Concrete should be dumped on the subgrade as nearly as practical in its final location in accordance with the good practices for handling concrete outlined in Chaps. 24-26. The mechanical spreader distributes the concrete between the forms. If reinforcing is used, the first spreader strikes off the concrete at the correct depth and the reinforcing steel is placed. The second layer of concrete is placed and spread by the same spreader making a second pass, or by a second spreader. The final spreader should leave a slight excess of concrete for the finisher.

FIGURE 26.23 Trailing edge of highway slipform paver showing inserter for polyethylene plane of weakness tape (about 2 in wide). Tape is carried on rolls, not visible, behind inserter. Note also electronic control riding guide wire to control elevation of trailing edge of slipform.

Consolidating the concrete is usually done by vibrating pans or screeds, or immersion spud or tube vibrators attached to the spreader or finisher or other independent equipment. Vibrating pans or screeds are suitable for paving thicknesses of 8 to 12 in. Immersion vibrators are required for greater thicknesses. Concrete must be thoroughly vibrated throughout its full depth with just enough mortar brought to the surface to finish satisfactorily. Excess soupy fines or mortar indicates overwet concrete, too high a sand content, or overfinishing.

Slipform pavers have largely replaced pavers which use side forms.

The slipform paver consists of equipment mounted on crawler tracks having moving side forms and incorporating the spreading, consolidation, finishing, and floating operation usually all in one piece of equipment (Fig. 26.23). As the machine moves forward, concrete at the trailing edge of the siding has stiffened sufficiently so that it does not slump as the side-form support is removed. The automatic machine is guided in both line and grade by sensors that follow guide wires set accurately in line and grade outside one or both edges of the slab. The nonautomatic machine is controlled in grade by the base on which it rides. Slipform pavers commonly pave 24, 36, or even 48 ft widths.

Slipform pavers incorporate most of the same features of side-form-type pavers. Concrete is delivered to the paver and spread laterally. A spreader operating ahead of the slipform is desirable to spread the concrete to a uniform thickness to produce more uniform conditions, helping the slipform paver to produce a smooth surface. Sometimes a spreader box is incorporated in the paver, or it may be separate. When separate, the paving mixer or delivery trucks discharge into the spreader box and pull it along, metering the concrete onto the base. Augers or screeds spread the concrete laterally. Discharge of concrete should be done so as to eliminate or minimize the effects of segregation. A strike-off or a conforming screed molds the concrete to the required cross section. Concrete is usually vibrated ahead of the screed both by spud vibrators and by a vibrating tube set just ahead of the leading edge of the screed to prevent surface tearing. The proper location of both types of vibrators is critical in order to prevent tearing of the concrete by the leading edge of the screed and to prevent the surge or boiling of concrete behind the screed. Variable-frequency vibrators are sometimes used so the operator can compensate for variations in the consistency of the concrete, reducing the surge.[12] Following the screed, the concrete is floated, utilizing a pressure pan, vibrating or oscillating screeds, vibrating pan floats, or a rotating screed plus a pan float. Final finishing consists of pipe floating, dragging a 6 or 7 in aluminum pipe forward and backward over the surface. The pipe is placed diagonally across the slab. The pipe can be moved either manually for narrow widths or by machine for wider paving. Straightedging may next be done so that irregularities can be detected and removed.

When reinforcing is required and slipform paving is used, the two-lift method of placing the reinforcing is usually followed. A stripped-down paver or other spreading equipment places the first lift to the required thickness for the reinforcing steel. The width is slightly narrower than the paving width. The mat of steel is placed and anchored and the second paver follows, placing the top layer of concrete. Internal spud vibrators, which would interfere with the reinforcing steel, cannot be used with this paver; so vibration must be obtained by vibrating screeds or pans or with tube vibrators.

There is normally a slight amount of edge slump with slipform pavers. This can be corrected by the edging operation or by addition of a small amount of mortar during the edging.

The final operation is to texture the concrete for skid resistance with a burlap or rug drag as mentioned for side-form pavers.

Good control of concrete deliveries and uniform consistency are essential to a smooth paving operation with slipform pavers. Variation in consistency will cause differences in surge of concrete behind the screed and consequent finishing difficulties. Overwet concrete causes edge slump that causes irregular edge surfaces. Sometimes concrete that is too wet or has a bleeding tendency will have "slip-outs" at the edge some time after the paver has passed. (Air entrainment will reduce bleedng and reduces this problem.) If concrete deliveries slow or cease for a short period, it is better to slow the slipform paver until deliveries resume rather than stop it completely.

Pavers for Small Jobs. Paving equipment discussed previously is suitable for large jobs. Some small jobs, such as city-street paving, are done by lightweight equipment. Such equipment is also sometimes used for large floor slabs or parking areas.

Dowels, tie bars, and reinforcing should be placed as required by the plans and specifications. Dowels should be held in place and in alignment so that they will function as intended. Tie bars are conveniently inserted in the concrete by equipment designed for this purpose. They may be inserted by hand or by a hydraulic inserter following the side-form paver, or they may be carried into the concrete by wires as the slipform paver moves ahead.

Joints should be made as required by the plans and specifications. Contraction joints are for the purpose of controlling the cracking of concrete due to shrinkage. These may be either preformed or sawed. Preformed joints may be made by a strip of sheet metal or plastic set and left in place during paving operations. Another type consists of inserting a wood or metal strip into a slot in the surface formed by vibrating a bar into the concrete. The final finishing produces a smooth surface over the strip which is removed the next day and the crack is later filled with joint sealant. Care should be taken that the strips do not cause irregularities in the concrete.

Joints may be sawed in the concrete by diamond saws. The time at which sawing is done is critical. It must be done late enough that the sawing does not ravel the concrete surface and before the concrete cracks. This may be from about 4 to 24 h after the concrete is placed. Sometimes such saw cuts are made diagonally at a slight angle across the slab to break the rhythmic effect on traffic. Sometimes it is desirable to saw every other joint initially when cracking is imminent, sawing the intermediate joints later. Joints in one lane should line up with joints in previously placed lanes.

Longitudinal joints (hinge or warping joints) can be sawed, formed with a strip of sheet metal left in place, or formed by a strip of polyethylene inserted in the concrete by an attachment on the slipform paver (Fig. 26.23).

Construction joints are installed at the end of a day's run. They should be placed to correspond to a regular contraction joint when possible. A transverse construction joint not intended as a contraction joint should have tie bars and keyways. Longitudinal construction joints should likewise have tie bars and keyways.

Expansion joints are not usually required except where the paving joins a bridge or other structure, curb, sidewalks, or at intersections. Preformed expansion-joint material such as cork, sponge rubber, or fiber treated with bitumen should be placed perpendicular to the concrete surface and held in place while concrete is placed against it. The expansion-joint filler should cover the entire surface of the joints; otherwise spalling will occur with expansion. The joint filler can be held to previously placed concrete with bitumen. See also Chap. 28. where joints are discussed in detail.

Curing paving can be accomplished by methods outlined in Chap. 29.[11] Water curing by ponding, sprinklers, wet earth, sand, or wet cotton or burlap mats is considered best. Covering with impermeable paper, plastic sheets, wet straw, or white-pigmented curing compound is also used. Curing should be started as soon as practical after the concrete is finished—as soon as it will not injure the surface—and should be continuous for the required period. Curing compound should be applied after the concrete surface has lost its shininess but before it has dried. Curing compounds should be thoroughly stirred before and during spraying, should not be diluted, and should be applied at the specified coverage to form a continuous uniform covering. Damaged curing compound should be resprayed after thoroughly wetting with water. The edges of slabs as well as the surface should be cured.

Traffic should be kept off the paving until it has attained its design strength, which is usually determined by flexural strength of beams cured alongside the slab.

Every effort should be made to construct smooth paving having satisfactory riding qualities as correction of roughness is expensive and not too satisfactory at best. Where satisfactory smoothness is not obtained as determined by measurements of the completed paving by a profilograph or other equipment, the bumps are required to be removed with a "bump cutter." One bump cutter is about 2 ft wide, consisting of about 100 diamond-saw blades mounted on a shaft. Another is a cylinder mounted on a shaft with diamonds set into the surface in a spiral pattern.

26.10 Canal Lining and Slope Paving

Concrete lining of canals, drainage channels, reservoir slopes, and other slopes such as around abutments on bridges and fills can be done in several ways. The method selected will depend primarily on the size of the job. The excavation work for the canal lining can likewise be done in several ways. Rough excavation is done by any of the excavation methods considered suitable. Fine grading is done with trimming machines, some of which are made particularly for the purpose. During excavation, the possibility of slides and unsuitable material such as expansive clays should be watched for. If overexcavation is required or occurs because of careless excavation, backfill should be with suitable material adequately compacted.

Shotcrete is suitable for small canals or areas. This method is discussed in Chap. 36.

Subgrade guided slipforms (Fig. 26.24) can be used for small canals. The canal section is excavated and trimmed to the required section. The form rides on sledlike runners, which are a part of the form and which space it away from the excavation to give the required lining thickness. The slipform is built so that it can be pulled along the canal with a winch or other suitable means. Concrete is delivered into the hopper of the form. As the slipform is pulled ahead slowly, the concrete in the hopper is spaded or vibrated, and fills into the space between the excavation and the pan or form which shapes the concrete. Vibration must be done with internal vibrators in the concrete in the hopper, not to the form.

Slope slipforms in which the concrete is placed in sections up the slope can be used on small jobs or jobs where other equipment is not suitable. A suitable slipform and its essential features are shown in Figs. 26.25 and 26.26. The slipform is supported on side forms or previously placed sections of concrete. The leading edge must be sharp, not rounded, where it rests on forms or concrete to keep concrete from getting between it and the form or concrete it rides on, thereby raising it. The form must be heavy—sufficient to resist the hydrostatic uplift of the concrete. As the form is slowly and uniformly moved up the slope by winch, air tuggers, or other means, the concrete in the "trough" at the leading edge is vibrated with internal vibrators. The vibrators must not be run under the form except at the start of a section before the slipform moves. The slipform must not be vibrated, as this would cause concrete to "boil" behind it, causing bulging and a wavy

FIGURE 26.24 Subgrade guided slipform placing concrete in a small canal. Canal section is excavated. Conforming screed slipform mounted on sledlike runners is pulled along as concrete deposited in its hopper is molded to the canal section. (*Portland Cement Association photograph.*)

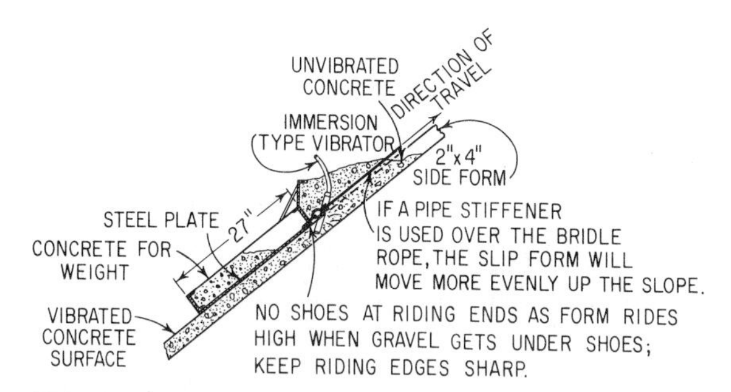

FIGURE 26.25 Slipform for placing concrete on a slope. Concrete is placed and vibrated as slipform is slowly pulled up the slope. For placing unformed concrete on slopes slipform screed should be steel-faced, weighted, and unvibrated. Concrete should be vibrated ahead of slipform by internal vibrators.

surface. The form must be long enough that concrete does not boil out below the form. When slipforms of this type are used, concrete is usually placed in alternate sections, the first sections being placed using side forms. The second sections are placed between the first sections. Inverts are placed using conventional flat-slab paving equipment.

Canal-lining paving trains (Figs. 26.27 through 26.30) used for paving larger and longer sections, where the size of the job justifies the large equipment expenditure, consist of the trimmer that fine-grades the section to required line and grade (Fig. 26.27); the lining machine, a longitudinal-operating slipform that places the concrete; the finishing jumbo where finishers use hand tools to obtain a better finish, where required, and to tool joints, place joint strips, and other handwork; and a curing jumbo or other equipment used for applying curing compound where membrane curing is permitted (Figs. 26.28 and 26.30).

The canal-lining machines may place the canal in one pass (Fig. 26.28) or in sections

FIGURE 26.26 Simple slope slipform in use. Concrete is placed in alternate panels. Larger and more elaborate ones are used.

FIGURE 26.27 Trimmer for fine-grading California Aqueduct described in Figs. 26.29 and 26.30.

(Figs. 26.29 and 26.30) and are supported and guided either by rails set to line and grade (this method is now obsolete) or by crawler tracks following guidelines set to line and grade. The equipment can also be controlled by sensors riding on the subgrade, which provides the most accurate control of the thickness of the lining.

Canal-lining machines consist of a concrete-distribution system utilizing chutes, drop chutes, conveyor belts, or drags.[8] The concrete is vibrated, usually by vibrating-tube internal vibrators in the hopper section ahead of the slipform portion forming the concrete to the required section as the machine moves slowly forward. Pressure plates at the rear of the machine impart a troweling to the concrete, eliminating most of the hand troweling. Equipment of this type was used in lining the California Aqueduct of the state water project (Figs. 26.29 and 26.30).

Control joints may be required at intervals as small as 12½ ft centers in canal lining. These can be conventional planes of weakness that are usually made by grooving the concrete. Such grooves are filled later with various types of joint-sealing compound. Longitudinal grooves can be made by the canal-lining machine with a projection on the underside of the pan that cuts the groove. These grooves may require hand tooling from the finishing jumbo. Longitudinal grooves can also be conveniently formed by a plastic strip that is intruded into the concrete through a tube from the front of the machine. The plastic strips are carried on reels on the front of the machine (Fig. 26.30).

Transverse grooves can be cut into the concrete by a cutter bar working off an eccentric which is actuated at the required groove interval. These grooves are tooled by hand from the finishing jumbo. In some cases plastic joint former is placed in the groove to hold it to the required dimension until the concrete stiffens enough that the plastic can be removed, or the plastic is removed after the concrete sets.

Water stops are sometimes placed in the concrete. One design of water stop includes vertical wings that serve as a plane of weakness to control the crack. Installation of this type of water stop, and water stop at construction joints, is satisfactorily accomplished longitudinally by intruding it into the concrete through a tube or other device from the front of the machine (Fig. 26.30) and transversely with special equipment carried on the finishing jumbo.

FIGURE 26.28 Front and rear views of a canal paving train placing concrete in South Bay Aqueduct of California Water Project. Canal has a bottom width and depth of about 6 ft and concrete is 3½ in thick. Train consists of a paving mixer, lining machine which places full section of concrete, a finishing jumbo, and spray equipment for applying curing compound. (*California Department of Water Resources photograph.*)

FIGURE 26.29 Lining one-half of canal of California Aqueduct with canal-lining train. See Fig. 26.30 for closeup (*California Department of Water Resources photograph.*)

Concrete mixes for canal lining are conventional mixes with additional concern for workability and finishing properties. Concrete with excessive slump is not used, as such concrete will not hold on the slope. For 4 in lining on a 1½:1 slope, 2½ in slump is about the maximum that can be used. Canal lining 4 in thick with 1½ in maximum aggregate size can be satisfactorily placed. One such mix satisfactorily used in the aqueduct of the California Water Project contains 40 percent sand and a cement content of 4.4 sacks per cubic yard, of which 70 lb is a pozzolan. The average 28 day strength is in excess of 3500 psi. A water-reducing retarding admixture is used. Entrained air is between 2 and 3 percent. It is common practice in canal-lining concrete to split the ¾ in 3 No. 4 size coarse aggregate into two sizes on the ⅜ in sieve. This allows the ⅜ in 3 No. 4 size to be kept at a small amount since an excess of this size sometimes causes finishing problems. The coarse-aggregate grading (in percent) being successfully used on two jobs is as follows:

Coarse-aggregate size	Job A	Job B
1½ × ¾ in	57	45
¾ × ⅜ in	33	35
⅜ in × No. 4	10	20

FIGURE 26.30 Closeup of canal paving train shown in Fig. 26.29 used to line California Aqueduct in two sections. Concrete is 4 in thick. Canal prism has a 40 ft bottom width, 32.6 ft depth, 1½:1 slopes, and 137.8 ft top width. Concrete mixed in 8 yd^3 batches in 10 yd^3 mixers in a central-mix plant is hauled to job in 8 yd^3 Agitor dump trucks. Paver places one slope and half the invert. Reels on front of paver carry plastic groove former or water stop inserted in concrete through tubes through front of machine. Note transverse groove cut into concrete to rear of paver and white plastic former being placed in groove to hold its shape. Finishers work from jumbos when inserting groove former and finishing concrete. Joints are at 12½ ft centers. Curing jumbo applies white-pigmented curing compound. (*California Department of Water Resources photograph.*) Similar canal of San Luis Project[32] has 110 ft bottom width, 32.8 ft depth, 2:1 slopes, top width of 250 ft, and 4½ in concrete thickness. Section was placed in four passes: one side slope and 10 ft of invert in one pass, and 90 ft of invert in two passes (45 ft width each). Same lining machine, with adjustments, was used for both side slopes and invert.

Sand-grading specifications sometimes require that the No. 50 to 100 mesh fraction be 15 to 20 percent instead of the 12 to 20 percent used for other concrete, to provide sufficient fines and added workability. Lining as thin as 3 in has been placed using 1½ in maximum-size aggregate. Table 23 of the "Concrete Manual"[40] gives details of mixes for a number of canal-lining jobs.

Figure 26.31 shows concrete facing for a rockfill dam being placed.

Subsequent to placing the concrete, curing should be started as soon as possible. On large jobs, curing with white-pigmented curing compound is almost universally used.

26.11 Curbs, Gutters, Sidewalks, Median Barriers, and Similar Structures

Construction of curbs, gutters, sidewalks, drainage ditches, and median barriers is auxiliary to highway or street and residential construction. Small amounts are usually done by hand methods. Forms are placed after excavation and fine grading are completed; concrete is placed in the forms and consolidated, then screeded, and hand-finished following

FIGURE 26.31 Placing concrete on sloping face of Spicer Meadows Rockfill Dam. Top photograph: concrete is delivered to top of dam in ready-mix concrete trucks, transferred down the dam slope by several flights of conveyors, then to side discharge conveyor on placing machine and discharged over the width of the placement. The placing machine is moved up the slope by winches at each end of the machine. Curing jumbo follows placing equipment. Bottom photograph: closeup of concrete placing equipment showing concrete being discharged from side discharge conveyor onto subgrade, vibrated with hydraulic vibrators inserted alternately into the concrete on each side of the operator, then screeded and finished by auger, roll, and drag at left of operator. One operator controls placement and movement of the equipment from side to side and up the slope.

conventional methods. On larger jobs, slipform or small paving equipment, which will accomplish such work, is available. Equipment such as that shown in Fig. 26.32 is capable of placing concrete in curbs; curbs and gutters; sidewalks; combined sidewalks, roll curbs, and gutters; drainage ditches; and median barriers. Attachments are available, or the machines can be adjusted to give a variety of cross sections.

Concrete work should be done with as much care as in any other work in order that the concrete will perform satisfactorily. Concrete with high slump (more than 4 in) should be avoided. Air entrainment should be used in climates where durability is a problem. Concrete should be carefully finished and properly cured.

TUNNELS AND CONDUITS

26.12 Tunnels

Tunnels for railroads, highways, sewers, water supply, and other utilities are sometimes lined with concrete.[20] This may be for the purpose of supporting the opening, improving flow conditions for water, containing water under pressure, sealing off the inflow of water, or usually a combination of these factors. Lining tunnels with concrete is one of the most difficult kinds of concrete work. Limited working room; difficult placing conditions due to confined spaces, reinforcing steel, steel or timber support, steel or timber lagging and blocking; water problems; and difficult cleanup conditions make unpleasant and discouraging conditions where it is difficult to get a good concrete job. For this reason, more than usual planning should be done to select the best procedures and equipment to accomplish the job.[7,33,34,35,40] Concrete is usually placed in tunnels using concrete pumps of either the piston or "air gun" types described in Chap. 36. Concrete is also sometimes placed using the preplaced aggregate method described in Chap. 38.[34]

Sequence of Tunnel Lining. The size of the tunnel, length, section, and sometimes available equipment determines the sequence of tunnel-lining operations. The usual method of lining is to place the concrete in sections between bulkheads. However, tunnel lining can be placed continuously (particularly applicable to long tunnels) by having sufficient length of forms so designed that they can be telescoped and leapfrogged ahead after concrete has hardened sufficiently that the last form section can be removed to obtain continuous placing.

Lining may be placed in a complete section, applicable particularly for small and circular tunnels, or placed in several placements in the section.

Several sections of tunnels are shown in Fig. 26.33. Sequences which have been or might be used are described as follows:

Tunnel section 1, a small circular tunnel, can be placed in a single placement in sections 40 or 50 ft long or more or in continuous placing (Figs. 26.34 through 26.37).

Tunnel sections 2, 3, 4, and 5 show circular or horseshoe tunnel sections placed in more than one placement. In these either the curb section or invert is placed first. The purpose of a curb section is to provide support for rails or for the invert screed or for supporting and anchoring forms. The curb or invert section is cast with appropriate embedded anchors for attaching rails if required, and for anchoring the arch or sidewall forms.

Tunnel section 5 is that for two diversion tunnels for Oroville Dam (Figs. 26.38 through 26.41). This was placed in four placements. Curbs A were placed first (Fig. 26.38).

FIGURE 26.32 Slipform concrete placing machines. Machines are guided by string set to line and grade. In one machine, concrete is elevated from delivery truck chute into the machine by an auger; the other machine uses a belt conveyor. The concrete is vibrated by hydraulic vibrators. These machines can go around curves. Top photograph: placing highway median barrier concrete. Bottom photograph: placing integral sidewalk and curb and gutter in a housing development. This machine also does final grading of subgrade.

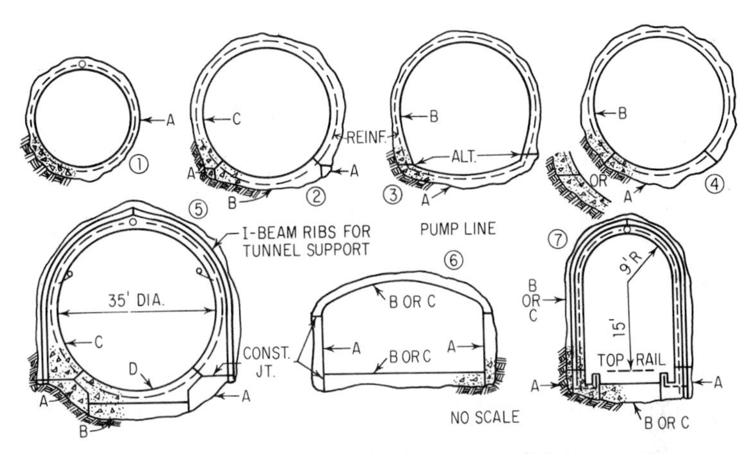

FIGURE 26.33 Several tunnel sections showing sequence of placing concrete.

FIGURE 26.34 Rail-mounted form in place ready to receive concrete for lining a 6 ft-diameter circular tunnel. Note bulkhead lumber at end of form held to angle bolted to steel form at one end and cut to fit rock and braced from rock at other end. Note also doors in side of form, slick line from concrete pump entering form at top, and rail-mounted carrier for slick line which is backed out as placing proceeds and connection for air line near bend in slick line for "slugging" concrete. See Figs. 26.35 through 26.37 for subsequent concreting operations. (*California Department of Water Resources photograph.*)

FIGURE 26.35 Form vibrators fastened to form shown in Fig. 26.34. Internal vibrators also used through door openings. Note hinge in form. (*California Department of Water Resources photograph.*)

FIGURE 26.36 Small-diameter special 6 yd^3 concrete car for delivering concrete from mix plant outside into 6 ft diameter tunnel. (*California Department of Water Resources photograph.*)

FIGURE 26.37 Finishing invert concrete after removing the invert form when concrete has stiffened sufficiently that it can be done.

FIGURE 26.38 Curb form being placed for lining one of the 35 ft diameter circular diversion tunnels for Oroville Dam. Top is tied and braced to rock. Bottom is anchored to rock. Note adjusting screws at bottom for elevation. Space between bottom of form and rock was filled with wood forming to hold concrete (water shown removed before concrete placed). Figures 26.39 through 26.42 show subsequent concreting operations. (*California Department of Water Resources photograph.*)

FIGURE 26.39 Front form is in place following completion of one 48 ft long placement, and rear form has been moved ahead by telescoping through front form and is being set. This permitted a 48 ft section to be placed each day. (Forms were required to be left on for 32 h.) Forms are hinged at two places on upper part of circle and are supported by curb section (Fig. 26.33, section 5 and Fig. 26.38). Note also working platforms on side of form, pump line located below upper platform to place concrete into openings to fill sidewalls. (*California Department of Water Resources photograph.*)

Next the subinvert *B* was placed, which was used to facilitate the remainder of the construction in providing a smooth surface for support and travel of the form carriers, concrete-delivery trucks, and the concrete pumps. The side-walls and arch *C* were placed in one operation in a section 48 ft long (Fig. 26.39). Finally the invert was placed using a slipform mounted on rails attached to the lower edge of placement *C* for one tunnel (Figs. 26.40 and 26.41). For the other tunnel, the invert section was formed and concrete was pumped into the section to fill it (Fig. 26.42). (This did not result in a satisfactory placement. The slipform produced much more satisfactory results.) In these diversion tunnels concrete with a maximum placing temperature of 50°F was specified to achieve a lower maximum temperature and resulted in relatively crack-free concrete. This resulted in strength gain in the concrete that would not permit removal of sidewall and arch forms and reuse on a 24 h cycle. Specifications required forms to be left in place for 32 h, and they were reused in a 48 h cycle. Two sets of forms were used and concrete was placed each day. The forms were designed so that one would telescope inside the other and could be leapfrogged ahead.

Tunnel section 6 is similar to a large double-track railroad or highway tunnel carrying two or more lanes. In the section shown, the sidewalls were placed first. The next placement is usually the arch, although the invert may next be placed to provide better working conditions. However, the invert is usually placed last.

Tunnel section 7 is for a single-track railroad relocation around the reservoir for Oroville Dam and was common for five tunnels (Figs. 26.43 and 26.44).

FIGURE 26.40 Slipform placing invert concrete in a diversion tunnel. Slipform supported on rails fastened to previous placement and slowly pulled ahead by air-operated tuggers or winches. Concrete delivered to top of chutes shown by conveyor receiving concrete from a delivery truck. Chutes shown delivered concrete to each side of invert. *(California Department of Water Resources photograph.)*

The same forms were used for all tunnels (with modification for the fifth tunnel, which was on a curve.) The curb sections were placed first. In these, the forms were anchored to and supported by steel tunnel support and rock. In the case of two tunnels, the tunnel-drilling jumbo and mucking equipment were rail-mounted. These same rails were used for the form carrier and concrete equipment. The sidewalls and arch were placed in the second placement (Fig. 26.44). These were placed in 50 ft sections. The invert was the third placement. In the other three tunnels, the tunneling equipment was rubber-tired wheel, not rail-mounted. The invert was placed following the curb placement to provide a smooth working surface for the form carrier and placing equipment. The invert was placed with paving equipment supported and operating off the curbs.

The curved tunnel utilized the same forms. Permission was granted to place the tunnel in tangents about 40 ft long. The sidewall and arch forms were modified by insertion of a wedge-shaped section to provide for the curvature.

Concrete in the sidewalls and arches of all the above railroad tunnels was placed on a 24 h cycle. Concrete had attained sufficient strength for the forms to be removed in 10 to 14 h.

Equipment Used in Tunnel Lining. Forms for tunnel lining vary considerably, depending on the section involved. Forms for curbs or invert sections can be either wood or steel. Forms for the curb section, *A* placement in section 7, were steel and were reused

FIGURE 26.41 Finishing invert concrete being placed in Fig. 26.40 from finishing platform at rear of slipform. (*California Department of Water Resources photograph.*)

many times in five railroad tunnels. Such forms usually position inserts used for aligning and supporting the sidewall and arch forms. The curb forms for this placement were supported and aligned from the steel tunnel-support ribs.

Forms for sidewalls and arch are usually specially designed for the job and are made of steel (Figs. 26.35, 26.39, and 26.43). Circular pressure tunnels with a steel liner use the liner as a form which remains in place. In such tunnels, the reinforcing steel may be placed on the liner outside the tunnel and moved into place in the tunnel. For small tun-

FIGURE 26.42 Form for invert concrete in one diversion tunnel. Form held in place by anchors in previous placement. Concrete was pumped under form. Placing was not nearly so satisfactory as with slipform in Fig. 26.40. (*California Department of Water Resources photograph.*)

FIGURE 26.43 Form for sidewalls and arch of railroad tunnel carried on truck. Note openings in form for access for vibration and for inspection. (*California Department of Water Resources photograph.*)

nels (such as section 1 in Fig. 26.33) the form may have a removable invert, which permits finishing the concrete at the time it can be satisfactorily finished where a smooth invert is desired (Fig. 26.37). Wall and arch forms are usually hinged at one or more places to facilitate placing and removal (Figs. 26.35, 26.39, and 26.44). Forms are usually carried and moved by a traveler or carrier mounted on rubber-tired wheels (Fig. 26.43) or on rails and are supported, positioned, aligned, braced, and removed by use of screw-jacks, hydraulic jacks, rams and cylinders, and steamboat jacks. Sidewall and arch forms have openings in the sides and arch to permit access for vibration and inspection (Figs. 26.35 and 26.43). In some cases concrete can be placed through such openings (Fig. 26.39). Form openings must be capable of being opened and closed easily and be tight. Openings at the crown should not be on the centerline but should be placed alternately on each side of the centerline to eliminate interference with the concrete pump line, which is usually placed on the centerline. Working platforms are built into the larger forms (Figs. 26.39 and 26.44).

Bulkhead forms in tunnels in rock may be built as shown in Fig. 26.34. Each tapered piece of wood is individually cut to fit the rock. Tunnels which are drilled with a mole may have reusable bulkheads which fit the outer more regular surface of the tunnel sufficiently closely so that individual fitting is not required.

Concrete-placing equipment for curb and invert varies considerably from conveyors, chutes, concrete pumps, or other means depending on the job (Fig. 26.40). Placing con-

FIGURE 26.44 Railroad-tunnel form in place. Note hinges at two points, working platforms at three levels corresponding to locations of access doors, and braces across forms at two levels. Form is supported by anchor bolts in curb section previously placed. *(California Department of Water Resources photograph.)*

crete in the sidewalls and arch is almost universally done with either piston or pneumatic ("air gun") concrete pumps (Chap. 36). These may be located outside the tunnel for short tunnels where pumping distances are short, or inside the tunnel. Concrete is usually consolidated by vibrators, internal type in invert and curb sections, and either internal or external (form) vibrators, or usually both, in the sidewalls and arch placement.

Concrete is mixed either outside in a central-mix plant or inside the tunnel in a variety of mixers, including transit and paving mixers or special mixers, materials being delivered in batch trucks or cars. Concrete is delivered in a variety of equipment ranging from special trucks or railroad cars to conventional concrete conveyance equipment such as transit mixers mounted on railroad cars or trucks, or tilting-type concrete trucks. On small tunnels special small-diameter cars or concrete mixers are frequently employed (Fig. 26.36). On long tunnels it is desirable to mix the concrete near the placing site to eliminate time delays, reduce the amount of concrete in transit, and minimize difficulties caused by slump loss, equipment delays, and breakdown. In long tunnels concrete or concrete batches may be delivered from the portals or concrete may be mixed above the tunnel and either pumped into the tunnel or dropped through a vertical shaft or a drilled hole into the tunnel. Concrete for one diversion-tunnel job was mixed in a central-mix plant above ground, moved horizontally about 100 ft on a conveyor belt, and dropped vertically about 150 ft through a drop chute placed in a hole drilled in the rock, thence into a collecting hopper. The concrete in the hopper discharged into Agitor trucks that transported

FIGURE 26.45 Concrete transfer for one diversion tunnel where access for concrete delivery was difficult. Concrete delivered outside to hopper and vertical pipe inserted through hole drilled in rock where it dropped 150 ft into hopper shown, then transferred into Agitor truck and hauled to delivery point. (*California Department of Water Resources photograph.*)

the concrete to the placing location (Fig. 26.45). Concrete for the adjacent diversion tunnel came from the same central-mix plant but was hauled in transit trucks into the tunnel, requiring a considerably longer time.

Some thought should be given to expediting concrete deliveries by providing passing tracks for trains or arrangements for trucks passing, or turnaround. For one of the diversion tunnels the width between curbs was insufficient for trucks to pass. A portable ramp and turntable were constructed to a elevate the trucks to provide additional clearance and to permit trucks to turn around near the placing site, which eliminated the necessity for trucks to back in slowly a considerable distance. In some of the railroad tunnels a portable ramp was constructed so that trucks were elevated sufficiently to clear the curb and pass. (These trucks were not able to turn around and had to back into the tunnel, however. The passing ramp permitted two trucks to be inside the tunnel at one time.)

Concrete Mixes for Tunnel Lining. Concrete pumps having 8 in lines are capable of placing concrete having maximum aggregate size of up to 2½ or 3 in. Pumps with 6 in lines usually limit the aggregate to about 1½ in maximum size. The mix used must be a workable cohesive mix without excessive bleeding suitable for pumping. Slumps as low as 3 in are pumpable, and air-entrained concrete is pumpable. Concrete with 21½ in maximum-size aggregate (natural sand and gravel) used in the railroad tunnel (Fig. 26.33), section 7 contained 5 sacks of cement per cubic yard, 33 percent sand, 5 to 7 in slump, 3 percent entrained air, and had an average 28 day compressive strength of 4000 psi. (This higher slump was required by the placing conditions to be discussed later, not by the concrete pumps.) In the lining of the diversion tunnels, using 8 in pumps, 1½ in maximum-size aggregate (natural sand and gravel) concrete was used because the smaller aggregate was more efficient for the higher-strength concrete required. The cement content was 5.5 sacks per cubic yard (including 70 lb pozzolan), 35 percent sand, 3 percent entrained air, 5 in slump, and average 28 day compressive strength was 4200 psi. (Required 1 year strength of 5500 psi was exceeded.) Concrete suitable for similar placements elsewhere from a strength, durability, and workability standpoint is suitable for curbs and inverts for tunnels. There are no special requirements otherwise.

Control of Seepage and Dripping Water. Control and effective removal of seepage and dripping water are some of the most difficult and ineffective parts of tunnel-lining work. Such water can cause damage to the concrete and should be controlled. Water entering the sidewalls and arch can be guided down to the invert and into drains by means of corrugated sheet metal placed so it fits closely against the rock. This is called "panning." if large amounts of water are entering, these may be sealed off at the time of driving the tunnel or later by grouting. Gravel or tile drains can be placed in the invert to carry water to a sump where it is pumped out, or drained outside the tunnel. Branches may be placed to pick up water from springs or that coming down the sides behind the pans.

In some cases holes are drilled into the rock and pipe is grouted into place with quick-setting cement to collect and drain away water.

Drains temporarily placed for controlling water during construction and not a part of the design are usually grouted later. Their location should be accurately detailed so they can be effectively grouted.

Plastic sheeting has been anchored to the rock surface prior to concrete placing. The purpose of this was to keep the concrete for a subway tunnel drier, preventing leakage through joints and cracks.

Cleanup. Cleanup of rock should be done by washing with water or air-water jets. In hard-rock tunnels, loose rock should be removed. In soft rock, cleanup should be consistent with the specified required results. Soft, loose material should be removed by pumping. Sandbags can sometimes help keep water out of the invert placing area.

Forms and Reinforcing Steel. Reinforcing steel is placed in accordance with the plans and specifications. Ties with No. 14 tie wire instead of No. 16 are recommended to avoid reinforcing steel displacement by the concrete-lining operations. Forms should be placed and set accurately in alignment and grade. Particular care should be taken to see that the forms are securely anchored and braced. Embedded items such as form anchors, inserts, drains, conduit, and grout pipe must be securely anchored so they will not become displaced.

Placing Concrete. Inverts can be placed with conventional methods. These include a heavy slipform or screed pulled by winches longitudinally along the tunnel on rails or other supports provided for support as concrete is placed and vibrated ahead of it. For smaller tunnels a transverse screed operated longitudinally may shape the surface. Wide curved inverts can be shaped by setting transverse screed guides and screeding with a straightedge parallel to the centerline. Concrete is delivered usually by transit mixers and chutes, conveyors, pumps, or other means.

Concrete must not be placed in water. If pools of water are still present and it is impractical to remove them or keep them dry by pumping, suction pumps, or blowing with air, concrete should be placed on previously placed concrete, and the toe of the concrete slope vibrated ahead, pushing water ahead of it, and subsequently out of the form or placing area.

Concrete in sidewalls and arches is placed behind the forms, using concrete pumps which are mentioned previously in this chapter and described more fully in Chap. 36. Construction joints against previously placed concrete should be clean prior to erecting the forms and recleaned if necessary prior to placing concrete. Small tunnels can be placed with a single pump line in the crown. Large tunnels are best placed with three pump lines, two located at or above the spring line through which concrete is delivered to both sides of the tunnel simultaneously. The pump line in the crown is used later to fill the upper or arch portion. For tunnels placed continuously, placing is done through a crown line only. The pump is moved as placing continues. The diversion tunnels section 5, (Fig. 26.33), were placed with three pump lines as shown. The lines on the side were located outside the form and concrete was chuted through the form openings (Fig. 26.39). Pneumatically operated gates at each side opening permitted concrete to be delivered through an opening (openings were at 8 ft centers). Location of steel rib support and reinforcing steel prevented placing the pump line through the bulkhead and behind the form as is sometimes done. When the line can be placed behind the form, concrete is placed first in the far end of the form and either the pump line is shortened or the pumps are backed away from the form as concrete rises to the level of the pump line. Concrete is vibrated with internal vibrators by workers working behind the forms or working through form openings. The level of concrete must be balanced on both sides of the form. Intermittent operation of form vibrators may also assist the placing. When the sidewalls are filled to the level of the side pump lines, pumping concrete is started through the crown line. This likewise starts at the far end of the form. As the concrete builds up and flows down the side its movement and consolidation can be assisted by form vibrators operated at proper locations. As the concrete level rises, it is more difficult to use internal vibrators, and form vibrators must be depended upon to move the concrete into the sidewalls and arch below the line. As the concrete builds up in the arch, the end of the line should become buried 5 to 10 ft in concrete. When the pumping resistance builds up, a section of line should be removed or the pump backed away, shortening the buried length. Keeping the line buried allows concrete to be forced up higher than the line, thus filling the space above the line. During this portion of the work, form vibrators near the area being filled should not be operated as this will pull the concrete down, losing the fill which has been achieved. The end of the pump line should be positioned between steel supports to prevent them from blocking the flow of concrete. With piston pumps, the air slugger should be used sparingly. It should be used in accordance with the manufacturer's recommendations that "Its sole purpose is to push concrete away from the discharge end of the pipe in order to fill the farthermost recesses of the form first—but it should be used sparingly if at all after the pipe is buried." Air slugging is used to move concrete away from the line, not scatter it. A skillful operation can use the air slugger to obtain this result.

Placing concrete with an air gun is likewise carelessly done many times, and only skilled and careful operation can achieve good results. The gun should be operated so that a small amount of concrete remains in the reservoir. Uncontrolled blasting of all concrete out of the gun into the forms results in dangerous conditions for workmen and much segregation of concrete. Air guns may create fog behind the forms, which makes it difficult to see. Uncontrolled use of air guns or air sluggers may cause loosening, dislodging, or moving of reinforcing steel or other embedded items.

Railroad tunnels (section 7, Fig. 26.33) are best placed similarly to the diversion tunnels using three pump lines as described above. However, some were placed fairly satisfactorily with only one crown line as shown. The end of the pump line was moved laterally to move concrete into one side or the other. The concrete was flowed from one end of the form to the other with vibrators (violating the principles of good practice outlined in Chap. 26; but as mentioned at the start of this section, tunnel lining is sometimes not so good as one likes). The considerable amount of steel ribs, lagging and blocking, and reinforcing steel required 5 to 6 in slump concrete to flow and fill satisfactorily. Experience showed the necessity of operating form vibrators near the end of the line in the crown to eliminate a premature buildup and hardening of concrete on the form at this point, which results in poor consolidation—evident when the forms were removed.

Sometimes construction joints in the first placement in a tunnel, conduit, or dam having sloping sides are detailed as shown in Fig. 26.46 with a short slope of concrete normal to the surface. Such sections are easily detailed but more difficult to place and secure adequate compaction, particularly if reinforced. Adequate vibration results in losing such a slope. It is usually much simpler to keep the entire construction joint horizontal as shown. The placement above can be adequately placed with careful placing, vibration, and inspection.

Keyways on horizontal construction joints are another detail which should be avoided if possible. Their use makes it more difficult to clean the construction joints and creates a depression that is a place for dirt to lodge and mortar used for joint preparation to fill. With good construction practices to obtain a good construction joint, many consider keyways unnecessary.

Curing. Curing of tunnels is best done by water sprinklers. Excessive drying conditions caused by movement of air through the tunnel can be prevented by closing one end where this is practical and can be done. Curing compounds, if used, must be of a type that will not be toxic to workmen in confined spaces.

26.13 Culverts and Cast-in-Place Pipe and Conduits

Structures such as single- or multiple-box culverts and cast-in-place pipe and conduits are usually constructed for water supply, sewers, or vehicular or pedestrian traffic. The principles outlined in Chaps. 24, 25, and 26 apply to these types of structures.

Culverts are usually placed in two or more placements (Fig. 26.47). The floor slab is placed as the first placement. The walls are placed next followed by the roof slab, or the walls and roof may be placed in one placement. If the latter is done, the placing of the

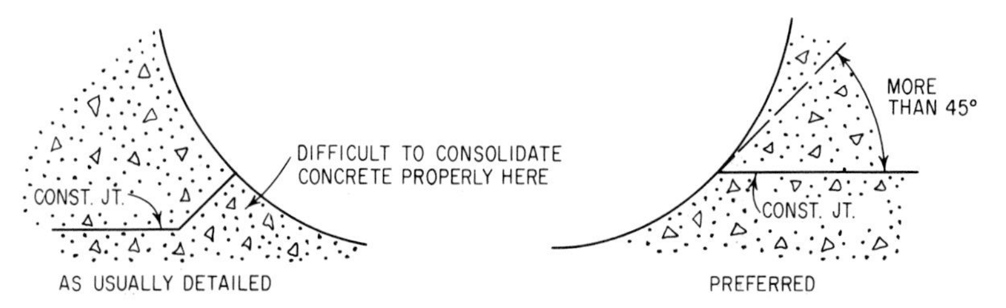

FIGURE 26.46 Sloping construction joint normal to surface in tunnel, conduit, or other construction is easily detailed but difficult to consolidate satisfactorily. Horizontal joint as shown in right sketch is satisfactory with good concrete-placing procedures.

FIGURE 26.47 Multiple-box culvert in various stages of construction. (*California Department of Water Resources photograph.*)

roof should be delayed until the wall concrete has settled, or the concrete at the junction of the wall and roof slab should be revibrated later to eliminate cracks which may form because of settlement of the wall concrete and restraint of the forms. Wing walls, if used, can be placed integrally with the walls or as a separate placement. Sloping surfaces on the wing walls should be covered with forms to retain the concrete during vibration and permit adequate consolidation. These forms are usually made in sections so they can be easily placed in position as concrete placing proceeds and to permit their removal shortly thereafter to permit finishing.

Cast-in-place pipe and conduit usually have a horseshoe or circular cross section. These may be cast in one placement, or usually the invert section is placed first followed by the sidewalls and arch in the second placement. Forms should be constructed so that concrete can be easily placed in the forms, vibrated, and inspected (Fig. 26.48). These structures may be heavily reinforced, which makes concrete placing difficult. The use of a drop chute outside the form, which spills concrete through the side of the form as shown in Fig. 24.14 may be satisfactorily used.

The upper portion of the outside form which slopes over the wall may be hinged to facilitate placing concrete in the sidewalls to the height of the hinge, the hinged section is then lowered and fastened in place, and placing continues (Fig. 26.48). Inspection openings may need to be placed in the side of the form to permit vibration access, observation, and inspection. Such openings should be located to provide convenient access and should be quickly and easily closed when concrete placing reaches their elevation.

Cast-in-place unreinforced-concrete pipe can be placed continuously by a slipform process.[17,18] Such pipe is used for drainage or irrigation-water supply. Its strength

(a)

(b)

FIGURE 26.48 Forms for conduits. (*a*) In this picture heavily reinforced section and overhanging portion of outer form made placing of concrete, vibration, and inspection very difficult. (*b*) Overhanging portion of form in this picture is hinged, which permits concrete to drop into the lower portion of the form more easily and allows better access for vibration and inspection. When concrete is placed to height of hinge, hinged portion is quickly lowered and fastened in place. (*California Department of Water Resources photograph.*)

depends on arch action. In one method (Fig. 26.49) a ditch having a semicircular bottom is excavated with an excavator specially built for this purpose. The lining machine consists of a steel sled or slipform that closely fits the sides and semicircular bottom of the ditch. A gasoline engine mounted on the forward end of the form operates a winch that moves the slipform forward by a cable attached to a deadman ahead. A generator provides power to operate a variable-frequency vibrator attached to the bottom of the inside form. Concrete placed in the hopper from a ready-mixed concrete truck alongside is vibrated and mechanically spaded into place. Collapsible aluminum-alloy semicircular forms about 4 ft long are fed into the front of the hopper assembly, hooked together, and support the top and sides of the pipe as the traveling form moves forward. The collapsible forms are held in place by wood struts or spreader bars or frames. These support forms are usually removed the following day. The invert of the pipe is finished by hand by a finisher working inside the slipform. This process produces pipe having inside diameters of from 24 to 84 in (Fig. 26.50). Concrete mixes suitable for structural concrete with slumps of 1½ to 3 in are suitable for this type of

FIGURE 26.49 Placing 72 in cast-in-place unreinforced pipe with a slipform. Trench is excavated to a semicircular cross section and slipform places concrete as it is pulled forward by a winch.

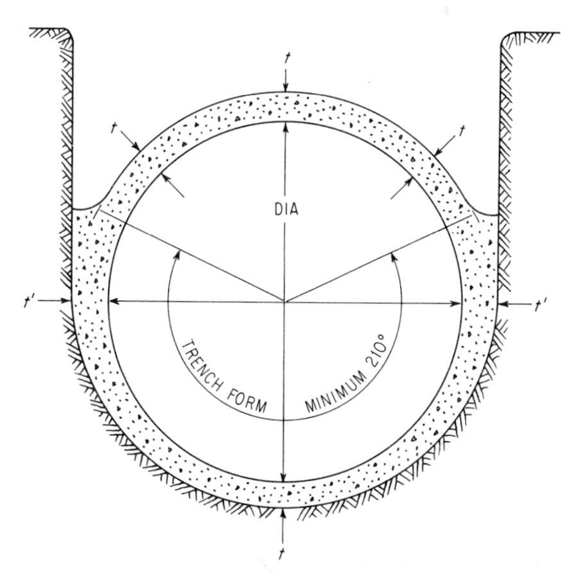

FIGURE 26.50 Typical section of 12 to 96 in cast-in-place nonreinforced concrete pipe. Minimum thickness t is 1/12 diameter plus 1/2 in. Thickness of sides t' is governed by clearance between pipe machine and trench wall. Thickness on horizontal diameter normally varies from t plus 3/4 in to t plus 2 in (t plus 2 to 5 cm) depending on size of pipe. (*From American Concrete Institute.*)[18]

concrete. The maximum size of aggregate should not exceed one-third to one-fourth the wall thickness. In another process, an inflated inner form of rubber and fabric is used to form the inner pipe circle.

PLACING CONCRETE UNDERWATER

Placing concrete underwater should be avoided whenever practical to do so. However, with care and attention to important details concrete can be placed underwater satisfactorily by four methods: underwater bucket, tremie, preplaced aggregate, and concrete pump, depending on the nature of the placement. References 36 through 39 describe several methods of concrete-placing operations in bridge foundations, graving docks, drydocks, wharfs, or caissons. Concrete should not be placed underwater in water colder than 35°F as strength gain is very slow at this level of temperature. Underwater concrete-placing methods require a very workable concrete with slumps as high as about 7 in. and cement contents of up to 650 $lb/yd.^3$ Large placements of this rich concrete have a considerable internal temperature rise.

Placing underwater concrete is usually within caissons, cofferdams, or forms. Foundation cleanup imposes problems but can be done by hydraulic jets and pumps. Concrete must not be placed in running water or allowed to fall through water. To do so would wash the cement from the concrete. However, recently developed antiwashout admixtures may be helpful in such cases.

26.14 Underwater Bucket

Underwater-bucket placing consists of lowering a special bucket containing concrete to the bottom of the foundation and opening it slowly to allow the concrete to flow out gently and without causing turbulence or mixing with the water. Subsequent buckets are landed on the previously placed concrete, and the gate should sink into contact with the previously placed concrete to prevent fall of concrete through water. Buckets should have drop-bottom or roller-gate openings. The gates obviously should have means for being opened from above water. If air is used to open the bucket, the air should discharge through a line to the surface to prevent disturbance. The top of the bucket must be covered in some way to prevent water from washing the surface of the concrete. One way is to cover with canvas or plastic sheets. Special buckets are made for underwater placing with a sloping top having a small opening to minimize water surge (Fig. 26.51).

Concrete for this type of concrete placing will have 6 or 7 sacks of cement per cubic yard, 5 or 6 in slump, 1½ to 2 in maximum-size aggregate, and sand contents higher than normal ≥40 percent. Concrete for foundations for the San Francisco-Oakland Bay Bridge was placed in water as deep as 240 ft with buckets.[43]

Bucket placement is a good way to start a placement by the tremie method; the latter is probably faster and possibly superior once it is well started without washing the first concrete placed. After a depth of at least 2 ft, preferably 3 ft, of concrete has been placed by bucket at the point the tremie pipe is to be operated, the empty watertight tremie pipe with its lower end covered with the pressure-seal plate shown in Fig. 26.52 is lowered and thrust deeply into the concrete already in place. Concrete is then placed in the tremie pipe. When it reaches a level somewhat above that sufficient to balance the water and concrete pressure, it will push off the lightly tied seal plate, and concrete will then extrude into and expand the mass already in place.

FIGURE 26.51 Special concrete bucket for placing concrete underwater. Note top, which convers concrete, preventing water surge from washing cement from concrete. (*GarBro. Manufacturing Co.*)

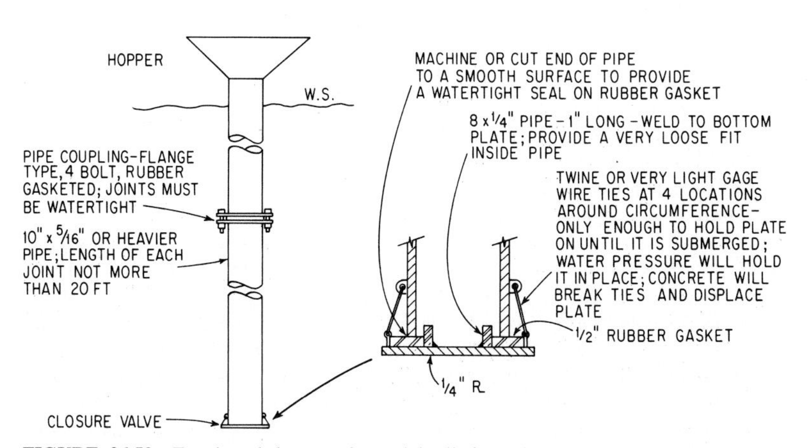

FIGURE 26.52 Tremie and closure-valve seal details for underwater concrete placement.

26.15 Tremie

Tremie placing consists of placing concrete through a vertical pipe, allowing the concrete to flow from the bottom (Fig. 26.52). The bottom of the tremie is kept submerged in the concrete at all times, and the concrete flows into the mass of previously placed concrete. Concrete placing with tremie is described elsewhere in the literature.[36–39]

A tremie consists of a pipe usually 10 or 12 in. in diameter with a hopper or funnel at the upper end. The equipment must be adequately supported and arranged so that it can be raised and lengths of pipe can be removed if necessary as the level of concrete rises in the forms. The spacing of tremies should not be more than 12 to 16 ft apart or 8 ft from the sides or ends of the enclosure, although greater spacing has been successfully used in placements of large vertical dimensions. Concrete is delivered to the hopper by bucket, truck mixer or transport, pump, or conveyor. This delivery must be at a good rate and without interruption. Delays for any reason impair the free-flowing mobility of the mass into which concrete is being placed and may endanger the success of the placement. Plugged pipe and cold joints can result from delays.

Starting a tremie placement is very important. Unless great care and proper methods are used, much of the first concrete placed will become badly diluted and washed with water. Initial discharges should be controlled to extrude very slowly into the water and should consist of cohesive starter mixes of high cement (and possibly pozzolan) content.

There are two basic methods of starting a tremie placement which might be called the "wet-pipe" and the "dry-pipe" methods. The wet-pipe method is practical in water depths under 100 ft and for large area work. The dry-pipe method is preferable for deeper work and for confined areas where nearly total freedom from any impairment of the first concrete on the foundation is important.

The wet-pipe method consists of placing a plug or a "go-devil" in the tremie pipe at about where the water level will be, with the pipe standing on the bottom in the water ready to start. This will be some distance below the concrete hopper on the upper end of the pipe, so there will be a sufficient weight of concrete in the pipe above the plug to overcome its friction and start it toward the bottom. These plugs may consist of a roll of burlap or cloth sacks, a cylinder of closed-cell PVC (polyvinyl chloride), or an inflated ball tight enough to stay well in place until the concrete pushes it down, forcing out the water in the pipe below it. The pipe is not lifted off the bottom until it is evident from the amount of concrete in the pipe and rising in the hopper that the plug is down to the bottom. The pipe is then slowly raised just enough to permit the plug to escape. This will be evident by a sinking of the concrete in the center of the hopper. The pipe is then quickly dropped to the bottom to shut off rapid flow of concrete into water; then it is repeatedly raised and lowered by small amounts as necessary to let concrete extrude rather than jet from the lower end of the pipe, until soundings show that the end of the pipe is embedded in concrete. Then the rate of flow can be increased because the concrete will be flowing into concrete and not into water.

The dry-pipe method requires a tremie pipe that is watertight, including gasketed, four-bolt flanges at the couplings. The pressure-seal plate shown in Fig. 26.52 is attached to the bottom of the tremie pipe in such a manner that the water pressure will make it completely watertight and the interior of the pipe will be empty and free of any water. With the pipe wall heavy enough to prevent flotation, the pipe is rested on the bottom to start, then filled with concrete, preferably a very rich, plastic, cohesive starter mix. When the pipe is lifted slightly, the weight of the column of concrete will break the strings holding the seal plate, and the concrete will extrude outward. With this arrangement, as in Fig. 26.52, it is evident that the pipe does not have to be lifted as high as for the plug to escape in the wet-pipe method. Accordingly, with careful manipulation of raising and lowering small distances, the concrete extrudes with a low velocity that causes a minimum of washing and dilution until sufficient concrete has been released to embed the end of the pipe; thereafter concrete will flow into concrete and not into water.

With either method of starting, once the discharge is well embedded in fresh concrete, greater concrete velocities and the resulting kinetic energy developed (which is achieved by a greater rate of concrete delivery) produce better results. Continuous delivery of concrete is desirable. Better results are obtained and there is less likelihood of a tremie plugging if concrete is discharged into the hopper in a manner to avoid segregation. If a tremie plugs, it can sometimes be freed by a quick lift of the pipe a few inches; quick dropping may plug it tighter. The end of the tremie must be kept buried in concrete at all times; from 3 to 5 ft is considered satisfactory. The pipe should not be lifted more than 6 to 12 in at any time. Soundings should be taken during placing, and the elevation of the concrete surface and position of the lower end of the pipe should be frequently noted from markings on the pipe and plotted to show continually their relation to each other. Placing should start at a tremie in the center of a large placement, and others should be started from the center outward. Air-lift pumps at the outside edge of the placement can remove mud pushed to the outer edge by the advancing concrete.

Very workable concrete is required for tremie placement. Seven-inch-slump, 7-sack concrete is considered standard. Maximum-size aggregate of 1½ to 2 in and sand contents of up to 45 percent are used. Anything that will improve the workability of the concrete will improve tremie placing. These include such things as use of air entrainment, pozzolans, natural gravel instead of crushed stone, reduction in amount of crushed material in coarse and fine aggregate, and increase in fines in sand if a coarse sand is being used.

26.16 Preplaced Aggregate

The preplaced aggregate method briefly consists of placing aggregate within the designated area and injecting grout into the mass. This method is described in Chap. 38.

26.17 Pumps

Concrete pumps have placed concrete satisfactorily underwater. Reference 44 describes placing concrete in a pier caisson of a wharf with a concrete pump. The caissons were 4½ ft diameter steel cylinders driven to bearing about 90 ft below pier top and 67 ft below water. Muck was removed from the caisson. The end of the concrete pump line was plugged and lowered to the bottom of the caisson. The line was filled with concrete, forcing the plug out, with the concrete at the bottom end sealing it. Concrete pumping continued, and the pump line was periodically raised as resistance increased. Sections of pipeline were removed above water as work progressed. The caisson was filled and allowed to overflow to remove laitance and excess water in the top concrete.

FARM AND HOME CONCRETE

26.18 Concrete Construction around the Farm and Home

Satisfactory work can be done by those unfamiliar with concrete work by observing the good practices pointed out in other chapters of this book. Much helpful information and literature can sometimes be obtained from service engineers of the Portland Cement

Association, cement companies, and local ready-mixed concrete producers. The following are some helpful hints for this type of work:

1. Plan the job. Have someone experienced supervise.
2. Prepare the forms and other details required before placing the concrete.
3. Visit jobs where professionals are at work.
4. Secure sufficient help. Divide a large job into smaller placements. Concrete work is hard work physically.
5. Use ready-mixed concrete if a quantity of concrete is involved.
6. Use air-entrained concrete for its superior durability and helpful effect on workability.
7. Use concrete with a consistency such that a little effort is required to place it—and see that such effort is used. The concrete should have a plastic, fatty, not soupy, appearance. Avoid overwet, soupy, runny concrete. Concrete which can be "poured" is "poor" concrete. Concrete should be placed, not poured. If vibrators are used, use a drier consistency. If vibrators are not available, work the concrete into place with spades, shovels, tampers, or the feet.
8. Use concrete with sufficient cement. The cost of an extra sack of cement per cubic yard is good insurance against some of the deficiencies of this type of work.
9. After placing and screeding the concrete, avoid further finishing operations, floating, and troweling until the concrete is ready. In other words, avoid overworking fresh concrete. Finish as outlined in Chap. 27.
10. After finishing, cure the concrete with frequent sprinkling with water for a few days (Chap. 29). Cover the concrete with burlap, old carpeting, sand, water-proof paper, or plastic to retain moisture. Concrete should be cured with water, not "dried." (Water is required for concrete to gain strength.) Curing compound may also be used, but water curing is better when possible.

PROBLEMS IN CONCRETE PLACING

26.19 Placing under Adverse Weather Conditions

Concrete placing during either cold or hot weather is covered in this chapter and Refs 9 and 10. The recommendations in this chapter and references should be followed.

Concrete placing during rain is, of course, not considered advisable. If rain is threatening at the time concrete placing is to start, its effect on the concrete and the advisability of canceling placing should be considered. If placing is allowed to proceed, plans should be made to take necessary precautions if rain starts. If rain starts after concrete placing starts, it is sometimes better to continue placing than to stop. The following precautions should be taken:

1. Cover the working area with tarpaulins or other protection to keep rain off the concrete, if possible.
2. Use a lower-slump concrete.
3. Dry up puddles of water collected on the foundation or on the construction joint before new concrete is placed.

4. Keep the surface of the concrete on a slight slope so water will run off, and provide a place for water to drain off.
5. Avoid working the surface of the fresh concrete.
6. If rain is so heavy that puddles cannot be removed and the surface of fresh concrete is being washed, it may be advisable to suspend concrete placing. If this is done, bulkheads, dowels, headers, or steps should be placed so that the placement is left in the best condition possible for continuing at a later time. The construction joint so formed because of discontinuing placing will, of course, be treated as a construction joint with suitable cleanup preparatory to placing the remainder of the concrete later.
7. During thunderstorms of short duration, cease placing concrete until the storm passes and cover the concrete with tarpaulins, plastic sheets, or other protection during the storm.

26.20 Avoidance of Cold Joints

Cold joints occur in a concrete placement when a layer of previously placed concrete hardens or sets to the extent that a newly placed layer does not bond to it. Cold joints are more likely to occur during hot or windy weather because concrete either dries or sets faster because of high temperatures and loss of moisture. In addition to lack of bond, the cold joint is visible and sometimes unsightly because of a lack of bond, the cold joint is visible and sometimes unslightly because of a lack of consolidation, or there is an offset due to the form's bulging from the pressure of the overlying concrete. Cold joints can be eliminated by planning placing procedures so that the concrete is not exposed too long before it is covered by subsequent layers. This can sometimes be accomplished by "stair-stepping" the layers or by speeding up placing. This speedup is achieved by more or different equipment, more workmen, or elimination of bottlenecks that cause slow placing.

Conditions that cause slower setting of the concrete such as the use of colder concrete, placing concrete at night during hot weather, shading the work area, maintaining high humidity at the concrete surface by fog sprays, or use of a retarder in the concrete are sometimes beneficial.

If the underlying concrete has hardened to the extent that a cold joint is likely to occur, the subsequent layer of concrete should be preceded by a layer of mortar, or the same procedure for starting concrete on a construction joint as discussed previously in this chapter should be followed. If cold joints cannot be prevented, the concrete placements should be reduced to a size that can be handled without cold joints.

26.21 Slump-Loss Problems

As stated previously, concrete stiffens or loses consistency or fluidity rapidly, commonly called slump loss. This is most likely to be troublesome during hot, dry, windy weather (Fig. 26.53), especially when the concrete is hauled long distances or transported on long chutes or conveyors or otherwise exposed to drying from wind and heat. The following can be helpful in reducing slump loss or its effects:

1. Organize and coordinate the job to transport and place concrete faster, and avoid delays. Select and use equipment that will expedite the work, not slow it. Schedule concrete delivery so that trucks can be unloaded and concrete used as soon as they

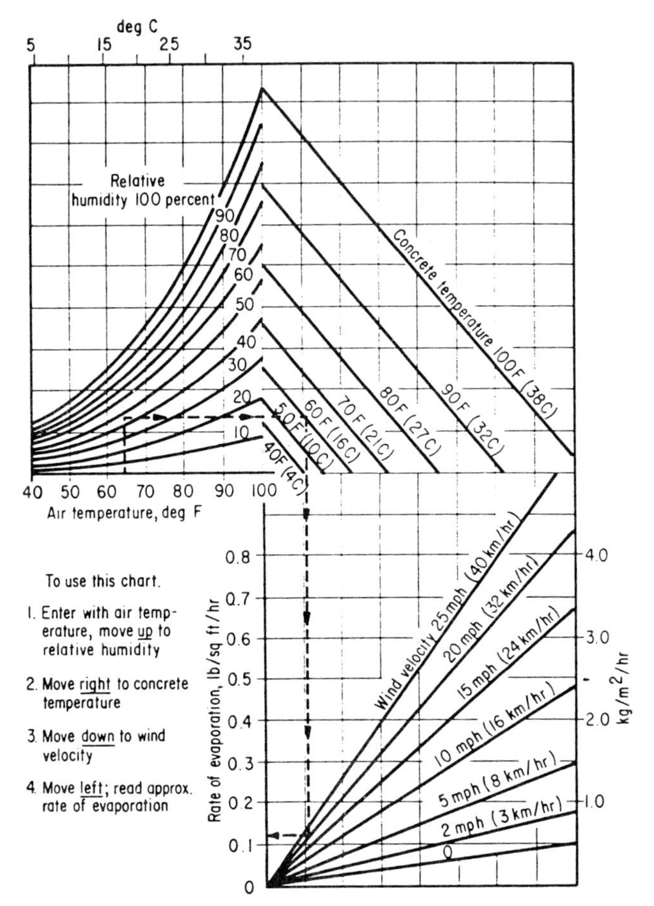

FIGURE 26.53 Nomograph for predicting moisture loss of concrete sufficient to cause plastic shrinkage cracking during hot, dry, windy weather conditions. Evaporation rate greater than 0.2 lb/ft^2/h will probably cause plastic shrinkage cracking. Evaporation rate greater than 0.1 lb/ft^{2}h cracking is possible.

arrive—do not have several trucks waiting to unload. Provide communications with the concrete plant so that deliveries can be changed to fit the progress at the placing site.

2. If using dry absorptive aggregate in concrete, dampen it before use.
3. If subgrade is dry, dampen it to lessen absorption.
4. Reduce temperature of concrete by
   a. Precooling aggregate with cold water, cold-air jets in the batcher bins, or vacuum evaporative cooling.
   b. Cooling cement.
   c. Cooling mix water by using chipped ice in mix water.
   d. Shading aggregate storage, water tanks, and lines.
   e. Painting truck mixers and water tanks and lines white.

5. Avoid overmixing.
6. Work at night.
7. Shade work and use windbreaks.
8. Use fog (not water) sprays in the concrete placing area to keep the humidity high and the air temperature low.
9. Investigate the possibility of false set or premature stiffening in the cement, and avoid cement having it.

Usually, the faster concrete can be handled and placed, the less difficulty will be encountered with slump loss and handling and placing problems resulting from such slump loss.

REFERENCES

The following references to American Concrete Institute Standards, Specifications, Guides, Recommended Practices, Committee Reports, and Special Publications are particularly pertinent. All may be found in the ACI "Manual of Concrete Practice," a five-volume set which is updated and published annually.

1. ACI 117, Standard Tolerances for Concrete Construction and Materials, 1990.
2. ACI 207.1R, Mass Concrete, 1987.
3. ACI 301, Specification for Structural Concrete for Buildings, 1989.
4. ACI 302.1R, Guide for Concrete Floor and Slab Construction, 1989.
5. ACI 303R, Guide to Cast-in-Place Architectural Concrete Practice, 1991.
6. ACI 304R, Guide for Measuring, Mixing, Transporting and Placing Concrete, 1989.
7. 304.2R, Placing Concrete by Pumping Methods, 1991.
8. ACI 304.4R, Placing Concrete with Belt Conveyors, 1989.
9. ACI 305R, Hot Weather Concreting, 1991.
10. ACI 306, Standard Specification for Cold Weather Concreting and Cold Weather Concreting, 1990.
11. ACI 308, Standard Practice for Curing Concrete, 1981 (Rev. 86).
12. ACI 309R, Guide for Consolidation of Concrete, 1987.
13. ACI 325.9R, Recommendations for Construction of Concrete Pavements and Bases, 1991.
14. ACI 318, Building Code Requirements for Reinforced Concrete, 1989.
15. ACI 318.1, Building Code Requirements for Structural Plain Concrete, 1989.
16. ACI 345, Standard Practice Concrete Highway Bridge Deck Construction, 1982.
17. ACI 346, Standard Specification for Cast-in-Place Non-reinforced Concrete Pipe, 1990.
18. ACI 346R, Recommendations for Cast-in-Place Non-reinforced Concrete Pipe, 1990.
19. ACI 347, Guide to Formwork for Concrete, 1988.
20. ACI 350R, Environmental Engineering Concrete Structures, 1989.
21. ACI Craftsman Series, Slabs on Grade.
22. ACI Craftsman Series, Cast-in-Place Walls.
23. ACI Craftsman Series, Supported Beams and Slabs.
24. ACI SP-4, Formwork for Concrete.
25. ACI SP-6, Symposium on Mass Concrete.

26. Townsend, C. L.: "Control of Temperature Cracking in Mass Concrete," paper presented at 62d annual meeting of American Concrete Institute, 1966.
27. Townsend, C. L.: "Control of Cracking in Mass Concrete Structures," U.S. Bureau of Reclamation Engineering Monograph 34, 1965.
28. Stodola, P. R., J. E. O'Rourke, and H. G. Schoon: Concrete Core Block for Oroville Dam, *Proc. ACI,* Vol. 62, p. 617, June 1965.
29. Price, W. H., L. P. Witte, and L. C. Porter: Concrete and Concrete Materials for Glen Canyon Dam, *Proc. ACI,* Vol. 57, p. 629, Dec. 1960.
30. Gillis, L. R., and L. S. Spickelmire: Slipform Paving, *Calif. Highways Pub. Works,* Vol. 44, nos. 1–2, 3– 4, pp. 63, 68, Jan.-Feb., March-April 1965.
31. Ray, G. K., and H. J. Holm: Fifteen Years of Slipform Paving, *Proc. ACI,* Vol. 62, p. 145, Feb. 1965.
32. Johnson, M. R.: Slipform Lining of the San Luis Canal, *Proc. ACI,* Vol. 62, p. 1313, Oct. 1965.
33. Wilson, R. J.: Lining of the Alva B. Adams Tunnel, *Proc. ACI,* Vol. 43, p. 209, Nov. 1946.
34. Davis, R. E., Jr., G. D. Johnson, and G.E. Wendell: Kemano Penstock Tunnel Liner Backfilled with Prepacked Concrete, *Proc. ACI,* Vol. 52, p. 287, Nov. 1955.
35. Tuthill, L. H.: Tunnel Lining with Pumped Concrete, *Proc. ACI,* Vol. 68, p. 252, April 1971.
36. Halloran, P. J., and K. H. Talbot: The Properties and Behavior Underwater of Plastic Concrete, *Proc. ACI,* Vol. 39, p. 461, June 1943.
37. Angas, W. M., E. M. Shanley, and J. A. Erickson: Problems in the Construction of Graving Docks by the Tremie Method, *Proc. ACI,* Vol. 40, p. 249, Feb. 1944.
38. Gerwick, B. C., Jr.: Placement of Tremie Concrete, Symposium on Concrete Construction in Aqueous Environments, *ACI Spec. Publ.* 8, 1964.
39. Khayat, K. H., B. C. Gerwick, Jr., and W. T. Hester: High Quality Tremie Concretes for Underwater Repairs, *ACI Spec. Publ.* 122, 1990.
40. "Concrete Manual," 8th ed., U.S. Bureau of Reclamation.
41. "Concrete Floors on Ground," 2nd ed., Portland Cement Assoc. Engineering Bulletin EB075.
42. "Building Movements and Joints," Portland Cement Assoc., Engineering Bulletin EB086.
43. Stanton, T. E., Jr.: Cement and Concrete Control—San Francisco–Oakland Bay Bridge, *Proc. ACI,* Vol. 32, p. 1, Sept-Oct, 1935.
44. Messina, R. F.: Pumpcrete Conveys Tremie Concrete to 90 Foot Loading Wharf Caisson, *Concrete,* Dec. 1943, p. 2.

PART • 7

FINISHING AND CURING

CHAPTER 27
FINISHING

Joseph A. Dobrowolski

Concrete surfaces are finished to the tolerances and textures as specified by the plans and specifications. The finish is the result of forms and liners used to form it and any subsequent treatment or finishing of the formed and unformed surfaces.

The plans and specifications may or may not specify procedures and materials to secure the desired finish with forms. Precast units will require only correction of defects acquired during shipment and erection or rejection of units not meeting the requirements, as they should have conformance to the plans and specifications prior to shipment.

Unformed surfaces are finished rough for subsequent toppings of concrete, mortar for paving tile or brick, asphalt, flooring, roofing, or other similar materials. Inset tile and linoleum require relatively level and smooth finishes, and any cracks should be repaired.

27.1 Tolerances

Tolerances for finished concrete surfaces are specified in two American Concrete Institute publications: Standard Specifications for Tolerances for Concrete Construction and Materials (ACI 117) and Specifications for Structural Concrete For Buildings (ACI 301). Table 27.1 presents the ACI tolerances for floor finishes in accordance with levelness and flatness factors. The flatness factor is a statistical analysis of elevation differences measured along a straight line consecutively at 12 in intervals. The levelness factor is determined from consecutive measurements of elevation differences at 10 ft intervals along a straight line. Procedures for measurement and calculation of the factors can be found in ASTM E1155. The measurements can be made by a proprietary profilometer or standard precision levels.

An alternate method uses a 10 ft straightedge, vertical and unleveled, to measure distances between the bottom surface of the straightedge and the finished concrete surface. The straightedge is placed anywhere on the floor and in any direction, and the gap between it and the floor should not exceed the following dimensions (in inch fractions):

Classification.	
Conventional	
Bullfloated	1/2
Straightedged	5/16
Flat	3/16
Very flat	1/8

TABLE 27.1 Floor Finish Tolerances Specified in the ACI 117 Standard*

	Minimum F_F F_L number required			
	Test area		Minimum local F no.	
Floor profile quality classification	Flatness F_F	Level F_L	Flatness F_F	Level F_L
Conventional				
Bullfloated	15	13	13	10
Straightedged	20	15	15	10
Flat	30	20	15	10
Very flat	50	30	25	15

The F_L levelness factor does not apply to slabs placed on unshored forms, previously shored form surfaces after removal of the shores, or inclined and cambered surfaces. The measurements must be made within 72 hs after concrete placement. No measurement should be made across a construction joint or within 2 ft of a wall.

27.2 Formed Surfaces

The finishes recommended by ACI Committee 301 in the Specifications for Structural Concrete are described in the following paragraphs.

Matching-Sample Finish. This type of finish requires the contractor to reproduce a surface finish, illustrated by a sample, on an area of at least 100 Ft^2 in a designated location. Acceptance must be obtained before proceeding with the finish in the specifically specified area.

As-cast Finishes. *Rough-form finish* requires tie holes and defects to be patched, fins exceeding ¼ in in height to be chipped or rubbed off, and to preserve the texture left by the forms.

Smooth-form finish requires tie holes and defects to be patched, all fins to be removed completely, and a rubbed final finish.

Architectural finish requirements can be found in Chap. 44.

Rubbed Finishes. Rubbed smooth form finishes consist of a smooth rubbed finish, a grout-cleaned finish, and a cork-floated finish.

Smooth rubbed finish requires the forms to be removed as early as can be permitted, followed by immediate patching, and then rubbing with a carborundum brick or other abrasive after wetting until a uniform color and texture is achieved. Grout to be used is that produced by the rubbing process.

Grout-cleaned finish requiring the entire surface to be available for cleaning prior to starting. The surface is wetted and a grout is applied which consists of 1 part portland cement and 1½ parts of fine sand. The grout is scrubbed into all the voids and all excess is removed. When the grout starts to turn white, the surface is rubbed and kept damp for 72 h. The grout mixture may require an addition of white cement to make colors uniform.

Cork-floated finish requires the forms to be removed as soon as possible, followed by any necessary patching, and removal of ties, burrs, and fins. A stiff grout, consisting of one part portland cement and one part sand, is applied to a wetted surface to fill all voids. The surface is then ground with a slow-speed grinder in order to compress the mortar into the voids. The final finish is produced with cork in a swirling motion.

Type of Finish Unspecified. When there is no specified finish, the smooth finish is expected in areas exposed to the public and the rough board finish where the surface is not exposed to the public.

27.3 Unformed Surfaces

Horizontal surfaces which have only side forms such as slabs on grade or both side and bottom when the slabs are suspended can be finished manually and/or mechanically. Choice of method is determined by the requirements of space, time, and economics, but should be adequate to result in a durable slab capable of excellent service. High-water-content grout mixtures and grout pumps have been used to produce concrete slabs which have poor finishes and poor durability.

Manual Finishing. The placement of concrete must be carefully monitored by supervisory personnel to prevent excessive water contents, to ensure proper consolidation around the accessories and at the base of the slab where lines of concrete placement meet. The placed concrete is then leveled to grade by screeds moved manually (Fig. 27.1), mechanically, or by a moving slipform.

After screeding additional leveling is done by a long-handled bullfloat (Fig. 27.2) having a magnesium or wooden blade which can cut high spots and fill low spots as it is pushed forward and pulled back over the concrete surface. Steel blades are not used for floating as they tend to seal off the surface, prevent the evaporation of bleed water, and result in surface blisters or soft surfaces. Stiff concrete mixes may require the use of a "jitterbug" (Fig. 27.3) or rollerbug to move the coarse aggregate down and provide sufficient mortar at the surface for finishing. A darby float (Fig. 27.4) is a small straghtedge used for smaller areas.

Floating is (Fig. 27.5*a*) is generally begun when the concrete will support a large person's weight but leave an imprint of a shoe to midsole depth. This process will bring further mortar to the surface, further filling of any voids, and prepare the surface for the final troweling (Fig. 27.5*b*) process. Premature floating creates a weak surface by mixing in

FIGURE 27.1 Hand-pulled vibrating screed.

FIGURE 27.2 Long-handled magnesium bullfloat.

the bleed water and also a thick nondurable mortar by driving the coarse aggregate too far below the surface. During this process edgers are used to round the upper corners of the concrete at the form face to prevent spalling of the edges and to improve the architectural appearance. Very flat surfaces require the use of bullfloats with 10 ft blades.

FIGURE 27.3 Jitterbug being used on surface of concrete.

Final troweling (Fig. 27.6) is done to densify and harden the surface for satisfactory service under severe weather conditions and expected traffic conditions. Except for small projects, most troweling is accomplished with power equipment. Manual troweling is still necessary for small areas, around utility openings, edges, and areas adjacent to walls and forms. This operation is begun when the surface is hard enough to leave only a fingerprint imprint. Weather conditions may require changes in procedures or use of monomolecular coatings to prevent excessive moisture loss.

Finishing with Mechanical Equipment

Slipform Pavers. Most highway pavements, city streets, airfield pavements, and parking aprons and commercial parking are being done by slipform equipment which is highly automated. Electronic feelers on one or both sides of the paver use

FIGURE 27.4 Darby float.

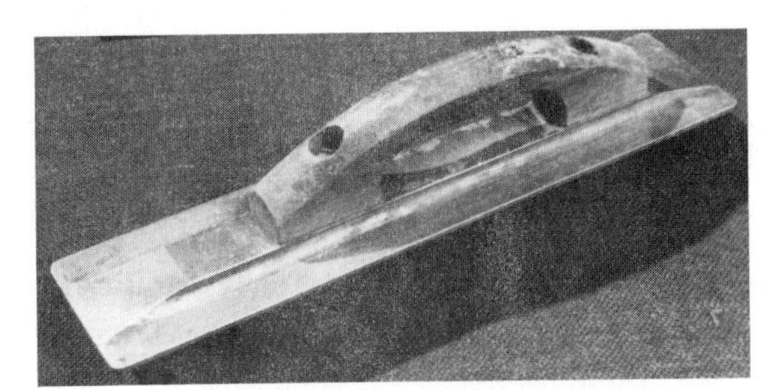

FIGURE 27.5*a* Hand float. (*Courtesy of Portland Cement Assoc.*)

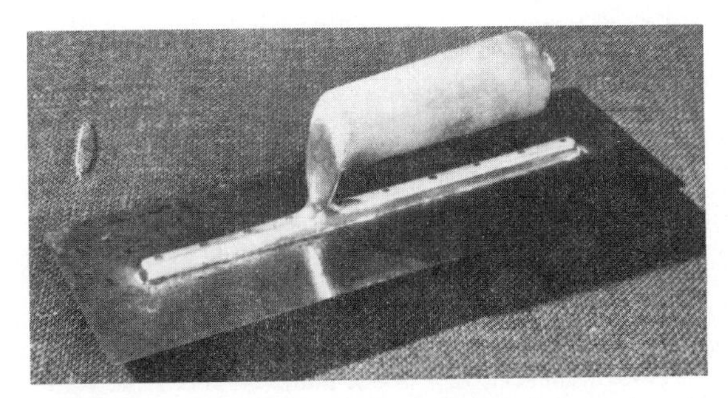

FIGURE 27.5*b* Trowel. (*Courtesy of Portland Cement Assoc.*)

FIGURE 27.6 Troweling in process.

string lines or piano wires set to grade and supported at 25 ft intervals to provide the machine signals which actuate hydraulic jacks on each of the four corners. The machine is propelled by tracks (Fig. 27.7) and carries its own screed in the form of an inverted flat channel which can be adjusted to the width and depth of pavement to be placed. Concrete is dumped ahead of the machine or in a self-contained hopper. Consolidation is accomplished by internal horizontal vibrators and the weight of the equipment on the screed form.

The concrete mixture must be of approximately 1½ to 2½ in slump to prevent edge slump and not slow down the forward movement. Most paving concrete is premixed and hauled to the site with agitation only. On smaller street projects and parking lots, the slump may be at the higher range because transit-mix trucks are used.

Additional finishing is required behind the slipform paver to meet highway smoothness requirements for rideability and to repair any surface blemishes, slight "holidays" of material, or low spots left in the surface caused by settlement of the slipform during changes of speeds or stops. This finishing is accomplished with a length of light irrigation pipe being pulled diagonally behind the paver and use of bullfloats. For smoothness requirements most highway bullfloats will have a blade that is 10 ft long. This procedure is followed by texturing and application of a curing seal.

Additional features required by the design may be the mechanical installation of edge and transverse dowel reinforcement.

Surface and rideability tolerances are specified by the individual highway departments and project specifications. For example, the California Highway Department requires a Profile Index of 7 in per mile or less for tangent alignment or horizontal curves having a radius of 2000 ft or more. The Profile Index may be 12 in per mile or less for pavements having a radius of 1000 ft or more but less than 2000 ft and any included superelevation transition.

The data is obtained by a manually pushed or truck-mounted profilometer (Figs. 27.8, 27.9) which measures the pavement surface 3 ft in from the edge and draws the surface on a graph having a horizontal scale of 1 in equal to 25 ft and a true vertical scale. A tem-

FIGURE 27.7 Slipform paving train followed by lightweight irrigation pipe for surface finishing, burlap drag, and final curing apparatus.

FIGURE 27.8 California Highway Department truck-mounted profilograph.

FIGURE 27.9 California manual profilograph.

plate having a scale length of 0.1 mile and a center horizontal band of 0.3 in is used to determine the Profile Index from the graph. The Index is a total of the inches measured which project above and below the 0.3 in band.

Individual high points in the pavement surface are ground down (Fig. 27.10) to the maximum of 0.3 in and other high points until the accumulated total is within the specified limit. Present equipment, used for grinding in the longitudinal direction, has approxi-

FIGURE 27.10 Bumpcutter used for pavement grinding.

mately 100 diamond-saw blades on a horizontal axle which are seperated by a 0.060 in space. The resultant surface has a serrated finish.

Nonslipform Mechanical Finishing

Screed Strike-off. Screeds used for striking off the surface concrete can be self-propelled or pulled by crank winches (Figs. 27.11 and 27.12). Each has an attached method of vibration and a metal or metal-covered straightedge which can be adjusted to compensate for sag or designed crown or slope. A continuous roll of mortar is required ahead of the straightedge to ensure against holidays of material. A uniform slump is necessary as excessively dry mixtures may cause the screed to run high and too wet mixtures may result in low spots.

Special equipment has been developed for screeding and finishing of sloped surfaces such as channel walls, auto raceway, velodromes, and for bridge decks (Figs. 27.13 and 27.14).

Power Floating and Troweling. The same power equipment is used to float and trowel the surface by varying the angle of the blades. Floating is done with the blades having a shallow angle, and troweling is done with a steeper angle. Equipment having a single

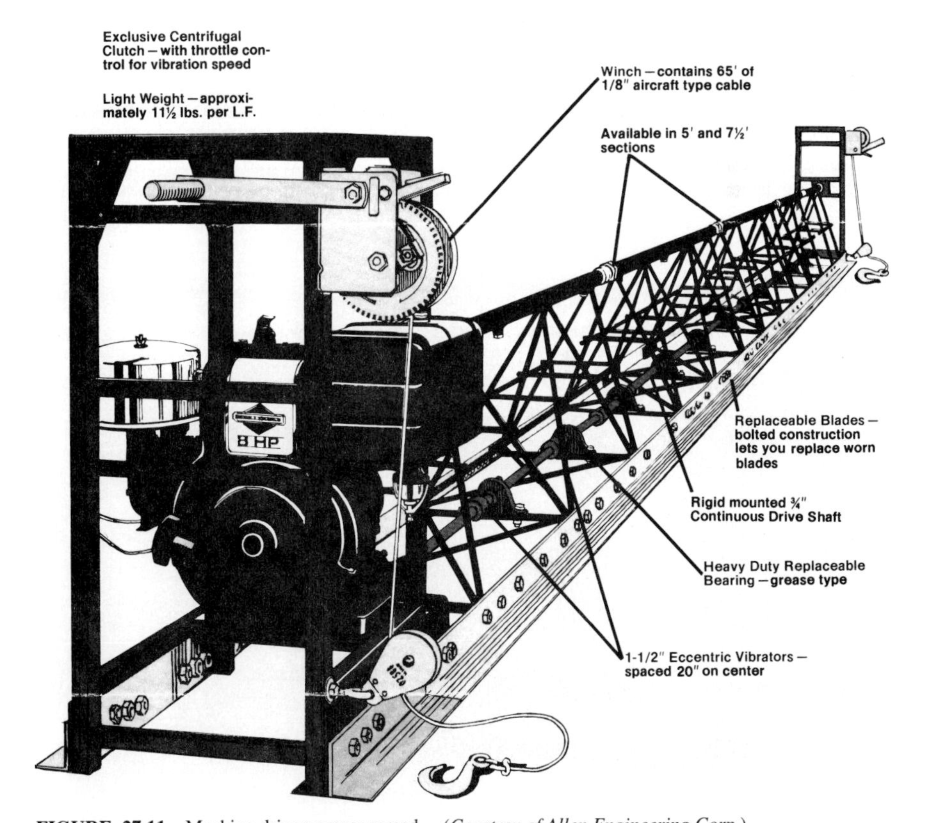

FIGURE 27.11 Machine-driven power screed. *(Courtesy of Allen Engineering Corp.)*

FIGURE 27.12 Self-propelled roller vibrating screed on forms.

rotary (Fig. 27.15) containing three blades and controlled by an operator immediately standing alongside is used on most projects, while equipment having up to three multiple rotaries (Fig. 27.16) with an astride operator have been used for projects with large expanses of concrete. Additional hand finishing is still required with the mechanical equipment in areas around utility openings and other obstructions which cannot be done mechanically.

FIGURE 27.13 Bidwell bridge deck finishing machine with float moving transversely across deck as it oscillates longitudinally.

FIGURE 27.14 Machine finishing topping slab of Olympic velodrome, California State University—Dominguez Hills.

Surface Textures. Various textures for the finished surface can be applied to meet service requirements. These can be listed as brushed, broomed, tinned, slotted, burlap drag, swirled, or floated. Architectural textures are discussed in Chap. 44.

Floated texture is used when an additional topping of mortar for paver tiles, asphalt, or concrete is to be applied.

A broomed or brushed texture (Fig. 27.17) is used to break up the slickness of the surface when wet to prevent falls. Depth of the brooming depends on the slope of the surface and the traffic.

FIGURE 27.15 Power float with single rotary, normally used on smaller projects.

FIGURE 27.16 Operator-driven power float with three rotaries. (*Courtesy of Koehring Construction Equipment Div.*)

FIGURE 27.17 Finisher brooming surface.

FIGURE 27.18 Swirled finish.

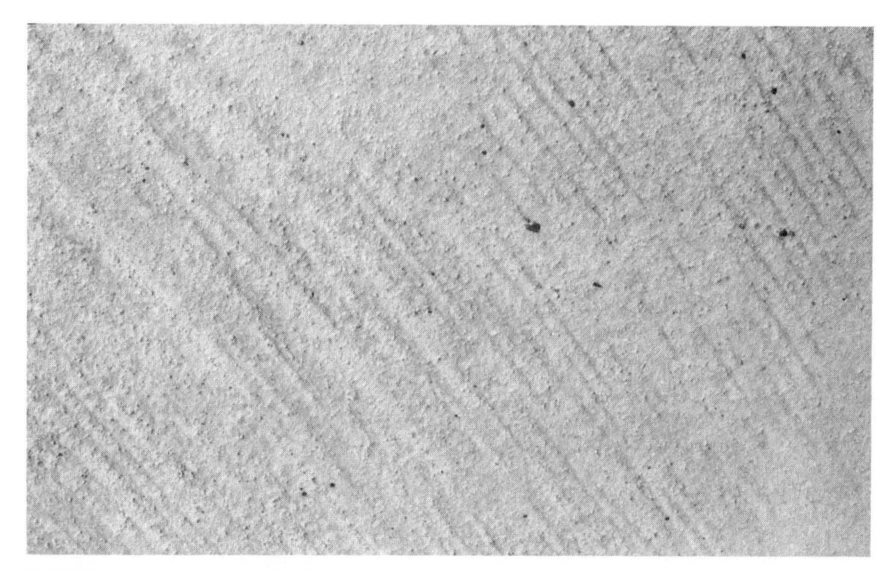

FIGURE 27.19 Burlap drag finish.

A tinned texture is the use of equipment having a number of stiff tines which are used to scratch the surface for animal pens or vehicular traffic.

Slotted textures are used on airfield runways to break up hydroplaning when rainwater is present. The orientation is transverse to the direction of travel.

Swirled textures (Fig. 27.18) are created by troweling in a circular fashion to produce circular ridges for traction. These ridges are approximately 1⁄16 in in height. Care should be taken as to timing of the troweling and to properly cure the ridges. Too early troweling will cause high, soft ridges and lack of curing, which will lead to excessive wear and early loss of traction. This texture is specified for most parking structures.

The burlap drag texture (Fig. 27.19) is extensively used for pavements having highway and street vehicular traffic. The burlap is dragged behind the screed or slipform, but only after the bullfloating is complete. Curing should follow immediately.

CHAPTER 28
JOINTS IN CONCRETE

Joseph A. Dobrowolski

TYPES AND INSTALLATION OF JOINTS

To minimize cracking, concrete needs control joints in various locations. The three types of recognized control joints are construction, contraction, and isolation joints. Type and installation are determined by design of the structure, method of construction, and expected use of the facility. If there are changes in usage during the service life, revisions in the joints may be required to prevent excessive maintenance.

28.1 Construction Joints

The construction joint is used to to limit the volume of placed concrete to the amount placeable during the chosen time period that can accommodate the shrinkage due to drying without uncontrolled cracking or to meet architectural requirements.

Vertical and horizontal construction joints are used in building and structural concrete. Most horizontal building joints will be occur at the bottom and top of each floor slab and be used to break up large floor areas into reasonable quantities. Initially, the contractor will form the columns and cast them. After column form removal, the shoring and forming for the floor slab will be constructed so that the floor slab placement can begin. On completion of the floor slab, the contractor begins to form the columns for the next floor and the sequence is repeated. Beams, girders, haunches, drop panels, and capitols are usually cast monolithically with the floor slab unless special arrangements for stress carryover are approved by the designer.

Horizontal construction joints in walls and other vertical structures are determined by economics, design limits of the forming, capability of the placing equipment such as pumps and hoses to place the concrete without segregation, and architectural requirements. Exterior vertical and horizontal construction joints are best when coordinated with rustication to diminish any contrast with the adjacent vertical wall surface by allowing tight fitting of the formwork to the existing concrete to prevent leakage and an accidental offset. For liquid-retaining structures, a water stop may be required across most construction joints to ensure watertightness. Most other exterior structural construction joint can be rendered weathertight by providing a rustication depth which can be caulked with color-matching sealant. Standard practice for construction joints is to have reinforcement projecting through the end bulkhead for tying in the next section. If practical, sections of

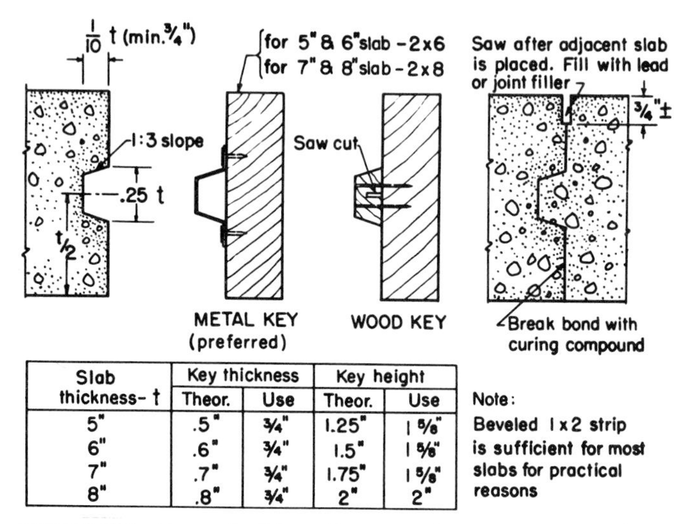

Slab thickness- t	Key thickness		Key height	
	Theor.	Use	Theor.	Use
5"	.5"	3/4"	1.25"	1 5/8"
6"	.6"	3/4"	1.5"	1 5/8"
7"	.7"	3/4"	1.75"	1 5/8"
8"	.8"	3/4"	2"	2"

FIGURE 28.1 Construction joints with tongue and groove. (*Courtesy of Portland Cement Assoc.*)

the building walls and slabs should be placed in alternate sections to allow volume change before the remaining areas are filled.

Slab construction joints (Fig. 28.1) are usually set along column lines in structures at grade and contain a tongue and groove design with dowels at midheight or a thickened edge to ensure against differential displacement. Construction joints for very flat floors have been located under planned shelving. This requires boxouts over column locations to allow unrestricted screed operations and rearrangement of other joints to concide with the column line diamonds. This method can also be used to accommodate mechanical screeds wider than the column spacing. Construction joints are required to be in the middle third of the span for suspended slabs, beams, and girders.

Recommendations no longer require a mortar bed in a horizontal construction joint prior to placing new concrete against existing construction for bonding purposes because of high placements and lack of insurance of the mortar placement. Sand or water jet blasting of the surface to remove any soft mortar or debris has been adequate. In some instances a concrete mixture containing only 50 percent of the coarse aggregate has been used for the first lift.

28.2 Contraction Joints

Contraction joints (Fig. 28.2) are used between construction joints to reduce the horizontal areas of concrete and to provide weakened planes for crack control. They are installed within the forms prior to concrete placement, in the surface of the concrete during concrete placement, formed during finishing procedures, or cut by sawing after sufficient hardening to prevent ravelling of the sawcut. Depths and distances between contraction joints will vary dependent on location and thickness of the slab.

Structural Walls and Slabs. Walls require a minimum contraction joint (Fig. 28.3) depth of 10 percent of the wall thickness in each face and at spacing of one times the wall

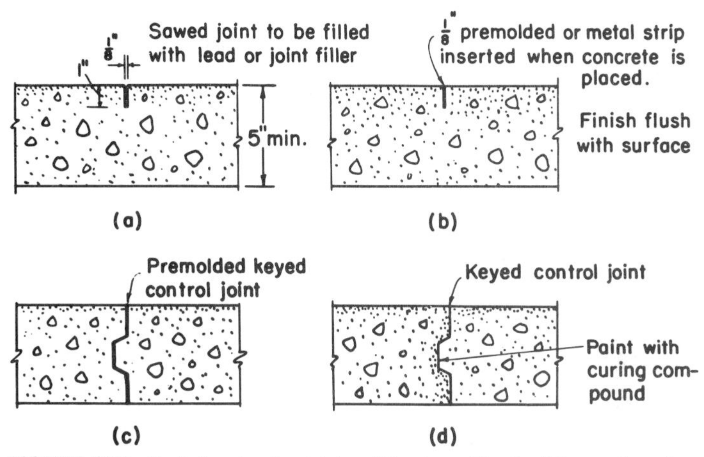

FIGURE 28.2 Typical contraction joints. (*Courtesy of Portland Cement Assoc.*)

height for high walls and three times the wall height for low walls. Recommended spacing for control joints in buildings is approximately 60 ft.

For suspended slabs, the designer may either (1) require sufficient reinforcement to keep any cracking to very fine widths and eliminate contraction joints or (2) specify their location and design. Most suspended concrete slabs have only construction joints to control cracking as deflection usually governs.

FIGURE 28.3 Formed vertical contraction joint in wall.

Slabs on Grade. Types of contraction joints in nonreinforced slabs on grade differ widely because of designer preferences, construction practices, and service requirements. Recommended depths are one-fourth to one-fifth of the thickness and a minimum of 1 in for all types of slabs. Distances between contraction joints have been recommended to be 15 to 20 ft, depending on the materials or whether reinforcement is used. In practice, good-quality concrete has performed adequately when the joint spacing does not exceed a distance in feet equivalent to four times the thickness of the slab in inches. For example, a 4 in slab containing excellent concrete and properly cured can have contraction joints at 16 ft.

Where the proposed use will not allow joints, the floor slab may be prestressed to contain volume change cracking, or placed on either polyethylene or sand to minimize subgrade friction in combination with reinforcement having 0.5 of 1 percent of the slab gross area, or by use of expansive cement with steel reinforcement. Reinforcing steel in the slab may also be used to increase distances between joints but should be terminated 2 in from the joint or have 50 percent of the reinforcement be cut at the joint.

FIGURE 28.4 Highway pavement with diagonal joints. Note center longitudinal crack due to lack of center longitudinal joint.

Highway paving contraction joints in nonreinforced concrete pavements were traditionally spaced on 15 ft centers. Because of modern requirements, spacing has been randomized in some states such as California to a repetitive series of 13, 19, 18, and 12 ft intervals in order to eliminate possible resonance in automobile suspension systems at high speeds. In addition, the transverse contraction or weakened plane joints are cut on a diagonal (Fig. 28.4) of 2 ft in 20 ft in order to lower the impact generated by each axle load.

Contraction joints may be made manually with a jointer and an edger, placement of plastic strips, sawing after sufficient hardening, or mechanically inserting a steel T-bar to form a groove or a strip of polyethylene during placement (Fig. 28.5*a* and *b*).

Manual tooling of contraction joints require jointers which are sufficiently deep to form the depth required by the thickness of slab. (Fig. 28.6*a* and *b.*) Many jointers do not have the required depth and need additional metal to be attached. Groovers are used to tool shallow depths of indentations for architectural reasons. Some specifications have required a full-depth tooled contraction joint to be backfilled with surface mortar until the joint is only ½ in deep. This has negated the effect of the original depth and caused uncontrolled cracking when additional volume change occurs. An edger (Fig. 28.7) may be used to round the edges of the joint or to leave a smooth line on each side of the joint known as a "shiner" (Fig. 28.8). If a shiner is not desired, additional troweling will uniform the surface to the joint.

Plastic inserts used for contraction joints are obtained as a T-shape extruded plastic strip (Fig. 28.9) with the prescribed depth and having a 10 ft length. The strips are generally inserted behind the screed after the concrete has been leveled. This should be done with care to ensure a a straight vertical joint. Prior to floating, the upper horizontal por-

FIGURE 28.5*a* Steel T-bar for forming joints automatically for slipform paver.

FIGURE 28.5*b* Attachment in paving train containing roll of plastic which unrolls into center of pavement to form longitudinal joint.

FIGURE 28.6a Manual jointer with 1 in depth

FIGURE 28.6b Groover for shallow architectural indentations.

FIGURE 28.7 Edger.

tion of the T is stripped off. The lower vertical portion of the T remains as the in-place joint. Misalignment and tipping of the plastic strips can occur if power floats are used prematurely on soft concrete. The spalls due to tipped plastic strips will occur under traffic and are expensive to correct.

Sawing the contraction joint should be accomplished as soon as the concrete is hard enough to prevent raveling of the edges by the saw blade, but still prevent random cracking. The timing of sawing is dependent on weather conditions and setting time of the concrete mixture. When drying conditions are severe, a combination of methods of installing joints may be required. The sawed joints should be vacummed clean to prevent future spalling as the slab moves as a result of thermal changes.

Newly developed lightweight saws (Fig. 28.10) with high speed have allowed sawing as soon as troweling is completed. Maximum depth of sawing is approximately ¾ in. Because of its lightness, the saw has a tendency to ride over coarse aggregate, especially when rushed. A few contractors have come back the next morning to deepen these shallow cuts and ensure crack control. Normal curing practices should prevail over the saw timing.

28.3 Isolation Joints

Isolation joints are used to prevent cracking during volume change due to restraint by full-depth seperation of rigid objects from the concrete slab. These objects would include footings, columns, walls, and utility penetrations. The seperation can be accomplished by using asphalt-impregnated fiber strips around the utility or adjacent to the structure, diamond column cutouts (Fig. 28.11), a sand blanket or flexible material between the bottom of the slab and the top of footing, and a delayed fill-in concrete strip used with tiltup construction.

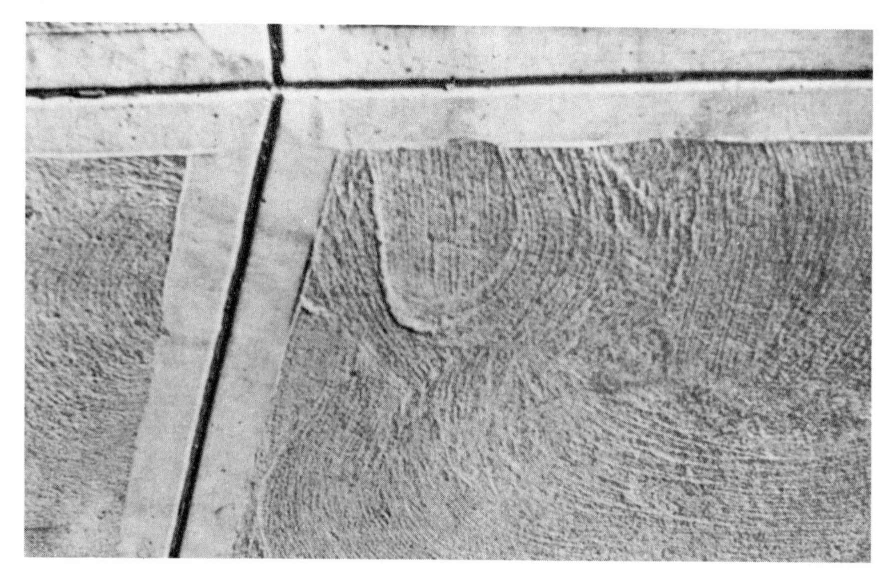

FIGURE 28.8 Joints with shiners.

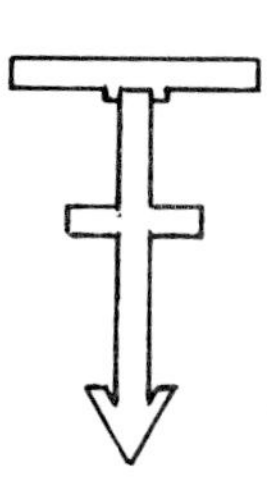

FIGURE 28.9 T-Shaped plastic strip used for forming contraction joint in plastic concrete.

FIGURE 28.10 Lightweight saw in operation.

FIGURE 28.11 Diamond blockout around column.

JOINT SEALANTS AND DOWELS

28.4 Joint Sealants

The use of sealants in joints of slabs on grade is determined by the expected traffic and environmental conditions. Normal conditions and traffic for residential concrete does not normally require joints to be sealed except where rain runoff would tend to carry sand or fine debris or sand is used to increase traction during ice conditions. The sand or fine debris tend to fill the joints while they are widest during cold temperatures, which results in edge spalling as the joints narrow during hot weather. In arid climates, pavement joints are not sealed as the joints are vacuumed clean by the high speed of high-volume traffic. Where traffic is low, blowing sand and dust requires joints to be sealed.

Use of forklift trucks in industrial warehouses has grown tremendously. This type of traffic will cause construction and contraction joints to spall due to high-point loadings and traffic volumes. Approximately 90 days after completion of slab placement but prior to occupancy, all cracks over $\frac{1}{16}$ in in width and joints located within the area traveled by the forklift should be sawed at least 1 in deep and filled with a low-modulus epoxy having a Shore A-scale hardness number of approximately 80 or a D-scale hardness number of 50. The sawed joint should be overfilled slightly and ground smooth after hardening with a sandpaper grinder. Tight cracks need not be repaired. An elastomeric sealant can be used in the remainder of the joints where floor debris is a problem. Daily vacuuming of joints in nontraffic areas can eliminate the need for sealants.

28.5 Dowels

Developments in dowel (Fig. 28.12) technology have shown that cracking due to horizontal dowel restraint has occurred in new slabs placed adjacent to an existing slab. New recommended details include square dowels cast into the concrete to allow horizontal movement but still maintain restraint to differential vertical slab movement.

FIGURE 28.12 Dowel assembly in formed pavement.

CHAPTER 29
CURING

Charles E. Proudley*

PURPOSE AND TYPES OF CURING

Curing is the procedure used to assure that there is enough water present in the concrete to provide for continuous hydration of the cement.[1] It is obvious that if the supply of water, usually present in ample quantity in the finished concrete product as mixing water, disappears as the result of evaporation, hydration of the cement will stop and there will be no further gain in strength and durability of the concrete.

As concrete dries, it shrinks, and if drying occurs when the concrete has little, if any, strength, cracks are sure to result. Furthermore, since drying occurs first on the surface, the cement will not be hydrated there but will be present as a dust coating having no strength to hold the aggregate particles together.

29.1 Temperature Effects

Temperature of the concrete during the chemical reactions of hydration has an important effect on the rate of strength gain. Temperatures near the freezing point of water retard the setting or hardening of cement to almost zero. Should the temperature fall below freezing, the free water in the concrete will turn into ice crystals, and since ice has much greater volume than the same water in a liquid state, the concrete is disrupted and on thawing will have no strength. On the other hand, should the temperature be above 90°F, there is great danger that the water will evaporate quickly and having once "boiled" off can be replaced only with considerable difficulty and much longer curing time for complete hydration. It is generally recognized that the problems of curing concrete during hot weather are much greater than during cold weather. Cracks in the surface of the concrete are likely to occur on a hot summer day because the setting reactions take place faster and the internal heat due to hydration causes expansion of the interior greater than at the surface.

In general, for the usual curing processes, special effort should be made to keep the concrete wet as evidenced by the continuous presence of moisture on the surface, and within a temperature of 50 to 80°F. Alternate wetting and drying can do more damage than no curing at all.

*Deceased. Chapter revised by Joseph J. Waddell and Joseph A. Dobrowolski.

There are two principal systems of curing: (1) application of water directly or through some material that holds a reservoir of water in contact with the surface, and (2) a seal to prevent or retard the escape of moisture from the concrete.

Temporary protection of flat slabs placed during hot and windy weather can be accomplished by the application of a sprayed-on monomolecular film (Confilm) that reduces rapid evaporation of water from freshly placed concrete. Available from some admixture producers, this material is sprayed on the concrete surface immediately after screeding. It is not a curing compound and is no substitute for proper curing, as its purpose is to reduce evaporation only while the concrete is in a plastic state. This material should not be used on concrete that bleeds excessively or under conditions of high humidity nor during high winds which will blow the film off the concrete surface.

29.2 Water Curing

One method of direct application of water is inundation of the concrete, as when horizontal slabs are kept flooded by means of an earth dam around the edge that retains about an inch, more or less, of water over the entire surface. The same effect is obtained by using a fine spray or mist from a number of nozzles arranged to cover the whole area. In either case, consideration should be given to water that runs off and whether it is objectionable in its effect on soil under or around the slab.

On dams and high piers it is sometimes the practice to attach spray pipes to the lower edge of the forms. As the forms are raised for successive lifts (Fig. 29.1), water curing is applied to the freshly exposed concrete. However, water running over the lower portions of the structure for a long period of time is apt to stain the concrete.

Burlap is used extensively as an absorbent cover for curing concrete. It should be closely woven and weigh at least 12 oz per 10 ft^2. Before it is used, burlap for curing concrete should be washed to remove soluble substances that are injurious to concrete or will mark the surface. Burlap is saturated with water before being placed over the concrete and successive strips are overlapped. The concrete must have set sufficiently so the wet burlap will not mar the surface; this will be before the concrete surface shows signs of drying. Later, additional layers of wet burlap can be added so that wetting will be required less often. Burlap must be kept continuously wet. Burlap curing is frequently followed after 12 h or more by some other method, such as membrane curing, that does not require further applications of water.

Sand, sawdust, or earth curing is used successfully if kept saturated and if there are no salts or substances present that will damage or stain the concrete. Granular materials must be watched to see that they do not dry to the point that they are absorbing moisture from the concrete instead of supplying it. Oak sawdust must never be used because of the tannic acid it contains.

Straw is acceptable as a curing medium, spread to a depth of 6 in or more, thoroughly saturated with water. It must be watched for fire hazard should it become dry between wetting, and also there is the possibility that the straw will be blown away unless it is held down by some other covering such as burlap or paper. The special advantage of straw is its insulating value against heat and cold.

Cotton mats or blankets, made of low-grade cotton bats quilted between two layers of 6½ oz burlap or osnaburg (cotton), and weighing a total of 25 oz/yd^2 when dry, are excellent for curing because of their high absorbency and thermal-insulating properties. Their weight makes it necessary to use a lighter burlap cover or a water mist until the concrete has set enough to withstand marring by the weight of the wet mats.

Used carpeting is an excellent curing medium as it holds large amounts of water and dries slowly. Such carpeting should be thoroughly rinsed with clean, clear water at least

FIGURE 29.1 Large areas of mass concrete are sometimes cured by means of water from spray pipes attached to the forms. Note that one block in the illustration is dry, the water having been shut off while the form is moved. (*U.S. Bureau of Reclamation photograph.*)

three times to eliminate any possible residue of organic cleaning compounds which may retard and soften the surface.

29.3 Impervious-Sheet Materials

Sealing the water in by means of waterproof, vaportight coverings has the advantage that it requires less attention subsequent to application of the curing, and also less water is needed, which can be a serious economic consideration where water is not readily available. One of the first waterproof or impervious-sheet curing materials was a paper blanket made by cementing sheets of kraft paper together by means of a bituminous adhesive, with a wide-mesh fiber fabric in between to give strength and durability. Sometimes the top or exposed surface is white to reflect the sun's heat. The kraft paper is treated to minimize dimensional changes as the paper comes in contact with water. The paper blankets come in widths up to 12 ft and are in rolls of 75 ft length. For wider blankets, sheets are cemented together to obtain a cover that can be sealed at the edges, usually by means of earth windrows on the paper to prevent escape of moisture, as shown in Fig. 29.2.

Repeated use of impervious-sheet curing materials is permissible, but the number of repetitions will depend on the condition of the sheets after removal from the concrete.

FIGURE 29.2 Applying waterproof curing paper. Ends and edges are held in place by small windrows of earth. (*Courtesy of Portland Cement Assoc.*)

Rips and tears, obviously, permit rapid loss of moisture and must be repaired to reseal the sheet material. Pinholes caused by walking or rolling equipment on the paper or impervious sheets are not so obvious but will permit the loss of moisture nonetheless. Pinholes are more difficult, if not impossible, to repair. It is necessary, therefore, to make a careful examination of the impervious-sheet curing covers between each use to determine if they will be effective. Sometimes an extra use or two may be made of slightly damaged or questionable covers by using them in double thickness. The careful worker may obtain as many as 20 reuses of a waterproof paper blanket for curing concrete, although under average conditions five or six uses are the limit for good curing.

The development of plastic-sheet materials has brought the use of polyethylene covering into prominence. It is available either clear or white-pigmented in rolls of varying widths up to 12 ft and in thicknesses from ½ to 4 mils. Clear polyethylene is generally used because it is less expensive than the white; however, in hot climates the white reflects heat from the sun and provides some thermal regulation. Black polyethylene should not be used for curing except where heat absorption is not a problem. The ½ mil sheet is extremely light and fragile, making it difficult to handle and keep in place. If used, it is generally discarded after one use. The 4 mil thickness is comparatively stiff but very durable. For most purposes the 2 mil polyethylene film is preferred.

There are also available combinations of polyethylene fused to burlap or containing glass reinforcing fibers and several other innovations that are designed to give better service or economy.

Impermeable sheets must be well lapped and weighted at joints and edges. Pieces of lumber or earth can be used for weights. Be sure corners and edges are well covered.

29.4 Liquid Membrane-Forming Curing Compounds

Liquid compounds to be sprayed on the freshly finished concrete that will dry quickly to an impervious membrane or film have been used extensively for vertical as well as horizontal surfaces. This material variously known as sealing, curing, or membrane-forming compound, is a paintlike liquid that, when sprayed on concrete, forms an impervious membrane over the surface. The formulation contains resins, waxes, gums, solvents, and other ingredients such as a fugitive dye or a pigment.

Compounds can be classified as white-pigmented, gray-pigmented, and clear, the latter containing a fugitive dye to aid in observing the coverage. The dye should render the film distinctly visible for at least 1 h after application and should fade completely in not more than 1 week. All compounds should be of a consistency suitable for spraying, should meet local environmental requirements, should adhere to a vertical or horizontal damp concrete surface when applied at the specified time and coverage, and should not react harmfully with the concrete. The clear compound should not darken the natural color of the concrete. Sometimes a thixotropic compound is added to hold the pigment in suspension, since the vehicle is usually so thin that pigments settle out rapidly, requiring special devices and constant care to keep the curing compound stirred and agitated for uniform application. The compound is formulated with either a wax-and-resin base or an all-resin base, some users specifying the all-resin type. One advantage claimed for the resin type is that it does not tend to separate as the wax-and-resin type does; hence there is less need for agitation.

Curing compounds are sprayed on at the rate of 1 gal per 200 sq ft^2 for horizontal surfaces or 1 gal per 150 ft^2 for vertical faces. Rough concrete surfaces require more compound than smooth surfaces. Recommendations of the manufacturer should be followed. For best results on all surfaces, the time of application should be when all surface water has disappeared, the surface is still dark with moisture but not shiny, and there are no dry areas. On a horizontal surface, one half of the curing compound should be applied in one direction and the other half in a direction perpendicular to the first application.

The compound should be applied to unformed surfaces as soon as the water shine disappears, but while the surface is still moist. If application of the compound is delayed, the surface should be kept wet with water until the membrane can be applied. Do not apply compounds to areas on which bleeding water is standing.

The compound should be applied to formed surfaces immediately after removal of forms, unless the surface is to be rubbed, in which case a moist covering should be used to protect the concrete until the surface has been rubbed. If the concrete surface is dry, it should be moistened and kept wet until no more water will be absorbed. As soon as the surface moisture film disappears, but while the concrete is still damp, the compound is applied, taking special care to cover edges and corners. Patching of the concrete is done after the compound has been applied. Compounds should be sprayed, not brushed, on patched areas.

When used solely for curing, it is necessary that the life of the membrane be only about 30 days or less, after which it will be either worn, washed, or blown away, or removed readily with a stiff bristle brush. Water retention of the membrane is very good for the first 3 days, still effective for another 4 days or more, but of little value for curing after 14 days. It is available in a variety of colors for aesthetic or practical purposes but if the concrete is to be rubbed, as on handrails, bridge piers, or architectural concrete, it should be clear without pigment or dye in order to avoid a mottled appearance.

Pressure-tank equipment, as shown in Figs. 29.3 and 29.4, should be used for spraying the compound, and the compound should be agitated continually during application. Apply in one coat, consisting of two passes of the spray nozzle at right angles to each other.

The compound should never be thinned. However, during cold weather, the compound can be heated if it becomes too viscous for application. Heating is done in a hot-water bath, never over an open flame, and the temperature should never exceed 100°F. When being heated, the container should be vented and there should be room in the container for expansion of the compound. The compound should be agitated during heating.

FIGURE 29.3 Automatic self-propelled spray machines are available for applying curing compound on large projects. (*Courtesy of Portland Cement Assoc.*)

29.5 High-Temperature Curing

Special methods of curing that accelerate, the curing process have been developed for concrete-products plants, permitting less storage space for finished units and quicker delivery on orders.[2, 3] The presence of adequate water for hydration is always of primary importance. More rapid hydration of the cement in the product is accomplished by elevated temperature. Systematic control is essential, with care taken to raise the temperature slowly to prevent the units from developing critical internal stresses through differential thermal expansion.

The normal schedule for atmospheric steam curing is about as follows:

1. *Preset:* Also referred to as "delay," "presteaming," or "holding." A delay period between fabrication of the product and the start of steaming. Hydration of the cement starts during this period, stabilizing the product. Products are stored in the steaming chambers at normal temperatures and prevented from drying out, if necessary, by the application of a light fog. During cold weather a small amount of heat might be necessary to keep the temperature above 60°F.

2. *Period of rising temperature:* During this time the products are enclosed in the curing chamber (sometimes called a "kiln") and heat is applied to raise the temperature to

FIGURE 29.4 Even on small jobs, manual pressure-spray equipment should be used. (*Courtesy of Portland Cement Assoc.*)

the maximum curing temperature. The temperature rise is controlled at a rate of 20 to 60°F/h, depending on the size and character of the products.

3. *Maximum-temperature period:* The temperature within the chamber is held at a constant temperature for whatever period of hours is necessary for the product to develop the necessary strength. In some plants this period approaches zero. See the following soaking period.

4. *Soaking:* After the product reaches maximum temperature the steam is shut off and the product "soaks" in the chamber as the temperature slowly drops.

5. *Cooling:* The product is permitted to cool to atmospheric temperature at a controlled rate, the rate again dependent on size and shape of product.

In a typical plant, after a preliminary curing at atmospheric temperature for 2 to 4 h, the temperature of the curing chamber or enclosure is raised at a rate of 40° per hour to a maximum of 140 to 175°F, holding at this temperature as a soaking period for 8 to 18 h or longer if tests show an advantage in a longer period, and reducing the temperature at not more than 1° per minute. Units having thin walls or small dimensions can be heated and cooled at a faster rate. Temperature recording or automatic temperature controls in the curing chamber are a necessity if uniformity of product is to be assured.

Some products plants supply moist curing at atmospheric temperatures by means of wet steam that maintains temperatures at about 135°F. Steam from the boiler is released through numerous holes in pipes laid along the floor between rows or racks of the units to be cured. If the temperature exceeds 135°, the steam is shut off or reduced to prevent units from becoming too hot, with consequent possibility of dehydration and undercuring. The units should always show a film of moisture while curing with low-pressure or atmospheric steam. Not more than 2 days of atmospheric steam is customary, provided tests show that adequate strength has been attained. When the chamber is opened there should be a period of several hours for the concrete to adjust to the temperature of the outside air; otherwise cracks may appear, especially in cold weather. Under high-temperature conditions both heat and moisture may be lost to the outside through the material used to cover the concrete; hence the importance of providing tight covers and kilns, insulated if possible. More moisture is required to maintain 100 percent relative humidity at higher temperatures than at low temperatures. At 70°F, 1.6 lb of water is required for 100 lb of air for 100 percent humidity, but 15 lb is required at 140°F and 66 lb at 180°F.

Under steam-curing conditions, the heat of hydration serves to raise the concrete temperature in the same way that it does under normal curing. Therefore, there comes a time when the concrete temperature will reach the ambient temperature within the enclosure, then rise above if control measures are not taken. After the concrete reaches the ambient temperature, the amount of heat should be reduced gradually in order to minimize the temperature differential between the concrete and the atmosphere within the enclosure and to permit a gradual cooling of the concrete.

When the concrete temperature exceeds the enclosure temperature, moisture will be lost from the concrete, unless wet steam is used as a source of heat. If heat is provided by other means than wet steam, there must be a source of additional moisture, such as hot-water sprays.

High-pressure steam curing can be used for any concrete product but is usually applied to masonry units in which, through choice of materials, there will be some reaction with the siliceous sand in the mix. An autoclave is necessary for high-pressure steam curing because the cycle of curing includes a period in dry steam at about 360°F. The cycle can be varied, but a typical sequence starts with a preset period at room temperature for 2 to 4 h; then the autoclave chamber charged with units is sealed and the temperature and pressure are raised over a period of about 3 h by means of saturated steam to about

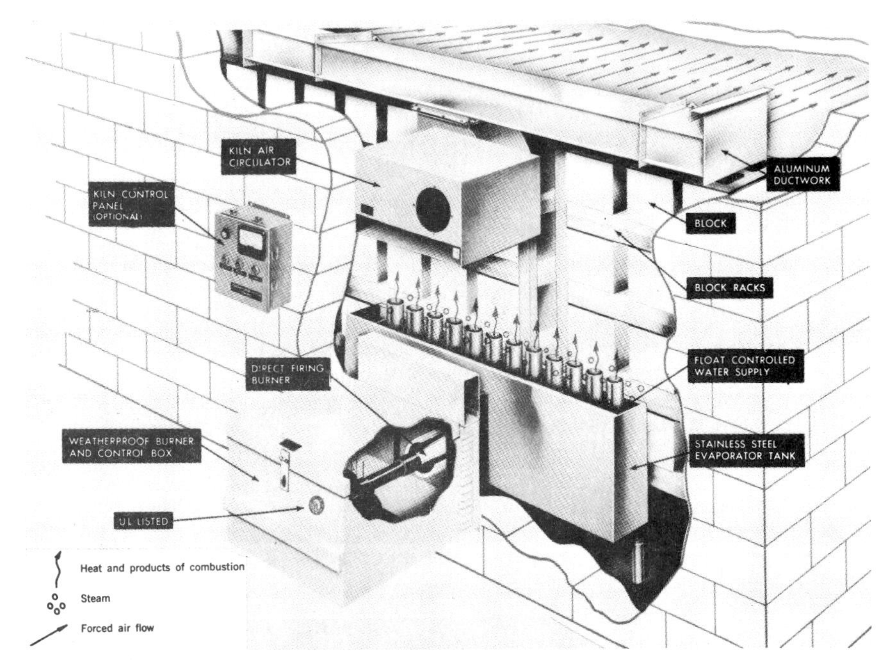

FIGURE 29.5 The Johnson direct-fired curing system, here applied to a masonry block-curing kiln, makes use of heat, steam, and carbon dioxide for curing the block. See Chap. 42. (*The Johnson Gas Appliance Co.*)

360°F and 140 to 150 psig. Pressure is released during the final 20 to 30 min, and cooling is rapid. Higher temperatures and pressures with shorter timing are often used, and also lower temperature and pressure with longer timing, depending on the manufacturer's facilities or choice based on materials and the quality or appearance of the concrete units resulting from the schedule.

Other curing methods involve the use of various means for applying heat while retaining moisture for hydration. Dry heat is applied where the concrete unit is sealed from loss of moisture that occurs when there is a vapor-pressure differential resulting from a difference between the temperature of the concrete and that of the surrounding atmosphere. The heat can be supplied from a heated casting bed, a removable enclosure in which there are heating units to heat the air as is the case when concrete is cured in the mold, or whatever arrangement is most convenient for the characteristics, dimensions, or design of the concrete product to be cured.

See Chap. 42 for a detailed discussion of high-temperature curing, including a direct-fired system that utilizes carbon dioxide during part of the cycle (Fig. 29.5).

SELECTION OF METHOD TO USE

29.6 Pavement, Roofs, Bridge Decks, and Exterior Floors

Horizontal slabs such as pavement, sidewalks, parking areas, bridge decks, runways, roofs, and all concrete exposed to climatic conditions should be cured in a manner to provide for temperature control insofar as possible. The first step under any condition of exposure is to prevent drying of the surface.[4, 5] Before the concrete has set enough to withstand marring, a lightweight sheet material such as polyethylene film should be used, sealing it along the edges to retard loss of moisture and to hold it in place by keeping wind from getting under it and ballooning the film. A fine fog to provide a mist that will envelop the area is excellent in hot weather if economically feasible since it will usually hold the temperature of the concrete uniformly within the recommended range while providing the required water for curing. The labor and facilities needed for water spray make this method more adaptable to curing large precast units at the manufacturing plant. Preliminary curing with a single layer of burlap is the most popular procedure for concrete pavement where construction moves along rapidly.

The most extensively used method of curing for surfaces that are outdoors is clear membrane-forming liquid compound since it can be applied almost immediately after the concrete is finished and no further attention is required. However, this is true only when temperature conditions are between 40 and 80°F. When the sun's rays and air temperature are likely to cause concrete temperatures to exceed 100°F, white-pigmented compounds should be used. When air temperatures are below 40°F and there is the possibility of subfreezing weather during the first 72 h, further temperature control is necessary. This is readily provided by placing a layer of straw about 12 in thick on the slab as soon as possible without damage to the surface, and covering the straw with a blanket material such as polyethylene, waterproof paper, or burlap to hold it in place. The straw insulation can be used without the liquid membrane-forming compound if straw curing alone is desired. In such cases only enough water should be applied to the straw to ensure that the concrete surface will not become dry during the curing period since excessive use of water may cause freezing in spots subject to greatest heat loss such as at edges and corners.

Sand, sawdust, soil, and similar granular materials are satisfactory for curing and provide considerable protection against high temperatures and a limited protection when temperatures dip below the freezing mark for short intervals during each 24 h. The sand, sawdust, etc. must be kept wet by sprinkling several times a day with a hose or a spray bar on a tank truck operating outside the paved area.

Bridge decks must be protected from below during cold weather unless the temperatures are not likely to be lower than about 15 or 20°F during the night, under which circumstances wood forms furnish ample insulation. To provide more protection and to accelerate curing, heaters placed at a safe distance below the formwork of the bridge floor will be helpful. The same recommendations apply to any slab above grade.

29.7 Interior Floors and Walls

In hot weather it is necessary only to prevent drying of the concrete inside buildings, and this can be done with liquid membrane-forming compounds, impervious-sheet materials such as paper or plastic, with a fine spray from a hose if the room is closed to maintain high humidity, or any of the absorbent covers such as burlap, cotton mats, sand, or sawdust. If workmen will be moving about on the concrete during the curing period, wet sand or sawdust is preferable to sheet materials or membrane curing because of the protection against abrasion; however, sand or sawdust must be kept wet and deep enough to be a protection until the curing period is over.

In cold weather the room or building should be closed and heated in addition to the application of curing materials as described above. Clear plastic sheets are used to cover windows and doors, and space heaters are employed to keep the temperature up to 60°F or better. Combustion gases from space heaters must be vented to the outside to avoid carbonation of the concrete surface and subsequent dusting. If the concrete slab is not protected by a building, a temporary enclosure of plywood, or lightweight frames or trusses covered with plastic or canvas, and heated by atmospheric steam, electric radiant heaters, or similar devices, can be constructed for protection against almost any cold-weather severity during concreting, finishing, and curing.

29.8 Exterior Walls, Columns, and Bridge Piers

Vertical surfaces to be cured after removal of forms can be protected from moisture loss, heat, and cold by enclosures, as has been described for horizontal slabs. Liquid membrane-forming compound sprayed or brushed on in one or two applications as may be necessary to obtain the required thickness of film is the least trouble for the degree of resulting moisture retention. If early rubbing is required, however, the removal of the film is troublesome, and unless it is completely removed the rubbed finish will not have a uniform appearance. If, however, the surface is rubbed immediately after the removal of forms, the sprayed-on curing can be applied after the rubbing, either clear without fugitive dye, white, or light gray, and the final appearance will be acceptable.

Where there is an abundance of fresh water available, good curing is obtainable by draping burlap against the piers or walls and keeping it saturated by a continuous trickle of water from pipes or hose at the top. Unless the water is free from minerals such as sulfides and iron, permanent stains may occur that cannot be removed satisfactorily by rubbing. Flowing water over a vertical surface under a covering of polyethylene film so attached as to be relatively airtight will cure satisfactorily but will not give much protection against cold or solar heat. In cold weather a framework around the walls or piers covered with impervious sheets of plastic or waterproof paper will give considerable thermal protection as long as the water flowing over the concrete does not freeze. Sometimes cotton mats are wrapped around columns or piers to preserve heat of hydration and prevent freezing (see Fig. 29.6).

29.9 Architectural Concrete

If the architectural concrete is cast in place in the structure and the forms must be removed before curing time has been completed, further curing is obtainable as described

FIGURE 29.6 A polyethylene enclosure, with heat supplied inside and straw on the deck, provides curing protection for this structure during cold weather.

FIGURE 29.7 Precast units in the casting yard are covered with heavy waterproof canvas for curing.

for vertical walls. Where the concrete is an overhang, strips of wood are attached by whatever means is most convenient to hold the blanket or sheet material as close to the concrete as feasible.

Architectural shapes, units, designs, or figures that are made in the concrete-products plant or shop will have been properly cured before being delivered to the job (Fig. 29.7). Job curing and plant curing must be coordinated to result in the same color and appearance of the finished concrete if it is desired to have them match in the structure. If curing results in differences that cannot be corrected otherwise, rubbing with a carefully chosen mortar mix sometimes improves the appearance.

29.10 Masonry Units, Precast Members, Concrete Pipe

Precast-concrete products will arrive on the job ready-cured for immediate use. The method of curing used by the manufacturer will depend upon his analysis of the economics of his operations, and these vary geographically and with the demands of the marketing area. The majority of plants use atmospheric steam curing since this can be controlled for changing weather conditions and does not require much financial investment. Water-fogging nozzles instead of steam are good but require supplementary heat during cold weather.

Large units such as culvert and sewer pipe that must be cast rather than machine-made are molded over a steam line that vents atmospheric steam to the interior of the mold. After casting and finishing the top end of the pipe, a jacket of dimensions to allow several inches of space around the outside and top of the mold is dropped over the entire assembly, leaving openings at the top and bottom so the warm, steam-laden air can circulate entirely around the mold. After 24 h the mold is removed and the cover replaced for more steam curing that can be continued for whatever curing period strength tests of the concrete in the pipe indicate is required.

Masonry units, sand-lime brick, concrete brick, cement-asbestos pipe, and other mass-production shapes are autoclaved with high-pressure steam when it is found by experience that the rapid production and shipment of units make it profitable, or when the characteristics of the autoclaved unit show marked improvement over any other means of production.

29.11 Tild-up and Lift Slab

For curing the concrete slab on which tilt-up or similar slabs are to be subsequently cast, there are available liquid curing compounds that act as a sealer to the concrete and as a parting compound to prevent the later concrete from bonding to the floor slab. The first coat is applied to the fresh concrete in the usual manner to serve as a curing compound. A second coat to serve as a bond breaker is applied after the forms have been placed but before the reinforcement and inserts are placed.

REFERENCES

1. ACI Standard 308-81, Standard Practice for Curing Concrete (Rev. 1986).
2. ACI 517.2R-87, Accelerated Curing of Concrete at Atmospheric Pressure.
3. Menzel Symposium on High Pressure Steam Curing, Special Publication 32, American Concrete Institute 1972.
4. "Current Road Problems, 1-2R, Curing Concrete Pavements," Highway Research Board, National Academy of Sciences, Washington, D.C., May, 1963.
5. Better Concrete Pavement Serviceability, Monograph No. 7, Edwin A. Finney, American Concrete Institute, 1973.

PART • 8

SPECIAL CONCRETES AND TECHNIQUES

CHAPTER 30
ROLLER-COMPACTED CONCRETE

Oswin Keifer, Jr.

DEFINITION

In Cement and Concrete Terminology (ACI 116R-90) the American Concrete Institute (ACI) defines roller-compacted concrete (RCC) as "Concrete compacted by roller compaction; concrete that, in its unhardened state, will support a roller while being compacted." In the past, some agencies have used the terms "rollcrete" and "rolled concrete," but these are now considered obsolete. The properties of hardened RCC are essentially the same as those of conventionally placed portland cement concrete and the in-place hardened product should simply be considered "concrete." However, the consistency of plastic RCC is different consistency from that of conventionally placed concrete; plastic RCC must be of sufficiently stiff consistency, less than zero slump, to support the rollers used for compaction. Equipment used for transporting, placing, and compacting RCC consists of high-capacity equipment used for earthwork construction and asphalt paving construction. Much less hand labor is required for RCC construction than for conventional concrete construction.

TYPES OF RCC

Two general types of RCC construction have been employed: (1) low-cement content "mass" RCC for construction of dams and other massive structures (retaining walls, stilling basins, levees, etc.) where high strength is not a requirement and (2) relative high-cement-content RCC for pavements and similar surfacing where strength is critical and where the surface must resist the abrasion of traffic. While both types of construction are referred to as "RCC," and use many of the same construction techniques, they are really no more related than conventional mass concrete and conventional concrete pavement. Both types do depend on high-capacity construction equipment and both types are compacted with heavy, dual-drum, vibratory steel rollers. However, since the mixtures used and the construction procedures used are so different, each type of RCC will be discussed separately herein.

30.1 Mass RCC

History and Development. Mass RCC construction was originally considered and developed as a means to design and construct lower-cost dams that could be constructed rapidly. It was originally considered as a lower-cost alternative to earth and rockfill dams. However, it soon was being considered and used as an alternative to conventional concrete dams. one of the earliest proposals for RCC dam construction[1] recommended that soil-cement principles be applied to the placement of an embankment with cement-enriched granular pit-run material using large earth-moving and compaction equipment. This cement-stabilized material would have increased shear strength, thus allowing a major reduction of the cross section of the dam as compared to a typical embankment dam.

Some of the early tests on and uses of RCC mass concrete were as follows:

1972—Cannon reported on the use of lean concrete with 75 mm maximum-size aggregate compacted with a vibratory roller at a Tennessee Valley Authority (TVA) project.[2]

1973—test sections of various RCC construction methods were studied near Lost Creek Dam in Oregon by the U.S. Army Corps of Engineers.[3]

1974 to 1982—more than 2.5 million m^3 of RCC were used at Tarbela Dam, Pakistan for rehabilitation work after the dam was subjected to reservoir filling.[4]

1976—at the TVA Bellefonte Nuclear Plant, 6000 m^3 of RCC were used to raise the turbine building base.[5]

The Japanese were among the first to use RCC for full-scale dam construction. RCC was used for the main body of Shimajigawa Dam started in 1978.[6] However, their approach was to use the RCC for a core constituting the main body of the dam, but totally encased with conventional concrete on both upstream and downstream surfaces.

The first full-scale RCC dam construction in the United States was Willow Creek Dam at Heppner, Oregon, constructed by the U.S. Army Corps of Engineers in 1982. Approximately 330,000 m^3 of RCC was placed in this 50-m-high dam in less than 5 months.[7]

In 1987 the U.S. Bureau of Reclamation (USBR) completed the 90 m high Upper Stillwater Dam in Utah, with 1.1 million m^3 of RCC.[8]

Since the construction of Willow Creek Dam, the use of mass RCC has grown rapidly and numerous RCC dams have been constructed in the United States and other countries throughout the world. RCC construction has also been used for strengthening existing older dams, which were either underdesigned or deteriorating.

Application and Advantages. Mass RCC is applicable to many locations where conventional mass concrete would be used. It is not applicable where any appreciable amount of reinforcing steel is required because of the difficulty of embedding the steel, although some use of reinforcing has taken place over gallery roofs and for anchors. Successful use of RCC does require a structure with sufficient space for the required heavy, earth-moving-type equipment to operate. The basic advantage of the use of mass RCC is economy. The overall cost of an RCC dam can be expected to be 15 to 30 percent lower than that for a conventional concrete dam. The lower cost is due primarily to the much shorter construction period required, the use of large earth-moving equipment, and the lower amount of labor required. Even large RCC dams can usually be constructed in one construction season. Spillways can be incorporated directly in the RCC dam. RCC dams will often be competitive in cost with embankment dams. Factors affecting this are the much smaller size of the dam, the use of an integral spillway, and the integral intake tower possible in lieu of the required freestanding intake in the reservoir for an embankment dam.

Design. The structural design of an RCC dam will essentially be similar to that of a conventional concrete dam. Items of particular concern to an RCC dam will be ensuring adequate bond and shear resistance at each lift joint and ensuring impermeability, especially at lift joints. Thermal studies similar to those used for conventional mass concrete are required for RCC mass construction. Usually thermal problems with the RCC will be slightly less than with conventional concrete. Early RCC dams were designed without contraction joints because of the lower shrinkage of the RCC. However, the tendency now is to install transverse construction joints at changes in the foundation shape and at regular intervals dictated by the shrinkage characteristics.

Upstream faces of RCC dams have been constructed with precast air-entrained concrete panels anchored to the mass RCC (e.g., Willow Creek Dam); with slipformed conventional air-entrained concrete "curbs" placed immediately ahead of the RCC (e.g., Upper Stillwater Dam); and with the use of a 0.6 to 1.2 m thick facing of conventional air-entrained cast-in-place concrete, formed and placed simultaneously with the RCC and integrally knit together with the RCC during consolidation (e.g., Elk Creek Dam[9]). A few RCC dams have used panels combined with some type of integral impervious membrane.

Downstream faces of RCC dams have typically been unprotected RCC placed at a slope between 0.75 H (horizontal) to 1.0 V (vertical) and 1.0 H to 1.0 V. Typically, this slope will be overbuilt by approximately 0.5 m with this area considered as sacrificial material. Since it is impractical to adequately compact the RCC at the outer edge, and since air entrainment has not yet become common in RCC use, this outer edge cannot be considered durable to withstand freeze-thaw weathering. However, service records to date show the deterioration to be quite slow. At Upper Stillwater Dam, the downstream face was constructed with a conventional slipformed concrete "curb" similar to the upstream face but of different shape. This same curb system was used for part of the downstream face of Elk Creek Dam, at the contractor's request.

Materials. Cementitious materials requirements for mass RCC are essentially the same as those for conventional mass concrete. Cement should be chosen to have low-heat properties and, where appropriate, low alkali content and sulfate resistance. Usually a pozzolan, most often fly ash, is used to help control heat gain, promote workability, and lower cost.

Aggregate requirements for mass RCC are essentially the same as those for conventional mass concrete. Early on, it was hoped that lower-quality aggregates, pit-run material, could be used. However, it soon became apparent that high-quality aggregates are just as important for mass RCC as for conventional mass concrete. A few RCC dams have employed aggregate gradations tailored specifically to the project. The tendency now is to use conventional ASTM C33 gradations with fine aggregate and with the coarse aggregate separated into individual sizes at the 19 and the 37.5 mm sizes. As with conventional mass concrete, the largest nominal maximum-size aggregate (NMSA) should be used for mass RCC. However, presently it is recommended that the NMSA not exceed 75 mm because of the increased tendency for segregation of RCC during handling when the aggregate size increases beyond this. It is common practice to require that at least half the aggregate needed for a project be produced and stockpiled near the mixing plant before RCC production commences, to help ensure that no interruption to RCC production occurs.

It presently is common to increase the fines in the fine aggregate slightly more than given by ASTM C33. A typical fine aggregate gradation used for mass RCC is as shown in the following table:

Typical Fine Aggregate Grading Limits

Sieve size	Cumulative % passing
9.5 mm	100
4.75 mm	95–100
2.36 mm	75–95
1.18 mm	55–80
600 μm	35–60
300 μm	24–40
150 μm	12–28
75 μm	8–18

Admixtures have not had as much use in RCC construction as with conventional concrete. While recent laboratory work has developed procedures to provide air entraining in RCC,[10] to date air entraining has seldom been used for mass RCC. A retarding admixture has been successfully used in some instances to keep the fresh RCC "live" and aid in placement and consolidation.

Mixture Proportioning. Mixture proportioning studies for RCC must consider required strength, workability, compactability, and durability. Included as part of durability for mass RCC is the limitation of heat rise due to hydration of the cementitious materials. Factors to be considered in attaining these criteria are use of the largest practical nominal maximum-size aggregate, use of the minimum required amount of cementitious material with part of this being a pozzolan, and use of cooling for the ingredient materials, particularly water and aggregates.

Different agencies have used different philosophies and approaches to the problem of determining mixture proportions. Three common proportioning methods as described in detail in Roller Compacted Mass Concrete, ACI 207.5 R89,[11] are the following:

1. *Proportioning RCC to meet specified limits of consistency.* Proportioning for optimum workability suitable for compaction uses the modified Vebe compactibility test, ASTM C1170, as the basis for determining workability and optimizing aggregate proportions. The modified Vebe apparatus consists of a vibrating table of fixed frequency and amplitude supporting an attached 0.01 m^3 container. A loose RCC sample is placed in the container under a surcharge and the sample is vibrated until consolidated. The vibration time for full consolidation is measured and compared with on-site compaction tests with vibratory rollers. The desired time is determined on the basis of density tests and evaluation of core samples. This vibration time is influenced by mixture proportions, particularly water content, overall aggregate grading, NMSA, fine-aggregate content, and fines content. RCC mixtures with paste contents in excess of aggregate voids will fully consolidate to approximately 98 percent of their air-free density. The optimum water content of a given mixture is that whose variability has the least effect on compactive effort for full consolidation. For a given cementitious material content, the strength of concrete at a given age will be maximized at the minimum paste volume that fills the aggregate voids.[11]

2. *Trial mixture proportioning for the most economical aggregate–cementitious material combination.* Mixtures for a number of RCC structures have been proportioned on

the basis of results of physical tests of samples made from trials using a fixed aggregate grading while varying the amount of cementitious material and comparing results. On the basis of these results, supplemental tests may be appropriate at a constant amount of cementitious material while adjusting the aggregate proportions. The most economical combination of cementitious materials and aggregate that provides the required strength and a field-usable mixture is then selected for the project. The proportioning of these mixtures has resulted in cementitious material paste contents that essentially fill the voids between aggregates. This procedure has been used with cementitious contents ranging from about 30 to 296 kg/m^3 with strengths at 1 year ranging from 4.00 to 36.5 MPa. The water content is adjusted to achieve satisfactory compaction. This corresponds to a moisture level just below the point where rutting or waving of the fresh mixture under construction traffic occurs, and just above the point where the dryness of the mixture causes a significant increase in segregation.[11]

3. *Proportioning using soils compaction concepts.* RCC mixtures have been proportioned using soils-compaction procedures. The soils-compaction procedure is more suited to smaller-aggregate mixtures and usually higher cementitious material contents. It involves determination of the maximum dry density of materials using modified Proctor compaction procedures and can be considered an extension of soil-cement technology. Optimum water contents are established using the same procedures to establish the optimum water content of embankment materials and soil cement. To accomplish this with RCC, a modification of the compaction equipment is required. RCC has been proportioned using soils compaction equipment similar to ASTM Standard D1557 (modified Proctor) test method. Compaction is dependent on the energy imparted to the specimen. The compactive effort of the modified Proctor test has been found to correspond closely to in-place density measurements of the smaller NMSA mixtures with which it is used. The optimum water content for use with soils compaction methods will depend on the aggregates used, the cementitious materials used, and the compactive effort applied. Loss of strength will occur with a water content either below or above optimum.[11]

All mixture proportioning methods require the use of trial mixtures to determine the suitability of the mixtures for construction. Hanson has listed many mixtures used in mass RCC construction.[12]

Tests and Test Section During construction the primary tests are the Modified Vebe test for consistency, and thus control of the water content, and determination of the density of the in-place compacted RCC by use of a Nuclear Density Meter, ASTM C1040. It has been found that the two-probe "strata gauge" density meters (Fig. 30.4) are particularly appropriate for the thicker-lift mass-concrete placements. While not extensively used for field control, test cylinders and beams are often fabricated and tested for record purposes. The cylinders and beams are fabricated and tested similarly to those for conventional concrete except that the stiff consistency of the RCC requires that they be made while being acted on by a vibrating table and with suitable surcharges on top.

Field test sections prior to start of mass RCC construction are almost mandatory. These test sections are necessary to check mixture preparations selected for construction and make any necessary adjustments and to prove out the contractor's proposed equipment and construction procedures. The test section should be sufficiently large to permit equipment and procedures to "settle down" to normal action. All materials, equipment, and procedures should be as proposed for use on the main project.

Batching and Mixing. Most mass RCC projects have used weigh batching equipment similar to that used for conventional mass concrete. Many projects have successfully used conventional concrete mixers, usually large tilting-drum mixers, for mixing mass RCC.

In recent years, several projects have very successfully used large twin-shaft pugmills for mixing mass RCC. Many engineers feel that the pugmills are really superior for mixing the relatively dry, harsh RCC because of their intense scrubbing action. While there should be radio communication between the placing site and the mixer operator, the mixer operator must nevertheless be very observant of the condition of the RCC as it is discharged from the mixer in order to detect any change in consistency even before it is reported from the placing site.

Transportation. There have been several different satisfactory methods used for transporting mass RCC from the mixers to the placing site. Regardless of the system used, it must be of sufficient capacity to support the project needs. Properly designed belt conveyor systems have proved to be very satisfactory. These conveyor systems can be designed to deliver the RCC directly to its in-place location or simply to feed trucks which operate only on the placing area and deliver the RCC to its in-place location. In either case, the use of a conveyor system prevents the tracking of mud and other foreign material which can occur when wheeled vehicles transport the RCC directly from the mixer. Wheeled vehicle transport from the mixer directly to the in-place location has been performed with rear-dump trucks, bottom-dump trucks, and scrapers. Some engineers prohibit the use of bottom-dump trucks because they feel these have more tendency to cause segregation at the edges of the dump. When wheeled transport directly from the mixer is used, great care should be taken to prevent tracking foreign material onto the placing site.

Lift-Surface Preparation. The preparation of the surface of each preceding lift of mass RCC is one of the most important considerations in mass RCC construction. Much of the requirements for bond strength and imperviousness depend on this preparation. The surface should be completely clean. With proper construction procedures and care, the surface should not need any specific cleaning unless placing has been delayed. Regardless of the age of the preceding lift, it is considered good policy to apply a mortar bonding coat immediately preceding the succeeding lift of RCC, particularly if the structure is a dam subjected to continual water storage (Fig. 30.1). The mortar should be a creamy cement-sand mortar and should be applied 7 to 9 mm thick. This is relatively inexpensive insurance to attain bond and impermeability of the joint. The mortar can be delivered with truck mixers or other equipment and can be spread with a saw-toothed squeegee mounted on the dozer blade of a small-wheeled garden tractor. Operations should be controlled so that the RCC is placed before the mortar dries or hardens (Figs. 30.2 and 30.3).

Spreading and Compacting. After mass RCC has been transported and dumped in place, various methods have been used to spread and compact the materials. One common method has been to simply dump the RCC in piles or windrows and then level these off with a dozer or motor patrol to a thickness which will give the required compacted thickness, usually 0.3 m. This lift is then compacted with a vibratory roller. A different method that has proved particularly effective is to use 0.6 m lifts made up of four 0.15 m thick layers. Each layer is dumped in place and then spread, worked back and forth to overcome any possible segregation, and the surface thoroughly "walked," all with a medium-sized tracked dozer. After all four layers are in place, the entire lift is compacted with a vibratory roller. With this method, it has been found that over 90 percent of the compaction is provided by the dozers.[9] Regardless of the method used, it is standard practice to require the use of double-drum vibratory steel wheel rollers of at least 9070 kg gross weight and producing a dynamic force between 6390 and 8030 kg per linear meter

FIGURE 30.1 Hauling and spreading mass RCC for dam construction. Mortar bedding layer in foreground.

FIGURE 30.2 Installation of sheet metal joint full depth in fresh lift of mass RCC to form transverse contraction joint in dam.

FIGURE 30.3 Mass RCC mix, spread but not yet compacted.

of drum width. Usually, requirements have been for four roller passes over the entire area of each lift. (A round trip of a double-drum roller across the same area constitutes two passes.)

Curing and Protection. At the completion of rolling, lift surfaces must be moistened and kept damp at all times until the next lift is placed on until the end of the required curing period, usually 7 days. It is imperative that the surface not be allowed to dry. RCC has essentially no bleed water whatever to help keep the surface damp at early age. Until the RCC has hardened, the curing water should be applied with fog spray nozzles. These nozzles which provide a very fine spray are readily available. After the surface has hardened, coarser sprays may be used. There is a common tendency to use the coarse sprays from the beginning, but this should not be permitted. Even though mass RCC will not develop laitence, such coarse sprays used too early can produce a surface scum much like laitance because of the overwetting and erosion. The surface of each lift should, of course, be protected from freezing, from heavy rain, and from abrasion from traffic. The aim should be to place mass RCC fast enough that each lift surface is exposed for the least time possible.

FIGURE 30.4 Two probe strata gauge, nuclear meter being used to test density of compacted mass RCC.

Properties. Like conventional mass concrete, mass RCC can be designed to cover a wide range of strength, thermal properties and volume change, durability, and costs. However, each of these factors must be balanced against the others to arrive at the overall optimum. Again, all these work down to the fact that the required structure with all its necessary properties can usually be constructed at a significantly lower cost when mass RCC is used instead of conventional mass concrete. While mass RCC construction requires critical care and uses construction procedures and equipment different from those used for conventional mass concrete, the procedures are no more complex and probably less complex, once learned.

30.2 RCC Pavements

History and Development. RCC pavement design and construction essentially evolved from the long-term use of soil cement and cement-treated base (CTB) for base course for pavements. Canadian engineers in British Columbia in the 1970s advanced to the use of coarse aggregate in the mix with relatively high cement contents to form a higher-grade base. The next step was to use this RCC as a full pavement without other surfacing course. The first use there was for a log sort yard at Caycuse on Vancouver Island in 1976.[13] This proved very successful in resisting the extremely heavy loads and abrasion of the logging equipment even though constructed on a poor subgrade. Since then RCC pavements have been used for numerous heavy-duty pavements in western Canada. The first modern use of RCC pavement in the United States was a successful test installation for a service road at the U. S Army Waterways Experiment Station at Vicksburg, Mississippi, in 1975.[14] The first significant RCC pavement in the U. S. was a large parking area for tanks and other tracked vehicles at Ft. Hood, Texas, in 1984. These were followed by a number of heavy-duty pavements at various military bases and numerous commercial heavy-duty pavements, particularly at port dock areas and intermodal freight yards being built to handle containerized freight cargo and the extremely heavy load equipment required to handle the containers. Two of the largest RCC paving projects to date were built during the 1989–1990 period: (1) roads and parking lots for tracked vehicles with a total area of 324,000 m^2 at Ft. Drum, near Watertown, New York and (2) roads and parking lots with a total area of 405,000 m^2 for the new General Motors Company, Saturn Division production plant near Spring Hill, Tennessee. All these RCC pavements have been performing quite successfully. Other countries throughout the world have been taking advantage of RCC pavement for numerous applications.

Application and Advantages. RCC pavement is particularly applicable to heavy-duty pavements, and in these applications performs as well as conventional concrete pavements. While the RCC pavement can be constructed to surface tolerances as tight as those for conventional concrete pavement, it has not received much use to date for high-speed traffic areas. The primary advantage of RCC pavement is lower cost. Experience has shown that RCC pavements can be constructed at a saving of 15 to 30 percent over similar conventional concrete pavements. Where a comparison was possible, RCC pavement has been shown competitive with asphalt pavement design to the same strength requirements.

Design. The structural design of RCC pavement essentially follows that of conventional concrete pavement. The exception is that normally the RCC pavement is not given any credit for any load transfer at joints or cracks. While RCC pavement can be expected to have a somewhat higher flexural strength than conventional concrete pavement with the same cementitious material content, this is seldom given credit in the structural design.

Materials. Cement used for RCC pavement will be similar to that used for conventional concrete pavement. Recognition must be given to any need for low alkali content and for sulfate resistance. While there have been projects using only portland cement, most RCC paving mixes have employed a pozzolan, usually fly ash. This has normally been at the rate of 20 to 30 percent of the cementitious material, primarily to increase workability.

While the Canadians originally started their RCC pavement work using pit-run aggregates, it soon became apparent that for most high-class RCC pavement, the aggregates needed to be of the same high quality required for conventional concrete pavements. A nominal maximum size of 37.5 mm has been tried, but the larger particles were found to cause excessive tearing of the surface as they passed the paver screed. Almost all projects have used a nominal maximum size of 19 mm. Many RCC paving projects now require two sizes of aggregates, 19.0-mm maximum, split on the 4.75 mm sieve such that they can be combined to produce a total grading within a band ranging from 83 to 100 percent by weight passing the 19.0 mm sieve to 2.8 percent passing the 75 μm sieve.[15]

Admixtures have been used in RCC paving mixtures but not to the extent used in conventional concrete paving. Several projects have successfully used a retarding admixture to provide longer working time for handling the RCC, particularly for aid in getting good bending at construction joints. No RCC paving projects to date have used air entraining other than on a trial basis. While the results of rapid freeze-thaw laboratory test, ASTM C666, on RCC paving mixtures have given mixed results, the actual service record of RCC durability in various types of severe winter climates has been excellent.[10, 16]

Mixture Proportioning. Mixture proportioning for pavement RCC will generally follow the guidelines previously given for mass RCC except the essential qualities will be flexural strength, workability, compactability, and durability, with no particular concern for heat rise due to hydration of cement. Since strength is a primary factor, the pavement RCC mix will have a much higher cementitious content than mass RCC, generally in the range of 282 to 356 kg/m^3. The most common method of mixture proportioning is method 3, under "mixing proportioning" earlier in this chapter (proportioning methods described in ACI 207.5 R89). Proportioning Using Soils Compaction Concept, as given under the discussion on mass RCC. Again, all mixture proportioning requires the use of trial mixtures to determine the suitability of the mixtures for construction. Ragan has listed typical proportions used for numerous paving RCC mixtures.[17]

Test and Test Section. During construction the primary control test is the field density test of the in-place compacted RCC pavement and, of course, straightedge testing of the pavement surface. Field density determination of the compacted RCC is performed with a nuclear density meter in accordance with ASTM C1040. Usually a single probe meter is used for RCC pavements. A common criterion for required density is that it meet at least 98 percent of the maximum density attained in the laboratory on the same material using ASTM Standard 01557. It is very important that the nuclear meter be calibrated daily on a test block of the RCC mixture of known dimensions and known weight. Frequent gradation tests of the aggregates must be made. Few agencies use the Vebe test for RCC pavement mixtures. Amount of water added to the mix must be constantly controlled by the mixer operator on the basis of visual observation of the mix going through the mixer and reports from the paving site. The operator must carefully control the water content to provide a RCC mixture that will place and compact properly; a plus or minus variation of only 0.1 percent in the water content will make a significant difference in placing and compaction. While test beams are seldom used for control, they are often fabricated and tested for record purposes. The beams are fabricated the same as for conventional concrete except that, because of the relatively dry, harsh mixture, they are normally supported on a vibrating table during fabrication, and a surcharge is used on top.

For RCC pavements, field test sections prior to start of paving should be used to check out the mixture proportions selected and make any necessary adjustments and for the contractor to determine the suitability of the equipment and construction procedures. The test section should be two paving lanes wide, with part of the joint between the lanes constructed as a fresh joint and part as a cold joint. The test section should be sufficiently long to give the equipment room to settle down to usual operating mode. All materials, equipment, and procedures should be as proposed for use on the main paving project.

Batching and Mixing. Batching and mixing for paving RCC can be performed in the same manner as previously described for mass RCC. However, almost all RCC paving projects have used continuous feed and continuous twin-shaft pugmill mixers, because of the capacity of this type of plant and the resultant economy. The use of pugmills (Fig. 30.5) is considered to be preferable because of their intense scrubbing action on the RCC mix, and the fact that paving engineers and contractors are familiar with pugmills. The size of the pugmill and the time the material is in it must be such that the mixture is uniformly intermingled and the aggregate particles thoroughly coated. This can be accomplished by adjusting the rate of feed and by adjusting the angle of the paddles in the pugmill, even to reversing some of the paddles. The aggregate materials should be fed from stockpiles into feed bins at the mixing plant. It is very important that the feed of each ingredient material from its bin be uniform and readily adjustable. Good plants will have adjustable speed on the feed belts for each aggregate with belt scales and electronic interlocks between the aggregate feed and the feeds for cementitious materials. This is important to ensure that the proportions in the mix remain constant. One necessary item when using a continuous plant is a transfer hopper at the end of the final feed to the hauling trucks (Fig. 30.6), to permit the mixing plant to operate continuously without stopping between trucks hauling the mixture. Stopping the mixing plant always causes an appreciable change in the mixture during restarting.

Transportation. The RCC mixture is hauled from the mixing plant to the paving site in rear-dump trucks. Unlike mass RCC which is dumped onto the work area, paving RCC RC is dumped directly into the hopper on the paving (lay-down) machines. Transportation, and every other operation, should be scheduled and controlled so that a uniform, continuous supply of RCC mixture comes to the paver, to prevent stopping the paver. It is

FIGURE 30.5 Twin-shaft pugmill used to mix RCC.

FIGURE 30.6 Automatic RCC continuous proportioning and mixing plant with pugmill erected on columns for truck drive-through. Project was RCC pavement construction.

impossible to stop the paver and restart it without causing an irregularity in the pavement surface.

Paving (Lay-Down). While mass RCC is spread and compacted to provide an only approximately smooth surface, paving RCC must be spread and compacted to provide a very smooth, uniform, dense surface similar to what is expected of conventional concrete and asphalt pavements. To do this, it is necessary to use a paving machine to spread, strike off, and partially consolidate the RCC. The machines used are the same as or similar to those used for asphalt paving. In the early stages, conventional American asphalt paving (lay-down) machines were used. The modern American asphalt paver has only vibrating screeds which do little to consolidate the RCC and the machines are not designed to successfully handle the thick lifts used with RCC. Soon, a German manufacturer brought into the market a heavy-duty paver with tamping screeds as well as vibrating screeds, built to handle RCC pavements (Fig. 30.7).

These machines made RCC paving a much more practical process, and there are a number of these machines in use in the United States as well as other countries. This machine will satisfactorily handle up to a 0.25 m compacted lift, and the tamping screed will provide 90 to 95 percent of the required total compaction, thus leaving less work for the rollers. Care must be taken to keep all phases of this machine and its operation properly adjusted to prevent small tears from appearing in the surface. The paver should be electronically controlled from stringlines on each side of the lane, or a stringline on one side and a ski on a previously placed lane. Slope control devices on the paver should not be used. In some instances, laser controls have worked well. Every effort should be made to keep the speed of the paver and the supply of RCC mixture coordinated so that the forward motion of the paver does not stop. Again, when the paver stops and restarts, an imperfection always occurs. During paving of succeeding lanes, the paver screed should

FIGURE 30.7 Heavy-duty, tamping-type paver for constructing RCC pavement.

be slightly higher to allow for compaction settlement. The screed should slightly overlap the edge of the other lane. Material deposited on the edge of the previous lane must be pushed back to the joint line with a lute before rolling occurs.

Compaction. After the RCC mixture has been spread and partially compacted by the paver (Fig. 30.8), the final compaction is provided by dual-drum, vibratory, steel-wheel rollers. Usual requirements have been the same as those for rollers used for mass RCC

FIGURE 30.8 Core from RCC pavement.

FIGURE 30.9 Vertical face of test beam sawed from RCC pavement.

construction: 9070 kg gross weight and producing a dynamic force between 6390 and 8030 kg per linear meter of drum width. Obviously, the vibrator on the roller must be used to provide required compaction (Fig. 30.9). As noted previously, usual requirements are that after compaction the field density must be at least 98.5 percent of the maximum density attained in the laboratory on the same material using ASTM Standard D1557. Usually four passes of the double-drum vibratory roller are required over the entire area, regardless, and more if the required density has not been attained. (A round trip of a double-drum roller across the same area constitutes two passes.) Density at longitudinal construction joints is usually specified at a lower number, often 96.5 percent. However, joint density is an extremely important item and must be given great care. Rolling of joints must be carefully controlled. Rollers must be kept operating closely behind the paver. Otherwise, particularly in hot weather, the mix will stiffen to the point where compaction cannot be completed.

Longitudinal Construction Joints (Lane Joints). Longitudinal construction joints are critical areas in RCC pavement construction. The primary problem involves the fact that it is nearly impossible to adequately compact the outer few inches of the free edge of a paving lane, that is, until an adjacent lane has been placed beside it to support the roller. That is why in multilane paving there is a time limit within which the adjacent lane must be placed before it is considered a cold joint. A usual time limit is 60 min in moderate weather and 30 min in hot weather. If the succeeding lane is placed within this time limit, the RCC in the first lane will still be sufficiently plastic that both sides of the construction joint can be rolled together, both sides will be adequately compacted, and the two sides will be knit together. If the time limit is exceeded, the outer 0.15 to 0.20 m of undercompacted material must be removed from the first lane before the second lane is placed. This is usually cut with a rolling cutter (disk) and the outer material removed and wasted. On large projects, this problem is often partially solved by having two pavers operating in echelon, the second immediately behind the first in the adjacent lane.

Transverse Joints. Originally RCC pavements were constructed with no transverse construction joints and the lanes allowed to develop transverse cracks where they would.

Because of the extreme dryness of the mix and its high strength, these cracks were spaced farther apart than would have been expected with conventional concrete, usually 15 to 20 m. These cracks had a minimal tendency to ravel or cause other distress. This method has proved suitable where appearance is no problem. In recent years, the thinking has been changing and many projects, particularly those where appearance is a factor, will be having sawed and sealed transverse joints similar to conventional concrete pavements. Spacing will be approximately 12 m. In other words, make it crack in neat lines where you want it. As yet, no one has done much in the way of sawing and sealing longitudinal joints. Generally, a well-constructed longitudinal construction joint between lanes opens very slightly and proves to be no problem.

Curing and Protection. Curing of RCC pavement is extremely critical. Curing must start immediately behind the rollers, and the surface must never be allowed to dry. Unlike conventional concrete which has bleed water, however scant, to keep the surface damp during the early minutes, RCC pavement has no bleed water and water must be added immediately and must be continuous. Otherwise, the surface can rapidly go dead, and the surface is the part that must take the abrasion due to traffic. Until the surface has hardened, water must be applied by fine-spray fog nozzles. After hardening, coarser sprays may be used and a common and very effective system is to use irrigation pipe and sprinklers. Curing should be continued for at least 7 days. While not as good, membrane-forming curing compound has been used. If used, it should be applied at twice the usual thickness and should be applied by a paving-type power applicator mounted in a self-propelled truss straddling the paving lane. The RCC pavement should be protected from freezing and from heavy traffic for the duration of the curing period. Some projects will use water trucks for curing. If necessary to be on the pavement, these should be controlled carefully and no sharp turns permitted, to avoid scuffing the surface. During paving, sufficient plastic sheeting should be at the site to protect fresh, unhardened pavement in case of heavy rainfall (Figs. 30.10 and 30.11).

FIGURE 30.10 Heavily loaded equipment handling containerized freight on RCC pavement.

FIGURE 30.11 Surface finish of completed RCC pavement. Knife for scale is 3½ in long.

REFERENCES

1. Raphael, J. M.: The Optimum Gravity Dam, in "Rapid Construction of Concrete Dams," American Society of Civil Engineers, New York, 1971, pp. 221–247.
2. Cannon, R. W.: Concrete Dam Construction Using Earth Compaction Methods, in "Economical Construction of Concrete Dams," American Society of Civil Engineers, New York, 1972, pp. 143–152.
3. Hall, D. J., and D. L. Houghton: "Roller Compacted Concrete Studies at Lost Creek Dam," U. S. Army Engineer District, Portland, Ore., June 1974, 56 pp.
4. Johnson, H. A., and P. C. Chao: Rollcrete Usage at Tarbela Dam, in *Concrete International: Design and Construction,* Vol. 1, No. 11, Nov. 1979, pp. 20–33.
5. Cannon, R. W.: Bellefonte Nuclear Plant—Roller Compacted Concrete—Summary of Concrete Placement and Evaluation of Core Recovery, *Report No. CEB-76-38,* Tennessee Valley Authority, Knoxville, Tenn., 1977, 10 pp.
6. Hirose, T., and T. Yanagida: Dam Construction in Japan: Burst of Growth Demands Speed, Economy, *Concrete International: Design and Construction,* Vol. 6, No. 5, May 1984, pp. 14–19.
7. Schrader, E., and R. McKinnon: Construction of Willow Creek Dam, *Concrete International: Design and Construction,* Vol. 6, No. 5, May 1984, pp. 38–45.
8. Oliverson, J. E., and A. T. Richardson: Upper Stillwater Dam—Design and Construction Concepts, *Concrete International: Design and Construction,* Vol. 6, No. 5, May 1984, pp. 20–28.
9. Hopman, D. R., O. Keifer, Jr., and F. Anderson: Current Corps of Engineers Concepts for Roller Compacted Concrete in Dams, in "Proceedings of American Society of Civil Engineers Symposium," Denver, Col., May 1–2, 1985.
10. Ragan, S. A., D. W. Pittman, and W. P. Grogan: An Investigation of the Frost Resistance of Air-Entrained and Nonair-Entrained Roller-Compacted Concrete (RCC) Mixtures for Pavement Applications, *Tech. Rept.* GL-90-18, U. S. Army Waterways Experiment Station, Vicksburg, Miss., 1990.
11. Roller Compacted Mass Concrete, ACI 207.5R-89, "Manual of Concrete Practice," Pt. 1, American Concrete Institute, Detroit, Mich., 1990.

12. Hansen, K. D., and G. W. Reinhardt: "Roller-Compacted Concrete Dams," McGraw-Hill, New York, 1991.

13. Keifer, O. Jr.: Paving with Roller Compacted Concrete, *Concrete Construction,* March 1986, pp. 287–297.

14. Burns, C. D.: Compaction Study of Zero-Slump Concrete, *Misc. Paper* S-76-16, U. S Army Engineer Waterways Experiment Station, Vicksburg, Miss., August. 1978.

15. Keifer, Jr., O.: Paving with RCC, *Civil Engineering,* Oct. 1987, pp. 65–68.

16. Ragan, S. A.: Evaluation of the Frost Resistance of Roller-Compacted Concrete Pavements, *Misc. Paper* SL-86-16, U. S. Army Corps of Engineers, Waterways Experiment Station, Vicksburg, Miss., Oct. 1986.

17. Ragan, S. A.: Proportioning RCC Pavement Mixtures, in "Roller Compacted Concrete II," Proceedings of the Conference, American Society of Civil Engineers, San Diego, Calif., March 1988, pp. 380–393.

BIBLIOGRAPHY

Keifer, O. Jr.: Corps of Engineers Experience with RCC Pavements, in "Roller Compacted Concrete II," Proceedings of the Conference, American Society of Civil Engineers, San Diego, Calif., March 1988, pp. 429–437.

Rollings, R. S., Design of Roller Compacted Concrete Pavements, in "Roller Compacted Concrete II," Proceedings of the Conference, American Society of Civil Engineers, San Diego, Calif., March 1988, pp. 454–466.

Saucier, K. L.: No-Slump Roller Compacted Concrete (RCC) for Use in Mass Concrete Construction, *Tech. Rept.* SL-84-17, U. S. Army Corps of Engineers, Waterways Experiment Station, Vicksburg, Miss., Oct. 1984.

U. S. Department of the Army, Corps of Engineers, Roller Compacted Concrete, in "Engineer Manual No. 1110-2-2006," May 31, 1991, Washington, D. C.

CHAPTER 31
LIGHTWEIGHT CONCRETE

John M. Scanlon

Lightweight concrete is defined in ACI 116R-87 as "concrete of substantially lower unit weight than that made using gravel or crushed stone aggregate." Lightweight-aggregate concrete is generally classified in three categories as low-density concretes, structural concretes, and moderate-strength concretes. Low-density concretes are used primarily as insulation and have unit weights below 50 pcf (800 kg/m^3). Although their unit weights are relatively low, compressive strengths normally range between 50 to 1000 psi (0.69 to 6.9 MPa). The two primary materials used as aggregates are perlite and vermiculite.

Structural lightweight concretes have ample strength and density, which permits their use in structural members. These concretes normally have unit weights between 90 and 120 pcf (1440 to 1900 kg/m^3). Minimum compressive strength as established by definition is 2500 psi (17.3 MPa). In many cases, strengths have exceeded 9000 psi (62.1 MPa). In seismic areas, codes limit the compressive strengths of lightweight concrete to 4000 psi.

Moderate-strength lightweight concretes bracket the low-density and structural lightweight aggregate concretes by ranging between 1000 psi (6.9 MPa) compressive strength and 2500 psi (17.3 MPa) with densities between 50 and 90 pcf (800 to 1440 kg/m^3).

Lightweight concrete is often used as a complete and suitable substitute for normal-weight concrete to decrease weight, although its ultimate compressive strength tends to be lower than that of normal-weight concrete. The higher cost per cubic yard of lightweight concrete is offset by the reduction in dead loads and increased fire resistance. The decreased dead load allows a reduction in the size of footings, number of foundation piles, and the size of foundation walls, columns, beams, and floor thickness. This reduction in mass of the concrete (bulk as well as density) will result in cost savings that far offset the increased cost of the lightweight concrete per unit volume. Furthermore, the heat insulating value of the lightweight concrete may be sufficient in itself to partially or completely eliminate the need for additional insulating material.

31.1 Low-Density Concrete

Low-density concretes weighing 50 pcf (800 kg/m^3) or less are most effective for insulation and sound deadening between floors. These can be used in combination with other materials in wall, roof, and floor systems where they are most advantageous in reducing heating and cooling costs and the transmission of sound between floors where lightweight

TABLE 31.1 Typical *R* values (thermal resistance) for Low-Density Concrete

Unit weight, pcf (kg/m^3)	Concrete thickness, in (m)			
	1 (0.03)	2 (0.05)	4 (0.10)	6 (0.15)
20 (320)	1.4 (0.24)	2.9 (0.50)	5.7 (0.97)	8.6 (1.46)
30 (480)	1.0 (0.17)	2.0 (0.35)	4.0 (0.70)	6.0 (1.06)
40 (640)	0.8 (0.14)	1.7 (0.29)	3.3 (0.56)	5.0 (0.85)
50 (800)	0.7 (0.12)	1.3 (0.22)	2.7 (0.46)	4.0 (0.68)
60 (960)	0.5 (0.09)	1.1 (0.19)	2.1 (0.36)	3.1 (0.53)
Structural lightweight 110 (1700)	0.2 (0.04)	0.4 (0.07)	0.8 (0.14)	1.1 (0.19)
Normal weight 145 (2320)	0.07 (0.01)	0.15 (0.03)	0.3 (0.05)	0.45 (0.08)

Source: Based on information in the PCA publication "Concrete Energy Conservation Guidelines."

concrete is a necessity. These low-density concretes can be divided into two groups according to composition.

Cellular concretes are made by incorporating air voids in a cement paste or cement-sand mortar through the use of preformed or formed-in-place foam. These concretes can have unit weights as low as 15 pcf (240 kg/m^3).

Aggregate low-density concretes are made with expanded perlite or vermiculite aggregate or expanded polystyrene beads. Oven-dry weights range from 15 to 60 pcf (240 to 960 kg/m^3).

At times, it is desired to include normal-weight fine aggregate in the low-density mixtures in order to improve creep properties. The use of normal-weight fine aggregate will increase unit weight and strength but is limited to weights of less than 50 pcf (800 kg/m^3) for maximum effectiveness in desired thermal insulation.

When dealing with low-density concretes, we are usually interested in thermal properties which increase as the weight decreases but must also recognize that strength values decrease as weight decreases. Optimization of the two is essential when setting the requirements for a particular application. Typical thermal resistance (*R* values) which can be anticipated from concretes of various unit weights are listed in Table 31.1.

Compressive strength and density relate very closely for the various types of low density concrete. Table 31.2 reflects this general relationship and is based on values presented in Ref. 7. A compressive strength of 100 psi (0.69 MPa) or less may be acceptible for insulating underground steam lines, although strengths of up to 500 psi (3.5 MPa) may be

TABLE 31.2 Typical Compressive Strength and Unit Weights of Low-Density Concretes

Cement type	Dry unit weight F^2 (kg/m^3)	Compressive strength, psi (MPa)
Perlite	20–40 (320–640)	80–450 (0.55–3.1)
Vermiculite	15–40 (240–640)	70–500 (0.48–3.5)
Cellular with sand	25–35 (400–560)	130–250 (0.9–1.7)
Cellular without sand	15–40 (240–640)	70–450 (0.48–3.1)

necessary to withstand the foot traffic of workers. Shrinkage due to drying is sometimes a problem with low-density concretes, especially when coarse aggregates are not used.

31.2 Moderate-Strength Concrete

Moderate-strength concretes are made with aggregates prepared by calcining, sintering, or expanding such products as slag, clay, fly ash, shale, slate, or aggregates processed from natural materials such as scoria, pumice, or tuff. The unit weights of concretes of this group normally range from 50 to 90 pcf (800 to 1440 kg/m^3).

By revising various ingredients in low density concrete, a moderate-strength concrete can be developed using materials such as perlite, polystyrene beads, and cellular-producing foam. Strengths normally change relative to density. The use of accelerating and water-reducing admixtures can drastically change the strength attainment by these various concretes. Figure 31.1 depicts the various types, principal ingredients, and unit weights of lightweight concretes.

31.3 Structural Lightweight Aggregates

This category includes aggregates that can be used in the making of concretes having strengths greater than 2500 psi and densities of less than 120 pcf but nominally greater

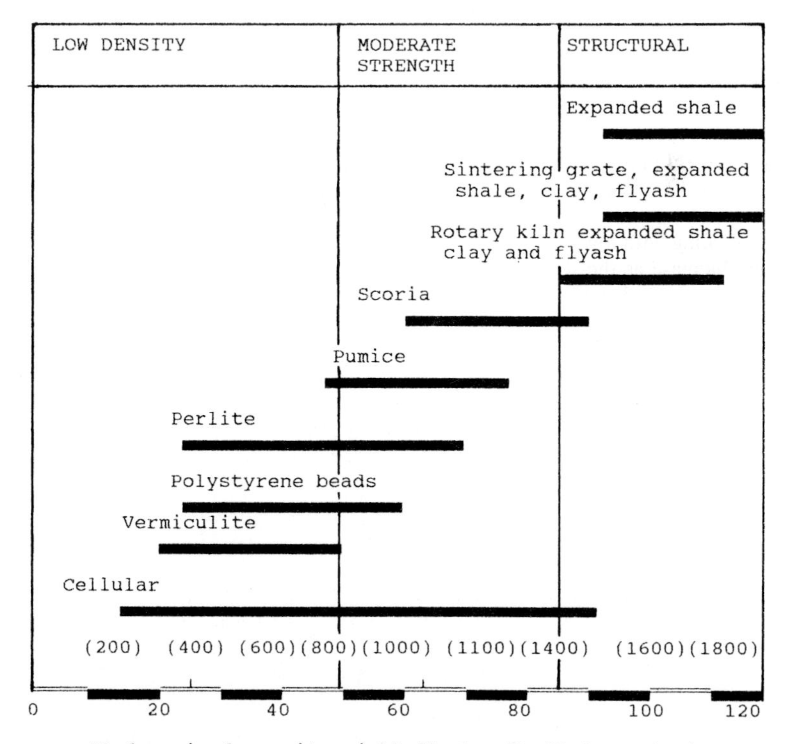

FIGURE 31.1 Lightweight concrete aggregates and unit weights.

than 90 pcf. Lightweight aggregates commonly meeting these requirements and those of ASTMC330, Standard Specifications for Lightweight Aggregates for Structural Concrete, include in general the following:

Expanded shale, clay, and slate
- Rotary-kiln process
- Sintered process
- Expanded slag
- Scoria
- Cinders
- Sintered fly ash

Compressive strengths of 4000 psi and higher are attained with some of these aggregates, whereas others may just make the 2,500 psi minimum applicable to this class of concrete. Incidentally, the strength of lightweight-aggregate concrete normally relates generally to density; density of concrete is more indicative of the density of the aggregate used in it, but concretes using the heavier lightweight aggregates are not necessarily the strongest. Actually, the highest concrete strengths are obtained with the rotary-kiln-processed expanded shales, clay, or slate aggregate. In the manufacture of this particular aggregate, the raw material must be of a type that bloats on heating to fusion or vitrification. The raw clay, etc. is introduced into the upper end of a rotary kiln (Fig. 31.2), and in its travel down the inside of the sloping cylinder, virtually all is subjected to high heat with essentially all being brought to fusion. Therefore, the probability that underburned or unburned particles will appear in the final product is remote. In preparing the raw clay, shale, or

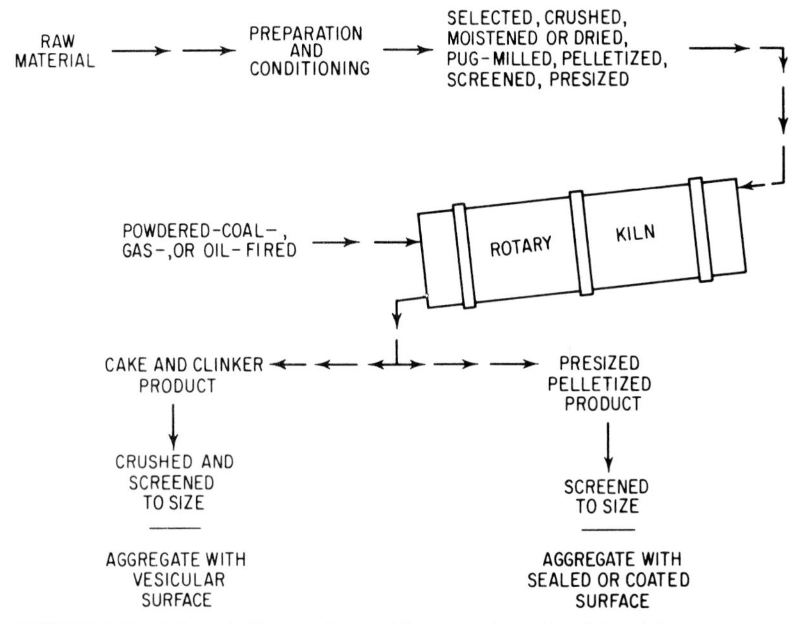

FIGURE 31.2 Schematic diagram of rotary-kiln process for making lightweight aggregate.

slate for firing, it may be mulled, mixed, or shredded before introduction into the kiln, but often it is pelletized into nuggets of predetermined size such that the subsequently fired or calcined particles conform to a prescribed gradation and maximum size. Usually, this firing of a pelletized particle results in a rounded or near-round particle having what appears to be a sealed surface enclosing a highly vesicular mass, the whole particle often having a density less than water.

In the sintering process (Fig. 31.3), as contrasted to the rotary-kiln process, the raw material is calcined on a fixed or moving grate, and it is therefore likely that even under perfect conditions, portions of the layered raw material will not reach the desired temperature. Thus the sintered product is normally rough-screened on discharge from the grate, the fine material (constituting the unburned portion) being returned to the raw-material stockpile. The acceptable clinkerlike material is then crushed to size as is also done to that production of the rotary kiln not employing the pelletizing step in its operation. This crushing and sizing after firing causes the aggregate particles to be harsh in appearance, open-textured, and showing outwardly their highly vesicular structure. Absorption of water by the crushed particle is much more rapid than that exhibited by the pelletized-to-size production. However, in a properly designed lightweight-aggregate concrete mix, the workability of that with the harsh-appearing crushed aggregate is not much different from that made with the pelletized aggregate; workability and ease of finishing are too much dependent as well on other factors such as proportioning, water content, air content, and slump.

The term *expanded slag* could be applied to air-cooled slags which are normally somewhat porous in nature although usually not enough so that they can be classified as lightweight aggregate. Thus, in order to make slag lighter in weight, either the molten material obtained directly from a blast furnace is poured into a vat of water, or the stream of molten slag and a stream of water are merged at a point; this combination is then spattered by mechanical means so that the slag becomes highly vesicular and porous. Although some of the expanded slag finds use in structural lightweight concrete, most of it is used in concrete block manufacture.

Sintered-fly-ash lightweight aggregate is now being produced by pelletizing moist-

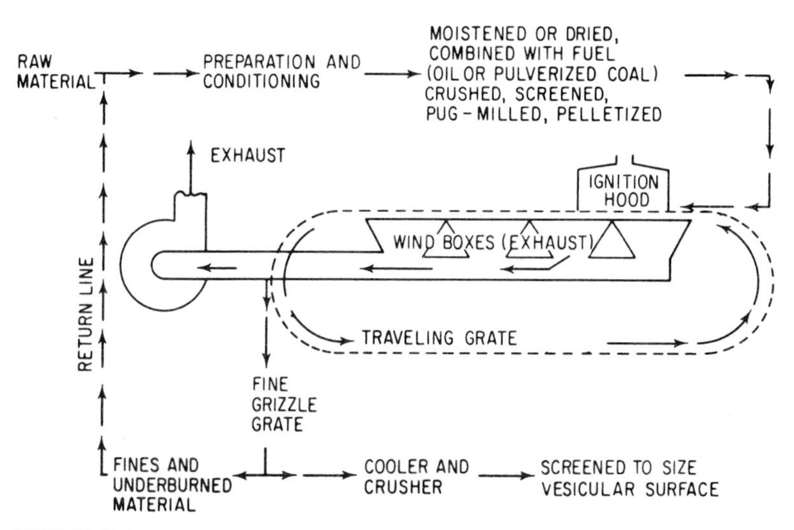

FIGURE 31.3 Schematic diagram of sintering process for making lightweight aggregate.

ened raw fly ash, partially drying the pellets, and then subjecting them to sintering much the same as that described above for sintered clay, shale, or slate. Since fly ash is a waste product of power plants employing pulverized coal as fuel, this sintering of fly ash may well be an economical way of disposing of this waste product, especially when the bulk of such material crops up in the larger metropolitan areas where disposal of anything is difficult and expensive. Cinders, on the other hand, appear to be diminishing in supply because of modern advances in combustion equipment; cinders are used for limited structural purposes and primarily in the manufacture of concrete block, although in this field the volume of cinder block is small compared with that of block made with expanded shale, clay, slate, and slag. Scoria and some grades of pumice are essentially the only naturally occurring lightweight aggregates found suitable for concrete; their use is still quite limited inasmuch as they are economically available in only a few areas.

31.4 Availability of Structural Lightweight-Aggregate Concretes

Few areas in the United States or in the world, for that matter, do not have locally available raw materials from which a lightweight aggregate may be made having a soundness, composition, and structure suitable for structural concrete. Also, whereas suitable clay, shale, or slate may be on hand in certain localities, normal-weight aggregates are often either lacking or in short supply there, or else they are deficient in some respect such as gradation, purity, and freedom from organics, freeze-thaw resistance, and the like. Of course, fuel to fire and calcine the raw material in the processing of lightweight aggregate must also be readily available and at economical cost. Even so, the production and use of structural lightweight aggregate is progressing rapidly, especially since it is now possible to make an aggregate from almost anything and yet design it to have certain attributes. These attributes might eventually be tailored so that certain required or desirable properties will be realized in concretes in which the aggregates are used. The quality of the aggregate itself certainly has an influence on thermal conductivity, compressive strength, flexural strength, bond strength, freeze-thaw resistance, sulfate resistance, modulus of elasticity, shrinkage and volume stability, creep properties, and the like.

31.5 Proportioning and Mixing of Structural Lightweight Concrete

Proportioning and mixing of lightweight concrete are discussed in detail in Chap. 11, but it might be mentioned here that properly proportioned and mixed lightweight-aggregate concrete can be handled by means conventionally employed with normal-weight-aggregate concrete. However, since it is light in weight, its slump for a workability equal to that of comparable normal-weight concrete is usually fixed at but one-half to two-thirds that of the normal-weight concrete. Whereas a 6 in slump may be used when sand-and-gravel or crushed-rock aggregates are employed in a particular placement of concrete, 3 to 4 in slump is used with lightweight-aggregate concrete. As to entrained air, normal-weight concrete might employ up to 4 percent for optimum workability, whereas lightweight-aggregate concrete would employ up to 6 or 8 percent depending on the aggregate texture and its vesicular structure. The harsher the aggregate, the greater will be the required air content. The use of a water-reducing admixture or water reducing and retarding admixture (ASTM C494, types A and D, respectively) is highly recommended here to effect a further reduction in water, to increase plasticity at a given slump, and to increase workability and ease of finishing.

Because of high absorption, lightweight aggregate is usually wetted before batching;

or else it is introduced into the mixer with an appreciable amount of mix water in order that the absorption may in part be satisfied before the cement is introduced into the mix. Pumping of lightweight-aggregate concrete appears to be greatly facilitated if the aggregate is saturated beforehand during manufacture through the use of a "hydrothermal" process or by a "vacuum" process;[1] high absorption is attained by controlling the aggregate's rate of cooling from its initial high temperature (2000°F or 1100°C) down to 300 to 400°F (150 to 230°C), whereupon it is sprayed with water and also inundated until it is relatively cool. In the vacuum process, the stocked aggregate is transferred to a tank and subjected to vacuum, whereupon water is introduced in copious amounts, thus causing the aggregate to be completely saturated. The technician or customer might benefit from suggestions of the producer of the aggregate or the local ready-mix supplier familiar with the particular lightweight aggregate being used.[2, 5] More attention is given in the mix design of lightweight concrete to cement content, slump, and yield than is given to the water/cement ratio.

Normally, the design of a lightweight-aggregate concrete mix[5] involves the making of trial batches using a range of three cement factors rather than a range of three water/cement ratios normally associated with the design of normal-weight-aggregate concrete. Proportioning of lightweight fines and lightweight coarse material by weight might seem unnatural to the newcomer in the concrete industry. The end result, however, is that the volume of the fines will amount to about 50 percent of the total aggregate. Approximately 32 ft^3 of lightweight aggregate (50 percent fine, 50 percent coarse) will be required to make 1 yd^3 of concrete. Since the fines have a specific gravity greater than that of the coarse material, the batch-weight ratios are much unbalanced as compared with those for normal-weight concrete; the fines in a batch will weight about 50 percent more than the coarse aggregate. This unbalance, or rather this revised balance as applied to lightweight aggregate, is similarly influenced when one begins to introduce into the mix a natural sand as a partial or whole substitute for lightweight fines. The substitution of natural sand for lightweight fines is often resorted to, although it is always and directly accompanied by an increase in dead weight of the lightweight-aggregate concrete and by a decrease in its heat-insulation value. The cost of shipping lightweight fines to some point quite distant from the aggregate plant might be uneconomical. However, certain benefits may be realized by the use of normal-weight fines of good quality with regard to strength, increased modulus of elasticity, and decreased volume change; these attributes might well be worth the accompanying increase in weight of the concrete.

31.6 Placing and Curing of Structural Lightweight-Aggregate Concrete

As mentioned above, the slump of lightweight-aggregate concrete should be about one-half that recommended for normal-weight aggregate concrete in any particular application. The use of entrained air and of lowest optimum water content is essential in obtaining assurance that these lightweight concretes are of a required workability for proper placing and finishing, especially those made with crushed, angular, and highly vesicular aggregates; bleeding, segregation, and undesirable floating to the surface of the larger less dense aggregate particles are each thereby reduced to a minimum. Vibration of lightweight-aggregate concrete should be done with extreme care so as to avoid segregation and consequent separation of aggregates into layers of variable density. As to the finishing of slabs or flat work, a jitterbug[4] consisting of a flat coarse wire mesh is often used to depress any large coarse aggregate down just below the finish elevation of the slab at a time when the concrete will hold the particle in that position. With regard to curing, the same methods and timing apply to lightweight concretes as are set forth for normal-weight concretes. While one should not condone any reduction in prescribed curing, it may be that a greater factor of safety in this regard is inherent in lightweight-aggregate concrete because it contains much more absorbed moisture than does normal-weight aggregate; since this moisture requires more time to evaporate, curing of the interior mass of concrete should be better. Conversely, this reservoir of moisture within the concrete is not of itself beneficial in the curing of surfaces like floors unless, of course, a curing compound is applied soon after they are finished.

31.7 Testing and Control of Structural Lightweight-Aggregate Concrete

Whereas control of slump, air content, and strength may be sufficient for normal-weight-aggregate concrete, this is not true for lightweight-aggregate concrete. A change in the unit weight of lightweight aggregate may have a marked effect on the inherent strength of the aggregate particle and thus on concrete made with it. Rather than wait until 28 day strengths are obtained, one should check the unit weight of the fresh concrete as well as slump and air content. Uniformity of unit weight within a 2 pcf range coupled with close adherence to mix-established limits for air content and slump will give better assurance that the density and quality of the aggregate and of the concrete are what they were purported to be. In making an air-content determination of lightweight-aggregate concrete, preference is always given to the volumetric method (ASTM C173), as described in Chap. 12, wherein the concrete is placed in a bowl, covered, and then carefully topped with water to a prescribed level. After the water and concrete are physically mixed by rolling the container on its side, the entrained air released from the concrete is measured as percent of entrained air simply by ascertaining the drop in the level of water. The pressure air meter (ASTM C231) is not generally used with lightweight-aggregate concrete inasmuch as it does not differentiate between entrained air and any entrapped air or voids in the lightweight-aggregate particles. This entrapped air in the aggregate voids has no bearing on the quality of the cement paste and mortar fraction of the concrete, and thus there is no merit in measuring it. At times, it may be feasible to use the pressure air meter if readings with it have been correlated with those of the volumetric meter beforehand, and if the moisture content of the aggregate in the freshly mixed concrete is known to be reasonably constant from test to test.

31.8 Nailing Concrete

Lightweight concretes in general incorporate an aggregate which will be deformed or crushed by the penetration of a nail so that the broken material will bind around the nail and make withdrawal difficult. Structural lightweight aggregates usually exhibit greater merit in this regard than do normal-weight aggregates. Natural-aggregate, crushed-stone, or sand-and-gravel concrete tends to fracture and break or split so that the nail may be easily withdrawn. Perlite and vermiculite heat-insulative concretes can be readily nailed, although their holding power is usually in direct proportion to their low compressive strength. Concrete made with sawdust[6] or with wood and other fibers as described below has evidenced good holding power. The Bureau of Reclamation finds that good nailing concrete is made by mixing equal parts by volume of portland cement, sand, and a pine sawdust with sufficient water to give a slump of 1 to 2 in. The proportion of sawdust is adjusted to result in the type of concrete suited for the purpose in hand, especially if some other source of sawdust is used (pine, fir, hickory, oak, birch, or cedar).

31.9 Conclusion

There are many uses for lightweight concrete, and as the production process improves, the demand for its use by owners and developers will drastically increase. Through use of admixtures, higher strengths will be attained on a routine basis.

REFERENCES

1. Cement and Concrete Terminology, (ACI 116R–85), American Concrete Institute, Detroit, Mich.
2. Guide for Structural Lightweight Aggregate Concrete (ACI 213R–87), American Concrete Institute."
3. Steiger, R. W.: Development of Lightweight Aggregate Concrete, *Concrete Construction,* June 1985, p. 519.
4. Structural Lightweight Concrete, *Concrete Construction,* March 1981, p. 247.
5. Standard Practice for Selecting Proportions for Structural Lightweight Concrete (ACI 211.2–81), American Concrete Institute.
6. Expanded Shale, Clay, and Slate Institute: *Information Sheet No. 7,* Lightweight Floor and Finishing.
7. Lightweight concrete acts as insulation, *Concrete Construction,* July 1983, p. 543.

CHAPTER 32
HEAVYWEIGHT CONCRETE

John M. Scanlon

Heavyweight or *high-density concrete* is defined in the American Concrete Institute Cement and Concrete Terminology Report (ACI 116R) as "Concrete of substantially higher density than that made using normal-weight aggregates, usually obtained by use of heavyweight aggregates and used especially for radiation shielding."[1] Although radiation shielding is the primary use of heavyweight concrete, it is also used in the fabrication of counterweights or simply as a means to increase the dead weight of some facility economically without increasing the bulk volume, as would be the case with normal-weight concrete. When we speak of heavyweight concrete, we normally refer to concrete having a density above 150 pcf and that, on the basis of size of aggregates and placing procedures, may attain a density as high as 400 pcf.

Heavyweight concrete is invariably more costly than normal weight concrete, even if compared per pound of mass, because more than normal care must be exercised in selecting an aggregate of adequate density and of a quality suitable for the purpose for which it is to be used, such as mining or quarrying the material, crushing and grading the aggregates, and mixing it into a concrete mix as well as in placing and finishing the concrete. Transportation cost for the necessary heavy aggregate will be relatively high when compared to normal-weight aggregates, which are normally available close to project sites. Most crushing and sizing equipment is geared to aggregates of normal weight: consequently, the wear and tear on such equipment would occur much more rapidly, and the volume of materials handled by such equipment should theoretically be inversely proportional to the densities of the aggregates. From a mixing standpoint, all concrete mixer capacities are based on volume of concrete, when the density of the concrete is around 150 pcf or around 400 lb/yd^3; therefore, the mixer capacities providing homogenous mixing may have to be reduced 40 percent or higher. As a result a 6 yd^3 mixer may be capable of mixing only 3.8 yd^3 of heavyweight concrete.

Although the heavyweight aggregates used in heavyweight concrete can be troublesome in crushing and handling during sizing, and result in costly problems in mixing, transporting, placing, and finishing, their use may be absolutely necessary or at least desirable in the design of many structures or facilities needing radiation shielding or dense counterweights, or where an increased density is required and especially where space is at a minimum. When design is based on density, the thickness of a wall or floor may be reduced 50 percent merely by doubling the density of the concrete used in its fabrication. There are many properties of concrete which are drastically increased as a result of increased density. One property which is becoming of greater importance is abrasion resistance, and with all other things being equal, the higher the concrete density, the higher the abrasion resistance.

Another area discussed below is the modern use of chemical admixtures; very little has been written about this, but basically with the use of some of these modern admixtures, the paste density can be increased by reducing the water/cement ratio while increasing workability and drastically increasing strength of the resulting concrete. In addition to chemical admixtures, condensed silica fume mineral admixtures permit greater paste density with decreased permeability and increased strength.

32.1 Nuclear or Radiation Shielding

Radiation shielding is provided and is necessary primarily for the protection of personnel operating in and about facilities which emit nuclear particles (neutron, proton, α, and β) and X or γ rays.[2] These particles or rays in general are stopped, deflected, transformed, or attenuated merely by mass, i.e., by the weight of concrete (pounds per square foot) which lies between the radiation source of energy and the persons being protected. Neutrons, however, require, in addition to mass, particular substances such as appreciable amounts of iron, hydrogen, and boron or cadmium. In some instances, the hydrogen contained in the moisture present in the concrete is sufficient for neutron attenuation; however, that this moisture present in the concrete may be dissipated during heat rise of the concrete during the curing process and if and when there is a subsequent rise in the temperature of the concrete in the shield. The area around the shield should be held in a relatively humid environment. The moisture in the concrete may be in the form of absorbed moisture, water of hydration chemically bound by the portland cement, or water of combination available in certain heavy aggregates (hydrous ores, limonite, and serpentine). Boron or cadmium, on the other hand, must be internationally introduced as an aggregate or as an admixture. A typical hot cell is shown in Fig. 32.1.

Strength of shielding concrete is very dependent on the aggregate quality and grading as well as on the water/cement ratio. Poor bonding of paste to aggregate seems to inhibit very high strengths for heavyweight concrete, but the denser the paste, the better, as long as the paste quantity is adequate for good workability. To the author's knowledge, very high strength (in excess of 12,000 psi) has not been attained with heavyweight concrete. It is felt that with the use of modern chemical admixtures which increase workability while reducing the water/cement ratio, and the possible use of condensed silica fume as a cementitious addition, much higher strengths would be attained. Normally, strength is not a criterion when requiring heavyweight concrete; consequently, the laboratories performing mixture proportioning are interested primarily in workability and density; relative low strength (1000 to 2000 psi) is normally acceptable. Shielding capacity is of primary importance. Furthermore, volume change and freedom from cracking are items of prime concern, especially where structural members form part or all of the shield. Thus it is concluded that the manufacture of shielding concrete is certainly more complicated than technology for producing normal-weight-aggregate concrete. First, one must know something about the energy source, nature, and intensity of the nuclear particles and rays which are to be stopped or attenuated at least to some acceptance limit. Second, the aforementioned choice must be made as to the aggregates which will permit the required density to be attained and to the admixtures which will result in the necessary workability and paste strength necessary for the particular structure. Normal Type I portland cement is regularly used as the binder, although in some instances Type II, IV, or V is chosen for its particular attributes. Likewise, low-alkali cement (less than 0.60 total alkalies as Na_2O) is sometimes required when the chosen aggregate is susceptible to alkali reactivity. In this same regard, an alkali-reactive aggregate should not be employed where the

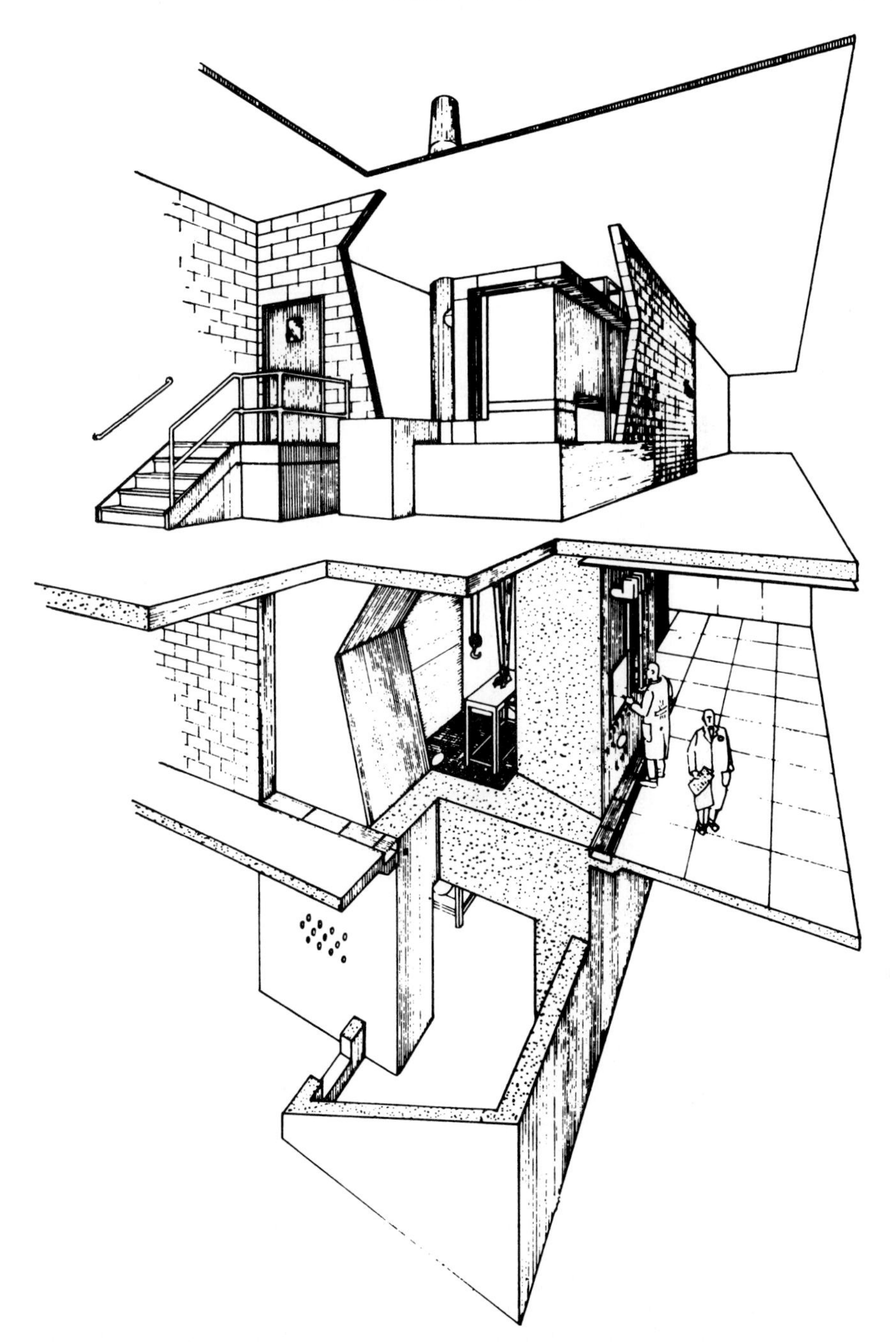

FIGURE 32.1 Cutaway drawing of hot cell shows all three levels that were surrounded by 66 in thick concrete walls. Isotope-handling room is located on second-floor cell, has 4 ft thick floor and ceiling. The concrete was placed in three separate lifts of one story each, the separate sections being joined by 3 by 10 in keys at each floor line. (*Courtesy of Master Builders Co.*)

resultant concrete will be exposed to liquors or other substances high in alkali regardless of the use of low-alkali cement in such concrete.

32.2 Counterweight Concrete

Heavyweight concrete is often employed in the fabrication of counterweights or simply as a means to increase the dead weight of some facility economically and yet without the bulk volume which normal-weight-aggregate concrete would occupy. The aggregates used for these purposes can be the same as those employed in radiation shielding concrete, except that the exposure of counterweight concrete to the environment may be even more critical from a different viewpoint. Thus additional stipulations may be required as to the quality of the concrete and the aggregate. If the concrete will be exposed to freezing and thawing in a saturated environment, the concrete should be made of a durable aggregate and should contain the usual recommended amount of entrained air commensurate with the maximum-size aggregate. Even though air entrainment results in a reduction in unit weight, it usually has less influence on weight than would a change in the aggregates, and air entrainment is absolutely necessary for concrete which will be exposed to freezing and thawing conditions in saturated environments.

If the concrete is to be exposed to salt water, deicing salts, or other chloride-ion exposures, care must be taken to avoid corrosion and subsequent detrimental expansion of the concrete made with steel punchings and the like as aggregate. In general, the use of high-cement factors, low water/cement ratios, and 3 to 4 percent entrained air is conducive to the making of an impermeable cement paste or mortar which should satisfactorily encase the iron aggregate in concrete subjected to almost any exposure. The use of condensed silica fume would greatly improve the impermeability. Materials containing excessive amounts of chlorides and other corrosive compounds should not be used. Submerged pipelines for gas, air, and even certain liquids are often counterweighted by attaching to them concrete saddles or encasing them in heavyweight concrete; they also use some normal-weight concrete. Some heavy-industrial wear-resistant floor surfacing mortars or concretes have densities of up to 222 pcf in as much as they employ specially graded and prepared metallic aggregate as one of their components; in this case, these metallic aggregates are not used because of their influence on weight but rather because of the resulting high resistance to abrasion of the metallic aggregate floor topping.

31.3 Heavy Aggregates

Normally, the exact chemical composition of heavy aggregates is not critical as long as they have the heavy density required to be used to attain the required density of the heavyweight concrete. In selecting aggregates for a specified density, the specific gravity (density) of the fine aggregate should be comparable to that of the coarse aggregate so that the density of the mortar will be closer to that of the coarse aggregate. Typical materials used in heavyweight concrete include those shown in Table 32.1. The ranges in specific gravity indicated for each coarse aggregate in the table are much dependent on the purity of the ore and how that ore was processed and manufactured into aggregates. These densities should be checked constantly, especially if the density of the concrete is critical to either the design or shielding capabilities. Ferrophosphorous is a slag and thus is subject to variations in density, as are the iron or ferrous aggregates. Serpentine and limonite are both hydrous ores and contain appreciable amounts of combined water over and above any absorbed moisture; this characteristic is of special merit in that this com-

TABLE 32.1 Typical Materials Used in Heavyweight Concrete

Material	Description	Specific-gravity range	Approximate resulting concrete density, (pcf)
Iron	Shot, pellets punching, etc.	6.5–7.8	310–380
Limonite	Hydrous iron	3.4–3.8	180–200
Goethite	ores		
Barite	Barium sulfate	4.0–4.5	205–230
Ilmenite	iron ores	4.2–4.8	210–250
Hematite			
Magnetite			
Ferrophosphorous	Slags	5.8–6.5	280–300
Crushed stone	Dolomite	2.6–3.0	150–165
Serpentine	Hydrous ores	2.4–2.6	—*
Boron frit		2.4–2.6	—*
Borates, colemanite		2.0–2.4	—*
Bauxite		1.8–2.2	—*

*Admixture-type materials

bined water will not be driven off until the temperature of the concretes in which they are used is equal to or higher than that to which the concrete would normally be exposed.

Magnetite and ilmenite are the most commonly used aggregates in the making of radiation-shielding concrete; barite, ferrophosphorous, and steel aggregate are used in much of the balance. Combinations of these aggregates are employed such as magnetite coarse aggregate and steel-shot fine aggregate. The borates, bauxite, and boron frits are usually introduced into the mix so as to better provide for neutron attenuation, especially when the concrete temperature in service is expected to be high, and when most of the water will be driven out. The borates (colemanite included) are quite fragile and are also somewhat soluble in concrete; the primary objection to this is the marked and uncontrollable retardation of setting time of the concrete which often accompanies their use with portland cement. The tendency is therefore to use the boron frits in lieu of the commoner borates.

32.4 Proportioning Heavyweight Concrete

Proportioning procedures for heavyweight concrete are very similar to procedures used to proportion normal-weight concrete. Appendix 4 of ACI 211.1 describes procedures and presents examples for developing trial mixtures of heavyweight concrete.[3] Additional considerations can be reviewed in ACI 304.3R, Heavyweight Concrete: Measuring, Mixing, Transporting, and Placing.[4] This author recommends that more trial mixtures are necessary to arrive at optimum amounts of coarse and fine aggregate inasmuch as the heavier and harsher aggregates behave somewhat differently than the normal-weight aggregate in normal-weight concretes. Additional recommendations, not found in either of the above ACI reports, are as follows:

1. The mortar should be proportioned to attain as high a density as possible; this can be done by using condensed silica fume and a high range water-reducing admixture

(HRWR), complying to the requirements of ASTM C494, Chemical Admixtures for Concrete. The condensed silica fume should contain at least 85 percent silica dioxide, a loss of ignition of 6 percent or less, and surface area (nitrogen absorption) of at least 15,000 m^2/kg; be noncrystalline within limits of detection by X-ray diffraction (XRD); and contain less than 5 percent oversize on a 45 μm sieve (wet). Also, with regard to workability the slump should never be greater than 2 to 3 in when HRWRs are not being used, and when placing conditions do not require a higher slump; even with the use of HRWR, the slump should be a compromise of around 5 to 6 in. The mixture should contain air entrainment (possibly 3 percent), even if not exposed to a freezing-thawing environment in a saturated condition; the air reduces the harshness.

2. The use of air entrainment and a minimum water content will aid measurably in reducing bleeding and separation of various aggregate sizes, and in the attainment of a more homogenous concrete.

3. When evaluating trial mixtures, the proportioner should establish families of mixtures so that rapid adjustments may be made during construction caused by nonuniformity of aggregates, such as variable gradings and breakage.

Essentially all the test methods stipulated for the control and evaluation of normal-weight concrete are likewise applicable to heavyweight concrete. Field inspection should include slump, air content, density, yield, and the making and curing of specimens (cylinders and beams) for strength testing. Tests made prior to concreting should include assurance testing to confirm that all the ingredients comply with the project specifications, particularly those pertaining to the density, grading, abrasion resistance, freezing-thawing resistance, and the potential alkali reactivity of the selected aggregates. Chemical admixtures as well as the aggregate-like mineral admixtures should also conform to established ASTM standards and to any particular project-specification requirements.

32.5. Construction Methods

There are primarily two construction methods which can be incorporated using heavyweight concrete: the conventional method and the preplaced aggregate method. When using the conventional method, we can incorporate many of the mixing, transporting, and placing requirements of normal-weight concrete, but we must always consider the increased density and its effect on equipment. The industrial concrete mixer capacities are designed primarily to volumetrically mix a particular volume of concrete having a density of approximately 150 pcf; consequently, we should not try to mix heavyweight concrete having a density of 300 pcf using the actual mixer volumetric capacity. In this case we should reduce the volume being mixed by at least 50 percent. Conventional heavyweight concrete should always be consolidated by vibration. These cautions in handling equipment applies also to the concrete chute supports, the capacity of cranes, size of concrete-conveying buckets, conveyor belts, and the strength of the forms, as well as other points of similar concern.

The preplaced-aggregate method of construction should always be considered, especially for heavyweight concrete. Its use almost always results in concrete having the greatest density. Use of this method permits the heavy coarse aggregates to be handled by more rugged equipment than that normally used to handle mixed concrete, and the mortar, although heavy, is normally batched and mixed close to the placement. Using this method, the coarse aggregate is distributed within the forms and the mortar is pumped into the base and forced upward around the coarse-aggregate particles. For heavyweight preplaced-aggregate concrete, it is essential that coarse-aggregate particles be thoroughly washed and be free of any undersized particles prior to placement in the forms to ensure unrestricted grout flow through the coarse-aggregate matrix. These aggregates are normally dropped into the form through flexible drop chutes in order to prevent breakage and separation into sizes. Frequently these coarse aggregates need to be handpacked around embedded items. After the coarse aggregates are in place, a rich grout containing fluidifier admixtures is injected into the aggregate matrix—of course, from the bottom up. A fluidifier must always be used in the grout to ensure complete filling of the voids between coarse-aggregate particles, to obtain controlled expansion of the grout during its plastic stage for optimum coarse-aggregate contact, minimize bleeding, and obtain satisfactory contact of the concrete with embedded items.

2.6 Conclusions

- Although heavyweight concrete technology is similar to that of normal-weight concrete, we need to be especially careful because of the effect of its density on equipment, forms, and employees.
- There are two primary methods of placing heavyweight concrete, either conventional method (mixing, transporting, and placing) or the preplaced-aggregate method (placing the coarse aggregate and grouting the matrix).
- The densities of existing heavyweight concrete can be slightly increased by using condensed silica fume as either a cementitious partial replacement or as supplemental materials, and by the use of high-range water-reducing admixtures to reduce the water content while increasing the workability.

REFERENCES

1. Cement and Concrete Terminology (ACI 116R-85), American Concrete Institute.
2. Concrete for Nuclear Reactors, SP-34, American Concrete Institute, Detroit, 1972.
3. Standard Practice for Selecting Proportions for Normal, Heavyweight, and Mass Concrete (ACI 211.1–89), American Concrete Institute.
4. Heavyweight Concrete: Measuring, Mixing, Transporting, and Placing (ACI 304.3R–89), American Concrete Institute.

PART • 9

BUILDING CONSTRUCTION SYSTEMS

CHAPTER 33
ON-SITE PRECASTING

Dean E. Stephan, Jr.

Individuality of a building results from a design developed by the architect, expressing the concepts of both the architect and the owner and incorporating necessary structural provisions evolved by the structural engineer. Careful preplanning and cooperation of the architect and builder can permit the contractor to explore the most economical and efficient methods of achieving the structural and aesthetic values sought by the architect.

BASIC PRECEPTS

For example, some types of building units lend themselves to precasting and prestressing on the site. Where there are many identical beams, panels, or similar units, the builder should investigate the possible economies resulting from on-site precasting. Beams and girders can be precast and prestressed, and wall panels can be precast in an efficient repetitive fashion in molds. Many other units can be cast on the site and lifted into position by crane or derrick.

In many cases, slight changes in structural design of conventional cast-in-place building units or initial coordination between the contractor and designers will permit precasting operations for the contractor. The economy of such a change is generally beneficial. However, each case must be considered individually in making a comparative cost analysis. Items for consideration on such a change, other than the costs of the members themselves, may include either more or less reinforcing steel, shoring and reshoring changes, cast-in-place joinery connections, sandblasting of beam tops, tensioning procedures, and creep, as well as design features, quality improvements due to precasting, erection operations, and savings or added costs in deck-forming methods that may be adaptable to the change. With respect to this last item, should deck-framing beams be changed from cast-in-place to precast, the cost of deck forming may be significantly reduced by using form members supported by the precast beams. Such a forming system can completely eliminate shoring requirements other than at the beams when composite action is desired.

A typical application of on-site precasting might be a multistory office building. The on-site precasting operation requires careful planning of the manufacturing operations, storage, transporting, and hoisting of the diverse building units. Preplanning and careful layout of the casting and storage areas are essential in maximizing automation and efficient work operations. The repetitive nature of on-site precasting places a premium on the efficiency of such planning. Factors such as forming and placing operations, number of pieces, unit weights, hoisting capacities and reaches, and project cycles and schedule require thoughtful consideration.

A typical 9 to 13 story office building will require as many as four, or even more, architectural panel molds in order to accomplish the proper cycle of production. If pretensioned, prestressed floor beams are being utilized, the stressing beds must be long enough to produce both manufacturing efficiencies and sufficient daily quantity of beams to meet the building schedule. Similarly, if posttensioned, prestressed beams are being precast, a sufficient quantity of form elements must likewise be used.

The selection of hoisting and handling equipment is a function of the character of the designed units. Conversely, the design of the units can be influenced by available equipment and equipment capacity. It can thus be seen that design and methods coordination are a large part of preplanning and can be an important factor in the final economic results.

Because of the many variations in shape, size, finish, and other properties of building units that can be precast, this discussion will dwell on generalities as they relate to a typical set of circumstances, rather than attempt to confine itself to any one form or unit.

In high-rise building construction, concrete curtain walls are frequently nonstructural. Even if they are structural, there should be no special problems in erection. The physical problems of making and handling precast units, on the other hand, are a function of the configuration. Long, narrow, and deep sections or highly detailed surfaces are literally impossible to efficiently cast in place; but such members can be efficiently built, utilizing precast-concrete forming elements, thereby giving the architect or designer the opportunity to incorporate very delicate lines and ornamentation on exterior architectural wall surfaces.

The molding of architectural panels generally requires removable form elements or parts, and the positive anchorage and seating of these elements are best obtained with either steel or fiberglass forms. These forms can be expensive as compared with more conventional wood or concrete forms. However, contractor-built fiberglass molds, although more costly than wood or concrete molds, should be considered because of better use factors, ease of duplicating additional molds, and configuration advantages, which are more limited in wood and concrete. A wood, concrete, or fiberglass mold built on the site by the contractor will probably be ready for use before a steel mold can be ordered and delivered—a factor that must be considered as lead time can often be limited.

Cost, timing, and the technicalities of efficient use and amount of reuse must all be considered when mold material is selected. Whatever the material, it is essential that work or artisanship be of the highest quality. Watertightness is essential, as are accuracy of lines affecting draft and alignment. One must remember that the finished units will be adjacent to each other on the building, both horizontally and vertically, thus requiring tolerances of a precise nature in dimensions and consistency in finish.

It is important that equipment and duplication of operations be employed to the fullest extent. Therefore, maximum capacity, reach, line speed, and other factors will serve to determine the type of equipment. Repetitive production-line methods will simplify educational and supervisory activities and produce product consistency and labor efficiency. Whether the molds are of steel, wood, fiberglass, concrete, or a combination of materials does not affect these decisions.

Most builders' new work opportunities represent "fixed plan" competitive-bid situations that generally do not lend themselves to the type of work discussed here. Some contracting techniques convey opportunities for builders to express themselves with their experience and knowledge. The willingness on the part of owners and architects, and the contractors' abilities to contribute to owners and designers with respect to matters of cost control, construction quality, and efficiency, will be truly represented when the factors discussed in this chapter can be realistically incorporated into the work during the preparation of the design.

Of fundamental importance is the experience of the contractor. Owners will probably not permit their jobs to serve as laboratories for experimenting with new and untried methods, nor are they likely to make them training grounds for inexperienced builders. Conferences between owners, architects, and contractors during the preliminary design stages will enable all to contribute their share to the successful and economical completion of the project.

To summarize, then, this discussion is based on three fundamental concepts: (1) that the innovation and imagination of both the designer and the builder constitute a dynamic mutual interest that can be used effectively in developing the design and execution of the building plans; (2) that builder has the necessary experience to engage successfully in the fabrication and erection of precast and sculptured concrete; and (3) that the owner or the designer feels that a limited budget makes it desirable to utilize the unique experience of the builder to take advantage of the economies of the on-site precasting.

33.1 Fiberglass Molds

Where the job schedule permits time to make fiberglass molds for production of units, it is generally economical to do so, especially when many forms are needed. Reinforced fiberglass is very durable, and forms made from it have an almost unlimited life. Fiberglass molds resist abrasion well and are not affected by weather. When forms of wood are not feasible because of high reusage, unusual architectural configurations, curved surfaces, or other special problems, the use of steel forms and fiberglass forms should be carefully considered. The use of steel forms is limited, however, because of the physical properties of steel itself. Architectural precast units that involve small-radius curves and very few flat surfaces may be extremely expensive, if not virtually impossible, to form in steel. In such cases, the use of reinforced fiberglass molds, as described in the following paragraphs and illustrated in Fig. 33.1, is recommended.

The master pattern must be dimensionally correct in every respect, as the production units will be exact copies. In very unusual architectural units where the design contains curvilinear contours, the master pattern will have to be made in sculptured plaster or a similar medium. Otherwise, straight-line contour patterns can easily be fabricated from prefinished plywoods. Extreme care must be exercised to ensure that the proper amounts of draft or taper are present on this master pattern, because in some cases the pattern will be used only once and will be destroyed during the process of stripping the first negative mold. If there is no draft, or very little draft, present for deep-sectioned units, the production units will not strip from the molds. Draft may be hardly measurable; but in deep sections it must be proportionately increased in order to overcome the increased friction and suction of the deeper form. In sections 24 in deep, a draft of ½ in has been satisfactory in the use of both hard-rock and lightweight concrete. The finished surfaces of the master pattern should have a smooth and high-gloss finish, for the fiberglass overlay will conform to these surfaces.

For illustration only, we have chosen a 12 × 12-in column showing a mold in simplest form. The master pattern as shown in step 1 in, Fig. 33.1, contains straight-line contours; so ½ in preglazed plywood has been selected as the material. The succeeding procedure will be to apply the releasing agents. To ensure a complete separation, a minimum of two coats of mold wax should be applied with one sprayed-on coat of polyvinyl alcohol. At this time it is important to mention that the correct resin materials be selected for this type of open-mold fabrication. Some resin manufacturers are designing their products to satisfy jobsite needs. Depending on climatic conditions, a choice should be made

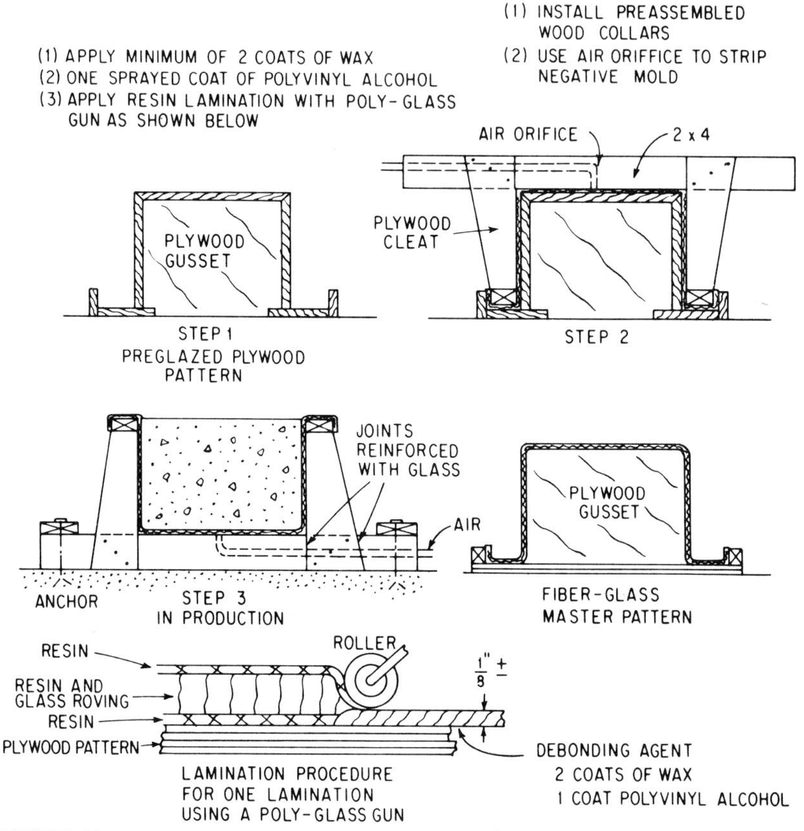

FIGURE 33.1 Sequence for making fiber-glass mold for repeated production of a simple 12 × 12 in column.

between winter or summer resins and for the proper degree of hardness. The hard resins are normally employed as outer skins for concrete resistance; the softer materials are more economical and can be used as backup laminations. Figure 33.1 illustrates the procedure of one lamination. This procedure can be repeated until the desired thickness is reached. Assuming that for this particular mold we have elected to apply three laminations of glass of ⅜ in thickness, then the wood reinforcing material, as shown in step 2 (in Fig. 33.1), will be installed after the second lamination. The last lamination, producing the required thickness of glass, will adhere to the wood, completely encasing it with glass. The preassembled wood collar, as shown in step 2, should be installed while the second lamination is still tacky, causing an adherence there, also. It is recommended that the wood joints be glassed for rigidity, as shown in step 3.

Air orifices are of great importance for stripping purposes. In some cases, air will be sufficient to create the initial separation between concrete and fiberglass. The crane will then make the final liftoff to the stacking area. In other cases, where unusually large and deep sections are being produced, a combination of air and hydraulic jacks is used for the initial liftoff. For large units, strongback systems are used. Steel I-beams as strongback members can be bolted to inserts cast in the units so that hydraulic hand jacks can be placed between the beam and the mold itself. For a positive stripping system, jacks must be placed to exert force directly between the unit and the mold. If the jacks bear against the ground of any place other than the mold, it is likely that they will pick up the unit and mold together.

Depending on the complexity of the pattern and the number of negative molds required, it is sometimes more economical to produce a fiberglass master pattern, as shown in Fig. 33.1, and then use this pattern to make negative molds. This is especially desirable when repeated stripping of negative molds causes extensive damage to the wood pattern. The procedure is to follow through with steps 1 and 2. Using the first negative mold, as shown in step 3, two laminations should then be applied. Plywood cross ribbing is added for rigidity. This type of pattern will give an infinite number of negative molds.

For equipment, it is necessary to have an air compressor capable of delivering 100 ft^3/min. The polyglass gun is a simple and reliable piece of equipment that is easy to maintain and use. However, its workings should be understood by all operators who are using it. It is also extremely important that the manufacturer's preventive maintenance program be carefully followed.

The fiberglass mold shown in Fig. 33.2 was used for forming the one-story-high panels shown in Fig. 33.3. Four of these molds were used to cast the precast concrete exterior for a 350,000 ft^2, seven-story office building. Figure 33.4 shows a fiberglass mold used in an off-site casting yard in the Midwest. The panels cast from molds of this type are illustrated in Fig. 33.5. Another use of fiberglass molds is shown in Fig. 14.8.

33.2 Steel Forms

Steel forms are most economical when many precast units can be cast from very few forms. They are extremely durable, and their rigidity produces high degrees of dimensional stability in the product. Also economically, their best use is for architectural units

FIGURE 33.2 Manufacturing a fiberglass mold.

FIGURE 33.3 Precast panels cast from fiberglass mold.

involving curved or straight surfaces which are regular. Another advantage is in having a rigid steel member for screeding the concrete, which results in exact alignment of interior edges of adjacent panels, eliminating the need for possible grinding, or similar to present a smooth interior finish at the joints.

In the construction of the forms themselves, it may be necessary to provide holes and fittings for air jets to aid in stripping the concrete castings from the molds, especially if the precast unit sections are very deep and thin or if parts of the form are surrounded by concrete on four sides as an inverted tub to form a recess in the unit. The placing of such jets, of course, depends on the shape of the form itself, but should any difficulty be antici-

FIGURE 33.4 Fiberglass mold for precast wall units.

FIGURE 33.5 Exterior of office building enclosed with wall units cast in the molds shown in Fig. 33.4.

pated in stripping the precast units, it is wise to provide these jets beforehand. Difficulty in stripping will also be increased should the forms be fabricated without the proper amounts of draft on all vertical surfaces, or with the steel surfaces or joints left rough. The joints between the steel plates should be welded and then ground off flush and checked with a straightedge to ensure that there are no small humps present. This will also eliminate the possibility of mechanically locking in the concrete unit and will also eliminate the joint lines showing on the finished concrete product. Caution is emphasized to make sure the forms have the proper draft and no mechanical-bind spots. If the concrete products are being cast and stripped each day, the product will normally be warm because of retained heat of hydration when stripping commences, whereas the steel form may be cool because of the air temperature at the same time. This will cause an additional tightness. The combination of this temperature differential with any lock of forms due to draft or bind spots mentioned above can prohibit stripping without damage to the concrete product.

In stripping the products, best results are usually obtained using the same strongback-type system and hand jacks as previously described. The jacks should be located so that the force to lift the product from the form is exerted between the product and the form itself. Figure 33.6 shows one type of panel that can be cast in steel forms, and in Fig. 33.7 are shown plan and sections of a form and panel. Positioning of stripping jacks shows clearly in Figs. 33.7 and 33.8. The panels shown here were cast so that two adjoining

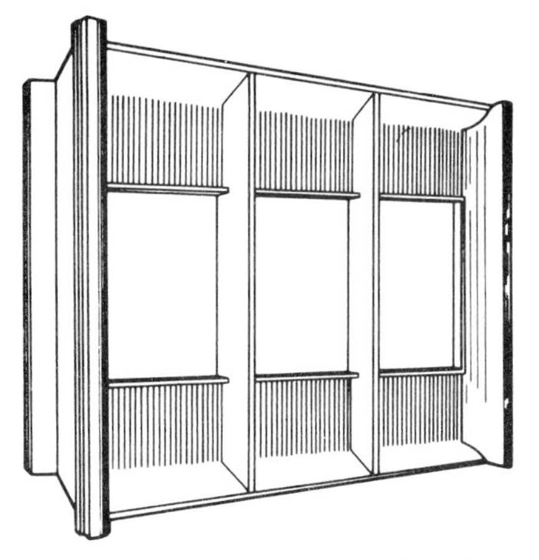

FIGURE 33.6 Type of precast wall panel made with steel form. Approximate weight 8200 lb. (See also Figs. 33.7 through 33.9.)

panels formed the closure for the concrete joiner column (Fig. 33.9). This eliminated any forming between the panels, and both exterior and interior elevations of the curtain wall presented the smooth surfaces that are obtained from the steel-form surfaces. In addition to the panels forming the column closure, the panels were cast so that the spandrel-beam soffit and sides were also formed by the panel, thereby eliminating any need for conventional beam forming and shoring.

33.3 Concrete Forms

Concrete molds are durable and rigid. The movable components of the mold are usually made of steel or wood. The negative mold is generally produced utilizing a wood and plaster master with subsequent negative molds made from a master pattern which was cast on the original negative.

Concrete molds do not have the "give" or elasticity of fiberglass and steel, and thus are not as suitable for deep relief or delicate patterns. The mass of the mold both enhances the removal of the product and makes it much less sensitive to temperature variations.

A wall panel and its form are shown in Figs. 33.10 through 33.12. After panels have been stripped, they can be surface-finished such as sandblasting and sealing, stored for curing, and then lifted and set in place (Figs. 33.13 through 33.15). On this particular job, erection of the panels was done with a tower crane, which proved to be the most economical piece of equipment for the purpose. The configuration of precast architectural panels are limited only by the designer's imagination (Fig. 33.16).

33.4 Precast Beams

On-site disposable-bed precast-beam forms for repeated usage are usually fabricated with the soffit and one side fixed and the other side either hinged or removable (see Fig. 33.17). With the fixed side being aligned and plumb before any units are cast, the removable side can easily be lined to it and spaced for each casting operation. In the stripping operation, the removable side is stripped first and the precast beams removed with a crane. This forming method, however, applies only to rectangular-sectioned beams. Should the beam section be an "I" configuration or another similar configuration, it will be necessary to have both form sides hinged or removable.

Portable steel or concrete beds are also used for jobsite precasting of pretensioned prestress beams. These beds offer the advantage of amortization of their capital cost over several projects. Portable beam beds can make the on-site manufacturer of prestressed beams economical on smaller projects.

Figure 33.18*a* and *b* shows an "E"-shaped portable concrete beam bed. The beams are cast within a channel created by the fixed concrete interior stem wall and a movable side form. The concrete bed takes the prestressing forces until they are transferred to the beam when the stressing cables are cut. The movable side form provides flexibility in beam width and facilitates removal of the beam from the bed. Past experience has

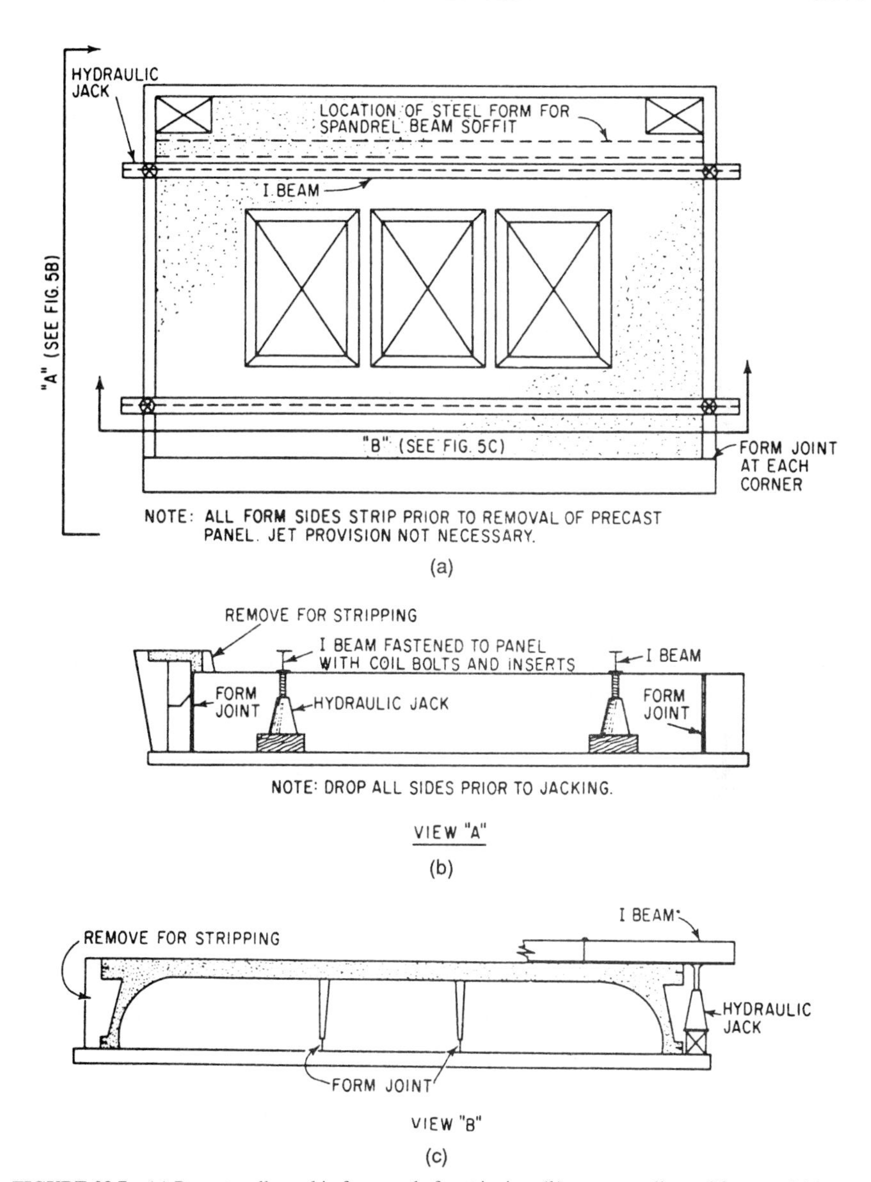

FIGURE 33.7 (*a*) Precast wall panel in form ready for stripping; (*b*) precast-wall-panel form; and (*c*) wall-panel form containing precast panel.

shown that casting within a concrete channel can create removal problems even when adequate draft of the concrete stem walls is provided. The stressing forces, when transferred to the beam, cause very high frictional forces between the beam and the stem wall along even the slightest imperfection which hinders the removal of the beam from the bed.

FIGURE 33.8 Strongbacks and jack used in stripping the panel from the form.

FIGURE 33.9 Precast panels in place in the building, showing method of forming joiner column and spandrel beam.

Types of precast beams fall into three categories: steel-reinforced beams, posttensioned beams, and pretensioned beams. For each type, the precasting sequence of form setting, placing concrete, and stripping varies only slightly. The reinforcing-steel cages are often fabricated and stockpiled ahead and then placed in the form as one unit. (Fig. 33.19) The sequence starts with cleaning and reoiling of the form in preparation for the next placement. Next the reinforcing is placed and then posttensioning cables or prestressing strands are inserted if the beams are of one of these types. Next the movable or hinged form side is set and the beam-end bulkheads set. In the case of pretensioned prestressed beams, the next operation is stressing of the strands. Then the concrete is placed. After the concrete has attained sufficient strength, the pretensioned prestressing strands are cut between the beams, transferring the stressing forces to the beam. In the case of posttensioning, the strands are stressed. The hinged or movable side form is stripped and the beam is lifted out of the form. The beams are usually stockpiled prior to erection. Planned storage time usually allows sufficient maturity for the beams to reach their specified strength. If they are of the posttensioned type, they can also be posttensioned in the stockpile or, where structural requirements dictate, be stressed in place. Usually the reinforcing steel and the pick-up points can be arranged so that the beams can be removed and stockpiled prior to posttensioning.

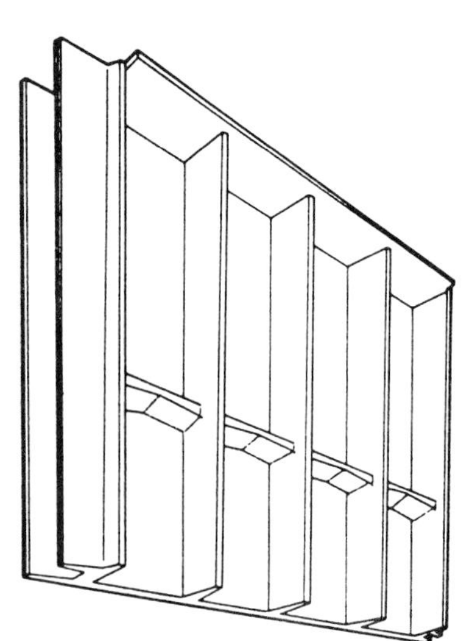

FIGURE 33.10 Another type of precast wall panel made with steel forms. Approximate weight 8400 lb. (See also Fig. 33.11.)

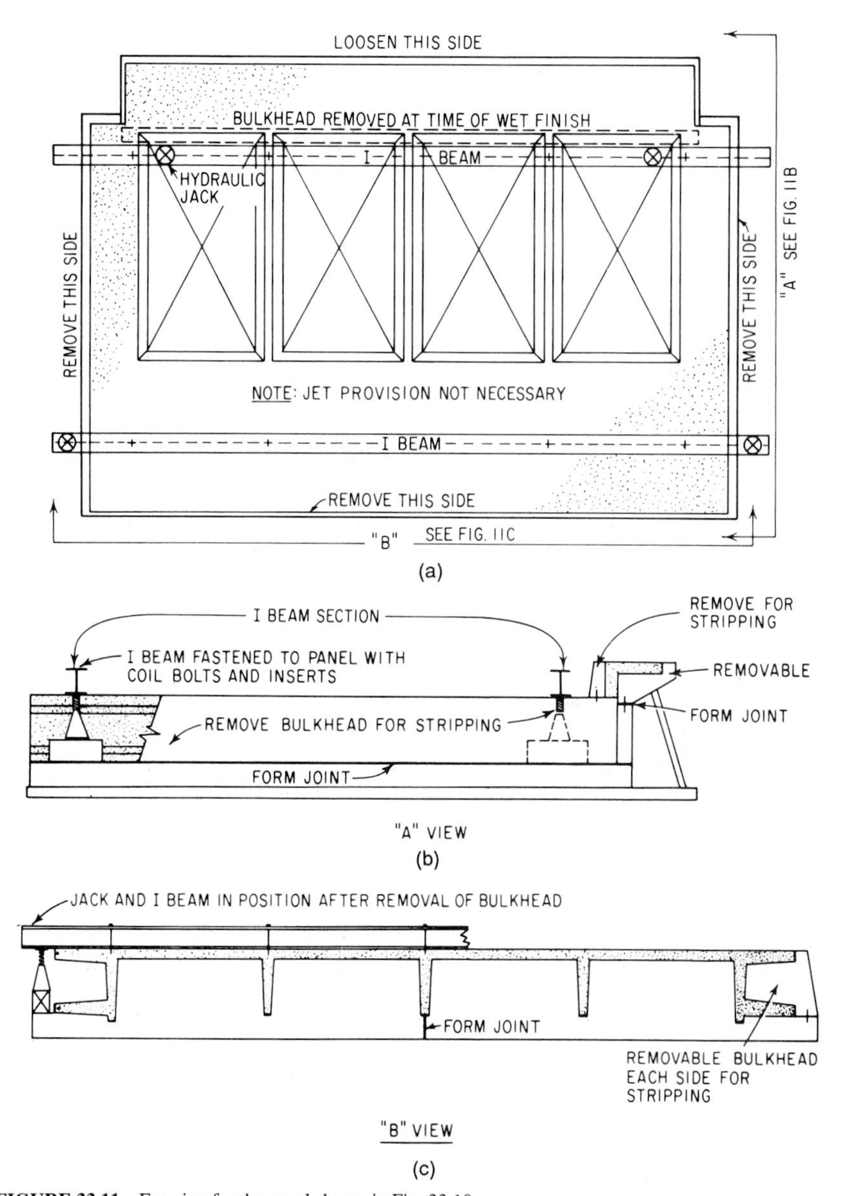

FIGURE 33.11 Forming for the panel shown in Fig. 33.10.

FIGURE 33.12 Photograph of the panel form shown in Fig. 33.11.

Details of prestressed concrete construction are discussed in Chap. 41. Methods of pretensioning and posttensioning that are used in central casting yards are, in general, applicable to on-site operations and need not be repeated here.

Placing and consolidating of the concrete, finishing, and curing all follow standard construction practices. Beams can be steam-cured in order to expedite strength development so that they can be moved sooner, thus making the casting beds available for the next round of girders. High-range water reducers are often used to facilitate placement of low-water/cement-ratio concrete mixes that are common to precast beam manufacturing. The low-water/cement-ratio mixes are used to achieve high strengths at an early age and improvement in concrete quality.

33.5 Precast Prestressed Floor Slab Elements (Planks)

In addition to precast beams (pretensioned or posttensioned), precast, prestressed floor planks can be job-cast. These, in conjunction with the beams, form the structural components of the floor slab over which a topping of 2 to 3 in of concrete is placed. The avail-

FIGURE 33.13 After the panel has been jacked loose from the mold, it is lifted out for surface finishing and storage.

FIGURE 33.14 Panel being positioned in the building.

ability of pretensioned planks and beams provides a great deal of flexibility to a building project, as this material can be easily installed, and, in conjunction with high-strength, lightweight concrete, it permits substantial reduction in the total weight of materials. Figure 33.20 illustrates a plan of an on-site casting bed for beams, planks, and wall panels. One traveling crane was used to service the casting operation, and a climbing crane was

FIGURE 33.15 After the panels are situated in the building, concrete is placed in the columns.

FIGURE 33.16 Panel configuration is limited more by the designer's imagination than the molds.

used in the building to erect the units. Figure 33.21 is a view of the precast beam and plank application in a 30-story high-rise office building.

33.6 Columns

Precast columns are generally cast on concrete casting stabs or portable wood platforms utilizing wood and sheet-metal side forms (Fig. 33.22). The length of precast column is controlled by weight limitations imposed by the equipment available for erecting the column and the flexural forces introduced during the process of raising the column from the horizontal position in which it is cast to its final vertical position. Because of the limitations imposed by column weights, it is generally advantageous to cast them close to their final erected position. This allows maximizing the handling capacity of the equipment utilized for column erection (Fig. 33.23).

When concrete casting slabs are utilized, it is advantageous to locate the top surface below the permanent building floor slab subbase elevation so that they can be abandoned in place rather than broken up and removed. The bases of precast columns are typically secured into position by the use of anchor bolts projecting through metal base plates that are cast into the column base. This connection technique is labor- and material-intensive. When conditions permit, a "wet" connection system is preferable. For columns connecting to footings, a socket receiving the column base which is grouted after the column is positioned is an economical solution (Fig. 33.24). Where multiple-tier columns are to be connected, column reinforcing steel projecting out of the column base which is inserted into grout-filled corrugated tubes in the column section below is a rapid connection technique. The important consideration for precast connections is to minimize the time on the "hook." The procedure should be one which can be readily accomplished so that the erection crew and equipment are not tied up with extensive positioning, alignment, and securing procedures while the piece is being erected.

Figures 33.25 and 33.26 are general overall views of an off-site yard used for casting elements for an office building and adjacent parking garage. Figures 33.27 and 33.28 are views of an on-site casting operation which occurred in the 40-ft fire lane between the office building and the parking structure. In both cases, the elements produced consisted of precast, prestressed beams, precast columns, and precast architectural wall panels for the office building and the parking structure.

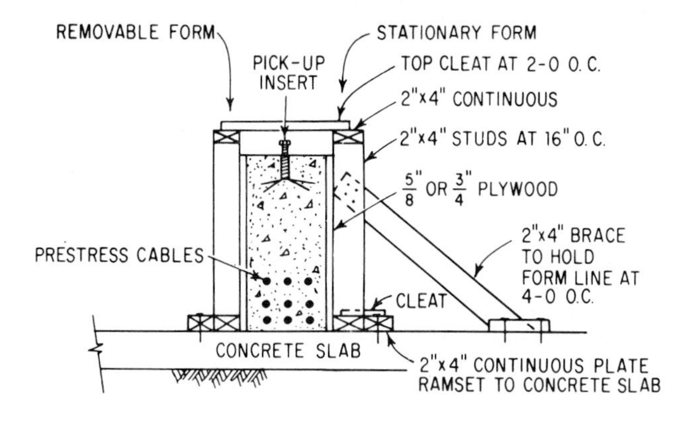

(A) STANDARD PRECAST BEAM FORM ON CONCRETE SLAB

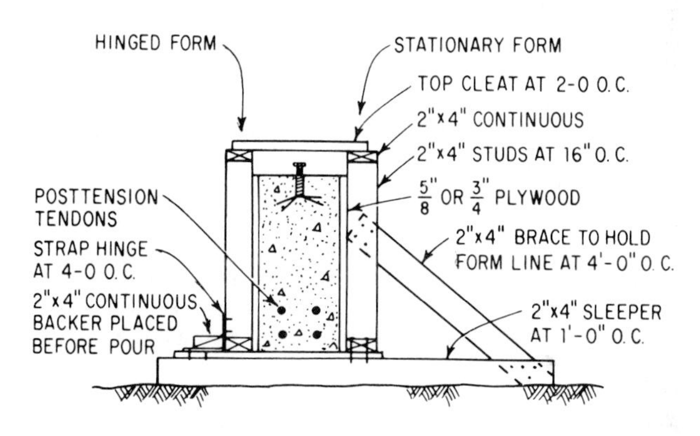

(B) STANDARD PRECAST BEAM FORM ON SOIL

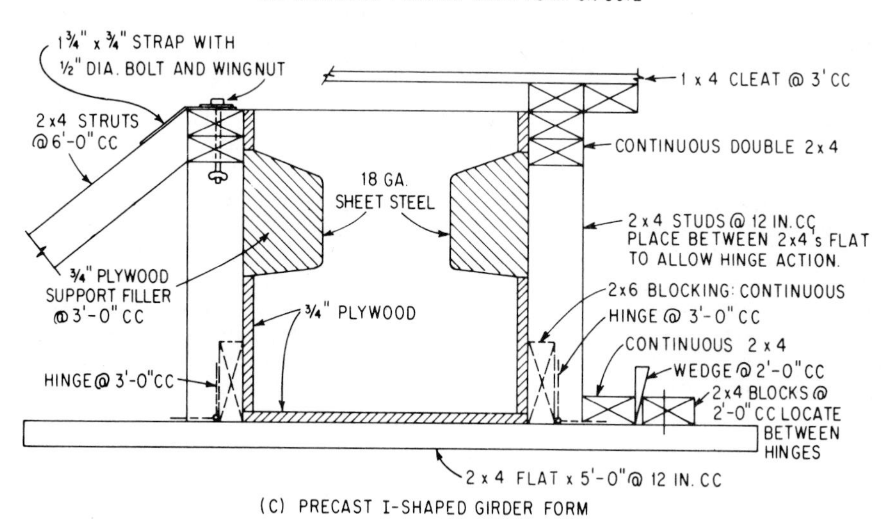

(C) PRECAST I-SHAPED GIRDER FORM

FIGURE 33.17 Beam forms for repeated use showing: (*a*) removable side; (*b*) hinged side; and (*c*) precast I-shaped girder form.

FIGURE 33.18 Portable beam bed.

FIGURE 33.19 Placing beam-reinforcing steel cage in the beam bed.

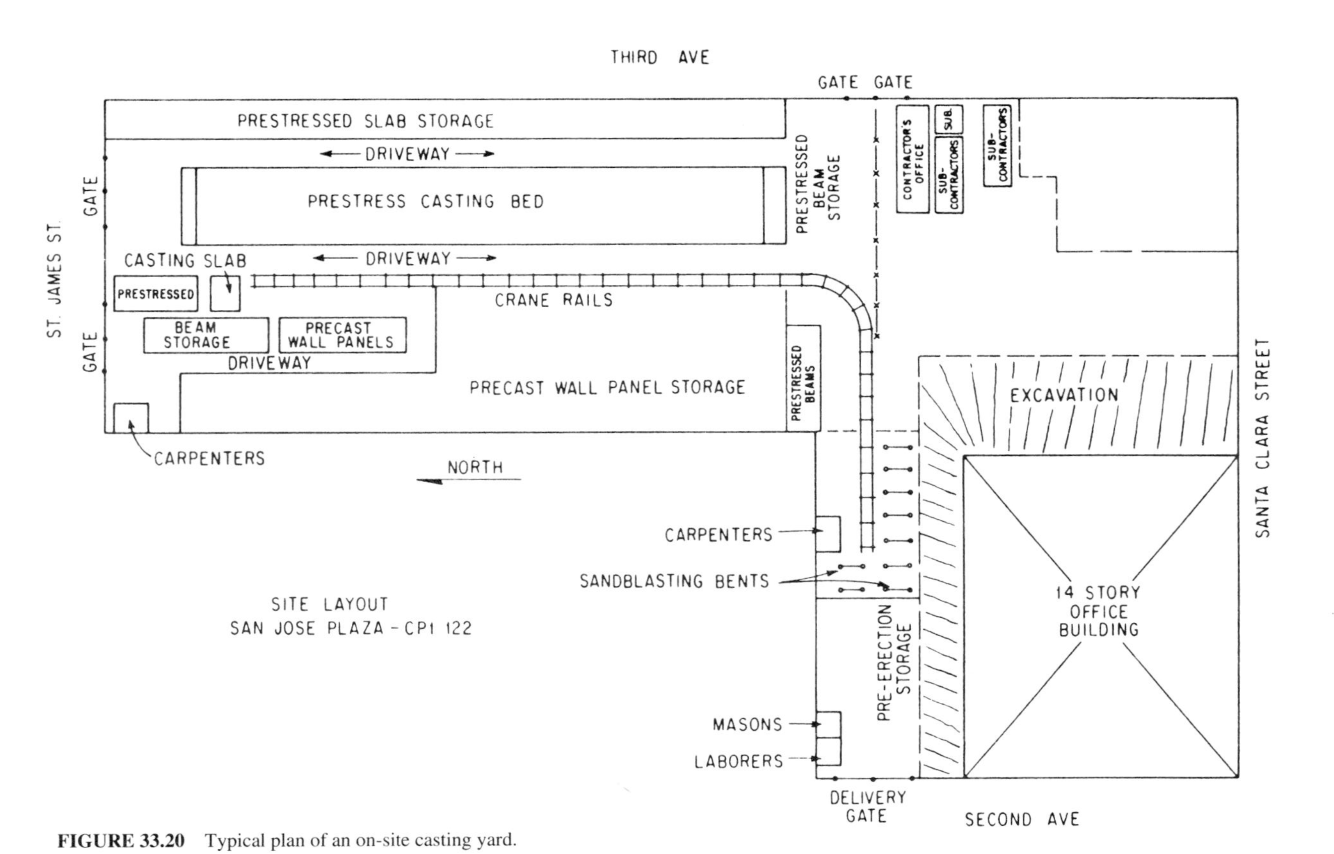

FIGURE 33.20 Typical plan of an on-site casting yard.

FIGURE 33.21 Installing precast planks on precast beams in a 30 story office building.

FIGURE 33.22 Precast column forming.

FIGURE 33.23 Precast column erection.

FIGURE 33.24 Precast column erection socket.

FIGURE 33.25 A bird's-eye view of an on-site casting yard. A portion of the 400 ft stressing bed is shown to the left of the crane rails. Prestressed beams are stored adjacent to the bed, with precast columns on the extreme left.

FIGURE 33.26 Another view of the yard shown in Figure 33.25. Crane rails and traveling tower crane are on extreme left. Two fiber-glass garage-wall forms are in the center of the figure; completed garage-wall panels are stored at the right. Forms and completed wall panels for the main building are in center background.

FIGURE 33.27 On-site precast fabrication.

FIGURE 33.28 On-site precasting.

CHAPTER 34
SLIPFORM CONSTRUCTION OF BUILDINGS

Dean E. Stephan, Jr.

The purpose of this chapter is to further the readers' understanding of slipform construction and bring to their attention methods by which the full efficiency of the slipform can be utilized. In the area of office building construction slipforming has been limited almost exclusively to the "core" portion of the building, or that portion which generally contains the elevators, stairs, bathrooms, and so on. This type of form is usually small in area and can be handled as a secondary operation in relation to construction of the balance of the building. In apartment, hotel, and similar residential-type multistory buildings, the slipform method of construction is highly effective, if constraints on architectural and structural design features are properly followed.

The ultimate approach in incorporating a slipform into building construction is to slipform all vertical concrete. This makes the slipform the primary operation controlling the other phases of construction. All vertical concrete structural elements are extruded by the slipform as shown in Fig. 34.1.

First, before discussing details of a slipform operation, we should evaluate the advantages and disadvantages of the basic system in order to determine whether a job should be slipformed. It is generally recognized that slipforming is a faster method of construction than conventional forming; so it follows that if the construction cycle is geared to the slipform as the controlling item, an earlier completion date can be obtained. From the owner's standpoint, an earlier completion date means an earlier return on invested capital. From the contractor's standpoint, an earlier completion date means lower overhead costs and more jobs that can be completed within the organization in a given period of time.

Under ideal conditions a slipform will provide substantially lower construction costs, which directly result in lower completed-building costs. This may make possible the building of projects that otherwise could never have been built because of limited financing.

Slipforming is one system of automation which can be more broadly included in building-design systems. The efficiency of its use depends on the building design and its adaptability to construction details that are necessary to slipforming and which are discussed later in this chapter.

In order to achieve this automation, work of all the various trades must be organized and coordinated so that they mesh with each other with a minimum of conflict. The materials deliveries, such as concrete and reinforcing steel, should be made on schedule in

FIGURE 34.1 All vertical concrete building elements being slipformed.

order to develop a maximum rate of slip. Specialized help is required for the duration of the slipform work, much more so than in conventional construction. It is imperative that some person thoroughly familiar with the overall operation be on the form at all times. If care is taken in the planning stage and carried through the complete operation, many of the normal pitfalls can be avoided and an efficient operation can be achieved.

Not all jobs are suitable for slipforming, and each should be evaluated to determine whether a slipform is feasible. This should be done in consultation with the architect and structural engineer, for sometimes limitations of structural or architectural design preclude the possibility of making the changes necessary to set up a slipform operation. If the design is in an early stage, it is sometimes possible to design around a slipform or, in other words, make the basic design for one that can be slipformed. Many owners have resorted to this method of architect-engineer-contractor coordination and cooperation in order to obtain the most for their money.

34.1 Design Requirement

Following is a discussion of the general design requirements that are necessary in order to develop an economical and efficient slipform. Details vary, but these requirements remain constant.

Number one on the requirement list is a layout that remains typical from floor to floor. This will enable a form to be built that will not have to be modified to any great extent as the slips progress. If possible, the wall thicknesses should remain constant throughout the full height of the walls, although walls can be narrowed down by the use of filler panels inserted in the form. Sometimes the saving in concrete is more than offset by the additional labor spent in modifying the form, not to mention the time lost. Minimum wall thicknesses may vary depending on whether lightweight or hard-rock concrete is used, but generally they should be at least 7 in to avoid "pullups" resulting from the friction between the form face and fresh concrete.

The reinforcing-steel design is the number-two item that will govern the efficiency of a slipform operation. Large concentrations of steel, such as found in spandrel beams and interior beams, should be kept to an absolute minimum, for these concentrations make it difficult, if not impossible, to place the steel while the slip is in progress. In some cases horizontal construction joints have to be located with this in mind. This also causes the workload of the ironworkers to vary greatly, thus making it difficult to maintain an efficient crew size. If the slip is not to be a continuous one, then the vertical steel should be detailed so that it can be placed while the form is at rest. If continuous, a pattern of splice points satisfactory to the engineer should be worked out so that the vertical steel can be placed while the form is moving. Again, consideration should be given to providing as even a work load as possible for the ironworkers by establishing the pattern of splices so that only part of the steel has to be spliced at any one elevation. Horizontal steel for slab doweling, No. 5 bars or smaller, may be bent to lay along the face of the form and straightened out after the form has slipped by; or, if large in diameter, the bars can be welded. The ends of the bars to be welded are normally encased in a blockout that is set in the form and therefore exposed after the slip. If bars are to be welded, only weldable steel should be used. This should be settled before construction starts (see Chap. 3).

Where a slab occurs on both sides of the slipformed wall, slots through the wall are a cost-effective way to accommodate horizontal slab reinforcing steel that penetrates the wall (Fig. 34.2). The slots are most easily formed by tying styrofoam blocks to the vertical reinforcing steel at blockout locations. The styrofoam is then easily removed from the wall by a combination of chipping it out and sandblasting. Placing a sand layer in the slipform is an economical manner of achieving a continuous horizontal slot in the wall when such a configuration is required.

In order to achieve a satisfactory finished product, in this case a completed structure incorporating all required architectural details, care must be taken to ensure that all the

FIGURE 34.2 Slots in slipform wall for future reinforcing steel and prestress strand penetrations.

items to be installed within or connecting to the slipformed walls have designed-in tolerances to ensure proper fit. This is no different from conventional construction, but the difficulty of setting work to an exact position from a moving deck accents this importance.

ACI Committee 117[1] specifies that translation and rotation from a fixed point at the base of the structure for heights of 100 ft or less not exceed 2 in and for heights greater than 100 ft, $\frac{1}{600}$ times the height but not more than 8 in. From this it can be seen that on tall buildings where the elevator shafts are of slipformed-concrete construction and the elevator door frame and sill remain in constant vertical alignment, the profile of the shafts in relation to plumb should be plotted so that the rail line can be set and fit properly within the shaft. In this manner, the proper architectural detail is maintained at each floor. Using a plaster skim coat on the lobby side of the elevator walls allows more flexibility in setting the elevator frames to the exactness required. Differences in projection of the frames from the concrete wall can be adjusted for appearance by varying the thickness of the skim coat.

When framed openings are required in slipformed walls, they can be handled in several ways. The first method, which can be used in the case of pressed steel or channel frames, is to set the frame in its final position similar to the way it would be in conventional construction, then place the concrete around it as the slip progresses. If this method is used, it is necessary to position the frame firmly by welding or some other means to make sure that it does not shift while the form is sliding. The second method is to provide a blockout in the wall slightly greater than the out-to-out frame dimension and install the frame after the slip has moved out of the way. The third method is to provide oversized blockouts in the walls and then set the frames after the slip has passed, and either grout them in or plaster around them. Any of the above-mentioned methods is satisfactory, but the method selected should be dependent on the architectural requirements of the job. The cast-in-place frame method is least desirable.

FIGURE 34.3 Finisher's scaffold suspended below the slipform.

34.2 Finishing

Another basic consideration in setting up a slipform is the type of concrete finish that is going to be required. The normal "slipform finish" is a rubber-float finish that is applied to the wet surface of the concrete as the slip progresses or applied dry after the slip is complete.

In the first method, a finisher's platform is suspended below the form and the finishing crew works with the freshly exposed concrete to give it the texture required. A membrane cure is sprayed on the surface after the finishing operation is complete, and it is usually unnecessary to go over the surface again once the slip is complete. Figure 34.3 shows a slipform. The finisher's platform can be seen below the form. Another slipform is shown in Fig. 34.4.

The second method, that of applying a "dry" finish, is to avoid touching the surface to be finished until after the slipping

FIGURE 34.4 Slipform has started up the core of the building.

operation is completed full height. When complete and the form has been removed, the walls are float-finished from a swinging stage by applying either a sand finish or a cement-plaster finish. If this method of finish is to be used, the finisher's scaffold is usually omitted except at critical areas where it is, or may be, necessary to have access under the form. Thus, the cost of fabricating and installing a finisher's platform is saved and can be applied to the higher unit cost of applying the finish in this manner. If a smooth finish is desired on the exterior of the building, a plaster skim coat can be applied over the rubber-float finish. Concrete finishing is discussed in Chap. 27, and plastering is covered in Chap. 39. Sandblasting of slipformed concrete is not recommended. The possibilities of uniform texture are practically nil. Furthermore, if sandblasting is attempted and found unacceptable, corrective procedures are costly in both time and money.

34.3 Elevators

Since we have been discussing basic considerations in order to determine whether a slipform is feasible, we should include the elevator requirement in arriving at a conclusion. One advantage of slipforming, as mentioned before, is that the elevators can be started before the building is topped out. This is especially true where the portion of the building containing the elevator shafts and equipment rooms is a stable structure within itself and can be slipped to the top and completed very early in the job. On tall buildings where the shafts cannot be slipped to the top ahead of the balance of the building, it is possible to get a head start with the elevator installation by establishing a rail line based on the first seven or eight floors and projecting it upward in similar increments as the shafts are completed. It is also necessary to block off the shaft at some point below the slipform to protect the elevator constructors. The most important thing to consider is the alignment of the shafts for, once the rail line for the first segment is established, it is necessary to continue this line for segments above. This makes the elevator-front design more important than ever.

34.4 Jacking Systems

After it has been decided to slipform a portion or all of the work, the details necessary to put the system into operation must be worked out. Of course, the initial decision must be that of selecting which jacking system to use. At present there are four primary systems: hydraulic, pneumatic, electrical and manual. Of these, the hydraulic and pneumatic systems seem to be the most common, and we shall discuss their details.

The hydraulic jacking system is a patented system employing a network of hydraulic jacks connected by oil lines to a central reservoir and powered by an electric pump. The jacks usually climb on pipe that has a diameter of 1 in and a wall thickness of ⅛ in. Each jack is calibrated to climb approximately 1 in each time the pump is activated. It is necessary to set the pump pressure high enough to make sure all jacks have raised before the pump is turned off. If this is not done, that portion of the form which is lightly loaded will progressively gain on the balance of the form and throw the deck out of level. The yokes that are used with this system are made of steel and are furnished by the slipform equipment supplier as part of the equipment leased to the job. The clearance from the working deck to the underside of the yoke is approximately 12 to 14 in, depending on the location of the bottom waler. Normal spacing between the yokes is 6 to 8 ft, with 7 ft considered to be optimum. If this spacing has to be increased because of architectural or structural requirements, special steps should be taken to ensure that the deflection of the walers between the yokes is kept within acceptable limits.

The pneumatic system of slipforming is also a patented system employing a network of jacks connected by air lines to an air compressor generally located near the base of the slipform. The control is simply a pressure exhaust valve located on the form and operated manually to raise the form in ½ in increments. The rods that these jacks climb on are most often a solid 1 in in diameter, drilled and tapped at each end for a coupling stud. The yokes are made of wood and fabricated on the job, which gives an added degree of flexibility in the form designing. Occasionally, jacking rods are withdrawn for reuse as the slip progresses. The expense of this process is seldom warranted by the material cost of the jacking rods.

As mentioned before, some systems employ manual and electric jacks, but these are not employed to any great extent in building-construction work. Figure 34.5 illustrates the wood yokes and shows the pneumatic jacks in place.

34.5 Form Design

After the jacking system has been selected, the next step is to design a form around this system making proper provisions for the jacks, yokes, and so on. Slipforms for building construction are usually built from 4 to 5 ft high (Fig. 34.6). Figure 34.7 illustrates a simple slipform layout and shows how a template is used to aid in the vertical reinforcing steel placement. Yoke spacing should be held to a maximum of 7 ft with "dummy" yokes (nonjacking) added as necessary to provide additional support. Where the load of the form is high, jacks will have to be spaced closer together or doubled up to provide the required lifting capacity. The weight of concentrated loads, such as a concrete hopper or stockpiled reinforcing steel, is an example (Fig. 34.8).

FIGURE 34.5 Operating deck of a slipform, showing yokes, steel template, and jacks.

Great care should be taken during the initial fabrication and setup period to see that design dimensions are maintained and that the form is strong enough to withstand the stresses of the slipping operation. The forms would be battered about ⅛ in in a 4 ft height so that they will tend to clear themselves as the slide progresses. In some cases, that batter is built into one side only, with the other side remaining vertical.

FIGURE 34.6 A slipform building core being set up.

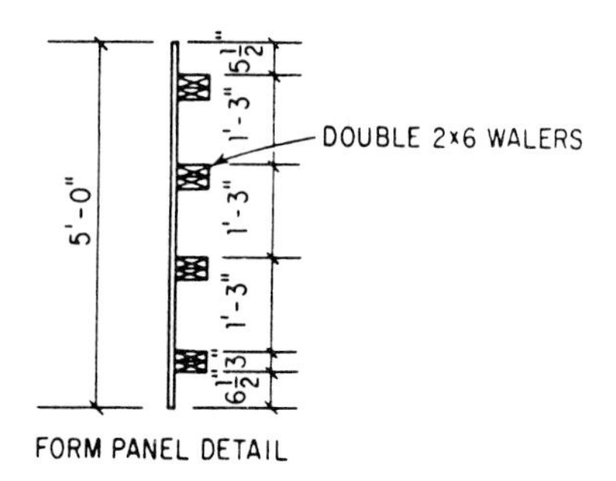

FIGURE 33.7 Timber yoke for a 5 ft slipform.

The form panels can be built from ¾ in plywood or some other material such as sheet steel or 1 in boards, and may utilize a two- or three-waler system. The size of the walers will depend on the spacing between yokes and may be 2 × 6 or 2 × 8 timber or, in some cases, structural steel. Blocking should be placed between the walers at each jacking point to transmit the load from the yoke into the form. The two-waler form usually will incorporate a herringbone bracing system between the walers continuous along the form. Form surfacing of 1 × 4 tongue and groove placed vertically permits slippage along the way between yoke locations where vertical adjustments are occasionally made. A 1 × 4 form board will not "cup" as much as wider boards. If plywood is used, a "plate girder" effect tends to occur, thereby inducing "corrective" action into adjacent jacking locations. This effect tends to self-align adjacent jacking points.

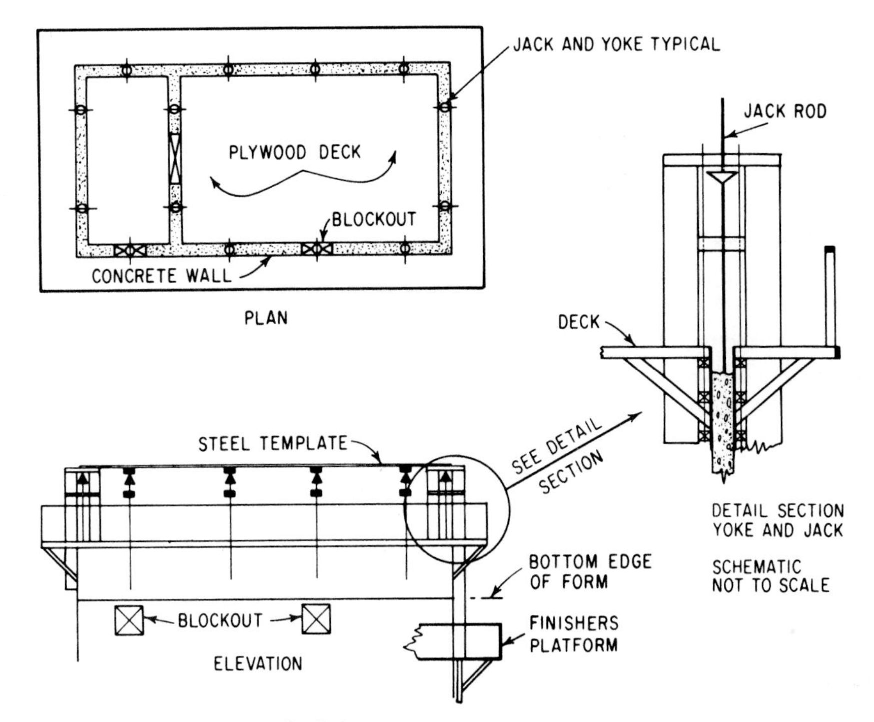

FIGURE 34.8 Typical layout of a slipform.

The yoke design is usually standardized to allow for walls of various thicknesses to be obtained from one size of yoke, and as much clearance as practical is left between the crosstie of the yoke and the working deck. This is the area in which the horizontal reinforcing steel must be placed; thus, a greater clearance between the yoke crosstie and the deck, the easier it is to place the horizontal reinforcing steel while the form is sliding. Large concentrations of horizontal steel (especially in earthquake zones) make it necessary to delay the slip or, in some cases, stop the slip in order to place the steel properly. There has been at least one attempt to increase the time available for placing this steel by increasing the clear distance between the yoke cross tie to the working platform from 12 to 36 in. The results of this high-yoke trial were quite satisfactory and can be helpful in critical areas of heavy reinforcing.

The deck-leveling method used for each of these jacking systems is a water level connected to a central reservoir with branch lines running to each jacking point. The working deck is leveled at the time of initial setup, and the water level is referenced so that as the slip progresses the level of the deck can be adjusted as desired. It is necessary to know relative elevations of the deck in order to control the lateral movement of the form. Plumb bobs and/or optical plummets are used and referenced to a line at the bottom of the slip. These should be checked continuously so that the slip can be adjusted quickly to compensate for deviations. Correction for deviation from line is done by changing the relative elevation of portions of the slipform deck. This adjustment is done by means of shutting down certain jacks during several lifting operations until the error has been compensated. This is a difficult operation and should be under the direction of a supervisor experienced in slipform work.

FIGURE 34.9 Skid used to correct form drift.

Drift and rotation of the slipform can also be adjusted by varying the location of vertical loads on the form or exerting cantilever forces on the form to pull it horizontally. An example of a cantilever force is a skid from which a 55 gal drum of water is suspended (Fig. 34.9). The water drum allows for ready variations in the amount of load applied to the form.

The slipform will tend to relax and grow over prolonged heights; thus the construction detailing of the working platform to allow it to work as a diaphragm between adjacent form units is essential. Rods with turnbuckles running between forming units are also useful. This helps maintain the desired dimensional and squareness relationships between units and allows for minor corrections when necessary.

The elevation of the form is registered by a steel tape, or similar device, attached to the base of the slipformed walls at a known elevation. The point where the tape is read at the form level is usually adjacent to the water level, therefore providing a direct method of establishing elevation anywhere on the form.

A good reference book giving design loads, methods of calculating pressures, and additional slipforming information is published by the American Concrete Institute.[2]

In order to slipform a portion or all of a building, a method must be selected for hoisting concrete, reinforcing steel, and the rest of the construction material required. This can be a material hoist, a mobile crane, a tower crane, a gin pole, or any other method that will satisfy the hoisting requirements of the slipform. Tower cranes have been used for many years to speed up the construction of high-rise buildings and are now very popular. In a slipform application the tower crane may be carried to the top with the form and used to satisfy the hoisting requirements, or a portion of the hoisting requirements, for the whole building after the slipform operation is complete. Whatever method is selected must be based on an objective view of the whole job with the ultimate goal of satisfying the needs of all phases of the construction. Figure 34.10 shows a typical slipform which was poured using a tower crane. Tight scheduling enabled the tower crane to service the slipform in addition to erecting the precast members.

A portion of another central core slipform, which was slipped over 400 ft, is shown in Fig. 34.11. All the precast structural elements and architectural wall panels on this job were hoisted by use of a climbing tower crane. Figure 34.12 shows the slipform on a 23 story apartment building in which all the vertical concrete shearwalls were slipformed. Figure 34.13 shows a slipform central building core surrounded by a structural steel perimeter building frame.

34.6 Slipping

The last major division of slipforming to be considered is the actual operation of the form during and between the slip. The quality of this portion of the work is dependent on the proper preparation of the details prior to starting to slip; scheduling of material to be delivered to the job, primarily concrete; coordination of subcontractors; and a thorough knowledge of the slipform work, including all details of all work to be incorporated into

FIGURE 34.10 Typical building core slipform. Material service to the form accomplished by a tower crane which climbs within the slipformed core.

FIGURE 34.11 A portion of a central-core slipform. Note the base of the tower crane in the center of the form.

FIGURE 34.12 Slipforming of building shearwalls.

FIGURE 34.13 Slipformed concrete core for composite steel and concrete framing.

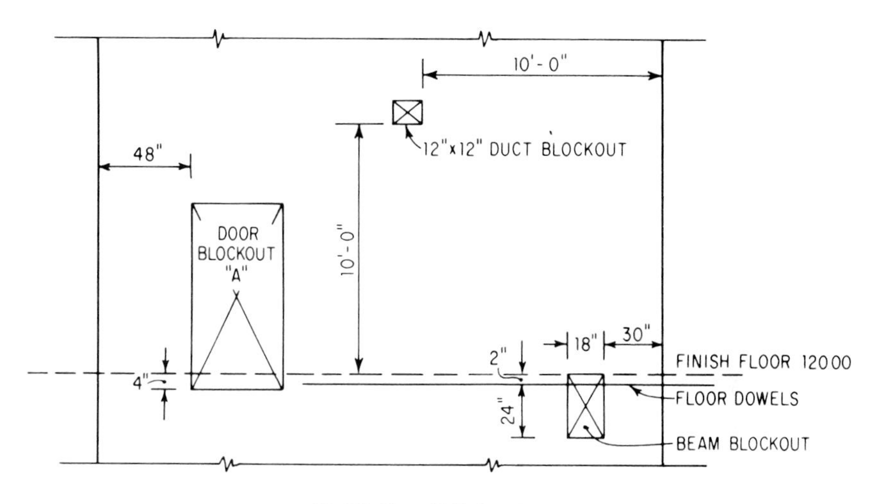

FIGURE 34.14 Detail sheet for a slipform operation.

the slip. Figure 34.14 shows a typical detail sheet which is made for the slipform operation. This type of drawing is made for each wall at each floor and shows all items to be incorporated into the slip. The responsibility for this should be given to one person who stays on the form at all times and has the authority to see that the necessary work is completed as the slip progresses. This person should keep an eye out for any indication of form trouble, which often shows up as a change in wall thickness, errant plumb-bob readings, form noise, or similar. Concrete should be placed in a definite pattern in lifts of 6 to 8 in and about 2 and 3 in slump with the vibrator penetrating just into the lift below. By keeping the form full, the maximum rate of slip can be obtained consistent with installation of other work. Rate of slipping will vary according to concrete slump, weather, and installation of other work, but the average is about 12 in/h with some slips approaching 24 in/h at certain times.

To summarize, the single most important thing necessary in order to obtain a satisfactory slipforming operation is thorough planning. From conception, through the structural design or modification thereof, through the form design and construction, to the actual operation of slipping, planning will prove the difference between a good low-cost job and one where troubles are prevalent and costs are high.

REFERENCES

1. ACI Committee 117 Specification, Standard Specification for Tolerances for Concrete Construction and Materials (ACI 117-90).
2. Hurd, M. K.: "Formwork for Concrete," ACI *Spec. Publ.* 4, American Concrete Institute, 1963.

CHAPTER 35

LIFT-SLAB AND PRECAST TILTUP CONSTRUCTION

Joseph J. Waddell
Joseph A. Dobrowolski

In the lift-slab technique, the building floor usually serves as the casting floor on which the roof and upper-floor slabs are cast at ground level. Several slabs may be involved, stacked one on top of the other. After the slabs have cured and have been posttensioned, they are raised, each to its required elevation, by means of jacks mounted on top of the preerected building columns, opening having been left in the slabs to enable them to rise along the columns.

Costly equipment, special experience, and the critical nature of many of the steps make lift-slab construction a job for the specialist. Preconstruction planning and consultation between the architect, engineer, builder, and lift-slab specialist will enable full advantage to be taken of this unique type of construction to effect the greatest time and money savings of which the system is potentially capable.

Many contractors have selected the lift-slab method to enable them to construct buildings in the most efficient manner to meet the requirements of safety and structural adequacy. Almost all types of buildings of medium to large size with an area of about 25,000 ft^2 or more are appropriate to this construction system, as shown in Fig. 35.1. Most structures have been three to five stories in height, but the method has been used in constructing a number of high-rise apartment and office buildings (Fig. 35.2). Advantages claimed are the elimination of costly formwork and shoring, as only edge forms are necessary; slabs can be cast, finished, and tensioned faster and more easily because the work can be done at ground level; electrical and plumbing services are easily installed before the concrete is placed; curing and weather protection are necessary only at ground-level area where the slabs are being cast; and tile, flooring, or other finishing materials can be applied directly to the slab concrete. Posttensioning the units minimizes cracking because shrinkage is restrained, and it is not necessary to compensate for deflection of the slab when installing partitions.

The foundation for a lift-slab structure is usually the ground-floor slab of the building. Special care is necessary to make sure that the foundation slab is smooth and free of blemishes, as the underside of the slab cast on it will reflect any such blemishes.

Before casting the lift slabs, it is necessary to erect the building columns, which may be of precast concrete, cast-in-place concrete, or steel. Columns are frequently spliced, the upper portions being installed after lifting of the slabs is commenced. For example,

FIGURE 35.1 Lifting is not confined to flat slabs, as this domed auditorium roof was lifted in one section. (*Vagtborg Lift Slab Corp.*)

slabs may be parked at the first three floor levels while the columns are being spliced; then the slabs are raised to their final elevations for attachment to the columns. Figure 35.3 shows slabs temporarily parked while columns are being extended. Column design must take into account the stresses imposed on them during slab lifting, considering among other things wind loads, stability, and stiffness during construction. Because of

FIGURE 35.2 Lift-slab project with jacks on concrete columns.

FIGURE 35.3 Slab being lifted from ground stack.

the design of the lifting equipment, there is small likelihood of eccentric loading on the columns. Steel lifting collars, cast into the slab later, are stacked on each column at the time the columns are erected. There should be clearance between the collar and column to ease lifting and permit relative movement caused by shrinkage and other factors.

Casting, finishing, and curing follow closely the procedures for any first-class slab production. Slabs may be up to 10 in thick, frequently of lightweight concrete. Conventional reinforcement, stressing cables, conduit piping, and inserts are all installed in the usual manner. Concrete with a slump not exceeding 4 in is placed and consolidated by vibration to eliminate all rock pockets, then is finished to a smooth finish with power tools. Curing is best done with a resin-base liquid material that also serves as a parting compound on which concrete for the succeeding slab can be placed, after the first slab has developed sufficient strength.

After a suitable curing period the slabs are posttensioned (see Fig. 35.6) usually in both directions; that is, tendons are positioned and cast in the concrete at right angles to each other. Tendons should not be deflected or diverted in plan around openings, as erratic friction losses are apt to be introduced. Stressing heads should be placed in wooden box forms recessed several inches along the edge of the form. Short lengths of heavy wire for pulling can be stapled or nailed in the boxes and, after removal of the boxes and completion of stressing, the cavities can be dry-packed, thus protecting the stressing heads from corrosion. Tendons less than about 70 ft long can be stressed from one end; those longer should be stressed from both ends. Usually adjacent short tendons are stressed alternatively from opposite ends, thus tending to equalize stresses in the slab.

After the columns have been erected, special hydraulic jacks are placed on top of the columns, one to each column (see Fig. 35.4). The jack illustrated has a large-diameter ram in the center that has a stroke of about ½ in. Follower nuts, following closely behind the jack movement, automatically hold the slab when the jack is retracted. A jack of this type may have a capacity as high as 100 tons under full hydraulic pressure of 2000 psi. When the slabs are ready for raising, suspension rods, reaching down from the jacks on top of the columns, are attached to the collars that were previously cast in the top slab (usually the roof slab), one collar encircling each column. All jacks are synchronized, operating from a central control console (Fig. 35.5). Should one of the jacks move ahead or lag behind by ½ in in elevation, a signal light on the console gives a warning. The slab

FIGURE 35.4 A typical hydraulic lifting jack atop a steel H-column. Follower nuts are rotated through the chain drives that are actuated by small hydraulic motors. (*Vagtborg Lift Slab Corp.*)

FIGURE 35.5 Regulation of all jacks on the slab is effected at the operating console. (*Vagtborg Lift Slab Corp.*)

is maintained level within a tolerance of about ¼ in. Usual jacking rate is between 5 and 15 ft/h, with 7 to 10 about normal. If no splicing of the columns is required, the roof slab is first raised to its final position, followed by the floor slabs, normally one at a time. On a small building more than one slab might be lifted simultaneously.

Height of the building and the number of slabs to be lifted determine the exact lifting sequence to be followed in each case. For a building requiring more than one column section (because of height) in the interest of providing column stability during construction, it is sometimes prudent to park several slabs at the top of the first section of columns. Part of the slabs can be lifted and braced, then columns extended, and finally the upper slabs lifted to their final elevations.

FIGURE 35.6 Lift slabs being posttensioned prior to lifting.

Slabs are connected to the columns by means of metal collars (Fig. 35.7) previously cast in place.

Connections between slabs and walls should not be made until the slab is at least 30 days old to reduce the possibility of horizontal movement of the slab relative to the shear walls resulting from creep of the slabs.

FIGURE 35.7 The cast-steel collar which is embedded in the slab concrete is lifted by two jack rods, one of which is shown just above the collar. (*Vagtborg Lift Slab Corp.*)

PRECAST TILTUP PANELS

This is a type of precast construction in which wall panels are cast in a horizontal position at the site, tilted to a vertical position, and moved into final location to become the building walls. Panels may be either solid concrete or of sandwich construction in which relatively thin, high-strength conventional concrete surface layers are separated by a core of low-density insulating material. Generally, tilt-up wall panels (Fig. 35.8) are cast directly on the floor slab of the building. Where there is not enough room available, they are usually stacked—that is, cast one on top of the other. Floor layout for the panels must be thorough to achieve a smooth economical lifting sequence. The layout should provide for efficient casting and lifting of the panels. These requirements are met if the area for the panels does not exceed 85 percent of the total usable floor space. Otherwise, stack casting will be required.

In considering a building for tilt-up construction, the floor must be designed to support the loads of cranes, ready-mix trucks, and similar equipment that may impose heavier loads than the proposed occupancy of the building. A well-compacted base to support the slab is essential to prevent movement and cracking of the slab. Sometimes it may even be desirable to increase the slab thickness, at least in areas where heavy equipment will be operating.

The casting floor should be perfectly smooth and uniform, as any imperfection in the floor will show on the panel cast against it. Imperfections in the floor can be smoothed off by filling them with stiff mortar, after thorough removal of all curing compound or dirt in the area being patched. Temporary fillings such as would be required in a joint or crack in the floor, and which are to be removed from the floor after the panels have been removed, should be made with nonabsorbent plastic filler.

Fabrication is accomplished by first placing a bond breaker (Fig. 35.9) on the casting floor. Liquids of various types are generally used, although sheets of plywood, metal, or

FIGURE 35.8 Tiltup panel with monolithic pilaster forming. Note plastic form liner on base slab.

FIGURE 35.9 Worker applying bond breaker to base slab.

paper can be used. Liquids consist of special formulations for this purpose, curing compound, and waxes, which are applied in two coats, the second coat being applied shortly before the panel concrete is placed. Uniformity of application is important. Paper and felt bond breakers nearly always wrinkle, and asphalt materials will stain the concrete and hence should be avoided.

A good type of bond breaker is one that combines curing with bond breaking. This material is applied to the floor as soon as final troweling is completed. A second coat is applied to the cleaned surface after forms have been placed but before the reinforcement and inserts have been set in place. Silicone-base materials have proved to be very satisfactory.

Side forms are usually of lumber. Forms for window and other openings may be metal or wood, with metal preferred because the swelling of wood frames makes them difficult to remove and might crack the concrete. Figure 35.10 shows a section of typical edge form for a flat wall panel. Dowels extending through the edges of panels fit into cast-in-place columns that are constructed after the panels have been braced in their final position. The presence of these dowels makes necessary the use of split forms on the vertical edges of the panels, as shown in Fig. 35.10.

Reinforcing steel may be fabricated in place, steel mats may be prefabricated on the site outside the forms, or heavy welded steel mesh can be used, depending on the relative cost at any one building site.

The panels are cast inside face up if they are to be erected from the inside, and outside face up if they are to be erected from the outside. However, should there be beam protrusions monolithic with the panels, or should a type of concrete finish on one face or the other dictate that that face be cast up, it may not be possible to erect in this manner. For

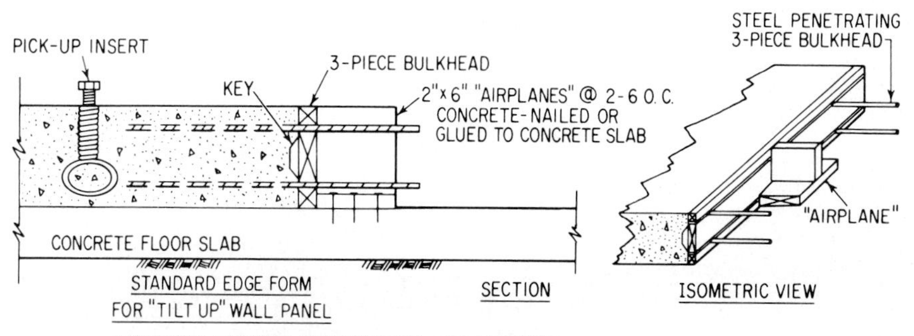

FIGURE 35.10 Details of edge form for tiltup panel.

this case, the erection of each panel is called a "suicide pick" as the panel would then fall into the crane. This situation is therefore to be avoided if at all possible.

Placing and consolidating the concrete should be done in accordance with good construction practices. Attachment fittings for lifting hardware are inserted in the form prior to concreting. Edge forms serve as screeds. Because of the presence of inserts and other hardware, it is frequently not possible to use a vibrating strike-off, and immersion vibrators must be used to consolidate the concrete thoroughly. Short (5 in long) internal vibrators have been developed to allow vertical vibration in thin tiltup panels which does not disturb any architectural treatment or the bonding compound on the bottom surface. Many special finishes can be applied while the concrete is still plastic, including the embedment of architectural details and ornamentation, such as exposed aggregate.

In finishing a tiltup panel, it should be remembered that there will be more bleed water rising to the surface than there would be for a slab on grade, as there is no absorption of water by the subgrade. This feature will probably make necessary somewhat more wood floating than is normally required.

Various types of cranes and gin poles have been used to lift the panels into place but motor cranes are normally used (Fig. 35.11). The panels are tilted onto a layer of mortar on the foundation and braced temporarily until the columns have been constructed.

Tilting places unusual stresses on the panels which should be considered in the design. The panel should reach the required design strength with a factor of safety, before the panel is tilted.[1] Field-cured cylinders will provide this information. Pickup points must be carefully located and lifting equipment designed so as to avoid high localized stresses in the panel that might cause cracking, splitting, or spalling of the concrete. Vacuum lifting attachments are sometimes used. Figure 35.12 shows two rigging arrangements for erecting panels. The location of the pickup points to erect the panels is very important. If they are not located correctly in relation to the center of gravity of the panel, the panel will not

FIGURE 35.11 Tilting panel into place with a motor crane.

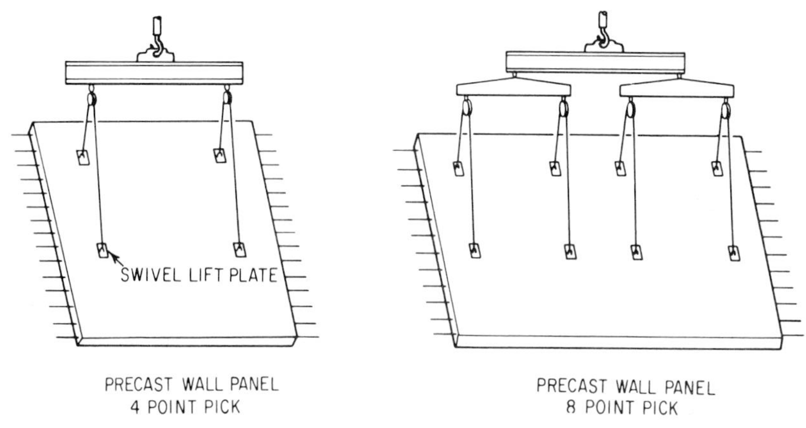

FIGURE 35.12 Two types of rigging for tilting panels.

hang in the proper position for erection. The normal position for easy erection is slightly out of plumb in a direction away from the crane. Computations for the pickup-point locations are usually done by the structural design engineer or the engineering staff of the concrete accessory companies.

Considerable more lifting force is required to break the panel loose from the casting floor than is necessary to lift the panel after movement has started (Fig. 35.13). For this reason, it is a good idea to move the panel slightly, if possible, before lifting. This can be accomplished with jacks operating in a horizontal motion to slide the panel a fraction of an inch to break bond. In some instances a wedge is used help to break bond at the start of lifting. Lifting stresses are usually limited to $5.5\sqrt{f'_c}$ to $6\sqrt{f'_c}$ for solid panels and $3.5\sqrt{f'_c}$ for sandwich panels. These stresses in the panels are usually greater than those due to full design loads. If these stresses are expected to be exceeded, strongbacks should be used to prevent cracking the panel.

Temporary bracing is best provided by tubular braces, including turnbuckles, for adjustment, that can be attached to the panel before it is erected (Fig. 35.14). After the panel is in place, the braces are attached to floor anchors and final adjustment is made to the panel location. Proper coordination of lifting and bracing will help to free the crane for other work.

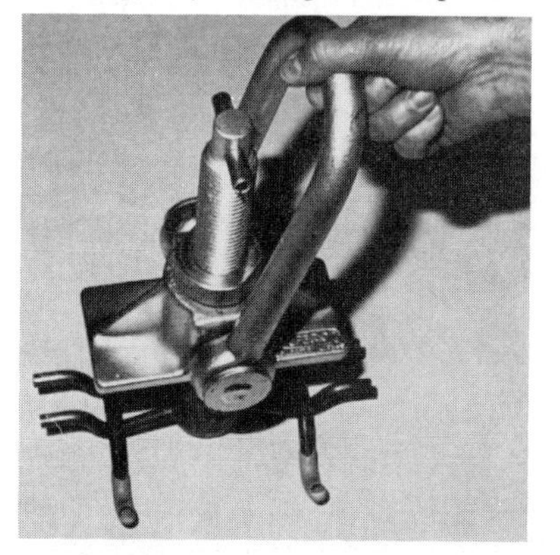

FIGURE 35.13 One type of lifting hardware which is cast into the concrete panel with turnbuckle attached.

After the panels have been erected and plumbed in their final positions, the joinery columns are formed and the concrete is placed in them. There has been some success in using shotcrete to place this concrete with properly trained operators. Where pilasters are cast monolithically with the panel, the connecting steel reinforcement is welded after a delay to allow shrinkage to occur. Figure 35.15 illustrates a connection with a precast column face

FIGURE 35.14 One type of lifting hardware which is cast into the concrete panel with turnbuckle attached.

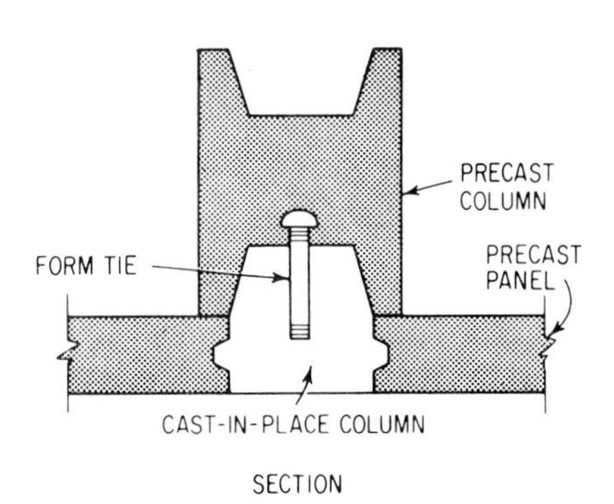

FIGURE 35.15 Precast panel and column joinery.

form which can have an architectural appearance to match or contrast with that of the adjacent panel. Lift hardware and braces are attached to fittings (Fig. 35.14) that are embedded in the concrete when the concrete is cast.

A general rule of thumb is that the crane capacity be at least twice that of a load from the heaviest lift. However, height of the panel, expected travel distance for the crane while carrying the panel, and reach distance of the crane are additional factors to be considered. Recommended safety factors for the lifting equipment are 2:1 for spreaders and 5:1 for the sling, snatch block, and shackles. Normally the lifting crew consists of the crane crew, rigger supervisor, two journeyworker riggers, and welders, if required. For stack-casted panels, two to three additional workers are required. In addition, the panel contractor should provide additional workers to install the bracing.[2, 3]

REFERENCES

1. Building Code Requirements for Reinforced Concrete (ACI 318-89), American Concrete Institute.
2. Tiltup Construction, *Concrete International*, Vol. 2, No. 4, April 1980, American Concrete Institute.
3. Tiltup Concrete Structures (ACI 551-92), American Concrete Institute.

PART • 10

SPECIALIZED PRACTICES

CHAPTER 36
PUMPED AND SPRAYED CONCRETE AND MORTAR

Joseph J. Waddell
Joseph A. Dobrowolski

Conveying of mortar and concrete through a pipeline has been a highly satisfactory method of transportation for many years. Two common methods are employed: pumping by mechanical pumps and conveying the material in a stream of air.

Pumped concrete or mortar is premixed, including the water, and conveyed through a pipeline or hose by the mechanical force exerted by a piston, pneumatic pressure, or a roller acting on the material. The material moves slowly through the conduit and is discharged by dropping into the form or a receiving hopper (see Figs. 36.1 and 36.2).

Shotcrete, or gunite, on the other hand, consists of either a dry mix, to which water is added at the nozzle, or a wet mix containing the necessary mixing water. Both are conveyed at relatively high velocity by pneumatic pressure through the conduit and are discharged at high velocity and forcibly deposited on the receiving surface.

There is some overlapping of methods and equipment, and some of the pneumatic machines perform as pumps as well as shotcrete guns; so no hard and fast rules govern the classifying of some of the equipment in one category or the other. It may fit in both.

PUMPED CONCRETE AND MORTAR

Conditions that lend themselves especially well to pumping exist on those sites and areas where access is limited and the site is crowded with materials and equipment, such as many city building sites.[1] Nearly all pumps, large and small, can pump 100 to 150 ft vertically, thus lending themselves very well to high-rise building construction. In recent times concrete was pumped vertically as much as 1038 ft in 1989. A pump takes small space and can be located any place that the ready-mix truck can reach. Conveying hose and pipe are easily placed out of the way and take little room. In locations difficult to reach with the ready-mix truck, such as on a steep hillside, a pump can easily move the concrete over obstructions that would be exceedingly difficult for the trucks to overcome. In many cases, the cost of pumping concrete is less than that of other methods of transporting. Each job, of course, must be analyzed on its own requirements and conditions to determine the most economical method of moving concrete.

FIGURE 36.1 Discharge of concrete from hose of small-line pump.

36.1 Equipment

Heavy Equipment. For use when relatively large quantities of concrete are to be placed, heavy mechanical pumps with a rated capacity of up to 138 yd^3/h are used. As shown in Table 36.1, these machines can pump 3 in slump concrete through a ≥5-in pipeline up to 4000 ft long, raising it as much as 1200 ft vertically. Concrete frequently contains aggregate as large as 2½ in.

FIGURE 36.2 A boom arrangement such as this enables the pipeline to clear obstacles and facilitates concrete placing. (*Transit Mixed Concrete Co.*)

TABLE 36.1 Specifications for Trailer Concrete Pumps

Model	750-15R series	2000-20R series	8000-18R series
Capacity, yd^3/h, theoretical maximum	42	117	138
Maximum strokes per minute	30	30	34
Maximum size aggregate, in	1½	2½	2½
Pump cylinder diameter, in	6	8	7
Stroke length, in	39	63	79
Maximum pumping distance, ft, straight line			
Horizontal	1000	1700	4000
Vertical	300	450	1200

Source: Schwing America, Inc., 1991 Technical Bulletin.

One of the largest of these machines, is shown in Fig. 36.3. Pistons and valves are mechanically operated as shown in Fig. 36.4. The receiving hopper is a pugmill-type mixer that helps to maintain uniformity of the concrete and minimize segregation.

The advent of truck mounted pumps with booms (Fig. 36.5), and seperate placing booms mounted on crane towers, fixed pedestals, or hydraulic self-climbing pedestals has revolutionized the placement of pumped concrete. The towers and pedestals have zero radius, which allows placement of the concrete adjacent to the tower or pedestal or out to over 150 ft within a full 360°. Supplied pipelines vary from 4 to 6 in in diameter.

Small-Line Pumps. A large variety of pumping equipment is available suitable for almost every concreting job. These pumps owe their name to the fact that the concrete is pumped through conduit 4 in or less in diameter, this being small when compared with the 5 and higher in. lines of the heavy pumps. Many of these rigs are small and portable enough to be hauled behind a pickup truck (Fig. 36.6). Others are truck-mounted. Capacities of up to 80 yd^3/hr are claimed, the actual rate depending on length of hose, vertical lift, maximum size of aggregate, mix proportions, and slump. In general, these small-line pumps evolved from grout and plaster pumps. There are several makes of piston pumps, either hydraulically or mechanically driven, most of them with two pistons, alternating on

FIGURE 36.3 Heavy-duty truck-mounted concrete pump with boom. *(Schwing America, Inc.)*

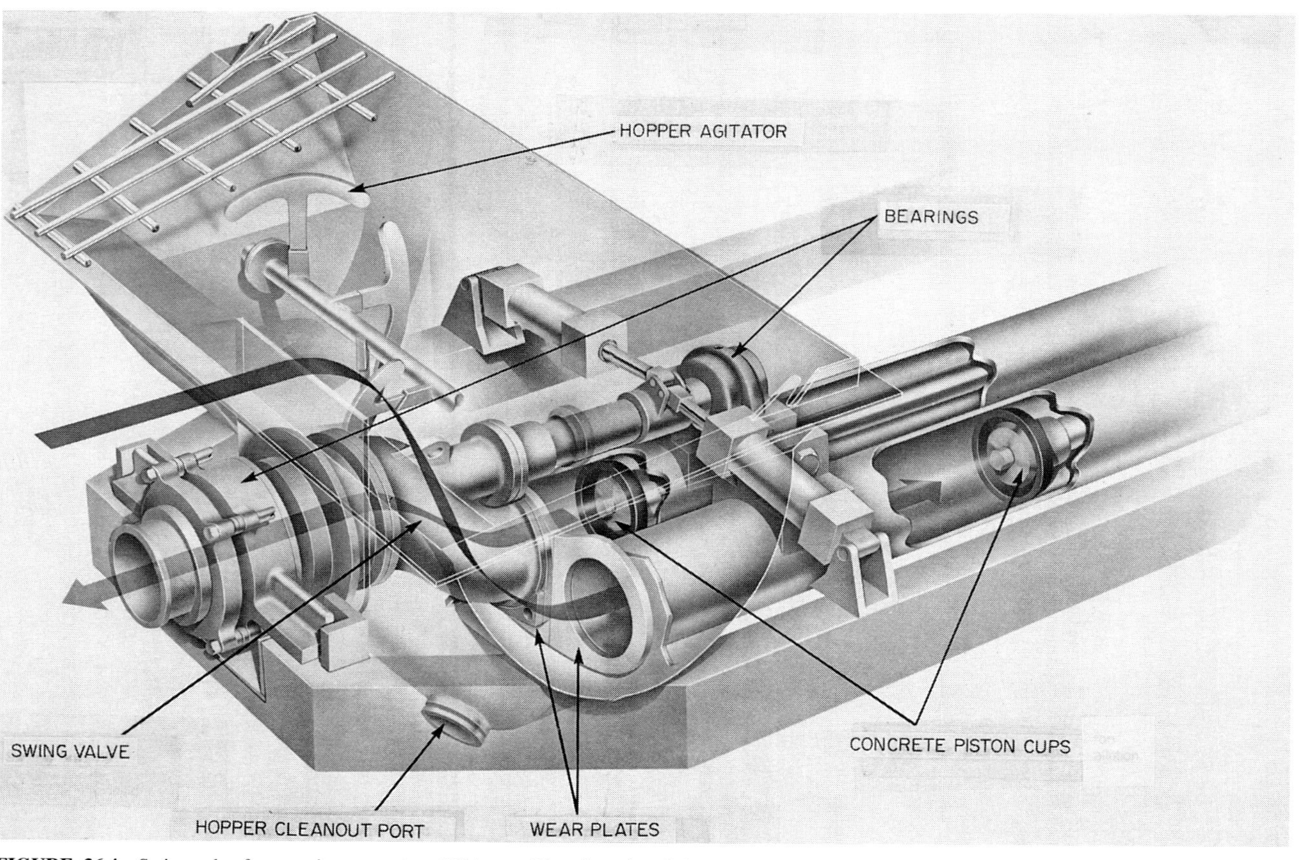

FIGURE 36.4 Swing valve for pumping concrete. *(Whiteman Manufacturing Co.)*

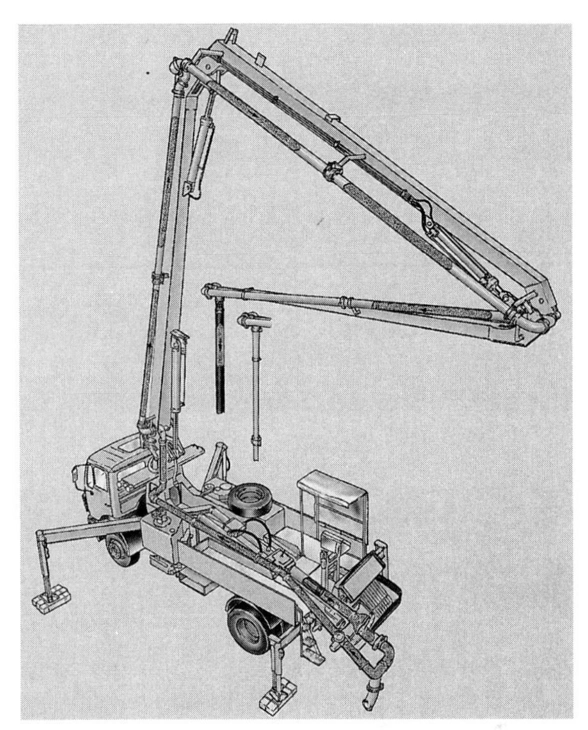

FIGURE 36.5 Operation of heavy-duty truck boom. (*Schwing America, Inc.*)

the power stroke. The large (6 to 8 in diameter) low-velocity pistons force concrete through reductions to the pipe or hose, which may be from 2 to 4 in in diameter.

Concrete from the ready-mix truck is deposited in the hopper leading directly to the loading chamber, passing through valves into the unloading chamber, where the piston forces it into the pipe or hose for delivery to the forms.

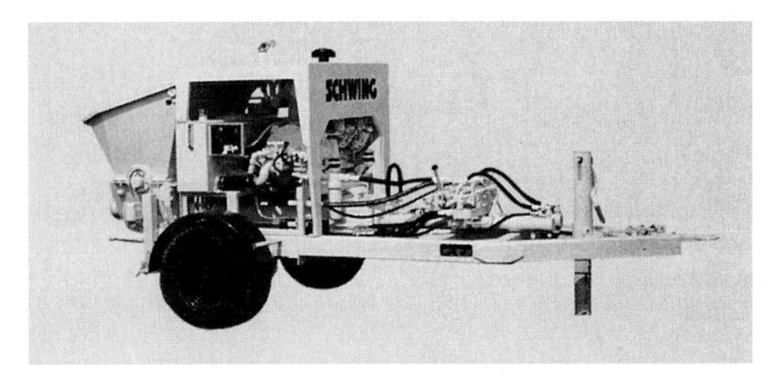

FIGURE 36.6 Small-line concrete pump with a rated capacity of 42 yd^3/h. This pump can handle concrete containing 1½ in aggregate. (*Schwing America, Inc.*)

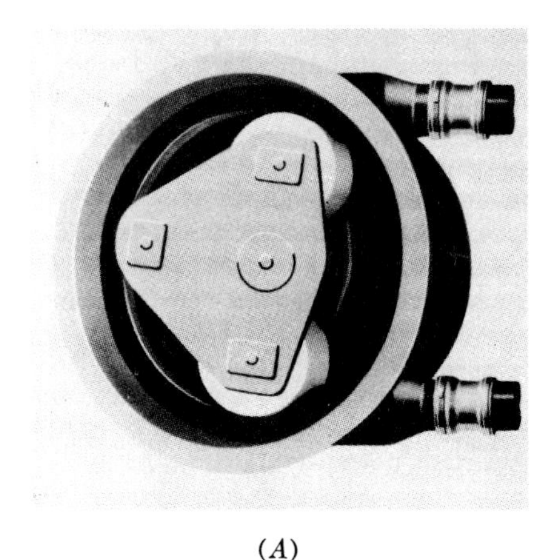

(*A*)

(*B*)

FIGURE 36.7 *(a)* Interior of the vacuum-squeeze pump; *(b)* a schematic cross section of the pump. A vacuum in the pumping chamber causes the pumping tube to open, drawing the concrete into the tube immediately behind the roller at the bottom, while the roller at the top is forcing the concrete out through the pressure line. (*Challenge-Cook Bros., Inc.*)

Another small-line pump consists essentially of a pair of rubber rollers that squeeze the concrete out of a rubber tube lining half of a drum-shaped vacuum chamber (Fig. 36.7). Pumping tube sizes, from 2 to 5 in, provide maximum pump capacities of 12 to 90 yd^3/h. By matching the pumping-tube diameter to that of the placing pipe line, there is no reduction in diameter of the concrete stream being pumped, and therefore no increase in pressure as the concrete enters the pipe. There are no valves, pistons, or other mechanisms in direct contact with the concrete.

A third type of small-line pump is the pneumatic type in which the concrete is carried through the pipe by air pressure in a manner similar to shotcrete, but discharge is at low velocity.

Remote control is sometimes provided, enabling the worker at the hose discharge to start or stop the pump. This feature is especially desirable when visual signals cannot be used because the hose discharge is not visible from the pump. Telephone or radio communication is sometimes used.

Pipes. Large-diameter pipe for heavy-duty machines may be about 8 in in diameter. It is quite apparent that a fairly large amount of concrete may be contained in a long line. Figure 36.7 will aid in determining the amount of concrete thus contained, and is of value in holding concrete loss to a minimum when nearing the end of a run. This chart can also be used to determine the amount of water required to clean out the line with the "go-devil", a plug that is inserted in the line at the pump and is forced through the pipe by air or water pressure.

In figuring pipeline for a job it is necessary to make an adjustment for vertical lifts and bends, converting them to equivalent horizontal pumping. The following equivalents are recommended:

1 ft of vertical pipe = 8 ft horizontal

One 90° bend = 40 ft horizontal

One 45° bend = 20 ft horizontal

One 30° bend = 13 ft horizontal

For example, assume that a line has an actual length of 360 ft which is made up of the following sections:

320 ft straight pipe

Two 90° bends

Four 45° bends

There is a 40-ft vertical lift at the end of the pipeline.

The equivalent length of straight horizontal pipe is determined as follows:

320-ft straight pipe equals	320 ft
Two 90° bends (each equivalent to 40 ft straight pipe)	80 ft
Four 45° bends (each equivalent to 20 ft straight pipe)	80 ft
40 ft vertical lift (1 ft vertical is equivalent to 8 ft horizontal)	320 ft
Total	800 ft

Layout of the pipeline (called the "slickline") for any size of pump is important, keeping in mind the fact that bends introduce additional frictional resistance. Alignment from the pump to discharge should be as direct as possible. Large areas of placement are most efficiently covered by adding sections of pipe as the work progresses. In some cases, however, such as on a large slab, it may be more desirable to work in the opposite direction, progressively removing sections of pipe. The sections removed should be cleaned immediately. Normally, a length of hose is used at the discharge end to facilitate placing the concrete exactly where required.

When it is necessary to run the pipe over a large area where a considerable amount of concrete is to be placed, it is sometimes feasible to run the pipe on steel supports that can be left in the concrete on completion of the placement. In this situation the steel supports are cut off just below finish grade before the concrete is finished.

In general, the larger the inside diameter of the pipe or hose, the less pressure is required to move a given quantity of concrete in a certain time interval. (Hose, however, requires more pressure than pipe because of higher frictional resistance.) Larger pipe requires more labor to handle the pipe sections and may require heavier bracing and supports. The total weight of a 3 in OD hose full of normal-weight concrete is 67 lb for a 10 ft length; a similar length of 5 in hose full of concrete weighs 204 lb.

Aluminum pipe must not be used because of a reaction of the aluminum with the cement in the concrete that produces an expansion of the concrete, causing the concrete to suffer a significant loss of strength. A number of cases have been reported in which this happened on the job.

36.2 Materials

Aggregates. The size, shape, grading, and proportions of aggregates are all important in obtaining a pumpable concrete. Some operators suggest that the maximum size of coarse aggregate should be no larger than about 40 percent of the conduit diameter, as shown in Table 36.2. The oversize should be eliminated by finish screening.

Rounded or subrounded aggregates make better mixes for pumping than aggregates containing a large proportion of crushed material, although the latter can be used satisfactorily. If it becomes necessary to use crushed-rock coarse aggregate, every measure should be taken to assure maximum plasticity of the concrete as described in the chapter on mixes.

Grading of the aggregates should conform to the requirements of the code or specifications under which the work is being performed. Sand must contain adequate fines, with 15 to 30 percent passing the No. 50 screen and at least 5 to 10 percent passing the 100 mesh. In 1 or 1½ in mixes, the total aggregate should contain about 10 to 15% pea gravel.

Poorly graded aggregates are apt to result in concrete that is difficult to pump, the friction of such concrete in the conduit offering so much resistance that maximum pumping efficiency is not achieved. Skip or gap gradings are not suitable.

Cement. Any of the usual types of portland cement can be used in concrete to be pumped.

Admixtures. Good construction practices govern the use of admixtures, and neither special limitations nor tolerances need to be applied so far as pumping is concerned. During hot weather a set retarder may be desirable if concrete is being pumped a long distance. Water reducers can be used to advantage. Pumping-aid admixtures are available and can be helpful in difficult pumping, such as long pipe line, high lift, harsh mix, or poor aggregate grading. Among the latter are finely ground minerals, such as hydrated lime, pulver-

TABLE 36.2 Maximum Aggregate Sizes for Different Conduit Sizes

Pipe or hose ID, in	Maximum aggregate size, in
1½	⅝
2	¾
2½	1
3	1¼
4	1½

ized quartz or limestone, and pozzolans. Some users report favorable results from certain water-soluble cellulose polymers.

36.3 Mixes

Concrete for pumping must be plastic and workable. Because of this many persons have felt that a very high percentage of sand, as much as 65 percent of total aggregate for a 1 in maximum-aggregate concrete, is necessary. However, the heavy pumps can handle concrete with far less sand, and the small-line pumps have now been improved to the point where they can handle mixes with a higher proportion of coarse aggregate. Heavy pumps handle concrete with 1½ in maximum aggregate containing 33 percent sand, 3 to 5 percent entrained air, and cement contents of about 580 lb per cu yd. While a slightly oversanded mix is more pumpable, other things remaining the same, excessive oversanding is not considered necessary. The percentage of sand in the mix should be based on the void content of the coarse aggregate, a properly proportioned mixture containing sufficient sand to fill the voids in the coarse aggregate and enough paste to coat the aggregate particles. If this condition is obtained, a pumpable, concrete can be realized even with relatively poorly graded coarse aggregate. However, the better the overall grading, the better the pumping results will be.

If harsh, angular coarse aggregate must be used, it is especially important to make sure that both the fine and coarse are well graded, as previously described. A somewhat higher cement content and percentage of sand than normal will be helpful, and the maximum size of coarse aggregate should be about ¾ in. Entrainment of about 4 percent air will be beneficial.

A plastic, workable mix of about 2½ to 5 in slump is best. Visual inspection of the concrete as it emerges from the line assists in evaluating plasticity. The above slump limits are broad enough to cover any normal material and job conditions but should not be used as limits on any one job. Once a plastic, workable mix has been developed every effort must be made to maintain all conditions within reasonably close limits so as to keep a supply of suitable concrete coming to the pump. A water-reducing admixture, if used under careful technical control, is desirable in order to keep the amount of water as low as possible. If the slump is too low, the friction of pumping is higher, resulting in less volume of concrete per unit of pump horsepower. However, segregation tendencies are lessened. Mixes that are too wet, on the other hand, are more apt to segregate, a condition that is liable to result in line blocks or plugs.

Cement content is usually around 470 or 520 lb/yd^3, with a lower limit in the neighborhood of 420 lb/yd^3. Richer mixes can be pumped successfully. Pumps can handle 1½ in aggregate mixes, but ¾ or 1 in mixes are commonly used.

In general, pumping difficulties that are caused by the concrete mix result from attempting to pump concrete with too much or too little slump; harsh, unworkable mixes caused by poorly graded or angular aggregates; concrete containing porous, highly absorptive aggregate, or aggregate too large for the machine or conduit.

A loss of slump during pumping is normal and should be taken into consideration when proportioning the concrete mixture. A loss of 1 in per 1000 ft of conduit is not unusual, the amount depending on ambient temperature, length of line, pressure used to move the concrete, and moisture content of aggregate at time of mixing. The loss is greater for hose than for pipe, and is sometimes as high as ¾ in per 100 ft.

Air-entrained mixes can be successfully pumped, although some operators recommend slightly less air than is normally desired for durability of the concrete. There is probably a slight loss of air during pumping, and evidence indicates that the concrete, while under pressure in the pipeline, suffers a temporary loss of workability because of compression of the entrained air.

36.4 Pumping

Before pumping of concrete is started, the conduit should be primed by pumping a batch of mortar through the line to lubricate it. A rule of thumb is to pump 5 gal of mortar for each 50 ft of 4 in hose, using smaller amounts for smaller sizes of hose or pipe. Dump concrete into the pump hopper before the last of the mortar disappears into the pump loading chamber, pump at slow speed until concrete comes out the end of the discharge hose, and then speed up to normal pumping speed. Once pumping has started, it should not be interrupted if at all possible as concrete standing idle in the line is liable to cause a plug. Of great importance is to keep concrete in the pump receiving hopper at all times, which makes necessary the careful dispatching and spacing of ready-mix trucks. From the standpoint of the pumper, it is better to have a ready-mix truck stand by for a few minutes rather than to shut down the pump to wait for concrete. However, having ready-mix trucks standing by waiting to discharge their concrete is neither efficient nor desirable.

Other causes of line blocks are slump too high; harsh, unworkable mix resulting from poor aggregate grading; mix too dry or undersanded; bleeding of the concrete; a long line exposed to the hot sun; improper adjustment of the pump valves; dirty and dented pipe sections; or a kinked hose.

With heavy equipment, a line plug can cause a delay of an hour or more while the pipe sections are separated and the plug is located. With a 10 ft section of 8 in pipe full of concrete weighing over 200 lb, the amount of labor involved can be appreciable. During an extended delay at the placing end, it is good practice to run the pump for a few strokes every few minutes, even though the concrete has to be wasted, in order to avoid a plug in the line. This is especially necessary in hot weather. If the pump can be reversed, thus reversing the flow of concrete, an attempt can be made to free the blockage in this manner. If the blockage cannot be freed after one or two attempts, the effort should be abandoned and the blockage cleaned from the system by other means. The plug in the pipe can be located by tapping on the pipe, starting at the pump. The dull thud of the tapping changes to a more hollow sound after the plug has been passed. Another method is to loosen but not release the couplings between pipe sections. Grout or mortar will spurt out if the coupling is between the pump and the blockage; after the blockage has been passed, the grout will not escape.

Concrete can be pumped upward, but downhill pumping has a few special problems, because the concrete is apt to separate or segregate in the pipe unless there is resistance to pump against. Resistance can be provided by a valve at the discharge end that can be adjusted to restrict the flow of concrete, or by inclining the final lengths of pipe upward.

If it is necessary to run the line up and over an obstruction, it is advisable to install an air-release valve at the highest point of the line to prevent an accumulation of air that could result in a line plug.

When nearing the end of a placement, with a heavy pump, the amount of concrete in the line should be computed, using Fig. 36.8 for this purpose, and the supply of concrete cut off in sufficient time to avoid excessive waste of a line full of concrete. Cleaning of the last of the concrete out of the line is accomplished with a "go-devil." When using air, there is no problem of disposal of the water following the go-devil, but care is necessary to prevent the go-devil from coming out of the end of the pipe like a bullet. As the go-devil nears the end of the line (determined by tapping on the pipe) the air pressure should be progressively lowered. Near the end, the resistance to the go-devil decreases to the point where the air can be shut off entirely, and the expanding air will force it out.

The use of water under pressure does not present the safety problem that compressed air does. However, arrangements must be made at the forms to dispose of the water outside the forms. This can sometimes be done by breaking the line before it comes over the forms.

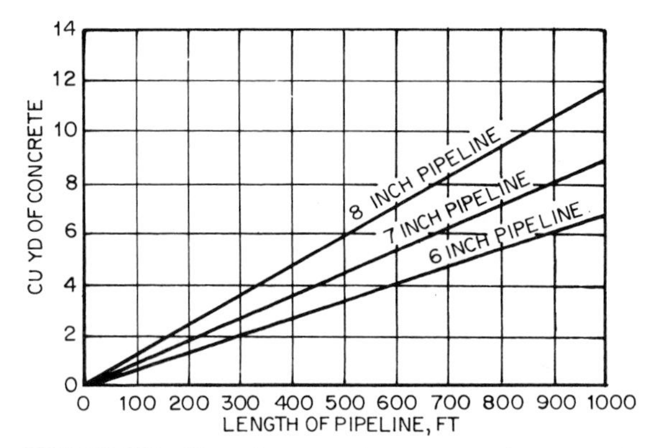

FIGURE 36.8 Chart for computing amount of concrete in pipelines of various diameters and lengths.

Small-line pumps are also cleaned by pumping clean water. This is accomplished by placing a large sponge or rubber flush plug in the feed end of the conduit. (The sponge is inserted in the pump section line of the vacuum roller machine.) After being washed out, the hopper is filled with water and the pump is started, forcing the remaining concrete out of the conduit followed by the sponge or plug. One type of pump has a separate washout pump that supplies water to force the plug through the conduit. All parts of the pump that come in contact with the concrete, including the hopper, heads, and cylinders, must be cleaned of concrete and mortar.

Users of the vacuum squeeze pump have developed a booster system for pumping concrete for high-rise construction in which the suction line of one pump is connected directly to the discharge line of the previous pump, all pumps in the system being controlled from one central panel. In this way, it is possible to maintain a steady maximum output of concrete regardless of height or distance. There are no hoppers on the booster pumps. The entire system can be reversed to simplify cleanup and discharge of wash water at ground level.

Lightweight structural concrete can be successfully pumped. The main problem in pumping lightweight concrete has been the loss of plasticity of the concrete resulting from absorption of water by the dry, porous aggregate particles when the concrete undergoes pressure in the line. Some operators report a slump loss of about 3 in between the time the concrete is dumped into the pump hopper and when it comes out of the hose. This has made necessary a slump of 7 in or more at the pump. Loss of slump can be minimized by prewetting the coarse aggregate, accomplished by some aggregate producers by vacuum and other treatments at the manufacturing plant. Success is more likely if natural sand can be used for the fine aggregate.

PNEUMATICALLY APPLIED CONCRETE AND MORTAR

Sprayed dry-mix mortar has been used for many years for both new construction and repair of old structures. Equipment is also available for the pneumatic delivery and application of wet concrete mixes containing coarse aggregate as large as 1 in. No formwork is necessary in many instances; yet intricate shapes and thin overlays can be successfully constructed, provided good materials are used and proper procedures are followed.

36.5 Dry-Mix Shotcrete

Description. When cement mortar is sprayed on a surface, the product is variously known as pneumatically applied mortar or shotcrete. The process consists essentially of mixing dry sand and cement in a mixer, then placing this mixture in the delivering equipment, the first part of which is a vertical double-chambered vessel (known as the "gun") wherein the mixture is placed under pneumatic pressure (Fig. 36.9). Under pressure, the mixture flows through a rubber hose to the nozzle, where water joins the material, wetting the mixture as it leaves the nozzle under high velocity. Compaction is achieved by the force exerted by the impact of the mortar on the receiving surface.

Shotcrete produces a high-quality material with the following desirable properties:

1. Low water-cement ratio, resulting in high strength and low permeability.
2. Dense concrete, as a consequence of low water content and high impact velocity.
3. Superior bonding ability, making it especially suitable for repair work on many types of structures.
4. Relatively simple work-area requirements. An air compressor and source of water are necessary, but buckets, cranes, elevators, trucks, and similar equipment are not necessary. Concrete (or mortar) is easily transported through a hose from the gun to the site of application.
5. Shotcrete lends itself to the production of many shapes and thin sections with a minimum of formwork, or no formwork at all.
6. Resistant to weathering and many types of chemical attack. Good abrasion resistance.
7. With proper aggregates, has good refractory properties.

FIGURE 36.9 A small mortar-gunning machine with a rated capacity of $2\frac{1}{2}$ yd^3/h.

Uses. Application of mortar by the pneumatic method is adaptable to either new construction, the application of a coating or covering, or repair of existing structures.

New construction includes linings for canals, reservoirs, tunnels, and pipe; thin slabs, walls, and domes; erosion control on earth slopes; and swimming pools.

Coatings may be applied to deteriorated concrete or masonry, after removal of unsound, deteriorated material; to rock surfaces to prevent scaling or disintegration of newly exposed surfaces; to steel and timber for fireproofing; to pilings for encasement. Coatings are used for refacing dams and for roof protection in mines and tunnels.

Materials. Any standard cement conforming to ASTM Designation C150 or

TABLE 36.3 Combined Grading for Fine-Aggregate Shotcrete*

	% passing	
Sieve size	Minimum	Maximum
⅜ in	100	
No. 4	95	100
No. 8	80	100
No. 16	50	85
No. 30	25	60
No. 50	10	30
No. 100	2	10

*Moisture content 3–6%; fineness modulus 2.15–3.38.

Source: Table 2.1, ACI 506R85, Gradation No. 1.

C595 may be used. Normally Type I or Type II is used, unless another type is specified. Aluminous cement is used for refractory applications.[8]

Aggregate should conform to the requirements of ASTM Designation C33, or C330, except for grading (see Table 36.3). (Some operators use plaster sand for the finish cover coat.) To prevent segregation and to ensure efficient mixing and application, the sand should have between 3 and 6 percent moisture.

Any potable water suitable for use in concrete can be used in shotcrete. A uniform supply at steady pressure is necessary.

Reinforcement consists of bars, welded mesh, or expanded metal lath, depending on the nature of the structure, thickness of the coating, and extent of the work. Mesh and lath may be self-furring, or used with chairs. Steel must be free of loose scale, heavy rust, oil, paint, or any material that will interfere with bond.

Mixes. Usual mixes consist of 1 part cement to 4 or 4½ parts of sand by dry loose volume. Strength of these mixes will normally be between 3000 and 5000 psi at 28 days, or perhaps slightly higher for the 1:4 mix. Leaner mixes are sometimes used, but the amount of rebound is considerably higher with the lean mixes. Water-cement ratio is from 0.30 to 0.50 by weight.

Preparation of Surface. Forms, when required, must be substantial, true to line and grade, and so constructed as to permit the escape of air and rebound. When constructing a wall, only one side needs to be formed. Columns are formed on two adjacent sides, and beam forms usually consist of one side and the soffit. Sufficient grounds must be provided to establish the surface and finish lines of the work. Ground wires should be taut and true to line and surface, and must be maintained in a taut condition until the finish coat has been applied.

In repairing old concrete or masonry, all the old unsound material must be removed so that the shotcrete can be applied to a sound surface. All weathered and unsound concrete or masonry should be chipped off with pneumatic tools, using sawtooth bits and gads instead of chisels or sharp points. The entire surface should be sandblasted in areas where chipping is not required. Finally, the surface is cleaned with compressed air and water. Heavily corroded steel should be sandblasted. About an hour before application of mortar, concrete and masonry should be wetted, but not saturated; otherwise suction will be reduced. Surface should be moist, but not wet, when the shotcrete is applied.

Earth surfaces to receive shotcrete should be well compacted, trimmed to line and grade, moist, and free of frost.

Water seepage or leakage must be removed by installation of drainpipes to take the water away from the surface being shot. Seeps in rock or old concrete can be sealed with a quick-setting mortar, such as a mixture of portland and aluminous cement with 2 parts sand, portland cement mortar with 5 to 10 percent (by weight of cement) of sodium carbonate, or certain proprietary compounds.

Reinforcement must be rigidly anchored in place so it will not move. In placing reinforcement for a repair, the steel can be wired or welded to existing reinforcement, or anchored to dowels secured in holes drilled into the concrete. Cement mortar or epoxy mortar can be used to grout the dowels in the concrete. Joints in reinforcing bars should be lapped 40 diameters, and mesh should be lapped one full mesh. Reinforcement should be furred at least ½ in from the surface and covered with a minimum of ¾ in of mortar on surfaces not subject to fire hazard. Fire-hazard exposure should have the same cover as cast-in-place concrete.

Equipment. Mixers are of either the rotating-drum or paddle type. Some rigs include a mixer as part of a self-contained unit. More elaborate rigs include mixer, elevator, air compressor, gun, and necessary auxiliary parts, such as a hose reel, all mounted on a truck or trailer. Placement capacities of 20 yd^3/h are common.

The commonest gun is the vertical double-chambered type (Fig. 36.9). The dry-mixed mortar or concrete is admitted to the upper pot while the gate between the two pots is closed. The charging port is then closed, the upper pot is pressurized, and the transfer gate is opened, allowing the material to flow by gravity into the lower chamber. By means of a feed wheel or similar device, the mix is fed into the delivery hose, where a stream of air carries the material to the nozzle. As soon as the upper pot is empty, the transfer gate is closed, pressure is released on the upper chamber, a new batch of material is admitted through the charging port, and the cycle is repeated. Meanwhile, pressure is maintained on the lower pot, which continues to deliver an uninterrupted flow of material to the hose.

Another type of gun makes use of a screw to feed the mixture to the delivery hose, where the air stream picks up the mixture and delivers it to the nozzle.

Special hose in 1¼ to 2 in inside diameter is available. Most manufacturers rate their machines with 150 or 200 ft hose. However, hose lengths of 500 ft are frequent, and lengths of 1000 ft have been used on occasion. The Gunite Contractors Association recommends[3] not less than 365 ft^3 of free air per minute at a minimum pressure of 45 psi in the gun chamber for proper placement and adequate blowout jet requirements. If more than 100 ft of hose is used, air pressure should be increased by 5 psi for each additional 50 ft of hose. However, a maximum of 75 psi is usually adequate.

Water is conveyed through a separate hose to the nozzle under a pressure of about 15 psi higher than the air pressure in the hose at the gun. Radial perforations in the water ring at the nozzle spray water into the cement-sand mixture as it passes through the nozzle.

Application. Before starting to shoot dry-mix shotcrete, precautions should be taken to protect property in the area. Adjacent construction, openings, shrubbery, and all areas that might be discolored or damaged by rebound, cement, water, or dust must be covered with tarpaulins or plastic sheets to protect them from damage.

The amount of water added at the nozzle should be a minimum but should be sufficient to prevent excessive rebound and ensure hydration of the cement.

The nozzle should be held as nearly perpendicular as possible to the surface being treated and should be held uniformily the same distance from the surface at all times. For a large nozzle, the distance is 2 to 6 ft, ranging down to as close as 12 in for a small nozzle in close quarters.

When enclosing reinforcing steel, the stream from the nozzle should be directed at an angle but never more than 45° so as to fill the space behind the bars, shooting from each side of each individual bar. The nozzle operator's helper should use an air jet to blow out rebound ahead of the application of shotcrete.

The flow of shotcrete out of the nozzle should be uniform, without slugs of wet or dry material. If the nozzle starts slugging, it must be turned away from the work until the flow becomes steady again. Slugs are usually caused by insufficient air for the amount of material being handled or the length of the hose being used.

Dry-mix shotcreting should not be done during a rainstorm, as the rain will wash cement out of the material, nor should it be done during high winds which blow against the nozzle spray and make control difficult. Work should be suspended during freezing weather, unless operations can be protected from the weather.

Shotcrete requires expert and conscientious work and skill and should be done under careful supervision and inspection with experienced workers. The ACI Guide to Shotcreting[2] points out the need for skill and experience in the recommendations described in the following paragraphs.

The crew should consist of a supervisor who can double as the nozzle operator, a finisher, and an assistant nozzle operator and a gun operator and mixer operator. If a project has more than one crew, a superintendent or project engineer may be required.

The supervisor, who is usually proficient in all positions and should have a minimum of 3000 h of experience, is responsible for the equipment, quality control, and ordering of the material.

The nozzle operator, who should be certified by the process enumerated in ACI 506.3 or have a minimum of 3000 h of shotcrete experience and have completed one project as a nozzle operator, is responsible for approving the area for shotcrete placement as to being clean, sound, and free of loose material, and that the anchors, reinforcing, and ground wires are properly spaced and secure. Control of water, air pressure, and removal of rebound and sag are also the nozzle operator's responsibility.

The finisher trims the shotcrete to line and grade; locates and corrects sand pockets, sags, and areas that are too dry; and brings attention to low areas. The finisher should have sufficient experience to substitute for the nozzle operator when required.

The assistant nozzle operator is responsible for dragging the hose, relays messages between the nozzle operator and the gun operator, blowing away the rebound ahead of the nozzle operator. This assistant should be an apprentice nozzle operator who has had sufficient experience to substitute in that capacity.

The gun operator must be familiar with the shotcrete-delivery system and preparation of the materials.

On vertical or overhanging surfaces, the mortar is usually applied in layers not exceeding 1 in in thickness. On horizontal or nearly horizontal surfaces (when shooting down), the thickness may be as much as 3 in. Excessively thick layers will cause the mortar to slough, or sag. Sagging may also occur if insufficient time elapses between layers. Time between successive layers should be at least 30 min but not so long as to permit the previous layer to set completely, as the surface is apt to glaze, with a resulting loss of bond between the layers. Light brooming of the surface will remove any rebound and assist in bonding the next layer of mortar.[3]

Rebound. A portion of the mortar bounces from the surface where it is being applied, the amount varying with air pressure, quality of sand, placement conditions, and the cement and water contents. The amount of rebound varies from 25 to 50 percent when shooting overhead, 15 percent or less when shooting down, and 15 to 25 percent on vertical or sloping surfaces. It is higher with high nozzle velocity and decreases with higher water content. This rebound material, consisting mainly of coarse sand particles, should not be reused because of its variable quality, low cement content and possible contamination.

Rebound is higher when a coarse sand is used, when a lean mix is used, or when the water content of the mix is too low. Frequent moving or "waving" of the nozzle, or shooting at an angle, increases rebound. Sometimes holding the nozzle close to the surface being shot decreases rebound, but this is apt to result in a rough or irregular surface. Use of a wetting agent or water-reducing admixture, introduced with the sand, sometimes reduces rebound.

Finishes. In many applications, the surface produced by the shooting is satisfactory. Tunnels, channel linings, ditches, and dam facings are a few of the structures frequently treated in this manner. Usual practice is to apply the finish coat by starting at the top and working down to avoid shooting over the finished work.

If finishing is required, the first operation after shooting is to rod (screed) off to the ground wires or other guides, then remove the grounds, after filling in low spots revealed by the rodding. The thin-edged rod is worked back and forth with a slicing motion, starting at the bottom of the area. Sometimes this is all the finishing that is required. Additional finishing might consist of brooming, sacking, floating with a rubber or wood float, or troweling. The amount of pressure required is normally less than that required for regular concrete finishing.

Curing. A fine water spray, or fog, should be applied to the surface of the shotcrete as soon as possible, to prevent drying out. The initial application of fog should be just enough to maintain the moisture in the mortar without wetting the surface. As soon as the surface has hardened, a heavier application of water is permissible, keeping the shotcrete wet for a period of 7 days, depending on weather conditions, or as required by the specifications.

Hot, dry weather, especially when accompanied by wind, causes the surface of shotcrete to dry very rapidly, leading to cracking, spalling, scaling, and low strength. Under these conditions the fresh shotcrete must be covered with burlap or similar material kept damp for the required curing period.

Liquid membrane-forming compounds are satisfactory for curing, and should be applied after first dampening the surface with water. The surface must be dark in color but not shiny-wet when the compound is applied.

Testing. Shotcrete can be tested by either of two methods.[6] One is to cut three cores from the completed shotcrete for at least each 50 yd^3 placed or once each shift. These cores are to be soaked in water a minimum of 40 h before testing. The other method is to make a test panel of 18 × 18 × 3 in, gunned in the same position as the structure, for each 50 yd^3 of shotcrete placed or at least once each shift. The panel is to cured as the structure, except that the cores should be cured for a minimum of 40 h prior to testing.

The designated testing agency will cut three 3 in diameter cores or 3 in cubes from each panel for testing in accordance with ASTM C42. The average compressive strength of the three cores from the structure or panel must equal or exceed $0.85f'_c$ with no one core of the three less than $0.75f'_c$. The average of three cubes from a panel must equal or exceed f'_c with no single cube out of the three less than $0.88\,f'_c$.

The ICBO Uniform Building Code[9] allows the shotcrete panels to be 12 × 12 in and the test samples from the structure or panels to be cores of 2 in diameter or 2 in cubes when the maximum size of the shotcrete aggregate is less than ⅜ in.

The presence of hollow spots or areas containing an excess of rebound included in the shotcrete can be detected by sounding the surface of the finished work with a hammer. Imperfections should be removed and replaced with new shotcrete.[5]

36.6 Refractory Applications

Shotcrete is sometimes used for lining smokestacks, breechings, and process equipment in furnaces, steel mills, oil refineries, and similar installations, making use of aluminous cement and certain selected aggregates (Table 36.4), where resistance to high temperature is an important factor. Mixes are usually about 1:3 or 1:4 by dry loose volume. A small amount of plastic fireclay (not over 12 lb per cwt of cement) may be added to improve workability. There are available on the market premixed dry-packaged castable refractories, consisting of aluminous cement and aggregate, suitable for gun application.

Steel must be clean, and a reinforcing mesh should be placed about 1 in away from the steel shell, covered with at least 1 in of shotcrete. Mesh is usually 3 × 3, 10−10, or 2 × 2, 12−12, rigidly supported.

After application, the shotcrete should be permitted to harden and followed immediately by water curing for the first 24 h in order to prevent the temperature of the refractory shotcrete from exceeding 77°F when high-alumina cement is used.

36.7 Wet-Mix Shotcrete

The previous discussion in this chapter has dealt with dry-mix shotcrete, a product that has been used for many years. Equipment for pneumatic application of premixed mortar and concrete is also commonly available. The principal advantage of wet-mix shotcrete over dry-mix is that the amount of water can be established beforehand and maintained at this value during the gunning operation. Gunning capacity is about the same as for dry-mix, a maximum of about 20 yd^3/h (Fig. 36.10). Pumps are of the auger type, in which the mixture is fed into the hose by means of a screw. A high-velocity flow of air conveys the mixture to the nozzle where it is shot onto the receiving surface. Another type consists of a pressurized tank in which rotating mixing paddles intermittently introduce air with the material into the hose or pipe. These machines handle concrete containing 470 lb of cement per cubic yard and ¾ in aggregate, with a zero slump. Recommended combined gradings for shotcrete can found in Table 36.5.

TABLE 36.4 Aggregates for High-Temperature Exposures*

Material	Max temp, °F	Strength	Conductivity	Abrasion resistance	Insulating value	Corrosion resistance	Volume stability
Siliceous sand	500	High	High	Good	Poor	Good	Poor
Calcined clay or shale	2000	High	Medium to low	Good	Good	Good	Fair to Good
Vermiculite and perlite	2000	Low	Low	Poor	Excellent	Good	Good
Traprock (igneous)	1800	High	High	Good	Poor	Good	Fair
Crushed firebrick	2500	Medium	Medium	Fair	Fair	Fair	Good
Granulated blast-furnace slag	1000	Low	Low	Good	Good	Good	Fair

*Ratings are comparative only and may vary widely for different materials.

FIGURE 36.10 Application of wet-mix shotcrete to a ditch lining.

Much of the previous discussion covering surface preparation, application, finishing, and curing of dry-mix shotcrete also applies to wet-mix shotcrete.

36.8 Fiber-Reinforced Shotcrete[4]

Fiber-reinforced shotcrete is essentially a conventional shotcrete with steel, glass, or polypropylene fibers added. Steel fibers have had the most use in shotcrete with the majority in dry-mix shotcrete. Originally, mixtures have incorporated a sand/cement ratio of 2.4:1 by weight (940 lb cement per cubic yard) but with the use of ⅜ and ¾ in aggre-

TABLE 36.5 Recommended Limits for Combined Aggregate Gradings

Sieve size, U.S. standard square mesh	Percent by weight passing individual sieves		
	Gradation 1	Gradation 2	Gradation 3
¾ in	—	—	100
½ in	—	100	80–95
⅜ in	100	90–100	70–90
No. 4	95–100	70–85	50–70
No. 8	80–100	50–70	35–55
No. 16	50–85	35–55	20–40
No. 30	25–60	20–35	10–30
No. 50	10–30	8–20	5–17
No. 100	2–10	2–10	2–10

Source: Table 2.1 ACI 506R-85, "Guide to Shotcreting."

gate, the cement content has been reduced. Fibers have been added in the amounts of 0.5 to 2 percent by volume. Sizes have been from ½ to 1½ in in length.

Batching and mixing has been accomplished by mixing the dry materials in a transit mixer and delivering the mixture to the shotcrete hopper. Fibers having a clumping capability should be added through a screen into the hopper. A good electrical grounding of the gun and nozzle has been found to greatly reduce the fiber clumping and plugging. The practice for handling wet-mix shotcrete fibers is similar to that for concrete. the fibers should be added to the concrete mixing in the transit mixer and not on the vanes. Mixing should be fast enough to keep the fibers from piling up.

The application of the shotcrete is similar to normal practices except for modifications needed to minimize clumping of the fibers. They are elimination of 90° elbows, abrupt changes in hose size and adding vibrators to the hopper. The hose diameter should be twice the fiber length. Generally, more fibers rebound than aggregate. To reduce rebound of the fiber, strategies such as reducing the air pressure, air velocity, or the amount of air at the nozzle; making the mix finer or using smaller aggregate; changing to shorter, thicker fibers, predampening the aggregate to get the proper moisture content; and wetting the mix without creating sagging or sloughing should be considered.

36.9 Polymer Portland Cement Shotcrete

The addition of certain latex additives to a normal portland cement shotcrete mix has improved flexural and tensile strengths, bond, and reduced absorption so that there is increased resistance to chloride penetration. These characteristics may increase the use of shotcrete for the repair of concrete in concrete bridge and marine environments, plus structures subject to chemical attack.

36.10 Gun Casting

In place of the standard nozzle, a casting head used with standard dry-mix equipment will slow down the velocity of the material at exit. The result is less rebound, minimal finishing, efficient use in close-confined quarters, minimal forming requirements, and a reduction in labor and material costs.

36.11 Certification of Shotcrete Nozzle Operator

The American Concrete Institute Committee 506 has developed a certification procedure for shotcrete nozzle operators.[7] The program was provided for use by specifiers, public agencies, shotcrete applicators, concrete users, and testing laboratories to increase their knowledge and as a means of establishing such a certification procedure. The program consists of a training and study period, followed by a certification examination having a written or oral examination and a field demonstration test.[7] The certification should not be used to eliminate preconstruction samples required by the specifications.

REFERENCES

1. Placing Concrete by Pumping Methods (ACI 304.2-71) (rev. 1982), American Concrete Institute.
2. Guide to Shotcreting (ACI 506-85), American Concrete Institute.

3. Gunite, *Brochure G76,* Gunite Contractors Association, Burbank, Calif.
4. State of the Art Report on Fiber Reinforced Shotcrete (ACI 506.1R-84), American Concrete Institute.
5. "Manual of Concrete Inspection," 7th ed., SP-2 (ACI 311.1-81) (reapproved 1985), American Concrete Institute.
6. Specification for Materials, Proportioning, and Application of Shotcrete (ACI 506.2-77) (rev. 1983).
7. Guide to Certification of Shotcrete Nozzlemen (ACI 506.3R-82), American Concrete Institute.
8. Robson, T. D.: "High-Alumina Cements and Concretes, Wiley," New York, 1962.
9. Uniform Building Code, 1991, International Conference of Building Officials, Whittier, Calif.

CHAPTER 37
VACUUM PROCESSING OF CONCRETE

Joseph J. Waddell
Joseph A. Dobrowolski

Vacuum processing of concrete is accomplished by applying a vacuum to surfaces of fresh concrete, either formed or unformed. This patented process removes as much as 40 percent of the water from a few inches of the concrete surface, producing in effect a "casehardened" concrete. At a depth of 6 in it is common to remove 20 percent of the water. In addition, air bubbles or "bug holes" are removed from the surface, as shown in Fig. 37.1.

The effect of vacuum processing is one of compaction; the cement paste is consolidated or densified on the removal of water from the mass. Removal of air is a surface phenomenon only, but the effect of vacuum in removing water is effective for a depth of as much as 12 in. This compaction and lowering of the water/cement ratio results in markedly greater compressive strength; up to an age of 7 days the strength of vacuumed concrete is as much as 100 percent higher than unprocessed concrete, and normal 28-day strength will be reached in 8 or 10 days. Final strength will also be higher. The curves in Fig. 37.2 compare 28 day strengths for vacuum and plain concrete at different water/cement ratios. Durability is improved, the resistance to erosion, abrasion (see Fig. 37.3), and cycles of freezing and thawing being greater than for plain concrete. Shrinkage and cracking tendencies are reduced.

Another advantage of vacuum concrete is an early solidification or apparent hardening of the concrete that in some instances permits removal of forms in a matter of minutes instead of hours. Figure 37.4 shows that vacuum processing reduces pressure on the form practically to zero for each lift processed, but the untreated concrete is still exerting pressure on the entire height of form after comple-

FIGURE 37.1 Surface finish of vacuum concrete cast against forms. The lower portion was cast against ordinary lumber forms, and the upper portion was vacuum-processed. Note absence of pits in the vacuumed surface. (*Courtesy of Aerovac Corp.*)

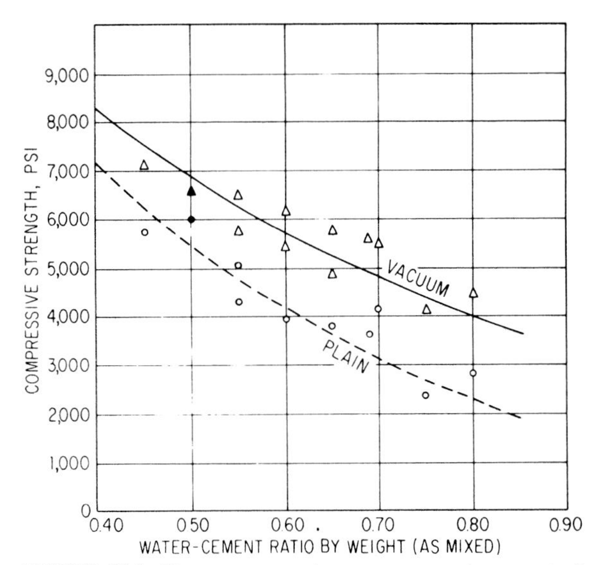

FIGURE 37.2 Vacuum treatment improves compressive strength of concrete at all water/cement ratios.

tion of placing. Similarly, finishing of slabs, as illustrated in Fig. 37.5, can start within a few minutes after placing the concrete and application of vacuum. Early strength is especially desirable in precast work, as it permits removal of forms as soon as processing has been completed, thus releasing the equipment for the next round of casting.

The value and extent of early form removal and early finishing must be considered in connection with any other advantages when determining whether to use the vacuum pro-

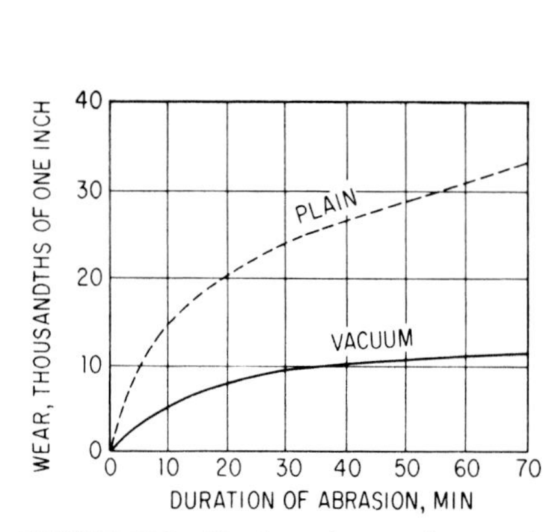

FIGURE 37.3 Abrasion resistance of concrete is improved by vacuum processing.

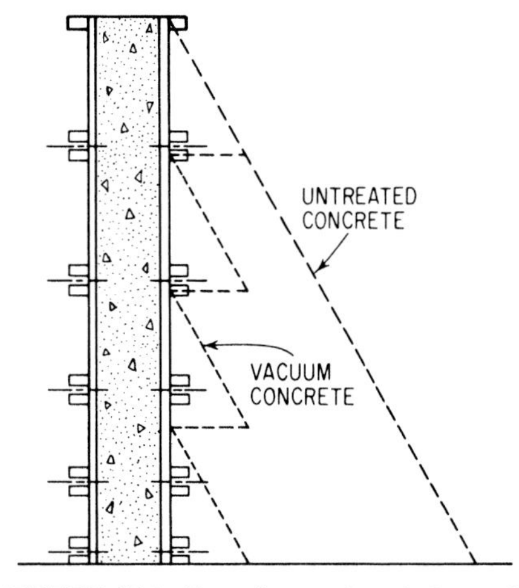

FIGURE 37.4 Ten minutes after placing each lift of concrete, vacuum processing reduces pressure against the form to zero by reducing slump to zero.

FIGURE 37.5 Vacuum pads, shown in the center, have been applied to the fresh concrete. After processing, the concrete is immediately finished. (*Courtesy of Kalman Floor Co.*)

cess, as there are some limitations. If careful formwork and proper vibration will do the job, there is no advantage in applying a vacuum treatment. It is difficult to justify vacuuming merely for improving the appearance of concrete, and usually proper air entrainment will provide all the durability that is required, although it has been reported that entrained air and vacuum processing together give better durability than either treatment alone. Because of the time involved, there is little benefit in attempting to process to a depth greater than 12 in.

Normal concrete mixes can be successfully vacuumed, although best results are obtained when the fines are at a minimum, that is, for relatively lean mixes containing the minimum practicable amount of sand, and with the coarser sands. Uniformity of slump from batch to batch is important, and variations should not exceed 1 in. Vacuum should be applied as soon as possible after the concrete has been placed, while it is still plastic. Vibration of the concrete during the first few minutes of processing is desirable, as this tends to close the small channels in the concrete that are formed as the vacuum draws the water out, thus providing improved watertightness. Efficiency of the vacuum system is somewhat higher at lower temperatures.

Because of the original vacuum processing equipment for slabs being cumbersome, expensive, and complicated, the use of vacuum processing declined in the United States during the 1960s and 1970s. Within the last few years, the equipment has been simplified and its use in Europe, especially in Scandinavia, increased in frequency. The vacuum process is now being used more frequently in the United States.

After the concrete slab is vibrated, the surface is immediately covered with a base filter pad and a suction mat connected to a vacuum pump. The pump (Fig. 37.6) is started to create a vacuum under the mat (Fig. 37.7). This causes the atmospheric pressure to compress the concrete and extract the water. The base pad is designed to distribute the vacuum over the entire surface evenly and allow passage of the water.

FIGURE 37.6 Vacuum pump for horizontal surface. *(Courtesy of Kalman Floor Co.)*

FIGURE 37.7 Vacuum mat for processing inclined or horizontal formed surface. Lining of lath, screen, and filter cloth is applied to the rectangular areas. Vertical and horizontal seal strips limit height and width of lifts subjected to vacuum processing.

FIGURE 37.8 Removing the filter cloth and vacuum bags in preparation for finishing procedures. *(Courtesy of Kalman Floor Co.)*

The vacuum is applied for about 3 to 5 min for each inch of slab thickness. The areas between mats is dried with burlap and cement. After the pads and mat are removed, the surface is firm enough to walk on and finishing procedures begin (Fig. 37.8) with specially designed low amplitude, high-frequency disk floats.

The slabs are water- and paper-cured for 7 to 10 days. Properly done, the process results in hard dustproof floors which require no sealers and little maintenance.

CHAPTER 38

PREPLACED-AGGREGATE CONCRETE

Joseph J. Waddell
Joseph A. Dobrowolski

Originally developed for structural concrete repairs, preplaced- or prepacked-aggregate concrete, or prepacked concrete, which is composed of graded coarse aggregate solidified with a special portland-cement grout, has been found to be suitable for use in certain types of new construction as well.[1] Preplaced-aggregate concrete is a system in which coarse aggregate is placed in the form and a special grout is then pumped into the form, starting at the bottom, to fill the voids and create a solid concrete unit. Some of the equipment, procedures, and materials are patented, and the user should check with authorized installers before using this method.

One of the main advantages claimed for preplaced-aggregate concrete is the ease with which the concrete can be placed in locations where conventional concrete could be placed only with extreme difficulty. In some cases on jobs where ready-mixed concrete is not readily available, preplaced-aggregate concrete can effect a saving in time and money. The point-to-point contact of the preplaced-aggregate particles and the special properties of the grout reduce shrinkage to a negligible amount. Drying shrinkage, following proper curing, is normally in the range of 200 to 400 millionths as compared with ordinary concrete, which ranges between 400 and 600 millionths.

Preplaced-aggregate concrete has been used for repairing damaged or deteriorated bridge piers, for refacing dams and spillways, for repairing structural concrete, for pile encasements and in areas heavily reinforced for seismic or other reasons. It is well suited to underwater construction and cofferdam seals. In some usages, such as in mines and tunnels, waste rock can serve as coarse aggregate. Special procedures for repair methods are described in Chap. 46.

A common usage of preplaced-aggregate concrete is in constructing high-density biological shielding of nuclear reactors and other equipment making use of radioactive substances. A portion of one of these structures, with preplaced aggregate in place, is shown in Fig. 38.1. Heavy aggregates such as barite or iron ore are required which, because of their heavy weight, angular shape, and friable nature, are extremely hard to handle. Hand placing of the aggregate is sometimes necessary because of the congestion of pipes, instruments, reinforcement, and similar gear within the form.

About 450,000 yd^3 of prepacked concrete was used in the substructure for the Mackinac Bridge. All 32 piers below water, and much of the above-water concrete in the prin-

FIGURE 38.1 Preplaced aggregate in an atomic-reactor shield, ready to receive grout.

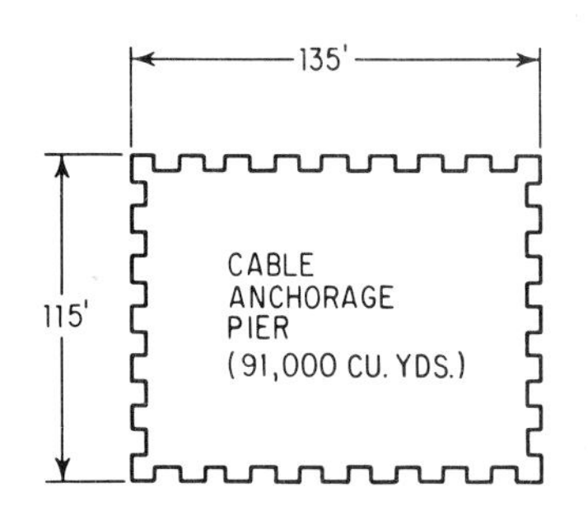

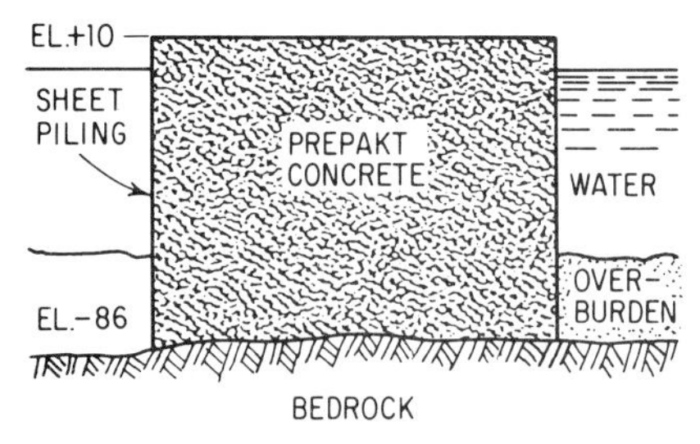

FIGURE 38.2 One of the large piers for the Mackinac Bridge which was placed in one continuous operation by the prepacked method.

cipal piers, employed prepacked concrete. Figures 38.2 through 38.4 show some of the details of this construction, which is typical of the large-scale work sometimes performed.

Figure 38.5 shows a cofferdam cell for the Columbia River Bridge at Astoria, Ore. This was typical cofferdam construction, in which washed coarse aggregate was placed below water inside the cofferdam, and grout was then injected from the bottom up through the pipes shown, displacing the water upward and out.

A typical repair job was the refacing of a railroad grade separation in which scaled and disintegrated concrete was first removed from the wing wall, forms were erected (Fig. 38.6), aggregate was placed, and finally grout was pumped into the bottom of the form. As the grout rose inside the form, it filled the voids in the aggregate, resulting in a solid concrete facing for the wall. Bond with the old concrete was assured by removing all old disintegrated concrete down to solid, sound concrete, and by virtue of the small amount of shrinkage of the preplaced aggregate facing.

FIGURE 38.3 Self-unloading boat depositing aggregate for the large pier. Note that grout pipes are in position. (*Courtesy of Intrusion-Prepakt, Inc.*)

FIGURE 38.4 Aggregate in place ready for injection of grout through the grout pipes extending above the aggregates. (*Courtesy of Intrusion-Prepakt, Inc.*)

FIGURE 38.5 Cofferdam for Pier 169 for the Columbia River Bridge. Grout will be injected through the long pipes that protrude above the river. The shorter pipes serve as gages to measure the elevation of grout as it is pumped upward in the aggregate. A floating batch plant is in the background to the right.

FIGURE 38.6 Forms erected for refacing of a deteriorated railroad wing wall.

Any of the types of portland cement complying with ASTM C150 or C595 can be used. Fine aggregate for the grout should conform to the requirements of ASTM Designation C33, except that the grading is somewhat finer, as shown in Table 38.1. A well-graded sand near the middle of the limits is best. Uniformity of grading is very important, and specifications usually place a limit on the amount of fineness modulus of a single sample taken at the mixer, which can vary from the average fineness modulus.

Coarse aggregate, which can be either gravel or crushed stone, must meet ASTM C33, and should be graded so as to have a minimum void content, usually specified to be between 35 and 40 percent after compaction, being least for a large-maximum-size smooth, well-graded, well-compacted gravel and highest for small-size, poorly graded crushed stone. The maximum size can be as large as convenient to handle, provided the maximum size does not exceed one-fourth of the minimum dimension of the portion of the structure in which the aggregate is to be used, or two-thirds of the minimum distance between reinforcing bars. The minimum size should be no smaller than ½ in when the section is not more than 12 in thick, or ¾ in for thicker sections, in order to provide ample space for the grout to flow into interstices and become distributed throughout the form. The specifications should allow a tolerance of 5 to 10 percent undersize passing the nominal minimum-size screen. Table 38.1 provides the grading limits for preplaced aggregates. Grading 1 or 2 is usually used in the Americas and the Orient, while Grading 3 is generally used in Europe. When labor cost is low, the larger aggregates are feasible because of low hand selection and placing costs.

Admixtures consist of a pozzolanic filler which imparts a degree of workability and contributes to ultimate strength of the grout. ASTM C618, Class F or N fly ash and simi-

TABLE 38.1 Grading Limits for Coarse and Fine Aggregates for Preplaced-Aggregate Concrete

Sieve size	Percentage passing		
	Grading 1 for ½ in minimum-size coarse aggregate	Grading 2 for ¾ in minimum-size coarse aggregate	Grading 3 for 1½ in minimum-size coarse aggregate
	Coarse aggregate		
1½ in	95–100	—	0.5
1 in	40–80	—*	—
¾ in	20–45	0–10	—
½ in	0–10	0–2	—
⅜ in	0–2	0–1	—
	Fine aggregate		
No. 4	—		100
No. 8	100		90–100
No. 16	95–100		80–90
No. 30	55–80		55–70
No. 50	30–55		25–50
No. 100	10–30		5–30
No. 200	0–10		0–10
Fineness modulus	1.30–2.10		1.60–2.45

*Grade for minimum void content in fractions above ¾

Source: Table 1, ACI 304.1R-92.

lar fine materials have been used for this purpose. A proprietary intrusion aid or fluidifier used in the grout to inhibit setting, lower water requirement, delay early stiffening, and entrain air usually contains a small amount of aluminum powder or similar material that reduces settlement shrinkage by causing a slight expansion of the grout just before initial set. They should meet the requirements of ASTM C937.[2]

Forms are constructed in much the same manner as good formwork for conventional structural concrete, but absorptive form liners cannot be used. Because of the fluid nature of the grout, joints between form panels must be sealed to prevent leakage. Forms should be adequately vented to allow air and water to escape.

Proportions of the grout vary, depending on the type of construction or repair being performed and the materials available. Homogeneity and uniformity from batch to batch are important. An actual mix which was used consisted of portland cement, 190 lb; sand, 3 ft^3; pozzolanic filler, 70 lb. Preplaced aggregate was ¾ to 6 in gravel and ¾ to 2½ in crushed granite. Other projects have used from 165 to 750 lb of cement per cubic yard. Standard procedures for determining proportions of the grout with the proposed aggregate are given in ASTM C938.[3] Tests to determine physical properties of the grout can be found in ASTM C939[4], C940[5], C942[6], and C943[7]. Initial ratios of cement to sand for trial are Grading 1, 1:1; Grading 2, 1:1.5; and Grading 3, 1:3.

The washed and drained coarse aggregate should be handled and placed in the forms in such a manner that segregation and breakage are minimized. Sometimes an 8 or 10 in pipe chute can be used, with rubber-belting baffles. Reinforcement and items to be embedded in the concrete are fixed in place, and vertical grout pipes, usually ¾ to 1¼ in in diameter, spaced about 5 ft apart, are installed as placement of the aggregate starts and are extended above the top of the form. Actual spacing of the grout-injection pipes is controlled by the characteristics of the section being constructed, such as size and depth of the section, size of aggregate, and proximity of forms or embedded items. In an extremely tight and congested location aggregate may be placed by hand to ensure closure around pipes and other items. The aggregate, when placed above water, can be lightly tamped or vibrated during placing to minimize voids. Vent pipes must be provided in certain types of construction where the aggregate is completely enclosed.

Grout is pumped through the injection pipes starting at the lowest point in the form. Pumping should be done slowly and at a uniform rate without interruption so the grout, as it rises in the aggregate-filled form, will completely fill all voids in the coarse aggregate. Vibration of the forms in the area of the grout surface during pumping helps to remove air bubbles and improve appearance of the completed concrete.

If the top surface is to be finished, grout is pumped into the form until the prepacked aggregate is flooded, any diluted grout is removed by brooming, and a thin layer of pea gravel is spread over the surface. After the pea gravel has been tamped down into the grout, the surface can be struck off and floated or troweled, as required.

Stripping of forms and curing are accomplished in accordance with good construction practices described elsewhere in this handbook.

REFERENCES

1. Guide for the use of Preplaced Aggregate Concrete for Structural and Mass Concrete Applications (ACI 304.1R–92), American Concrete Institute.
2. Standard Specifications for Grout Fluidifier for Preplaced Aggregate Concrete, ASTM C937–80 (reapproved 1985).
3. Standard Practice for Proportioning Grout Mixtures for Preplaced Aggregate Concrete, ASTM C938–80 (reapproved 1985).

4. Standard Test Method for Flow of Grout for Preplaced Aggregate Concrete (Flow Cone Method), ASTM C939–87.
5. Standard Test Method for Expansion and Bleeding of Freshly Mixed Grouts for Preplaced Aggregate Concrete in the Laboratory, ASTM C940–80.
6. Standard Test Method for Compressive Strength of Grouts for Preplaced Aggregate Concretes in the Laboratory, ASTM C942–86.
7. Standard Practice for Making Test Cylinders and Prisms for Determining Strength and Density of Preplaced Aggregate Concrete in the Laboratory, ASTM C943–80 (reapproved 1985).

CHAPTER 39
PORTLAND CEMENT PLASTER

Joseph J. Waddell
Joseph A. Dobrowolski

Portland cement plaster is a facing material for either interior or exterior application to almost any building surface. When properly prepared and applied, it has properties akin to those of good concrete. Stucco is sometimes meant to designate exterior plaster. In the western United States and throughout the majority of the United States, stucco is understood to mean the final coat of exterior plaster in which the color and texture of the plaster are obtained.

Plaster is mixed on the job by combining plastering sand, portland cement, and water to produce a mixture with a consistency suitable for either hand or machine application.

The chief advantage of portland cement plaster, in addition to strength and durability that make it especially suitable for exterior exposure or for interiors to be exposed to wetting or severe dampness, is its versatility. It can be used to construct free-form curved and irregular surfaces without formwork by using the reinforcement as a form. In addition to form and texture, plaster lends itself to such artistic treatments as fresco and graffito. High-rise buildings in many areas now are clad with curtain wall panels covered with portland cement plaster reinforced with glass fibers.

39.1 Materials

Cement may be blended cements meeting ASTM C595 or Type I, II, or III conforming to the requirements of ASTM C150, except in respect to the limitations on insoluble residue, air entrainment, and additions subsequent to calcination. Plasticizing agents in an amount not exceeding 12 percent by volume may be added to Type I, II, or III cements in the manufacturing process. This cement is then known as "plastic cement" and is used for hand- or gun-applied plaster. In some locations masonry cement, conforming to ASTM C91, is used. When plastic cement is not available, regular Type I or Type II cement is employed, requiring addition of a plasticizer such as hydrated lime. When lime is used with regular cement, its amount should not exceed 20 lb per sack of cement for first and second coats. Lime, conforming to ASTM C206,[5] may be added up to an amount equal to the amount of portland cement for the finish-coat plaster.

Aggregate may be normal-weight sand (ASTM C33 and C897)[4] or, where weight is of prime importance, expanded shale meeting ASTM C330 or perlite meeting ASTM C35 specifications. Sand should be washed clean, free of organic matter, clay, and loam and

TABLE 39.1 Sand Grading, Cumulative Percents Retained

Sieve size	UBC[a] and ASTM C35		ACI[b] Committee 524		Stucco Manufacturers Association[c]		PCA[d] and ASTM C897	
	Min.	Max.	Min.	Max.	Min.	Max.	Min.	Max.
No. 4	—	—	—	—	—	—	—	—
No. 8	0	5	0	10	0[e]	5[e]	0	10
No. 16	5	30	10	40	10	35	10	40
No. 30	30	65	30	65	30	65	30	65
No. 50	65	95	70	90	70	90	70	90
No. 100	90	100	95	100	90	95	95	100
No. 200	—	—	—	—	—	—	97	100

[a]Uniform Building Code (Finish coat grading to meet above except all aggregates shall pass No. 8 sieve).[1]
[b]Guide for Portland Cement Plaster, ACI 524-92.[11]
[c]Specifications for Finish Coat Stucco.
[d]"Portland Cement Plaster (Stucco) Manual," Portland Cement Association, 1980. (Fineness modulus must be between 2.05 and 3.05.)[2]
[e]No. 12 sieve.

be well graded, conforming to the specified grading, as shown in Table 39.1. The portion retained on (or passing) each sieve should be near the midpoint of the specifications. Especially to be avoided is an excess of fines. A well-graded sand has a minimum of voids, making a plaster with a minimum water requirement, other things being the same.

Quality and type of aggregate for finish-coat stucco are determined by the stucco manufacturer. In most cases color of the aggregate is important. Sometimes special materials, such as perlite fines or mica flakes, are added for special effects.

Perlite aggregate is a volcanic glass that has been expanded by heating to 1400°F. The dry weight per cubic foot is less than 15 lb, and plasters containing the aggregate will weigh approximately 60 pcf. By recommendation, it is limited to interior use. The aggregate can be used where insulating and lightweight properties are desired. Where available and economical, a natural lightweight pumice aggregate has been used. Its weight averages about 30 pcf.

Water should have a quality suitable for concrete. Municipal water is usually suitable. Where questionable, the water should meet the requirements of ASTM C191 for setting time and ASTM C109 for strength.

Stucco finish coat is a factory produced material consisting of standard portland cement having a very light gray or white color, fine aggregate, hydrated lime, pigment, and plasticizing agent, all mixed and sacked ready for use, requiring only addition of water at jobsite. The sacked mix conforms to local building codes and due to more rigid quality standards of packaging and blending allow a more uniform color in the finish coat than from jobsite mixing and proportioning.

Color finish coats are usually a factory mixed product. However, if additions of color are to be made jobsite, use white cement for the light-pastel colors and light-gray cements for the deeper colors to minimize color cost. Pigments should consist of metallic oxides or pigmented admixtures for their stable and nonfading properties. Iron oxides are used for black, gray, brown, red, and yellow; chromium oxide for greens; and cobalt oxide for blues. Liquid dyes are also available.

Admixtures are rarely added during mixing as plastic hand or gun cement, masonry cement, and the manufactured stucco mix contain all the necessary plasticizers for workability and pumpability. The addition of waterproofers or damproofers is not recommended. Additional air-entraining admixtures are not required as these have already been added.

Calcium chloride should not be used because of possible corrosion of the metal reinforcement and accessories. If acceleration of setting time and strength gain is desired, heating the aggregate and water or use of Type III cement is recommended. Accelerators other than calcium chloride are available.

Lime is necessary only when regular cement is used. When added at the mixer, it should conform to the specifications for special finishing hydrated lime, ASTM C206, Type S.[5]

Asbestos additions as a plasticizer have been prohibited for environmental and health reasons.

Lath consists of numerous types and forms of metal reinforcement as wood types of lath have become uneconomical. One common form of reinforcement is expanded-metal lath, which is made by cutting parallel, staggered slits in sheet metal, then expanding or stretching the sheet to form diamond-shaped openings (Fig. 39.1). The sheet becomes self-furring when the dimples keep the main sheet at least ¼ in from the backing or smooth surface. V-rib lath is expanded metal lath having metal ribs every 6 in and is available in standard sheet sizes of 24 or 27 in wide × 96 in long. The V ribs are ⅜ in high to provide stiffness to the lath and allow wider spacing for the supports. This property makes them more desirable for use on ceilings and soffits. Use should be limited to the interior and dry areas as it is impossible to provide proper protective plaster cover for the ribs. There are such lath made from copper-bearing steel sheets, painted with rust-resistant paint, or from zinc-coated steel sheets. Weights are 2.5 lb/yd^2 for use on vertical surfaces or 3.4 lb/yd^2 for horizontal surfaces. Requirements for metal lath and installation requirements can be found in ASTM C1063.[9]

FIGURE 39.1 Diamond-mesh lath attached over waterproof paper with control joint.

Stucco mesh is similar to diamond mesh, except that it has openings about 3 × 1½ in. This reinforcement is used for exterior plaster and should weigh at least 1.8 lb/yd^2 to conform to QQ-L-101c and ANSI-A42.3.

Paper-backed wire fabric lath consists of square or rectangular mesh of No. 16 gauge (minimum) galvanized wire with a maximum spacing of 2 in, welded at the intersections. The paper backing is integrally woven with the wire mesh, thus providing a backing for machine-applied plaster. Furring crimps are also woven into the mesh.[8]

Stucco netting is a hexagonal woven wire mesh (chicken wire) of 17 or 18 gauge galvanized wire. It is fastened with furring nails over waterproof building paper. Mesh is available in 1 or 1½ in size, depending on the wire gauge, and with paper backing.

Miscellaneous materials include waterproof paper or felt meeting the requirements of Federal Specifications UU-B-790, Type II, Class D. The paper should be free of holes and breaks, and weigh at least 14 lb per 108 ft^2 roll. Flameproof paper specified for certain locations should comply with ASTM D777.

There are an assortment of base screeds, casing beads, corner reinforcements, clips, furring nails, tie wires and others, for special uses.

Grounds and screeds are devices for establishing the proper thickness. Grounds are narrow strips of metal or wood attached to the base adjacent to edges and openings in the area to be plastered. Screeds are narrow strips of initially placed plaster within large open areas to be plastered which provide elevation points for obtaining the full thickness throughout.

39.2 Bases for Plaster

Any base for plaster must be rigidly secured to prevent movement during or after plastering. Since it is a thin membrane, the plaster should not be used in the design to handle a significant portion of the design loads by ensuring that the connections to the plaster do not transfer the structural loads. Table 39.2 lists typical approved methods of attaching lath and appurtenant fixtures.[10]

Wood frame structures are either sheathed or open-frame. In open-frame construction, soft annealed wires of No. 18 gauge or heavier, called "line wire" or "string wire," are stretched horizontally across the studs at a vertical spacing of about 6 in and attached to every fourth stud. The wire is drawn taut by raising or lowering attachments to intermediate studs. Waterproof felt or paper is next nailed in place, with edges lapped at least 2 in. In sheathed construction, the felt is applied directly to the sheathing without the string wires.

Metal lath or reinforcement is applied over the felt.[7] For exterior plaster, this consists of hexagonal stucco netting, applied with furring nails so that the netting clears the paper by ¼ in. On wood frame structures, paper-backed wire-fabric lath can be applied directly to the studs or sheathing without first attaching string wire and felt.

Metal reinforcement must not be jointed at corners but returned 6 in on wood sheathing and to the first stud on open construction. Installation should be with the long dimension across the studs and fastening with galvanized or rust-resistant nails and staples. Aluminum nails should not be used because of the potential for chemical reaction with the fresh plaster. Spacing of fasteners should not exceed 6 in with 1½ in barbed roofing nails on horizontal supports and 1 in roofing nails or 1 in, 4-gauge staples for the vertical supports.

Steel frame construction is similar to wood frame except that the lath is attached to steel studs by means of 18-gauge wire ties spaced on 6 inch centers.

Masonry and concrete surfaces must be cleansed free of paint, oil, dirt, or other material which could affect bond. Bond can be obtained mechanically if the surface is rough

TABLE 39.2 Attachments for Exterior Plaster Reinforcement, String Wires, and Paper Backing

	Attachments for[a]		
Type of construction	String wires 6 in apart vertically[e]	Paper backing[b]	Reinforcement (furred out ½ in min.)[c]
Wood frame sheathed	—	Nails and approved staples[d]	Furring nails, approved staples, or other furring device, 6 in apart vertically on supports
Wood frame open	32 in on center horizontally	Nails or approved staples[d]	Furring nails, approved staples, or other furring device, 6 in apart vertically on supports
Steel frame open	—	Clip or other attachment	Clip or other attachment 6 in apart vertically on supports
Masonry or concrete (when reinforcement is used)	—	—	Furring device 6 in apart vertically and 16 in apart horizontally

[a]All nails, staples, or other metal attachments of lath and reinforcement should be corrosion-resistant.

[b]Paper backing may be omitted if (1) exterior covering is of approved waterproof panels, (2) construction is backplastered, (3) there is no human occupancy, (4) backing is water-repellant panel sheathing, (5) approved paper-backed metal or wire-fabric lath is used, or (6) metal lath, wire lath, or wire-fabric lath is used for noncombustible construction.

[c]Self-furring metal lath or self-furring wire-fabric lath meets requirements. Furring not required on steel members with flange width of 1 in or less.

[d]Attachment for paper backing should not tear paper.

[e]String wires must be securely fastened and rigid by stretching taut through staggering the attachments up or down (herringbone fashion) or by wrapping the wire around the attachments.

enough to provide a key or by suction caused by absorption of moisture into the surface being plastered. Lightweight concrete block, common brick, and porous tile have high suction while dense materials such as granite, hard burned brick, and glazed tile have low suction. Cleaning of old masonry surfaces can be accomplished by sandblasting, brushing, or washing depending on the thickness and nature of the coating which needs to be removed. Efflloresence can be removed by washing with (1) a 10 percent solution of muriatic acid and (2) an ample clean-water rinse. As the acid is caustic, protective clothing and eyeglasses are necessary. Grease can be removed by scraping and washing with a strong detergent or by burning with a blowtorch. Disintegrated or weathered masonry must be removed to a sound surface. Because of environmental restrictions, sandblasting with sand containing 8 percent moisture to prevent dust may be required. Local regulations may require that special precautions are used to collect and dispose of any effluent from the cleaning and washing procedures.

Where bond is questionable, the application of a dash coat of plaster or bonding agent is recommended. The dash bond coat consists of 1 part cement to 1 or 2 parts of sand combined with water. Application is done by manually casting the mortar on the surface from a brush or small broom or by machine. The dash coat is allowed to set sufficiently, kept damp for a day or two, and then allowed to dry before application of the second coat. Bonding agents should conform to ASTM C932,[6] and application should follow manufacturer's recommendations. An excess of bonding agent on the surface may cause the first coat to slough.

Where bond is impossible, the surface should be covered with felt and metal reinforcement attached over it so that the reinforcement is furred out a minimum of ½ inch. Paper-backed wire fabric can be applied directly to the surface.

Chimneys are covered with properly furred metal lath without waterproof paper or felt prior to application of plaster.

Old plaster or stucco is treated in the same manner as masonry and concrete. The unsound material must be removed.

Gypsum products such as gypsum lath, plaster, or block, and magnesite plaster should not be used as a base for portland cement plaster because of a possible chemical reaction leading to bond failure.

39.3 Proportioning and Mixing

Mixes. Codes and specifications permit volumetric proportioning (Table 39.3). Measurement of sand is accomplished by counting the number of shovels of sand per sack of cement which is done by equating seven No. 2 shovels to 1 ft^3 of sand. The amount of water is controlled by the appearance of the plaster in the mixer. In some instances slump measured with a 6 in high cone, having a 4 in diameter at bottom and a 2 in diameter at the top, has been required to be 1½ to 3 in for either hand- or gun-applied plaster at point of application.

The use of plastic or masonry cement simplifies jobsite mixing and proportioning because no additional plasticizers are required. When portland cement is used, 20 percent by weight of the cement hydrated lime may be added to the mix as a plasticizer. Other plasticizers may be used when approved by the engineer or architect. The plasticizer should not reduce the strength of the plaster by more than 15 percent.

Mixing should start by adding (in the following order) (1) the water, (2) 50 percent of the sand, (3) the cement, (4) any admixtures, and (5) the remainder of the sand. The mixture should be mixed in a power mixer for at least 3 to 5 min. If the mixer is progressively charged manually while mixing, allow a minimum of 3 min of mixing after all the ingredients are in.

TABLE 39.3 Exterior Portland Cement Plaster

Coat	Maximum volume of sand per volume of cement	Minimum thickness, in	Minimum period of moist curing, h	Minimum interval before applying succeeding coat
First, or scratch	4	⅜*	48	48 h†,‡
Second, or brown	5	1st and 2d coats, ¾	48	7 days‡
Third, or finish	3§	1st, 2d, and 3d coats, ⅞		

*Measured from backing to crest of scored plaster.

†To ensure suction, there should be a drying period.

‡When applied over gypsum lath backing, the brown coat of reinforced exterior plaster may be applied as soon as the first coat has become sufficiently hard. In some areas when approved by the building official, the brown coat may be applied the day after scratch coat has been applied. Shrinkage cracks in the brown coat should be filled in with the finish coat.

§If proprietary product is not used.

Avoid sloppy and overwatered mixes as they can cause segregation, glazing, shrinkage cracking, slow takeup, sloughing, effloresence, or crazing. Batches that are delayed and become stiff may be retempered once to restore consistency. Plaster must be discarded when not used within 2½ h of start of initial mixing. Incomplete mixing results in nonuniform workability.

Better results are experienced with manufactured finish-coat mixes. However, jobsite proportioned mixes can be more successful when 1 part white cement is used with ½ by volume of cement of hydrated lime and 2 to 3 parts of sand by volume. Pigments should be weighed for each batch to achieve uniform color.

39.4 Application of Plaster

Hand Application. On many projects, plaster is applied manually with the familiar hawk and trowel. These methods are most suitable for the small isolated job such as a residential dwelling.

Plaster is machine mixed (Fig. 39.2), and whether it is hand- or machine-applied is determined on an economics basis.

Machine Application. Gun plastering machines are either the piston- or worm-drive type. The piston-type machine has the mixed plaster flow by gravity or suction from the hopper to the pump cylinder. Large piston machines are capable of lifting the material to the upper floors of a building where the gun applies the material directly to the wall. The worm drive has a screw in a rubber-lined stator which forces the plaster mix through the hose to the gun where air is applied to spray the plaster mix onto the wall (Fig. 39.3).

Inadequate maintenance of equipment, improper operation, or use of unsuitable material can cause operating and plastering difficulties as shown in Table 39.4.

FIGURE 39.2 Jobsite plaster system containing sacked cement, a pile of moist sand, and a rotary-paddle mixer. The pump is not visible.

FIGURE 39.3 Gun used to spray plaster onto the wall.

General. Most plaster coats are applied in three applications known as scratch (initial coat), followed by the brown coat and the final finish coat. Prior to application of plaster to masonry or concrete, the surface should be damp but not saturated.

The application of the scratch coat should have sufficient pressure to force the material through and completely embed the metal reinforcement so that no more than 10 percent of the reinforcement is left exposed. After application the surface of the scratch coat is scratched or scored horizontally with a special tool to form ridges for mechanical bonding of the brown coat (Fig. 39.4).

The second or "brown," coat is applied as soon as permitted by the local building code (Table 39.3). The scratch should be evenly damp but not saturated for application of the brown coat if there is a waiting period. Where allowed, the brown coat is applied as soon as the scratch is hard enough to hold the brown coat without damage (Fig. 39.5).

Thickness of the plaster after application of the brown coat should be ¾ in. The brown coat is brought to an even, flat surface by smoothing or straightening with the rod or darby and is left in a rough floated condition for application of the finish coat. Defects such as rough spots and deep scratches need to be removed as they will show in the finish coat. Application should be continuous between corners, columns, pilasters, doorways, downspouts, belt courses, and control joints where joints in the plaster can be hidden as the appearance of the plaster will be different from one side of the joint to the other. This should also apply to joints at scaffold levels. Horizontal joints in the brown coat should be made at least 6 in below the scratch-coat joint.

After curing the brown coat, it is good practice to allow the plaster to take its shrinkage in order to have the finish coat fill any resulting cracks. The required brown coat 7 day curing period also results in a more uniform suction. The brown coat should also be dampened without excess moisture prior to application of the finish coat. The application

TABLE 39.4 Sand Blocks

Cause	Cure
Dirty pump, hose, or nozzle; gravel or sand left in hose after inadequate cleanup; cement buildup in pump, hose, or gun	Pump, hose, and gun should be thoroughly cleaned after every use or long delay; cleaning is done by filling the hopper with clean water and flushing the hose until clear; disconnect hose at the feed end, insert a rolled-up sponge, reconnect hose, and flush sponge through hose
Lumps clogging hose or gun from aggregate or hardened plaster breaking off from hopper, hose, or mixer; insufficient mixing	Use screen on pump hopper to catch lumps; keep equipment clean.
Loose or leaky hose connections which permit separation of water or cement paste from sand	Provide regular preventative maintenance; repair or discard worn hose or fittings; keep all joints tight
Hot, dry hose at start	Cool hose prior to start with water
Mortar variability due to insufficient mixing	Mix dry materials to uniform color before adding water, then mix for 5 min
Mortar variability due to irregularities in batching	Manual batching is variable and can be minimized by careful attention to batching, measuring, and sand moisture; keep sand pile close to mixer and use standard shovel
Oversanded mix	Add cement
Stiff mix	Check whether slump with 6 in cone needs water

FIGURE 39.4 Typical scratch coat texture.

FIGURE 39.5 Typical texture of brown coat on left as it is being applied over scratch coat. Note taping of metal joint opening.

thickness is usually a minimum of ⅛ in. Total thickness of the plaster is required to be a minimum of ⅞ in.

39.5 Special Finishes

A great variety of finishes are obtainable by using different tools during finishing. Highly textured finishes are common. Special effects are obtained by the application brushes, brooms, or sponges to the surface while plastic. An even-sanded texture is obtained by floating with a carpet or sponge rubber float. Other finishes are obtained by spraying the color coat with bits of mica, colored metallic tinsel, or ground glass or plastic. Patterns may be obtained by various tooling methods such as combing or scoring. A rough coating surface may be obtained by dashing a final thin layer and partially floating the mounds level.

Marblecrete (Fig. 39.6) consists of exposed natural or integrally colored aggregate, partially embedded in a natural or colored bedding coat of plaster. It is frequently used for making colored and textured murals, either interior or exterior. Aggregates consist of chips or pebbles of marble, quartz, crushed glass, crushed china, seashells, cinders, or other material having a hardness of at least 3 Mohs' scale. They must be durable, alkali-resistant, and inert, and must be free of all dust and dirt at time of application.

Marblecrete can be applied to the following:

1. Unset plaster base that has been allowed to take up until sufficiently firm to carry the marblecrete
2. Hardened plaster that has been moistened to create even suction, or has been treated with a liquid bonding agent
3. Concrete that has been treated with a bonding agent (or a leveling coat may first be applied to the treated concrete before bedding coat is applied)
4. Concrete masonry that has been coated with a plaster dash coat and allowed to cure for 48 h

FIGURE 39.6 A small hand gun is used to lay the marblecrete aggregate on the still plastic bedding coat.

Thickness of the bedding coat depends on the size of aggregate used, and should be at least 80 percent of the nominal aggregate dimension (see Table 39.5). After the bedding coat has taken up to the proper consistency, the aggregate is laid on by throwing it onto the bedding coat by hand or by mechanical hand-hopper machine that embeds the aggregate slightly; then the aggregate is tamped gently and evenly so as to embed the aggregate from one-half to three-fourths of the aggregate dimension. Aggregate smaller than No. 4 can be applied by machine, but larger particles must be hand-placed. After 24 h, if weather conditions make it necessary, the marblecrete can be cured by keeping it damp with a light fog of water. After the surface dries out, a clear nonstaining, nonpenetrating liquid glaze coating may be applied, or a clear nonstaining waterproofing sealer can be used. Either material should resist deterioration and discoloration from weathering.

39.6 Curing

Scratch and brown coats must be cured, as shown in Table 39.3, by careful spraying with water. Most codes are silent on curing of the finish coat, there being some objection because of the danger of causing streaks or color and texture variations if the surface becomes too wet. However, curing improves the durability and strength of plaster, especially during hot, dry weather, and if properly done will not cause discoloration. After the finish plaster has hardened for 24 h a light fog is applied, taking care to apply just suffi-

TABLE 39.5 Bedding-Coat and Aggregate Sizes

Chip* No.	Sieve size, in		Minimum bedding-coat thickness, in
	Passing	Retained on	
0	1⁄8	1⁄16	3⁄8
1	1⁄4	1⁄8	3⁄8
2	3⁄8	1⁄4	3⁄8
3	1⁄2	3⁄8	3⁄8
4	5⁄8	1⁄2	1⁄2
5	3⁄4	5⁄8	5⁄8
6	7⁄8	3⁄4	3⁄4
7	1	7⁄8	7⁄8
8	1 1⁄8	1	1

*Chip sizes conform to National Terrazzo and Mosaic Association Standards.

cient fog so the plaster does not dry out, keeping the surface damp for about a day, but avoiding a heavy application of water that will wash the surface and cause streaks.

Do not spray water on any coat of dry, hot plaster, and do not permit the plaster to dry out between applications of water; otherwise cracking and discoloration will develop. A fogging nozzle of the type used in hothouses or nurseries is especially well adapted to curing plaster.

39.7 Plaster Failures and Distress

The proper choice and use of materials and methods will result in a good job of plastering. Problems with plaster can invariably be traced to the failure to follow good plastering practices.

Discoloration. Causes of nonuniform appearance are variations in mix proportions, especially the water content; too much water in the mix; color coat spread too thin; incomplete mixing of job-mixed finish coat; inferior pigments; dirt, clay, or organic material in the sand; dirty tools, including pump and accessories, mortar boards, and hand tools; retempering batches that have dried out; variations in surface, suction, or moisture content of base coat; careless workmanship, including variations in trowel or float angle and pressure; overtroweling; dry floating; starting and stopping joints or joinings; rain on fresh color coat; and variations in curing.

Discoloration can occur at any time after application of the stucco because of a number of causes. Rusting of metal brackets, trim, etc., will cause staining. Improper design or installation of flashings, drips, sills, etc., permits water to enter the hardened stucco. Staining material originating in the backing material is brought to the surface by migrating water. The obvious remedy is proper and adequate design and installation of all appurtenances to stucco as well as proper application of the plaster and stucco.

Efflorescence. A whitish bloom, film, or crystal formation resulting from the deposition of mineral salts on the surface by water migrating from the interior of the wall and evaporating from the surface is called efflorescence, for example, mineral salts present in the sand or mixing water; organic matter in aggregate or mixing water; soluble calcium hydroxide leached out of the plaster; efflorescence already present on concrete, brick, or masonry base.

Efflorescence can be prevented by attention to certain details. Use only clean materials conforming to applicable specifications. Do not use unwashed sand or brackish, stagnant water. Do not apply plaster directly to masonry or concrete surfaces showing evidence of efflorescence. Prevent water from entering the wall by proper design, such as the use of an impervious membrane. Stop plaster above the soil line to prevent movement of salt-carrying water from the soil into the plaster. Present codes require a horizontal noncorrosive metal weep screed at or below the foundation plate line on exterior stud walls. The weep screed is to be fastened to the stud wall a minimum of 4 in above the earth or 2 in above paved areas and be of a type which allows trapped water to drain to the exterior. The plaster and lath should cover and end on the attachment flange of the weep screed.

Cracking. In general, cracking is caused by structural movement of the building resulting from settlement or swelling of foundation soil; failure to provide control joints about every 25 ft on large areas; inadequate joining of lath to studs or backing; failure to lap and fasten lath to each other; shrinkage or warping of wood framing; excessive loading of structure; weak sections such as at changes in cross section or at openings; continuous plaster membrane over two different backing materials; too much suction resulting from application of plaster to a dry base (including previous coat of plaster); shrinkage of plaster caused by wet mixes, inferior, poorly graded, or dirty aggregate; mix too rich; rapid drying of the fresh plaster; plaster membrane too thin, especially over studs; inadequate curing.

Other conditions that contribute to the formation of cracks are hot, dry, windy weather; overfinishing, or finishing too soon; loose line wires in open-frame construction; bond failure between the plaster coats or between plaster and substrate; and thermal shock, or sudden and large differences in temperature.

Low Strength. Properly applied plaster possesses sufficient strength for any exposure condition normally encountered. Conditions that lead to poor strength performance are much the same as those affecting concrete. Low strength may be caused by inferior aggregates, poor aggregate grading (especially excessive fines), insufficient cement in the mix, too much water (sloppy mix), incomplete mixing, retempering batches retained too long after mixing and before using, lack of curing, or hot, dry weather.

Softness of the finished plaster is evidence of low strength. Improper use of such admixtures as detergents and soap, low temperatures that retard hydration, freezing, and impurities in the water or aggregate all lead to low strength and soft plaster.

39.8 Plastering in Cold Weather

Because exterior plaster is a thin membrane on an exposed structure that is difficult or impossible to protect from low temperature, especially if accompanied by wind, plastering during cold weather should be avoided. During cold weather setting and strength gain of mortar are slower than at more moderate temperatures, causing long delays between application of the plaster and floating or finishing. Particularly risky is plastering when nighttime temperatures are below freezing as there is great danger that the fresh plaster will suffer irreparable damage if it freezes before it has hardened, resulting in low strength, softness, and unsoundness.

If it is necessary to plaster during cold weather when the minimum temperature is above freezing, setting of the cement should be accelerated so the plaster will take up faster and floating can be accomplished within a reasonable time. Sometimes all that is necessary is to heat the mixing water to about body temperature, easily accomplished by directing the heat from a salamander or other heater against the water barrel. Water should be warm but not hot, as hot water might cause a flash set.

REFERENCES

1. Uniform Building Code, 1991, International Conference of Building Officials, Whittier, Calif.
2. "Portland Cement Plaster (Stucco) Manual," 1980, Portland Cement Association
3. Standard Specification for Application of Portland Cement-Based Plaster, ASTM C926-81.
4. Aggregate for Job-Mixed Cement-Based Plasters, ASTM C897-83.
5. Standard Specification for Finishing Hydrated Lime, ASTM C206-84.
6. Standard Specification for Surface-Applied Bonding Agents for Exterior Plastering, ASTM C932-80 (reapproved 1985)
7. Standard Specification for Metal Lath, ASTM C847-83.
8. Standard Specification for Welded Wire Lath, ASTM C933-80.
9. Standard Specification for Installation of Lathing and Furring of Portland Cement-Based Plaster, ASTM C1063-86.
10. "Plaster and Drywall Systems Manual," 3d ed., BNI Books and McGraw-Hill, New York, 1988.
11. Guide for Portland Cement Plaster (ACI 524-92), American Concrete Institute.

CHAPTER 40
CONCRETE MASONRY

Joseph J. Waddell

Masonry is used in all types of buildings: dwellings, schools, churches, industrial, commercial, and farm buildings; from small single garages to high-rise apartments. Retaining walls are frequently built of concrete masonry.

Concrete masonry is used extensively as a backup for other material (stone, brick, tile, plaster), for strictly utilitarian walls without adornment, as a finish wall material for both exterior and interior applications, and for ornamental grilles, solar screens, and garden walls. Masonry building walls may support part of the structural load of the building, in which case they are called load-bearing, or they may be non-load-bearing space dividers.

40.1 Masonry Building Construction

The six types of masonry building construction are hollow-unit masonry, solid masonry, grouted masonry, reinforced hollow-unit masonry, reinforced grouted masonry, and cavity wall masonry. A seventh type, used for architectural and decorative purposes, is the screen wall used for grilles, garden walls, and solar screens.

Hollow-unit masonry consists of hollow units all laid and set in mortar. When two or more units are used to make up the thickness of the wall, header courses or metal ties are required.

Solid masonry is made of concrete brick or solid load-bearing concrete-masonry units laid contiguously in mortar. All head, bed, and wall joints are required to be solidly filled with mortar. Adjacent withes must be bonded by means of header courses or metal ties.

Grouted masonry is made with solid brick units or hollow units in which interior masonry joints and all cells or cavities are filled by pouring grout therein as the work progresses. Type S mortar is specified. Wall may be of either low-lift or high-lift construction. High-lift construction requires special wire wall ties and permits grouting in 4 ft lifts.

Reinforced hollow-unit masonry is made with hollow masonry units in which some of the cells, containing vertical reinforcement, are continuously filled with concrete or grout. Type S mortar is specified. A cleanout opening is required at the bottom of each cell to be filled where each lift exceeds 4 ft in height.

Reinforced grouted masonry is the same as grouted masonry except that it is reinforced and the thickness of grout or mortar between masonry units and reinforcement must not be less than ¼ in. However, ¼ in bars can be laid in horizontal joints at least

½ in thick and steel-wire reinforcement can be laid in horizontal joints at least twice the thickness of the wire diameter.

Cavity wall is made with units in which facing and backing are completely separated except for metal ties that serve to bond the two tiers together. The Uniform Building Code[1] requires a minimum net thickness of 3½ in for both the facing and the backing, and a net width of the cavity between 1 and 3 in.

Screen walls are made of special patterned, pierced, or open block. Besides their ornamental value, they are used to separate space, both interior and exterior, as solar screens to shade windows and other areas, and to provide privacy yet permit passage of air and light.

40.2 Masonry Units

Dimensions. Concrete-masonry units are usually referred to by their nominal dimensions. Actual dimensions are smaller to allow for the thickness of mortar joints. For example, an 8 × 8 × 16 in unit actually measures 7⅝ × 7⅝ × 15⅝ in, allowing for ⅜ in mortar joint. Common sizes and types are shown in Fig. 40.1.

Pierced or screen blocks are available in a variety of sizes that should be ascertained locally before use.

Modular Construction. If advantage is taken of the nominal sizes of units, construction can be greatly simplified and cutting or breaking of units is minimized. Half-length blocks are available for further simplification. Tables 40.1 and 40.2 show the nominal lengths and heights of walls built with the two commonest sizes of blocks. All vertical dimensions should be in multiples of nominal full-height units, and horizontal dimensions in multiples of half-length units. Sizes and location of doorways and window openings should be planned to accommodate these dimensions, making proper allowance for sills, jambs, lintels, and frames.

Kinds of Block. Blocks are designated as hollow load-bearing, solid load-bearing, hollow non-load-bearing, and brick. These may be either normal-weight concrete or lightweight concrete (about 105 pcf), with the latter usually preferred because they weigh about one-third less than the heavy ones. Lightweight blocks weigh 24 to 30 lb each.

Many types of building units are ornamental by virtue of special textures, patterns, or colors incorporated in them at the time of manufacture. Other units are specifically designed for aesthetic value such as *grille* and *screen* block cast in special molds that give a pierced or open effect, especially desirable for screen walls, garden enclosures, and fences; *slump* block, a rustic-finish block resembling adobe brick in appearance, made with a mix of such consistency that it slumps or sags when the form is removed, resulting in blocks of variable appearance, texture, and height; *split* block that looks like rough stone, made by splitting a hardened block; *shadow* block, with embossed or recessed face patterns that give a three-dimensional effect; and *faced* or *glazed* blocks that are coated with a ceramic or plastic material giving a smooth, sanitary surface for use in bathrooms, kitchens, and similar locations.

Construction Requirements. Blocks must be dry when they are laid in the wall; otherwise the units will shrink and the wall will tend to crack. For this reason blocks must be permitted to age for about 4 weeks before they are used. Aging periods may be shorter in a hot, dry climate, or when autoclaved blocks are used. Blocks are classified by ASTM as follows:

ASTM C55	Concrete building brick
ASTM C90	Hollow load-bearing concrete masonry units
ASTM C129	Hollow nonload-bearing concrete masonry units
ASTM C-145	Solid load-bearing concrete masonry units

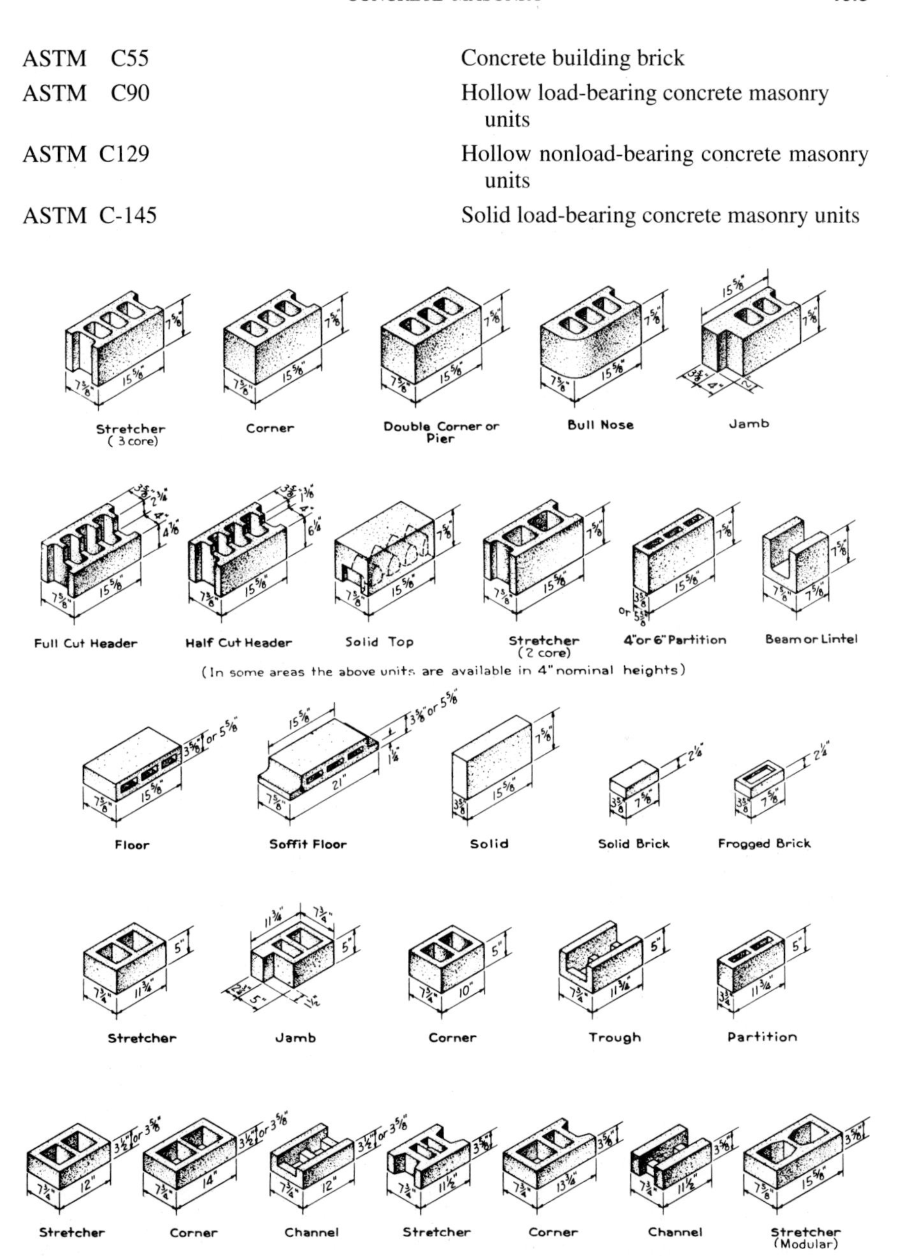

FIGURE 40.1 Typical sizes and shapes of masonry units. Dimensions shown are actual unit sizes. A 7⅝ × 7⅝ × 15⅝-in unit is commonly known as an 8 × 8 × 16 in concrete block. Half-length units are usually available for most of the units shown below. See concrete-products manufacturer for shapes and sizes of units locally available. *(From "PCA Concrete Masonry Handbook," Portland Cement Assoc., Skokie.)*

Blocks in each of the classifications (except C139) are further designated as Type I, moisture-controlled units; and Type II, nonmoisture-controlled units. Blocks conforming to

TABLE 40.1 Nominal Length of Concrete-Masonry Walls* by Stretcher

No. of stretchers	Nominal length of concrete-masonry walls	
	Units 15⅝ in long and half units 7⅝ in long with ⅜ in thick head joints	Units 11⅝ in long and half units 5⅝ in long with ⅜ in thick head joints
1	1 ft 4 in	1 ft 0 in
1½	2 ft 0 in	1 ft 6 in
2	2 ft 8 in	2 ft 0 in
2½	3 ft 4 in	2 ft 6 in
3	4 ft 0 in	3 ft 0 in
3½	4 ft 8 in	3 ft 6 in
4	5 ft 4 in	4 ft 0 in
4½	6 ft 0 in	4 ft 6 in
5	6 ft 8 in	5 ft 0 in
5½	7 ft 4 in	5 ft 6 in
6	8 ft 0 in	6 ft 0 in
6½	8 ft 8 in	6 ft 6 in
7	9 ft 4 in	7 ft 0 in
7½	10 ft 0 in	7 ft 6 in
8	10 ft 8 in	8 ft 0 in
8½	11 ft 4 in	8 ft 6 in
9	12 ft 0 in	9 ft 0 in
9½	12 ft 8 in	9 ft 6 in
10	13 ft 4 in	10 ft 0 in
10½	14 ft 0 in	10 ft 6 in
11	14 ft 8 in	11 ft 0 in
11½	15 ft 4 in	11 ft 6 in
12	16 ft 0 in	12 ft 0 in
12½	16 ft 8 in	12 ft 6 in
13	17 ft 4 in	13 ft 0 in
13½	18 ft 0 in	13 ft 6 in
14	18 ft 8 in	14 ft 0 in
14½	19 ft 4 in	14 ft 6 in
15	20 ft 0 in	15 ft 0 in
20	26 ft 8 in	20 ft 0 in

*Actual length of wall is measured from outside edge to outside edge of units and is equal to the nominal length minus ⅜ in (one mortar joint).

Source: PCA "Concrete Masonry Handbook."[3]

Type I, at the time of delivery to the site, should have a maximum moisture content ranging in value from 25 to 45 percent of the total absorption, depending on the average annual relative humidity of the geographical area in which the blocks are to be used and the allowable shrinkage of the blocks. Blocks are further classified as to grades as shown in Table 40.3.

Strength, absorption, density, and allowable moisture contents vary considerably for different classifications of blocks. For these reasons, the producer and user should check

TABLE 40.2 Nominal Height of Concrete-masonry Walls by Courses*

No. of courses	Nominal height of concrete-masonry walls	
	Units 7⅝ in high and ⅜ in thick bed joint	Units 3⅝ in high and ⅜ in thick bed joint
1	8 in	4 in
2	1 ft 4 in	8 in
3	2 ft 0 in	1 ft 0 in
4	2 ft 8 in	1 ft 4 in
5	3 ft 4 in	1 ft 8 in
6	4 ft 0 in	2 ft 0 in
7	4 ft 8 in	2 ft 4 in
8	5 ft 4 in	2 ft 8 in
9	6 ft 0 in	3 ft 0 in
10	6 ft 8 in	3 ft 4 in
15	10 ft 0 in	5 ft 0 in
20	13 ft 4 in	6 ft 8 in
25	16 ft 8 in	8 ft 4 in
30	20 ft 0 in	10 ft 0 in
35	23 ft 4 in	11 ft 8 in
40	26 ft 8 in	13 ft 4 in
45	30 ft 0 in	15 ft 0 in
50	33 ft 4 in	16 ft 8 in

*For concrete-masonry units 7⅝ and 3⅝ in in height laid with ⅜ in mortar joints. Height is measured from center to center of mortar joints.
Source: PCA "Concrete Masonry Handbook."[3]

TABLE 40.3 ASTM Requirements for Concrete-Masonry Units

ASTM designation	Grades
C55	Grade U: For use as architectural veneer and facing units in exterior walls and for use where high strength and resistance to moisture penetration and frost action are desired
	Grade P: For general use where moderate strength and resistance to frost action and moisture penetration are desired
	Grade G: For use in backup or interior masonry, or where effectively protected against moisture penetration
C90 C145	Grade N: For general use, as in exterior walls below and above grade that may or may not be exposed to moisture penetration or the weather, and for interior walls and backup
	Grade S: Limited to use above grade in exterior walls with weather-protective coatings and in walls not exposed to the weather
C129	No requirements

local code provisions concerning requirements for blocks and should also be sure that the latest applicable ASTM specifications are being used, since these are subject to revision. Materials, dimensions, and appearance are all covered in the standard specifications.

Block, on delivery to the site, must be stored on planks or other supports to keep them off the ground, and should be covered with canvas, plastic, or waterproof paper to protect them from moisture, if atmospheric conditions require.

Estimating Quantities. The number of block and the amount of mortar required for a plain masonry wall can be estimated from Table 40.4.

40.3 Foundations

Concrete-block walls and structures must be constructed on concrete footings of adequate width and thickness to carry the expected loads, situated on undisturbed or well-compacted earth, in accordance with the design and local building requirements. The foundation should be constructed in accordance with good construction practices and screeded level. At the time blocks are laid on it the concrete should be at least 3 days old, clean, and free of all laitance and dirt.

Size and depth of footings is a function of design, and the builder is concerned with making the footing in accordance with the plans. A footing width twice the wall thickness is usually adequate for light buildings. Base of the footing should be at least 12 in below ground line or below the frost line, whichever is deeper. In areas of expansive or unstable soil, the designer may impose additional requirements. Exterior doorsills can frequently be cast as part of the footing. Footings for garden walls are made at least 12 in wide and 8 in deep.

TABLE 40.4 Weights and Quantities of Materials for Concrete-Masonry Walls*

		For 100 ft² of wall				
			Average weight of finished wall			
Actual unit sizes (width × height × length), in	Nominal wall thickness, in	No. of units	Heavy-weight aggregate, lb†	Light-weight aggregate, lb‡	Mortar,§ ft³	For 100 concrete units, mortar,§ ft³
3⅝ × 3⅝ × 15⅝	4	225	3050	2150	13.5	6.0
5⅝ × 3⅝ × 15⅝	6	225	4550	3050	13.5	6.0
7⅝ × 3⅝ × 15⅝	8	225	5700	3700	13.5	6.0
3⅝ × 7⅝ × 15⅝	4	112.5	2850	2050	8.5	7.5
5⅝ × 7⅝ × 15⅝	6	112.5	4350	2950	8.5	7.5
7⅝ × 7⅝ × 15⅝	8	112.5	5500	3600	8.5	7.5
11⅝ × 7⅝ × 15⅝	12	112.5	7950	4900	8.5	7.5

*Table based on ⅜ in mortar joints.
†Actual weight within ± 7 percent of average weight.
‡Actual weight within ± 17 percent of average weight.
§With face-shell mortar bedding. Mortar quantities include 10 percent allowance for waste. Actual weight of 100 ft² of wall can be computed by formula $WN + 150M$, where W = actual weight of a single unit, N = number of units for 100 ft² of wall, M = ft³ of mortar for 100 ft² of wall.

Source: PCA "Concrete Masonry Handbook."[3]

40.4 Reinforcement

The amount and location of reinforcement is a design consideration, and the builder must be governed by the plans and specifications in this respect. The following comments are offered to assist in placing reinforcement in accordance with the design requirements.

Footings. Usually footings for light buildings and walls are unreinforced, but the plans and local code should be consulted in this respect. Heavy loads, poor soil conditions, and areas subject to high winds or earthquakes may make reinforcement necessary. For example, in earthquake zones, it is customary to place two horizontal ⅝ in bars in the footing, one 3 in from the bottom and one a like distance down from the top. Minimum steel to tie walls to footings may consist of ½-in vertical dowels adjacent to door openings. In earthquake areas two vertical dowels may be required adjacent to door openings in exterior and bearing walls and at corners and intersections, with additional dowels in the footing.

Wall and Joint Reinforcement. The amount and location of vertical steel is usually determined by design considerations for vertical structural loads, and seismic or wind conditions. Vertical bars, when required, are placed in the cores of the blocks at each foundation dowel, lapping the dowel 30 bar diameters. Bars should be wired to the dowels or spaced one bar diameter, at splices, and placed at the centerline of the wall, at least ¼ in clear of the masonry. Bars must be held at sufficient points to prohibit movement of the bars while placing masonry and grouting, usually at intervals not exceeding 192 bar diameters. Where two bars are called for at a jamb, the individual bars can be placed in adjacent cores. If the cores are to be grouted in lifts exceeding 4 ft, a cleanout hole should be provided at the bottom of the lift in the cell or core containing the steel.

Where horizontal reinforcement is required, it usually consists of ¼ in rods or commercially available fabricated assemblies of longitudinal wires joined by means of diagonal cross wires, laid on the webs of the masonry, and placed in the face-shell mortar. Splices in wire-reinforcement assemblies should be lapped at least 8 in, with one cross wire in each assembly included in the lap. If cross ties are not in the same plane as the longitudinal reinforcing wires, the assembly should be laid with the cross ties down, thus providing better embedment by holding the wires up from the surface of the lower block.

Many walls are built without horizontal joint steel. However, the steel strengthens the wall and distributes stresses that might occur in the wall. If the wall shrinks and cracks, the cracks will be much finer and closer together than in a similar wall without the steel. Horizontal steel is necessary in bond beams at roof and floor levels, at the top of wall openings, in parapets, and in walls of stacked bond. See Fig. 40.2 for typical details.

Special fabricated wire assemblies are available for use as horizontal steel to tie together the two withes or tiers of cavity walls. Cross rods are provided with a drip to prevent migration of moisture.

40.5 Mortars for Concrete Masonry

Materials. Cementitious material for masonry mortar consists of masonry cement, plastic cement, or common portland cement with hydrated lime or lime putty. Cements are described in Chap. 1. Hydrated lime should conform to ASTM Designation C207, Type S, and quicklime to ASTM C5.

Aggregate consists of sand conforming to the grading shown in Table 40.5; this is the grading shown in ASTM C144. Neither ASTM nor the Uniform Building Code has a limitation on the amount of natural sand passing the 200 mesh sieve. However, the fines should be kept below 5 percent, as a sand that is too fine requires more water for worka-

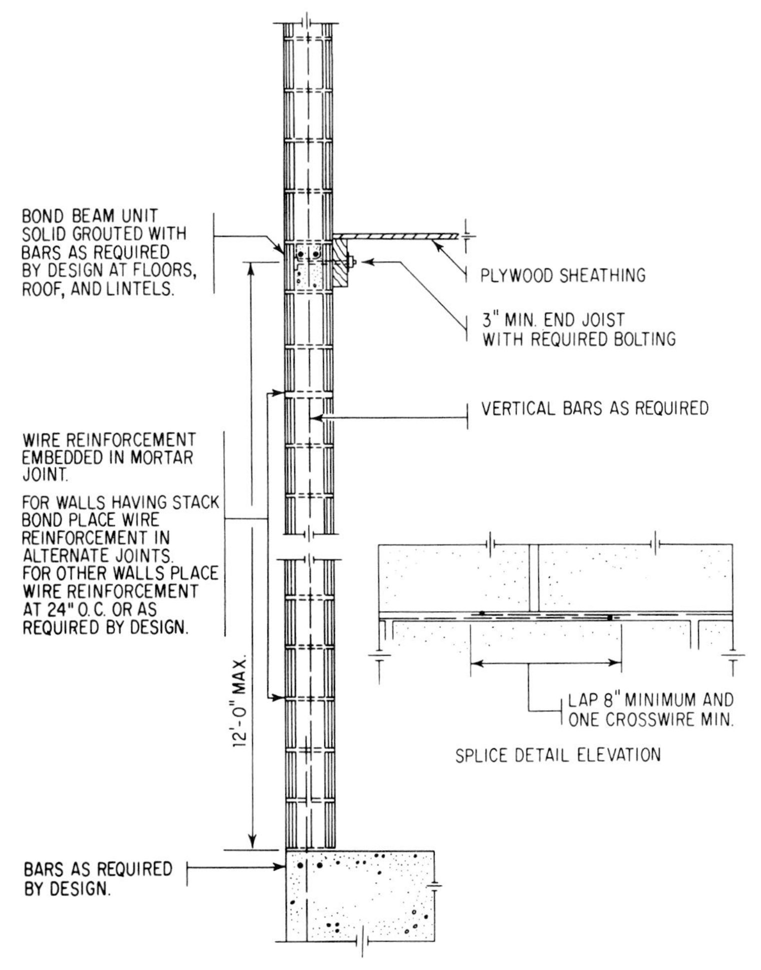

FIGURE 40.2 Wire reinforcement in mortar joints. Wire conforming to ASTM A82 may be used as temperature steel or to replace running bond and may be added to bond beam and foundation bars for required minimum reinforcement. Spacer wires welded to longitudinal wires should be spaced at 16 in on centers maximum. Wire reinforcement is usually specified as No. 9 wire or 3⁄16 in diameter wire. *(From "Reinforced Masonry Engineering Handbook, Masonry Institute of America, 1978.)*[4]

bility and is apt to produce weak mortar. On the other hand, a sand deficient in fines passing the No. 50 and 100 sieves will probably be harsh and unworkable, making it difficult to produce weathertight joints. Sand should be washed and free of clay, silt, and organic matter.

Cement and aggregates should be stored where they will not become contaminated but should be close to the mixer.

Properties of Mortar. Mortar should be workable, or "buttery," have good water retentivity, and be capable of developing the specified strength and bond. These properties are

TABLE 40.5 Sand Grading for Masonry Mortar, ASTM C 144

Sieve size No.	Percent passing	
	Natural sand	Manufactured sand
4	100	100
8	95–100	95–100
16	70–100	70–100
30	40–75	40–75
50	10–35	20–40
100	2–15	10–25
200		0–10

achieved by the use of well-graded sand, proper mix proportions, and the minimum amount of water to give the required plasticity.

Because masonry units are dry when laid in the wall, their suction draws water out of the mortar, resulting in a loss of plasticity of the mortar and sometimes affecting the bond. Water retentivity is the property of the mortar that resists rapid loss of water to the masonry as determined by the method of ASTM C91.[2] Mortar with a low water-retention value (less than 75 percent) undergoes a rapid loss of water when placed in the wall, resulting in loss of plasticity and the danger of serious loss of bond.

The property of mortar that unites it to the masonry is the bond. Bond is affected by the amount and kind of cementitious material, workability and plasticity of the mortar, texture and suction of the masonry units, and the skill used in laying the masonry.

The specifications provide for several grades of mortar as shown in Table 40.6, and the class of mortar used for any structure depends upon design and exposure conditions. The Uniform Building Code designates the types of construction and allowable unit stresses for the different classes of mortar. Code requirements should always be ascertained before starting construction.

Mixes. The strength of masonry mortar depends on the quantity and kind of cementitious material, amount of water in the mix, quality of aggregate, water retentivity, plasticity, and work or artisanship. Mortar should have good plasticity but overly rich mixes should be avoided as they are subject to volume change. Fireclay and similar materials should not be used.

Mortar should be mixed in a power mixer. Half of the water and sand are put in the mixer, then the cement (and lime, if any), followed by the balance of the water and sand. Mixing is continued for 3 min. Mortar should be used within 2½ h after mixing when the air temperature is 80°F or higher, or 3½ h when the air temperature is below 80°F. Water can be added to the mortar within these time limits to maintain plasticity if the mortar has stiffened because of evaporation of water.

Color. In joint mortar, color can be obtained by the use of pigments as described in Chap. 4. Experiment with the mix and materials to be used on the job by making small test panels and storing them on the job for about a week. Brilliant and dark colors are difficult to obtain. What is sometimes thought to be fading is actually a deposit of efflorescence on the surface that hides the color. A faded appearance also results from weathering in which the aggregates are exposed by erosion of the surface.

TABLE 40.6 Mortar Proportions Parts by Volume

Mortar	Type	Portland Cement or Blended Cement*	Masonry cement†			Mortar cement‡			Hydrated lime or lime putty*	Aggregate measured in a damp, loose condition
			M	S	N	M	S	N		
Cementlime	M	1	–	–	–	–	–	–	¼	≥2.5 to ≤3 times the sum of the separate volumes of cementitious materials
	S	1	–	–	–	–	–	–	+¼–½	
	N	1	–	–	–	–	–	–	+½–1¼	
	O	1	–	–	–	–	–	–	+1¼–2½	
Mortar cement	M	1	–	–	–	–	–	1		
	M	–	–	–	–	1	–	–	—	
	S	½	–	–	–	–	–	1	—	
	S	–	–	–	–	–	1	–	—	
	N	–	–	–	–	–	–	1	—	
Masonry cement	M	1	–	–	1	–	–	–	—	
	M	–	1	–	–	–	–	–	—	
	S	½	–	–	1	–	–	–	—	
	S	–	–	1	–	–	–	–	—	
	N	–	–	–	1	–	–	–	—	
	N	–	–	–	1	–	–	–	—	

*When plastic cement is used in lieu of portland cement, hydraulic lime or putty may be added, but not in excess of 1/10 of the volume of cement.

†Masonry cement conforming to the requirements of ASTM C91-67.

‡Mortar cement conforming to the requirements of UBC Standard No. 24-19.

Source: Table 24-A, 1991 Uniform Building Code.[5]

40.6 Laying Masonry

The First Course. Alignment of the first course is extremely important in order that the finished wall will be true and straight. The concrete footing should be swept clean of all sand, dirt, and other debris. Block laying starts from the corners, using a chalked snap line to mark the footing. Lay blocks with the thick end of face shells up, using full bedding; that is, the webs of the units as well as the face shells are embedded in mortar. Mortar should not extend into the cores more than ½ in. If the wall is more than 4 ft high, provide cleanouts of about 12 in^2 area at the bottom cell containing vertical reinforcement. Frequent use of the straightedge and level ensures a straight course of masonry.

Laying the Wall. Corners are kept higher than the rest of the wall. Ends must not be toothed, but racking a maximum of five tiers is permitted. Unless specified otherwise, the blocks are laid in common or running bond.

The method of bedding depends on the type of wall under construction. Load-bearing building walls, columns, piers, and pilasters, especially those for heavy loads, and the first or starting course on the footing or foundation wall should have full bedding; that is, mortar should be applied to the webs as well as the face shells of the units. All other hollow concrete masonry can be laid in face-shell bedding.

Use a mason's line stretched between the corner blocks to keep the units on the proper line and grade. Place the bed-joint mortar on two or three blocks in the previous course of blocks just ahead of the unit being placed, but not so far ahead that it loses moisture and plasticity before the block is placed on it. For vertical joints, mortar is usually applied only to the face shells. Some masons butter the end of the block before setting it in the wall, while others apply mortar to the end of the block already in place. Either method is satisfactory. When the block is placed, shove it against the previously laid block so as to compact the vertical mortar joint. Once a block has been set into the wall it should not be moved after the mortar has started to stiffen, as bond will be broken and a leaky wall will result. Instead remove the block and mortar and apply fresh mortar; then reset the block.

The mason should use the level, straightedge, and story pole to maintain line and grade and to assist him in keeping alternate head joints (in common bond) in vertical alignment. Cores containing vertical reinforcement must be carefully kept in alignment. It is usually best to lay up pilasters at the same time as the wall.

When building up the corners first and working from both ends toward the middle of the wall, the last block laid in each course is the closure block. The closure block is laid by buttering all edges of the opening and all four vertical edges of the block. If any of the mortar falls out before the block is tapped into position, remove the block, apply new mortar, and reset it. When work is interrupted for any reason, cover the top of the wall with canvas, building paper, or polyethylene to keep out rain and snow. Be sure the covering is anchored so it will not blow away.

Solid grouting for bond beams, as shown in Fig. 40.2, is accomplished by using special bond-beam blocks that are filled with concrete or mortar after installation of specified reinforcing steel.

Openings and Intersections. Depending on personal preference and site conditions, either of two methods can be used for door openings. One method is to square and brace the door frame in its final position before laying blocks. Anchor bolts or ties are then attached and extended into the joint mortar as the wall is laid up. Another method is to set in a temporary wood frame of the exact size and shape and lay the blocks around it, removing the frame later. The latter method is normally used when special jamb blocks are laid, and is the preferred method. Window openings are constructed in the same manner. Metal sash requires special attention, as metal-sash blocks, grooved to accept the metal frame, are used, and it is sometimes necessary to set the metal frame before the lintel is laid.

Doorsills are sometimes cast in the concrete of the foundation. Windowsills are of two types, both of which have sloping tops and project 1½ in outside the wall, with a drip groove on the bottom to prevent rainwater from running back into the wall. The lug sill is set in place as the wall is built up, and the slip sill is mortared in place after the wall has been completed.

Lintels may be precast or cast-in-place concrete, steel angles supporting concrete block, or lintel block. Precast-concrete lintels may be either one-piece or split, i.e., consisting of two separate beams side by side. One-piece lintels are preferred. Lintel blocks are U-shaped blocks, placed with the open end up. They must be well supported during construction. After the blocks are placed, two reinforcing bars are placed in the bottom of the channel formed by the blocks and the channel is filled with mortar or concrete. The use of lintel blocks avoids the difference in texture that results when precast is used.

Usual practice is to tie corners of bearing walls with a masonry bond, but other intersections are not bonded; instead the intersecting wall terminates and is tied to the other by means of metal ties. Nonbearing wall intersections are tied by means of strips of metal lath or hardware cloth in the mortar joint in alternate courses. The vertical joint at the intersection should be raked and caulked. Metal ties of this type, used for tying interior partitions to exterior walls, are slightly flexible and permit slight differential movement.

Anchors and Attachments. Anchors and inserts must be set in mortar or grout. Plate anchor bolts can be set in the mortar or concrete in the bond beam. Where an anchor or insert extends into the core of a block, the core is filled with mortar after first placing a piece of metal lath in the horizontal joint just below. If the anchor extends into a core containing vertical steel it should be solidly wired to the steel to prevent displacement when the cores are grouted.

When interior finish requires the use of furring strips, these can be attached to masonry by the use of special nails designed for this purpose.

Flashings over door and window openings, on parapets, and on roofs are shown on the drawings. Where the flashing extends into the mortar joint, it should not penetrate into the joint more than 1 in.

Screen Walls. Laying block in screen walls is done in virtually the same manner as in other types of masonry. Many of these walls are laid in stacked bond, requiring horizontal steel in every joint.

Cold Weather. During cold weather the work area should be enclosed, and the newly laid wall protected from low temperatures for at least 48 h. When the air temperature drops to 40°F the water should be heated to not over 160°F in order that the mortar temperature will be between 40 and 120°F. When the air temperature drops below freezing the sand should be heated also, taking special care that frozen lumps are thawed out. Masonry units should be stored indoors where they can be protected from frost and snow. If it is necessary to lay masonry when the air temperature is below 20° the units should be warmed to about 40 to 50°F.

Block should not be laid on a frozen foundation or on frozen block. When cold weather makes protection necessary, both sides of the wall must be protected.

40.7 Mortar Joints

The success of a masonry wall depends on the quality of the joints, hence the importance of good work or artisanship and materials. The properties of mortar that lead to good joints are clean sand of the proper grading, adequate cement of the proper type, minimum water to give the necessary plasticity and workability to the mortar, and adequate water retentivity of the mortar.

Making the Joint. Mortar joints should be ⅜ in thick, plus or minus ⅛ in. The starting bed joint on the foundation should be laid with full mortar coverage on webs as well as face shells. All other joints are laid in face-shell bedding except in pilasters, columns, cores at vertical steel, and as shown on the plans. Bedding mortar is spread on the previously laid blocks ahead of the one being laid but should not be permitted to dry out or stiffen before the block is laid. Joints must be full and level. Head joints must be carefully buttered, laid up to the full width of the face shell. Excess mortar is struck off the face of the blocks as soon as it is squeezed out of the joint by the block being shoved into place.

When laying a cavity wall, droppings can be kept out of the cavity by laying a board across the metal ties. When the masonry reaches the level of the next row of ties the board is removed by wires attached to it, cleaned, and reset on the new level of ties.

Tooling Joints. Customary practice is to tool the joints to dress them up and make them watertight. This is accomplished with a jointing tool, going over the joint to compact the mortar and seal it tightly against the masonry units. Excess mortar is then trimmed off. Tooling should be done when the mortar has stiffened somewhat but before it sets. All exterior joints must be tooled if a watertight wall is to be obtained. Joint finishes of the type shown in Nos. 1 through 5 in Fig. 40.3 are necessary for exposed walls, as they are watertight when properly made. Unexposed and interior walls, or exterior walls in a mild dry climate, can be of the types shown in Nos. 6, 7, and 8 of Fig. 40.3.

Tooling makes a dense surface texture that sheds water, gives better bond by forcing the mortar against the masonry units, and reveals places where mortar is thin or lacking. An untooled joint, which is made by striking off the fresh mortar with a trowel, is apt to be porous and permeable, with the mortar torn and drawn away from the units.

When a smooth, uniform texture is desired, as on a wall to be painted, the joints can be struck flush and then sack-rubbed (No. 5 in Fig. 40.3).

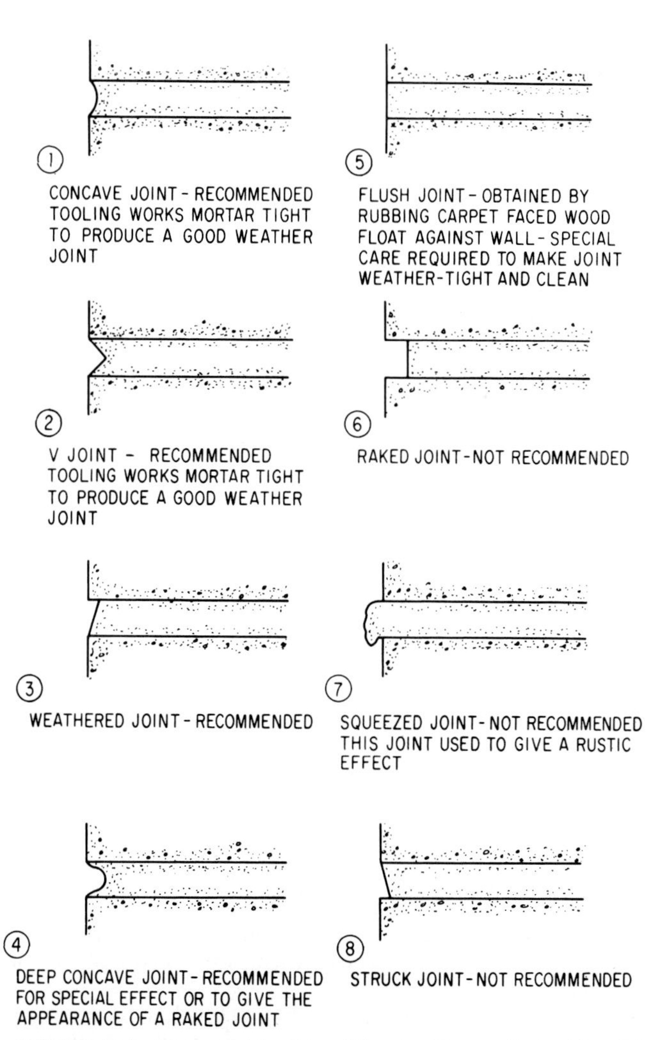

FIGURE 40.3 Masonry joints. (*From "Concrete Masonry Design Manual," Concrete Masonry Association of California, Los Angeles, 1964.*)

40.8 Grouting

Mixes. Grout for grouted masonry, unless otherwise specified, consists of 1 part cement, 3 sand, and 2 pea gravel mixed to a consistency as fluid as possible for pouring without segregation. Where the grout space is less than 3 in in its least dimension the grout consists of 1 part cement with 2¼ to 3 parts sand. All proportions are by damp loose volume. Mixing should be continued for at least 3 min after all ingredients including water are in the mixer.

Grout to be pumped is usually delivered to the job in ready-mix trucks and should contain at least 660 lb of cement per cubic yard and sufficient water to be of fluid consistency, as described above. Minimum compressive strength is 2000 psi at 28 days.

Grouting. Vertical cells to be filled must be aligned so as to have a continuous vertical cell measuring at least 2 × 3 in in area. Cells or cores not to be grouted should be covered with wire lath, a broken piece of block, or similar material to keep grout from entering. All mortar droppings and other foreign material should be cleaned out of the cell and off the reinforcement; then the cleanout hole is closed.

All bolts and anchors should be tightly grouted, as well as spaces around metal door frames and similar items. Cells containing reinforcement should be carefully filled with grout. When it is necessary to stop grouting, a key is formed by stopping 1½ in below the top of the course of masonry. Beams over doors and other openings should be grouted in one continuous operation. All grout must be consolidated by vibration or puddling.

High-Lift Grouting. In high-lift grouted masonry vertical grout barriers of solid masonry are built across the grout space between the withes at a horizontal spacing of not more than 25 ft. Walls of the two withes of masonry should be permitted to cure for a minimum of 3 days. Hollow-unit masonry should cure for a minimum of 24 h. Grout barriers or dams are, of course, not necessary when the grouting consists only of filling the cells in hollow units.

Grout should be puddled or vibrated at the time it is poured, and reconsolidated a second time just before it loses its plasticity.

40.9 Cracking of Masonry

Causes of Cracking. In most instances, cracking of the wall can be traced to shrinkage of the blocks. Other causes of cracking are improper design or work in the footing that leads to settlement, or movement caused by frost, or soils that expand when wet. Temperature changes may cause movements. Expansion or contraction of floor or roof elements can cause abnormal distortion. Overloads on the wall might result from improper design, vibration, earthquakes, or extremely high winds. Substandard units, such as thin webs and shells, or weak concrete, or poor work in laying the units can cause cracking.

Crack Prevention. Of greatest importance is to use blocks that have dried out adequately. Specifications for Type I blocks limit the total moisture content of blocks at the time of delivery. These are maximum values, and lower ones are preferred. Blocks that have been stockpiled and permitted to dry out for 28 days will satisfy these requirements. In very dry climates, more rapid drying permits using the block sooner. Autoclaved blocks dry out more rapidly than water- or steam-cured ones. Moisture content and absorption tests should be made to determine when the blocks have dried out sufficiently. The drier the block, the less shrinkage it will undergo after construction of the wall, and the less cracking the wall will suffer. Examples of concrete-masonry construction are shown in Fig. 40.4.

Mortar must be made of sound materials, of the correct proportions, properly mixed and applied. The use of reinforcement in the joints in accordance with good design practices is desirable.

Contraction joints should be installed at intervals of about 25 ft in long walls. The joint should be entirely through the wall, plumb, and of the same thickness as the other mortar joints so it will not be conspicuous. Sometimes it can be hidden behind a downspout or other feature of the building..

The designer has a wide range of joint designs available, including special rubber strips that are inserted in grooves in the block, the use of building paper to break bond,

FIGURE 40.4 Examples of concrete-masonry construction, showing structural walls of split block, typical one-story construction, screen walls, and fences.

FIGURE 40.4 *(Continued)*

installing Z-bar reinforcing on weakened joints, and others. All joints must be caulked with a flexible or plastic caulking compound.

40.10 Testing

Methods of sampling, inspecting, and testing masonry units are described in the ASTM specifications referred to in Table 40.3.

Grout and mortar should be sampled twice a day, or whenever there is a change in materials or mix. Specimens are handled in the standard manner and are tested in compression.

Grout. The method prescribed in the Uniform Building Code[1] is as follows: On a flat nonabsorbent base, form a space approximately $3 \times 3 \times 6$ in high, i.e., twice as high as it is wide, using masonry units having the same moisture condition as those being laid. Line the space with a permeable paper or porous separator, such as paper toweling, so that water may pass through the liner into the masonry units. Thoroughly mix or agitate grout to obtain a fully representative sample and place into mold in two layers, and puddle each layer with a 1×2 in puddling stick to eliminate air bubbles. Level off, immediately cover mold, and keep it damp until taken to the laboratory. After 48 h, carefully remove the masonry units and place the sample in the fog room until tested in a damp condition.

Mortar. Again quoting the Uniform Building Code, the recommended method is to spread mortar on the masonry units ½ to ⅝ in thick, and allow to stand for 1 min, then remove mortar and place in a 2×4 in cylinder in two layers, compressing the mortar into the cylinder using a flat-end stick or fingers. Lightly tap mold on opposite sides, level off, and immediately cover mold and keep it damp until taken to the laboratory. After 48 h, remove the mold and place the sample in the fog room until tested in a damp condition.

REFERENCES

1. Uniform Building Code, International Conference of Building Officials, Whittier, Calif., 1991.
2. ASTM Specifications C91, Specifications for Masonry Cement, American Society for Testing and Materials, Philadelphia, 1983.
3. "Concrete Masonry Handbook," Portland Cement Association, Chicago, Ill., 1991.
4. "Reinforced Masonry Engineering Handbook," Masonry Institute of America, 1978.
5. Building Code Requirements for Concrete Masonry Structures (ACI 531–79) (rev. 1983), American Concrete Institute.

PART • 11

PRECAST AND PRESTRESSED CONCRETE

CHAPTER 41
PRESTRESSED CONCRETE

Joseph J. Waddell
Joseph A. Dobrowolski

Prestressing is accomplished when two stressed materials are joined in such a manner that a force acting in one is balanced by an opposite force in the other. Thus in prestressed concrete a tensile force applied to steel tendons embedded in the concrete generates a compressive force in the concrete. The reason for prestressing concrete is to enable it to withstand tensile stresses, as concrete is weak in tension, its tensile strength being roughly one-tenth of its compressive strength. This permits the use of smaller, lighter concrete members. Prestressing imparts greater stiffness to the member and enables the designer to take advantage of high-strength steel. Becasue the concrete is in compression, applied loads can be greater than for reinforced-concrete members of the same size before actual tensile stresses are set up in the concrete.

In any prestressing operation, the procedure follows good concreting practices in regard to materials, mixing, placing, curing, and other phases of concrete making. Additional work involves placing and tensioning strands or wires and, in the case of pretensioning, release of tension at the proper time.

Control of prestressed concrete includes the usual inspection and supervisory procedures for any good concrete.[2] Uniformity of the concrete is especially important, as nonuniform concrete results in variations in camber and other features of the finished units. Concrete strength as high as 8000 psi is frequently specified.

Steel forms are almost universally used, although wood is acceptable for a limited number of reuses. They should be stoutly built and should close tightly to prevent leakage of mortar when the concrete is placed in them. Because forms are used repeatedly, they are apt to warp and bend, resulting in mortar leaks and uneven surfaces.

Prestressed units require conventional reinforcing steel, the same as nonprestressed concrete. Stirrups in girders and beams should be accurately placed as shown on the drawings. Frequently special steel is installed to control certain types of cracking. A relatively small amount of extra stirrup reinforcement near the end of a pretensioned girder is effective in minimizing the occurrence of cracks.

GENERAL PROCEDURES

41.1 Basic Methods

There are two basic methods of prestressing concrete: pretensioning and posttensioning. Pretensioning lends itself particularly well to mass production of members in a central casting yard. In pretensioning, the stressing strands are tensioned in long beds before the concrete is placed in the forms. After the concrete hardens and has cured, the tensile load on the strands is relieved, transferring the stress to the concrete by means of bond between the concrete and steel. Forms are usually continuous, separated by bulkheads to give the requisite length of members. In some yards, concrete is placed continuously in a long form and the hardened concrete section is sawed into appropriate lengths.

In posttensioning, which is especially suitable for on-site construction, ducts are formed by embedding hollow conduits in the concrete when the member is cast. After the concrete has reached the required compressive strength, the strands or wires in the ducts are tensioned, thus putting the concrete in compression, and the ducts are (usually) grouted.

Whether to use one method or the other is largely a matter of economics. Pretensioning usually presents greater economy when there are many identical or near identical units. However, if units exceed about 80 ft in length, there are apt to be problems in transporting them to the jobsite, unless rail or barge transportation can be used. Pretensioning beds require heavy abutments to withstand the loads imposed by the stressed strands, but posttensioning loads are carried by the concrete of the member being stressed. Posttensioning can sometimes increase the effective spans of cast-in-place slabs or beam and slab systems, at the same time offering good deflection control, and aids in developing continuity in precast construction.

41.2 Materials

Prestressing Steel. Steel for prestressed concrete consists of stress-relieved high-tensile-strength wire or strand or, less frequently, high-strength rods, as described in Chap. 3. The following types are regularly used:

1. Small-diameter strand (½ in or less in diameter) made up of six uncoated wires wrapped helically about a center wire. Used for pretensioned and posttensioned concrete. ASTM Designation A416.
2. Cold-drawn single wire, uncoated and stress-relieved, or hot-dip-galvanized, used in groups of two or more essentially parallel wires for posttensioning. ASTM Designation A421. Type BA has button anchorage and Type WA has wedge anchorage.
3. Cold-stretched alloy steel bars. Used mostly for posttensioning.
4. Large-diameter strand consisting of 37 or more wires. For posttensioning.
5. Low-relaxation strand which meets ASTM A421 and A416 has a higher yield strength. It must meet certain relaxation loss requirements as measured by ASTM E328 and a minimum yield strength at 1 percent extension of not less than 90 percent of minimum breaking strength. Its specifications are based on yield strength rather than ultimate strength. The significantly low loss of initial tension in low-relaxation strand can result in an improved and more predictable service performance and a higher load-carrying capacity.[3]

Detail requirements differ for different users, but the reference ASTM specifications provide sufficient detail for the applicable types. Strength of the wire and strand is usually specified as 250,000 psi or more. The elastic modulus is about 28 million psi.

Many of the comments relative to reinforcing steel apply to stressing steel also. For example, a small amount of rust has been found to be beneficial to bond, but severe corrosion should not be permitted. Avoidance of corrosion requires more care for stressing steel than for reinforcing steel, as high-strength steel is more susceptible to corrosion. Strand should be protected from the weather during storage, preferably by keeping it inside a building. Severe corrosion may occur if the steel is exposed to galvanic action while in storage.

In pretensioned concrete, and some types of posttensioned concrete, the prestress forces are maintained exclusively by bond between the steel and the hardened concrete, hence the importance of maintaining the steel free of deleterious coatings and contamination.

Concrete. Materials and mixes are covered elsewhere in this handbook. Only clean, well-graded aggregate of high quality should be used. Natural or artificial lightweight aggregates can be used to reduce the weight of members. Any of the standard types of cement, including Type III, are satisfactory. Admixtures that can be employed are air-entraining agents and water-reducing retarders. Admixtures containing calcium chloride must be avoided, because of the subsequent danger of serious corrosion of the stressing strands. Air entrainment has a tendency to lower strength values for rich mixes but is desirable from the standpoint of improved workability and durability. A water reducer permits improved strength development by reducing the amount of water for the same slump.

Compressive strength for detensioning (transfer of stress from the steel to the concrete) and ultimate strength are specified by the designer. Detensioning strength must be adequate for the requirements of the anchorages or transfer of stress through bond and must meet camber or deflection requirements.

The ACI Building Code[1] specifies that concrete stresses immediately after stressing (before time-dependent prestress losses) do not exceed the following limits:

1. Extreme fiber stress in compression $= 0.60\, f'_c$
2. Extreme fiber stress in tension except as permitted in 3 below $= 3\sqrt{f'_{ci}}$
3. Extreme fiber stress in tension at ends of simply supported beams $= 6\sqrt{f'_{ci}}$ where f'_c = specified compressive strength and f'_{ci} = compressive stress at time of initial prestress

The maximum stress allowed in the prestressing steel is 30,000 psi.

Mixes are usually quite rich, containing as many as 8 sacks (752 lb) of cement per cubic yard. Type III high-early-strength cement is frequently specified. Slump is low, in some plants less than 1 in. Consolidation is always by vibration, either internal, external, or both.

41.3 Camber and Cracks

Camber. Prestressed-concrete members have a natural tendency to bend in a vertical plane so the midpoint is higher than the ends. This curvature is called *camber* and results from the eccentric application of the prestressed load. Both pretensioned and posttensioned members will camber.

It is difficult to forecast exactly how much a member will camber, since camber depends largely on the modulus of elasticity of the concrete, and increases with time because of creep. The main quality control consideration is the difference in camber between adjacent members in the structure. For example, box beams in a bridge deck, to be covered with a cast-in-place composite slab, could tolerate a difference of 1 in, while adjacent tees in a building could accommodate perhaps half as much.

Camber can be measured while the member is still on the stressing bed, just after detensioning of pretensioned members, or stressing of posttensioned members, using the soffit plate or bed as a line of reference.

Cracking. One of the principal reasons for prestressing is the lessened occurrence of diagonal tension and flexure cracks under load. However, some cracking does occur during the manufacturing process which, in moderation, has no effect on the structural integrity of the member.

Some minor cracking results from shrinkage of the concrete. Because of the very low-slump concrete normally used, shrinkage cracking is minor and of little or no consequence. Short discontinuous cracks, commonly not over 3 in long, have been observed occurring horizontally at the angle of the web and bottom flange of I-beam girders. These are shallow shrinkage cracks and are not serious. Cracking may occur as a result of failure to follow proper curing and detensioning procedures. Vertical cracks can result from shrinkage and cooling before the member is detensioned. Once formal steam curing is discontinued, holddowns and strand anchorages should be released immediately.

Some cracks occur at the ends of pretensioned members that can be controlled by proper conventional reinforcement. This will not prevent the cracking, but it will minimize the size and frequency of cracks. Stress concentrations occur at the ends of posttensioned members that can lead to cracking. Transverse reinforcement will reduce the size of such cracks.

PRETENSIONING

In pretensioned concrete, the steel tendons are placed in the forms and a tensile load is applied to them by means of jacks. The concrete is then placed, and after it reaches a certain specified strength, the tendons are cut loose at the ends. This transfers the load to the concrete as a compressive force, and it is held by means of the bond between the concrete and the steel.

Steel for pretensioning consists of twisted strands that are anchored in grillages at the ends of the casting bed and are elongated by means of hydraulic jacks. All the strands in a member may be stressed simultaneously, or the strands may be stressed one at a time. The trend is now toward single-strand tensioning.

41.4 Casting Beds

Casting, or stressing, beds may be less than 100 ft long, ranging up to 650 ft. The reason for the long beds is to permit several forms to be set up in a line when cross section and tendon pattern in the units permit. One typical plant had four beds ranging from 20 to 450 ft in length. In this plant, the bed consisted of a raised soffit plate supported on two pilot liners, or I-beam rails set in the concrete base. The raised soffit plate provided access to the underside of the form for installation and release of holddowns for draped, or deflected, strands. Additional support for the soffit plate was provided by short lengths of I-beam spaced at 30 in. intervals and set at right angles to the long axis of the bed. This

FIGURE 41.1 The soffit plate rests on pilot liners and steel beam sections.

was a specialized bed, designed for large bridge beams of either I-section or hollow-box section (see Fig. 41.1).

The bed itself should be of concrete, finished to a smooth surface to support the steel forms, with end anchorages built into each end. Horizontal loads imposed on the end anchorages by the fully stressed strands in a bridge girder may approach 2 million lb, requiring extremely heavy construction to resist the moments induced by such loads. End anchorage may be constructed of reinforced concrete or structural steel, extending down into the ground so as to transmit the force into the ground or into the concrete of the casting bed. The strands pass through openings in the end anchorage to the stressing equipment beyond, as shown in Figs. 41.2 through 41.4. Instead of the heavy end anchorage, the bed can sometimes be designed so the forms take the reaction to the stressing load.

FIGURE 41.2 End anchorage for a small duplex bed designed for a maximum of twelve ⅜ in strands on each side.

FIGURE 41.3 The end anchorage for a large bed for heavy bridge girders. Twenty ½ in strands are shown. Some girders have 60 or more strands.

Many prestressed units are designed with deflected or draped strands, sometimes called "harped" strands, in which part of the strands are raised at each end of the girder. This construction makes necessary the introduction of holddowns in the forms to guide the strands into the proper pattern.

Members in a long bed are separated by means of bulkheads installed in the forms prior to placing the concrete (Fig. 41.5). A number of bulkheads are available on the market, and some precasters make their own. For a limited number of uses plywood can be used, but most bulkheads are of steel. One type is segmented so it can be fitted around the strands at the proper location in the form. Another type can be threaded onto the strands in a bundle, then separated and each bulkhead moved to its proper location. Usual practice is to anchor the bulkhead to the strands to prevent movement during concrete placing. When draped strands are used the bulkheads serve also as a "holdup" for the strands.

Curing facilities must also be provided and may consist of steam, hot water, or hot-oil piping (see "Curing" in Sec. 41.7).

FIGURE 41.4 Another view of the anchorage shown in Fig. 41.3. The single-strand jack is shown in a movable framework.

FIGURE 41.5 Tendons have been pulled through the bed and bulkheads set, indicating the ends of the girders.

41.5 Forming

Usually only steel side forms and steel or concrete bottom forms are recommended. Forms should be of sufficiently thick material, rigidly braced, and anchored, to withstand vibratory placing of fairly stiff concrete without bulging or deflecting.

Forms must be easily cleaned and should be cleaned after each usage.

Joints in forms must be smooth and tight to prevent unsightly offsets, sand streaks, or other blemishes on the concrete. Joints in the surface of a form must be ground smooth. Joints between form sections, soffits, and bulkheads must be tight, and gasketed if necessary. Certain types of chamfer strips of rubber or plastic can double as gaskets. All edges should be chamfered.

Most forms are externally supported, and no form ties are necessary. If form ties are necessary, they must be snap ties or she-bolts that break away beneath the surface of the concrete, leaving no pieces of metal in the surface of the concrete.

A form oil or parting compound must be applied to the forms to prevent adhesion of the concrete. There are many good materials on the market especially compounded for high-temperature curing in steel forms. Special care must be exercised, when applying the form oil, to avoid getting any of it on the prestressing strands. If the bond breaker is applied to the form before the strand is placed, the form must be covered with paper, plastic, or other protective materials so the strands will not be contaminated when they are installed. If the form oil is applied after the strands are in place, special care is likewise needed to keep the strands clean. Any form oil that gets on the stressing wires should be removed immediately with a solvent.

Most forms include provisions for the attachment of heating pipes for curing.

Some designs, especially heavy bridge girders, require bearing plates at the girder ends. These bearing plates must be placed accurately on the casting bed before the forms are set. Other inserts may be required, and the plans should be checked to determine type and location.

Prestressed units contain conventional reinforcement also (Figs. 41.6 and 41.7). Stirrups are especially important in controlling some types of cracking. Reinforcement is usually prefabricated into cages that are inserted in the forms either before or after stressing, depending on the type of unit. Some forms, especially those for I-beams, are so arranged that the entire assembly of stressed tendons and conventional reinforcement is in place before the side forms are set up. Bulkheads and deflecting hardware can be placed also before setting forms.

FIGURE 41.6 The tendons have been stressed and conventional reinforcement, consisting of longitudinal bars and stirrups, has been placed. Note that some of the tendons are draped or deflected.

FIGURE 41.7 The bed is now ready for the forms to be set in place. Note the end anchorages, stressing strands, and reinforcement.

FIGURE 41.8 The same bed shown in Fig. 41.7. Forms, spreaders, and bulkheads have been set, part of the concrete placed, and void formers installed. Balance of the concrete will be placed immediately, before a cold joint develops.

Void formers, discussed in Chap. 14, are placed in the forms to displace concrete near the neutral axis of a member to reduce weight with no loss in load-carrying capacity. Small, cylindrical voids are formed in roofing and deck units and large voids of square or rectangular section are used in making hollow box girders for bridges. Installation of formers is shown in Figs. 41.8 and 41.9.

41.6 Tensioning

When a tensile load is applied to a piece of steel, the steel starts to stretch or elongate. This elongation (called *strain*) is proportional to the applied load (called *stress*) up to a point called the proportional elastic limit. If the load is now released, the wire will resume its former length. If the load is carried beyond the elastic limit, the wire is permanently stretched to a degree depending on the load. This is called *permanent set.* In prestressed concrete, it is desirable to elongate the strands to a stress just below the elastic limit, by Code not to exceed 0.94 yield stress or 0.80 ultimate stress of prestressing steel. (see Fig. 41.10).

The modulus of elasticity E is a measure of the elasticity of a material. It is equal to the unit stress divided by the unit deformation or strain. The modulus of elasticity of steel for prestressing is in the neighborhood of 28 million psi. This is apt to vary by as much as 7 or 8 percent from lot to lot, thus introducing a possible error in stress values computed from strain measurements. By using jack pressure for computing stress, a check is obtained. As long as the two are within reasonably close agreement, there is no question of the accuracy of the stress measurement. Each job will have to set up its own standards and tolerances, but a difference of 5 percent between stress computed from jack pressure and that computed from measured elongation is acceptable for pretension-

FIGURE 41.9 A form for a hollow-box girder. All steel is in place and a large void former tied in place.

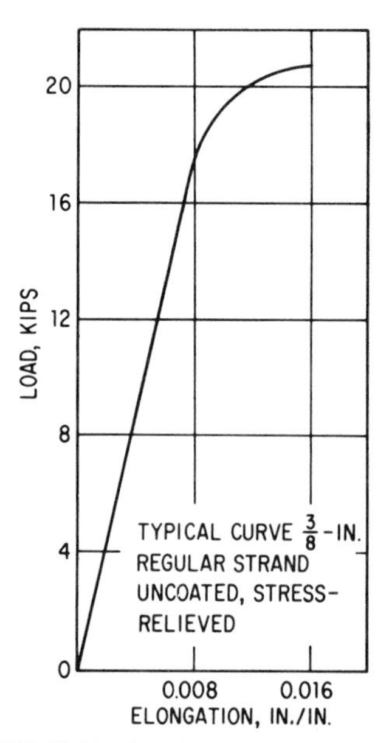

FIGURE 41.10 A typical stress-strain curve for prestressing strand.

ing and 7 percent for posttensioning. The ACI Building Code states that the prestressing force shall be determined by measuring tendon elongation and also either by checking jack pressure on a recently calibrated gage or by the use of a recently calibrated dynamometer.

Knowing the properties of the steel, tensioning requirements, and dimensions of the stressing bed, it is not difficult to compute the required pressure to apply to the stressing jacks, or the elongation to be obtained. Differences in the modulus of elasticity of different lots of steel, or inaccuracies in pressure gages on the stressing jacks, are sources of error in measuring stressing loads. These measurements are complicated when deflected or draped strands are used. All equipment should be calibrated at regular intervals, preferably under operating conditions. Any measurements should be checked occasionally with a dial gauge extensometer applied to the wire during the stressing operation.

As a check on stressing, for either straight or deflected strands, the occasional use of load cells is recommended at the beginning of a job. A load cell makes use of electric strain gages (SR-4) for measurement of loads.

Strain gauges applied directly to the strand are useful to determine the uniformity of tension along a deflected strand. This procedure is used when checking a new tensioning system but is impractical as a routine check.

The sudden release of energy that occurs when a strand breaks or an anchor fails causes rapid recoil of the strand and other hardware in the stressing bed. For this reason, inspectors and workmen must be alert for any unusual occurrences during stressing.

Tensioning Methods. In pretensioning, strands are anchored at one end of the bed and are attached to the tensioning device at the other end. In both multiple-strand and single-strand stressing, special hydraulic jacks with long-ram travel are used.

It is common practice to apply a small preload to the tendons (strands) by tensioning them to a low value which is the minimum force that will take the slack out of the tendons and equalize stresses. The load required depends on bed length, size of strands and whether draped strands are used. Values commonly used are between 5 and 20 percent of final stressing force. Preload can be applied by using the stressing jack or other means to apply a measured load. After application of the preload, reference points must be established from which total elongation can be measured. Elongation must be measured within a tolerance of ¼ in, which corresponds to 3 percent for 50 ft bed, 1 percent for 150 ft bed, or ½ percent for 300 ft bed.

In multiple-strand tensioning, the jacks push a crosshead that pulls the template to which the strands are attached. When the proper tension is reached, as indicated by jack pressure and elongation, the strands are anchored in the grillage and the jacks are removed to the next bed. When a large number of strands are being stressed, more than one jack might be used.

Single-strand tensioning permits the use of smaller, more portable jacks. A center-hole jack may be used, in which the strand to be tensioned is fed through the center hole of the jack and anchored on the outboard end. After stressing, another anchor on the strand holds the strand against the grillage, and the jack is released and moved to the next strand.

Temperature Effect. Normally, the difference in temperature between the steel at time of tensioning and that of the concrete is not significant in causing stress change in the tendons. If steel is stressed at a low temperature and warm concrete is placed around it, the steel will expand and reduce the tension. Of course, the opposite effect results if temperatures are reversed.

Total length change in the steel resulting from temperature change is expressed by the equation (all length measurements in inches)

$$\Delta_t = etl \tag{41.1}$$

in which Δ_t = length change of steel due to temperature change
e = linear coefficient of expansion of steel = 0.0000065 in/in/°F
t = temperature difference
l = length of strand subject to temperature change

Unless there is a large temperature difference, the effect can usually be neglected.

Discrepancies and Losses. Small discrepancies can be expected between calculated and measured elongations. As long as these are recognized and evaluated, there is no cause for concern. One source of discrepancy is slippage in the gripping devices or anchors on the ends of the strand. This can easily be determined and included in the computations, as shown in the paragraph on computing stresses. At the start of any prestressing job, the movement of anchoring abutments and elongation of anchor bolts should be evaluated. Usually this can be ignored, but it should be checked if there is an otherwise significant discrepancy.

In single-strand stressing, the strand will have a tendency to untwist, causing the jacking ram to rotate. A rotation of one turn is permissible per 100 ft of strand.

Sometimes a wire will break during stressing. If the number of broken wires is less than 2 percent of the total area of wires, the stressing is still acceptable.

Computing Prestress. The size, number, and distribution of tendons, and the stress to be applied, will be computed by the designer, but this information has to be translated into pressure and elongation values for field use.

The following computation shows how to compute the tensioning load in a group of strands being stressed simultaneously. If single-strand stressing is being used, the equations are still valid, in which case $N = 1$ and $J = 1$.

Let A = area of one strand, in^2
a = net area of jack ram, in^2
d = unit deformation or strain, in/in
E_s = modulus of elasticity of strand, psi
Δ = total change in length l, in (elongation)
J = number of jacks
L = net length of stressing bed, ft
l = length, in
M = measured movement of jack-end anchors, in
N = number of strands in one group
P = jack pressure, psi
S = unit stress applied to strand or required, psi
T = total slippage, in
W = total load on N strands, lb

The number of strands N is the group of strands being stressed at one time by a group J of jacks. This will vary with different plants but will probably be the number of strands in one casting, assuming that the bed consists of several castings in tandem. The unit stress S is determined on the basis of the physical properties of the steel being used, and this will be the basis for the field computations. The modulus of elasticity E_s is determined from laboratory test and is the basis for determining the change in length of the strands in the bed.

To obtain the jack pressure reading P required to produce the unit stress S in the strand, we have

$$W = ANS$$

then
$$P = \frac{W}{Ja} = \frac{ANS}{Ja} \tag{41.2}$$

In using E_s to determine the total elongation of the strands, it is necessary to consider the amount of slippage of the strands in the anchors, both at the dead end of the bed and at the jack end. As the load is applied, there will be a slight movement in the anchors at each end as they grip the strand. This is normally of the magnitude of ¼ in. To find what it is, mark each strand, under no load, just where it emerges from the anchor, using crayon, soapstone, or similar material. After the bed has been stressed, measure the distance the marks have moved away from the ends of the anchors. The sum of the slippages at each end is the total slippage T.

From the relationships,

$$d = \frac{S}{E} \text{ and } d = \frac{\Delta}{l}$$

we get
$$\Delta = \frac{Sl}{E_s} = \frac{12SL}{E_s} \tag{41.3}$$

where Δ is the total required elongation in a bed of length L. Movement of the ram

$$M = \frac{12SL}{E_s} + T \tag{41.4}$$

The slippage T should be checked for each operation. Once the pressure and elongation have been set up, there will be no change unless a change is made in the other constants. In some plants it will be possible to measure the elongation Δ directly, without measuring slippage. In this case, use Eq. (41.3).

The following example illustrates the computations.

A certain plant has a stressing bed 400 ft long using one jack with a net ram area of 32 in^2. It is planned to stress a string of beams using 16 strands of 0.035 in^2 area. The steel has an elastic modulus of 28 million psi, and the strands will be stressed to 175,000 psi. Slippage in the anchors is ¼ in at the jack end and ¼ in at the dead end. Compute required jack pressure and movement of jack-end anchors.

From the above,

$A = 0.035$ $\quad\quad$ $N = 16$

$a = 32$ $\quad\quad$ $J = 1$

$E_s = 28{,}000{,}000$ $\quad\quad$ $S = 175{,}000$

$L = 400$ $\quad\quad$ $T = \frac{1}{2}$

$$P = \frac{ANS}{Ja} = \frac{0.035 \times 16 \times 175,000}{1 \times 32} = 3062\ psi$$

$$M = \frac{12SL}{E_s} + T = \frac{12 \times 175,000 \times 400}{28,000,000} + \frac{1}{2} = 30.5 \text{ in}$$

In practice it is necessary to apply a slight load to the strands in order to take up all the slack and give a reliable starting joint for measuring elongation. This is done by applying a small load, say 200 psi, to the jack which will take the catenary or droop out of the strands through the forms. The stress in the strands can be computed from Eq. (41.2) and the initial elongation by proportion.

Using values from the above example, for a 200 psi initial jack load, from Eq. (41.2),

$$S = \frac{PJa}{AN} = \frac{200 \times 1 \times 32}{0.035 \times 16} = 11,428 \, psi$$

$$\frac{11,428}{175,000} = 0.0653$$

$$\text{Elongation at 200 psi} = (M - T) \times 0.0653 = 30 \times 0.0653 = 1.96 \text{ in.}$$

Deflected Strands. The foregoing computations apply to a bed with straight strands. Many prestressed units, on the other hand, contain deflected, or draped, strands. A draped strand is one that extends a predetermined distance near the bottom of the beam or unit being made, equal distances each side of the center of the span, then rises and emerges from the ends of the beam near the top. The purpose of deflecting strands is to provide a better stress distribution in the prestressed member and to reduce eccentricity of the prestress force at the ends of the member.

Elongation and load computations are made as for a straight strand. However, it is sometimes not possible to stress a deflected strand from one end only, because of friction in the hardware holding the strand down and up at several points through the length of the bed. In this case, the full load is applied to one end of the strand and the elongation noted. The jack is then moved to the other end of the bed and the full load applied at that end, the elongation there being noted. The sum of the two elongations should equal the computed total elongation within 5 percent.

There are two common methods of stressing deflected strands, both of which appear to give satisfactory results. In one system, the strands are tensioned the same as straight tendons, as described above. If the bed contains more than one or two members in tandem, it is usually necessary to stress from both ends. In the second system, all the tendons, running straight through the forms, are partially stressed a predetermined amount with the jack. The balance of the stress is applied by pulling the strands up or pushing them down to final deflected position.

A great variety of hardware is available for holddown and holdup units. Holddowns in some cases extend through the bottom form and are held in the anchor plates by means of pins that can be easily and quickly removed. Holdups consist of various assemblies, some of them with small grooved wheels to carry the strands to minimize friction.

An effect similar to that obtained with deflected strands can be achieved by breaking the bond between the concrete and the end of the tendons. The distance to be unbonded is determined in the design office. Bond is prevented by enclosing the tendon in a paper or plastic sleeve, or by coating the strand with a bond-inhibiting compound.

41.7 Concreting

After the tendons have been tensioned, the reinforcement placed, and the forms buttoned up (Fig. 41.11), concrete is placed in the conventional manner, following good construction and control practices.

Some plants use ready-mixed concrete, but most have their own batching and mixing plants, frequently with a turbine-type mixer because of the low-slump concrete used. Distribution of concrete in the yard is by practically any kind of handling equipment as shown in Fig. 41.12. Concrete should be distributed in the forms in lifts not over about

FIGURE 41.11 A small bed for thin concrete planks, ready to receive concrete.

16 in in depth, thoroughly consolidated by vibration, both internal and external. Cold joints must be avoided.

Concrete test cylinders should be made during the time that concrete is being placed. The exact number will depend on the manufacturing schedule, but a minimum of six cylinders should be field-cured with the concrete members, and the time of releasing stress on the tendons determined by the strength developed in these specimens. Cylinders should be made for standard curing also. Specimens for standard curing may be cast in either cardboard or metal molds, but the field-cured ones should be cast in metal molds.

Curing. Some sort of accelerated curing is practiced in nearly every casting yard, because of the requirement for 4000 psi compressive strength, or more, in about 16 h.

FIGURE 41.12 Various methods of transporting concrete are used. In this plant, they hauled the concrete in a front-end loader.

Many of the forms have incorporated in them pipes for steam, hot water, or hot oil. Means must be provided to keep the units moist during the curing period. Beds can be covered with heavy tarpaulins or boxes to prevent loss of moisture and heat. When heat is supplied by wet steam, sufficient moisture is usually present. If heat is supplied by hot water or hot oil in pipes, moisture must be supplied to the top exposed areas of concrete by soaker hoses, fog spray, wet burlap, or other means. Water must be hot. In general, the best curing is obtained with wet steam under tarpaulins except in exposed outdoor beds in cold weather.

Start of high-temperature curing should be delayed 2 to 4 h after the last concrete is placed, and the temperature should not exceed 160°F. Temperature rise in enclosure should be limited to a maximum of 40°F/h. After the expiration of the curing period (determined by field cured cylinders) the concrete is permitted to cool to ambient temperature. When the air temperature is below 50°F, the concrete should be cooled at a rate not exceeding 20° per h.

Curing methods are depicted in Figs. 41.13 and 41.14.

41.8 Detensioning

As soon as the concrete reaches the specified strength for transfer of stress to the concrete, the strands are released, or detensioned. This may be accomplished by releasing the jacks when multiple-strand tensioning has been applied to straight strands. Single-strand tensioning, and the presence of draped strands, make other methods necessary. One method is to cut the strands individually with an acetylene torch, using a sequence or pattern that has to be developed for each type of unit and stressing bed.

The relationship between curing, release of tension in the strands, stripping of forms, and release of strand holddowns is important in the control of cracking in the units. Forms should be loosened as early as possible in the curing cycle, but in any event while the concrete is warm and before the detensioning is done.

This should be done as rapidly as possible to minimize cooling of the concrete before the bed has been completely released; otherwise undesirable cracking may result.

FIGURE 41.13 Heavy canvas is spread over the finished girder, and then steam is admitted through perforated pipes. The canvas is on frames so there is circulation of steam around the beam.

FIGURE 41.14 This beam, exposed to normal atmospheric temperatures, was covered with burlap kept wet with soaker hoses. Such a procedure could not be used during cold weather.

Multiple-Strand Release. The entire stressing load is picked up by the jack, or jacks, and is then gradually released. There will be some sliding of the stressed members on the bed as stress is released, the amount depending on the length of the tendon exposed between members and between the last member and the dead end. For this reason, there should be no restraint of movement of the members, except friction on the bed.

Single-Strand Release. Using a sequence that keeps the stresses nearly symmetrical about the axis of the members, the strands are heated with a low-oxygen flame until the metal softens and loses its strength, causing the strands to part gradually, in a matter of several seconds for each one. Best results are obtained if heating can be done at both ends of the bed simultaneously.

Deflected Strands. For heavy members, in which the weight of each member is at least twice as much as the holddown force within the member, the holddown devices are released; then the strands, both deflected and straight, can be detensioned by either multiple- or single-strand release method.

In the case of light members weighing less than twice the holddown force, the draped tendons are first released by heating the tendons at each uplift point (between the bulkheads), and then the holddowns are released. Finally the straight strands are released by either single- or multiple-strand method. Light members can also be detensioned in the same manner as heavy ones if sufficient weight or restraint is applied to the members directly over the holddown points.

41.9 Handling and Transporting

Types of handling equipment include cranes, lift trucks, and straddle carriers of various types. Rubber-tired or rail-mounted gantry cranes are sometimes used. In enclosed plants, overhead-cab-operated bridge cranes are suitable.

Transportation of girders to the job is usually accomplished with motor trucks. Long girders are hauled on a tractor-and-bogie combination. Frequently the bogie is equipped with a steering device similar to those used on long fire trucks.

Prestressed members must be maintained in an upright position at all times, and supported or picked up only at designated pickup points. Beams and girders must be picked up at the ends, over the bearings. Disregard of these precautions will result in a damaged member.

Figures 41.15 through 41.18 illustrate handling of members.

POSTTENSIONING

Prestressing by the posttensioning process is accomplished by forming ducts through the concrete at the time it is cast. When the concrete has reached the required strength, as indicated by field-cured cylinders, the stressing wires are inserted in ducts, tensioned, anchored, and grouted. Whereas pretensioned members are manufactured at a central casting yard or plant, posttensioned members are frequently made at the site. One reason for posttensioning is that there is no need to transport the units over highways and streets; hence the logistics of getting the units in place does not impose a limitation on size. Pretensioned girders are rarely over 80 ft long for this reason.

FIGURE 41.15 Straddle carriers are frequently used for moving girders short distances.

FIGURE 41.16 On the Lake Pontchartrain Bridge job, a special 200 ton gantry picked the precast deck unit out of the stressing bed and carried it to a barge for transportation to the construction site.

FIGURE 41.17 Cranes are used to move large girders about the yard and to erect them in the structure. Pickup must be made at the bearings, with the member in an upright position at all times. (*Illinois Tollway photograph.*)

FIGURE 41.18 Massive precast pretensioned "I" ready to be lifted by gantry cranes to top of pier for bridge over a dry wadi near Riyadh, Saudi Arabia; Owner, Department of Communications.

41.10 General

The same care in production as for pretensioning is necessary. Frequently the ducts are curved or draped, a condition that is apt to lead to friction on the wires while they are being stressed. For this reason, it is especially important to check jack pressure and elongation at both ends. The ACI Building Code states that friction losses in posttensioned steel should be based on experimentally determined wobble and curvature coefficient, verified during stressing operations. The values of coefficients assumed for design and the acceptable ranges of jacking forces and steel elongations should be shown on the plans. These friction losses can be calculated by the equation

$$P_s = P_x \epsilon^{Kl} + \mu\alpha \tag{41.5}$$

When $KL + \mu\alpha$ is not greater than 0.3, Eq. (41.6) may be used.

$$P_s = P_x(1 + Kl + \mu\alpha) \tag{41.6}$$

in which K = wobble friction coefficient per ft of prestressing steel
l = length of prestressing steel element from jacking end to any point x
P_s = steel force at jacking end
P_x = steel force at any point x
ϵ = base of Naperian logarithms
α = total angular change of prestressing steel profile, radians from jacking end to any point x (Fig. 41.19)
μ = curvature friction coefficient

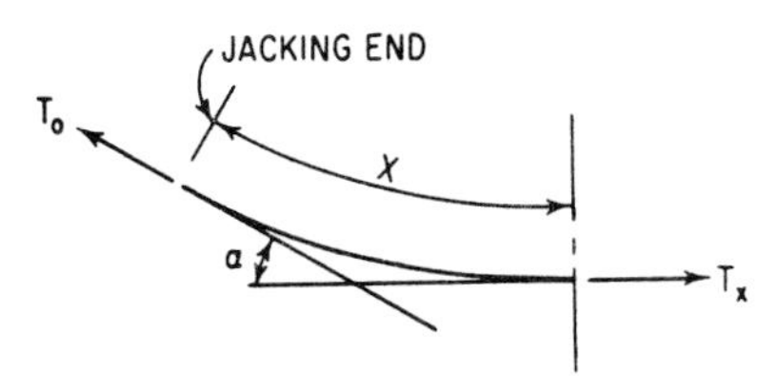

FIGURE 41.19 Angle α, expressed in radians and measured as shown here, is a maximum at the midpoint of the draped strand.

Values of K and μ can be taken from Table 41.1, which shows values for metal sheathing that can be used as a guide. Values of K (per lineal foot) and μ vary appreciably with duct material and type of construction.

Ducts or core holes for wire can be formed in the concrete by casting in flexible metallic tubing which is left in the concrete, or by flexible rubber tubing, hose, or fiber conduit that can be removed. Conduit or tubing should be securely tied to the reinforcing cage at intervals not exceeding one-tenth of the length of the member, with maximum intervals of 10 ft. In making straight ducts in lengths of hollow piles that are subsequently joined together, the ducts are formed by inserting steel rods, covered with rubber hose, in the form. Removal of the rods loosens the rubber, which can then be pulled out.

Individual wires are usually used in posttensioning, instead of the strand commonly used in pretensioning. By means of special equipment illustrated in Fig. 41.20, a group of wires is forced through the duct and anchored by means of special wedges. In another type of posttensioning, the wires are accurately measured and assembled into groups in a length of flexible metallic tubing. Each wire is then upset on each end, forming small buttons that bear against a special perforated plate. The entire tube and wire assembly is located in the forms and the concrete is placed. After the concrete has developed the required strength, a special jack grips the perforated plate and elongates the wire, which is then anchored, and the tube is grouted.

41.11 Stressing and Grouting

A small preload is usually applied to take the slack out of the tendons, as described under "Pretensioning," and then the design stress is applied, with proper allowances for

TABLE 41.1 Values of K (per Lineal Foot) and for μ Metal Sheathing*

			Wobble coefficient, K	Curvature coefficient μ
Grouted Tendons	in metal sheathing	Wire tendons	0.0010–0.0015	0.15–0.25
		High strength bars	0.0001–0.0006	0.08–0.30
		7-wire strand	0.0005–0.0020	0.15–0.25
Unbonded tendons	Mastic coated	Wire tendons	0.001–0.002	0.05–0.15
		7-wire strand	0.001–0.002	0.05–0.15
	pre-greased	Wire tendons	0.0003–0.002	0.05–0.15
		7-wire strand	0.0003–0.002	0.05–0.15

*From ACI 318-89, Table 18.6.2

FIGURE 41.20 Around the edge of the 54 in concrete cylinder (right) are 12 holes extending the length of the section. One person operating the machine collects 12 wires from turning spools, right background, and shoves the wires through each hole to join six 16 ft sections into one 96 ft pile.

friction. The tendons are next fixed in the on housing devices, and the elongation is accurately measured, as shown in Figs. 41.21 and 41.22. It is important that both elongation and jack pressure be measured and the discrepancy noted (agreement must be within 5 percent).

Grouting. The purpose of grouting is to protect the tendon from corrosion and to provide bond between the tendon and the concrete. Grouting should be done as soon after stressing as possible, but in no event more than 48 h later. Ducts to be grouted must be provided with ports for entrance and discharge of the grout and may require air bleeders at the high points of the profile.

Ducts must be clean when the grout is admitted, which may require flushing with water, after which the duct is blown out with air. Grout may be admitted in one or more

FIGURE 41.21 Two posttensioning jacks on a 36 in pile. The wires in part of the ducts have been stressed, anchored, and burned off. As soon as wires in all ducts have been tensioned, grout will be pumped in. After a suitable curing time, the wires will be burned off, releasing the tensioning ring.

FIGURE 41.22 Marking reference points for measuring elongation.

points, depending on the length and configuration of the duct. Grout is pumped continuously until it flows steadily out the discharge port. The discharge port is then closed and the grout pressure gradually increased to a pressure of 75 to 125 psi and held for 10 or 15 s. The entrance vent is then closed. Pressures at the inlet end should not exceed 250 psi.

Mixing, agitating, and pumping units are available for handling grout. The grout used for this purpose is quite fluid, sometimes containing a small amount of fine sand. Materials for grout must be accurately measured. After mixing, the grout is passed through a screen before it is admitted to the pump. Some specifications will allow a small amount of unpolished aluminum powder (about 1 teaspoonful per sack of cement) to provide 5 to 10 percent expansion in the grout.

CIRCUMFERENTIAL PRESTRESSING

Another type of posttensioning is applied to cylindrical objects, such as large concrete pipe and concrete tanks. After the pipe or tank has been cast in a conventional manner and has attained the necessary strength, special machines wrap high-tensile wire around the object, stressing the wire to the required tension. The entire outer surface is then coated with shotcrete.

Manufacture of concrete pipe by this process is briefly described in Chap. 42.

Tanks are made by first constructing a conventional cylindrical tank on a concrete slab. The joint between the slab and the tank walls is usually flexible; that is, some movement is possible. Leakage is prevented by installing a rubber of plastic water stop in the joint. A special machine, suspended from the top of the tank and traveling in a circular course, wraps high-tensile-strength wire around the tank under the proper degree of prestress. The machine is operated so as to space the wire wrapping properly from bottom to top of the tank. After the prestressed wire has been installed, the entire tank is coated with shotcrete to protect the wire.

CHEMICAL PRESTRESSING

Portland cement can be formulated in such a way that an expansive force is generated during the hydration process. Called expansive, shrinkage-compensating, or self-stressing, this is a calcium sulfoaluminate cement.

Experimental use has been made of expansive cement to produce a self-stressing concrete. It has been found that, under favorable circumstances, when concrete is properly restrained with adequate reinforcement, a prestressing load can be induced in the concrete and steel similar to the stresses induced by mechanical prestressing. Although still in an experimental stage, the process shows promise. Dependent to a large extent on the amounts of normal cement and expansive cement in the mix, results depend also on the particular type of cement being used, the amount and distribution of reinforcing steel, and curing conditions.

POSTTENSIONED SEGMENTAL CONSTRUCTION

By the early 1980s over 50 posttensioned segmental bridges had been built in North America (e.g., see Fig. 41.23). Their construction could be classified in four categories as balanced cantilever, span by span, progressive placing, and incremental launching or pushout.[4]

1. The balanced cantilever consists of segments cast, bonded, and posttensioned to the previous section on each side of a pier so that balance is maintained until the midspan is reached and a closure segment is cast to fill in the space at midspan from segments cast

FIGURE 41.23 Cantilever posttensioned segmental bridge under construction on Interstate Route I-8 at Pine Valley, Calif., for the California Department of Transportation.

and stressed from an opposite pier. This type has the least interference with the environment and surface traffic flow and requires a small labor force and an excellent speed of erection.

2. The span-by-span category is constructed from end of the structure by spans which are supported by a movable truss supported on each pier. The segments can be cast in place on the trusswork or cast on the ground and lifted into place.

3. Progressive placing is similar to the cantilever, except the segments are placed on the same side of the piers rather than on each side. As the erection stresses become excessive, temporary stays are added either by temporary falsework or cables from a center tower.

4. Incremental launching or pushout construction consists of segments being positioned and posttensioned together behind an abutment. The assembly of the units is done by pushing out one unit at a time, which allows the placing of another unit behind the abutment. Hydraulic jacks are used to move the units forward on special sliding bearings. To reduce large negative moments, a fabricated structural steel launching nose is attached to the lead segment and temporary piers may be used to subdivide long spans.

REFERENCES

1. ACI Building Code Requirements for Reinforced Concrete (ACI 318–89) and Commentary (ACI 318R-89), American Concrete Institute.
2. "PCI Manual for Quality Control for Plants and Production of Precast Prestressed Concrete Products," 1985.
3. Martin, L. D., and D. L. Pellow: Low-Relaxation Strand, Practical Applications in Precast Prestressed Concrete, *PCI J.,* July-Aug. 1983.
4. Recommended Practice for Precast Post-tensioned Segmental Construction, *PCI J.,* Jan.-Feb. 1982.

CHAPTER 42
PRECAST CONCRETE

Joseph J. Waddell
Joseph A. Dobrowolski

Precast concrete, as discussed in this chapter, includes precast building units, concrete masonry, pipe, and various specialty items such as railroad ties and poles.

PRECAST BUILDING UNITS

Precasting of building units can be divided into two general categories: (1) precasting on the site, using tiltup, lift-slab, or other on-site construction methods (see Chaps. 33 and 35), and (2) plant precasting, using manufacturing assembly-line techniques (Fig. 42.1). Units made by either method may or may not be prestressed. Prestressed units may be either pretensioned or posttensioned (Chap. 41). This chapter is concerned with off-site plant precasting.

Concrete lends itself well to prefabrication of building units, not only because of its structural qualities but also because it can be cast into intricate shapes and details (Figs. 42.2 and 42.3). Off-site production-line manufacture of building components is an important part of the "system building" concept, which is merely a term to describe a system of construction in which design, production, erection, and overall administration of the building and its construction are coordinated, or preengineered (see also Chap. 33).

The same basic design and construction requirements that apply to any on-site concrete work apply also to precast work.[1,3] In addition, there are several special requirements peculiar to precast units. Concrete wall panels, for example, may be of regular weight or lightweight concrete. Some are solid; some consist of a "sandwich" of concrete on the outside with a center layer of foamed, lightweight insulating material.

42.1 Shop Drawings

The manufacturer of precast concrete is usually required to furnish shop drawings showing complete information for making and installing units. Drawings should include an erection plan with all pieces marked, showing the necessary bracing that will be required during erection. Shop drawings should also give all dimensions, including tolerances. If units are to be prestressed, the allowance for elastic shortening should be designated.

FIGURE 42.1 Modern precasting plants require adequate space for storage of finished members.

Drawings should show size, amount, and location of reinforcement; details and location of connections and method of adjusting; location and details of erection devices and special reinforcement for same; details on all inserts, reglets, attachments, etc., and method of anchoring; joint details and materials; concrete finishes; and methods of protecting exposed surfaces during storage and erection.

42.2 Forms

Steel, wood, fiberglass-reinforced polyurethane, concrete, and plaster can be used for forming material, depending on the configuration of the units, number to be made, and preference of the manufacturer (see also Chap. 14).

Steel forms are desirable when many reuses of the form are required and where complicated irregular and reentrant surfaces are not necessary. Rivet heads and joints between sheets must be carefully ground smooth. Steel forms must be opened up or disassembled to permit removal of casting. Sheet-steel contact areas must be checked occasionally to detect buckling or dimpling, and must be coated with a light form oil before each usage. Steel forms are especially desirable when the product is to be steamed.

Fiberglass can be molded to complicated shapes on a mast or pattern and can be used many times (Chap. 39). The material is strong but requires support to give it rigidity. Concrete surfaces cast against new fiber glass are apt to be very smooth and glossy.

Wood forms are easily fabricated into intricate details but are not so durable as steel or fiber glass for many reuses. Wood forms should be treated with a sealer to protect the wood and should be coated with a parting compound when used.

FIGURE 42.2 The contrast between the precast white concrete and gray concrete lends interest to this office building. White-concrete units were manufactured by the Schokbeton process. (*Rockwin Prestressed Corp., Clairmont, Calif.*)

A concrete form is made by casting against pattern of concrete, wood, or plaster, making an allowance for shrinkage. Waxes and sealers should be applied to the concrete.

42.3 Fabrication

Many times a casting yard is combined with a prestressing yard, thus permitting maximum utilization of such facilities as batching plant, mixers, and cranes. Forming, casting, curing, and handling are similar in the two plants. Mass-production assembly-line techniques are common. Individual reinforcing bars are cut, bent, bundled, and tagged, and bar mats and mesh can be cut to size. By setting aside an area of the yard for steel fabrication, jigs can be set up on benches at a convenient working height, and the individual steel components assembled into complete reinforcing units, or cages, for the precast units. The completed cages are stockpiled for later installation in the forms. Frequently inserts, such as lifting eyes, connecting hardware, and anchor bolts, can be attached to the steel cage, and electrical conduit can be tied in. Where similar but different units are being produced, a system of numbering must be established before manufacturing commences, and the reinforcing cages identified with these numbers.

FIGURE 42.3 Precast panels may be two or more stories in height. Large units require fewer joints than small ones but usually entail little or no additional work in handling. Instead, there being fewer units to handle, erection costs are less.

While the steel crew is making reinforcing cages, another crew prepares the forms for concreting. The previous castings are stripped out and moved to storage for additional curing, using a motor crane, gantry, or straddle buggy. Forms are then cleaned of all loose or adhering concrete and other material, oiled, buttoned up, and the steel cage set in place. Small and light cages can be handled manually; heavier ones require the use of a crane. Cages must be accurately positioned in the forms and tied in such a manner as to prevent any movement during concrete placing. The type of concrete surface specified determines the type of chairs to be used, if any, or other means of fixing the steel in place.

Before placing concrete, the supervisor and the inspector should inspect the forms carefully to make sure that reinforcement and inserts are accurately located, well braced, and of the correct size, type, and quantity; that the forms are clean, well braced, and properly "buttoned up"; that the parting compound or form oil has been applied to the form; that void formers, if required, are well tied down to prevent movement or floating. Void formers will float in the plastic concrete during vibration, carrying reinforcement with them, unless they are well tied down to the form. Many of these units are prestressed, and the reader is referred to Chap. 41 covering this phase.

Concreting follows conventional methods. Slump of concrete is 2 in or less. Immersion vibrators consolidate the concrete, frequently assisted by form vibrators. The desire for high-early-strength concrete to permit rapid turnover of the forms and casting beds makes necessary the use of rich mixes, steam curing, and sometimes Type III cement. Compressive strength of 5000 psi is commonly specified for facing mixes, and 4000 psi for other mixes. Air entrainment is recommended for attainment of maximum durability and to minimize permeability.

Curing, discussed in detail in Chap. 29, can be by moist curing at a temperature above 50°F for 5 days if regular cement is used, or 3 days for Type III cement. Many plants steam-cure at temperatures up to 160°F. Curing compounds are rarely used.

After curing, units are sometimes stored for a period of time. Storage should be in such a manner as to minimize warping or distortion. Panels can be stored flat if they are fully supported on a flat surface; however, there is less danger of warping if they can be stored on edge, provided the panel edge is fully supported and the panels rest against a rigid nonstaining frame. Beams should be supported at their normal support points. If storage is for more than a few days, units must be covered with polyethylene to protect them from atmospheric staining.

All repair to minor spalls and similar defects must be made as early as possible.

In handling and transporting units, care is necessary to avoid staining or marring. Edges and corners are especially vulnerable and should be protected with timber lagging if it is necessary to lift the unit with a sling or tie it down to a truck or car bed. Ropes are less apt to damage surfaces than cables or chains.

42.4 Sandwich Panels

In this type of construction the wall panel, cast in a horizontal or flat position, consists of a layer of structural concrete, a layer of lightweight insulating material, then another layer of structural concrete. By this method, the panel has the advantage of relatively high structural strength, at the same time possessing fairly high insulating value.

Besides possessing good thermal-insulating properties, the insulating material must be resistant to the absorption of moisture, as the insulating ability diminishes when moisture is absorbed. A desirable material is one having low absorption, no capillarity, and capable of little or no vapor transmission. In addition, the material must be capable of developing good bond with the enveloping concrete, and must develop a reasonably high modulus of elasticity and compressive strength. These strength properties are necessary to provide a satisfactory degree of interaction between the concrete and the insulation, and to give the unit the required load-carrying capacity.

Materials that meet these requirements are portland-cement concretes using expanded mineral aggregates (perlite, vermiculite); foamed concrete, foamed glass, certain foamed or expanded plastics such as polystyrene and polyurethane, and paper honeycomb. Preformed insulation must be protected from damage during handling and storage, and protected from absorption of moisture. Densities of these materials range from less than 15 pcf to a little over 50 pcf.

The outside concrete layers of sandwich panels are interconnected by means of mechanical shear ties, or by concrete spacers or ties. In general, properly located concrete ribs are more efficient than metal ties.

42.5 The Schokbeton Process

In this plant precasting process, zero-slump concrete is placed in the mold and the entire assembly is subjected to a low-frequency vertical shock or impact that consolidates the concrete without segregating it. Speed of vibration is of the magnitude of 250 high-amplitude blows per minute. Since the entire element is subjected to the same processing simultaneously, there are no conflicting impulses.

Molds must be accurately made, rigid, and rugged to withstand the casting process. Details must be accurate, with precise corners and arises, so as to assure excellence in the finished product. Mold dimensions are usually required to maintain a tolerance of plus 0, minus $3/32$ in for 10 ft or over. Resulting concrete possesses high strength, high density, and low water absorption, making members with exact tolerances and sharp arises.

Reinforcing steel must be accurately placed and rigidly tied. Manufacturer's specifications require a minimum coverage of $3/4$ in of concrete for smooth surfaces and 1 in for exposed-aggregate surfaces, unless the local code specifies other requirements.

The concrete is specified to have a compressive strength of not less than 6000 psi at 28 days.

42.6 Extrusion Process

Hollow-core prestressed precast slabs are made by a macnine (e.g., see Fig. 42.4) which extrudes concrete as it moves along the casting bed. These machines generate a motive force as a reaction to the force that extrudes the concrete. These extruders are particularly well adapted to long-line continuous manufacture of prestressed concrete.

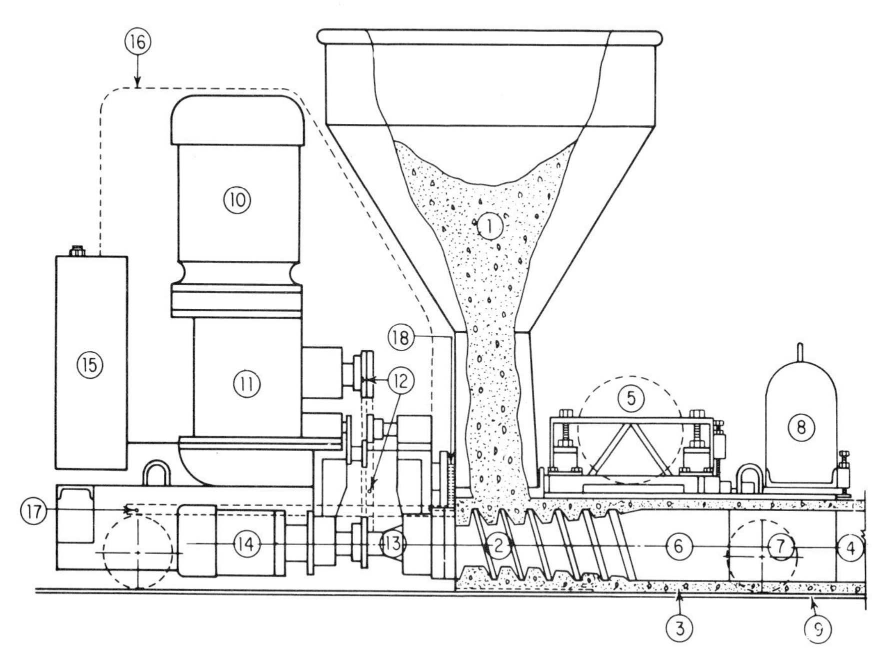

FIGURE 42.4 Operation of the Spiroll extruder starts with introduction of the concrete mix into the machine hopper (1). The process continues as follows: (2) a set of rotating augers collects the concrete mix from the discharge end of the hopper and forces it toward the rear of the extruder; (3) the concrete, which has been forced to the rear by the auger flights, enters a molding chamber where the bottom, top, sides, and cores of the slab are finally finished before it is extruded from the machine; (4) the extruded slab, as soon as it leaves the extruder, is strong enough to support a large person's standing weight; (5) in order to assist the pressure generated by the augers to consolidate the concrete mix, vibrators are affixed to the top forming plate adjacent to the hopper; (6) for the same reason, a vibrator is mounted inside the hollow core of each auger; (7) the final finish is applied by the separate follower tubes which are rubber-mounted to the ends of the augers so that extreme vibration is not transmitted this far—vibration at this point of the extrusion tends to deteriorate rather than improve slab compaction; (8) a counterweight is mounted on the rear of the machine just over the final troweling plate to prevent it and the machine from rising—naturally, there will be a tendency for this to occur as the resultant force of extrusion is downward as well as rearward; (9) the extruder travels on, and the slab is cast into polished steel pans—these, of course, are laid between the stressing abutments, and the extruder starts at one abutment and continues until it reaches the other; (10) the electric-drive motors impart their torque through (11) a gearbox and (12) chain-and-sprocket drive to (13) the auger drive shaft, which is mounted in the main bearing housing—this bearing housing is welded solidly to the extruder's frame and forms the very core of the machine; (14) the vibrator, which is mounted inside each auger, is driven through the hollow core of the auger and its drive shaft by a high-speed electric motor mounted on the front side of the bearing housing; (15) the extruder controls are electrical and are operated from a box at the front of the machine by using push buttons on the top of the box; (16) the drive motors and other working parts are covered by a lightweight fiber-glass housing; (17) hydraulically powered rods may be used to insert transverse steel in the top of the slab when required; (18) the transverse steel can be stacked at the front of the hopper where it is picked up by the plunger rods.

FIGURE 42.5 Typical Spancrete panel extruder containing three hoppers for premixed material. Material is compacted around metal tubes by tampers and vibration. Note distributed strand on adjacent beds.

Casting surfaces are metal pallets, permanently attached to the bed or a concrete bed. The first operation is to clean the steel pallets and apply a parting compound or lightly sand the concrete bed. Stressing strands are then placed on the bed and tensioned. Tracks attached to the pallets or angles guided by the edges of the casting bed provide guidance for the extrusion machine while zero-slump concrete feeds by gravity from the receiving hopper or hoppers to the rotating augers or moving tampers in the molding machine. To allow design variability, the Spancrete system (Figs. 42.5 and 42.6) has three hoppers for supplying bottom, middle, and top concrete-mix layers. This allows an adjustable amount of cover below the prestressed strands and use of lightweight concrete in the bottom layer only.

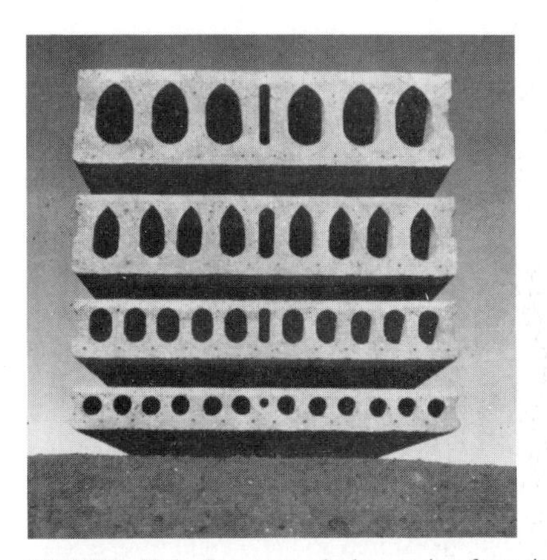

FIGURE 42.6 Spancrete planks varying from 4 to 10 in in thickness and a 1 m in width. The lower prestressing strands are located between the voids at the required height from the bottom. The sides are notched and the bottom of the plank is wider than the top to allow the placement of grout.

As the slab emerges from the machine, it is covered with wet burlap or curing compound to prevent drying during the preset time prior to application of steam. After steaming, the Spiroll slab is allowed to cool, cut to length and removed to storage. The Spancrete system allows the slabs to be stacked in separate castings up to five in height before cutting to size and removal to storage.

Speed of the machine along the bed varies in accordance with the mix proportions, type of aggregates, moisture content of the mix, size of slabs being extruded, and thickness of the slab.

Conventional methods of mixing and handling zero-slump concrete are used. Some plants use a gantry spanning several beds to feed the extruder hopper.

42.7 Finishes

Materials for precast units are covered in detail in Part 1 of this handbook. Because precast units, especially wall panels, are frequently exposed, special architectural finishes may be specified requiring special aggregates such as selected gravel or crushed rock, marble, white limestone, quartzite, granite, or artificial aggregates, such as expanded shales and clays, ceramics, and crushed vitreous aggregate. If a proposed aggregate does not have a satisfactory service history in a similar usage and exposure a careful investigation of that aggregate must be made before it is used. White cement may be specified, sometimes with color added. Finishes may be smooth formed surfaces, or aggregates exposed by retarded surface, acid etching, sandblasting, or bushhammering (see Chap. 43). Sometimes individual aggregate particles are hand-placed on the surface of the fresh concrete.

Uniformity of materials and finish is very important. Besides color of materials, the aggregate must be unvarying in size and grading if it is to be exposed.

Liners of many varieties are available to be used in combination with the forms to provide special finishes. Rubber and plastic sheets are commonly used. One technique that is used on panels for a deep reveal is to prepare an absorbent bed of sand within the form, hand-place the surface aggregate, and then place the matrix mortar, taking care that the mortar fills all interstices between the aggregate particles. Finally the reinforcing steel, inserts, and the backup concrete are placed. See Chap. 43 for finishing techniques.

42.8 Erection

Handling of precast members should be kept to a minimum. Working loads on lifting insets should be based on a factor of safety of at least 3, and the design engineer of record should specify or approve location and type of inserts. Many precast units are subjected to greater loads, and sometimes different loadings, during erection, from those to be experienced after the units are in place in the building, and this fact should be taken into consideration when designing the structure. Lift points should be located so the crane hook will be directly above the center of gravity of the unit while it is being lifted, thus keeping the bottom edge level. Figures 42.3 and 42.7 illustrate good rigging for wall panels.

Large wall members should not be laid flat after they have been removed from the molds. If they must be stored, it should be in a vertical position, with adequate support under the bottom edge and corners. Units should not be stored so the weight of one bears against one previously stored.

As soon as a unit has been erected in place in the structure, shores and braces must be installed while temporary connections are being made. Whether the temporary connections are adequate without bracing must be given very careful engineering analysis. Serious accidents have been caused by failures of temporary connections and bracing, and a

FIGURE 42.7 Rigging for a wall panel with a horizontal axis much longer than vertical.

little extra bracing is good protection against failure. In addition, good rigging practices must be followed when handling units.

42.9 Joints

As a rule, joints involving precast concrete units are not intended to be weathertight until they have been caulked. Usual construction is to insert a backup filler in the joint to restrict the depth of sealant and prevent the sealant from adhering to the back of the joint. Foamed polymers that change volume with changes in joint width are commonly specified. These include certain urethanes, ethylenes, vinyl chloride, and others.

After insertion of the backup filler, the sealant is applied, preferably with a power-actuated gun, in accordance with the manufacturers' instructions. Thermosetting plastics, such as a polysulfide, are preferred as joint sealants, although some thermoplastics and mastics have been found suitable. Oil-base compounds eventually dry out and are not suitable. Cement mortars do not possess the extensibility necessary for a good joint.

The minimum amount of sealant to make a satisfactory joint should be used. Usually most of the joint is filled with the premolded elastic filler, with the sealant used on the exterior ½ in of the joint. Sealant should be applied to a clean and dry joint. Do not apply during freezing weather.

42.10 Connections

Connections are defined as the hardware used for joining the precast elements to the building frame and to each other. Design of connections, a function of the structural engineer, should be taken care of when the members are designed, as both members and connections share the responsibility for structural adequacy with the building frame. Analyses of building failures resulting from earthquakes have shown that structural weaknesses leading to the failure were the result of improperly designed or installed connections, rather than any weakness of the building units.

Reference 1 is a good source of information on design of connections. This article points out that structural adequacy is of greatest importance, followed by architectural function, which requires that connections be neat and clean, essentially invisible, and compact. Finally, economy requires both ease of erection and ease of fabrication. For example, it is well to design a connection so it can be temporarily fastened with erection bolts, thus freeing the crane for other work while the permanent connection is made. Permanent fastening is accomplished by welding, bolting, or riveting.

CONCRETE BLOCK

Concrete blocks, or concrete-masonry units, are made in a variety of machines that vibrate, jolt, press, and otherwise consolidate a concrete of barely moist consistency.[2]

The manufacture of concrete masonry units involves six basic steps: (1) selection of raw materials of a type and quality, and of proper gradation in the case of aggregates, that will permit production of finished units meeting all pertinent specifications; (2) accurate batching of the raw materials to meet the requirements of selected mix designs; (3) thorough mixing of the batched materials to produce concrete of suitable workability; (4) molding in a machine properly adjusted to produce units of consistent quality from an adequate range of mix designs; (5) proper curing to develop product qualities that will minimize problems in initial handling and ultimately meet pertinent specifications with respect to strength and other characteristics; and (6) storage of finished units by methods that will promote the development of ultimate strengths and moisture content requirements.

42.11 Materials

All masonry units contain portland cement, aggregate, and water. In addition certain admixtures are frequently used, such as coloring pigments, air entrainers, pozzolans, accelerators, and retarders (see Part 1 of this handbook).

Cement can be standard Type I or II, depending mainly on location of usage. Where early strength is desirable, or to reduce curing time, Type III high-early-strength cement may be used. Most cement companies now produce a block cement which is of a lighter color than the standard or normal types, and approaches Type III in strength development. White cement is commonly used in such specialty block as split block, where the mortar color is especially important, and in decorative grille and screen units.

Aggregate should (1) be clean, and free of particles or substances that are unsound, dirty, or would interfere with strength and durability, or would cause surface imperfections; (2) have sufficient strength, toughness, and hardness to resist the expected loading; (3) be durable so as to resist freezing and thawing, temperature fluctuations, and changes in moisture content; and (4) have a uniform grading from finest to coarsest size to produce a workable, economical, and uniform mix.

Normal or heavy aggregates include sand, gravel, crushed stone, and blast-furnace slag. All sand should pass the No. 4 mesh screen, have 15 to 25 percent pass the 50-mesh and 5 percent minus 100-mesh, with a fineness modulus between 2.60 and 3.00. A maximum of 5 percent silt is usually not objectionable. Coarse aggregate passes the ⅜ in screen and is retained on the No. 8 with a tolerance of not more than 5 percent passing the No. 8. About 25 to 40 percent should pass the No. 4 screen.

Many block manufacturers use lightweight aggregates. Natural lightweight aggregates are pumice, diatomite, and scoria. Manufactured lightweight aggregates are expanded shale, expanded clay, water-quenched blast-furnace slag, and cinders. Grading for lightweight aggregate should be about the same as for regular aggregate. Other important properties are unit weight and absorption, which should be maintained at uniform values. Many lightweight aggregates are quite angular and harsh, and some, especially the very lightweight ones, are apt to be somewhat fragile. The desirable properties listed above for normal-weight aggregates apply also to lightweight. Aggregate should be free of sulfur compounds, excessive lime, iron particles, pyrites, and any material that might cause staining and popping.

Control of segregation is important for all types of aggregates, as segregation results in variations in grading. Grading variations cause variations in workability of the mix, water demand, texture and strength of the finished block, and yield (number of blocks) per yard of aggregate.

Water should conform to the requirements for mixing water for plastic concrete (Chap. 4).

Admixtures play an important part in block production. A block producer who is considering the use of any admixture should carefully evaluate it in his own plant, using his normal materials and methods, observing the effect of the admixture on all manufacturing operations and all properties of the block. The amount to use and the conditions under which it can be used depend on the mix, materials, equipment, weather, and many other factors.

An air-entraining agent causes the formation of many microscopic air bubbles in the mix, greatly improving durability of the block. It improves workability of the mix, especially with aggregates lacking fines, and improves surface texture and density of the block. Absorption of the block is usually decreased, and breakage or damage to green block is lessened.

The use of calcium chloride as an accelerator in reinforced precast concrete or masonry block is not recommended, as research and field experience has shown that it also accelerates or induces corrosion of the adjoining or encased reinforcement in the presence of moisture.

Some of the metallic stearates perform as waterproofing or capillarity-reducing agents, but their effectiveness is limited.

Pozzolans, such as fly ash, burned and pulverized shale, or pumicite, can be successfully used with some aggregates to improve the quality of the block. Pozzolans should be carefully evaluated by laboratory and plant tests before they are used in production.

42.12 Materials Handling

Most plants use bulk cement, delivered in special tank rail cars or trucks equipped with pneumatic pumps that convey the cement through a pipe to the top of the silo or bin. Some plants receive the cement in a ground-level hopper from which screw conveyors or bucket elevators carry it to the silo. Sacked cement is rarely used except in very small plants.

Cars or trucks discharge aggregates into ground-level hoppers from which inclined-belt conveyors or vertical bucket elevators lift the aggregate to overhead storage bins. Occasionally a clam shell unloads cars into stockpiles or moves aggregate from cars and

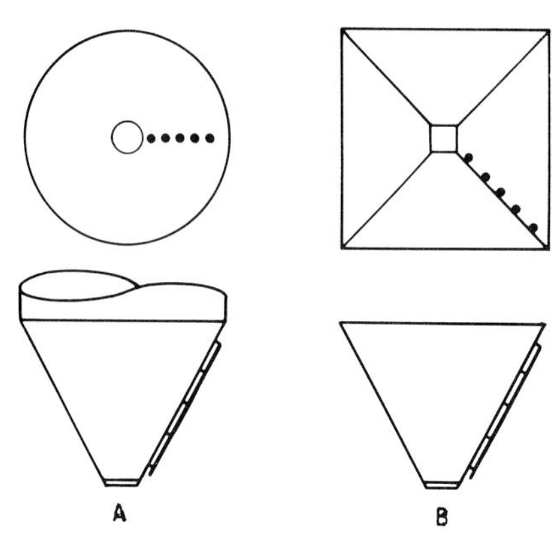

FIGURE 42.8 Installation of low-pressure air jets in the bottom of a cement silo fluffs the cement and keeps it in a freeflowing condition. Air pressure should not exceed 5 psi. The aerator fits on the outside of the silo and admits a large volume of air through a fabric held between a circular housing and a grill. (*a*) In a conical-bottom silo, one row of jets is usually adequate. (*b*) In a pyramidal-bottom square bin, a row of jets should be installed adjacent to each of the valleys. One row of jets is shown.

stockpiles into bins, but this system is seldom used any more because it is costly and inefficient.

Movement of materials, especially cement and other fine bulk material, is aided considerably by the proper installation of external vibrators on hoppers and bins. Low-pressure air jets, installed in a conical silo bottom just above the gate opening, aid in preventing arching and sticking of cement (Fig. 42.8).

42.13 Mixes

Combined aggregate, that is, minus ⅜ in gravel or other coarse aggregate plus the sand, must be so proportioned that the fineness modulus is between 3.5 and 4.0. Usually a mix will consist of 55 to 70 percent sand with 45 to 30 percent gravel. These proportions might be changed to 40 to 75 percent sand and 60 to 25 percent of crushed rock. Strength of the block for any given cement content is higher when the aggregate grading is coarser. However, sufficient fines are always necessary to produce a workable mix that will give regular and uniform surfaces, and minimize breakage.

Proportions of aggregate and the cement content have to be determined by trial mixes in the plant. Workability in the machine, cracking, surface texture, green strength, density, absorption, strength of the cured block, economy, and general appearance of the block all have to be considered. The maximum proportion of coarse aggregate that the mix will tolerate should be used. Sometimes this may be as little as 25 percent of the total aggregate but usually it is around 40 to 50 percent. Trial mixes should consist of varying proportions of coarse to fine aggregate, combined with different amounts of cement. Richest mix will probably be about 1 part cement to 6 parts of damp, loose mixed aggregate, ranging down to 1:10. Because of the very dry nature of the concrete, the water-cement ratio is not critical from a strength and durability standpoint.

Behavior of the mixture in the machine, appearance of the green block, and the ease of handling the green block all give clues as to the final quality of the block. Once a mix has been decided on, every effort should be made to keep all operations and materials the same and equipment in good operating condition.

When using lightweight aggregate it may be necessary to use less fine material in the mix than for normal-weight aggregate in order to obtain the required light weight, texture, insulating value, and acoustical properties. This is accomplished at a sacrifice of some of the strength.

Appearance of the blocks as they come off the machine is important in judging whether the amount of water in the mix is correct. Differences in machines and materials affect the amount of water. It is best to use the maximum amount permissible without causing the block to slump. If insufficient water is used, the block will have a dull gray color, with a "dry"-appearing texture. Too much water will cause the block to slump. The correct amount produces a water web or sheen on the molded faces of the block when it is demolded.

42.14 Batching and Mixing

Part 5 of this handbook covers descriptions of most types of batching and mixing equipment. In most modern plants, batching is done in automatic or semiautomatic batchers. The weigh batcher is positioned under the storage bins and a measured amount of each ingredient except water is discharged into the batcher. The batcher then moves to a position over the mixer and discharges the batch into the mixer while water is measured into the mixer. The admixture, if any, is introduced with the water. One mixer installation is shown in Fig. 42.9. Other mixers, both large and small, are frequently horizontal-shaft pugmill units.

Control of mix water is best accomplished by means of some type of moisture meter. One type has probes in the mixer that sense the moisture content of the batch, usually by electrical-resistance methods. A dial gauge shows the moisture content and enables the operator to adjust the amount of water. The meter can be made to actuate automatic controls that regulate the amounts of water and sand in the batch. Probes should be cleaned regularly and the equipment kept in adjustment to ensure proper operation.

Adequate mixing is essential and results in better block than short mixing. Adequate mixing results when the mixer is not overloaded, the time period is sufficiently long, and the proper consistency of the batch is obtained through control of the water. The mixer is overloaded if the batch in the mixer covers the mixer shaft; length of time period depends on the type of machine, batch proportions, and materials. It must be determined by trial.

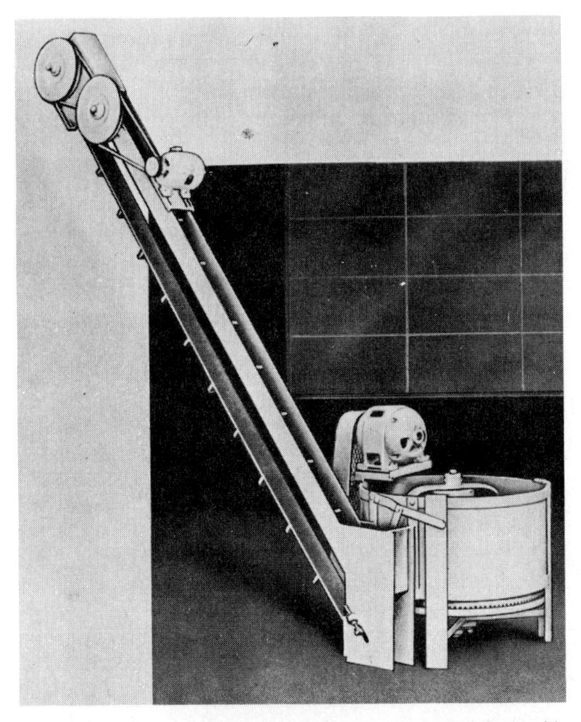

FIGURE 42.9 A small turbine mixer can be installed flush with the floor to facilitate charging. The inclined conveyor elevates mixed concrete to the block machine.

With lightweight aggregates, it may run about 7½ min, including about 1 min of prewetting of the aggregate before cement is added. Water control is best achieved by use of automatic equipment.

42.15 Making the Block

After mixing, the concrete batch is discharged into the block-machine hopper, from which controlled quantities of concrete are fed into the molds. Vibration and pressure consolidate the concrete in the molds, which rest on a steel or wood pallet during molding of the block. Output of the machine is based on the number of 8 × 8 × 16 in blocks, or equivalent amount of concrete, produced per machine cycle, the machine speed usually being about 5 cycles per minute. Figures 42.10 through 42.12 are views of block machines and plant layouts.

The pallet, usually of steel, but occasionally of wood, measures 18½ × 26 in, the 26 in length permitting working space to mold three 8 in blocks, two 12 in, four 6 in, six 4 in, or 12 bricks. Thus the machine, by changing the mold box, can produce any of the common modular units.

As soon as the concrete is compacted and struck off, the mold box lifts and the pallet holding the green blocks emerges from the machine, while a new pallet is pushed into the other side of the machine. The pallet containing the green blocks then travels to the curing area (Fig. 42.12).

Many of the steps are automated, even on some of the small machines. Push-button controls permit one-person operation of machines with a capacity exceeding a thousand 8-in blocks per hr. Block machines are ruggedly built, as they are subject to moisture, vibration, and exposure to abrasive materials, and proper maintenance is essential to keep them operating efficiently. Thorough cleaning and lubrication every shift are very important. As an aid in this respect, compounds are available that can be sprayed or brushed on

FIGURE 42.10 This semiautomatic block-making machine makes two 8 in high blocks at a time and a speed of up to 8 Hz. (*The Besser Co.*)

FIGURE 42.11 A large automatic block-making machine of this type can make 1400 8 × 8 × 16 blocks per hour. (*The Besser Co.*)

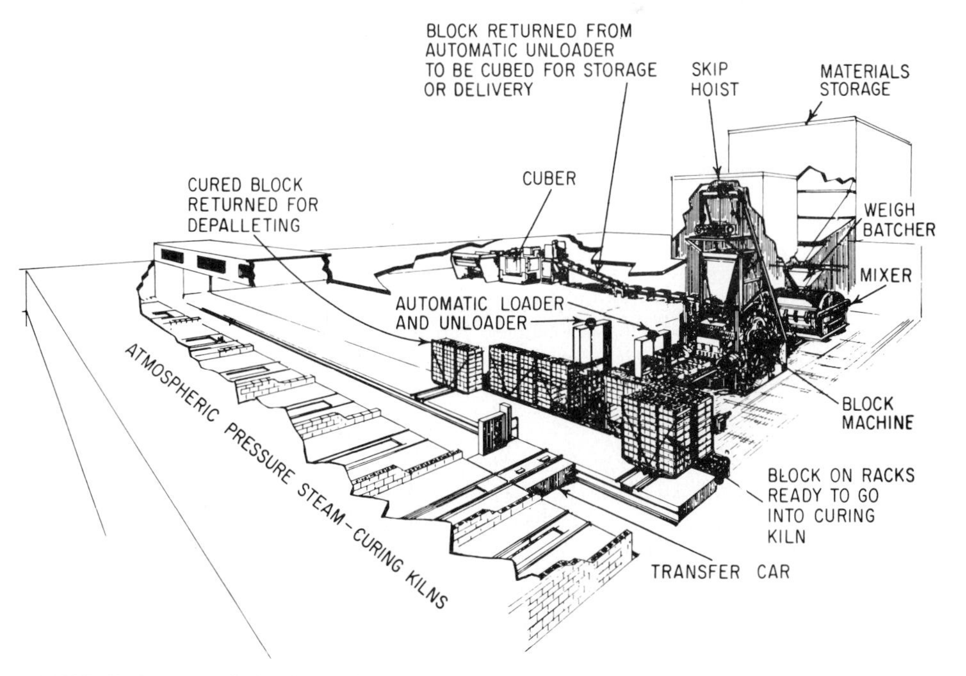

FIGURE 42.12 Mechanical equipment in the several stages of block manufacturing serves to automate this one-car automatic plant with atmospheric curing. Many combinations of equipment are available for automation of almost any type of plant. (*The Besser Co.*)

surfaces to prevent adhesion of concrete. Pallets must be kept clean and mold boxes checked for wear; those showing excessive wear should be discarded. Blocks should be checked for density and dimensions occasionally to be sure they are maintaining the proper quality.

Offbearing, the process of moving the block off the machine, in many plants is either semiautomatic or automatic. In some plants this operation is performed manually. The loaded pallets are placed in steel curing racks which are now taken to the curing kiln or autoclave. Transporting of racks is accomplished with lift trucks, casters, small rail cars, or monorail.

42.16 Curing

High-temperature steam curing, almost universally practiced, can be accomplished with saturated steam at atmospheric pressure, or can be accomplished by means of high-temperature high-pressure steam in a pressure vessel called an *autoclave.* Curing is discussed in Chap. 29.

Atmospheric-pressure steam–curing rooms, called *kilns,* should be tight so as to minimize heat loss. Either hinged or vertical sliding doors can be used, or sometimes heavy drop curtains. Walls and roofs should be made of insulating material, such as lightweight-aggregate concrete or block, and the floor should be concrete. Adequate drains must be provided. The ideal situation is one in which every block in the room is exposed to exactly the same curing cycle. This ideal is seldom achieved, but adequate space around all racks of block, to permit free movement of the saturated warm air, will help to approach the ideal. There should be a clearance of 18 in above the highest block in the racks, and 6 in. on all sides of each rack.

Most kilns and steam-curing equipment are improvisations developed by the block maker personally. There apparently is no general standard for the construction of kilns. They are made from all types of materials, from a canvas-covered frame and canvas flap door to a concrete or concrete-block building with seal-tight quick-opening doors. However, a good kiln has walls, top, and door correctly insulated against heat losses and has a door that will give a tight seal, preventing cold air from entering the kiln and steam from escaping.

The best method in block curing is to have a set of kilns in which the various operations of loading, steaming, soaking, and unloading can be carried on separately. In this way, production is in chainlike manner; that is, loading one kiln, unloading another, steaming another, and soaking still another, all at the same time.

Some block makers load an entire 8 h output of blocks in one large curing room. Then they apply steam to the entire lot at one time. This apparently is not the ideal method, as it requires the continuous-chain production method without any increase in output.

The most important part of steam curing is to know the temperature of the block throughout the entire curing cycle. If heat is controlled during the entire curing cycle, then the block maker knows that each batch of blocks is of precisely the same quality and color as the preceding one. Since a temperature rise of 60° per hr is permissible in a block, it is possible with the proper steam equipment to carry out the steaming operation in a matter of 3 or 4 h.

Steam is evenly distributed in the kiln by means of a perforated steam header. Condensate can be allowed to drain out the door of the kiln, or through special drains. Thermometers are usually installed in the kiln so that they can be read from outside the kiln.

FIGURE 42.13 Autoclaves are used for high-pressure high-temperature steam curing of block, pipe, and similar precast materials.

High-pressure steam curing in an autoclave produces blocks which are lighter in color, which exhibit greater dimensional stability, that is, less shrinkage, and which are considerably drier when removed from the kiln, compared with blocks cured under atmospheric conditions. One-day strength of autoclaved block is about equal to standard 28 day strength.

The autoclave, shown in Fig. 42.13, is a long, tanklike steel cylinder of diameter large enough to accept racks of blocks. Blocks from the machine are held from 1 to 4 h before placing in the autoclave to permit initial hydration of the cement and hardening of the blocks. Next the blocks are heated gradually in the autoclave so that full pressure is reached in not less than 3 h. Units are then steamed for 5 or 6 h at a temperature of 325 to 375°F. Thick units require slightly more time. After the steaming, or soaking, period, pressure is rapidly released over a period of ½ h or less so the blocks will dry as much as possible. This rapid pressure release removes sufficient moisture from normal-weight blocks to bring them to a relatively stable air-dry condition. Additional drying is usually necessary for lightweight block.

Heat Computations. Augmenting the discussion of steam curing in Chap. 29, the following computations illustrate the steps to determine the necessary boiler capacity for a high-temperature steam-curing system in a block plant.

In any high-temperature curing system, the following information is required:

1. Mean average low outside air temperature (if autoclave or kiln is set outside in the weather).
2. Mean average low air temperature of the room (if autoclave or kiln is set inside building).
3. Desired curing temperature of the blocks.
4. Time required to raise temperature to curing temperature.
5. Total weight of blocks to fill autoclave or kiln.

6. Total weight of racks for holding blocks.
7. Total weight of pallets for holding blocks.
8. If more than one autoclave or kiln is used, determine cycle for peak steam load.
9. Operating pressure and temperature of autoclave. Operating temperature of kiln.
10. If autoclave or kiln is insulated, determine K factor for radiation loss.
11. If autoclave or kiln is not insulated, determine K factor for radiation loss.
12. Complete dimensions and weight of autoclave (diameter, length, and thickness). Dimensions of kiln.
13. Other steam requirements—such as water heating, heating aggregate, and space heating.

Kiln Curing. By far the largest steam requirement occurs during the time the loaded kiln is brought up to the curing temperature. This includes bringing up to temperature the kiln, the concrete blocks, the concrete-block racks, the pallets, and any other handling equipment that is contained within the kiln during curing.

The general formulas used are as follows:

$$\text{hp} = \frac{\text{weight of all items} \times \text{specific heat} \times \text{temp. rise(h}^{-1})}{33,500}$$

(temperature rise should be at the rate of 60°F/h)

$$+ \text{ steam load to fill kiln} = \frac{\text{ft}^3 \text{ of kiln} \times \text{Btu/lb of steam}}{33,500 \times \text{volume/lb of steam}} + \text{radiation loss of kiln}$$

$$= \frac{\text{heat - transfer coefficient} \times \text{surface area} \times \text{temp. difference}}{33,500}$$

Because there is a difference of only 2 or 3 hp between heatup and holding rates, the full radiation rate has been used, to simplify computations in the following general formula:

$$\text{hp} = \frac{W(60°) \text{ specific heat}}{33,500} + \frac{\text{volume (Btu/lb)}}{\text{ft}^3/\text{lb} \times 33,500} + \frac{(K \text{ factor})(\text{surface area})(T_1 - T_2)}{33,500}$$

Example—Kiln Steam-Curing Concrete Blocks. Assume the following:

1. Kiln located outside building
2. Minimum low outside temperature +20°F
3. Initial temperature of all materials including concrete blocks, racks, etc., +40°F
4. Final temperature of above material +165°F
5. Size of kiln, 10 ft wide × 7 ft 6 in high × 30 ft long; 2 doors 10 × 7 = 140 ft^2; exposed walls and roof = 780 ft^2; cubical contents 2250 ft^3
6. Racks (14) for holding concrete blocks, 700 lb each × 14 = 9800 lb
7. 350 pallets at 50 lb each = 17,500 lb
8. Weight of steam and condensate distributing pipe, say, 700 lb
9. Weight and number of concrete blocks, 1000 at 44 lb each = 44,000 lb
10. Specific heat of concrete blocks, 0.25

11. K factor for exposed walls for kiln 0.25; for doors for kiln, 1.00; for cubical contents of kiln, 0.02
12. Temperature rise $(T_1 - T_2) = 165 - 40 = 125°$
13. Heatup time at 60°F/h = 125⁄60 = 2.08 h (some authorities recommend a maximum rate of temperature rise not to exceed 40°F/h)

Estimate:

Weight of racks	9800 lb
Weight of pallets	17,500 lb
Weight of pipe	700 lb
Total	28,000 lb

$$\frac{28,000 \times 0.12 \times 60° \text{ temp. rise } (\text{h}^{-1})}{33,500} = \text{say, } 6 \text{ bhp}$$

Concrete blocks:

$$\frac{44,000 \times 0.25 \times 60° \text{ temp. rise } (\text{h}^{-1})}{33,500} = \text{say, } 19.7 \text{ bhp}$$

Steam to fill kiln:

Operating pressure = 75 psi
Latent heat, Btu/lb of steam at 75 psi = 894.7 Btu
Volume of steam at 75 psi = 4.896 ft^3/lb of steam
cu ft of kiln = 2250 less load, say, 50% = 1125 ft^3

$$\frac{1125 \times 894.7}{33,500 \times 4.896} = \text{say, } 6.2 \text{ hp}$$

Radiation loss kiln:

$$T_1 - T_2 = 165°\text{F} - 20°\text{F} = 145° = \text{temp. rise}$$

Doors

$$\frac{140 \text{ ft}^2 \times 145° \times 1.0}{33,500} = \text{say, } 0.61 \text{ hp}$$

Exposed walls

$$\frac{780 \text{ ft}^2 \times 145° \times 0.25}{33,500} = \text{say, } 0.85 \text{ hp}$$

Cubic contents

$$\frac{1125 \text{ ft}^3 \times 145° \times 0.02}{33,500} = \text{say, } 1.10 \text{ hp}$$

	1.56 hp
Plus 10% safety factor	0.16 hp
	1.72 hp

Weight racks	6 hp
Concrete blocks	19.7 hp
Steam to fill kiln	6.2 hp
Radiation loss	1.7 hp
	33.6 bhp/h required during heatup
Say 10% for transmission-line losses	3.4 hp
Total	37.0 bhp/1000 block/h during heatup

After 3 h heating time, soaking time requires

Radiation loss	1.7 hp
Transmission-line loss	2 hp
	3.7 hp

To these estimates, any additional steam load such as space heating, heating stockpiles, heating water, steam jetting hoppers or railroad cars, or steam cleaning must be added to the total horsepower load. Also, future plant expansion must be considered; the steam generator should be large enough, or provisions should be made for the installation of additional units.

Autoclave Curing. As in the kiln method, the largest steam requirement occurs during the time the loaded autoclave is brought up to curing temperature. This includes bringing up to temperature the autoclave, the concrete blocks, concrete-block racks, pallets, and any other handling equipment that is contained within the autoclave during curing. The general formulas used are as follows:

$$\text{hp} = \frac{\text{weight of all items including autoclave} \times \text{specific heat} \times \text{temp. rise}}{33,500 \times \text{min temp. raising time, h}}$$

$$+ \text{ steam load to fill autoclave} = \frac{\text{ft}^3 \text{ of autoclave} \times \text{Btu/lb of steam}}{33,500 \times \text{volume/lb of steam}}$$

$$+ \text{ radiation loss of autoclave}$$

$$= \frac{\text{heat - transfer coefficient} \times \text{surface area} \times \text{temp. difference}}{33,500}$$

The general formula is as follows:

$$\text{hp} = \underset{\text{Items}}{\frac{W(T_1 - T_2)\text{ specific heat}}{33,500 \times \text{h}}} + \underset{\text{Autoclave}}{\frac{\text{volume (Btu/lb)}}{(\text{ft}^3/\text{lb}) \times 33,500}} + \underset{\text{Radiation loss}}{\frac{(K\text{ factor})(\text{surface area})(T_1 - T_2)}{33,500}}$$

Note. To simplify estimate, 33,500 Btu/hp was used in place of 33,475 Btu per hp.

Example—Autoclave Steam-Curing Concrete Blocks. Assuming the following:

1. Autoclave located outside building
2. Minimum low temperature +40°F
3. Initial temperature all materials including blocks, racks, etc., +40°F

4. Final temperature of above material +350°F
5. Size and weight of autoclave, 8 ft 0 in diameter × 90 ft 0 in long, ⅞ in thick, specific heat 0.12, weight of steel per ft^2 = 35 lb; two heads 8 ft 0 in diameter (πR^2) × 2 = 100.5 ft^2; shell, 90 ft 0 in. long ($\pi D \times 90$) − 25.12 × 90 = 2261 ft^2 + 101 = 2362 ft^2; 2362 × 35 = 82,670 lb
6. Racks (28) for holding concrete blocks 700 lb each (28 × 700 = 19,600 lb)
7. 700 pallets at 50 lb each = 35,000 lb
8. Weight steam and condensate distributing pipe (assume 1000 lb)
9. Weight and number of blocks, 2000 at 44 lb each
10. K factor for insulated autoclave = 0.29 (0.29 Btu/h/ft^2/temp. difference = 0.29 × 310 = 90 Btu/loss/ft^2)
11. Specific heat of concrete block −0.25
12. Heatup time 3 h
13. $T_1 - T_2 = 350 - 40 = 310$°F

Estimate:

Weight autoclave	82,670 lb
Weight racks	19,600 lb
Weight pallets	35,000 lb
Weight piping	1000 lb
Total	138,270 lb

$$\text{hp} = \frac{138,270 \text{ lb} \times 0.12 \times 310}{33,500 \times 3} = \text{say, } 51 \text{ hp}$$

Concrete blocks:

$$\text{hp} = \frac{2000 \times 44 \times 0.25 \times 310}{33,500 \times 3} = \text{say, } 68 \text{ hp}$$

Radiation loss:

$$\frac{90 \times 2362}{33,500} = \text{say, } 6 \text{ hp}$$

Steam to load autoclaves:

Latent heat of steam at 150 psi = 857.0 Btu
Volume of steam at 150 psi = 2.752 ft^3/lb

Ft^3 of autoclave = area × length = 50 × 90 = 4500 cu ft (deduct for block load) = say, ca. 50% or 2250 ft^3

$$\text{hp} = \frac{\text{volume (Btu/lb)}}{(\text{ft}^3/\text{lb})(\text{h})\ 33,500} = \frac{2250 \times 857.0}{2.752 \times 3 \times 33,500} = \text{say, } 7 \text{ hp}$$

Total hp required to heat autoclave in 3 h:

Heat-handling material, say	51 hp
Heat product, say	68 hp
Radiation loss, say	6 hp
Load autoclave	7 hp
Total	132 hp
Say plus 10% for transmission-line losses	13 hp
Total hp required per hour for first 3 h	145 hp

After 3 h, steam required for heating time, soaking time requires, per hour

Radiation loss	6 hp
Transmission-line loss	13 hp
Total	19 hp

In total, 145 hp − 19 hp = 126 hp available for the second autoclave or other steam requirements.

42.17 Carbonation

It has been demonstrated that exposure of concrete to carbon dioxide in a normal atmosphere will result in shrinkage of the concrete. Tests by the Portland Cement Association and the National Concrete Masonry Association have shown that precarbonation of concrete-masonry units reduces shrinkage by as much as 50 percent under subsequent exposures to carbon dioxide or cycles of wetting and drying. Conclusions based on the PCA-NCMA tests[4] were as follows:

1. The relative humidity in the kiln should be between 15 and 35 percent.
2. The CO_2 content should be as high as possible but not less than 1.5 percent.
3. The temperature should be between 150 and 212°F.
4. Generally a 24-h treatment will be required, but under optimum conditions a shorter period may provide considerable reduction in shrinkage.
5. Potential shrinkage reduction of more than 30 percent may be obtained under favorable conditions.

Practice in the industry has been to introduce waste flue gas into the kiln after the blocks have been cured and are in the drying period. During this period, the blocks must be exposed to conditions that dry the blocks to a favorable moisture level, maintaining that level during treatment, and that provide the correct CO_2 concentration to effect rapid carbonation.

Equipment is available that provides an automatic cycle to cure the block in this manner. The system includes equipment that steams, dries, and carbonates the concrete in accordance with the following program:

Preset. Required of all concrete to be cured at any temperature above normal atmospheric temperature. During this period, air is circulated through the kiln to maintain a uniform atmosphere.

Steam. A gas-burning heater is activated to raise the kiln temperature, evaporate water for steaming, and introduce the products of combustion into the kiln to provide carbon dioxide.

Drying. Steam generation is stopped, but dry heat and carbon dioxide continue to enter the kiln, drying and carbonizing the block.

Exhaust. The kiln is now vented to remove moisture and carbon dioxide.

An installation of equipment of this type is shown in Fig. 29.5.

42.18 Handling and Storage

Cubing is the process of assembling blocks into convenient groupings for further handling. On completion of curing, racks of block are conveyed to the cubing station, where the cubes are assembled either by hand or by automatic machinery. A cube consists of five or six layers of 15 to 18 blocks, 8 × 8 × 16 in, or an equivalent volume of other size of block (Fig. 42.14). Other sizes of cubes may be used, depending on local custom and facilities. It usually requires four laborers to cube the production of one three-at-a-time block machine, but a semiautomatic cuber can do the job with two workers. After cubing, the blocks are moved to storage.

Lift trucks, or forklifts, are extensively used for moving racks and cubes of block in the yard. These machines, normally operating on pneumatic tires, are very versatile and can operate in unpaved yards. By cubing the blocks on pallets, or cubing them so the tines of the fork lift can slide through the cores of the bottom course, the entire cube can be lifted into the stockpile or onto a truck.

42.19 Properties of Masonry Units

Types of block are briefly discussed in Chap. 40, and common types are shown in Fig. 40.1. Screen and grille blocks offer an almost infinite number of interesting and artistic designs, and the introduction of textures and color (including pure white) has made available a wide array of specialty precast products including bas-relief surfaces, garden accessories, and other special-use units. Different textures can be obtained by use of different aggregate gradations, by changing the mix consistency, or by mechanical means.

While concrete blocks are similar to structural concrete in many respects, there are important differences. In general, quality of the product depends on the qualities of materials, mix, and work, the same as structural concrete. However, block mixes contain less cement (260 to 375 lb/yd^3) and less water. Aggregate rarely is larger than ⅜ in and is frequently lightweight. Curing is nearly always at elevated temperatures.

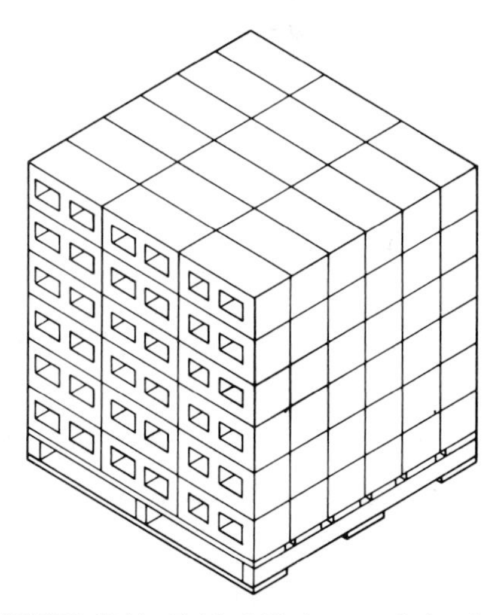

FIGURE 42.14 Finished blocks are cubed and palletized for ease in handling. This pallet contains 108 8 × 8 × 16 in blocks. The pallet shown here is not the type of pallet that is used in the block machine, but constructed of 2 × 4's and 1-in boards; it is commonly used for palletizing many kinds of merchandise.

Strength. Compressive strength is determined on the entire block and is computed on the gross area of the block, that is, overall dimensions, including core space. Compressive strength based on net area, excluding core space, is about 1.8 times the gross-area value. Compressive strength is influenced principally by the type and amount of cement per unit, type of aggregate, grading of aggregate, degree of compaction attained in the molding machine, age of the specimen, curing, and moisture content at time of test. Lightweight blocks are usually not as strong as heavy ones.

Water Absorption. A good indication of the density of block is obtained from the absorption value. Absorption varies widely, from about 10 to 20 pcf depending on type of aggregate and amount of compaction of the block. Permeability, thermal conductivity, and acoustical properties are also influenced by absorption. Although high absorption is not desirable, it of necessity accompanies lightness in weight and desirable thermal and acoustical properties. Block for exterior exposed walls should have low absorption.

Volume Changes. Masonry, the same as any concrete, undergoes changes in dimensions due to changes in temperatures and moisture content. One of these changes is the original drying shrinkage of green concrete. If blocks are laid up in a wall before they have taken their drying shrinkage, tensile stresses are developed that are apt to lead to cracks in the wall. By proper curing and drying of block in the yard, the moisture content of the block can be reduced to an equilibrium condition with the surrounding exposure and cracking tendencies are minimized.

Texture and Color. Texture is the disposition of particles and voids in the surface of the units. The desired texture can be obtained by adjusting (1) the aggregate gradation, (2) the amount of mixing water, and (3) the degree of compaction in the mold. In addition, special face molds and rubber or plastic mold liners are available for special finishes, and the units can be ground with an abrasive, or the aggregate can be exposed by chemical means. The desire for color and texture has led to the use of bonded facings, such as those obtained with thermosetting resin binders with silica sand, and vitreous glazed surfaces.

Color depends mainly on the materials, although autoclaving results in a lighter color. Different cements are of different shades of gray. Block cement is lighter in color than regular cement, and white cement results in white units. Pigments, available from reputable producers, make possible a veritable rainbow of colors. The use of white cement with pigments results in clean, pastel colors, and gray cement gives the dark colors. The choice of cement, aggregate, and colors is especially critical for split block. The many types of standard and specialty blocks are described and illustrated in Chap. 40.

42.20 Specifications and Quality

The most commonly accepted specifications are those of the American Society for Testing and Materials (Table 40.3). The federal government and other agencies also have their own specifications. Most building codes, including the Uniform Building Code, refer to ASTM specifications.[6] Testing of block is covered by ASTM Designation C140. The number of samples required for testing is shown in Table 42.1.

An industry program is the "Q block quality-control program."[5] sponsored by the National Concrete Masonry Association. Manufacturers participating in this program are required to have their products tested by an independent testing laboratory. The NCMA reviews the laboratory reports and issues a certificate of quality to the qualifying producer, who is then authorized to market the product as "Q block."

TABLE 42.1 Number of Concrete-Masonry Samples

	No. of specimens	
No. of units in lot	For strength test only	For strength, absorption, and moisture content
≤10,000	3	6
10,000–100,000	6	12
>100,000	3/50,000	6/50,000

CONCRETE PIPE

Methods for making precast-concrete pipe include tamping, compression by means of a revolving packer head, centrifugally spun, wet cast, and dry cast. The pipe can be broadly classified into pressure pipe and nonpressure pipe, for which there are a number of specifications, as shown in Table 42.2.[7]

Nonreinforced pipe of 4 to 24 in diameter, classed as agricultural pipe, is used for irrigation and drainage of farmland. Irrigation pipe may have tongue-and-groove joints sealed with cement mortar or some type of rubber coupling. Drain tile may have plain butt ends or tongue-and-groove ends that are left open to admit drainage water.

Culverts and storm sewers, 4 to 36 in in diameter, are usually nonreinforced; large pipe is reinforced. Culvert joints may be cement mortar or rubber gasket. Storm sewers usually have rubber-gasketed joints.

Nearly all pressure pipe is laid with a gasket joint of some type. All pressure pipe is reinforced, some is prestressed circumferentially, and some consists of a steel cylinder with mortar on both the inside and outside. Pressure pipe is used for municipal water supply, pressure lines for irrigation, pressure sewer mains and outfalls, siphons, and in any installation where there is internal hydraulic pressure on the pipe.

42.21 Packer-Head Process

In the packer-head process a no-slump, moist concrete mix is fed into a stationary split cylindrical form, or jacket, and is packed or compacted radially outward against the inside of the form by the revolving and rising shoe, called a "packer head" or "roller head." Starting at the bottom of the form, the packer head slowly rises as it rotates at high speed, compacting the concrete that is fed into the open top of the mold as the packer head rises. Speed of revolution is about 250 r/min. This method is restricted to cylindrical pipe, and is adaptable to making both reinforced and nonreinforced pipe. Range of sizes produced is from 4 to 48 in in diameter with lengths up to 8 ft. Pipe can be made in either the bell-up or bell-down position. Figures 42.15 through 42.17 are views of three sizes of packer-head machines.

As soon as the packer head travels its course, the jacket containing the pipe is taken to the curing area and the jacket is immediately removed. Because of the dryness of the mix and degree of compaction, the freshly made pipe has sufficient "green strength" to stand without support. In some processes a vibrating core follows the packer head through the pipe-making sequence. It is retracted into a pit below the machine before moving the pipe.

TABLE 42.2 Listing of Standard Specifications and Sizes of Concrete Pipe*

Specification designation	Type of pipe	Diameter range, in
ASTM C412	Concrete drain tile	
Standard quality		4–12
Extra quality, special quality and heavy-duty extra quality		4–36
ASTM C14	Nonreinforced concrete	
Class 1	Sewer, storm drain, and culvert pipe	4–36
Class 2		4–36
Class 3		4–36
ASTM C76	Reinforced-concrete culvert, storm, drain, and sewer pipe	
Class I		60–144
Class II, III, IV, V		12–144
ASTM C118	Concrete irrigation and drainage pipe	4–24
ASTM C361	Reinforced concrete, low-head pressure pipe	12–108
ASTM C444	Perforated-concrete pipe	
Type 1		4–27 and larger
Type 2		4–27 and larger
ASTM C443	Joints for circular concrete sewer and culvert pipe using rubber type gaskets	
ASTM C478	Precast concrete manhole sections	48 & larger by design
ASTM C505	Nonreinforced irrigation pipe with rubber-type gasket joints	6–24
ASTM C506	Reinforced arch culvert storm drain and sewer pipe	Equivalent round 15–132
ASTM C507	Reinforced elliptical culvert storm drain and sewer pipe	Equivalent round 18–144 36–144
ASTM C655	Reinforced-concrete D-load culvert, storm, drain and sewer pipe	
ASTM C497	Method of tests	
ASTM C789	Precast reinforced-concrete box sections for	36 × 24 to 144 × 144
Table 1		
Table 2		
Table 3		
ASTM C822	Definitions of terms relating to concrete pipe and related products	
ASTM C850	Precast reinforced-concrete box sections for culverts, storm drains, and sewers with <2 ft of cover subjected to highway loadings	24 × 36 to 144 × 144
Table 1		
Table 2		
Table 3		
ASTM C877	External sealing bands for noncircular concrete sewer, storm drain, and culvert pipe	18–144
ASTM C923	Resilient connectors between concrete manhole structures and pipes	

FIGURE 42.15 Short lengths of small drain tile can be made on this machine, shown making 6 in drain tile with butt joints.

42.22 Tamped Process

Both reinforced and nonreinforced pipe can be made by the tamped process, in which a split jack and base plate, resting on a turntable, rotate about a stationary core. Dry-mix concrete is fed into the top of the annular space and is tamped by metal- or steel-shod wooden tamping bars striking at a rate of 430 to 600 blows per minute. As the pipe wall is built up, the tampers move upward also, at the same time striking the surface of the concrete with a constant amount of energy to compact the concrete uniformly from bottom to top of the pipe being made. Rotation of the jacket rotates the concrete about the stationary core, providing a troweling action that smooths the interior of the pipe. On completion of tamping, the pipe is removed from the machine with either the inner or outer form and moved to the curing area where the form is removed in Fig. 42.18.

42.23 Centrifugal Process

In the centrifugal process, low-slump concrete is placed in a horizontal mold rotating at slow speed, just fast enough so centrifugal force holds the concrete against the mold. The rotational speed is slowly increased, causing centrifugal force to compact the concrete and drive out the excess water, which, being lighter than the aggregate and cement, migrates to the interior of the pipe where it can be removed by light troweling or brushing.

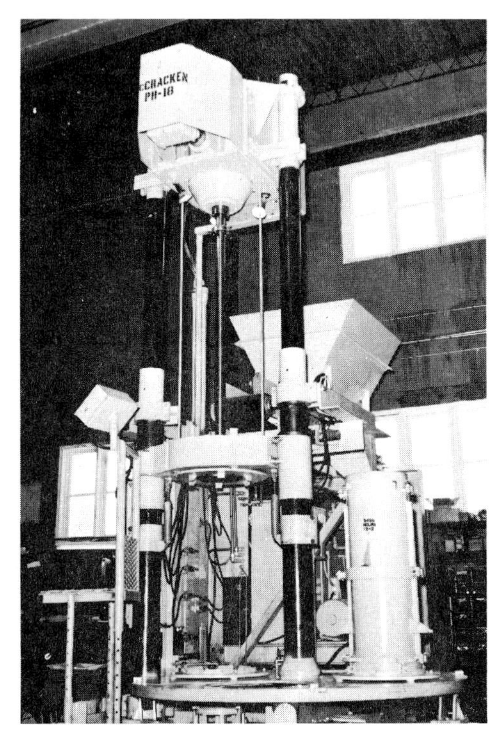

FIGURE 42.16 A medium-sized high-capacity packerhead machine with a size range of 4 to 18 in in 4 ft lengths, for producing gasket-joint pipe using machined socket curing pallets or mortar-joint pipe using socket formers. The complete pipe-making cycle can be automatic or manual with an operator.

High-strength dense concrete results from the centrifugal process. Slump of the concrete does not exceed 2 in when it is placed in the form. During spinning a large portion of the water is removed, leaving a concrete with a very low water-cement ratio. Pipes from 12 to 78 in in diameter and from 8 to 16 ft long are made by centrifugation. Pipe may be either reinforced or nonreinforced.

The so-called cylinder pipe consists of a steel cylinder in which the pipe concrete is spun, the cylinder thus becoming a part of the pipe. After the interior concrete or mortar lining has been made, a cement-mortar covering is shot onto the exterior of the steel shell. For high-pressure service (up to 350 psi) the exterior of the shell is wrapped with high-tensile-strength steel wire which is then stressed to about 135,000 psi, in effect applying a circumferential prestressing to the pipe. After the wire is stressed, the mortar coating is applied.

Steel rings for the bell-and-spigot joint are welded to the steel cylinder prior to spinning. To ensure accuracy of fit of the steel joint rings, they are stretched on a mandrel just beyond their elastic limit. Rubber gaskets are used in the joints.

A variation of the centrifugal process is one in which dry-mix concrete is placed in the rotating mold, compaction being achieved by means of a combination of centrifugation,

FIGURE 42.17 One of the largest packer-head machines available, this one makes 72 in pipe in 8 ft lengths. (*Concrete Pipe Machinery Co.*)

FIGURE 42.18 While the worker with the dolly picks up the empty mold and returns it to the pipe machine, the other worker carefully removes the split jacket, setting it aside to be "buttoned up, after removing and ragged pieces of concrete and smooths the groove end of the pipe.

FIGURE 42.19 A Cenviro machine for making pipe by a combination of centrifugal force, vibration, and pressure.

vibration, and compaction with a roller as shown in Fig. 42.19. After the requisite amount of concrete has been deposited in the form, vibration is applied to the form, and a heavy roller is brought to bear on the concrete in the mold.

Defects in spun pipe are rare when proper manufacturing techniques are followed. When they do occur, the manufacturer can eliminate them, as in any other process, by eliminating the cause. Occasionally a seam crack may occur, caused by leakage of mortar through the jacket gate during spinning. Other defects might include exposed steel, usually due to faulty alignment of the cage, rather than to faulty cage dimensions; inside diameters not within allowable tolerances; rough interior, frequently caused by attempting to fill an underfilled pipe after it has spun several minutes, or lightweight-aggregate particles; crooked gasket ring in spigot, misalignment of the ring ends, or ring not perpendicular to axis of pipe; and rough, sandy spots resulting from excessive form oil.

At times blisters, or drummy areas, in which the interior concrete separates from the main body of the pipe wall, occur in spun pipe. These blisters are attributed to a number of conditions, including concrete too soft or too wet (high slump), insufficient spinning, starting steam curing too soon after spinning, and steaming at too high a temperature (above 155°F).

Blisters are usually worse during wet, cold weather than at other times. Probably all the above factors contribute to their formation at times. It has been found that, by giving the pipe its initial steam curing in a vertical position, or at least steeply inclined, the blisters are largely eliminated. The effect of lower-slump concrete is definite in reducing the incidence of blisters.

42.24 Cast Pipe

Large pipe, usually larger than 48 in in diameter, is sometimes cast in vertical forms, using plastic concrete mixes and following normal good practices for structural concrete.

Forms must be tight. Joints and gates in the form jackets and inside liners, and the joint with the base ring must be sealed with gaskets or tape to prevent the loss of mortar that will result in sand streaks. To facilitate placing concrete in the form, a cone or plate should be placed on top of the form so concrete will feed into the annular space between the form jacket and inner liner. Usually several forms are set up at one time to give maximum production.

Concrete should be placed in shallow lifts and vibrated with high-speed form vibrators operating at speeds in excess of 8000 r/min. Speeds of this magnitude dictate the use of pneumatic or high-cycle electric vibrators. Vibrators should not be operated on empty areas of forms but should run while concrete is being placed in their vicinity and until the level of the concrete is well above them; then they should be stopped. Usually pipe is cast spigot end up. As the level of concrete rises in the form and approaches the top, particular care is necessary to avoid overvibration that produces weak concrete in the spigot. Slump should be kept low and the amount of vibration reduced.

Revibration is sometimes helpful in reducing the incidence of horizontal settlement cracks occasionally found near the top of the pipe. Revibration should be delayed as long as possible. The best time is just before the concrete loses its plasticity, and depends on temperature, slump, presence of admixtures, and other factors.

Soft spots in the end of the pipe cast against the base ring will result if an excess of form oil is allowed to accumulate and is not removed before placing concrete.

42.25 Mixes

Most manufacturers are unwilling to divulge information relative to the mixes they use, and it is necessary to proportion mixes based on commonly known principles within the limitations of specifications and pipe-making processes.

Some of the ASTM specifications set forth certain limits for some mix properties, and these limits can be used as a starting point. Trial mixes are necessary in order to achieve the most economical mix that will perform satisfactorily. Maximum size of aggregate depends on the size of pipe and method of manufacture. Packer-head and tamped pipes rarely contain aggregate larger than ⅜ or ½ in. Although ASTM C76 and similar specifications require a minimum cement content of 5 sacks (470 lb) per cubic yard, the need for plasticity, especially of packer-head and tamped pipe, makes more cement necessary. Some producers claim that the use of an air-entraining agent improves plasticity. The amount of air-entraining agent required, in the case of tamped pipe, is about half the amount normally used in concrete, but for packer-head pipe the amount may be twice as much as normal. Normal air entrainment would be of benefit for cast pipe, but there is no value in using air-entraining agents in centrifugal pipe.

As previously pointed out, pipe concrete is dense concrete. Density is achieved mechanically by the several methods of compacting and forming the pipe. In all methods, the mix is very dry. Cast and centrifuged pipe mixes have a slump of 2 in, ranging down to almost zero. Packer-head and tampered mixes are barely damp, sometimes called less than zero slump because considerably more water can be used in the mix before any indication of slump is obtained in a slump test. Water content may be as low as 3½ gal (30 lb) per cwt (hundredweight) of cement.

Materials should conform to applicable ASTM or local code requirements, should be measured accurately, preferably by weight, and should be well mixed. Plasticity of the fresh concrete is essential in all processes, but too much plasticity can lead to trouble. It might create a mushy mix which leads to broken tamping sticks, or it might cause the freshly stripped pipes to slough. The maximum proportion of coarse aggregate gives the best strength, but this has to be balanced against density and permeability.

42.26 Reinforcement

All pipe of 36 in diameter and over is reinforced with longitudinal bars and heavy wire wound into a cage. Pipe as small as 12 in, when subjected to internal hydrostatic pressure or heavy fill loading, is reinforced.

Reinforcing steel consists of longitudinal bars around which the circumferential steel is wrapped in a continuous spiral as shown in Fig. 42.20. Resistance welds are made at each crossing of the longitudinal bar and circumferential steel. In pipe designed to bear especially heavy fill loads, the circumferential steel may consist of two concentric cages, or a single cage may be formed into an elliptical shape. When elliptical cages are used, it is essential that some positive means of identifying the axes of the ellipse on the finished pipe be provided, as the long axis of the ellipse must be horizontal whenever the pipe is laid. Cases have been reported in which large pipe cracked in storage because it was laid down with the long axis of the cage in a vertical orientation.

Reinforced-concrete pipe may crack if the reinforcing steel is too close to the surface. The amount of cover must be determined by the manufacturing process, type and size of pipe, type of reinforcing cage, and experience. As little as ½ in of cover is adequate in some cases. (Note that this cover is considerably less than the 2 to 3 in recommended for structural concrete.) However, careless placement of the cage may result in exposure of the steel on either the outside or the inside of the pipe.

42.27 Curing and Handling

Curing. Machine-made pipe is handled and cured in a manner similar to concrete block, practically all pipe being low-pressure steam-cured in kilns as shown in Fig. 42.21 (see also Fig. 42.20). Large cast pipe is frequently steam-cured by covering individual

FIGURE 42.20 Reinforcing cages are made on some sort of mandrel. Shown is a large automatic cage machine that automatically wraps circumferential steel over the previously placed longitudinal bars, spacing the wraps correctly and welding each intersection. As the mandrel rotates on its longitudinal axis, the carriage moves along its track to space the circumferential spiral.

FIGURE 42.21 Pipe 36 in in diameter by 6 ft long in curing kiln awaiting the normal delay before the kiln is closed and steam admitted.

pipes, or groups of pipes, with an enclosure of wood, metal, or canvas and admitting steam into the enclosure.

Some producers provide fog nozzles in the curing kilns. If the fog nozzles are turned on progressively across the room as the kiln is filled with pipe, the pipe does not dry out before steaming.

Pipe sections, when steam-cured, should have both the exterior and interior exposed to the elevated temperature. Admitting steam to the interior only, especially if the ambient-air temperature is relatively low, might cause fine longitudinal cracks on the wall of the pipe.

Handling Pipe. Nearly as varied as the plants making pipe are the methods and equipment for handling. Figure 42.22 shows a forklift with a special hydraulically powered fitting for gripping large pipe. Dollies of numerous types, one of which appears in Fig. 42.18, are frequently used, especially for small sizes, with crawler and motor cranes to handle the large ones.

FIGURE 42.22 Special grips or fittings on forklifts can be adapted to handle heavy sections of pipe.

42.28 Properties and Tests

Most specifications require a load test, in which an exterior crushing load is applied to a pipe and the strength is determined in pounds per lineal foot of pipe when a 0.01 in crack appears along a length of 1 ft or more, or at ultimate failure. Figure 42.23 depicts a "D-load," as it is called, being applied to a piece of pipe, and Fig. 42.24 indicates the method of supporting the pipe and applying the load. In the case of large pipe, it is usually specified to withstand a certain load, and the 0.01 in crack load is not specified.

Absorption tests are made on pieces of pipe broken during a load test.

Some pipe, especially for pressure lines, is required to pass a hydrostatic test,

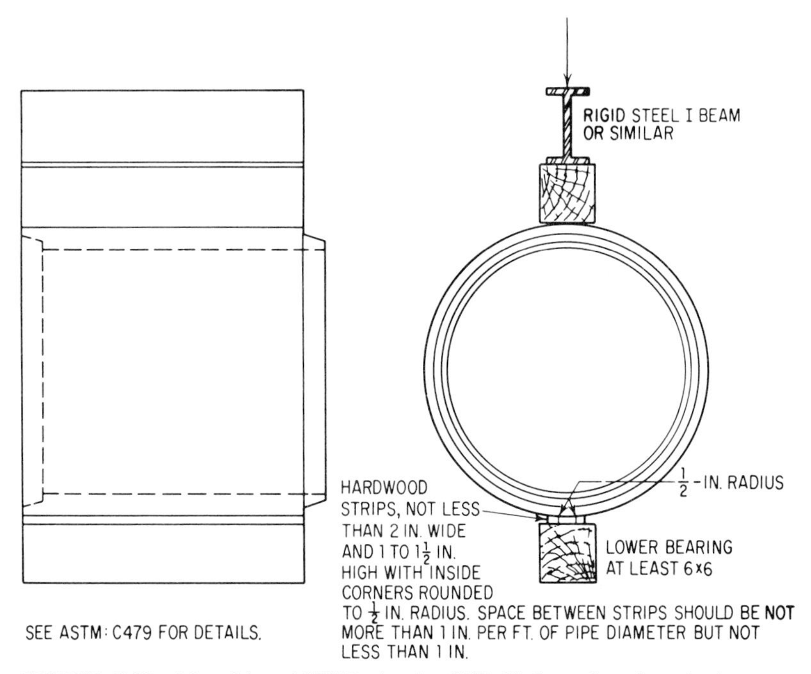

FIGURE 42.23 Adapted from ASTM Designation C497, this figure shows how pipe is supported and the load applied. Properly called the crushing test, this procedure is frequently called the D-load, or three-edge bearing test. Figure shows a length of tongue-and-groove pipe under test. The same arrangement is used for bell-and-spigot pipe except that load is not applied to the bell.

in which ends of the pipe are bulkheaded off (frequently two pieces of pipe are joined together so as to test the joint) and an internal hydrostatic load is applied.

Inasmuch as pipe is covered by standard ASTM specifications, structural analysis is not normally necessary. The D-load, or crushing, test is relied on for design purposes based on determination of pipe bedding and trench load factors in the installation.

In making tamped and packer-head reinforced pipe, reasonable care must be exercised to avoid displacement of the reinforcement, especially in the bells and spigots where it might be exposed. Displacement of the steel weakens the pipe and interferes with adequate compaction of the concrete.

Pipe may be rejected because of circumferential cracks, longitudinal cracks, ragged tongues, or rough interiors.

Besides cracks and spalls due to rough handling, other defects are circumferential cracking of reinforced tamped pipe. Sometimes these cracks are in the form of a helix, caused by reinforcing cage hoops poorly welded to longitudinals, inexperienced or careless strippers, poorly graded mix containing insufficient gravel, or twisting of cage during tamping released when the pipe is stripped. Cracks in general may be caused by drying of pipe before initial steaming, careless stripping, mix too wet, or length-diameter ratio too great (as 4 ft lengths of 15 in pipe).

Rough interiors result from poor aggregate grading or using a mix that is too wet or too dry. Pinholes, causing hydrostatic failures, sometimes result from using sand lacking in fines.

FIGURE 42.24 In the three-edge bearing test, a crushing load is applied to the pipe by means of hydraulic jacks and the load per lineal foot of pipe is computed. Note the rejected reinforcing cages serving as wells for the counterweights, and the two hydraulic jacks for applying the load.

42.29 Joints

Joints in pipe are either tongue-and-groove or bell-and-spigot (Fig. 42.25). Sealing can be either with cement mortar or by means of a rubber gasket as shown in Fig. 42.26. There are many types of gasket joints, some of which are patented. The advantage of gasketed joints is a slight degree of flexibility.

Concrete drain-tile lines have open joints to admit drainage water. In order to prevent infiltration of soil into the pipe, the pipe is laid in a specially proportioned bed of filter gravel.

CONCRETE PILES

42.30 Precast Piles

Piles can be manufactured in a casting yard adjacent to the construction site for a large project, or they may be made in a central plant and transported to the job by rail, barge, or truck.[8] An example of a jobsite plant is the casting yard for the Chesapeake Bay Bridge and Tunnel. A completed pile structure is shown in Fig. 42.27.

Square and hexagonal (or octagonal) piles are the commonest cross sections, although cylindrical spun piles, usually hollow, are frequently used also.

Square and Octagonal Piles. Except for a taper at the point, piles are normally of uniform section throughout their length. Whether to make square or octagonal piles depends mainly on personal preference. (In this discussion, "octagonal" is meant to include hexag-

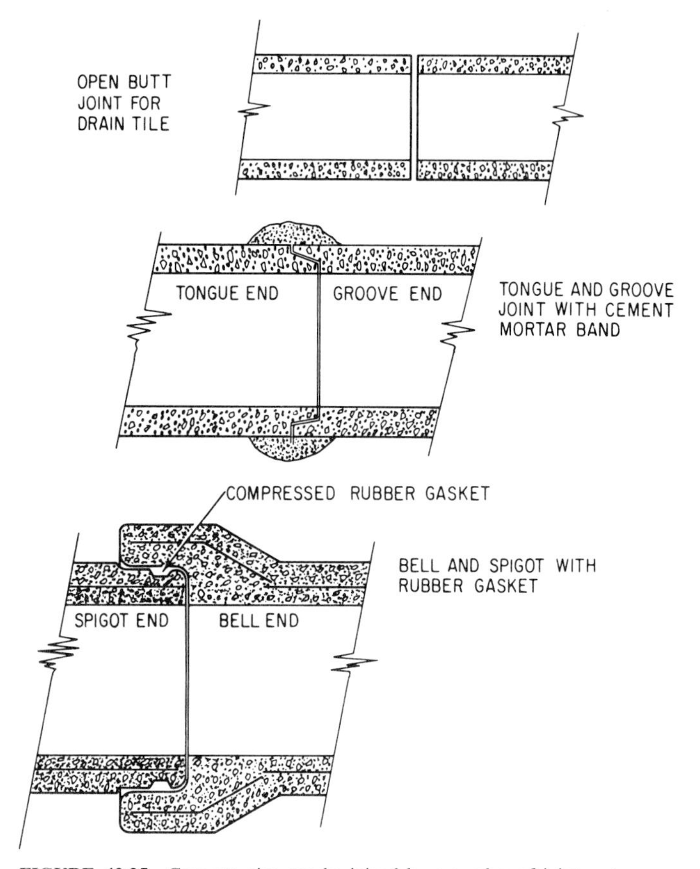

FIGURE 42.25 Concrete pipe can be joined by a number of joint systems, as shown. There are a number of modifications available, especially of the gasketed joints.

onal also. Octagonal section is more prevalent than hexagonal.) One advantage of an octagonal pile is that its flexural strength is the same in all directions. Octagonal piles in a group present a pleasing appearance because rotation of any of them during driving is not noticeable. Other advantages are that the lateral ties can easily be made in the form of a continuous spiral. They can be made in wood or metal forms and edge chamfering is not necessary.

On the other hand, longitudinal steel in a square pile is better located to resist flexure. Square piles are easy to form, especially in banks or tiers, and concrete placing is easier than in octagonal ones. Also, there is more surface area per volume of concrete.

Size of piles can vary to suit almost any conditions. Piles 24 in square over 100 ft long are fairly common and piles as small as 6 in square are also made. Reinforcement, consisting of longitudinal bars in combination with spiral winding or hoops as detailed on the plans, is designed to resist handling and driving loads as well as service loads. Coverage over steel should be at least 2 in except for small piles of special dense concrete when slightly less coverage is permissible. For piles exposed to seawater or freezing and thaw-

FIGURE 42.26 The workers are attaching a rubber joint gasket to the spigot end of a pipe. Besides being made of a ring of circular cross section as shown here, gaskets may be of different cross sections.

ing cycles while wet, coverage should be 3 in. Splices in longitudinal steel, if required, should be staggered.

The ends of the pile are called the head and the point (Fig. 42.28). The head should be carefully made, smooth, and at right angles to the axis of the pile. It is deeply chamfered on all sides, and extra lateral reinforcement is provided for a distance at least equal to the diameter of the pile. Shape of the point depends on expected soil and driving conditions. For hard strata and cohesionless sand and gravel, the tip may be about one-fourth the pile diameter with a length three times the pile diameter. In plastic soil the diameter may be $\frac{1}{4}D$ and the length $1\frac{1}{2}D$, or there may be no taper at all. In particularly difficult driving a metal shoe is sometimes provided. Longitudinal bars should follow the taper by drawing together in the center of the pile. Avoid bunching them together at one side of the point. Extra lateral reinforcement should be inserted in the tip.

FIGURE 42.27 Vertical and batter piles have been driven to support a waterfront warehouse and berth. (*Los Angeles Harbor Department photograph.*)

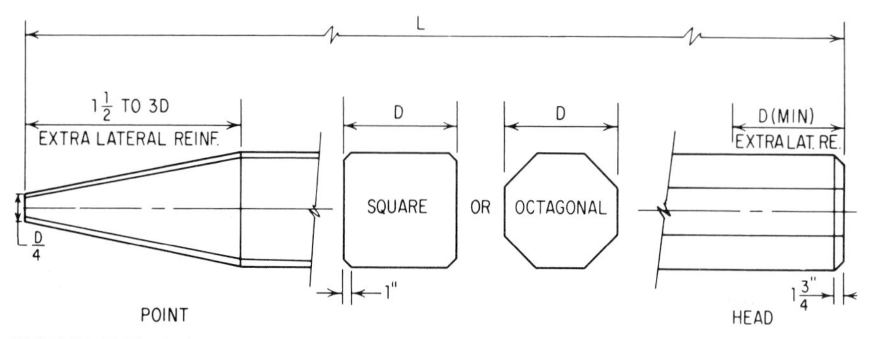

FIGURE 42.28 Solid concrete piles can be square, octagonal, or hexagonal (ELR = extra lateral reinforcement).

Manufacture of piles should follow the fundamentals of good concrete construction explained elsewhere in this handbook. Casting-yard facilities can range from very simple hand operations using ready-mixed concrete, when only a few piles are required, to elaborate casting yards of the type shown in Fig. 42.29. Regardless of the size of the casting yard, it should be designed to permit logical movement of materials and finished piles with a minimum of overlapping.

Square forms can be arranged on centers twice the width of the piles. Sides of the forms are covered with heavy building paper or sheet plastic which becomes the form for the second group of piles after the first ones have been cast and stripped.

Common practice is to assemble the reinforcement into cages that can be handled with a crane and set into the forms so as to preclude any movement during placing and vibration of the concrete. Small concrete or mortar blocks can be wired to the steel to space the steel on the sides as well as on the bottom of the form, or metal or plastic chairs can be used. The finished cage should be 6 in shorter than the pile, to give 3 in clearance at the head and point.

Mixes are usually fairly rich, especially for piles to be exposed to seawater, containing 6 or 7 sacks of cement (564 to 658 lb) per cubic yard. Low-alkali low-C_3A cement is desirable. Slump need not exceed 2 in for vibrated concrete containing entrained air. After the water sheen disappears from the top surface of the fresh concrete in the form a

FIGURE 42.29 Precasting yard for precasting 54 in prestressed cylinder and other kinds of piles. This yard is used for precasting pile caps and bridge-deck units.

wood float finish is applied. Finally, good curing practices must be followed. Concrete should normally have a compressive strength of not less than 4000 psi when the piles are driven, but the plans should be checked to determine what is actually specified for any certain job.

Handling imposes high stress in a pile and should be done carefully. The pile should be picked up at the proper designated points to avoid bending stresses and shocks and jolts should be avoided. Field-cured cylinders, exposed to the same conditions as the piles, give a reliable indication of strength of the concrete in the piles to determine when the piles can be moved. Straightness of the piles is important, and specifications require that the maximum allowable deviation of the longitudinal axis from a straight line drawn from the center of the tip to the center of the head shall not exceed a certain amount, usually ⅛ in per 10 ft of length of pile.

Hollow Cylindrical Piles. These piles were used on the Lake Pontchartrain causeway, numerous bridges on the Northern Illinois Tollway, the Chesapeake Bay Bridge, and other structures. This "cylinder pile," as it is called, is a hollow, cylindrical, precast, posttensioned concrete pile made up of a series of sections placed end to end and held together by posttensioned cables. The pile sections are manufactured by centrifugal spinning, similar to the manner in which concrete pipe is made. Each section is reinforced with a small amount of longitudinal and spiral steel to facilitate handling. Longitudinal holes for the prestressing wires are formed in the wall of the section by means of a mandrel enclosed in rubber hose.

After curing, the sections are assembled end to end on the stressing racks with the ducts lined up, and a high-strength polyester resin is applied to the joint surfaces. Posttensioning wires are inserted and tensioned, ducts grouted, and the sections allowed to rest while the grout cures. Next the stressing wires are burned off and the pile is ready for driving.

Cylinder piles are generally made in standard diameters of 36 and 54 in with wall thickness and number of stressing cables varied, depending on job requirements. The 36-in piles have wall thickness from 4 to 5 in containing 8, 12, or 16 prestressing cables; the 54 in piles are 4½, 5, or 6 in thick with 12, 16, or 24 cables.

There are variations in methods of spinning, form stripping, and curing, but the variations are minor and the piles made are essentially identical. These large piles require a hammer capable of delivering a large amount of energy, of the type shown in Fig. 42.30. Figures 42.31 through 42.33 show typical construction details.

Sheet Piles. Made in the shape of interlocking planks, sheet piles are a special type of precast piles. Planks vary in thickness from 6 to 12 in and in width from 15 to 24 in. Interlocking is provided by a tongue-and-groove design, as shown in Fig. 42.34. The point of the pile is beveled so that it will be forced against the previously driven adjacent pile, thus assuring continued alignment of the wall under construction.

Handling Piles. Piles must be rolled over, picked up, loaded, transported, unloaded, picked up, turned to a vertical position, and finally driven. All this handling imposes loads on the pile which must be considered when the pile is designed. Pickup points should be indicated on the finished pile, if they are critical, and the pile should be picked up at these points. Short piles (less than 25 ft long) can usually be handled with one pickup, but longer piles may require a two-, three-, or even four-point pickup, except that most prestressed piles can be handled with one pickup, as depicted in Fig. 42.35. Each suspension system must be designed for each situation.

FIGURE 42.30 This pile hammer, used for driving 54 in prestressed cylinder piles, is capable of delivering 50,000 ft-lb of energy with each blow. Most pile hammers are steam-operated, but many are also air-operated.

42.31 Cast-in-Place Piles

These members are constructed either by drilling a shaft in the earth and filling the shaft with concrete, or by driving a hollow sheet-metal shell by means of a mandrel. After driving, the mandrel is withdrawn and the shell is filled with concrete. Another type consists of a heavy shell (7 gauge or heavier) driven without a mandrel as shown in Fig. 42.36. A special type of cast-in-place pile is the caisson footing in which a shaft is drilled and the bottom is belled to a larger diameter to provide an extended bearing area for the concrete.

While these piles may come in the class of "rough" concrete, they are vitally important to the safety and success of the structure they support, and there is no reason to abuse usual good construction practices. They are not subject to weathering exposure such as may be the case with precast piles, but they may be subject to attack by aggressive waters in the soil. It is usually permissible to use a slightly leaner mix in cast-in-place piles than in precast piles because they receive vertical loading only. However, there are uncertainties in soils and foundation work, and the extra cost for good concrete is negligible. Water-cement ratio should not exceed 6½ gal per sack of cement (0.58 by weight). The hole should not contain free water when the concrete is placed in it. Tops of piles, if exposed, should be given proper curing and protection.

Shell piles may be either reinforced or unreinforced. If reinforcement is used, it is usually assembled into a cage that is lowered into the shell after the shell has been inspected. Inspection is accomplished by lowering a light into the shell. A shell that is not watertight, or that shows kinks, bends, or other deformation resulting from driving that would impair the quality of the pile, should be repaired or removed. The reinforcing cage should be provided with chairs or other devices to assure clearance from the sides of the shell. It is usual to place a foot or two of mortar in the bottom of the shell just before filling with concrete. The concrete should be vibrated. If practicable, all the piles in one construction unit, such as a pier, bent, or abutment, should be driven before concrete is placed in any of them. This provides better uniformity in construction, avoids damaging completed piles, and permits removal of a shell if it becomes displaced or damaged because of subsequent driving operations.

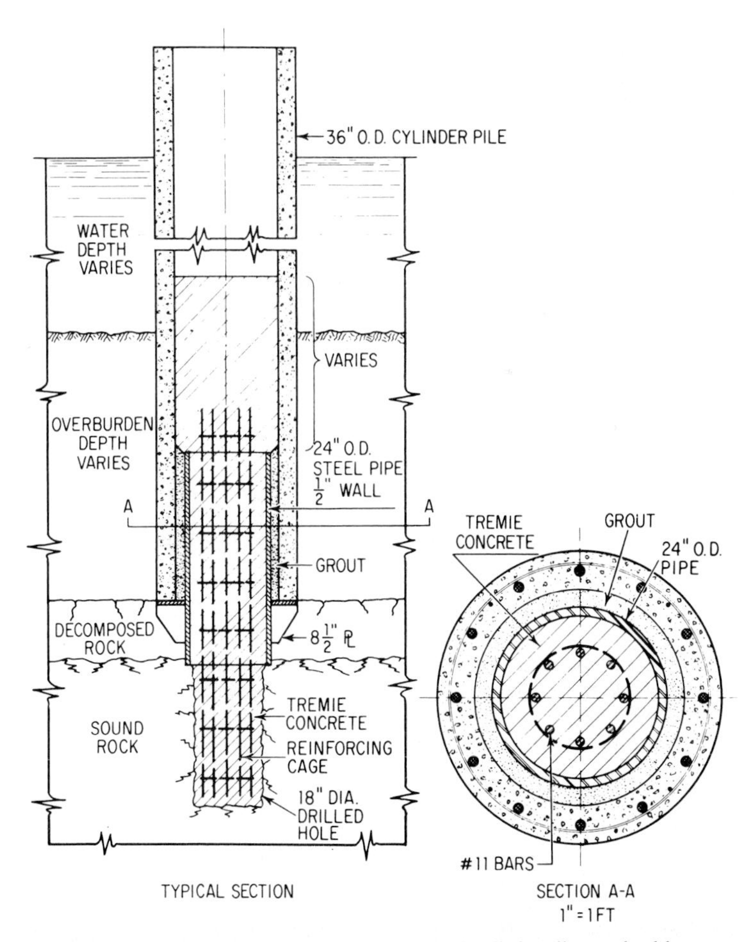

FIGURE 42.31 A method for securing a Raymond cylinder pile to a hard bottom where there may be insufficient overburden or where the pile may be subject to tensile or pullout forces. By concreting a reinforcing cage into a socket, or by driving a steel pin into soft rock and concreting the annular space between pin and pile, an effective method for anchoring the pile can be achieved. (*Raymond International, Inc.*)

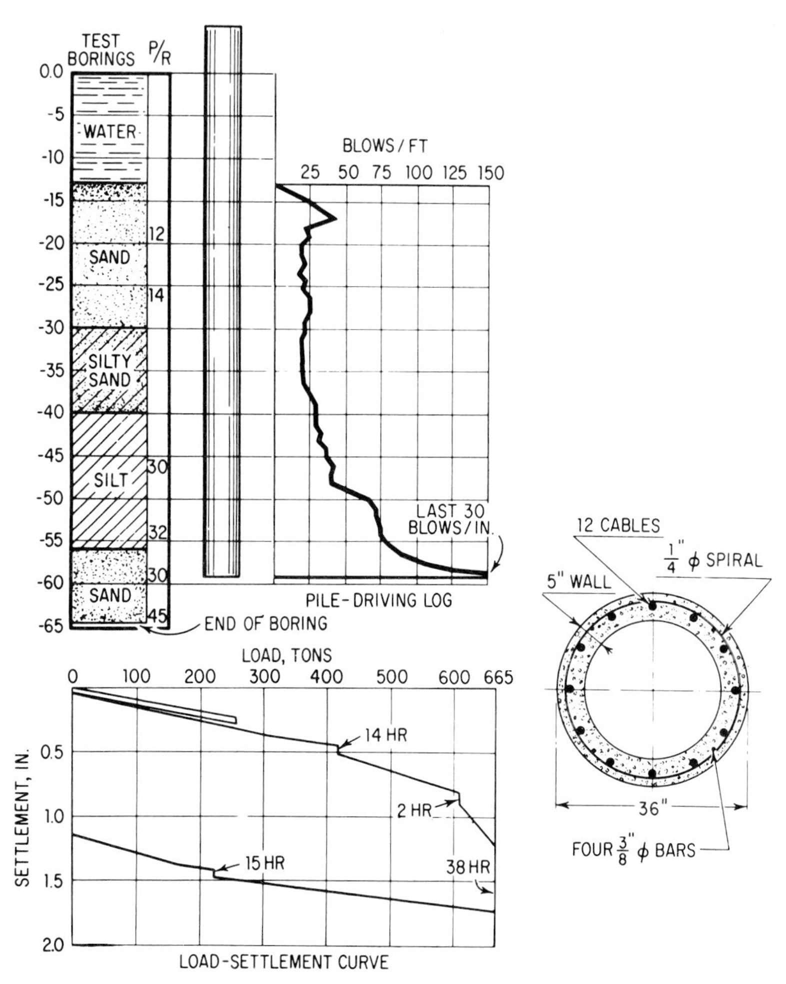

FIGURE 42.32 Pile-driving log and load-settlement curve of a 36 in OD Raymond cylinder pile driven in Lake Maracaibo in 1959. The 80 ft pile was fitted with a flat point by filling the lower 4 ft with concrete. Driving was done with a Raymond 5/0 hammer delivering 57,000 ft-lb of energy. (*Raymond International, Inc.*)

MISCELLANEOUS PRECAST ITEMS

Concrete roofing tiles are made on a variety of machines that extrude, press, or vibrate the concrete mixture. In one machine, mixed material in a hopper feeds by gravity onto a series of pallets moving forward on an endless chain, passing under vibrating tampers and trowels that densify and smooth the tiles. Color is achieved either through the use of integral coloring material that colors all the concrete and extends entirely through the tile, or by a cementitious glaze that is applied to the surface of the tile with a pneumatic gun. Each tile is 10½ × 15 in with an exposed area 8 × 12 in.

Railroad cross ties have been installed on a number of test installations in the United States and Canada, after favorable reception in Europe. The ties that have been used in North America have been prestressed, using high-quality low-slump concrete; most are steam-cured.

Other precast items include manhole rings for constructing manholes for sewers and utility installations underground; septic tanks; parking bumpers; meter boxes and covers; burial vaults; garden furniture; laundry trays; fireplaces; silo staves; cribbing for retaining walls; fence posts and rails; steps; multiple-duct electrical conduit; precast paving; and many others, apparently limited only by the ingenuity of the precaster.

Many of these items use rich mixes, low slump, or no slump at all, and compaction methods to give a very dense concrete. Precast roof joists are shown in Fig. 42.37.

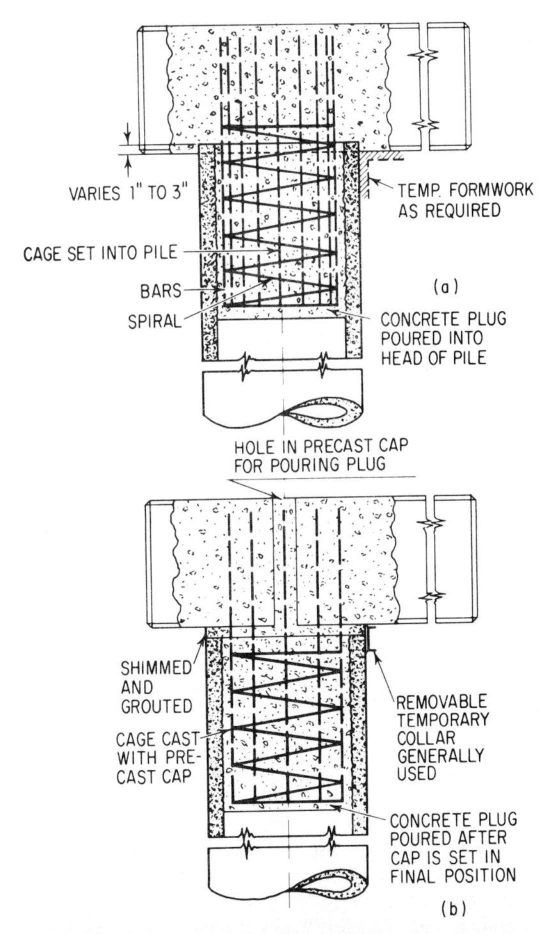

FIGURE 42.33 Two methods of attaching a pile cap are illustrated. A cast-in-place pile cap is applied to a cylinder pile following method *a*. For a structure utilizing a precast cap, method *b* is followed. In each case an expendable platform is inserted just below the cage to support the fresh concrete inside the pile. (*Raymond International, Inc.*)

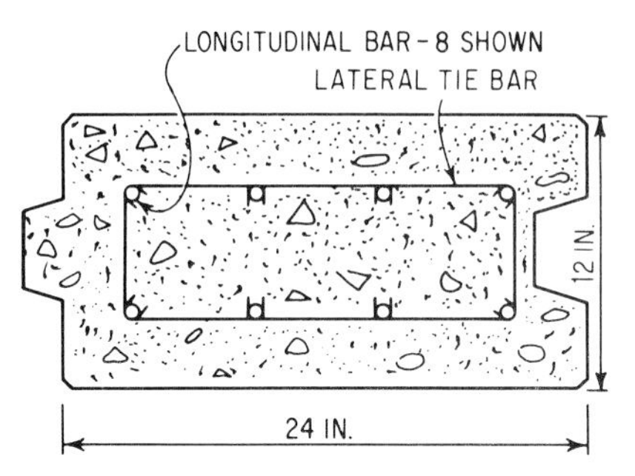

FIGURE 42.34 Sheet piling is available in a number of sizes and shapes. This one is typical. The amount of reinforcing depends on the length of the pile and anticipated lateral loading.

FIGURE 42.35 Special heavy equipment, including a self-elevating driver platform with elaborate leads, jets, and whirler crane, is used for handling of 54 in prestressed cylinder piles. Note single-point pickup. (*Raymond International, Inc.*)

FIGURE 42.36 Vertical leads, to hold the pile in a vertical position while it is being driven, can be mounted on skids, a barge, or railroad car or, as shown here, can be attached to the boom of a crane. Hollow steel shells for cast-in-place piles are being driven here.

FIGURE 42.37 Precast-concrete roof joist units in storage in the casting yard. On the trailer are four dome units of precast concrete.

REFERENCES

1. "Design Handbook," Prestressed Concrete Institute, 1991.
2. "Concrete Masonry Handbook," 5th ed., Portland Cement Association, 1991.
3. Building Code Requirements for Reinforced Concrete (ACI 318-89), American Concrete Institute.
4. Toennies, H. T., and J. J. Shideler: Plant Drying and Carbonation of Concrete Block-NCMA-PCA Cooperative Program, *J. ACI,* May 1963, pp. 617–633.
5. Q Block Specifications, National Concrete Masonry Association, Washington D.C.
6. Building Code Requirements for Concrete Masonry Structures (ACI 531-79)(rev. 1983), American Concrete Institute.
7. "Concrete Pipe Handbook," American Concrete Pipe Association, 1981.
8. ACI Recommendations for Design, Manufacture, and Installation for Concrete Piles (ACI 543-74)(reapproved 1980).

P A R T • 12

ARCHITECTURAL CONCRETE

Architectural concrete is defined by the American Concrete Institute[1,2] as concrete exposed as an interior or exterior surface in the completed structure, which definitely contributes to its visual character, and is specifically designated as such in the contract drawing and specifications. The materials, procedures, and finishes of architectural concrete will usually differ from those for structural concrete. Chapters 43 and 44 describe those differences and how to avoid problems.

CHAPTER 43
MATERIALS FOR ARCHITECTURAL CONCRETE

Joseph A. Dobrowolski

43.1 Cement

Most attributes of the cement are set by the structural and durability requirements of the location and use of the structure. However, because of possible changes in color of cements from different sources and of different types, cements for architectural concrete must be restricted to one source and type until completion. No particular cement should be specified for its color contribution to the concrete, until a guarantee is given that the cement will be available in ample and timely supply.

Times of shortage, use of foreign cement (see examples in Fig. 43.1) and variable cement delivery schedules may not allow ready-mix plants to guarantee a sole source. This can be solved by an on-site batch plant if space is available or using ready-mix plants having a storage silo available for storing only one type of cement for the project.

FIGURE 43.1 Precast panels containing different cements and contrasting colors. Top panel contains Saudia Arabia Type 5; middle panel, Saudia Arabia Type 2; bottom panel, a Spanish cement.[3]

43.2 Aggregates

In addition to the requirements of ASTM C33, an architect may have additional requirements that specify color, size, shape, and source. On some projects, architectural considerations such as color may override ASTM C33 requirements. Aggregates have been used which have had a history of reactivity or excessive wear when tested by the Los Angeles Rattler Test (ASTM C131)[4]. Such exceptions should be made with extreme caution and proper proportioning to reduce any risk.

As an example, Ventura County, California aggregates with a long history of

FIGURE 43.2 St. Basil's Church bushhammered exterior without any visible reactivity problems.

cement–alkali reactivity were used in a bushhammered exposed-aggregate exterior finish for St. Basil's Church in Los Angeles, California (Fig. 43.2). The concrete mixture contained a pozzolanic admixture and a low-alkali portland cement. After 25 years of service, the aggregate is still performing satisfactorily without any signs of reactivity.

The Equitable Life Building in Los Angeles was clad with precast units having exposed aggregate from Texas which had been chosen for color although a relatively soft limestone. Manual washing was used for exposure and a protective sealer applied prior to erection. The units have been performing satisfactorily since 1974 in the mild southern California climate, but might have proved unsatisfactory in a more severe climate.

Aggregates such as chert and shale may cause surface popouts (Fig. 43.3) in hardened concrete slabs due to an expansion caused by reaction with the cement or absorption of moisture. These aggregates are usually in the $\frac{3}{8}$ in to the No. 8 size and of lower specific gravity. Concrete mixtures containing such material in oversanded proportions, designed for a maximum size of $\frac{3}{8}$ in aggregate, and with a high water content will have a potential for a large number of popouts. The proportioning provides a larger amount of the material for flotation to the surface by the high water content. If the con-

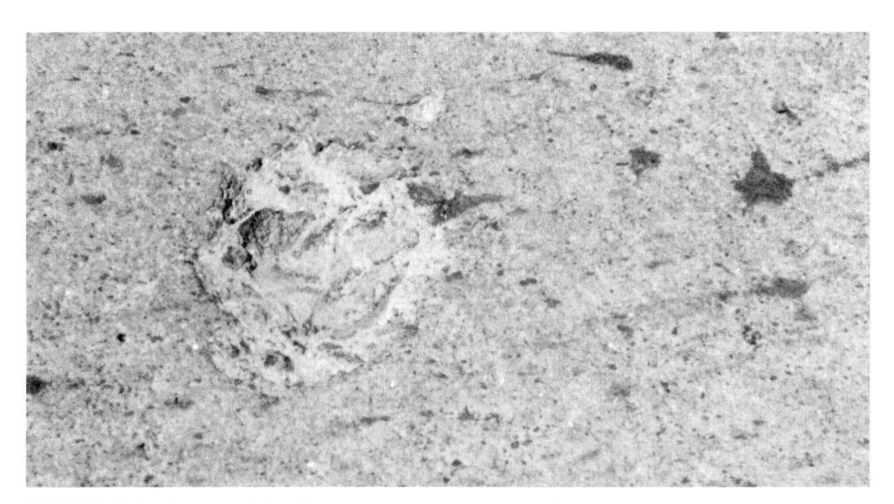

FIGURE 43.3 Concrete slab with visual popout due to expansion of aggregate near to surface. Note characteristic portion of aggregate left in bottom.

crete mixture is designed with minimum quantities of the suspect material and minimum slumps (3 to 4 in, 75 to 100 mm), the final product will have minimum popouts.

Aggregates chosen for color and exposure by sandblasting should have a sample checked for color after sandblasting, which may abrade and lighten the aggregate. A check should be made for iron content which could cause possible staining after exposure.

Cubical or rounded aggregate will provide better distribution and self-cleaning ability on exposure to weather conditions. Crushed aggregates will tend to accumulate and hold airborne contaminants.

43.3 Admixtures

General recommendations regarding the use of admixtures are as follows:

1. All admixtures to be used in the architectural concrete should be combined in a sample with other proposed concrete materials to determine its effect on color.

2. Air-entraining agents may be required to meet maximum durability conditions. If bugholes are a problem with high-density types of architectural forming due to the stickiness of concrete mixtures containing maximum percentages of entrained air, the lower range values may be substituted in mild climates.

3. Calcium chloride may cause discoloration of concrete and adversely affect surface retarders.

4. Superplasticizers and high-range water reducers have been used beneficially for workability and high-strength concrete. Their effect on the architectural concrete color needs to be checked prior to use.

5. Mineral admixtures have been used. As the color of mineral admixtures differ by source, each should be checked with the proposed concrete mixture to determine effect on color. Some have been used to give a warming color to light-colored cements. The use of microsilica has been known to heavily darken an architectural concrete. In some cases of colored concrete, a mineral admixture cannot be used due to a detrimental darkening. Other than color, mineral admixtures should not have a detrimental effect.

6. Color oxides and coloring admixtures may be used. Pigments used to color concrete mixtures are finely ground natural or synthetic mineral oxides. The field mockup sample should be used to determine whether the color has any reaction with compounds used on the forms or to clean the surface. The intensity of color is not increased by additions over 5 percent by weight of cement, and quality of the concrete is affected by additions of color over 10 percent by weight of cement. Coloring admixtures, developed to provide a more uniform color through the usual specified range of slump, are a blended mixture of pigment, a mineral filler, and a water-reducing admixture. Lampblacks may not be as durable or long-lasting as black metal oxides and may have a detrimental effect on air content. Local environmental requirements may require metropolitan ready-mix firms to recycle wash water in accordance to NRMCA standards. Experience has shown that color is not affected until the total amount of wash water from 30 yd^3 (22.9 m^3) is added to the concrete. If problems are occurring with the color of the field sample concrete, check the recycled wash water.

43.4 Water

Most water is satisfactory for architectural concrete unless it contains sufficient iron, rust, or other chemicals to cause staining in light, white, or colored concrete. Water proposed for the project should be checked in the field mockup. The requirement would also include water for curing.

43.5 Reinforcement

Structural requirements for reinforcement may not be sufficient for architectural concrete where normal deflections may create an appearance of sag in long-span, smooth concrete surfaces, cracking is to be minimized by increasing ACI 318 horizontal wall reinforcement by 50 percent, and minimal cover requires the use of epoxy-coated, stainless, or galvanized steel.

Plastic chairs or spreaders (Fig. 43.4) are best for architectural concrete to inhibit rusting. Proposed use should be investigated in the field mockup where further treatment, such as sandblasting or bushhammering, is specified.

43.6 Forms, Release Agents, Bolts and Ties, Rustication Strips, and Chamfers

Forms. Forms can be all wood, wood backing faced with thin plastic liners, a reinforced plastic, metal, or cast concrete or plaster. Type is determined by texture desired, number of expected reuses, method of form construction, installation, removal, form ties, and proposed storage.

Wood forms are, generally, rough boards, and plywood with smooth, striated, or plastic coated plywood of medium and high-density grades (Fig. 43.5). Rough boards are the most economical but require initial treatment with a lime or cement slurry to prevent the wood sugars from creating a soft, dusty surface on initial use. For uniformity of surface color, all form lumber for one project should be obtained from the same source. Plain B-B plywood can be used comfortably four times with proper care and high-density coated plywood can be expected to provide at least 20 uses. If grain raise is desired for the concrete surface, plain or sandblasted plywood should be used.

Plastic liner economics begin with the one use styrofoam and progress through the use

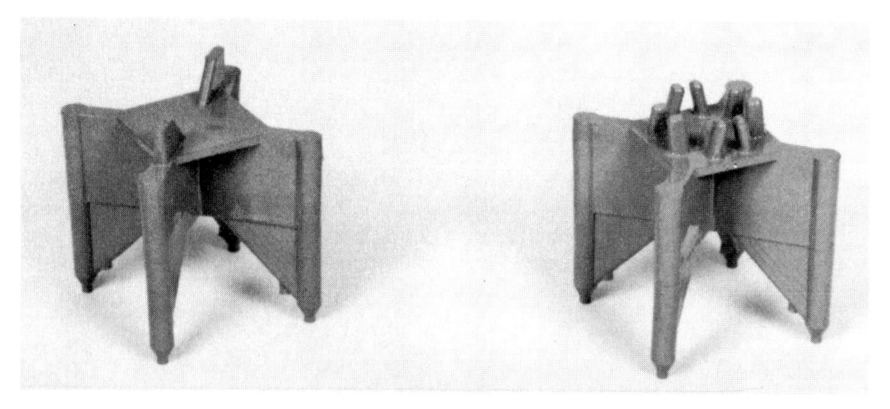

FIGURE 43.4 Plastic accessories used to support or separate reinforcement bars from each other or from forms. (*Courtesy of Fosroc-Preco Co.*)

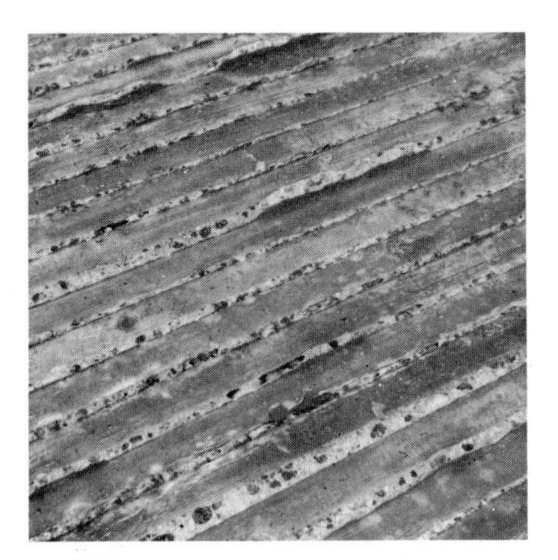

FIGURE 43.5*a* Rustic surface left by rough board forming.

FIGURE 43.5*b* Portion of resin-coated plywood form after 20 reuses.

of rigid extruded plastic, to the elastomeric (rubbery flexible plastic) capable of 50 or more reuses (Fig. 43.6).

Gaining some popularity is the new thin, flexible vinyl liner, which does not have a large experience record but after a high initial cost has the possibility of great economies in reuse, ease of form construction, gang-form handling, and excellent results. The thin vinyl liner (50 to 60 thousandths of an inch) is obtained in 3 ft wide rolls which are 50 to 60 ft long.

Fiberglass-reinforced forms (Fig. 43.7) have a high initial cost and are used for intricate details, which requires multiple reuse.

FIGURE 43.6*a* Expanded polystyrene form manufactured to simulate rough board finish.

FIGURE 43.6*b* Typical rigid plastic form liners of various configurations.

Metal forms are used for large, smooth, repetitive surfaces and require special cleaning procedures to preserve uniformity of reuse appearance. To prevent sticking of mortar to galvanized-steel forms (Fig. 43.8), they should be treated with a 5 percent solution of chromic trioxide solution.

Concrete molds (Fig. 43.9) or forms are used to precast the architectural face which is then incorporated into the structural portion of the building by additional placement of reinforcement and concrete. They require a need for many reuses and a definite maintenance procedure.

FIGURE 43.6*c* Various configurations of elastomeric form liners. Note flexibility and thickness.

FIGURE 43.6*d* Thin vinyl form liner used to cover plywood backing.

Plaster molds are used occasionally but have given way to Styrofoam forming, which can be sculptured to produce intricate figures or shapes at an economical one-use cost.

The designer should consider the potential effects of absorptive and impervious forming on the surface of the concrete. Impervious forming such as steel, coated plywood, and plastic liners will usually result in a lighter and more uniform color.

Release Agents. Release agents are applied to the form face to help prevent forms and form liners from sticking to the concrete, facilitate form cleaning after removal, and achieve successful architectural surfaces. Criteria for selection include compatibility with the form or liner; concrete admixtures and any existing form sealer or coating; and future applications to the hardened concrete such as sealers, coatings, or curing compound. Selection also depends on any resultant discoloration, size, an amount of surface bugholes, effect on ease and time of stripping, residue buildup on f rm face with each reuse,

FIGURE 43.7 Cross sections of fiberglass forms used for architectural forming.

FIGURE 43.8 Steel forming used to cast piers for Pine Valley bridge near San Diego, Calif.

FIGURE 43.9 Concrete mold used to precast architectural facings for columns of high-rise building.

temperature changes, and uniformity of appearance. Each should meet local environmental requirements as to contained solvents and be checked prior to use on the field mockup sample.

The main classes of release agents are described as the chemically active (most popular) and the barrier types. The barrier class includes various kinds of fuel oil, paraffin wax in a carrier oil, and chemical compounds containing silicon. Problems with the use of these agents have included form release and resulting blemishes and adhesion of other materials to the hardened concrete. They are not recommended for architectural concrete. Chemically active release agents contain fatty acids, obtained from animal fat, fish, and plants, which combine chemically with the fresh concrete to form a grease or calcium soap for the form release. An additional ingredient, a petroleum-based oil, is the solvent used to maintain uniform distribution of the more expensive fatty acids. The chemically active release agents can be either neutralized to prevent the ingredients from settling out or not neutralized. Most water-based release agents, including emulsions, are also considered to be chemically reactive.

Bolts and Ties. The spacing of bolt and tie holes may be a design feature and a factor in form design as the size of the tie is dependent on the spacing and expected form pressures. They should be capable of easy removal without spalling of the surface and leave no metal susceptible to rusting within 1½ in (40 mm) of the surface (Fig. 43.10).

On removal, ties will leave holes identifiable with their type. Holes approximately 1 in (25 mm) deep and ¼ in (6 mm) in diameter are left by wire snap ties (Fig. 43.11). To satisfy additional architectural requirements, wood or tapered plastic cones are available to increase the size of the hole to 1 in (25 mm) and to provide deeper break backs to 2 in (50 mm) and additional protection against mortar leakage. She-bolts range in size from ¾ to

FIGURE 43.10 Typical bolt-type form tie after form removal in preparation for bolt removal.

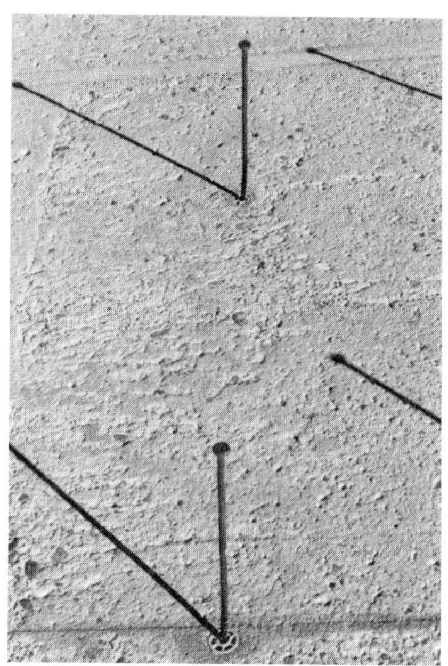

FIGURE 43.11 Snap ties after form removal and prior to breakoff.

1½ in (20 to 40 mm) in diameter, while he-bolts have a 1½ to 2 in (25 to 50 mm) diameter cone which will allow room for application of mortar cover for the steel left in the wall. Pull ties leave holes of ½ to 1½ in diameter, similar to diameters of holes left by rods. Snap ties without cones or special seals should not be used for architectural concrete.

Rustication Strips and Chamfers. Rustication and joint patterns are needed to break up large, flat areas. Right-angle corners can be obtained if the forms are designed for easy removal without damage. If allowed, chamfers can be used at internal form corners. They can be of wood or plastic. Wooden chamfers (Fig. 43.12) should have a minimum width of 1 in (25 mm). All chamfers should be securely fastened at uniformly close intervals for straight alignment. Joint depths for rustication strips are ¾ in (20 mm) for small grooves, are 1½ in (40 mm) for control joints and panel division, and contain a minimum draft of 15° to allow removal from the hardened concrete.

43.7 Design Reference Sample

Prior to completion of the project contract documents, casting of a design reference sample (Fig. 43.13) should be made to illustrate the desired concrete color and texture. Source of materials and texture treatment are usually identified in the specifications. Minimum size is 18 × 18 in and thickness, depending on the texture treatment and risk of breakage. Placement of the concrete should be similar to that expected on the project.

FIGURE 43.12 Wooden chamfer and rustication strips fastened to high-density forming.

43.8 Sealants, Sealers, and Coatings

Sealants are used to seal joints from moisture or debris penetration. They should be impermeable and be capable of deforming with the amount and rate of joint-width movement, and recovery without failure in adhesion or tension after each cycle. Ageing or low temperatures should not materially affect the resilient qualities of the sealant.

The majority of present-day sealants are thermosetting, which cure chemically or by the release of solvents. The chemically cured types are polysulfide-, silicone-, urethane-, or epoxy-based. Those that cure by releasing solvent have bases of chlorosulfonated polyethylene, certain butyls, and neoprene and are less affected by temperature changes.

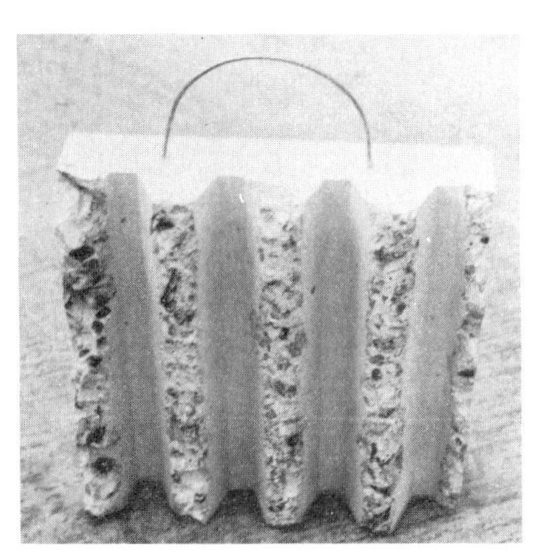

FIGURE 43.13 Example of design reference sample.

The terms *sealer* and *coating* have been used interchangeably with a common meaning connoting surface protection for the concrete. Generally, the two terms can be distinguished with respect to whether the application penetrates the concrete (sealer) or forms a film over the surface (coating). Most quality architectural concretes are relatively impermeable and do not require sealers for additional waterproofing. Primary reasons for sealers are to slow attack of airborne chemicals found in urban areas, seal a sandblasted or bush-hammered surface in freeze-thaw or coastal areas, and reduce absorption of moisture to minimize effloresence and possible reaction of marginal concrete aggregates. A sealer should not discolor, stain, or darken the finish surface; should allow water vapor to escape from the interior; and should be chemically resistant to the concrete.

Developments in sealer technology have produced the silanes and siloxanes, which can penetrate the surface by over a ¼ in.[5] The resulting reaction will result in a concrete surface which will repel water. There are two types: monomeric alkyloxysilane and oligomeric alkyloxysilane. Both have a small molecule. The monomeric is extremely volatile and has a high rate of evaporation which increases with the increase of drying conditions and increased concentrations of silane. The oligomeric has little vapor pressure during application, which holds down the evaporation rate and allows lower concentrations of silane. Although more expensive than most sealers, they are longer-lasting, are less affected by ultraviolet rays, and have little effect on the initial appearance.

Methyl methacrylate, an acrylic resin formulated with a low viscosity and containing a high volume of solids, is also used to protect concrete surfaces. The resulting film or coating left on the surface is durable, is generally glossy, and will usually darken or deepen the color of the concrete.

Any sealer or coating should be tested on the job field mockup for its effect on the architectural surface (Fig. 43.14) and weathering capability (Tables 43.1 and 43.2).[7]

FIGURE 43.14 Darkening effect of sealer applied to architectural concrete surface.

REFERENCES

1. Guide to Cast-In-Place Architectural Concrete Practice (ACI 303R-91), American Concrete Institute, 1991.
2. Engineering and Design Architectural Concrete (EM 1110-1-2009) US Army Corps of Engineers, Sept., 1987.
3. Dobrowolski, J. A., and Scanlon, J. M., How to Avoid Deficiencies in Architectural Concrete Construction, Dept. of the Army, Waterways Experiment Station, Dept. of the Army, Vicksburg, Miss., June 1984.
4. Standard Test Method for Resistance to Degradation of Small-size Coarse Aggregate in the Los Angeles Machine, ASTM C131-81 (reapproved 1987).
5. Concrete Sealers for Protection of Bridge Structures, National Cooperative Highway Research Program, Report 244, Dec. 1981.
6. A Guide to the Use of Waterproofing, Protective, and Decorative Barrier Systems for Concrete, (ACI 515.1R-79) (rev. 1985), American Concrete Institute, Detroit, Mich.
7. Architectural Precast Concrete, Precast/Prestressed Institute, 2nd ed., 1989.

TABLE 43.1 Clear Sealers

Type	Color-change resistance	Surface carbonation resistance	Water resistance	Vapor passage
Oil	Poor	Average	Above average	Excellent
Siliconate	Excellent	Poor	Above average	Excellent
Silicone	Excellent	Poor	Above average	Excellent
Silane	Excellent	Poor	Excellent	Excellent
Siloxane	Excellent	Poor	Excellent	Excellent
Stearate	Above average	Poor	Above average	Excellent

TABLE 43.2 Clear Coatings

Type	Color Change Resistance	Surface carbonation resistance	Water resistance	Vapor passage
Acrylic	Excellent	Excellent	Excellent	Above average
Butadienestyrene	Poor	Excellent	Excellent	Above average
Chlorinated rubber	Poor	Excellent	Excellent	Above average
Epoxy	Poor	Below average	Excellent	Poor
Polyurethane				
Aliphatic	Above average	Excellent	Excellent	Poor
Aromatic	Poor	Excellent	Excellent	Poor
Polyester	Poor	Excellent	Excellent	Poor
Vinyl	Above average	Excellent	Excellent	Above average

CHAPTER 44
CONSTRUCTION OF ARCHITECTURAL CONCRETE

Joseph A. Dobrowolski

44.1 General

The construction of architectural concrete requires additional precautions and care in proportioning, mixing, transporting, forming and form removal, placing, consolidation, and curing to result in uniformity of surface texture and color. Guidance will be given to avoid problems and procedures to correct any possible surface deficiencies.

Construction specifications can be of the performance type, which requires the contractor to provide a finished product having the same texture and color as the design reference sample without significant abnormalities, or of the prescription type, where procedures and materials are specified with final acceptance guaranteed if the specifications are followed. Generally, most architectural concrete specifications are a combination of both.

44.2 Field Mockup Sample

Prior to startup of operations for the production of architectural concrete, a field mockup sample (Fig. 44.1) should be constructed. This sample should be large enough to include full panel form joints, one lift joint, typical structural reinforcement, and planned rustication. Materials, personnel procedures, and equipment planned for the production architectural concrete should be used for construction of the field mockup sample. The sample should be protected from subsequent construction operations and maintained until acceptance of all of the production architectural concrete. Acceptance of the work or artisanship on the sample should be received prior to the construction of any production architectural forming.

On some projects, contractors have trained workers on nonarchitectural structural walls, to be subsequently covered by other materials or buried, to ensure excellent work or artisanship on the production architectural concrete. The Inglewood, California, City Hall (Fig. 44.2) had a five-story subterranean control center, specified to have an as-cast finish with normal concrete ingredients, was used by Swinerton Walberg, Contractor, to train personnel to produce a heavy sandblasted exposed aggregate concrete finish of excellent quality for use on the high-rise above-grade portion.

FIGURE 44.1 Field mockup sample constructed by Bomel Construction Co. for Skidmore, Owings and Merril, Architects, at the Hughes Corporation corporate headquarters, Marina Del Rey, Calif.

FIGURE 44.2 Appearance of concrete form joints sealed with plastic tape over joints after form removal. Aggregate exposure method to be heavy sandblast. Inglewood City Hall, Calif.

44.3 Forming

Tolerances. To prevent pillowing on plane surfaces, form surface deflections between supports should be limited to $L/400$. Tolerances such as the ⅛ in abrupt vertical offset and others listed in the American Concrete Institute ACI 117, Standard Tolerances for Materials and Construction, may require lower values in close-contact areas but values less than half those listed are impractical.

Joints, Bolts, and Ties. Leakage through form openings will leave surface blemishes such as hard, dark lines and circles which contrast heavily and are difficult to remove during any subsequent scheduled abrasive blasting. For as-cast finishes, the joints must be sealed with nonabsorbent plastic compressible tape or sealant. The sealant should be replaced with each use of the forms. Thin plastic tape can be used on the surface over the joint, if subsequent removal of the mortar by any proposed abrasive blasting is deep enough to remove the tape impression (Fig. 44.2). Such tapes must be secured with additional glue above and underneath to prevent slippage and wrinkling when the forming has a slick surface such as coated and metal forms.

Bolts and ties can be sealed by wrapping with tape or caulking to prevent leakage of the mortar.

Liners. To prevent bulging of plastic form liners as temperatures rise, they should be attached to the forms during warm parts of the day. Spraying the form faces with cold water just before placement will help diminish any bulging. Attach a rigid plastic type form liner with 3⁄16 in staples on approximately 3 in centers. Elastomeric liners cannot be fastened with nails because of a tendency to sag but must be glued entirely to the form backing with a mucilage type of glue. Because of the flexibility, joints become mortar-tight and do not need further sealing. Fiberglass liners should be fastened with ¾ inch long screws on 1 ft centers. A thin (0.06 in thick) plastic liner obtained in 3 to 5 ft wide and 50 to 60 ft long rolls has been successfully used to line gang forms by gluing the plastic sheeting entirely to the backing. The mucilage-type glue is squeezed up through the contact joints and around the ties to effectively seal them against leakage. Hard plastic liners and fiberglass types require tapes behind each joint to prevent leakage of mortar into the space behind the liner.

Release Agents. Because of the many types available, any form release agent should be tried on the field mockup sample for incompatibility with the concrete or any subsequent material application. Application should be with a fan-type nozzle and in very thin films. The use of cloths soaked with form release agent (Fig. 44.3) to apply the agent to high-density-type forming has proved beneficial in decreasing the amount of surface bugholes in as-cast surfaces. Any runs should be wiped out to uniform the application. Form release agents should be used for each use of wood or liner forming, except elastomeric types, to prevent differences in the cast architectural surface due to form wear. A partially full drum of form release agent stored for a long period may have a detrimental effect on the surface resulting from concentration of its reactive ingredients.

Storage and Reuse. To prevent warping, panel forming (either wood or steel) is stored flat. Coated or plastic lined form panels should be protected from the sun in order to prevent hairchecking of the coating which will transfer to the concrete and deterioration of the liners. Care during removal, storage, and reuse of the forming will prevent changes in the concrete surface from one use to another, and allow more reuses.

FIGURE 44.3 Workers manually applying form release agent to thin vinyl-lined curved form with agent-soaked cloths. The Museum of Contemporary Arts, Los Angeles, Calif.; forming contractor, Peck-Fager.

44.4 Batching

To prevent intermingling and contamination of the architectural concrete with other concrete materials, special bins or weigh hoppers may be required. On some projects the architectural materials have been batched prior to beginning of the normal concrete batching to ensure clean equipment and prevent comingling. When required, coloring admixtures are also added with the other ingredients. Stockpiles of special aggregates should be reblended on delivery to ensure uniformity in gradation. As the fines tend to accumulate in the bottom foot of a stockpile, this should be wasted as it will adversely affect the uniformity of an exposed-aggregate finish.

For colored architectural concrete, a minimum batch of 4 yd is recommended to ensure color consistency. To ensure color uniformity on a project, full loads are recommended for all architectural concrete. A "wet check" test furnished by the color manufacturer can be used at the jobsite to confirm color consistency of a delivered load of concrete.

44.5 Mixing

Architectural concrete may be mixed in transit, premixed and agitated during transit, or mixed at a jobsite mixer. If nonuniform color streaking is seen, additional mixing should be required to make the color uniform. With the use of coloring admixtures, 1 in slump variations apparently do not materially affect the color uniformity. Larger increases in slump tend to lighten the color.

44.6 Placement

Most equipment can be used to place the architectural concrete as long as it is clean from handling other types of concrete. Belted concrete should be covered from the sun to prevent excessive drying. Pumping architectural concrete may require revisions to the mix to ensure pumpability. Mixing of placement methods should be avoided as this may affect the uniformity of the exposed-aggregate surface. Changes in mix design to allow placement into steel-congested areas should be evaluated on the field mockup sample for their effect on an exposed-aggregate finish. Concrete trucks should have prescribed arrival times to allow uniform and consistent placement procedures. Protection of the form face against mortar splatter by using polyethylene or metal sheets withdrawn as the concrete rises in the form will prevent dark and hard mortar streaks in the surface of a sandblasted concrete (Figs. 44.4).

If foamed styrene is used to line the bottom of tiltup forming, walking within the form during placement or heaping of the concrete may damage the liner (Fig. 44.5).

44.7 Consolidation

Architectural concrete requires additional care in consolidation procedures to diminish or eliminate air bugholes on the surface of as-cast finishes, prevent holidays of aggregate with exposed-aggregate finishes and satisfactorily meld lift lines. Insertions should be spaced so that the radius of action will overlap by at least 50 percent. After 10 to 30 s of vibration, the internal vibrator should be withdrawn slowly at 3 in/s to allow the form surface air sufficient time to migrate upward. To prevent surface holidays in exposed aggregate or changes in density, the vibrators should be kept at least 2 in away from the form face for as-cast finishes and 3 in away for exposed-aggregate finishes.

FIGURE 44.4 Use of twin metal sheets to prevent form splatter.

Thick walls may require vibration adjacent to both mats of steel to ensure adequate consolidation and prevent subsequent settlement and horizontal cracking. Amount of vibration necessary for adequate consolidation can be determined on the field mockup sample by using the designed reinforcement and width. For harsh, gap-graded mixtures, the vibration at the lift line should stop while aggregates are still visible on the surface. Such mixtures tend to lock up after vibration ceases and cannot be penetrated when the next lift is placed, so that a mortar lense is visible on exposure.

Precast plants generally use form vibration to consolidate concrete except around accessories which require additional internal vibration. Special short stubby 2½ in diameter and 4½ in long internal vibrators have been developed for use with tiltup panel construction. Use of the normal-length internal vibrator placed horizontally on the tiltup reinforcement mat has not always sufficiently consolidated the bottom surface of concrete placed in circular arcs or lines. Sandblasting of the surface to expose the aggregate will accentuate the rocky appearance of these areas and have the appearance of irregular lift lines when erected (Fig. 44.6). They are difficult to repair. Any form liner used on the bottom surface of tiltup panels should be adequately fastened by mucilage-type glue to prevent dislodgement during placement and consolidation of the concrete.

FIGURE 44.5 Steel workers using shoes with foam-rubber soles to prevent damage to foamed polystrene liner in tiltup panel form.

FIGURE 44.6 Psuedo-lift lines accentuated when sandblasting is used to expose aggregate in tiltup panels.

External vibration requires formwork to have additional reinforcement for the external forces. Depth of concrete placements, when form liners are used, should follow manufacturer's recommendations as excessive heights may collapse rib-type liners. Normal lifts should be 18 to 24 in in height. Higher lifts to decrease bugholes by increased depth pressure or with high-range water reducers should be tested on the field mockup sample.

Revibration can be used after most of the bleeding has taken place, but before initial set has begun, to complete consolidation and remove additional surface bugholes from the upper portion of the concrete placement. This will require some additional concrete to relevel the top.

44.8 Curing and Form Removal

As with all previous procedures, consistency of curing procedures will assist in obtaining a uniform surface appearance to the architectural concrete. The most common type of cure is to leave the forms on for a certain period and have a follow-up procedure after form removal. Membrane curing compounds have been used where further exposure of the aggregate by bushhammering or sandblasting is to be used. Any membrane used on an as-cast finish should be compatible with further use of sealants, sealers, or coatings.

High-density forming has a tendency to produce dark concrete due to the locked-in surface moisture. This concrete will gradually lighten as it dries as can be seen in Fig. 44.7*a* and *b*. Polyethylene or waterproof paper sheeting should not be used with colored slabs as they tend to mottle the surface through nonuniform contact. A clear, colored resin or wax-base membrane is recommended.

Form removal for both precast and cast-in-place architectural concrete is usually set

FIGURE 44.7*a* Concrete surface immediately after high-density-type form removal. Note dark mottled appearance.

by selecting a period for hardening. Precast plants may use heat to accelerate the process but must maintain the same procedure for all units. Cast-in-place concrete form removal depends on specified architectural treatment and safety. Normal times are 12 to 24 h after concrete placement when ≥50°F air temperatures are present. When retarders are used on the form face, form removal must be done at uniform ages to obtain uniform exposure. Some accommodation will be required when the concrete operations continue through different seasons. Drops in surface temperature should be limited to 40°F per day in order to prevent hairchecking from thermal shock. Cracking of the forms to allow a gradual drop in concrete surface temperature has been advantageous as long as the top of the form is shielded to prevent rapid drying of the top portion (Fig. 44.8).

Removed forming should be cleaned, repaired when allowed, and stored horizontally to prevent warping. Steel forms should be stored after cleaning and a light application of

FIGURE 44.7*b* Same concrete surface after 2 weeks of drying. Note lighter appearance as surface has dried.

FIGURE 44.8 Lighter color at top of concrete placement due to differential drying.

oil. Plastic-coated forming and liners should be stored undercover to prevent deterioration by sunlight.

44.9 Exposed-Aggregate Surfaces

Placing Aggregate. There are numerous methods of placing the aggregate for an exposed-aggregate architectural surface (e.g., see Fig. 44.9). For horizontal surfaces, this may be done by seeding (spreading) special aggregate, with a normal or special grading, manually or with a hopper onto the surface of a freshly floated concrete, or preplacement in a bed of sand for casting into the lower face of tiltup panels. For vertical surfaces, temporarily gluing the aggregate to a plywood form face, or preplacement within vertical form walls is another method. In all cases, integral use as the concrete aggregate for monolithic placement is a final choice.

Aggregate Exposure. The various methods of aggregate exposure, which are detailed further below, consist of manually

FIGURE 44.9 Seeding (spreading) an imported aggregate on the horizontal surface of a colored concrete.

water brushing away the mortar, high-pressure water jet, sandblasting, bushhammering, acid etch, grinding, and fractured rib. Surface retarders may be used to assist the exposure of the aggregate by retarding the surface mortar.

An acid wash is used to enhance and clean the exposed aggregate after exposure. Because of the removal of a portion of the surface during aggregate exposure, additional compensating cover must be provided initially to end up with the prescribed minimum required cover for reinforcement.

Final appearance of the exposure achieved by the proposed personnel, methods, timing, and materials should be accepted on the field mockup sample prior to proceeding with the production architectural concrete. For tiltup construction and plant precast, a full-scale sample (Fig. 44.10) should be constructed for acceptance prior to full production.

High-Pressure Water Jet. High-pressure water jets (Fig. 44.11) with pressures of 3000 to 10,000 psi have been used to expose aggregates in concretes with 5000 psi compressive strengths. Assistance with abrasion by including 10 percent sand in suspension has resulted in insignificant alteration of the aggregate appearance. Attention must be given to the disposal of the wastewater.

Manual Water Brush. To preserve the natural color of exposed aggregate, water brushing (Fig. 44.12) is required for exposure. The brushing is begun as soon as the concrete surface has hardened sufficiently to retain the aggregate. Water is used to soften the surface mortar and assist in its removal. Cast-in-place vertical walls must be brushed within 4 h to allow exposure. Retarders may be used on the form surface for assistance, but care is needed to remove the forming and commence exposure at the same concrete age for overall uniform exposure. If exposure is delayed and sandblasting is required for exposure, a different texture and appearance is to be expected.

FIGURE 44.10 Full-scale tiltup sample erected and being sandblasted to expose aggregate in portions. Note design reference sample in lower left-hand corner of photograph.

FIGURE 44.11 High-pressure water-jet exposure of aggregate: 10 percent sand included with water. Natural color of aggregate preserved.

FIGURE 44.12 Water brushing used to expose soft aggregate in precast plant.

Acid Etching. Acid etching is generally confined to precast plants where safety precautions are rigidly enforced. The aggregates are exposed with a 5 to 35 percent hydrochloric solution by spray, brush, or immersion. This method is usually limited to siliceous or granite aggregates which are more resistant to the acid- rather than carbonate-type aggregates, which can be discolored or damaged. Use on vertical cast-in-place walls is not recommended because of safety problems and possible nonuniformity of aggregate exposure caused by nonuniform application of the acid.

An acid wash containing 2 to 5 percent hydrochloric acid is used to restore color and luster to exposed-aggregate surfaces which have been exposed by other means or coated by contaminants. Neutralization with an alkaline bath is required to prevent further softening of the concrete matrix. Minimum requirements for acid etching are 14 days of age or a compressive strength of 4500 psi.

Bushhammering. This method of aggregate exposure (Fig. 44.13) by a pneumatically driven tool with metal teeth results in a fractured rugged surface retaining the original color of the aggregate. Minimum compressive strength and age is required to be 4000 to 5000 psi and 14 days. If desired, more uniformity of appearance can be achieved by additional aging and drying of the concrete before bushhammering. If more than one worker is bushhammering, wear of the equipment should be kept uniform to maintain uniformity of the texture.

Sandblasting. Exposure of aggregate by sandblasting will tend to polish and lighten its color, and round edges of the aggregate dependent on its hardness and the hardness of the concrete matrix. The depth of exposure can be specified in degrees of reveal or projection of the aggregate from the matrix of the concrete. There are four degrees of sandblasted exposed aggregate: (1) brush having no reveal and used to remove sheen left by high-density forming with no guarantee that color will be uniform, (2) light with 1⁄16 in maximum reveal and expectations of color uniformity with an occasional exposed coarse aggregate, (3) medium with 1⁄4 in maximum reveal of the coarse aggregate, and (4) heavy with maximum reveal of 1⁄2 in or one-third of the coarse aggregate diameter having a predominance of visible aggregate (Fig. 44.14).

However, both brush and light degrees of sandblasting tend to accentuate any surface

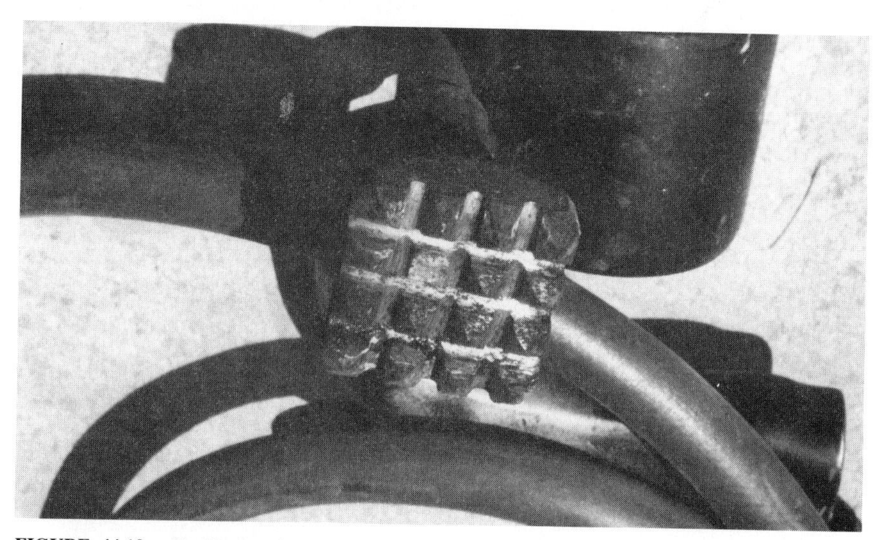

FIGURE 44.13*a* Bushhammering tool.

blemish. Fine cracks become wider as a result of rounded surface edges and bugholes increase in size and quantity. For those degrees, special care in consolidation, application of form release agents, sealing of form openings, and use of lower air-entrainment ranges for durability will result in a more satisfactory surface. As medium and heavy degrees have the coarse aggregate exposed, the contrast of the blemishes tends to be overshadowed by the nonuniformity of aggregate color, randomness of distribution, and the fact that any existing cracks can be seen only between the aggregates which now predomi-

FIGURE 44.13*b* Typical texture from bushhammering.

FIGURE 44.14*a* Brush sandblasting on left; light sandblasting on right.

nate. Age of concrete at sandblasting is not critical for brush and light degrees of sandblasting. Medium and heavy degrees of sandblasting require approximately uniform concrete ages of sandblasting to ensure uniformity in appearance of the exposed aggregate. As the concrete ages, the mortar strength increases and more energy will be needed to achieve the specified reveal. The additional polishing and rounding (Fig. 44.15) of the aggregate may result in contrasting differences in the exposed-aggregate appearance on the same project. Minimum concrete compressive strength for sandblasting to prevent dislodging the aggregate is 2000 psi.

FIGURE 44.14*b* Medium sandblasting on left; heavy sandblasting on right.

FIGURE 44.15 Heavily sandblasted surface resulting in rounding of sharp edges and polishing of the aggregate.

High-density forming is recommended to provide a more uniform surface density for sandblasting. Medium-density forming can be used when one use only is planned.

Grinding. Grinding of the surface of the concrete to expose the aggregate is expensive because more labor is required than in the other methods of exposure. The surface has a polished characteristic. The work is generally accomplished in precast plants where the required facilities are readily available.

Fractured Rib. This type of exposure requires the casting of an as-cast ribbed surface. The fracturing of the rib is accomplished manually with a hammer or with mechanical equipment by striking the side of the rib on 9 in centers vertically and on alternate sides for each consecutive blow. Testing for proper time of fracturing and training of personnel can be effectively accomplished on the jobsite mockup sample. The method and resultant texture are illustrated in Fig. 44.16.

44.10 Final Finishing

Acceptance. In order to eliminate controversy during the final stages of a project, final acceptance should be definitively spelled out in the specifications. In addition, proposed methods and materials for final finishing and repairs should be demonstrated on the jobsite mockup to allow early acceptance and examples for the production architectural concrete.

As-Cast Surfaces. Finishing of as-cast concrete surfaces is determined by the requirements of the specifications and desires of the designer. This can be a surface left as cast or a need for improvement by sack rubbing. Any proposed formulation for sacking

FIGURE 44.16*a* Workers manually fracturing ribs on as-cast finish. Loma Linda (Riverside County), Calif. Veterans Hospital; Robert E. McKee, Contractor.

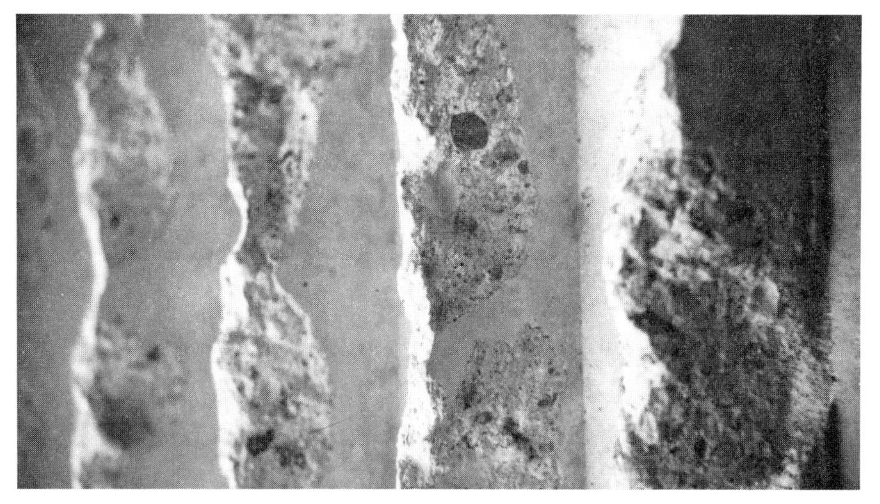

FIGURE 44.16*b* Resulting fractured-rib texture.

should be demonstrated on the field mockup. Typically, the mixes contain 1 part of cement to 2 parts of a well-graded sand passing through a No. 30 sieve and containing sufficient water to produce a consistency of thick paint. A portion of the cement should be white cement for color matching. The mixture is applied in a uniform thickness to a dampened surface by means of clean burlap so that all holes and depressions are filled.

Any excess mortar is scraped off with a trowel. When the surface has become sufficiently firm to prevent removal of the mortar from the holes, the surface is rubbed with dry burlap. No visible film of mortar should be left on the surface.

Sacking should be scheduled so that an entire portion of a structure such as a column, wall, or pier be done at one time to ensure uniformity of color. Air temperatures should remain above 50°F during sacking and for the required 48 h curing period to ensure that the surface is not damaged.

Form Tie Holes. Tie holes can be left as visible voids, partially filled to provide some shadowing or filled flush with the surface. Any remaining metal must be covered with a long-lasting substance to prevent rust staining. Holes should be filled with a wood rod having a rough end to ensure bonding of the various applied layers. Filling of the tie holes in exposed-aggregate concrete surfaces should be delayed until texturing is complete in order to allow proper matching of color and texture. If a bonder is to be used integrally with the patching compound, a test sample should be checked for excessive darkening and susceptibility to ultraviolet rays of the sun.

Repairs

As-cast Surfaces. As-cast surface repairs should be completed as soon as the forms are removed to allow the repairs to age with the adjacent surface. Integral acrylic bonders should not exceed 30 percent of the mix water to prevent excessive darkening or contrast with the adjacent concrete. As acrylics tend to gloss over within 20 to 30 min, any mixture containing them or being bonded should be finished within that time limitation. A repair should have a minimum edge depth of ¼ in to ensure proper compaction and bonding. Patch layers should be limited to a 1 in thickness per application to ensure satisfactory compaction.

Exposed-Aggregate Surfaces. Minor surface repairs should be delayed until after aggregate exposure to ensure proper matching of color and texture. The repair mixture should be checked on the field mockup. Major repairs which require replacement with the architectural concrete must be repaired as soon as possible to allow drying and aging for color uniformity.

Needle Gun. A pneumatic-type apparatus used to scab steel plate prior to painting has been found to effectively remove form leakage lines, tie-bolt leakage, minor lift lines, and surface mortar streaks. The use of chisel-pointed rods in the "needle gun" will diminish the contrast of such blemishes and allow acceptance (Figs. 44.17 and 44.18). Applica-

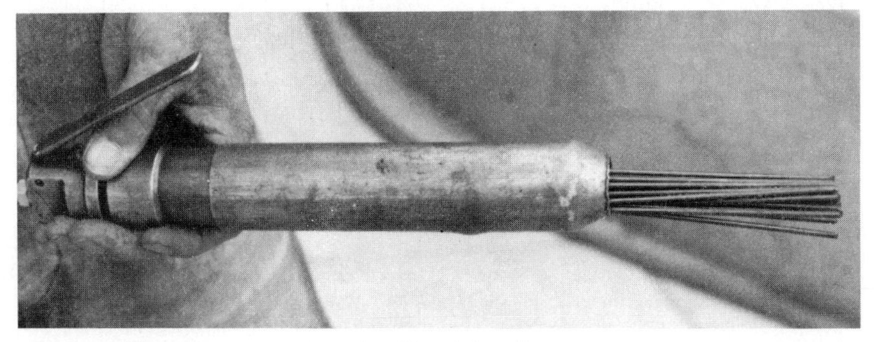

FIGURE 44.17 Needle gun used for repairs of form leakage lines.

FIGURE 44.18*a* Surface with form leakage lines prior to use of needle gun.

FIGURE 44.18*b* Surface after removal of contrasting form leakage line.

tion pressure of the rods should be gentle to prevent a bushhammer appearance or a depression below the surface which appears as a shadow in oblique sunlight.

Stain Removal. All precautions should be taken to protect finished architectural concrete surface from subsequent construction operations which may damage or stain. Rust stains can be removed by further sandblasting with sandblasted exposures. For as-cast surfaces, use a solution of 1 oz of sodium citrate per 6 oz of water applied for 10 min, followed by crystals of sodium hyposulfate and covered with a paste of Fuller's earth. After drying for 10 min, the surface is cleaned. The process may have to be repeated. Efflorescence and calcium carbonate may be removed with applications of mild solutions of muriatic acid. Care must be exercised with muriatic acid because of its toxicity and burning capability.

Coatings and Sealers

General. Coatings and sealers are materials used to protect the architectural concrete surface from airborne contaminants and weathering.[1,2] A sealer is expected to penetrate the surface up to ¼ in and leave little material on the surface, while a coating will penetrate slightly but also leave a visible film on the surface. A coating may be clear or opaque. Silane- and siloxane-type sealers have a proven service life of 5 to 10 years and do not materially affect the existing color of the architectural concrete.

Application. Prior to application, any sealer or coating should be tested on the mockup sample for compatibility with the surface color and further treatments. Many sealers and coatings tend to return the contrast of repaired or needle-gunned areas. Unless exposed-aggregate areas require a sealer or coating for freeze-thaw, a sulfate environment, coastal conditions, corrosive smog, or prevention of graffiti, their use should not be considered. Adhesion of such materials should also be checked on the field mockup sample especially when high-density forming is used for as-cast surfaces.

44.11 Quality Assurance

A quality control plan should be requested from the contractor to ensure that proper planning is initiated. A copy should be given to the inspector assigned to this portion of the work for personal use. In addition, the specifications and postbid meetings can be used to prescribe the duties and responsibility of the inspector in charge. Remember that contrast is the key to the resolution of most architectural concrete problems. If the contrast is reduced, acceptability becomes a reality.

REFERENCES

1. A Guide to the Use of Waterproofing, Protective, and Decorative Barrier Systems for Concrete, (ACI 515.1R-79) (rev. 1985), American Concrete Institute, Detroit, Mich.
2. Architectural Precast Concrete, Precast/Prestressed Institute, 2nd ed., 1989.

P A R T • 13

REPAIR AND GROUTING OF CONCRETE

CHAPTER 45
REPAIR OF CONCRETE

Joseph A. Dobrowolski

Concrete is a versatile building material that is cast inexpensively and is durable and strong and takes the shape of the form in which it is cast. Most concrete members and shapes are cast on site and in their final location. Marginal work and construction procedures cause imperfections that require repair. Natural forces and use in service cause deterioration or damage which must be repaired or replaced during maintenance of the structure.[1]

Most small repairs are made to make the structure appearance acceptable and when completed should blend with adjacent untouched surfaces by matching its color and texture. In addition, the repair must be

1. Thoroughly and permanently bonded to adjacent concrete.
2. Sufficiently impermeable to protect the original concrete.
3. Free of shrinkage cracks or checking.
4. Resistant to freezing and thawing where required.

Repairs to old concrete require analysis and planning as deteriorated concrete must be isolated and removed. Some dams have had complete refacing with shotcrete and prepacked-aggregate methods. In earthquake country, retrofitting old buildings and repair of seismic damage have become normal practice.

This chapter discusses recommended materials and methods for making most types of concrete repairs.

MATERIALS FOR CONCRETE REPAIR

The choice of a repair material must consider thermal compatibility with the base concrete, durability requirements, service conditions, chemical and electrical environment, shrinkage characteristics, modulus of elasticity, and placement conditions.[2]

45.1 Portland Cement

Most types of portland cement may be used. However, to ensure compatibility, the type of repair cement should be similar to the original. Type II or V is used for resistance to moderate or severe sulfate attack, respectively. Type III high-early-strength cement can be used for rapid strength gain. Shrinkage-compensating and high-alumina cements have been used in special situations. Further discussion can be found in Chap. 1.

There are many packaged proprietary quick-setting cements on the market which must conform to ASTM C928. Each must be investigated prior to use as some contain gypsum, which is not durable in an exterior or moist exposure, or chlorides, which may create corrosion difficulties. Their rapid setting and hardening properties restrict use to small quantities. Any material which has hardened or becomes stiff should be wasted.

45.2 Aggregates

Requirements for aggregates are similar to those for concrete and mortar. Special aggregates required by the specifications may have to be used in the repair of an exposed-aggregate finish to match the adjacent color and texture.

45.3 Admixtures

Admixtures used in concrete can also be used in repairs. These include water reducers, air-entraining agents, chemical accelerators, steel, alkali-resistant glass or polypropylene fibers, silica fume, fly ash, natural pozzolans, and integrally mixed polymers.

45.3 Bonding Agents

Bonding agents can be placed in three groups, such as epoxy-based, latex-based, and cement-based.

Epoxy bonders are required to conform to ASTM C881. Because of greater thermal coefficients and tensile strengths, thin repairs may pull loose from the base concrete during sharp temperature variations. Moisture trapped under an impermeable epoxy-based repair may cause failure under freeze-thaw conditions. As epoxies have a short pot life, quantities that can be mixed are limited in size and may be more limited during high ambient temperatures, which shorten the setting time. Once hardened, epoxies are not affected by moisture and most chemicals, but will soften with high temperatures and melt and char in temperatures above 450°F.[3]

Latex bonders should conform to ASTM C1059. They are of two types: Type I—reemulsifiable and Type II—nonreemulsifiable. Type I can be applied to the surface to be bonded a few days ahead of time but cannot be used where moisture or high humidity is present. It is not recommended for structural bonding. Type II has a higher bond strength and can be used in areas where moisture is present.

Cement grout or mortar consisting of fine sand and cement proportioned to a thick-cream consistency have been used for years to bond clean surfaces together.

Polyesters, acrylics, and methyl methacrylates are resin bonders similar to epoxy bonders but are more economical. They tend to have a higher shrinkage than epoxies. Polyesters do not adhere well to nonporous surfaces such as metal or glass, are not compatible with wet or damp surfaces such as fresh concrete, and have a short shelf life.

45.4 Sealants and Coatings

Repairs may require a sealant or coating for protection against severe weather or chemicals. Sealants are clear and are expected to penetrate the surface without leaving a visible film. Coatings are clear, opaque with or without color, and may have some penetration but will leave a visible film on the surface. Both should be capable of allowing vapor emission from the concrete but be impermeable to moisture penetration after curing. Silane and siloxane sealants have become popular because of their long service life (5 to 10 years), effective penetration (⅛ to ¼ in), and little effect on concrete color. Because of environmental requirements, solvent formulations have been revised to water emulsions, which has adversely affected certain characteristics of the sealants. Any planned use in areas which will be covered during further construction should be checked.

Coatings such as chlorinated rubber epoxy, colored epoxies, and acrylic latex paints may be used but should be checked for compatibility with expected thermal changes, color of the adjacent surface, and other service requirements.

PREPARATION OF CONCRETE FOR REPAIR

Most concrete repairs involve the removal of concrete which may be either deteriorated or damaged. Choice of the various methods for removal are determined by extent and location of the repair, economy, safety, effect on the concrete to remain and delay, or effect on present use of structure. Specifications should outline the expected result of the repair and may in some cases outline the methods of concrete removal which may be used.[4] Large repairs which require replacement of structural concrete should be done as soon as feasible after form removal. Where reinforcement is involved (Fig. 45.1), exposure of a reinforcing bar past its midpoint requires additional removal of concrete to provide at least one inch of space behind the bar.

FIGURE 45.1 Void left in concrete shearwall because of insufficient vibration and reinforcement configuration.

FIGURE 45.2 Cleaning surface of panel by waterblasting prior to repair.

FIGURE 45.3 Worker removing excess material from surface of wall with chipping gun.

45.5 Removal Methods

Methods of removing damaged or deteriorated concrete during repairs can be classified as abrasive, cutting, impacting, presplitting, and blasting.

Abrasive methods include sandblasting, shotblasting, and high-pressure waterblasting. They are used to remove thin layers of concrete which may be discolored, deteriorated, or uneven and to roughen the surface in preparation for the repair. Sandblasting may be done on vertical and horizontal surfaces. Shotblasting is limited to horizontal surfaces such as floors. High-pressure waterblasting can be used similar to sandblasting (Fig. 45.2).

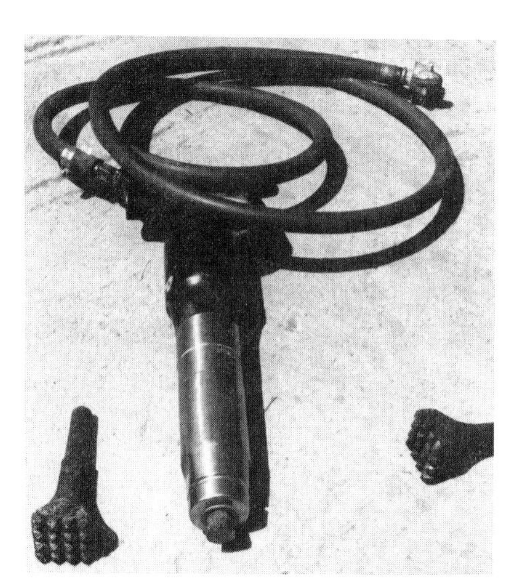

FIGURE 45.4 Bushhammer equipment with head and air gun.

Impacting methods consist of using a handheld chipping gun or bushhammer, scarifier, scabbler, or boom-mounted breaker. The chipping gun consists of a chisel-type bit operated pneumatically (Fig. 45.3). A bushhammer is also operated pneumatically but has a head containing a number of projecting teeth which is applied at right angles to the surface and used to uniform plumbness and remove deteriorated material (Fig. 45.4).

In the "scabbler," a machine traveling on wheels, a number of heads are mounted on the bottom of cylinders which are operated by air. The machine can be set to remove a predetermined depth of material.

A typical scarifier has up to 108 cutting bits on a rotating head and cuts with a rotating action. It can be used on vertical or horizontal surfaces which contain form ties and wire mesh. Dust and debris control can be easily handled with this machine.

Presplitting methods consist of mechanical wedging, water-pressure pulses, and expansive chemicals. A type of mechanical wedging uses a hydraulic splitter placed in predrilled holes for removal of large volumes of concrete in massive structures. Cracks in the concrete may prevent the buildup of hydraulic pressure in the bore holes. A secondary method may be required if reinforcing steel is present. The expansive chemical method involves mixing the chemical with water and pouring the mixture into a predetermined pattern of bore holes. The mixture undergoes a large increase of volume in a short time, which tends to split the concrete. Other methods may be more economical, but the advantages of the presplitting method are that large volumes of material can be removed from a concrete structure by placing the expansive material in deep holes with little effect in the remaining concrete.

Cutting methods include high-pressure water jet, diamond or carborundum saws, diamond wire, and thermal. The high-pressure water jet uses a small jet of water driven by pressures of 10,000 to 45,000 psi (Fig. 45.5). It is an excellent method to remove concrete when reuse of the reinforcement and minimal damage to the remaining concrete is desired. The diamond and carborundum saws are available in many sizes so that selection can be made for any project, but cutting is a slow and costly process (Fig. 45.6). Cutting with a diamond wire involves looping a wire, impregnated with diamonds, around the portion of the structure to be removed. The wire loop is spun by power as it is drawn out through the concrete.

Removal by blasting consists of using explosives in predetermined holes to rupture the concrete so that it can be removed. This method is used to remove large quantities of concrete.

45.6 Cleaning and Curing

Once the deteriorated or damaged concrete has been removed, any dust, dirt or loose material must be removed by waterblasting followed by blowing out any excess moisture.

FIGURE 45.5 Power of high-pressure water jet illustrated by cuts in concrete surface.

FIGURE 45.6 Machine developed for leveling floors and pavements. Diamond saws, spaced 0.060 in apart, are on horizontal axis for cutting adjustable depth.

If the repair material is cementitious, the base concrete to be repaired should be kept moist at least 48 h prior to the application of the repair material. Presence of oil or other contaminants will require washing with detergents.

REPAIR METHODS

A thorough study of the underlying reasons for the deterioration and damage to be repaired is necessary as it will determine the type of repair method. Epoxy injection of cracks still subject to thermal or load deflections will not be permanent. Once the cause or causes have been determined, a permanent repair solution can be found. Maintenance schedules and planning may also dictate the expected life of the repair. Use of the structure may require a method of repair that will have a minimal effect on operations during the repair application. Weather conditions will also influence the choice of repair. Modifications may be required to present environmental conditions to prevent future deterioration or damage.[9]

Methods of repair include concrete replacement which can be the same as the base concrete or modified with silica fume, acrylic, styrene butidiene latex or epoxy, drypack replacement, shotcrete application, preplaced aggregate with pumped or formed concrete, injection of cement grout or chemical formulations, and manually applied mortar.[6]

45.7 Concrete Replacement

Conventional concrete similar to the original concrete may be used for the repair if forming, placement, and consolidation are feasible. This method is used where large voids extend through the section (Fig. 45.1), holes in unreinforced concrete having an area of more than 1 ft^2 and over 4 in in depth, and holes in reinforced concrete over ½ ft^2 in over 4 in in depth, and holes in reinforced concrete over ½ ft^2 in area and deeper than the reinforcing steel (Fig. 45.7). Admixtures may be used to entrain air, accelerate or delay the setting time, increase strength, improve placement workability, or reduce water requirements or other properties. If the original problem was caused by alkali–aggregate reaction, the use of a pozzolan material would not only increase the resistance to such an attack but would also improve long-term strength and impermeability. Proportions should be similar to those in the original concrete except for changes reflecting new technology or space requirements demanding a smaller maximum-size aggregate. The use of shrinkage-compensating cement has been successfully used in repair concrete. A low-slump concrete will also minimize shrinkage.

The conventional concrete repair consists of forming on both sides of the concrete section so that the form is completely sealed around the repair. A chimney or hopper is constructed on top and integrally with the form so that a hydraulic head will assist in maintaining pressure onto the concrete (Fig. 45.8). The top of the concrete void should have a slight upward incline to the outside of 10° minimum to allow air to escape during consolidation. In addition, the form hopper should be as wide as the void to be repaired and have sufficient opening for concrete placement and insertion of internal vibrators.

Prior to placing concrete, the existing area of concrete to be repaired should be dampened so as to be saturated surface-dry. Lifts should be no greater than 12 in, and openings should be provided at sufficient levels to allow placement of concrete and consolidation. The form should be constructed to allow stripping the next day so that the excess concrete can be removed, if required. As soon as the removal is complete, the repair should be cured by application of a curing membrane or a taped-on impervious covering.

Silica-fume additions are made in either liquid or powder form in quantities of 5 to 15 percent by weight of the cement. Compressive strengths of silica-fume concrete can attain strengths of 12,000 to 15,000 psi.

FIGURE 45.7 Typical void requiring regular concrete replacement.

FIGURE 45.8 Forming for structural repair with conventional concrete. Note hopper at top.

Epoxy-modified concretes result in high flexural, compressive, and tensile strengths. Durability in freeze-thaw climates is satisfactory. Impermeability, chemical resistance, and bond are improved. Typical water/cement ratios are 0.25 to 0.35. Levels of epoxy in concrete range from 15 to 20 percent polymer solids by weight of cement. A typical epoxy-modified mix design would be as follows:

Cement	752 lb (8 sacks)
Water	250 lb (30 gal)
Sand	1532 lb
⅜ in × No. 4	1296 lb
Epoxy resin	122 lb (13.19 gal)
Epoxy curing agent	19 lb (2.29 gal)

One day of moist curing and several additional days of air drying at 60 to 80°F prior to application are required.

Styrene butadiene and acrylic latexes have been most effective and predictable in concrete restoration. Effective ratios of levels of polymer solids to weight of cement are similar to those of epoxies, and water/cement ratios have ranged between 0.30 and 0.40. Premature surface thickening and shrinkage cracking have been experienced during field placement. As the modulus of elasticity of the latex-modified mixes is lower, its use to repair high-load-bearing structural members should be investigated ahead of time.[12,13] Acrylics are useful in exterior repairs with white or colored concretes where color retention is required.

45.8 Topping Repair of Unformed Concrete

The repair of damaged or deteriorated concrete slabs and decks involves the placement of toppings or complete replacement. The complete replacement is the same as normal good concrete placement and finishing. Only slabs less than 2½ in in thickness need to be bonded to the base slab. Maximum size of aggregate is determined by the thickness of the topping and included reinforcement. Any cracks in the base slab should be repaired by filling with a low-modulus epoxy or covered with a strip of nonbonding sheathing. All joints in the base slab should be duplicated in the topping. Curing membrane application should quickly follow the finishing operations to prevent cracking. Toppings 2½ in and over should have a definite bond breaker between the topping and the base slab.

Bonding of thin toppings is accomplished by brooming into the existing surface a thick creamy mortar of 1 part cement and 1 part sand, similar to that in the topping, just before application of the topping. Acrylic latex emulsions and epoxies have also been successfully used as bonders for toppings.[8] Proportions should follow manufacturer's instructions. Because of the short pot life of the synthetic bonders, their proportioning and application should be limited to a total of 20 to 30 min. To prevent excessive air entrapment, the mixing of the mortars containing the synthetic materials is limited to a maximum of 2 min. Epoxies that need to cure in humid environments and bond to damp surfaces must conform to ASTM C881.

To minimize early shrinkage cracking, polypropylene, steel, and glass fibers have been added to topping mixes. Rusting of the steel fibers has been objectionable on some projects. The initial surface appearance of the toppings containing glass and polypropylene fibers is not as smooth as a normal steel-troweled finish, but the fibers disappear with time as a result of wear and weathering. A light small saw has been developed which allows concrete to be sawed as soon as finished. Depth of cut is limited to ¾ in, which is sufficient for thin toppings.

45.9 Shotcrete

Shotcrete is an excellent method of repairing vertical and overhead surfaces. Details on proportioning and materials can be found in Chap. 36.[5] Where repair areas are deep, shotcrete should be built up in layers which are allowed to take some initial set before an additional layer is placed to prevent sloughing. For repairs of over 2 in depth, wire mesh having a minimum size of 2 × 2 is used. For large areas, the mesh should be fastened to the base surface with dowels. If possible, the wire mesh is attached to existing reinforcement. Shotcreting around existing reinforcement is done by angling the nozzle to a maximum of 45° so that shotcrete can be placed in back of the reinforcement.

Either the dry-mix or wet-mix shotcrete can be used for repair. The water for the dry mix of blended damp (3 to 6 percent moisture by weight of sand) sand is added at the nozzle while water for the wet mix is added to the sand and cement at the batch plant and mixed in a transit-mix truck. The mix is chuted from the truck into a pump hopper which pumps the concrete or mortar to the nozzle. Air, under pressure, is used to impell the wet mix against the form or base concrete. The applied shotcrete is finished to match the adjacent concrete. Finishing should be done carefully to avoid disturbing the bond of the finished product. Curing is done with a curing membrane or impervious sheeting. Further details are given in Chap. 36.

Fibers have been added to shotcrete on some projects. The material provides added strength to the shotcrete repair, but there is insufficient data regarding benefits and costs.

45.10 Prepacked Aggregate with Mortar Intrusion

The prepacked-aggregate method was developed for large repairs of piers, abutments, and walls. The deteriorated concrete is removed and coarse aggregate preplaced between forms and the area to be repaired. A cement-sand grout is then injected to fill the voids starting at the bottom. As the grout surface rises, the grout pipes are gradually pulled up to follow but remain below the surface. The process is suitable for underwater repairs as it replaces the water as the voids are filled with grout. Repairs have been successful for water tanks, beams, and columns in industrial plants. Details for this method may be found in Chap. 38.[11] Since the technique involves higher form pressures, form joints must be sealed to prevent leakage of the mortar.

45.11 Injection Methods

Various injection materials are used to repair cracked slabs that are on grade or suspended, fill voids under slabs and structures to restore support and grades, repair cracked structural members, and fill voids in structural members. These include cement grout, epoxies, acrylamides, polyurethanes, and methyl methacrylates which are injected by pressure or gravity.

Cement grouts can be jobsite formulated or purchased as proprietary packaged material.[7,10] They are economical and easy to obtain and install. Admixtures are available to minimize shrinkage and mineral fillers to make them more economical when large quantities are required. Since cement grouts are composed of solid particles in suspension, the minimum crack width feasible for injection is about ⅛ in. Support and restoration to grade of subsided slabs on grade and footings can be accomplished by injecting cement grout under pressure. Continuous monitoring of the changing elevations is required to prevent excessive uplift. Cement grouts have been used to fill large voids and cracks in mass structures. Such grouts should conform to ASTM C1107.

Epoxies have wide ranges of viscosity and setting time. Cracks as narrow as 0.002 in have been filled successfully. They have excellent bond to clean and dry substrates, and some types will bond to plastic concrete. For instance, the structural integrity of a cracked structural member can be restored by epoxy injection (Fig. 45.9). Cracked slabs on grade have been restored to monolithic and load-supporting capacity. The short pot life and working time is further shortened during high ambient temperatures. In some cases it may be more economical to remove and replace a badly cracked on-grade slab rather than repair it by epoxy injection (Figs. 45.10, 45.11).

FIGURE 45.9 Cracked prestressed double-tee wall panel with structural integrity restored by epoxy injection.

Gel-type formulations such as the acrylamides and polyurethanes are excellent for repair of water-containing structures as some formulations have viscosities allowing injection into any crack or opening that water can flow through. They cannot be used to repair a structural member.

Methyl methacrylates have been used to repair cracked structural and on-grade slabs in three stages. First, the cracks are gravity-filled by pot with a spigot. Then

FIGURE 45.10 Crack in on-grade slab with surface sealed by epoxy and openings for injection at intervals.

the entire surface is flooded with the methyl methacrylate. The final stage requires the flooded material to be scrubbed into the surface with a stiff-fiber broom. This method does not restore structural integrity but will improve impermeability and rigidity of the floor. Protective clothing is required for this method. For confined areas with poor ventilation, a breathing mask is also required.

Pressure injection requires the sealing of cracks at the surfaces with epoxy or a heavy wax and leaving injection openings or placing injection ports at 10 to 12 in intervals. The injection nozzle has two lines converging which blend the epoxy and its catalyst just before ejection. The nozzle has a slit for openings or a round tip for injection into ports. The epoxy or gel is injected at one end of the crack (Fig. 45.12) until it is seen at the next adjacent port. The injection device is then transferred to that adjacent port or opening, and the process is continued until the entire crack is filled. Cracks in slabs on grade cannot be sealed on the bottom. This requires some judgment as to penetration of the epoxy before moving to the next port. In some cases, a higher viscosity can eliminate most of the waste into the subvoids. Because of this waste, costs for on-grade slab repairs are higher per foot than those for structural crack repairs unless scaffolding costs are encountered. Scheduling of night and weekend

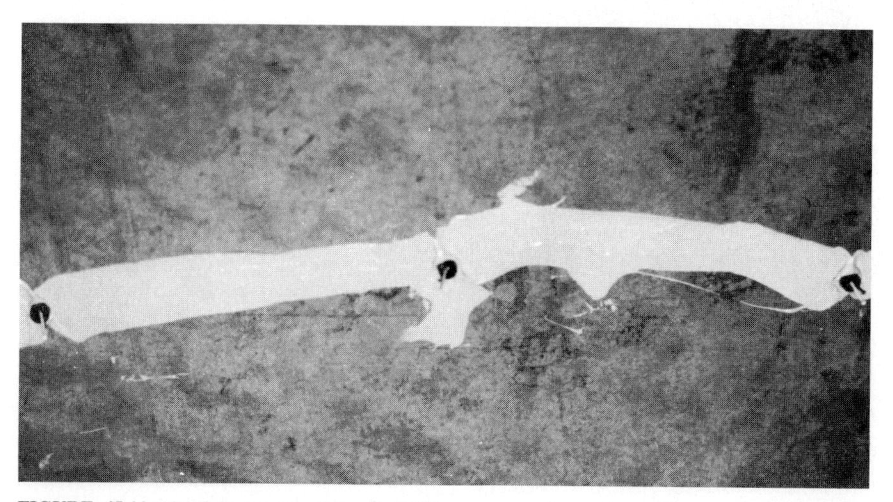

FIGURE 45.11 Crack in on-grade slab sealed with epoxy at surface and containing ports at intervals for epoxy injection.

repair operations is preferable, to avoid interruption of weekday industrial schedules during these repair periods, as the repaired floors can be used the next morning (Fig. 45.12).

If only light traffic is anticipated to continue and the slab cracks have spalled edges, continued deterioration can be prevented by grooving out the cracks and gravity-filling the grooved cracks with a low-modulus epoxy (Fig. 45.13). Typical low-modulus epoxies have a hardness Shore A of 80 and Shore D of 50, tensile strength of 400 to 500 psi, and adhesion to concrete of 150 to 250 psi.

45.13 Drypack Repairs

Drypack mortars consist of 1 part cement, 2½ to 3 parts passing through a No. 16 sieve, and sufficient water to mold the material into a ball with slight pressure and no evidence left of water on the hand. Because of its low water/cement ratio, shrinkage is minor, and the repair has good durability, watertightness, and strength. Some white cement is combined with the gray cement to match the adjacent existing concrete surface. This method cannot be used for shallow patches but is excellent for small holes with greater depth than area, patching of rock pockets, and form tie holes. The repair material should be applied in 1 in layers so that proper consolidation is accomplished by tamping. Care should be taken to use a tamper with roughness rather a smooth metal surface as the 1 in sections may not bond because the interface was polished by a metal tamper. Tampers for form tie holes should be round wood dowels rather than steel rods.

Acrylic bonders have been added integrally in ratios of 20 to 25 percent of the mix water, which has aided adhesion and durability without affecting color. Curing is done by taping waterproof paper over the patch or applying a membrane cure (Fig. 45.14). Integral use of epoxies should be investigated as some patches containing the epoxies have changed color after exposure to sunlight.

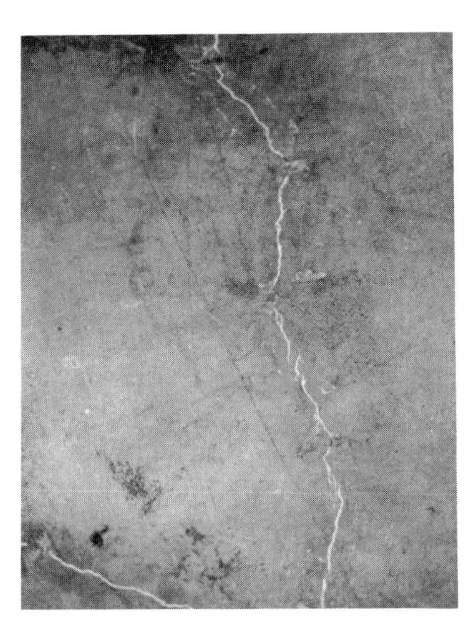

FIGURE 45.12 Repaired crack after removal of top seal.

FIGURE 45.13 Crack grooved out and filled with epoxy by gravity.

FIGURE 45.14 Curing of drypack repair on building.

45.14 Manually Applied Mortar

Mortar applied manually can be mixed at the jobsite from regular portland cement and sand, prepackaged cementitious material, or polymer-modified mortars. These materials should not be used to repair areas where reinforcing steel is exposed and undercut as it is difficult to compact these materials behind the reinforcing bar.

The surface must be roughened and cleaned to achieve bond. Curing is critical for this repair and should be initiated as soon as a portion is available for large areas and as soon as application has been completed in small areas. Application of a curing membrane is probably the most efficient. Care must be taken on incremental curing that the curing membrane does not drift over on uncompleted work. Examples of manual application of mortar are shown in Figs. 45.15 through 45.17.

45.15 Spall Repairs

A common problem with contraction and construction joints in industrial buildings is spalling of the edges in forklift traffic lanes. The spalling tends to widen and deepen under traffic with time as the spalling debris drops further into the spalled area when the joint widens during lower temperatures. The depression may become deep enough to affect the stability of forklift loads and cause stock to fall off the pallet and become damaged. The problem can be prevented by the installation, during construction, of a minimum 1 in depth of low-modulus epoxy in sawed construction and contraction joints. The sealant is finished flush with the surface for smoothness. Spalls may also develop when plastic strip inserts are tipped during the finishing operations (Fig. 45.18). Invariably the upper wedge of concrete will break off under the pressure of traffic.

The size and location of a individual spalled area will determine the method of repair. Small surface spalls resulting from impact of dropped objects need to be cleaned before

FIGURE 45.15 Thin surface patch being applied to deck of parking structure. Minimum depth is $\frac{1}{4}$ in to provide edge for compaction.

FIGURE 45.16 Undesirable surface removed by chipping gun.

FIGURE 45.17 Surface repaired with applied mortar.

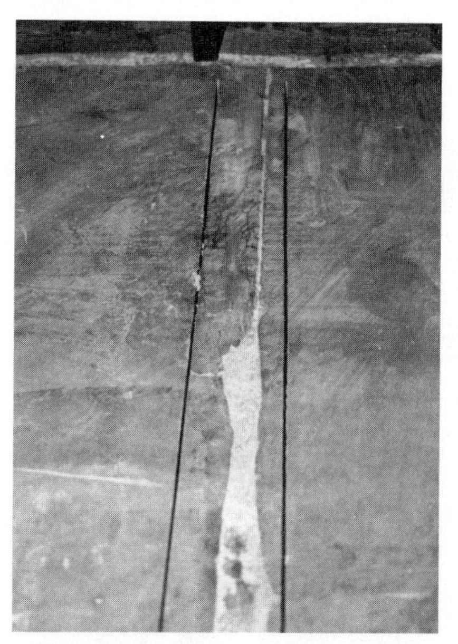

FIGURE 45.18 Contraction joint spalled as a result of tipping of plastic strip. The joint has been sawed on both sides in preparation for removal of concrete.

the addition of fresh concrete and dampened prior to repair. The repair can consist of a mortar having 1 part portland cement to 2 to 3 parts clean sand. Mix water should contain 25 to 30 percent acrylic bonder and be added to blended dry materials in a quantity to be workable just for placement. Too much water causes excessive shrinkage and haircheck-ing. Curing consists of application of a curing membrane or taped-over impervious cover such as polyethylene or building paper or water. After 48 h, the repair should be allowed to dry. Pot life for the mix is approximately 20 to 30 min. Mixing should be limited to 2 min to prevent entrapment of excessive air.

If joint spalling of both edges is no more than ¾ in total, repair consists of cleaning the joint of all debris, loose material, and contaminants, followed by gravity-filling with a low-modulus epoxy. The joint should be overfilled so that it can be cut or ground with a sandpaper grinder to meet the adjoining surfaces.

When a spalled joint is over ¾ in in width, the repair consists of sawing approximately 1 in clear of the spall edge. The saw cut should be 1½ in deep (Fig. 45.18). The concrete between the saw cut and the joint is chipped out to the 1½ inch depth. After the spalled section has been removed, a separation strip of ⅛ in thick pressed board is placed in the joint and the chipped-out areas on one or both sides are slightly overfilled with a mixture of 1 part epoxy and 3 parts dry silica sand. After hardening, the repair is ground flush with the surface with a sandpaper grinder which will also grind down the pressed-board separator strip which maintains the joint (Fig. 45.19). An alternate procedure is to omit the pressed board and saw-cut the joint after hardening. In any case, the joint opening needs to be preserved as it will cause failure of the repair by cracking over the joint opening or separating at one of the repair edges.

Spalls due to popouts, where a chert or reactive aggregate just below the surface has expanded and popped out a piece of the concrete surface, require repairs that involve more than merely filling of the hole. To prevent additional reaction, the remaining portion of the aggregate should be drilled out. The rest of the repair is similar to that for other spalls. Some experimentation should be done to ensure color matching.

FIGURE 45.19 Completed spalled joint repair. Note ground-pressed board strip in center.

REFERENCES

1. Guide to Durable Concrete (ACI 201.2R-92), American Concrete Institute.
2. Causes, Evaluation, and Repair of Cracks in Concrete Structures (ACI 224.1R-89), American Concrete Institute.
3. Use of Epoxy Compounds with Concrete (ACI 503R-80), American Concrete Institute.
4. Standard Specification for Repairing Concrete with Epoxy Mortars (ACI 503.4-92) (rev. 1986), American Concrete Institute.
5. Guide to Shotcrete (ACI 506R-85), American Concrete Institute.
6. Guide for Use of Polymers in Concrete (ACI 548.1R-86), American Concrete Institute.
7. Specification for Packaged, Dry Combined Materials for Mortar and Concrete, ASTM C387.
8. Specification for Epoxy Resin Base Bonding Systems for Concrete, ASTM C881.
9. Thermal Compatibility between Concrete and Epoxy-Resin Overlay, ASTM C884.
10. Specification for Packaged, Dry, Rapid-Hardening Cementitious Materials for Concrete Repairs, ASTM C928.
11. Practice for Proportioning Grout Mixtures for Preplaced-Aggregate Concrete, ASTM C938.
12. Specification for Latex Agents for Bonding Fresh to Hardened Concrete, ASTM C1059.
13. Bonding Strength of Latex Systems Used with Concrete, ASTM C1042.

CHAPTER 46
GROUTING OF CONCRETE

Robert E. Philleo

Grout is normally considered to be a mixture of portland cement and water or a mixture of portland cement, sand, and water to which chemical admixtures may or may not be added. In recent years the term has also been applied to chemical formulations, entirely unrelated to portland cement, which are used for the same purposes as portland cement grout.[4]

The principal uses of grout are the setting of machine bases, sealing fissures in foundations under hydraulic structures, filling joints in concrete structures, filling cavities behind tunnel linings, and filling the voids between preplaced particles of coarse aggregate in the "prepacked" method of construction. This chapter is concerned only with the grouting of joints and tunnels. Since the use of chemical grouts is confined primarily to foundation grouting or soil stabilization, only cement grouts are discussed here.[7]

GROUTING OF MASS CONCRETE

There are two principal types of joints in mass concrete which must be grouted: (1) vertical radial contraction joints in arch dams and (2) longitudinal vertical joints in high-gravity and arch dams. In addition some organizations grout contraction joints in gravity dams, although this is by no means universal practice. Gravity dams are divided into individual monoliths by contraction joints perpendicular to the axis. Each monolith is designed to be stable. Therefore, there is no load transfer across the contraction joints. The only reason for grouting these joints is to prevent leakage. Since it is customary to place water stops in these joints, many authorities consider grouting unnecessary. Some organizations also used keyed joints to increase the path length for water and to increase the probability of the joint's becoming plugged with silt if a leak should occur. When longitudinal joints (joints parallel to the axis) are included in a gravity dam, however, the joints cannot be left open. Shear stresses must be transferred across the joint. The separation of the monolith into two or more individual monoliths by longitudinal joints produces a stress distribution entirely different from that assumed in design and might seriously affect the safety of the structure. Every effort is made to avoid longitudinal joints in gravity dams. As the height of the dam is increased, the length of the lower portion of each monolith is increased and the problem of temperature cracking brought about by the interaction of thermal volume changes in the concrete and foundation restraint increases. Longitudinal joints are introduced to prevent the accidental separation into individual monoliths by cracking and to produce the separation in a controlled location where it can be grouted. In the past longitudinal joints were common in gravity dams. In recent years precooling of materials and postcooling of the concrete have so well controlled the temperature that monoliths over 700 ft high have been built successfully without longitudinal joints.

In arch dams radial contraction joints will close and transmit compression when water pressure is present behind the dam. Although little or no shear stress needs to be transferred, the joints are grouted to ensure a uniform transmission of compressive stress across the joint. Nonuniform stress produces effects not contemplated in design. In very-high-arch dams longitudinal joints are introduced for the same reason that they are used in high-gravity dams.[5]

Grouting is always postponed until the concrete has cooled. Where artificial cooling is employed, the grouting can be done quite early, in some cases within 2 months after placing the concrete. Otherwise, a long delay is necessary. Waiting until the completion of cooling achieves the double objective of opening the joint to its maximum width to facilitate grouting and ensuring that no further contraction will occur following grouting. Such contraction could reopen the joint. Where pipe cooling of the concrete is employed, it is common to cool the concrete to a temperature below its final stable temperature so that, following grouting, the attempt of the concrete to expand on warming will place the concrete in compression and positively prevent reopening of the joint.

46.1 Design of Layout for Grouting

The objective in planning a grouting program is to achieve complete filling of the joints without damaging the structure. The latter is a definite possibility which must be guarded against. Whereas in foundation grouting it is common to use pressures of several hundred pounds per square inch to fill small remote voids in the underlying rock, such pressure cannot be tolerated within the joints of a structure. Excessive pressure in a joint may force the joint open to an undesirable extent so that it closes parts of adjacent joints or even moves adjacent monoliths out of line. In extreme cases it may cause shear failures along horizontal construction joints and thus increase rather than reduce cracking. Grout pressure should be restricted to about 50 psi. The area of joint to be grouted at one time should be restricted to that area which can be covered with a pressure of 50 psi. As a result of this requirement it has become common practice to grout a height of 50 ft at one time. Means should be provided to ensure that no harm is being done to the structures. These include pressure measurements, overall displacement measurements by theodolite, and joint-opening measurements by dial gauges near the top of each grouting lift.

Each area to be grouted at one time should be isolated by metal sealing strips, commonly called "grout stops," around the periphery of the joint. The material normally used is 24 oz soft-tempered copper, although annealed No. 20 gauge stainless steel has been used. The strips should include an accordion fold to accommodate changes in joint opening after installation. Connections should be carefully brazed or welded since the grout stops must resist pressure during grouting.

Experience has indicated that it is not advisable to attempt to grout more than 100,000 ft^2 of joint area a day. This limitation may influence the grouting procedure. In small arch dams and small gravity dams in which contraction joints are to be grouted, it is desirable to grout all the joints from abutment to abutment in one operation in order to avoid the problem of excessive bending and joint opening. When the size of the dam becomes too large, the normal procedure is to work from each abutment toward the center, although sometimes construction contingencies require some deviation from this plan. Even when it is possible to grout an arch dam from abutment to abutment in one operation, the grout pressure may produce a problem, as illustrated in Fig. 46.1. The resultant of the pressure

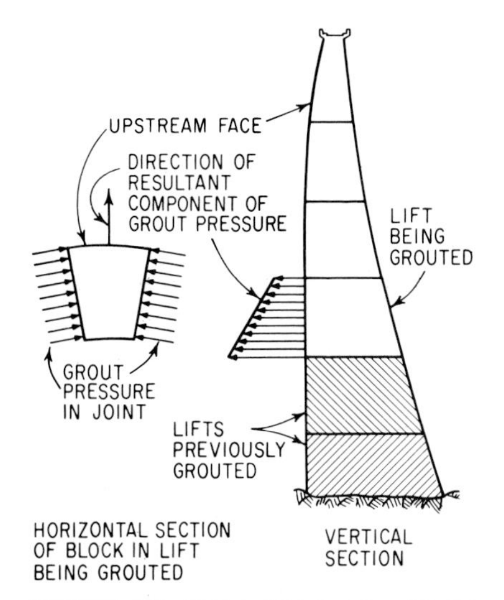

FIGURE 46.1 Forces acting on a monolith of an arch dam during joint grouting.[1]

on the two sides of the trapezoidal blocks is a force tending to bend the monolith as a cantilever in the upstream direction. This condition is most effectively counteracted by partially filling the reservoir prior to grouting.

Joints may be grouted from a gallery or from the exterior. In large gravity dams it is common to include galleries at 50 ft intervals of height for this purpose. Thinner arch dams are usually grouted from temporary catwalks on the downstream face. Figure 46.2 shows the piping arrangement for an arch dam; and Fig. 46.3, for a large gravity dam with a gallery, from which contraction joints are to be grouted. The grout is introduced at the bottom of the lift through a supply header, which in turn feeds grout to the riser pipes. Along the risers grout outlets are spaced uniformly over the face of the joint. The outlets commonly consist of two modified electrical conduit boxes connected by tees to the riser pipes. The supply header makes a complete circuit returning to the gallery or downstream face. During grouting if the supply header should become plugged, grouting can be continued through the return header. At the top of each grouting lift a horizontal groove acts as a vent to collect air, water, and thin grout that must escape as they are displaced by thick grout. The vent is equipped with a pipeline similar to the header at the bottom of the lift. When it is impossible to complete the grouting through the bottom supply pipe, grout may be introduced through the vent pipe. The end of the vent pipe must be equipped with a stopcock and pressure gauge so that during the final stages of grouting the joint can be placed under pressure and the pressure may be monitored.

All the piping shown in the figures is embedded during construction. In addition, temporary supply lines must be provided. Except on small structures which can be grouted from abutment to abutment in a single operation, two such temporary systems must be provided. One is for grout, and the other is for water. When the joint adjacent to a joint being grouted is not to be grouted at the same time, it is maintained under water pressure during the grouting operation. This procedure minimizes the pressure difference on the two sides of the block and thus prevents excessive deflection and shear failure. The joint above the one being grouted is also kept filled with water during the grouting operation. The water helps to keep the joint open, minimizes the loss of grout if a leak should exist in the upper grout stop, and provides a means for checking for such leaks during grouting.

46.2 Installation of Embedded Pipes

The piping is installed just inside the form of the leading monolith prior to placing of the concrete. Pipes and outlets must be securely anchored to prevent damage during placing of the concrete. The supply header should be mounted on metal supports. When electrical conduit boxes are used for outlets, the covers are removed and the open boxes attached directly to the forms so that after stripping the covers may be replaced. All plumbing connections should be inspected and tightened, and just before concrete placement the sys-

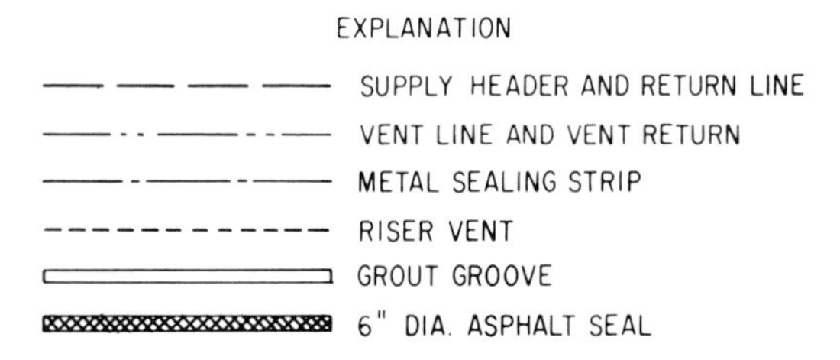

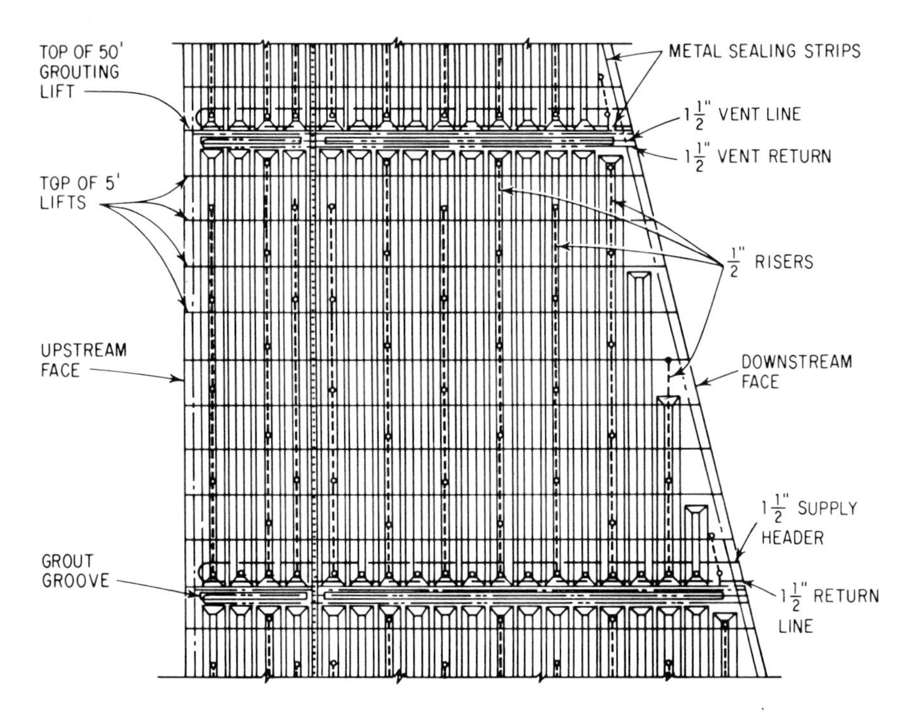

FIGURE 46.2 Radial contraction joint in arch dam showing header and riser-pipe layout.[1]

tem should be pressure-tested. The metal sealing strips should be carefully inspected and joints or welds tested by the soap-bubble method. During placing of concrete special attention should be given to the pipes and seals to ensure that they are not damaged or displaced. The same care is required during placing of the adjacent following lift since half the grout stop is contained in that lift. Most troubles that occur during grouting are attributable to damage that took place during the embedment.

46.3 Grouting Operation

There is a considerable amount of preparation to be done the day before the actual grouting. All pipefitting must be in place both for grout and for filling adjacent joints with water. A communications system must be installed linking mixer, pump, vent pipes, and leak inspectors. Dial gages for measuring joint openings and pressure gauges for measuring vent pressure must be in place. Both the pipe system and the seal system must be checked. The pipes are checked and cleaned by running water through them until the effluent from the vent pipe appears clean. Then, the vent stopcock is closed, and the joint is placed under pressure. If the pressure cannot be maintained, there is a leak in the system which must be located and caulked. Wetting the joints in advance of grouting also lubricates the surfaces and causes the grout to flow more readily.

46.4 Composition of Grout

While both neat-cement grout and sanded grouts have been used in grouting joints, the practice now is to use only neat-cement grout. The use of sand increases the minimum width of joint that may be grouted and introduces the problem of keeping the coarser material in suspension prior to pumping. While it is desirable to have joint openings of at least ¹⁄₁₆ in at the time of grouting, neat-cement grout has successfully penetrated local areas where the opening was only 0.02 in. The cement should be free from lumps or coarse particles. Some organizations require that 100 percent of the cement should pass a No. 100 screen and 98 percent should pass a No. 200 screen. Most cement companies can supply a product meeting this specification so that no processing of cement is required on the job. It is important, however, that the cement be properly stored at the site so that contact with moisture does not produce lumps.

Type II portland cement is normally used for joint grouting. Conditions requiring low-heat or oil-well grouting cements usually are not present in joint grouting. The water/cement ratio varies as the grouting operation proceeds. During the course of the operation the water/cement ratio by weight may be reduced from an initial value of 1.5 to a final value 0.5.

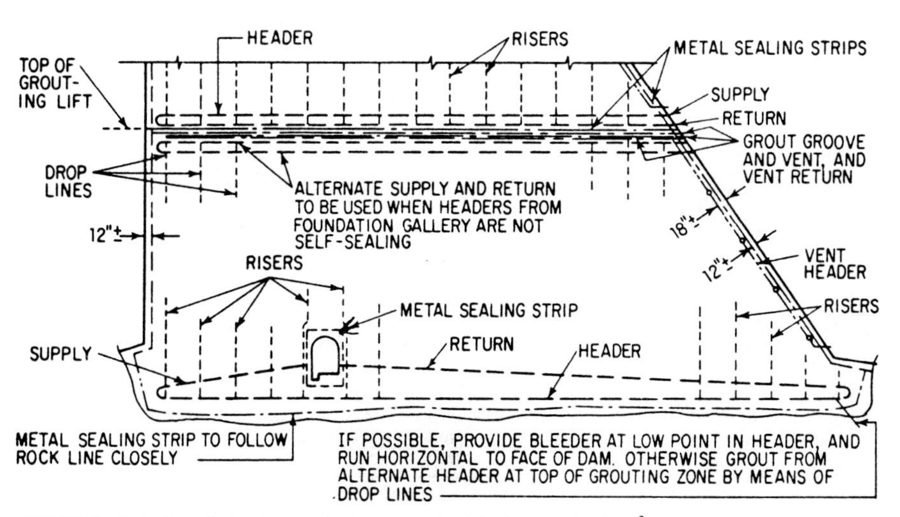

FIGURE 46.3 Installation for grouting a contraction joint in a gravity dam.[2]

46.5 Mixing and Pumping Equipment

The same type of equipment as is used in foundation grouting is suitable for joint grouting.[5,6] To grout 100,000 ft^2 in an 8 h shift a mixer and agitator tank with a capacity of 20 ft^3 and an air-driven $10 \times 3 \times 10$ in grout pump are adequate. The mixer size must be large enough in relation to the pump capacity to maintain a steady flow of grout, as an interrupted flow may produce areas of set grout which later impede the flow. The agitator tank must provide sufficient motion to prevent settling of the cement. The pump must be capable of close control of pressure and a continuous uninterrupted flow. A complete set of standby equipment should be available in the event of a breakdown.

46.6 Pumping Procedure

A few batches of very thin grout (water/cement ratio of 1.5 by weight) should be placed in each joint to lubricate the lower portion of the joint. This is followed by a thicker grout with a water/cement ratio of about 0.7. The grout supply should be rotated among joints so that the level is maintained approximately level throughout the grout zone and the grout has an opportunity to settle in each joint. This can be accomplished by placing a measured amount in each increment. The schedule must be adjusted so that the time lapses are not so great that the pipes become plugged. When the joints are three-quarters filled, the water/cement ratio may be reduced to 0.5 or 0.55. The material coming out of the vent pipe should be observed. When grout nearly as thick as that being pumped is flowing from the vent, it may be assumed that the joint is filled. At this time the stopcock at the end of the vent line should be closed. Pumping should be continued until the pressure at the vent reaches the allowable maximum figure or until the opening indicated by the dial gage at the top of the joint reaches the maximum permitted value. Normally for a 50 ft high grouting lift the increase in joint opening is restricted to 0.02 in as measured by a dial gauge having a least reading of 0.0001 in. During this period leaks in the top grout stop may be detected by drawing off water from the supply header in the lift above that being grouted. If any grout leaks into a joint which has not yet been grouted, water must be circulated through the joint to prevent setting of the grout. During the early stages of the period that the grout is maintained under pressure it usually is necessary to continue intermittent pumping as the pressure drops because of the forcing of water into the concrete. When the pressure remains essentially constant for 30 min, grouting can be considered completed.

Following the grouting, immediate cleanup is important before the remaining grout sets. Not only does the equipment contain grout, but excess grout is always spilled in the galleries or on the face of the structure. If it is not removed promptly, cleanup becomes difficult and unsightly discolorations result. Ordinarily 2 days are required to dismantle one setup and prepare for the next grouting operation.

GROUTING OF TUNNELS

The grouting of tunnel linings differs from the grouting of joints in mass concrete principally in the sizes of the voids to be filled. Whereas the joints in structures have a thickness of the order of $\frac{1}{16}$ in. the space between tunnel lining and the adjacent rock may be several inches. The existence of such a cavity can be a serious problem in a tunnel designed to carry water since it creates an undesirable water passage outside the lining. Because of the volumes to be grouted it is customary to use sanded grouts rather than neat-cement grouts. Sanded grouts are cheaper and have less shrinkage in place. The larger required openings for sanded grouts are not a disadvantage. Where very large openings are involved, the primary grouting with sanded grout may be followed by high-pressure grouting with neat-cement grout to fill the voids left by contraction of the sanded grout. Expansive admixtures, such as aluminum powder, may be added to the sanded grout to minimize the contraction problem.

46.7 Preparation for Grouting

In long tunnels some means must be found to separate the tunnel into zones for grouting. This can be done by hand-packing a mortar dam at intervals around the periphery of the lining when the forms are stripped. The spacing must be such that the volume between adjacent dams may be grouted in a continuous operation. The period of continuous grouting is usually somewhat longer for tunnel grouting than for joint grouting. With the use of retarding admixtures in the grout, this period is sometimes as long as 48 h. Both grout pipes and vent pipes must be in place prior to grouting. Sometimes holes for the pipes are drilled. In other cases they are cast in the concrete. The vent pipes are placed at or near the crown and are extended clear through the void to the overlying rock or soil (Fig. 46.4). This permits the escape of air but prevents loss of grout until the final stages of grouting. The grout pipes are removed some distance from the crown and extend through the lining and partially through the void. Rows of holes are spaced about 20 ft apart.

46.8 Grouting Operation

The procedure and equipment for tunnel grouting are less standardized than for joint grouting because conditions are much more variable. Figure 46.4 illustrates a jumbo-mounted grouting apparatus that was used successfully on a large tunnel project. Grout was mixed outside the tunnel and hauled to the jumbo by rail.

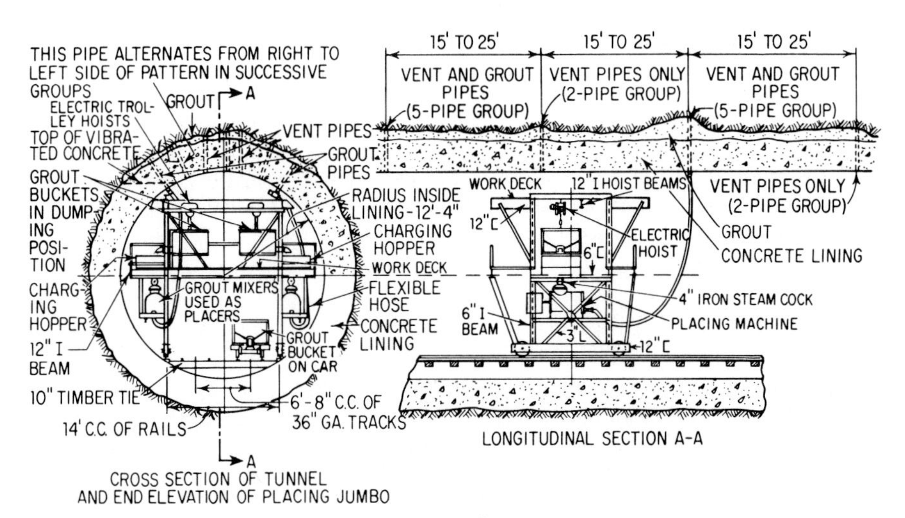

FIGURE 46.4 Jumbo-mounted rig for tunnel grouting.[3]

A typical grout mixture contains equal quantities of cement and sand and a water/cement ratio of 0.45 by weight.

It is normal to use a well-graded sand, having a fineness modulus within the range of 1.50 to 2.00, with not more than 5 percent retained on the No. 16 sieve. For filling large cavities a coarser sand is sometimes used. However, the coarseness of the sand is limited by the difficulty of maintaining it in suspension and of pumping the grout rather than by the size of the cavity.

Pressures of the order of 50 psi are used. Grouting begins at the end of the zone and proceeds in one direction. Initially, connection is made to all the grout pipes at the first location. Pumping is continued to refusal at the specified pressure. If grout starts to flow through the vent pipe, it is plugged. No grout may appear in the vent until grout is pumped at the second location. The operation is continued throughout the zone. Each vent pipe is plugged when it starts to pass grout. If any vent pipe should fail to vent, a grout delivery hose is placed in the vent, and pumping continues until refusal.

REFERENCES

1. Simonds, A. W.: Arch Dams: Theory; Methods and Details of Joint Grouting, *Proc. ASCE, J. Power Div.,* Vol. 82, p. 991–1, 1956.
2. Elston, J. P.: Grouting Contraction Joints in Dams Controls Cracking and Seepage, *Civil Eng.,* Vol. 19, pp. 624–628, 1949.
3. Jacobs, J. D.: Grouting the Tunnels, *Eng. News-Record,* Vol. 117, pp. 274–277, Aug. 20, 1936.
4. Kneuer, R. L., and Meyers, M.: Strengths and Limitations of Epoxy Grouts, *Concrete International,* March 1991, American Concrete Institute.
5. "Concrete Manual," 8th ed., 1974, Bureau of Reclamation.
6. Grouting Methods and Equipment, February 1970, AFM 88-32, Department of the Air Force.
7. Innovative Cement Grouting, *ACI Spec. Publ.* SP-83, 1984, American Concrete Institute.

INDEX

Abrasion:
of aggregate, test of, **2.**19
of concrete, **7.**18
Absolute volume of aggregate, **11.**16
Absorption:
of aggregate, **2.**20
test of, **13.**2
of concrete: definition of, **8.**1
tests, **8.**2
Absorptive form lining, **7.**19, **37.**1
Accelerator:
chloride (*see* Calcium chloride)
(*See also* Admixture, chemical)
Accident prevention, **15.**15, **22.**2
Acid:
attack on concrete, **7.**15
etching, **44.**10
Acoustic properties:
sound absorption, **10.**8
coefficient, **10.**8
noise reduction coefficient, **10.**8
sound transmission, **10.**8
loss, **10.**8
Adhesives, **4.**21
bonding, **4.**22
surface preparation, **4.**22
Admixtures, **4.**1 to **4.**13, **11.**10
antifreeze, **30.**7
architectural concrete, **43.**5
chemical, **4.**3 to **4.**9
accelerators, **4.**5, **6.**6, **11.**10, **30.**11
in plaster, **39.**2
air-entraining agents, **4.**3, **11.**10
(*see* Air-entrained concrete; Entrained Air)
release agents, **35.**7, **43.**9
set retarding, **4.**6, **11.**11, **30.**5
for tunnel grouting, **46.**7
Admixtures, chemical (*Cont.*):
water reducing, **4.**6, **11.**11, **30.**5
high range, **4.**7, **11.**6
cold-weather concreting, **20.**1, **20.**10, **29.**1
for dampproofing, **4.**13, **11.**12
dispensing, **4.**2, **17.**5
hot weather use, **11.**24,
in masonry block, **42.**11
in mass concrete, **26.**11
mineral, **4.**9 to **4.**12, **6.**7
coloring, **4.**12, **43.**5
pozzolan, **4.**10, **6.**8, **11.**11
silica fume, **4.**11, **43.**5
in plaster, **39.**2
in prestressed concrete, **41.**5
in roller-compacted concrete, **30.**6, **30.**12
storage, **4.**3
testing, **4.**1, **4.**4, **4.**12
Adverse weather, placing concrete in, **26.**52
Age, influence on strength, **6.**3, **6.**11 to **6.**14
Aggregate:
abrasion test, Los Angeles method, **2.**19, **7.**18, **12.**2
absorption, **2.**19, **13.**2
absorptive effect on cracking, **9.**1
adherent clay coating, **2.**27
aerial surveys, **2.**10
batching, **17.**3, **17.**7 to **17.**10
moisture compensation, **17.**7
beneficiation:
cage mill **2.**46
heavy media (HMS), **2.**48
jig, **2.**47
blending for gradation, **2.**26
clay in, **2.**27, **12.**27
coarse:
in concrete (test), **12.**16

Aggregate, coarse (*Cont.*):
definition of, **2.**2
deleterious substances (table), **2.**29
properties, **11.**8
requirements, simplified practice recommendations (table), **2.**25
volumes for mix proportioning, **11.**16
crushers (*see* Crusher)
definition of, **2.**2
deleterious substances, **2.**27
exploration:
resistivity method, **2.**16,
seismic method, **2.**12
exposed, **44.**8
feeders, **2.**37
fine:
definition of, **2.**3
deleterious substances (table), **2.**28
(*see* Sand)
fineness modulus, **2.**26, **11.**9, **12.**2
freeze-thaw tests, **2.**30, **2.**52
geophysical exploration, **2.**11
glacial origin, **2.**4
grading, **2.**22, **2.**27, **12.**2
handling in batch plant, **19.**5
heavy, **11.**10, **11.**40, **32.**4
iron or steel, **32.**5
lightweight, **11.**9, **11.**30, **31.**3 to **31.**6
log washer, **2.**42
maximum size, **2.**24, **20.**2
mineral, definition of, **2.**2
moisture content, **13.**2
organic impurities, **2.**30
particle shape, **2.**17
processed, sampling of, **2.**13
production:
rock quarries, blasting of, **2.**32
screening, **2.**39
storage areas, **2.**15
stripping overburden, **2.**31
water supply, **2.**36
prospecting and exploration, **2.**7
purpose in hardened concrete, **2.**1
in plastic concrete, **2.**1
reactive, **1.**8, **2.**27, **2.**28, **7.**15 to **7.**17, **9.**8, **12.**2, **43.**3
for repairs, **44.**15, **45.**4
roller compacted, **30.**5, **30.**12
sample identification, **2.**16
sampling, **2.**13, **12.**2, **13.**4
processed, **2.**13

Aggregate, sampling (*Cont.*):
during production, **2.**14
prospective sources, **2.**13
stockpiles, **2.**15, **13.**1
screens:
deck-type, **2.**41
revolving, **2.**40
(*see* Screens)
scrubber, **2.**42
segregation, **19.**2, **19.**5, **19.**6
sieve analysis, **2.**22 to **2.**27, **12.**2
size:
effect on strength, **11.**8
selection of: **11.**4, **11.**15
soundness, sulfate test, **2.**30, **2.**52
sources of, **2.**14
sampling: auger boring, **2.**12
quarry sites, **2.**13
specific gravity, **2.**20, **11.**16, **11.**33, **11.**36, **32.**4
factor, lightweight, **12.**20
specifications, **2.**49
stockpiling, **2.**15, **19.**1
structural strength, **2.**18
surface texture, **2.**17, **11.**8
temperature of, **20.**4, **20.**8, **20.**18, **46.**8
tests, **2.**49, **12.**2
(*see* Specific property concerned)
transfer finish, **44.**8
transportation: belt conveyors, **2.**37, **19.**2
front end loader, **2.**15, **19.**2
motor trucks, **2.**36
railways, **2.**36
unconfined, alcohol-water freeze-thaw test, **2.**30
voids, **2.**22, **11.**15
Agitators, **26.**18, **23.**5
Agitor, **26.**18
Air cooling, aggregate, **26.**17
Air-entrained concrete:
in canal lining, **26.**26
in mass concrete, **26.**11
measuring air content, **12.**6
errors in, **12.**5
gravimetric, **12.**5
pressure, **12.**6
volumetric, **12.**5
for lightweight concrete, **12.**5
mix design, **11.**12 to **11.**16, **11.**19,
architectural concrete, **43.**5
pavement, **26.**18

Air-entrained concrete, mix design (*Cont.*):
  pumping, **36.**11
  roller-compacted concrete, **30.**6, **30.**12
  tests, **4.**4, **12.**4, **12.**5
  (*see* Entrained air)
Air-entraining agent, addition, **17.**5
  (*see* Admixture, chemical)
Air-entraining cement, **1.**18, **11.**8
Air guns, **26.**41
Air meter, types of, **4.**4, **12.**5
Air slaking rock, protection with shotcrete, **25.**2
Air slugger, **26.**30, **30.**41
Airports (*see* Pavements)
Alkali-aggregate reaction, **1.**15, **2.**28, **7.**17, **43.**1
  locations where reported (map), **2.**29
Alkali-carbonate reaction, **2.**28, **7.**17
Alkali-silica reaction, **7.**15
Alluvial fan, **2.**5,
Aluminum powder, **9.**8
  reaction with calcium chloride, **4.**6
Ammonium salts, effect on durability, **7.**4, **7.**10
Andesite, **2.**28, **7.**15
Arches, domes, paraboloids, and folded plates, concrete placement, **26.**5
Architectural concrete:
  admixtures, **43.**5
  aggregates, **43.**3
  batching, **44.**4
  bolts and ties, **43.**10, **44.**2
  cement, **43.**3
  chamfer and rustication strips, **43.**11, **44.**2
  consolidation, **44.**4
  curing of, **44.**6
  definition of, **43.**1
  Design Reference Sample, **43.**11
  exposed aggregate, **44.**8 to **44.**13
    methods:
      acid wash and etch, **44.**10
      bushhammering, **44.**10
      fractured rib, **44.**13
      grinding, **44.**13
      high-pressure jet, **44.**9
      manual water brush, **44.**10
      sandblasting, **44.**10 to **44.**13
        degrees of, **44.**12
  Field Mockup sample, **44.**1
  final finishing, **44.**13
  forming, **43.**6 to **43.**9, **44.**6 to **44.**8
    coated, **43.**6, **44.**6
Architectural concrete, forming (*Cont.*):
    elastomeric, **43.**7, **44.**2
    joints, **44.**2
    liners, **43.**6, **44.**2
    removal, **44.**6
    storage and reuse, **44.**3
    tolerances, **44.**2
  mixing, **44.**4
  placement, **44.**4
  quality assurance, **44.**17
  reinforcement, **43.**6
    spacers, **43.**6
  release agents, **43.**9, **44.**3
  repairs, **44.**15
  sealers, sealants, coatings, **43.**12
Argillaceous limestone:
  effect on popouts, **45.**18
  rock, **2.**27
Autoclaving, **6.**21, **29.**7, **42.**16, **42.**19
  (*See also* Curing, high temperature)

Backfill concrete, lean, **25.**4
Ball penetration test, **5.**5, **12.**4
  slump conversion, **12.**5
Bar:
  geological term, **2.**5
  reinforcing (*see* Steel, reinforcing)
Barite, heavy aggregate, **32.**4
Basalt, **2.**7, **2.**8, **2.**9
Batch plant, **16.**1 to **16.**16, **17.**1 to **17.**14
  inspection, **13.**1
  maintenance, **21.**2
  operation, **21.**1, **21.**2
Batching:
  accuracy, **13.**2
  admixture, **17.**5
    capacity of equipment, **17.**5
    dispensers, **17.**5
  aggregate, **17.**6, to **17.**9,
  cement, **13.**2, **17.**3
  equipment, **16.**1 to **16.**16, **17.**1 to **17.**14
  recorder, **17.**25
  tolerances, **13.**2, **17.**2
Beam:
  properties of, **15.**3
  test, **12.**12
Belt:
  conveyors, **2.**37, **19.**3, **24.**24
  inundation, **46.**7
  spraying, **46.**8

Beneficiation, aggregate (*see* Aggregate, beneficiation)
Bins:
  aggregate, **19.**5
  cement, **19.**7
Biological shielding, **32.**2, **38.**1
Bituminous coatings, **7.**4
Blasting rock quarries, **2.**32
Bleeding, **5.**6, **9.**1, **24.**16, **45.**4
  effect on cracking, **44.**4
Blemishes, surface, **44.**13 to **44.**17
Block, concrete, **42.**10 to **42.**25
  admixtures in, **42.**11
  carbonation, **42.**22
  cubing, **42.**23
  curing, **42.**16 to **42.**23
    atmospheric steam, **29.**7, **42.**16
    autoclave, **29.**7, **42.**17
    heat computations, **42.**18 to **42.**22
  manufacturing, **42.**14, **42.**17
  materials for, **42.**10
  mixes for, **42.**12
  properties, **42.**23
  Q-block program, **42.**24
  sampling, **42.**25
  shrinkage, **42.**17, **42.**24
  specifications for, **42.**24
  types of, **40.**2, **40.**3, **42.**24
  (*See also* Masonry)
Blockouts, **25.**21
Boilers, **20.**10
  boiler horsepower, **20.**8
Bolt holes, **44.**2
  repair, **44.**15
Bond:
  adhesive for, **4.**21, **45.**4
    application, **45.**13, **45.**18
  breaker, tiltup, **35.**6
  between lifts of concrete, **8.**6, **25.**5, **28.**2
  in masonry, **40.**9
  in plaster, **39.**5
  strength, **6.**2, **6.**13
Bonding agents, **4.**21, **45.**4, **45.**14
Breccia, **2.**7 to **2.**10
British thermal unit, **20.**6
Brushed finish, **27.**13
Bucket:
  clamshell, **19.**3
  elevator, **19.**3
Buckets for concrete, **23.**2 to **23.**5, **24.**3
"Bugholes, elimination of, **24.**25, **44.**4, **44.**6
Building construction, **26.**1, **33.**3 to **33.**25, **35.**1
Bullfloat, **27.**5
"Bumpcutter," **26.**21
Bushhammering, **44.**13

Caisson footing, **42.**41
Calcareous rock, **2.**7, **2.**8
Calcium chloride:
  amount used, **4.**4, **11.**11
  batching, **4.**5
  for deicing, **7.**4
  effect in concrete, **4.**5, **6.**5
  in plaster, **39.**3
  in prestressed concrete, **41.**5
  reaction with aluminum, **4.**6
  (*See also* Admixtures, chemical)
Canal lining:
  control joints, **26.**24
    water stop, **26.**24
  curing, **26.**25
  methods:
    shotcret, **26.**22
    slope slipform, **26.**22
    subgrade guided slipform, **26.**22
  placing concrete, **26.**21 to **26.**27
Capacity, batching plant (*see* Production rate, concrete plant)
Capillarity, **8.**1
Capping test cylinder, **6.**17, **12.**9
Capital, column, **15.**9
Carbon dioxide:
  effect on dusting, **6.**26, **29.**9
Carbonation, **6.**21, **7.**8, **44.**7
  concrete block, **42.**22
  effect on volume change, **9.**7
Cart for concrete, **24.**3
Cast-in-place unreinforced pipe, **26.**42 to **26.**46
Casting yard, on-site, **33.**20
Cavitation, **7.**18
Cellular concrete, **31.**2
Cement-aggregate reaction (*see* Alkali-aggregate reaction; Carbonate-aggregate reaction)
Cement:
  air-entraining, **1.**18, **11.**8
  alkali content, **1.**15, **44.**10
  aluminous, **1.**19, **36.**19
  antibacterial, **7.**5
  approval of, **12.**1, **13.**8

Cement (*Cont.*):
architectural concrete, for, **43.**3
bacteria, effect of, **7.**5
batching, **13.**1, **17.**3
composition, **1.**3, **1.**12
Concrete Reference Laboratory, **13.**7
content, computation, **11.**14
conveying, **1.**17
cooling, **1.**15, **20.**27, **26.**16
expansive, **1.**19, **9.**8, **41.**25
factor, computing, **11.**14
false set, **1.**15, **5.**8, **12.**2
fineness, **1.**16, **5.**8
effect on strength, **6.**5
flash set, **20.**10
gun plastic, **1.**19, **39.**2
handling, **1.**17, **19.**7
air-pollution control, **1.**8
heat of hydration, **1.**15, **4.**9
pozzolan effect, **4.**12
high alumina, **1.**19
inspection of, **12.**1
low alkali, **1.**15, **7.**15, **43.**4
low heat, **1.**5, **1.**18, **6.**4
magnesite, **1.**17
for mass concrete, **1.**18, **10.**5
oil well, **1.**19
packset, **12.**2
paste, shrinkage, **2.**2, **9.**4
physical properties, **1.**16
plastic, **1.**19, **39.**2
portland blast-furnace slag, **1.**19
prehydration, **1.**15
production, **1.**3 to **1.**12
dry process, **1.**6
wet process, **1.**7
raw materials for, **1.**3
in repairs, **45.**3
for roller-compacted concrete, **30.**5, **30.**11
sampling, **12.**2
silo, **1.**12, **19.**7
Sorel, **1.**17
specifications for, **1.**17 to **1.**19
sticky, **1.**17, **42.**11
storage, **1.**17
sulfate resistant, **1.**18, **7.**6
temperature, **20.**8
testing, **12.**1
transporation, **1.**12, **19.**8
for trial batch, **11.**12
types of, **1.**18,
Cement, types of (*Cont.*):
relative strengths, **1.**17
warehouse set, **1.**17
white, **1.**19
Centering for formwork, **25.**13,
(*See also* Shoring)
Central-mixed concrete, **16.**11
(*See also* Ready-mixed concrete)
Certification, manufacturer's, **13.**8
Chace air meter, **4.**5, **12.**6
Chalcedony, **2.**28, **7.**15, **44.**11
Chemical attack:
acids, effect of, **7.**4
carbonation, **7.**8
leaching, **7.**5
sea water, **7.**7
sewage, **7.**4
sulfates, **7.**6
cement-type, **7.**6
uncombined lime and magnesia, **7.**8
Chert, **2.**10
popouts, **43.**4
Chesapeake Bay Bridge and Tunnel, **42.**33
Chord modulus of elasticity, **6.**23
Chutes:
for concrete, **24.**3
Cinders as aggregate, **11.**9, **31.**1, **42.**4
Classification of finishes, **27.**2
Classifier for aggregate:
hydraulic, **2.**42
rising current, **2.**43
Clay, **2.**27
coating on aggregate, **2.**27
expanded, aggregate, **11.**9
Cleaning surfaces:
acid wash, **25.**7, **44.**10
grout cleaning, **27.**4
sack rubbing, **27.**4, **44.**13
sandblasting, **25.**5, **25.**8, **44.**10
stain removal, **44.**16
Cleanup:
construction joint (*see* Construction joint, preparation)
in form, **25.**15
Clinker, cement, **1.**12
Coal and lignite in aggregates, **2.**29
Coarse aggregate, definition, **2.**2
Coatings:
on aggregate, **2.**27
form, **14.**4, **43.**6, **44.**17
on reinforcement, **3.**11

Coefficient:
of expansion, epoxy resin, **48.**4
surface conductance, **10.**2
thermal conductance, **10.**2
thermal conductivity, **10.**1, **10.**3
thermal expansion, **9.**3, **10.**5
thermal resistance, **10.**2
variation **12.**27, **12.**31
Cohesiveness, **5.**5, **11.**4
Cold concrete, **5.**9, **25.**16
Cold joints, avoidance of, **26.**51, **30.**8
Cold weather:
concreting in, **5.**9, **30.**17, **44.**7
curing in, **29.**9, **30.**5, **44.**7
insulation, **4.**15
masonry construction in, **40.**12
plastering in, **39.**13
Color in concrete, **4.**12, **43.**5, **44.**4
"wet check" test, **44.**4
Colorimetric test, sand, **2.**30
Compaction, roller-compacted concrete, **30.**8
Compressive strength (*see* Strength of Concrete)
Conchoidal fracture, **2.**17
Concrete-Mobile, **23.**7
Concrete piles (*see* Piles)
Concrete pipes (*see* Pipes)
Concrete Plant Manufacturer's Bureau, **16.**3
Concreting (*see* Placing)
Condensation, **10.**3
prevention of, **10.**3
Conductivity, thermal, **10.**1, **20.**14, **46.**14
of aggregates, **10.**3
Conduits, placing concrete in, **26.**42 to **26.**46
Consistency (*see* Slump)
Consistency meter (mixer), **18.**4
Consolidation of concrete, **24.**17
hand methods, **24.**26
with vibrators, **24.**17, **24.**19 to **24.**26
(*See also* Vibration)
Construction joint, **25.**5 to **25.**10, **28.**1
bond between lifts, **8.**6, **28.**9
cleanup, **25.**5 to **25.**9
dowels, **28.**8
laitance on, **5.**7, **25.**5
mortar layer, **25.**9, **28.**2, **44.**5
in paving, **26.**21
preparation: acid-etch method, **25.**7
green cutting, **25.**8
hand methods, **25.**7
retarder on, **25.**8
Construction joint, preparation (*Cont.*):
wet sandblast, **25.**8
protection of, **25.**9
in tunnel, **26.**42
water leakage through, **8.**6, **25.**5
(*See also* Joints)
Contraction joints, **28.**2
Control charts, strength, **12.**30
Control systems:
batching automatic, **17.**9
electronic, **17.**9
manual, **17.**8
with power assist, **17.**9
semi-automatic, **17.**9
slump, **17.**7
Convection of heat, **20.**21
Conversion tables:
cement, weight-volume, **11.**32
density-volume, **11.**31
water, weight-volume, **11.**30
water-cement ratio, **11.**33
Conveyor:
bucket elevator, **19.**7
for cement:
for concrete, **24.**5, **24.**16
pneumatic, **19.**8
Cool concrete, **20.**16, **29.**7
Cooling:
of aggregates, **20.**17 to **20.**25, **26.**17
of cement, **20.**27, **26.**17
of concrete, **16.**5, **20.**16 to **20.**27, **26.**16
time required, **20.**1
water, **20.**22, **26.**17
Core from hardened concrete, **12.**23, **12.**30
Corrosion:
aluminum, **7.**19
copper, **7.**19
galvanic, **7.**19
lead, **7.**20
steel, **3.**11, **7.**19, **25.**33
zinc, **7.**20
Cracking, **9.**7
caused by:
alkali-aggregate reaction, **7.**15
(*See also* Aggregate, reactive)
differential temperature, **9.**3
evaporation, **9.**2
galvanic action, **7.**19
joints, **28.**2
rusting of reinforcement, **7.**19
settlement, **26.**3, **44.**5

Cracking, caused by (*Cont.*):
temperature, **9.**3
fatigue, **6.**25
in hardened concrete, **6.**1, **9.**8
of masonry, **40.**14
prevention of, **40.**14
of plaster, **39.**13
plastic shrinkage, **4.**14, **9.**2, **26.**52
prevention of, **4.**14, **26.**52
of prestressed concrete, **41.**6, **41.**9
repair of, **45.**12 to **45.**14
structural, **6.**1, **26.**3, **28.**1
tiltup, **35.**6
(*See also* Crazing)
Crazing, **4.**14, **44.**7
Creep, **9.**9 to **9.**13
cracking, **45.**12
effect of:
age at initial load, **9.**12
constituents, **9.**10
exposure conditions, **9.**12
high-performance concrete, **9.**12
load intensity, **9.**12
manufacturing methods, **9.**10
proportions, **9.**10
reinforcement, **9.**12
time at sustained load, **9.**12
equation, **9.**12
in lift slab, **35.**1
test method, **9.**12
Crushed gravel, definition, **2.**3
Crushed stone:
definition of, **2.**3
sources of, **2.**5
Crusher for aggregates, **2.**38
gyratory, **2.**38
hammer mill, **2.**38
impact, **2.**38
jaw, **2.**38
primary, **2.**39
roll, **2.**39
secondary, **2.**39
Crushing:
test, pipe, **42.**33
theory, **2.**39
Cubing of concrete block, **42.**23
Culverts, placing concrete in, **26.**42 to **26.**46
Curbs, gutters, and sidewalks, placing concrete in, **26.**27
Curing, **4.**19, **29.**1 to **29.**12
architectural surfaces, **44.**6
Curing (*Cont.*):
autoclave (*see* Curing, high pressure)
block, **29.**7, **40.**12, **42.**16
canal lining, **26.**23
cold weather, **29.**9, **30.**17
compounds (*see* Curing materials)
cotton mats, **29.**9
effect on:
crazing, **44.**7
strength, **6.**9 to **6.**12
exterior, **29.**10
floor, **27.**31
with forms, **25.**16
heat computations, **42.**17 to **42.**20
high-pressure, **6.**10, **29.**7, **42.**16
block, **42.**16, **42.**19, **42.**20
high-temperature, **6.**10, **29.**6 to **29.**8, **42.**16
schedule for, **29.**7
hot weather, **29.**1, **30.**10, **30.**17, **36.**18
impermeable sheets, **29.**3
interior, **29.**9
interrupted, **29.**1
liftslab, **29.**12
low-pressure steam, **6.**10, **29.**7
block, **42.**15, **42.**17
mass concrete, **26.**17
materials for, **4.**14 to **4.**15
liquid-membrane forming, **4.**11, **29.**4, **29.**5
mats and blankets, **4.**15, **29.**3
paper and plastic sheets, **4.**15, **29.**3, **29.**4
water, **4.**14, **29.**2
paving concrete, **26.**21
pipe, **29.**11
precast units, **29.**11
purpose of, **29.**1
roller-compacted concrete, **30.**17
selection of method, **29.**8
shotcrete, **36.**18
slabs, **29.**9
temporary, **29.**2
tiltup concrete, **29.**12
tunnels, **26.**42
water, **29.**2
inundation (ponding), **29.**2
spray or mist, **29.**2
Cyclone, use of, for sand, **2.**44

Dacite, **2.**28

Dampproofing:
  definition of, **8.**7
  materials, **4.**13, **8.**8
  (*See also* Waterproofing)
Dams (*see* Mass concrete)
Darby, **27.**7
Darcy's law, **8.**3
D-cracks, **7.**17
Deadman, dragline, **2.**34
Deformation:
  elastic, **9.**9
  inelastic, **9.**9
Deicing agents, **7.**4
Density:
  of concrete, **10.**5, **12.**13
  computation, **11.**16, **11.**33
  heavy, **11.**41
  lightweight, **11.**33
Deposit, river, as source of aggregate, **2.**5
Design Reference Sample, **43.**11
Dew point, **20.**8, **20.**19
Diabase, **2.**9, **2.**10
Diatomaceous earth, pozzolan, **9.**5
Diatomite, **11.**9, **45.**4
Dicalcium silicate, **1.**14
Diffusivity, **10.**1, **10.**5
Diorite, **2.**9, **2.**10
Dispatching mixer trucks, **21.**1, **22.**2
D-load, pipe, **42.**33
Dolomite, **2.**9, **2.**10
Dome pans, **15.**11, **15.**12
Dowels:
  in pavement, **26.**20
  in slabs, **28.**8
  in structure, **28.**2
Drag, burlap, **26.**20, **27.**15
Dragline excavation, **2.**34
Drain tile, **42.**25
Drainage of foundation, **25.**4
Dredge excavation, **2.**36
Drop chutes for concrete, **24.**6, **24.**11, **44.**6
Dropping concrete, **24.**7, **44.**6
Dry pack:
  filling holes, **44.**15, **45.**13, **45.**14
  rockpockets, **45.**13
Drying shrinkage, **9.**4 to **9.**6, **26.**51, **28.**2
  affected by:
    exposure conditions, **9.**6
    materials, **9.**4
    proportions, **9.**5
    size and shape, **9.**5
Drying shrinkage (*Cont.*):
  of lightweght concrete, **9.**4
  restraint by aggregates, **2.**2
  unfavorable factors, **9.**6
  (*See also* Cracking, Volume change)
Dumpcrete, **23.**5, **26.**18
Durability, **7.**1 to **7.**20, **11.**6
  factor, **7.**3
  influence of aggregate, **7.**15 to **7.**18
    bleeding, **5.**7
    curing **7.**1
    proportions, **7.**2, **24.**23
  tests of, **7.**3, **12.**22
Dusting, **27.**5
Dynamic modulus of elasticity, **6.**23
  (*see* Modulus of elasticity)

Efflorescence, **39.**12
  on plaster, **39.**13
Elasticity:
  modulus of (*see* Modulus of Elasticity)
  steel, **41.**11, **41.**14
Electrical properties, **10.**10
Elephant truncks for concrete, **24.**6, **24.**11
Elevator:
  bucket, **19.**3
  in building, **34.**5
Embedded grout lines:
  installation, **46.**3
  materials, **25.**22
Entrained air:
  effect on:
    bleeding, **4.**3, **5.**9
    durability, **7.**1, **11.**10
    freezing and thawing, **7.**2
    segregation, **5.**9
    shrinkage, **9.**4
    strength, **5.**9, **6.**7, **6.**8, **11.**5
    workability, **4.**3, **5.**9
  in fresh concrete, **4.**4, **5.**9, **6.**7, **7.**2, **11.**5
  in hardened concrete, **5.**9, **6.**7, **7.**2
  mix design for, **11.**19 to **11.**23
  recommended amount, **11.**6, **43.**5
Environmental protection, **1.**8, **16.**10
Epoxy resin:
  adhesive, **4.**21
  mortar, **45.**5, **45.**11
  for repairs, **45.**10, **45.**12, **45.**18
Equipment for handling and moving concrete
    (*see* Handling concrete, equipment for)

Esker, **2.**4
Evaporation, **20.**18, **30.**17
  effect on cracking, **26.**51
  eduction of, **26.**52, **29.**2, **30.**10
Excavation:
  dragline, **2.**34
  hydraulic, **2.**35
  power shovel, **2.**33
Expansion, thermal, coefficient of, **9.**3, **10.**5
Expansive cement, **1.**19, **9.**8, **41.**25
Expansive concrete, **9.**8
Expansive materials, **9.**8
  aluminum powder, **9.**8, **41.**25
  hydrated calcium sulfoaluminates, **9.**9
  iron (finely divided), **9.**9
Exposed aggregate finish, **44.**8 to **44.**13, **44.**15
  acceptance, **44.**13
Extrusion precast concrete, **42.**6

False set, **1.**5, **5.**8
Falsework, form, **14.**20, **15.**9, **25.**13
  (*See also* Shoring)
Farm concrete, **26.**19
Fatigue strength, **6.**25
Federal Housing Authority, **2.**53
Feeders for aggregate, **2.**37
Fiber-glass forms, **14.**10, **33.**5 to **33.**7
Field mockup sample, **44.**1
Fine aggregate (*see* Aggregate, fine; sand)
Fineness modulus, **2.**22
  computation of, **2.**22, **11.**7
  in mix proportioning, **11.**16
Finish:
  aggregate transfer, **44.**8
  architectural, **42.**7, **44.**8, **44.**13
  cork-floated, **27.**4
  exposed aggregate, **44.**8 to **44.**13
  floor, **26.**49, **27.**5 to **27.**14, **28.**3 to **28.**9
    cold weather, **26.**50, **29.**1
    hot weather, **26.**51, **29.**1
  form liners for, **43.**6 to **43.**8
  formed surface, **27.**4
  pavement, **27.**6 to **27.**11
    roller-compacted, **30.**2
  precast concrete, **42.**7
  retarded, **44.**9
  rough form, **27.**4
  rubbed, **27.**4
  sandblasted, **44.**10
  shotcrete, **36.**18
Finish (*Cont.*):
  stoned, **27.**4
  textured, **27.**13, **44.**8
  tiltup panels, **35.**8
  tolerances, **27.**3, **44.**2
  unformed, **27.**5
  vertical slipform, **34.**4
Finishes, classification of, **27.**4
Finishing, **27.**3 to **27.**16
  effect on crazing, **27.**5
  machines, paving, **26.**17
  mass concrete lifts, **26.**13
Fire endurance of wall:
  aggregate-type, effect of, **10.**3
  concrete quality, **10.**5
  construction methods, **10.**5
  resistance, **10.**1, **10.**6
    tests, **10.**7
  wall thickness, **10.**2
Flash set, **5.**8
Flat work (*see* Finishing)
Flexural strength, **6.**2, **6.**12, **11.**24, **12.**12
Float, **27.**11
  longitudinal, **27.**12
  paving, **26.**17
Floor:
  armored, **7.**19
  base for, **25.**1
  base course, **26.**17
  bonded, **45.**11
  construction, **27.**5 to **27.**16, **28.**1 to **28.**9
  coverings, **7.**9
  hardener, **4.**23, **7.**9, **45.**2
  heavy-duty, **7.**19
  jointing and edging, **27.**5, **28.**1 to **28.**9
  repairs, **45.**11, **45.**12, **45.**15
  skidproof, **27.**13
  for tiltup, **35.**6
  tolerances, **27.**3
  vacuum-processed, **37.**3
Fluosilicate:
  floor hardener, **4.**23, **7.**10
  protective sealer, **7.**9
Flyash, **4.**10, **6.**8, **46.**5
Foamed concret, **31.**1
Fog spray to prevent cracks, **26.**52
Foliated rock, **2.**6
Form:
  accessories for, **14.**6, **14.**14, **14.**18, **15.**2, **43.**6
    chamfer strip, **43.**11

Form, accessories for (*Cont.*):
grade strip, **25.**12
release agent, **15.**13, **43.9**
rustication strip, **25.**12, **43.**11
she-bolt, **25.**11, **44.2**
snap-tie, **14.6**, **43.**20
spreader, **25.**11, **14.6**
ties, **14.6**, **25.**12, **25.**19, **43.**10, **44.2**
void formers, **15.2**, **25.**22
in prestressed concrete, **41.9**
anchorage, **25.**12
mass concrete, **26.**12
architectural, **43.6**
bracing, **14.**5, **15.**2, **25.**11
coating, **14.**4, **25.**14, **41.9**,**43.9**
construction details, **15.**1 to **15.**14, **25.**11
damages from vibrators, **24.**24
liner, **43.6**, **44.2**
fastening of, **44.2**
plastic, **43.6**
lumber loads on: compression, **15.6**
deflection **15.4**
shear, **15.8**
stress in, allowable, **15.**5
oils, **15.**13, **25.**14
in prestressed concrete, **41.**8
related defects, **43.**6, **44.**3, **44.**11, **44.**15
removal, **25.**15, **44.6**
plastic liners, **43.6**
precast units, **33.**10
prestressed units, **41.**18
for repair work, **45.9**
side openings in, **25.**13
Formed surfaces, **43.6** to **43.9**
acid etching, **44.**8, **44.**10
bushhammering, **44.**10
grinding, **44.**13
grout cleaning, **44.**13 to **44.**15
rubbed, **27.**4, **44.**13
sandblasting, **44.**10
stoning, **27.4**
(*See also* Finish)
Forming systems, **14.**7 to **14.**22
selection of, **14.4**
Forms:
absorptive liner, **7.**19, **43.9**
application, **15.**9 to **15.**14
care of, **44.**3
checking during concrete placing, **25.**15
column, **14.**10, **15.9**
concrete, **33.**10, **43.**8
Forms (*Cont.*):
corrugated paper, **14.**10
dam, **26.**12, **26.**15
deflections, **15.**3 to **15.**5, **44.2**
design, **15.**3 to **15.**14
basic formulas, **15.2** to **15.5**
dome pans, **15.**11
factory built, **14.7** to **14.**22
fiber-glass, **14.**10, **33.5** to **33.**10
flying deck, **14.**25
handling, **14.**4, **15.9** to **15.**14
high density, **43.**13
medium density, **43.**13
layout of, **15.**1
materials for:
aluminum, **14.9**, **14.**11
applications, **14.4**
effect of concrete finish, **14.4**
fiber-glass, **14.**10, **33.**5, **43.7**
plaster, **43.9**
plywood, **14.**4, **43.6**
steel, **14.7**, **27.**12, **33.**7 to **33.**10, **43.6**
styrofoam, **43.6**, **44.5**
thin vinyl, **43.7**, **44.**3
pan joist, **15.**11, **15.**13
for precast concrete, **33.**5 to **33.**10, **42.**3
prefabricated, **14.**11 to **14.**17
pressure on, **15.2**, **15.5**
for prestressing, **41.9**
requirements for, **14.4**
reuse, **14.**5, **25.**.17
safety provisions, **15.**15, **25.**14
slab and beam, **15.**9 to **15.**14
specifications for, **14.5**
steel, for precasting, **33.7** to **33.**10
temporary holding, **24.**24, **26.**15
textured, **43.6**
tightness, **25.**10, **44.2**
for tiltup, **35.6**
tolerances, **25.**14, **44.2**
tunnel, **26.**29 to **26.**42
wall, **15.2** to **15.6**
Foundation:
cleanup, **25.**1 to **25.5**
placing concrete in, **26.**2
seal, **25.5**
Free fall of concrete, **24.**16
Free lime and magnesia, **1.**15, **7.8**
Freezing:
effect on popouts, **45.**18
and thawing; factors involved, **7.2**

Freezing, and thawing (*Cont.*):
tests:
field, **7.**3
laboratory, **7.**2, **12.**22
Frequency distribution of strength tests, **12.**29
curve, **12.**29
tabulation, **12.**33
Fresco, **39.**1
Fresh concrete test (*see* Tests of fresh concrete)
Frost resistance, **7.**1, **7.**2

Gabbro, **2.**7
Galvanic corrosion, **7.**19
Gas concrete, **9.**8, **31.**2
Gates, batcher, **17.**3
Glaciation, areas (map), **2.**4
Glass, **2.**8
Glen Canyon Dam, **26.**11
Gneiss, **2.**7
Gradation test, aggregates, **2.**22 to **2.**27, **12.**2
Grade strips for forms, **25.**12
Grading:
of aggregate, **2.**22 to **2.**27
requirements, coarse aggregates (table), **2.**25
Graffito, **39.**1
Granite, **2.**8
Gravel:
crushed, definition of, **2.**4
definition of, **2.**2
Gravimetric, method of air content, **12.**5
Green cutting, construction joint, **25.**8
Grinding surfaces, **27.**4, **44.**9
Grizzlies for aggregates, **2.**40
Groove joint, **26.**24, **28.**2, **28.**4
Grout:
cleaning, **27.**4
neat cement, **46.**5
sanded, **46.**5
stop, **46.**4
Grouting, **46.**1 to **46.**6
mass concrete, **46.**1 to **46.**6
tunnels, **46.**6 to **46.**8
Gunite (*see* Shotcrete)
Gunite Contractors Assn., **36.**16

Handling concrete:
correct and incorrect methods, **24.**7 to **24.**17
equipment for:
buckets, **23.**2 to **23.**5, **24.**3

Handling concrete, equipment for (*Cont.*):
carts, **24.**3
chutes, **24.**3
conveyors, **24.**4
drop chutes, **24.**6, **24.**10, **24.**16
elephant trunks, **24.**6, **24.**16
pumps, **24.**4, **36.**3 to **36.**13
receiving hoppers, **23.**1
selection of, **24.**3, **24.**16
underwater, **26.**46 to **26.**49
wheelbarrows, **24.**3
Hardened concrete:
entrained air in, **5.**9, **6.**7, **7.**1
tests of cores, **12.**23
compared with test hammer, **12.**24
Hardener, surface (*see* Floor hardener)
Hardness, of concrete surface, **7.**18
Hauling units, number required, **23.**6
Heat:
of fusion, **20.**6
of hydration, **1.**15, **4.**12, **46.**5
insulating concrete, **31.**1
loss through wall, **10.**2
required, **20.**8
of vaporization, **20.**6
(*See also* Temperature; Thermal properties)
Heat-balance, **20.**8, **20.**15
Heat computations, curing block, **42.**17 to **42.**22
Heat transfer, **10.**1 to **10.**5
Heating:
aggregates, **20.**14 to **20.**16
coils, area required, **20.**15
concrete, **20.**1 to **20.**26
time required, **20.**1
water, **20.**10
Heavy concrete, **11.**40, **32.**1 to **32.**7, **38.**1
adjustments, **11.**43
aggregate for, **11.**8, **11.**40, **32.**4
counterweights, **32.**4
mixing and placing, **32.**6
nuclear or radiation shielding, **32.**2 to **32.**4, **38.**1
proportioning, **11.**41, **32.**5
Heavy-media seperation, **2.**48
Hematite, **11.**41
High-pressure steam curing (*see* Curing, high pressure)
High-temperature exposure, **26.**8, **29.**6
aggregates for, **36.**19
shotcrete, **36.**19

Highways (*see* Pavements)
Histogram for strength test analysis, **12.**30
Home concrete, **26.**49
Honeycomb (*see* Rock pockets)
Hoppers:
  aggregate, **17.**3
  cement, **17.**3
  receiving, for concrete, **23.**1, **24.**6
Hot concrete, **26.**8
Hot weather concreting, **5.**10, **26.**17
  curing, **26.**17, **26.**52
Humidity, **20.**7, **26.**52
  effect on cracking, **26.**52
Hydration:
  effect on volume change, **9.**4
  heat of, **1.**15, **4.**12, **26.**17
Hydraulic structure permeability, **8.**1
Hydroxylated carboxylic acid, **4.**6, **11.**11

Ice, in mixing water, **16.**6, **20.**22
Insoluble residue, **1.**14
Inspection:
  ACI recommendations for, **13.**1, **13.**9
  of aggregate, **13.**2
  of batching and mixing, **13.**2, **22.**1
  of materials, **13.**8
  off-site, **13.**8
  placing concrete, **25.**24 to **25.**26
  ready-mixed concrete, **22.**1
  reports, **13.**1, **13.**8
  service, employment of, **13.**1
Insulating concrete, **31.**1
  cold weather use, **4.**15
  of forms, **25.**16
  winter concreting, **4.**15, **25.**16
Inundation for cooling aggregates, **16.**1, **20.**18
Iron:
  heavy aggregate, **11.**10, **11.**40
  stains, **44.**16
  sulfide popouts, **42.**11

Jigging, hydraukic, **2.**47
Jitterbug, **27.**5, **31.**7
Joint(s), **28.**1 to **28.**7, **44.**2
  construction, **8.**6, **28.**1
    cleanup, **25.**5 to **25.**9
  contraction, **28.**2
    in dams, **46.**1
Joint(s) (*Cont.*):
  control, **28.**1
    in canal lining, **26.**24
  expansion, **26.**21, **28.**6
  filler (*see* Sealant)
  isolation, **28.**6
  in masonry, **40.**7, **40.**12
  in paving:
    contraction, **26.**20
    expansion, **26.**21, **28.**3
    isolation, **28.**3
    sawing, **26.**20, **28.**6
  in pipe, **42.**35
  in precast units, **42.**9
    field-molded, **4.**15 to **4.**19
    premolded, **4.**19, **28.**6
  in roller-compacted concrete, **30.**8, **30.**16
  (*See also* Construction joints)

Kame, **2.**5
Kelly ball, **5.**5, **12.**4
Key, shear, **28.**2
Kiln:
  cement, **1.**5
  curing, **42.**16

Laboratory:
  commercial, **13.**6
  field equipment, **13.**6
Laitance, **5.**7, **25.**9
Latex, **45.**4
Lath, **39.**3
Leaching, **7.**5
Leakage through forms (sand streaks), **25.**11, **44.**11, **44.**15
Lift-slab, **35.**1 to **35.**5
  post tensioning, **35.**3
Lightweight aggregate (*see* Aggregate, lightweight)
Lightweight concrete, **11.**30 to **11.**40, **31.**1 to **31.**9
  adjustments, **11.**27
  fill, **31.**1
  heat insulation, **31.**2
  nailing, **31.**8
  placing and curing, **31.**7
  proportioning, **11.**30 to **11.**33, **31.**6
  strength of, **31.**33, **31.**1, **31.**2
  testing, **31.**8

Lightweight concrete (*Cont.*):
  trial mixes, **11.**34 to **11.**36, **31.7**
Lignosulfonate acid, **4.**5, **11.**11
Lime, uncombined, **7.8**
Limestone, definition of, **2.**13
Limonite, heavy aggregate, **32.5**
Line block:
  concrete pump, **36.**12
  plaster pump, **39.9**
Liner:
  form, **43.6**, **44.2**
  plastic, **43.6**, **44.2**
Linseed oil:
  protective coating, **7.**3
  stains, **44.7**
Log washer for aggregate, **2.**42
Los Angeles abrasion test, **2.**18, **7.**19, **12.**2
Low-temperature concreting, **5.9**, **29.**1, **29.**5, **29.9**, **39.**13
Lumber:
  form: effect on appearance, **14.**4, **43.6**
    effect on dusting, **43.6**

Mackinac Bridge, **38.**1
Magnesium fluosilicate hardenr, **4.**23, **7.**10
Magnetite, heavy aggregate, **32.5**
Maintenance, plant, **21.2**
Manufacture's certification, **13.9**
Map-cracking, **44.7**
Maps:
  geological survey, **2.**10
  soil survey, **2.**10
Marblecrete, **39.**10
Masonry, **40.**1 to **40.**17
  anchor bolts in, **40.**12
  cold weather, **40.**12
  cracking of, **40.**14
  estimating, **40.6**
  foundation for, **40.6**
  grouted, **40.**14
  high-lift, **40.**14
  joints, **40.7**, **40.**12
  laying, **40.**10
    bedding, **40.**10
  modular construction, **40.2**
  mortar for, **40.7** to **40.**16
    color, **40.9**
  reinforcement, **40.7**
  testing, **40.**16
  types of construction, **40.**1
Masonry (*Cont.*):
  units, **40.2**
  (*See also* Block, concrete)
Mass concrete, **26.8** to **26.**17, **46.**3
  cleanup and joint treatment, **8.6**, **25.**5 to **25.9**, **26.**12
  curing, **26.**17
  definition of, **26.8**
  final stable temperature, **26.**16
  forms for, **26.**12
    temporary holding, **26.**15
  lift thickness for temperature control, **26.**17
  mixes for, **17.**13, **26.9**
  pipe cooling system, **26.**17
  placing, **26.**11
    preparations for, **26.**12
    sequence, **26.**13
  plant for, **16.**3
  size of placements, **26.9**
  temperature control, **10.**5, **26.**17
    required, **26.**17
  thermal changes, **10.**5
  time interval between lifts, **26.**17
Materials:
  architectural concrete, **43.**1 to **43.**13
  at batch plant, inspection of, **13.**1
  sampling, **13.2**
  storage, **16.**3
Mechanical analysis, aggregates (*see* Grading; Sieve analysis)
Metamorphic rock, **2.6**
Meter:
  air, **4.**4, **12.**5
  consistency, moisture, **13.**2, **18.**4
  water, **17.**4, **18.**4
Mill scale on reinforcing steel, **3.9**, **25.**21
Mill test report, cement, **13.8**
Mineral, definition of, **2.6**
Mineral admixture (*see* Admixtures, mineral)
Mineral aggregate, definition of, **2.2**
Mineralalogical terms, **2.6**
Mix:
  adjustments, **11.**24, **11.**27, **11.**43
  design (*see* Proportioning below)
  prepared mixes, charts, **11.**25 to **11.**29
  proportioning, **11.**3 to **11.**43
    data required for, **11.**5
    examples, **11.**17 to **11.**25
    for heavy concrete, **11.**40 to **11.**43, **32.5**
    for lightweight concrete, **11.**36 to **11.**40, **31.6**

Mix (*Cont.*):
  selection of, **11.**12
Mixer Manufacturer's Bureau, **18.**1
Mixers, **18.**1 to **18.**8
  design, **18.**1
  efficiency test, **12.**15
  motor-load indicator, **18.**4
  nontilting, **18.**5
  pan, **18.**5
  paving, **18.**5
  sizes of, **18.**1
  tilting, **18.**1
  truck, **18.**6
    number required, **23.**7
  turbine, **16.**12, **18.**5
  uniformity tests, **12.**5, **12.**6
  vertical shaft, **18.**5
Mixes:
  for block, **42.**12
  canal lining, **26.**21
  mass concrete, **17.**14, **26.**9
  paving, **26.**17
  piles, **42.**38
  pipe, **42.**31
  prestressed concrete, **41.**5
  pumping, **36.**11
  radiation shielding, **32.**2
  tunnels, **26.**39
Mixing:
  architectural concrete, **44.**4
  in block plant, **42.**13
  cycle time, **16.**12
  heavyweight concrete, **32.**5
  lightweight concrete, **31.**6
  long time, effect of, **6.**16
  plant location, mass concrete, **16.**11
  (*See also* Batch plant)
  roller-compacted concrete, **30.**7, **30.**13
  trial batch, **11.**20
Modified cube test, **6.**18
Modulus of elasticity, **6.**23, **12.**18
  determination of:
    compression test, **6.**16 to **6.**20, **12.**10
    dynamic test, **6.**24
    flexural test, **6.**19
    torsion test, **6.**24
  effect: on creep, **9.**10
    of temperature, **10.**6, **6.**16
    of variables, **6.**15, **6.**24
  relationships, **6.**24
  stressing steel, **41.**6, **41.**14
Modulus of rigidity, **6.**24
Modulus of rupture, **6.**2, **6.**12, **12.**12
Moisture in aggregate, **13.**2
  compensation for, **17.**7
  determining:
    drying method, **13.**2, **17.**7
      flask method, **13.**2
      measuring meter, **13.**2, **17.**7, **18.**4
  tests for, **13.**2
Moisture content of concrete, **16.**9
Molds,elastomeric, **43.**7, **44.**2
  (*See also* Forms)
Mortar:
  content, test for in concrete, **12.**17
  dry patching, **45.**4
  layers between lifts of concrete, **25.**9
  masonry, **40.**7 to **40.**10
Moving concrete, equipment for (*see* Handling concrete)

National Concrete Masonry Association, **42.**24
National Stone Association, **2.**54
National Ready-mixed oncrete Association, **18.**1
National Aggregates Association, **2.**54, **18.**1
National Slag Association, **2.**54
Noise-reduction coefficient, **10.**8
Non-destructive test(Test, non-destructive)
Non-shrink or expansive concrete, **9.**8
Normal curve, strength variations of, **12.**26 to **12.**33
Nuclear reactor, shielding, **32.**2, **38.**1

Ocrate concrete, **7.**5
Opal, **2.**8, **2.**28, **7.**15
Openings, cracking near, **26.**3
Operation, batch plant, **21.**1
Organic impurities in fine aggregate, **2.**30
Oroville Dam, **25.**9
Outwash plain, **2.**5
Overmixing, **9.**7, **26.**53
Overvibration, **24.**23

Paint, **44.**17, **45.**5
  stains, **44.**16
Pan joist forms, **15.**14, **15.**18
Parting compounds, **15.**13, **25.**14, **41.**9, **43.**6

Pavement:
composition of, **17.**10
curing, **26.**21
finisher, **27.**5
finishing, **27.**5 to **27.**16
foundation for, **26.**17
placing concrete in, **26.**18
subgrade trimmer, **26.**18
texture of, **27.**13
tolerances for, **27.**8
Paver:
consolidating concrete, **26.**19
side form, type of, **26.**1, **26.**18
slipform, **26.**17 to **26.**20, **27.**6
canal lining, **26.**21
subgrade guided, **26.**19
for small jobs, **26.**20
Penetration ball, **5.**5, **12.**4
Penetration probe guage, **12.**25
Perlite, **11.**9, **31.**2, **39.**2
Permeability:
air and vapor, **8.**8
bleeding, effect of, **5.**6
coefficients (table), **8.**3
definition of, **8.**1
tests, **8.**2
(*See also* Watertightness)
Phyllite, **7.**15
Pigments, **14.**12, **43.**5, **44.**4
Piles, **42.**35 to **42.**42
cast-in-place, **26.**2, **42.**40
caisson, **26.**46, **46.**49, **42.**40
shell, **42.**41
precast, **42.**35 to **42.**39
hollow cylindrical, **42.**39
mixes for, **42.**38
pickup, **42.**39
prestressed, **42.**39
reinforcement, **42.**38, **42.**41
sheet **42.**39
sizes, **42.**36
square and octagonal, **42.**35
Pipe:
cast-in-place, unreinforced, **26.**42 to **26.**46
precast, **42.**25 to **42.**35
air entrainment in, **42.**31
cast, **42.**31
centrifugal, **42.**27
curing, **42.**32
cylinder, **42.**28
defects, in, **42.**30
Pipe, precast (*Cont.*):
joints in, **42.**35
mixes for, **42.**31
packerhead, **42.**25
plant for, **16.**12
reinforcement, **42.**32
specifications for, **42.**26
tamped, **42.**27
testing, **42.**33
Pipe cooling system, mass concrete:
advantages and disadvantages of, **26.**17
design of, **26.**17
Placing concrete:
in adverse weather, **26.**51, **29.**9
in arches, domes, paraboloids, and folded plates, **26.**5
in architectural concrete, **44.**4
in bridges, **26.**6
during cold weather, **29.**9
correct and incorrect methods, **24.**7 to **24.**17
in culverts and cast-in-place pipe conduits, **26.**42 to **26.**46
in curbs, gutters, and sidewalks, **26.**27
depositing in forms, **26.**13, **26.**41, **44.**4
effect of bleeding on, **26.**20
equipment, selection of, **24.**3, **24.**17, **44.**4
in floor, **26.**2, **27.**5
in foundations, **26.**2
heavy, **32.**6
in heavy piers, **26.**6
during hot weather, **5.**10, **26.**51
in layers, **26.**13, **26.**17
lightweight, 316
in mass concrete structures, **26.**13 to **26.**17
in pavements and canal linings, **26.**17 to **26.**27
in precast concrete, **42.**3
preparations for:
blockouts, **25.**22
cleaning within forms, **25.**15
construction joints, **25.**5 to **25.**10
embedded materials, **25.**21
final checkout sheet for placing, **25.**4
final inspection, **25.**24
forms, **25.**10 to **25.**17
foundations, **25.**1
reinforcement, **25.**17 to **25.**24
in prestressed members, **41.**16
in retaining walls, **26.**6
in slabs, **26.**2, **26.**20, **26.**27

Placing concrete (*Cont.*):
on slopes, **26.**5, **26.**21
in tunnels and subways, **26.**20
underwater, **26.**46
in walls, columns, and beams, **26.**2
Placing temperature, reduction of, **26.**8
Planning, construction method, **33.**4, **34.**1, **35.**1, **35.**6
Plant:
central-mix paving, **16.**11
commercial ready-mis, **16.**8, **16.**10
dry-batch paving, **16.**8, **18.**1
mass concrete, **16.**11
mobile, **16.**10, **17.**27
precast products, **16.**12
Plaster, portland cement, **39.**1 to **39.**14
admixtures in, **39.**2
application: hand, **39.**7
machine, **39.**7
sandblock, **39.**9
attachments, **39.**4
bases for, **39.**4
concrete, **39.**5
masonry, **39.**5
steel frame, **39.**5
wood frame, **39.**4
bonding, **39.**5
coats, **39.**8
in cold weather, **39.**13
cracking, **39.**13
discoloration of, **39.**12
failures, **39.**12
finishing and curing, **39.**10, **39.**11
line block, **39.**9
materials for, **39.**1
mixes for, **39.**6
reinforcement, **39.**4
strength, **39.**11
Plastic flow, (*see* Creep)
Plastic shrinkage, **9.**2, **26.**52
(*See also* Cracking; Volume change)
Pneumatic trough conveyor, **1.**17, **19.**8
Pneumatically applied concrete and mortar (*see* Shotcrete)
Poisson's ratio, **6.**25
Pollution control, **1.**8, **16.**12
Polyester resin, **45.**8
Polymerizing, **6.**26
Popouts, **43.**4
repair, **45.**18
Pore structure of concrete, **8.**1
Porosity:
of aggregate, **8.**2
of concrete, **8.**1
Postcooling, **26.**8
Pot life, epoxy resin, **45.**1
Power shovel excavation, **2.**33
Pozzolans, **4.**11
alkali-aggregate reaction, **4.**8, **45.**9
in architectural concrete, **43.**5
handling, **19.**7
in masonry block, **42.**11
in mass concrete, **4.**11
(*See also* Admixtures, mineral)
Precast concrete, **42.**1 to **42.**10
beams, **33.**10
columns, **33.**16
connections, **33.**16, **42.**9
erection, **33.**16, **42.**8
extrusion, **42.**6
floor slabs, **33.**14
forms, **33.**5 to **33.**17, **42.**2
miscellaneous items, **42.**42
panels, **33.**4, **33.**7, **33.**10
placing concrete, **42.**3
sandwich panels, **42.**5
Schokbeton process, **42.**5
shop drawings, **42.**1
Spancrete extruder, **42.**7
Spiroll extruder, **42.**6
Preconstruction planning, **33.**4, **34.**1, **35.**1, **35.**6
Precooling, **17.**7, **20.**16, **26.**8
Premature stiffening, **1.**8, **5.**8
Prepacked concrete (*see* Preplaced aggregated concrete)
Preplaced aggregate concrete, **9.**8, **38.**1 to **38.**7
admixtures in, **38.**5
advantages of, **38.**1
aggregates for, **38.**5
biological shielding, **38.**1
cofferdam construction, **38.**1
forms, **38.**6
mortar for, **38.**6
for repair, **38.**2, **45.**12
underwater, **26.**46, **38.**2
Prestress:
computation of, **41.**13 to **41.**16
loss of, **9.**13, **41.**14, **41.**22
Prestressed concrete, **41.**3 to **41.**27
admixtures in, **41.**5
camber, **41.**5
casting beds, **41.**6

Prestressed concrete (*Cont.*):
  cracking of, **41.**1, **41.**6
    reinforcement, **41.**6
  curing, **41.**17
  detensioning, **41.**18
    strength for, **41.**5
  forms, **41.**1, **41.**9
  handling units, **41.**19
  materials for, **41.**4
  mixes, **41.**5
  placing concrete, **41.**16
  posttensioned, **3.**21, **3.**24, **41.**3, **41.**20 to **41.**27
    anchorages, **3.**24
    ducts for wire, **41.**26
    friction losses, **41.**22
    grouting, **41.**23
    lift-slab, **35.**3
    stressing, **41.**23
    tendons, bonded, **3.**21, **41.**24
  reinforcing, **3.**29, **41.**4
  segmental, **41.**26
  steel for, **3.**21 to **3.**24
    corrosion of, **41.**5
  strand for, **3.**20 to **3.**24, **41.**4
    harped, or draped, **41.**16
    small diameter, **3.**22
  wire for, **3.**23, **41.**4
    stress-relieving, **3.**22
Prestressing:
  chemical, **41.**25
  circumferential, **41.**25
Probability of low strength, **12.**29
Production rate, concrete plant, **18.**2
  central-mix paving, **16.**11
  dry-batch paving, **16.**5
  mass concrete, **16.**1
  products plant, **16.**12, **18.**5
  ready mix, **16.**5, **18.**7
Properties of concrete (*see* Specific property)
Proportioning mixes (*see* Mix, proportioning)
Protective treatments against aggressive substances, **7.**8 to **7.**14
Pullout test, **12.**25
Pumice, **2.**17, **11.**9, **39.**2, **42.**11
Pump(s):
  concrete, cleaning, **36.**12
  heavy, **36.**4
    specifications for, **36.**5
  line block, **36.**12
  operation, **36.**12

Pump(s) (*Cont.*):
  pipes for, **36.**9
    friction in, **36.**9
    go-devil, **36.**12
    volume of, **36.**9
  slickline, **36.**9
  small line, **36.**5
  vacuum squeeze, **36.**8
  water, for aggregate production, **2.**41
Pumpcrete, **26.**32, **36.**8, **36.**13
Pumped concrete, **36.**3 to **36.**13
  air-entrained concrete, **36.**11
  difficulties, **36.**12
  downhill, **36.**12
  equipment for, **36.**4 to **36.**10
  lightweight concrete, **36.**13
  materials and mixes for, **36.**10 to **36.**11
  slump loss, **36.**11
  in tunnel lining, **26.**39

Q-block specification, **42.**24
Quality control, **12.**28
  charts, **12.**30
Quarry sampling:
  diamond drilling, **2.**32
  shot drilling, **2.**32
Quartz, **2.**6
Quartzite, **2.**18

Radiation:
  of heat, **20.**20
  nuclear, shielding, **32.**2, **38.**1
Railroad ties, **42.**43
  electrical properties, **10.**10
Rain, placing concrete during, **26.**50, **30.**10
Reactive aggregate (*see* Aggregate, reactive)
Ready-mixed concrete:
  sampling, **12.**3
  typical compositions, **17.**10 to **17.**14
  uniformity tests, **12.**15
  (*See also* Mixers)
Receiving hoppers for concrete, **23.**1
Recorder, batching, **17.**10
Refractory concrete, **36.**19
Refrigeration, **20.**8
  absorption, **20.**25
  vacuum, **20.**24 to **20.**27
Reinforcing steel (*see* Steel, reinforcing)
Relaxation due to creep, **9.**10

Relief patterns, **43.**11
Repair of concrete, **45.**3 to **45.**19
  architectural concrete, **44.**15
  bridge deck, **45.**11
  concrete replacement, **45.**8 to **45.**10
  cracks, **45.**12 to **45.**14
  curing, **45.**7
  dry pack, **45.**13
  needle gun, **44.**15
  popouts, **45.**18
  preparation for, **45.**5 to **45.**8
  preplaced aggregate concrete, **38.**1, **45.**12
  qualifications for, **45.**3
  shotcrete, **45.**11
  synthetic patches, **45.**14
  tools for, **44.**15, **45.**6
Required average strength, **12.**29
Rescreening, **19.**6
Resin, epoxy, **45.**4, **45.**10, **45.**11
Resistivity, electrical, **10.**10
Resistivity surveys for aggregates, **2.**11
Retaining walls, placing concrete in, **26.**6
Retarders (*see* Admixtures, chemical)
Revibration, **24.**27
Rhyolite, **2.**28, **7.**14
Rock(s):
  classification, **2.**7
  compression strength of, **2.**18
  definition of, **2.**7
    extrusive, **2.**6
    igneous, **2.**6
    intrusive, **2.**6
    metamorphic, **2.**6
    sedimentary, **2.**6
  summary of engineering properties, **2.**10
Rock ladder for aggregates, **19.**6
Rock pockets, **24.**17, **24.**23, **25.**10, **45.**5
  prevention of, **24.**13
  repair, **45.**5 to **45.**10
Roll-a-meter, **4.**4
Roller-compacted concrete, **30.**3 to **30.**19
  application and advantages, **30.**4, **30.**11
  curing and protection, **30.**8, **30.**17
  definition, **30.**3
  design, **30.**5, **30.**11
  history and development, **30.**4, **30.**11
  joints, paving, **30.**16
  lift preparation, **30.**8
  mass concrete, **30.**4, **30.**19
  materials, **30.**5, **30.**11
  paving, **30.**14
Roller-compacted concrete (*Cont.*):
  proportioning, **30.**6, **30.**12
  spreading and compacting, **30.**8, **30.**14, **30.**15
  test, **30.**7, **30.**12
  ypes, **30.**3
  transportation, **30.**8, **30.**13
Roof tiles, **42.**42
Rubbed surfaces, **27.**4, **44.**13
  curing compound on, **26.**21, **29.**4
Rupture, modulus of, **6.**2, **6.**12, **12.**12
Rust:
  on prestressing steel, **41.**5
  on reinforcing steel, **3.**9, **25.**24, **43.**6
  stains, **44.**16
Rustication strips, **25.**11, **43.**11, **44.**1

Sack rubbing, **27.**4, **44.**13
  to eliminate surface voids, **27.**4, **44.**15
Safety, **15.**15, **22.**2, **25.**14
Salamanders, effect:
  on dusting, **29.**9
Salt scaling, **7.**4
Samples, preparing and shipping, **13.**11
Sampling:
  aggregates, **2.**14, **12.**2
    sample size, **12.**2
  concrete, **12.**7
  materials, **12.**2
Sand:
  definition of, **2.**3
  grading, **2.**22, **11.**9
    canal lining, **26.**27
    roller-compacted concrete, **30.**6
  manufactured, **11.**9, **11.**25
  moist, bulking of, **2.**30
  production, **2.**31
    drag, **2.**34
    dry classification, **2.**44
    hydraulic size classification, **2.**42
    rake, **2.**43
    screw classifier, **2.**43
    washing, **2.**41
    water scalping, **2.**43
  streaking, **44.**2, **44.**15
  (*See also* Fine aggregate, Architectural concrete)
Sand block, plaster, **39.**9
  (*See also* Line blocks)
Sand-gravel, **7.**15

Sandblasting:
finish, **44.**10
construction joints (*see* Wet sandblast)
to remove laitance, **25.**5
Sawing joints, **26.**21, **28.**2, **28.**6
Scaffolding, **14.**20, **15.**9
Scales, batching, **16.**7, **17.**2
accuracy of, **16.**7, **17.**2
central mix, **16.**11
checking, **13.**2, **17.**10
dry batch, **16.**8
mass concrete, **16.**6
precast products plant, **16.**12
truck-mix plant, **16.**8
Schist, **2.**7
Schmidt hammer (*see* Test hammer)
Schokbeton, **42.**5
Scoria, **2.**7, **11.**9, **31.**4, **42.**11
Screed, **24.**21, **27.**11
Screens:
at batch plant, **19.**6
capacity of, **19.**7
deck, **2.**41
revolving, **2.**44
Scrubber for aggregate, **2.**42
Sealant:
joint: field-molded, **4.**15 to **4.**19
premolded, **4.**19 to **4.**21
Sealer, penetrating, **43.**12
Seawater (*see* Water, sea)
Secant modulus of elasticity, **6.**23, **12.**18
Segregation:
of aggregate, **19.**3, **19.**6
of concrete, **5.**10, **24.**6 to **24.**17
cause of, **24.**7
effect of: on manufacture, **5.**10
on placement, **5.**10
on proportions, **5.**10
on strength, **6.**15
mass concrete, **26.**13
prevention of, **24.**2, **30.**6
Seismic surveys for aggregates, **2.**12
Seperation of concrete (*see* Segregation of concrete)
Setting time:
accelerated, **4.**6, **11.**10
retarded, **4.**6, **11.**11
Settlement, cracking, **26.**43
Sewage attack, **7.**4
Shale, **2.**9
effects on popouts, **43.**4
Shale (*Cont.*):
expanded, **11.**9, **11.**30
Shear failures due to grouting, **46.**2, **46.**3
epoxy repair, **45.**12
Shear key, **28.**2
Shearing strength, **6.**14
Shielding concrete, **32.**2, **38.**1
Shoring, **14.**20, **15.**9, **25.**13, **26.**7
safety rules, **15.**15
scaffold, **14.**20, **15.**9
vertical, **15.**6
capacities, **15.**6
Shotcrete, **36.**3, **36.**13 to **36.**22
ACI Recommended Practice, **36.**22
application of, **36.**16
canal lining, **26.**22
curing, **36.**18
dry-mix, **36.**5
equipment, **36.**16, **36.**21
fiber-reinforced, **36.**20
finishing, **36.**18
materials and mixes, **36.**15
properties of, **36.**14
rebound, **36.**17
refractory, **36.**19
for repairing concrete, **45.**11
for rock protection, **25.**2
polymer, **36.**21
surface preparation, **36.**15
testing, **36.**18
uses, **36.**14
wet mix, **36.**19
Shrinkage (*see* Drying shrinkage; Volume changes)
Sieve analysis, aggregate, **2.**22 to **2.**27, **12.**2
Silo, cement, **1.**13, **19.**7, **43.**1
Skidproof floor, **27.**15
Slab:
finishing, **27.**5 to **27.**16
and floor:
hot weather finishing, **25.**53
placing concrete in, **26.**2
weight of, **15.**12
Slackline technique, dragline, **2.**34
Slag:
blast-furnace: air-cooled:
definition of, **2.**3
production of, **2.**5
chemical composition, **2.**6
expanded, **31.**4, **42.**11

Slag, expanded (*Cont.*):
properties, **11**.8
open-hearth, **11**.8
Slate, **2**.6, **2**.8, **2**.10, **11**.8
Slipform:
for building construction (*see* Slipform, vertical)
canal lining, **26**.18, **27**.8
subgrade guided, **26**.20
vertical, **34**.1 to **34**.13
building design requirements, **34**.2
design of, **34**.6
evaluation, **34**.1
finish, **34**.4
jacking, **34**.6
operation of, **34**.10
Slope paving, **26**.21
Slump, **5**.4, **18**.2
adjusting, **11**.4
canal lining, **26**.26
cone, **5**.4
factors affecting, **18**.2
heavy concrete, **32**.1
lightweight, **31**.7
loss placing problems, **26**.50
in pumping, **26**.39, **36**.11
mass concrete, **26**.8
meters (mixer), **18**.4
pavement, **26**.17
permissible, **11**.4
roller-compacted concrete, **30**.6
specified, **11**.4
temperature effect on, **26**.52
test, **5**.4, **11**.4, **12**.4, **30**.6
in truck mixer, **18**.7
tunnel lining, **26**.29
water required for, **11**.14
(*See also* Workability)
Snowshoes, **26**.15
Sodium silicate:
protective coating, **7**.13
used as a hardener, **29**.34
Sound absorption, **10**.7
Sound insulation, **10**.9
Sound transmission, **10**.8
Spall, repair of, **45**.15
Spancrete extruder, **42**.7
Specific gravity:
of aggregate, **2**.20
factor, lightweight, **12**.20
heavy, **11**.40, **32**.5
Specific gravity, of aggregate (*Cont.*):
saturated surface dry, **11**.17
volume relationship (table) **11**.31
(*See also* Aggregate, specific gravity)
of cement, **1**.14, **11**.16
of concrete, (*see* Density)
Specific heat, **10**.5, **20**.3, **20**.6
Specifications:
for aggregates, **2**.50, **43**.3
for cement, **1**.16 to **1**.19
performance, **1**.16 to **1**.19
prescription, **11**.3
strength, **12**.29
Spiroll extruder, **12**.6
Splitting tensile (*see* Strength of concrete, splitting tensile)
Spreaders, paving, **26**.18
Stains, **43**.6, **44**.16
Standard deviaiton:
strength, **12**.26
within text, **12**.26
Statistical analysis of compressive strength, **12**.26 to **12**.36
sample computation, **12**.33
simplified evaluation, **12**.36
Steam curing:
atmospheric, **6**.10, **29**.6
boiler capacity for, **42**.17, **42**.20
of conrete block, **42**.16 to **42**.22
high-pressure, **6**.22, **29**.8, **42**.17
low-Pressure, (*see* Atmospheric above)
of pipe, **42**.32
(*See also* Curing)
Steel reinforcing, **3**.1 to **3**.19
accessories, **3**.13, **43**.6
bar supports, **3**.13, **25**.27
beam bolsters, **3**.14
chairs, continuous high, **3**.14
joist, **3**.14
slab bolster, **3**.14
tie wire, **3**.15, **25**.20
bar, **3**.2
axle steel, **3**.5
billet steel, **3**.5
deformed, **3**.2
lists, **25**.17
numbers, **3**.3
plain **3**.3
rail steel, **3**.5
spacing, **25**.18
bendability, **3**.7

Steel reinforcing, accessories (*Cont.*):
bending:
heavy, **3.**7
light, **3.**7
radial, **25.**22
bends, typical, **25.**19
branding, **3.**6
cleaning, **3.**9, **25.**21
coated, **3.**11, **43.**6
compressive, **9.**13
concrete cover, **25.**19
in architectural concrete, **43.**6
in piling, **42.**34
in pipe, **42.**38
connection details, **25.**19
corner and wall intersection details, **25.**19
deformations, **3.**2
elongation, **3.**4
epoxy coated, **25.**18, **43.**6
fabric, one way, **3.**9
fabrication, **3.**7, **25.**20
tolerances, **25.**18
grades, **3.**4
hooks and bends, **25.**19
identification, **3.**3
inspection check list, **25.**25
lateral reinforcement, **25.**21
mechanical property, **3.**5
metal mesh, expanded, **39.**3
in paving, **26.**20
placing, **3.**15, **25.**20 to **25.**24
tolerances, **25.**18
for prestressed concrete (*see* Prestressed Concrete)
rusting, **3.**9, **25.**21
splices, **3.**15, **25.**18
storage, **3.**9
strength, tensile, **3.**5
tack welds, **3.**18
welded-wire fabric, **3.**9 to **3.**11
deformed, **3.**10
"two-way", **3.**10
welding, **3.**17 to **3.**19, **25.**18
chemical composition for, **3.**17
inert-gas shielded-electric-arc, **3.**17
joints, **3.**18
butt, **3.**18
cross, **3.**18
lap, **3.**18
processes, **3.**17
resistance, **3.**17

Steel reinforcing,accessories, welding (*Cont.*):
shielded-metal-arc, **3.**17
wire: deformed, **3.**10
sizes, **3.**11
yield point, **3.**4
Stockpiling aggregates, **19.**1, **19.**2, **30.**5
Stone, crushed, definition of, **2.**3
Stone ladder (*see* Rock ladder)
Stoned finish, **27.**4
Storage at batch plant, **16.**3 to **16.**5, **19.**6, **30.**5
Storage tank, hot water, **20.**12
Straightedging pavement, **26.**20, **27.**3
Streets (*see* Pavements)
Strength of concrete:
bond, **6.**2, **6.**13
compressive: analysis of, **12.**26 to **12.**36
of cores, **12.**30
design, **12.**26
effects: entrained air, **5.**8, **6.**7, **11.**5
moisture content, **12.**23
equired average, **12.**29
factors affecting. **6.**2
admixtures, **6.**5
age, **6.**3, **6.**11 to **6.**12
aggregate size, **6.**5, **11.**4
aggregate type, **6.**4, **11.**8
cement content, **6.**3
fineness, **6.**3
type, **6.**3, **6.**4, **11.**8
curing, **6.**9, **12.**8
lateral restraint, **6.**19
loading rate, **6.**19
moisture content, **6.**19
proportions, **6.**2
temperature, **6.**9, **6.**19
water-cement ratio, **6.**2, **11.**12
fatigue, **6.**25
flexural, **6.**12, **11.**24
test, **6.**2, **12.**12
gain, **6.**11
accelerated, **6.**5, **6.**10, **6.**21
normal, **6.**11
retarded, **6.**20
general considerations, **6.**1 to **6.**22
impact, **6.**26
moist curing vs. non-standard, **29.**7
prediction of, 611, **6.**22
relationshipd, **6.**4, **6.**12
shear, **6.**14
specimen: capping, **6.**17, **12.**9

Strength of concrete, specimen (*Cont.*):
curing, **12**.8
field-cured, **12**.7
flexural, **12**.12
making, **12**.7
molds for, **12**.7
number required, **12**.29
testing, **6**.16 to **6**.20
transporting, **12**.8
trial mix, **11**.20
splitting tensile, **6**.1, **6**.12
test, **6**.1, **12**.11
temperature effect, **6**.9, **6**.19
tensile, **6**.1, **6**.12
tests, **6**.1, **12**.7 to **12**.13
bond, **6**.2
compression, **6**.1, **12**.7 t0 **12**.10
flexural, **6**.1, **12**.12
tension, **6**.1
variations, **6**.15 to **6**.20, **12**.26
causes of, form and size of specimens, **6**.17
loading rate, **6**.19
materials, **6**.15
moisture content, **6**.19
production, **6**.16
sampling, **6**.16
testing procedures, **6**.16, **6**.19
Stress-strain relationship, **6**.23
Strike off, **27**.11
Stucco, **39**.1
netting, **39**.4
(*See also* Plaster)
Substances, effect of, protective treatments, **7**.10 to **7**.14
Subways, placing concrete in, **26**.29, **26**.39
(*See also* Tunnels)
Sulfate attack, **7**.6
resistance to, **7**.6
Sunshades to prevent cracking, **26**.53
Surface retarded, **44**.9
Surface hardeners
(*See also* Floor hardeners)
Surface treatments, protective, **7**.9 to **7**.14, **44**.17
Swiss hammer (*see* Test hammer)
System concept, building, **33**.3, **42**.1

Tamper, **27**.5
Tangent modulus of elasticity, **6**.22
Temperature:
calculation of:
aggregate, **20**.1, **20**.17
concrete, **20**.8, **20**.16
of cement, **20**.4, **26**.52
drop to control cracking, **26**.21
dry-bulb, **20**.4
effect of: on cracking, **9**.3, **26**.52
on crazing, **26**.52
on properties, **5**.9, **10**.6
on strength, **6**.9, **6**.19
on volume change, **9**.3, **26**.52
rise on hydration, **1**.15, **9**.3
of concrete, **1**.14, **9**.3, **10**.5, **20**.1 to **20**.27, **30**.6
wet-bulb, **20**.7
Tennessee Valley Authority, **30**.4
Tests:
aggregate, **12**.2
(*See also* Specific property concerned)
batch, **11**.20 to **11**.25, **30**.7, **30**.12
cement, **12**.1
field, **12**.3 to **12**.7
of fresh concrete:
air content, **4**.4, **12**.5
ball penetration, **5**.5, **12**.4
bleeding, **5**.6
compaction, **30**.7, **30**.16
flow, **5**.3
slump, **5**.4, **12**.4
trowel, **5**.6
unit weight, **12**.14, **12**.17
hammer, for compressive strength, **6**.27, **12**.24
impact, **7**.18
modified cube, **6**.18
non-destructive, **6**.26
indentation (*see* Hammer above)
pulse transmission, **6**.26, **7**.3, **12**.26
radioactive, **6**.26
sonic, **6**.26
pullout test, **12**.25
pulse velocity, **6**.26, **7**.3, **12**.26
shrinkage, **12**.3
slump, **5**.4, **12**.4
sonic, **6**.26
volumetric for air content, **4**.4
Windsor probe, **12**.25
of water, **12**.2
(*See also* Specific property concerned)
Testing agency (*see* Laboratory)

Tetracalcium aluminoferrite, **1.**15
Texture:
  architectural concrete, **44.**8
  pavement, **27.**13
  slab, **27.**13
Thermal expansion, **9.**3
Thermal properties:
  conductivity, **10.**1, **20.**3, **20.**14
  specific heat, **10.**5, **20.**3,**20.**6, **20.**17
  thermal diffusivity, **10.**5, **20.**3
  thermal expansion, **9.**3, **10.**5
Tie bar, pavement, **26.**20
Tilt-up, **27.**5, **29.**11, **29.**12, **36.**6 to **35.**10, **42.**1
Tolerances:
  in batching, **17.**1
  building, **14.**5, **25.**14
  concrete, **25.**14
  forms, **25.**14
  pavments, **27.**8
  placing reinforcing steel, **25.**18
  slabs, **27.**3
Tools, finishing, **27.**5 to **27.**8
Toughness, **6.**26
Transit mix, **16.**8
  concrete (*see* Ready-mixed concrete)
  trucks, **23.**5
    capacities, **18.**7
    loading time, **16.**8
    number required, **16.**10, **23.**6
Transporting concrete, **23.**1 to **23.**8, **30.**8, **30.**13
Traprock, **2.**7
Tremie for placing concrete, **26.**46
Trial mix:
  adjusting, **11.**24
  lightweight, **11.**38
  proportioning, **11.**17, **11.**34, **30.**6, **30.**12
  tables for, **11.**26 to **11.**29
Tricalcium aluminate, **1.**7
Tricalcium silicate, **1.**7
Tridymite, **2.**28, **7.**15
Trowel, **27.**6, **27.**8
Truck-mixed concrete (*see* Ready-mixed concrete)
Tuff, **2.**7, **11.**9
Tunnels, **26.**29 to **26.**42
  cleanup, **26.**40
  concrete delivery equipment, **26.**37
  consolidation of concrete, **26.**41
  control of seepage, **26.**40
  curing, **26.**42
  forms, **26.**35 to **26.**37
Tunnels (*Cont.*):
  grouting, **46.**6 to **46.**8
  mixing equipment, **26.**38
  placing concrete, **26.**29, **26.**40
    pumps for, **26.**37 to **26.**42
    sequence, **26.**29
  reinforcing steel, **26.**41

Ultrasonic pulse, test, **6.**26, **7.**3
Underdrain, **25.**4
Underwater concrete, **25.**5
  placing, **26.**46 to **26.**49
    bucket, **26.**46
    pumping, **26.**49
  preplaced aggregate method, **26.**49
Uniform Building Code, **39.**3, **40.**7, **40.**9, **40.**16
Unformed surfaces, **27.**3, **27.**15 to **27.**16, **30.**11
  (*See also* Floor; Slab)
Unit weight:
  aggregate, **11.**16
  concrete (*see* Density of concrete)
U.S. Government, Army Corps of Engineers, **2.**53, **12.**1, **30.**11, **38.**5
  Bureau of Reclamation, **2.**66, **3.**9, **12.**1, **30.**4
Unsound aggregate particles, **2.**26, **7.**15 to **7.**18

Vacuum cooling of aggregate, **20.**23, **26.**52
Vacuum heating, **20.**16
Vacuum processing of concrete, **7.**19, **37.**1 to **37.**4
Vapor pressure, **20.**6, **20.**7
Variation, coefficient of, **12.**27
Vent pipe in grouting operations, **46.**5, **46.**6
Vermiculite, **11.**9, **31.**1
Vibra-Tak, **24.**17
Vibration of concrete:
  amplitude, **24.**17
  effect of:
    on air void system, **24.**23
    frequency, **24.**17
  to prevent surface voids, **24.**25, **44.**4
    of reinforcement, **24.**20
    (*See also* Consolidation of concrete)
Vibrators:
  internal, characteristics and application of (table), **24.**18

Vibrators (*Cont.*):
  measuring frequency of, **24.**17
  types of, **24.**20 to **24.**23
Voids:
  in aggregates, **2.**20, **11.**8
  in concrete, **8.**1
  surface defect, **24.**6, **24.**24, **45.**5
Volcanic glass, **45.**18
Volume of dry-rodded coarse aggregate in mix proportioning, **11.**15
Volume changes:
  autigenous, **9.**2
  bleeding and setting shrinkage, **9.**1
  in hardened concrete, due to:
    alternate wetting and drying, **9.**6
    carbonation, **6.**21, **9.**6, **29.**9
    cement hydration, **9.**2
    drying, **9.**4
    moist storage, **9.**3
    thermal changes, **9.**2
    undesirable chemical and mechanical attack, **9.**8
  measurement, **9.**1
  plastic shrinkage, **9.**2
  (*See also* Drying shrinkage)

Wale, **14.**5
Walls, columns, and beams, placing concrete in, **26.**2
Waste mold, **43.**9
Water:
  absorption, aggregates, **2.**20
  for curing, **4.**13, **29.**2, **43.**5
  equivalent (calorimeter), **20.**5
  gain (*see* Bleeding)
  for mixing, **4.**13, **12.**2
    adjusting, **11.**14
    amount required, **11.**10, **11.**14
    conversion table, **11.**30
    dangerous substances in, **4.**13
    measuring, **17.**4
    quality, **4.**13, **12.**2
    temperature, **17.**4, **26.**52
  sea:
    attack, **7.**7
    for mixing, **4.**13, **12.**2
  stop, **4.**19, **8.**7, **28.**1
    canal lining, **26.**24
Water-cement ratio:
  adjusting, **11.**14
Water-cement ratio (*Cont.*):
  conversion chart, **11.**33
  curves, **11.**24
  determination of, **11.**12
  effect of:
    on creep, **9.**10
    on permeability, **8.**4, **8.**5
    on strength, **6.**2, **11.**12
  permissible: for exposure, **11.**6
    for strength, **11.**12
Water chillers, **20.**22, **26.**52
Water-reducing admixtures, **4.**7, **6.**5, **11.**10, **43.**5
  (*See also* Admixtures, chemical)
Water proofing:
  definition of, **8.**7
  methods for, **8.**8, **42.**11
Watertightness, factors affecting:
  admixtures, integral, **8.**3, **11.**12
  aggregate size, **8.**4
    type of, **8.**4
  cement, **8.**4
  construction practices, **8.**6
  curing, **8.**6
  entrained air, **8.**4
  proportions, **8.**4
  surface application, **8.**8
  water-cement ratio, **8.**5
Wear, **7.**18, **30.**11
Weathering, **2.**12, **7.**1, **44.**17
Weighing (*see* Batching)
Welding reinforcing steel, **3.**17 to **3.**19, **25.**18
Wet-sand blast:
  construction joints, **25.**5, **25.**8
  to expose aggregate, **44.**10
Wheelbarrows for concrete, **24.**3
Wind velocity, effect of, on cracking, **26.**52
Windbreaks to prevent cracking, **26.**52
Winter concreting (*see* Adverse weather; Cold weather)
Wood stain, **43.**6
Workability, **5.**4 to **5.**7
  agent, **5.**8
  definition of, **5.**3
  effect of admixtures on, **5.**8, **30.**6, **30.**12
    aggregates, **5.**8, **11.**9, **30.**6, **30.**12
    cement, **5.**8, **30.**5, **30.**11
    temperature, **5.**9, **30.**10
    time, **5.**8, **30.**10, **30.**17
    water reducer, **5.**9

Workability (*Cont.*):
  measurement of, **5**.4 to **5**.7
  trowel test, **5**.6
  (*See also* Slump)

Yield computation, **12**.13
Yield point, **3**.5
Young's modulus of elasticity (*see* Modulus of elasticity)

Zeolite, **7**.15
Zinc fluosilicate hardener, **7**.9

ABOUT THE EDITORS

Joseph J. Waddell, formerly a consulting engineer of construction materials and methods at Riverside, Calif., is now in retirement at Los Osos, Calif. He directed, previously, the materials inspection and control program for the Northern Illinois Tollway and administered the construction quality control and inspection services at Soil Testing Services, Inc. Mr. Waddell is also the author of McGraw-Hill's *Practical Quality Control for Concrete*.

Joseph A. Dobrowolski who is a consulting engineer with offices in Altadena, Calif., specializes in concrete materials and concrete construction technology. He is a fellow of the American Society of Civil Engineers and the American Concrete Institute. Mr. Dobrowolski is the author of the U.S. Corps of Engineer's *Engineering Manual EM -1110-1-2009, Engineering and Design—Architectural Concrete* (1987), and during 1979-1986 served on the American Concrete Institute's Technical Activities Committee which reviews all committee reports for technical content prior to publication.